Biological Control Programmes in Canada, 1981–2000

Dedication

This book is dedicated to biological control specialists who have, over the years, shown that Canada is a world leader in this field. One of them, Don Wallace (1929–1995), who worked for the Canadian Forest Service, will always be remembered for his selfless leadership to biological control. His untimely death ended an outstanding career.

Biological Control Programmes in Canada, 1981–2000

Edited by

P.G. Mason

Agriculture and Agri-Food Canada,
Eastern Cereal and Oilseed Research Centre,
Ottawa, Ontario, Canada

and

J.T. Huber

Natural Resources Canada,
Canadian Forest Service,
Ottawa, Ontario, Canada

CABI *Publishing*

CABI *Publishing* is a division of CAB *International*

CABI Publishing
CAB International
Wallingford
Oxon OX10 8DE
UK

Tel: +44 (0)1491 832111
Fax: +44 (0)1491 833508
Email: cabi@cabi.org
Web site: www.cabi-publishing.org

CABI Publishing
10 E 40th Street
Suite 3203
New York, NY 10016
USA

Tel: +1 212 481 7018
Fax: +1 212 686 7993
Email: cabi-nao@cabi.org

A catalogue record for this book is available from the Library of Congress, Washington DC, USA.

Library of Congress Cataloging-in-Publication Data

Biological Control Programmes in Canada 1981–2000 / edited by P.G. Mason and J.T. Huber.
p. cm.
Includes bibliographical references (p.).
ISBN 0-85199-527-6 (alk. paper)
1. Insect pests--Biological control--Canada. 2. Weeds--Biological control--Canada. 3. Phytopathogenic microorganisms--Biological control--Canada. I. Mason, P. G. (Peter G.) II. Huber, John T. (John Theodore)

SB933.32.C2 B56 2001
632′.96′0971--dc21

2001029512

ISBN 0 85199 527 6

Typeset by Columns Design Limited, Reading.
Printed and bound in the UK by Biddles Ltd, Guildford and King's Lynn.

Contents

Introduction

P.G. Mason, J.T. Huber and S.M. Boyetchko

This book summarizes biological control programmes in Canada since 1981. Previous volumes in the series were published in 1962, 1969 and 1984. While similar in format to the previous books, this one departs in important ways.

First, it includes much more on pathogens (viruses, bacteria, fungi and nematodes), either as targets for control or as biological control agents themselves, acting either directly as hyperparasites and/or pathogens or indirectly as antagonists that compete successfully for the same resources as the target pest. The emphasis on introducing insects for classical biological control against insect pests has been relatively reduced, particularly in forestry. In contrast, before 1980, relatively few pathogens were used as biological control agents, e.g. *Bacillus thuringiensis* Berliner and some viruses, and none were targeted for biological control (Fig. I.1). The number of plant diseases targeted for biological control here, and the list of potential biological control agents evaluated on each target disease, underscores the amount of research by plant pathologists that has been undertaken during the past 20 years.

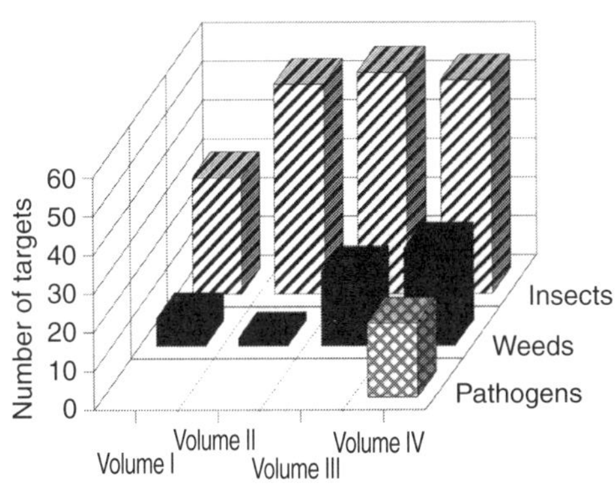

Fig. I.1. Comparison of numbers of insect, weed and pathogen targets in Volumes I–IV of *Biological Control Programmes in Canada.*

Second, chapters were added that address the important issues of invasive species and the change in philosophy of pesticide use, to provide a context for continued pursuit of biological control. Another chapter re-emphasizes the link between biological control and that most basic science, taxonomy.

Third, the treatments of pests are grouped by target host, i.e. weeds, insects and plant pathogens, because pests and biological control agents cross sectors, e.g. forestry, agriculture, health, and environment and commodity groups. For example, *Lygus* bugs are important pests of forest tree seedlings as well as many agricultural crops, and some parasitoids of the spruce budworm also attack various leaf rollers on apple trees. *Sclerotinia sclerotiorum* causes diseases in several crops, such as canola and bean, while powdery mildews, which show host-specificity based on the fungal genus and species, affect a variety of host plants, from roses to cucumber to cereals.

Apart from producing a comprehensive update of biological control programmes in Canada, a primary motivation for the project was the need to capture the collective knowledge of people who have made important contributions to biological control, before this knowledge is lost. Several projects were not completed because of the retirement or untimely death of the principal investigator. Moreover, this book illustrates the dedication of several researchers from government and universities to write up those unfinished biological control projects, despite the lack of long-term funding. Additionally, changes to the way project funding is allocated have impacted heavily on what biological control research is undertaken in Canada, and it is important to provide an updated scientific summary of the discipline as a basis for future allocation of resources. Frequently, biological control research has been conducted on a project-by-project basis, often dependent on external funding, and under conditions where infrastructure and/or resources were limited. In comparison, research on chemical pesticides has enjoyed optimum financial resources and well-coordinated research efforts.

Factors contributing to the changing emphasis in biological control include the following.

1. Significantly increased global trade, resulting in increased spread of pests (Chapter 1). Some of these are highly invasive and preventing their introduction is essential. If they do manage to establish, immediate action is essential to reduce their impact.

2. Environmental concerns, emphasizing sustainable development and biological diversity, e.g. doing adequate surveys for native natural control agents of either native or introduced pests. These surveys are essential because a pest may already be partially controlled by another organism in certain situations and it is important not to introduce exotic agents needlessly.

3. The number of candidates for introduction as biological control agents has apparently declined because the most obvious choices have been tried. Also, more detailed, basic biological studies in countries of origin are required before introductions are permitted. More biological control agents are likely to be required because of continued new pest introductions, e.g. various wood-boring beetles.

4. Changes in use patterns of chemical pesticides have encouraged development of biological control. Pressure on growers from the general public to further reduce pesticide use has increased demand for development of biological control agents, e.g. for pests of greenhouse crops, market vegetables, small fruits, ornamentals and medicinal crops. Deregistration or loss of chemical pesticides due to fewer being registered for minor use, and serious consideration by several urban municipalities to ban the use of chemical pesticides for cosmetic reasons, is prompting the necessity to consider biological controls as alternatives to chemicals.

5. Increased governmental emphasis to share research costs by establishing links with industry to develop biological control products. One limitation, however, is that industry

does not necessarily want to become engaged in developing permanent controls through release of classical agents. Instead, industry's goal is to develop products that can be sold yearly in sufficient quantities to guarantee a profit. For inundative agents, the private sector may be a very worthwhile partner, e.g. BioMal (Chapter 75), and *Trichogramma* (Chapter 12), but for classical biological control it is not. However, the industry most likely to invest in biological control products represents the small to medium-sized enterprises, not the large multinationals. The former often lack the resources and/or capital to invest in the research and development of biologically based products until they are near-market, often resulting in 'orphaned' biopesticides that may be highly effective but do not reach the marketplace.
6. Molecular biology and genetic engineering replacing organismal biology. More emphasis is placed on tinkering with known organisms, rather than studying the biology of new ones in preparation for their eventual use as biological control agents.

Worldwide, the decline in classical biological control since the 1970s has largely occurred due to a reduction in the number of specialists working in this field. In 1972, the biological control laboratory at Belleville, Ontario, closed. This laboratory had one of the largest concentrations of biological control specialists in the world. Its scientists either retired or found employment in universities and other government laboratories, not necessarily all in Canada. The decline is also partly due to a shift in emphasis to different methods of doing biological control, described above. Throughout the history of classical biological control in Canada a close link has existed between CAB *International* (formerly IIBC or CIBC) in Delémont, Switzerland, and Agriculture and Agri-Food Canada (AAFC, formerly Agriculture Canada) and Canadian Forest Service laboratories. Although funding has declined over the past two decades this close cooperation with CAB *International* continues, particularly with AAFC. Interestingly, since 1980, many general books on biological control have been published (Appendix I) but, in contrast, fewer actual field projects have been undertaken, at least in Canada.

The past two decades have experienced a greater level of activity in the evaluation and development of fungal and bacterial pathogens for inundative biological control of weeds and plant diseases. This has also stimulated new approaches to biological control, e.g. soil amendments (Chapter 102). Factors that have generated interest include organic crop production and low/no pesticide agriculture, development of resistance to chemical pesticides, e.g. resistance of grass weeds to herbicides, and de-registration of chemical pesticides by Canadian regulators. Biological controls represent the next generation of pest-control products, with potentially new modes of action aimed at controlling pesticide-resistant insects, weeds and plant pathogens. The lack of perceived success has often not been the result of poor biological control candidates, but has been most likely attributed to the inability of industry to capture the technology to bring these biological control agents to implementation. Researchers working in this area have concentrated mainly on continued screening and testing of yet another 'potential' biological control agent, while neglecting the tools required to evaluate their efficacy in the field. Significant advances in fermentation and formulation technology are now facilitating the development of biological control agents towards the product-development phase. Although biological control must be evaluated on its own merits, in reality, producers make similar comparisons in efficacy and cost to chemical pesticides. It will be necessary to educate the public, producer, industry and pesticide regulators on the merits of biological control and the tangible and intangible benefits that biological control technologies can offer the consumer.

Notable biological control successes over the past 20 years are purple loosestrife (Chapter 74), mountain-ash sawfly (Chapter 46), birch leafminer (Chapter 25), greenhouse aphids (Chapter 9), *Sphaerotheca* and *Erysiphe* powdery mildews using Sporodex® (Chapter 100). These projects resulted in complete control in some areas. Others have resulted in successful establishment of agents, e.g. birch leafminer (Chapter 25), European apple sawfly (Chapter 28), hemlock looper (Chapter 30), wheat midge (Chapter 50), leafy spurge (Chapter 69) and dalmation toadflax (Chapter 72). Yet others have resulted in development of cost-effective inundative products, e.g. *Chondrostereum purpureum* is being registered in Canada and the USA under the tradename Chontrol®, by MycoLogic (Victoria, British Columbia) for control of stump sprouting and re-growth of alder, birch and poplar in utility rights-of-way and forest vegetation management (Dr W.E. Hintz, MycoLogic Inc., personal communication; Chapter 59).

Important spin-offs resulting from practical biological control include a better knowledge of Canada's fauna and flora, and an overall increase in our level of biological knowledge of target species and their natural enemies. This information can be used in other ways, e.g. in integrated pest management, conservation and environmental studies. Further, suppliers of biological control agents have become well established in Canada (Appendix II), providing safe, effective agents. The challenge now is to develop multidisciplinary teams of researchers, e.g. entomologists, pathologists, weed scientists, taxonomists, ecologists and agrologists, plant and microbial physiologists, etc., working on similar targets/biological control agents to advance some of the more promising projects. Researchers must also be diligent in selecting the most appropriate biological control approach, e.g. classical versus inundative, based on the target pest and the needs of the farmer. For example, inundative biological control agents may be more appropriate when the level and speed of pest control are critical for minimizing yield loss, while classical control approaches may be utilized when ecologically sound pest-management options are more appropriate and economical.

Biological control in Canada is thriving, albeit in ways different from the past. Although some projects will be completely successful, there are failures in the sense of lack of pest control below economic levels. When all projects are taken together, however, there are clear economic benefits that justify continued support for the science. Biological control is also a pest-control option that has important environmental benefits. The future for biological control in Canada is therefore promising.

Acknowledgements

Preparation of this book was only possible through the hard work of the many authors who contributed to it. Their willing and patient cooperation is greatly appreciated and they deserve any accolades. Any errors or omissions are the responsibility of the editors. We especially thank John Bissett, Stephen Darbyshire and Michael Sarazin for their careful checking of scientific names in the index and reference citations throughout the text. The readily available taxonomic expertise at the Eastern Cereal and Oilseed Research Centre greatly facilitated the process of verifying scientific names in a diversity of taxa, and their willingness to help is greatly appreciated.

Publication costs were shared between the Canadian Forest Service and Agriculture and Agri-Food Canada. The directors of AAFC Research Centres at Summerland, Lethbridge, Saskatoon, Harrow, London and Ottawa, and CFS Forestry Centres at Fredericton, Sainte Foy, Sault Ste Marie, Edmonton and Victoria, and CFS Science Branch (Catherine Carmody, in particular) in Ottawa are thanked for their support.

1 Invasive Species and Biological Control

D.J. Parker and B.D. Gill

Introduction

Invasive alien species are those organisms that, when accidentally or intentionally introduced into a new region or continent, rapidly expand their ranges and exert a noticeable impact upon the resident flora or fauna of their new environment. From a plant quarantine perspective, invasive species are typically pests that cause problems after entering a country undetected in commercial goods or in the personal baggage of travellers. Under the International Plant Protection Convention (IPPC), 'pests' are defined as 'any species, strain or biotype of plant, animal or pathogenic agent injurious to plants or plant products', while 'quarantine pests' are 'pests of economic importance to the area endangered thereby and not yet present there, or present but not widely distributed and being officially controlled' (FAO, 1999).

Origins

Traditionally, most invasive pests in North America came from Europe, reflecting trading patterns of the past 500 years (Mattson *et al.*, 1994; Niemela and Mattson, 1996). Vast numbers of weeds, phytophagous insects and stored products pests arrived as stowaways in cargo or on horticultural products exported from Europe. In Canada, 881 exotic plants have become established, representing 28% of the total flora (Heywood, 1989). A diversity of soil-dwelling insects arrived in the ballast of ships (Lindroth, 1957; Sadler, 1991), until this pathway was inadvertently curtailed when soil ballast was replaced by water ballast, favouring aquatic invaders (Bright, 1999). While the rate of introductions has increased greatly over the past 100 years (Sailer, 1983), the period 1981–2000 has seen political and technological changes that may unleash an even greater wave of invasive species. The collapse of the former Soviet Union and China's interest in joining world trade have opened up new markets in Asia. These vast areas, once isolated, can now serve as source populations for additional cold-tolerant pests, e.g. the Asian longhorned beetle, *Anoplophora glabripennis* (Motschulsky), and the lesser Japanese tsugi borer, *Callidiellum rufipenne* (Motschulsky). Examples of a few insects introduced to Canada since 1981 include apple ermine moth, *Yponomeuta malinellus* Zeller, European pine shoot beetle, *Tomicus piniperda* (L.), leek moth, *Acrolepiopsis assectella* (Zeller), cherry bark tortrix, *Enarmonia formosana* (Scopoli), and the yellow underwings, *Noctua pronuba* (L.) and *Noctua comes* (Hübner).

Canada is no longer susceptible to invasion of pests from temperate locations only. Cultivation under glass, currently about 1470 ha (K. Fry, Vegreville, 2000, personal communication), is expanding rapidly and there is growing concern about possible introductions from tropical and subtropical regions that may adversely affect horticultural plants and greenhouse vegetable production. Recent introductions have included western flower thrips, *Frankliniella occidentalis* (Pergande), to eastern Canada, sweetpotato whitefly, *Bemisia tabaci* (Gennadius), and leafminers, *Liriomyza* spp. Other technological advances that facilitate the movement of pests are the greatly increased volume of traffic and increased speed of transport of commodities and people around the world. Container ships cross the oceans in record time, offloading their sealed containers directly on to railcars that are promptly delivered to the heart of the continent. Many hitch-hiking species now arrive alive instead of dying in transit. Finally, establishment of the World Trade Organization, which promotes expansion of global trade, both in volume and extent, is sure to increase the problem. All of these factors point towards invasive species or 'biological pollution' as being a major threat to the biodiversity and the economic health of North America (Office of Technology Assessment, 1993; Wallner, 1996; Bright, 1999).

Costs

The costs of invasive organisms are difficult to estimate. A report on harmful, non-indigenous species in the USA estimated that losses from invasive pests between 1906 and 1991 amounted to US$97 billion (Office of Technology Assessment, 1993). Insects accounted for $92 billion of this amount. Pimentel *et al.* (2000) have estimated that the economic and environmental losses due to non-indigenous species in the USA, combined with their control costs, amount to US$137 billion per year. While the values of control costs and economic losses can be estimated with a fair degree of precision, the damaging cost to the environment through habitat loss or species extirpation (even extinction) due to invasive organisms cannot be estimated in monetary terms. In the words of Pimentel *et al.* (2000), 'the true challenge for the public lies not in determining the precise costs of the impacts of exotic species but in preventing further damage to natural and managed ecosystems caused by non-indigenous species'.

Regulations

Alien species may cause economic damage to plants or plant products and are therefore regulated as quarantine pests. Biological control agents can also be considered as invasive species. In this case, the invasive species are intentionally introduced to reduce problems caused by foreign or native pests.

Since the enactment of the Destructive Insect and Pest Act (DIP 1912), 'an act to prevent the introduction or spreading of insects, pests and diseases destructive to vegetation', the Federal government has been charged with protecting Canada's plant resources from invasive plant pests. Under the current Plant Protection Act, the Plant Health and Production Division of the Canadian Food Inspection Agency regulates the importation of plants. In the past, plants were regulated on the basis of their role as carriers of diseases and pests and not in terms of their potential invasiveness or weediness. Although most of the weeds causing problems in agriculture and natural environments today were introduced into Canada well before the Destructive Insect and Pest Act of 1912, some sanctioned introductions of exotic (non-indigenous) agricultural, horticultural and ornamental plants have indeed become invasive (e.g. purple loosestrife, *Lythrum salicaria* L.; European buckthorn, *Rhamnus cathartica* L.; Norway maple, *Acer platanoides* L.; and Russian olive, *Elaeagnus angustifolia* L.). All importations of exotic plants should undergo a risk assessment, both for their potential to harbour pests and diseases, and to determine their potential invasiveness in natural and disturbed habitats.

The same legislation that is used to exclude exotic plant pests has also been used to regulate the importation of plant pests for biological control. The Act has been amended several times (DIP, 1954; Plant Quarantine Act, 1969; Plant Protection Act, 1990) and the regulations have been modified to reflect changes in pest and disease conditions in Canada and throughout the world. While the definition of a pest in the legislation has changed over the years, permits for the introduction of foreign biological control agents have been issued under the authority of the

Plant Protection Act and its predecessors for about 90 years. Biological control organisms that attack plants are strictly regulated and cannot be released into the environment until they have successfully passed a pest risk assessment. This is carried out by Canadian Food Inspection Agency (CFIA) entomologists, assisted by the Biological Control Review Committee (BCRC) of Agriculture and Agri-Food Canada in consultation with the United States Department of Agriculture, Animal and Plant Health Inspection Service (USDA-APHIS) and their review panel, the Technical Advisory Group (TAG). Most releases have been of phytophagous agents for the control of exotic weeds (classical biological control). Entomophagous biological control organisms are regulated with regard to their potential to be indirectly injurious to plants, because plant pests are loosely defined under the Act. Recently, attempts have been made to formalize the review of entomophagous insect petitions for biological control by developing guidelines and protocols for import and release. The North American Plant Protection Organization (NAPPO) has developed information guidelines, i.e. standards for the import and release of phytophagous and entomophagous biological control organisms. Since intentional introductions have the potential to affect ecosystems in Mexico, USA and Canada, it is important that all three countries are aware of planned introductions and participate in the petition review process. Commercial entomophagous biological control organisms are regulated in much the same way as classical agents. Species that have a history of importation without negative effects, e.g. predacious mites, are admitted under permit (see Appendix II). Random audits of commercial agents are made to determine species purity. New, exotic commercial agents for inundative release in greenhouses and interior landscapes must be reviewed by the BCRC and the regulatory entomologists of the CFIA. Microbial biological control agents are regulated by the Pest Management Regulatory Agency (PMRA).

Exotic Introductions and Classical Biological Control

As regulators, it is our responsibility to review import applications and to issue permits and conditions for all insects, mites and terrestrial molluscs entering Canada. Our legislative mandate is to prevent the introduction and spread of exotic plant pests. We also assess petitions for the importation and release of non-indigenous agents for the classical biological control of introduced weeds and plant pests. Balancing these two, often contradictory, viewpoints is difficult. Classical biological control is only one technique of integrated pest management. Augmentation of numbers of existing natural enemies, conservation of habitats for predators and parasites, crop rotation, diversification, as well as the more conventional chemical methods may be as important to successful farming and forestry as is classical biological control. The challenge facing scientists and regulators alike will be to ensure that classical biological control is safe for non-target organisms. This will require more effort and research in host-specificity testing and in trophic-level interactions, particularly with entomophagous agents. Through guidelines, research and review committees, the few classical introductions that occur each year in Canada are being assessed more thoroughly than ever before. But problems are fast approaching. The continued erosion of taxonomic support in Canada will make the practice of classical biological control very dangerous. Without accurate names on organisms or access to taxonomists who can authoritatively identify them, the science of classical biological control will cease to be a safe and effective component of integrated pest management. In this case, regulators will be given little choice but to deny introductions.

References

Bright, C. (1999) Invasive species: pathogens of globalization. *Foreign Policy* 116, 50–64.

FAO (Food and Agriculture Organization of the United Nations) (1999) *Glossary of Phytosanitary Terms.* Secretariat of the International Plant Protection Convention. International Standards for Phytosanitary Measures, Rome, Publication No. 5.

Heywood, V.H. (1989) Patterns, extents and modes of invasions of terrestrial plants. In: Drake, J.A., Mooney, H.A., di Castri, F., Groves, R.H., Kruger, F.J., Rejmanek, M., and Williamson, M. (eds) *Biological Invasions: a Global Perspective.* John Wiley and Sons, New York, New York, pp. 31–60.

Lindroth, C.H. (1957) *The Faunal Connections Between Europe and North America.* John Wiley and Sons, New York, New York.

Mattson, W.J., Niemela, P., Millers, I. and Inguanzo, Y. (1994) *Immigrant Phytophagous Insects on Woody Plants in the United States and Canada: An Annotated List.* United States Department of Agriculture-Forest Service, North Central Forest Experiment Station, General Technical Report NC-169.

Niemela, P. and Mattson, W.J. (1996) Invasion of North American forests by European phytophagous insects – legacy of the European crucible? *BioScience* 46, 741–753.

Office of Technology Assessment (1993) *Harmful Nonindigenous Species in the United States.* OTA-F-565, United States Congress, Washington, DC.

Pimentel, D., Lach, L., Zuniga, R. and Morrison, D. (2000) Environmental and economic costs of non-indigenous species in the United States. *BioScience* 50, 53–65.

Sadler, J. (1991) Beetles, boats and biogeography: insect invaders of the North Atlantic. *Acta Archaeologica* 61, 199–211.

Sailer, R. I. (1983) History of insect introductions. In: Wilson, L. and Graham, C.L. (eds) *Exotic Plant Pests and North American Agriculture.* Academic Press, New York, New York, pp. 15–38.

Wallner, W.E. (1996) Invasive pests ('biological pollutants') and US forests: whose problem, who pays? *European Plant Protection Organization Bulletin* 26, 167–180.

2 Pesticides and Biological Control

K.D. Floate, J. Bérubé, G. Boiteau, L.M. Dosdall, K. van Frankenhuyzen, D.R. Gillespie, J. Moyer, H.G. Philip and S. Shamoun

Introduction

Synthetic organic pesticides are the primary method of control for weeds, insects and pathogens. In Canada, sales of these products exceeded Can$1.4 billion in 1998, primarily for herbicides applied to cereal and oilseed crops (Figs 2.1 and 2.2; Anonymous, 1998). Historically, use of these pesticides has been marked by constant change. For example, the discovery of the insecticidal properties of DDT in 1939 was followed by the development of organochlorine-, carbamate- and organophosphorus-based insecticides in the 1940s and 1950s. Use of synthetic pyrethroids and macrocyclic lactones became widespread in the 1980s and 1990s. Most

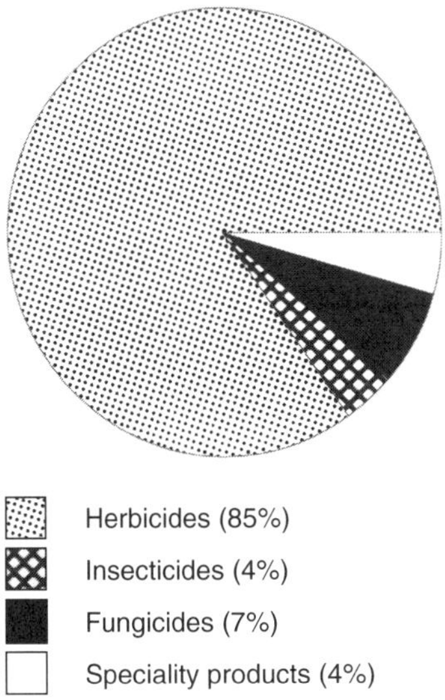

Fig. 2.1. Percentage sales in 1998 by product group.

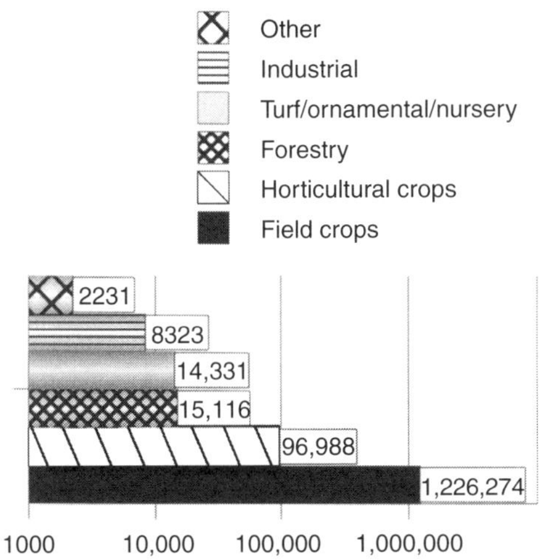

Fig. 2.2. Pesticide sales ($1000s) in 1998.

recently, pesticidal proteins have been genetically engineered into crop varieties.

The continuous development of new pesticides reflects two main factors: firstly, a desire to replace existing products with products having greater target specificity, reduced environmental persistence and lower mammalian toxicity; and secondly, the need to find alternatives to products that are no longer effective due to the development of pesticide resistance. Such resistance has been reported for target populations of weeds, plant pathogens and, in particular, insects and mites. This problem is compounded when resistance to one product confers resistance to other products in the same chemical class and/or to products in a different chemical class.

While recognizing the tremendous benefits of pesticides in modern agriculture, concerns of non-target effects and efficacy will continue to affect usage patterns. The Food Quality Protection Act (1996) in the USA (Anonymous, 1999a) requires the reassessment of all carbamate and organphosphate insecticides by 2006 for compliance with a new standard: reasonable certainty that no harm will result from aggregate exposure to each pesticide from dietary and other sources. The Pest Management Regulatory Agency in Canada is reviewing all pesticides registered prior to 31 December, 1994 (74% of the 550 currently registered active ingredients) to stay current with the reassessment under way in the USA.

Because chemical and biological control are frequently, but mistakenly, viewed as competing methods of pest control, historical emphasis on developing new pesticides has hampered the growth of the biological control industry. We review briefly how changes in pesticide use during the past 20 years have affected biological control research and implementation in Canada.

Herbicides

The first herbicides, 2,4-D (2,4-dichlorophenoxyacetic acid) and MCPA (4-chloro-2-methylphenoxyacetic), were marketed in 1946. By 1995, more than 300 herbicides were listed in *Weed Abstracts* with global sales exceeding US$12 billion (Casely, 1996). Their widespread adoption provided farmers with a degree of weed control that increased crop yields to levels not previously possible.

However, use of herbicides is not without problems. In western Canada there is intensive, continuous cropping, primarily with rotations of wheat, *Triticum aestivum* L.; barley, *Hordeum vulgare* L.; and canola, *Brassica napus* L. and *B. rapa* L. These cropping systems typically rely on frequent applications that select for herbicide resistance. For populations of wild oat, *Avena fatua* L., in Alberta, Saskatchewan and Manitoba, Beckie *et al.* (1999) reported resistance to acetyl-CoA carboxylase inhibitor herbicides (Group 1) in more than half of the fields surveyed, the frequent occurrence of multiple-group resistance, and discovery of four populations resistant to all herbicides registered for use in wheat. Resistance to one or more herbicide classes has been reported for populations of seven broadleaf weed species on the Canadian prairies in the past decade (Beckie *et al.*, 1999).

Applications of herbicides also introduce chemical residues into the environment, with undetermined consequences. Harker and Hill (1997) reported low levels of herbicide residues in a majority of shallow groundwater samples recovered in Alberta, with concentrations in some samples exceeding the guidelines for drinking water. Herbicides such as 2,4-D, bromoxynil and dicamba frequently are present in rainfall at concentrations that may have adverse effects on sensitive species of plants and on the quality of surface water (Hill *et al.*, 1999). Partially because of these concerns, 2,4-D and other herbicides are being re-evaluated (Anonymous, 1999b). The potential removal of 2,4-D from the market is of particular concern, because it remains efficacious at a time when weeds are becoming resistant to newer herbicides with narrower modes of action.

The most significant development in recent years has been the release of crop varieties genetically engineered for herbicide tolerance. These varieties are very attractive to industry, because they increase the versatility of existing products, i.e. popular, non-selective herbicides can be now used in major crops. This technology provides both benefits and detriments to the farmer (Marshall, 1998). The use of non-selective herbicides such as glyphosate for in-crop weed control reduces the use of residual herbicides for weed control in crops such as corn, *Zea mays* L., and canola. However, the 'volunteer' offspring of herbicide-tolerant versus conventional varieties are more difficult to control. In addition, cross-fertilization can transfer traits for herbicide tolerance to conventional varieties or to closely related species of weeds, to produce populations of plants resistant to one or more groups of herbicides. In Alberta, cross-fertilization among transgenic varieties has been implicated in the discovery of canola plants with resistance to imidazolinone, glufosinate and glyphosate (Hall *et al.*, 2000).

Biological methods of control most frequently target weeds of rangeland and permanent pastures, where widescale application of herbicides is not cost effective and where there is a greater risk of adversely affecting non-target species than in intensively managed cropland. More than 70 exotic arthropod species have been released in Canada since 1952 as biological agents to control 21 weed species. Mycoherbicides are another method of biological control against weeds, particularly in forests being managed for desirable species (Wagner, 1993; Shamoun, 2000).

Vegetable Crops

The history of control for Colorado potato beetle, *Leptinotarsa decemlineata* (Say), on potato, *Solanum tuberosum* L., illustrates the general pattern of pesticide use for control of vegetable pests. Demand for high quality, abundant and inexpensive potatoes has promoted use of insecticides despite repeated development of insecticide resistance by *L. decemlineata*. Hence, control of the beetle is possible only because new insecticides are being registered as current products become ineffective. One positive consequence of this process is the development of insecticides that are kinder to the environment, to the applicator, and to the consumer. Nevertheless, declining efficacy of registered products and the de-registration of still effective insecticides for environmental reasons (e.g. aldicarb in 1991) stimulated research on non-chemical control methods.

Foliar sprays of the bacterium *Bacillus thuringiensis* (Berliner) (*B.t.*) were introduced in 1991 using existing spray technologies. Although very effective, adoption of *B.t.* was hindered by its high cost, an effectiveness limited to early instar larvae, the need for repeated applications, low residual toxicity and an inability to stick to plants during rain. Predators and parasitoids showed promise for control of *L. decemlineata* in small-scale field studies, but problems associated with handling, storage and application prevented their commercialization. Further efforts to develop biological, cultural and mechanical methods of control were stymied by the registration of the insecticide imidacloprid in 1994. Imidacloprid was immediately adopted by potato growers, which greatly reduced demand for alternative control methods.

The most recent development for control of *L. decemlineata* has been transgenic potatoes that express insecticidal proteins, e.g. NewLeaf, first registered in 1996. Initially well received, subsequent demand has slowed because the varieties are expensive and growers must sign agreements that restrict farming practices. In addition, ongoing controversy regarding potential risks of transgenic varieties to human health and to the environment has increased market uncertainty (Dean, 2000).

Ultimately, insect pest control in vegetable crops requires a strategy of integrated pest management (IPM), including biological control. Boiteau and Osborn (1999) showed that IPM was effective in preventing economic losses to potato by *L. decemlineata*, at a cost only 1.6–3.9 times higher than that of the conventional insecticide-based strategy. Non-chemical methods of control at field perimeters are already used, e.g. plastic-lined trenches, flaming, vacuuming and border spraying of biological insecticides. Other vegetable crops, particularly root crops where availability of synthetic insecticides is negligible, provide an even stronger rationale for IPM use.

Greenhouse Crops

The Canadian greenhouse vegetable industry once relied heavily on pesticides, but now an estimated 30 biological control agents are used to manage about 15 pest species. Pesticide resistance in greenhouse whitefly, *Trialeurodes vaporariorum* Westwood, and twospotted spider mite, *Tetranychus urticae* (Koch), forced adoption of biological control in greenhouses around the world in the late 1970s (van Lenteren and van Woets, 1988).

Support for biological control was reinforced following a pesticide-related food safety case in British Columbia. Misapplication of aldicarb to a cucumber crop caused serious illness in consumers of the treated produce (*The Vancouver Sun*, 3 June, 5 June, and 6 June, 1985). The grower was convicted under the Pest Control Products Act and the Food and Drug Act (*MacLean's*, 27 April 1987, p. 34). The negative publicity forced the industry to re-examine its reliance on chemical pesticides. It now promotes biological control and IPM standards as components of produce quality and enforces compliance among growers.

Another factor favouring adoption of biological control is that resistance to new insecticides and miticides usually has developed in pest populations elsewhere before these new products receive registration for use by the Canadian greenhouse vegetable industry. The pyrethroid insecticide permethrin was registered for greenhouse use in 1982 but resistance among *T. vaporariorum* populations was universal in British Columbia by 1985. The implication is that resistance was already present in *T. vaporariorum* populations that had been imported on plant stock from elsewhere.

Differences in pesticide registrations between Canada and the USA further strengthen support for biological control. Fenbutatin-oxide is registered in Canada to control *T. urticae* on tomato, pepper and cucumber, but is not registered for use in the USA. Hence, produce with fenbutatin-oxide residues cannot be sold in the USA. To retain this major market, the British

Columbia greenhouse industry now relies almost exclusively on biological agents to control *T. urticae* on tomato, *Lycopersicon esculentum* L.

Adoption of bumble bees, *Bombus* spp., in the late 1980s to pollinate greenhouse tomatoes also increased reliance on biological control. Bumble bees are cheaper and more effective than hand pollination of flowers, but they are sensitive to many insecticides. Hence, when *Bombus* spp. are present, either pesticide applications in greenhouses must be avoided or the bees removed prior to applications. The latter is expensive and is possible only for pesticides with short residual toxicities. The industry has responded by promoting the use of selective pesticides with short residual times and by maintaining its reliance on biological control.

A steady increase in the number of pest species attacking tomato, pepper and cucumber has occurred since 1980. Because biological controls and IPM programmes are not available for new pests when they first occur, control may rely initially on registered broad-spectrum pesticides that are generally incompatible with use of biological control agents. Hence, the industry promotes the registration of pesticides having minimum impact on natural enemies to supplement ongoing efforts to develop biological controls for new pests.

Field Crops

Chemical control of insect pests in field crops during the past 20 years has shifted from reliance on products with relatively low activity, e.g. azinphos-methyl, methomyl and methamidophos, to compounds with high activity requiring less product per unit area, e.g. cyhalothrin-lambda and deltamethrin, but that nevertheless have broad-spectrum activity on contact with both target and non-target species. There is little pressure to reduce reliance on chemical controls, because resistance development is uncommon for insect pests of field crops. With relatively short growing seasons and sporadic occurrence of pests, repeated application of insecticides, even against multivoltine species, is rare within years, and few pest species are repeatedly treated with insecticides in consecutive years.

Improved delivery of insecticides, e.g. by adjusting spray angle and nozzle type for low volume, uniform coverage of the crop canopy, has reduced drift and minimized harmful effects on beneficial species (e.g. Elliott and Mann, 1997). Insecticidal seed coatings rather than foliar sprays for controlling pests such as flea beetles, *Phyllotreta* spp., and wireworms (Elateridae) have been adopted. Seed treatments deliver less insecticide per unit area, specifically target the pest species and are generally safer to apply. With Canada's participation in an international protocol to restrict or eliminate persistent organic pollutants that contribute to transboundary pollution, the most widely used insecticidal seed treatment for insect pest control in Canada (lindane) is being replaced by compounds considered less environmentally damaging.

Attempts to implement classical biological control for insect pests of field crops in Canada are hindered by the instability of annual cropping systems (Turnock, 1991). Perhaps the greatest innovations in biological control have been achieved with microbial agents, especially the entomopathogens *Nosema locustae* Canning and *Beauvaria bassiana* (Balsamo) Vuillemin for grasshopper control (Johnson, 1997). The efficacy of these agents has improved steadily, but adoption has been hindered by low infection rates, environmental constraints and the availability of cheaper chemical products. Foliar sprays of *B.t.* have not been used extensively in field crops. Bertha armyworm, *Mamestra configurata* Walker, larvae are naturally resistant to commercial formulations of *B.t.* (Morris, 1986) so control has relied on chemical sprays. Transgenic varieties of canola that express the gene for producing *B.t.* delta endotoxin are being developed to control diamondback moth, *Plutella xylostella* (L.), and flea beetles. Transgenic *B.t.* corn resistant to

European corn borer, *Ostrinia nubilalis* Hübner, is available commercially. Spray formulations of *Nucleopolyhedrovirus* and synthetic sex pheromones are being developed to control lepidopteran pests.

Implementation of the Food Quality Protection Act in the USA will have a major impact on insecticide use in Canadian field crops because so many of our field crop commodities are exported to the USA. The de-registration of some insecticides currently used will likely increase demand for new biological control agents for use in IPM programmes.

Tree Fruits

Before 1980, insect and mite pests of tree fruits were managed primarily by four synthetic pyrethroids, eight organophosphates, six carbamates, three organochlorines and four miscellaneous chemistries for mites. Development of resistance to these products by tentiform leafminers, *Phyllonorycter blancardella* (Fabricius) and *P. mespilella* (Hübner), Oriental fruit moth, *Grapholita molesta* (Busck), obliquebanded leafroller, *Choristoneura rosaceana* (Harrison), and pear psylla, *Cacopsylla pyricola* Förster (Croft *et al*, 1989; Anonymous, 1999c, d), reinforced the already active promotion of IPM to reduce reliance on insecticides.

The successful implementation of biological control of resistant mites in the late 1960s and 1970s demonstrated that preservation of natural enemies can maintain pest populations below action thresholds. Research and extension efforts in British Columbia and Ontario emphasized the preservation of natural enemies to manage pear psylla by reducing application rates or by replacing existing products with products less harmful to natural enemies. It was during this period that the use of *B. thuringiensis* serovar *kurstaki* (*B.t.k.*) increased to control lepidopteran pests resistant to organophosphate and synthetic pyrethroid insecticides. New products were developed with acceptable or no impact on important insect and mite predators and parasitoids – avermectin, pyridaben and amitraz (pear psylla and mites), clofentezine (mites), tebufenozide (Lepidoptera) and imidacloprid (leafminers and sap-sucking insects). Other, new, active ingredients expected to be registered include spinosad (Lepidoptera, leafminers, thrips), indoxacarb (Lepidoptera), thiamathoxam (sap-sucking insects), bifenazate (mites) and acetamiprid (sap-sucking insects).

The use of sex pheromones to disrupt mating in Lepidoptera is increasing (Evenden *et al.*, 1999a, b). The combination of a sex pheromone with an insecticide (a formulation termed an attracticide) is being developed to attract and kill male codling moths, *Cydia pomonella* (L.) (Charmillot *et al.*, 2000). Expanded research and development of host-derived semiochemicals will allow monitoring of females and improve the usefulness and performance of current semiochemical-based control tactics. The adoption of these tactics in combination with 'softer' control products will greatly enhance the opportunity to develop and implement more biological control-based pest management programmes.

Biological control of fruit tree diseases promises to reduce the need for multiple weekly applications of chemical fungicides (Bernier *et al.*, 1996).

Forests

Prior to the North American commercialization of *B.t.k.*, forest protection programmes were characterized by extensive use of synthetic insecticides to control defoliating Lepidoptera. In 1960, the Canadian Forest Service conducted the first experimental aerial applications of *B.t.k.* (Thuricide®, Bioferm Corporation) against spruce budworm, *Choristoneura fumiferana* (Clemens). New formulations based on the HD-1 *kurstaki* isolate generally improved field efficacy during the 1970s. Cost effectiveness improved in the late 1970s with advances in production and formulation technologies. By the end of that decade, *B.t.k.* was generally considered an operational alternative for control of *C. fumiferana*. However, its use was limited because of inconsistent results and treatment costs that exceeded those of synthetic insecticides (Smirnoff and Morris, 1984).

The past two decades have been characterized by the rapid replacement of synthetic insecticides with commercial *B.t.k.* products to control *C. fumiferana* (van Frankenhuyzen, 1990). Operational use for control of this pest increased from less than 5% of the total area sprayed in the early 1980s to 50–65% by the mid-1980s. This increase was due primarily to a political decision by various provinces to curb aerial application of synthetic insecticides in public forests in response to growing public opposition and environmental concerns. However, limited operational use prior to that decision had catalysed significant cost reductions and critical improvements in the formulation and application of *B.t.k.* The trend of the late 1970s to increase product potency continued in the 1980s. New high-potency formulations were designed for undiluted (neat) application in ultra-low volumes (ULV). By the mid 1980s, it was possible to apply the recommended dosage rate of 30 billion (10^9) international units (BIU) ha^{-1} in application volumes as low as 2.4 litres. Low spray volumes increased spray plane work rates, while the higher product potency increased efficacy and reliability of control operations. By the mid-1980s, these improvements, together with the shift in political climate that favoured the use of biological control, resulted in the widespread acceptance of *B.t.k.* as a fully operational, and often the only available, option for control of *C. fumiferana* and of gypsy moth, *Lymantria dispar* L., and other forest defoliators.

Recent developments also promise a role for biological control in the management of tree pathogens. Already operational for foresters in Europe, a formulation of the fungus *Phlebiopsis gigantea* (Fries) Julich is being developed in Canada to control Annosus root rot (Bussières *et al.*, 1996). Mycoviruses, *Hypovirus* spp., are being studied as a biological control for the causative agent of chestnut blight (Baoshan *et al.*, 1994). Foliar fungal endophytes show promise for control of tree pathogens, including white pine blister rust, *Cronartium ribicola* Fischer (Bérubé *et al.*, 1998). Biological control of pathogens, e.g. Scleroderris canker, *Gremmeniella abietina* (Lagerberg) Morelet, stem rusts, *Cronartium comandrae* Peck, Dutch elm disease, *Ophiostoma ulmi* (Buisman) Nannfeldt, beech bark disease, *Nectria coccinea* (Persoon: Fries) Fries var. *faginata* Lohman, Watson and Ayers, and Septoria canker, *Mycosphaerella populorum* G.E. Thompson, may be attainable in the coming decades.

Livestock

Since 1980, changes in the livestock industry have reflected the introduction of synthetic pyrethroid and macrocyclic lactones into the Canadian market. In 1978, 12 of the 21 chemicals available to livestock producers were organophosphates with the remainder being carbamates, organochlorines, botanicals and sulphur (WCLP, 1978). In 1999, 27 chemicals were available to producers, of which 11 were organophosphates, four were synthetic pyrethroids and five were macrocyclic lactones (WCLP, 1999). The newer insecticides provided alternatives to organophosphates, carbamates and organochlorines at a time when resistance to these compounds was becoming a problem. Harris *et al.* (1982) reported multiple resistance within populations of house fly, *Musca domestica* L., to more than 20 carbamate, organochlorine and organophosphate insecticides, and showed that this pest quickly developed resistance to synthetic pyrethroids.

Insecticidal ear tags, first registered in Canada in 1981, combine a plastic matrix impregnated with active ingredients, usually an organophosphate and/or synthetic pyrethroids, that are slowly released on to the treated animal. Ear tags provided an effective method of season-long control of horn fly, *Haematobia irritans* (L.), with a single application. By 1987, however, resistance of *H. irritans* to synthetic pyrethroid ear tags had been reported in Manitoba (Mwangala and Galloway, 1993), Alberta and British Columbia (D. Colwell, Lethbridge, 2000, personal communication), and to both synthetic pyrethroid and organophosphate ear tags in Ontario (Surgeoner *et al.*, 1996). Ear tags with both synthetic pyrethroid and organophosphate components are used to manage *H. irritans* populations resistant to one, but not both, insecticide types.

Ivermectin, the first macrocyclic lactone registered in Canada, was quickly adopted by producers because a single application controls both internal parasites, e.g. nematodes (Nematoda) and cattle grubs, *Hypoderma* spp., and external parasites, e.g. lice (Anoplura) and ticks (Ixodoidea), providing a significant advantage over other products. Four additional macrocyclic lactones have been registered since 1995. Macrocyclic lactones are effective for control of several arthropods affecting livestock, but there is at least one report of ivermectin resistance in house fly populations (Pap and Farkas, 1994).

Black fly (Simuliidae) control illustrates how reliance on insecticides has hindered implementation of biological controls. Initially, biological controls were not considered because cheap and effective insecticides were available. Hence, although the insecticidal properties of *B.t.* were reported in 1902, isolation of a strain, *B. thuringiensis* serovar *israelensis* (*B.t.i.*), toxic to Simuliidae did not occur until 1978 (Lacey and Undeen, 1986). DDT (dichlorodiphenyltrichloroethane) was used as a larvicide until banned in Canada in 1970 because of its environmental persistence. Its replacement, methoxychlor, was used as a larvicide in western Canada from 1969 to 1988 (P. Mason, Ottawa, 2000, personal communication), when its use was banned because of its broad-spectrum activity (Dosdall and Lehmkuhl, 1989). These and other concerns rekindled interest in *B.t.i.*, which waned with the introduction of synthetic pyrethroids as adulticides, e.g. in ear tags and self-application devices. When synthetic pyrethroid resistance occurred, development of biological controls was again emphasized (Watkinson, 1994). Control of Simuliidae now relies exclusively on treating rivers and streams with *B.t.i.*, now the larvicide of choice.

Biological Control in the New Millennium

This synopsis of pesticide use identifies a common theme. Over-reliance on synthetic chemicals leads to development of pesticide resistance by the target species. Pesticide resistance generates support for biological controls that wanes when new synthetic pesticides become available. This cycle of chemical dependency exists until external factors force consideration of alternative control methods. When sustained support for biological control has been provided, researchers either have developed economically viable methods or have made significant progress towards this objective. Based on changes in pesticide use during the past 20 years, we forecast the following for biological control.

Demand by consumers for inexpensive food coupled with a drive to maximize profit margins for producers and manufacturers will ensure that pesticides remain the primary method of pest control in large-scale crop production in the early part of the new millennium. Synthetic chemicals provide the most economical method of pest control, particularly in large-scale agricultural settings where they are easy to apply, effective, fast-acting and relatively inexpensive. However, the realization that pesticides have 'hidden' costs to the environment and to human health will continue to pressure private industry to develop safer pest control methods.

Private industry – not necessarily producers or the general public – will increasingly dictate the direction of biological control research. Government laboratories traditionally have developed biological methods of control to benefit producers and, indirectly, the general public. Adoption by industry of methods that could be commercialized will increase. This framework shifted in the 1980s when government laboratories became more reliant on industry for research support. The shift has accelerated commercialization of biological agents, e.g. mycoherbicides, *B.t.* formulations, that benefit industry, producers and the general public. However, this emphasis has reduced funds for research on biological controls that primarily benefit producers and the general public, e.g. classical biological control of weeds, by providing longer-term pest suppression while reducing control costs. Nevertheless, classical biological control will remain a strong option for control of exotic species of weeds.

Cultivation of transgenic crops will promote research on biological controls of arthropod pests. Transgenic crops with insecticidal proteins only affect insects that feed on plant tissues. Hence, use of biological agents is more compatible with transgenic versus conventional varieties where broad-spectrum insecticides are applied. Development of resistance by the target pest to the insecticidal protein(s) in transgenic host tissue is predicted. Hence, there will be continued support for biological methods of control. Industry is likely to incorporate additional types of insecticidal proteins into transgenic varieties, rather than fund research on biological controls.

The 'organic' food industry will be a major advocate for implementation of biological control in agriculture. Fuelled by controversy regarding the safety of transgenic crops and pesticide residues, sales of organic products have increased by 25–30% per annum during the past 5 years and will continue to increase. Because national guidelines being developed for 'organic' agriculture in Canada and the USA specifically reject use of transgenic varieties and synthetic pesticides, there will be a large demand for continued research on biological controls.

Historically, the trend by industry and government researchers has been to develop pesticides and application methods of higher pest specificity and fewer adverse environmental effects. This has culminated in the development of pathogens (e.g. bacteria, fungi, and viruses) as microbial pesticides (e.g. Morris *et al.* 1986), the use of which conserves predators and parasitoids of pest species. Biological controls have been incorporated into IPM programmes to a degree that varies among commodities. The role of biological controls in IPM programmes will continue to increase in future years.

References

Anonymous (1998) *Crop Protection Institute 1998 Sales survey pest control product in Canada: report and discussion.* http://www.cropro.org/sales/sales97.htm (25 February 2000).

Anonymous (1999a) The Food Quality Protection Act (FQPA) of 1996. http://www.epa.gov/oppsps1/fqpa/index.html (18 May 1999).

Anonymous (1999b) *Discussion Paper – A New Approach to Re-evaluation.* Pesticide Regulatory Agency, Ottawa, Ontario.

Anonymous (1999c) *Fruit Production Recommendations 1998/99.* Ontario Ministry of Agriculture, Food and Rural Affairs, Toronto, Ontario.

Anonymous (1999d) *Tree Fruit Production Guide for Commercial Growers Interior Districts 1998/99.* British Columbia Ministry of Agriculture and Food, British Columbia Fruit Growers' Association, Victoria, British Columbia.

Baoshan, C., Choi, G.H. and Nuss, D.L. (1994) Attenuation of fungal virulence by synthetic infectious hypovirus transcripts. *Science* 264, 1762–1764.

Beckie, H.J., Thomas, A.G., Legere, A., Kelner, D.J., Van Acker, R.C. and Meers, S. (1999) Nature, occurrence, and cost of herbicide resistant wild oat in small grain production areas. *Weed Technology* 13, 612–625.

Bernier, J., Carisse, O. and Paulitz, T.C. (1996) Fungal communities isolated from dead apple leaves from orchards in Québec. *Phytoprotection* 77, 129–134.

Bérubé, J.A., Trudelle, J.G., Carisse, O. and Dessureault, M. (1998) Endophytic fungal flora from eastern white pine needles and apple tree leaves as a means of biological control for white pine blister rust. In: Jalkanen, R., Crane, P.E., Walla, J.A. and Aalto, T. (eds) *Proceedings of the First IUFRO Rusts of Forest Trees WP Conf., 2–7 Aug. 1998, Saariselka, Finland.* Finnish Forest Research Institute, Research Papers 712, pp. 305–309.

Boiteau, G. and Osborn, W.P.L. (1999) Conventional and IPM control of the Colorado potato beetle: summary of a three year project. In: Boiteau, G., Leblanc, J.-P.R., Osborn, W.P.L., Parsons, A.J. and Sandeson, P.D. (eds) *Assessment of Long-term Pesticide Based and Biorational Based Colorado Potato Beetle Control Programs on Potatoes 1996–1998.* Potato Research Centre, Agriculture and Agri-Food Canada, Fredericton, NB, Final Report, Potato Insect Ecology, pp. 3–13.

Bussières, G., Dansereau, A., Dessureault, M., Roy, G., Laflamme, G. and Blais, R. (1996*) Lutte contre la maladie du rond dans l'ouest du Québec. Projet No.4023, Essais, expérimentations et transfert technologique en foresterie.* Service Canadien des Forêts, Ressources naturelles Canada, Ottawa, Ontario.

Casely, J.C. (1996) The progress and development of herbicides for weed management in the tropics. *Planter* 72, 323–346.

Charmillot, P.J., Hofer, D. and Pasquier, D. (2000) Attract and kill: a new method for control of the codling moth *Cydia pomonella. Entomologia Experimentalis et Applicata* 94, 211–216.

Croft, B.A., Burts, E.C., van de Baan, H.E., Westigard, P.H. and Riedl, H. (1989) Local and regional resistance to fenvalerate in *Psylla pyricola* Foerster (Homoptera: Psyllidae) in western North America. *The Canadian Entomologist* 121, 121–129.

Dean, L. (2000) GMO at crossroads. *Spudman* 38, 34–36.

Dosdall, L.M. and Lehmkuhl, D.M. (1989) Drift of aquatic insects following methoxychlor treatment of the Saskatchewan River system. *The Canadian Entomologist* 121, 1077–1096.

Elliott, R.H. and Mann, L.W. (1997) Control of wheat midge, *Sitodiplosis mosellana* (Gehin), at lower chemical rates with small-capacity sprayer nozzles. *Crop Protection* 16, 235–242.

Evenden, M.L., Judd, G.J.R. and Borden, J.H. (1999a) Simultaneous disruption of pheromone communication in *Choristoneura rosaceana* and *Pandemis limitata* with pheromone and antagonist blends. *Journal of Chemical Ecology* 25, 501–517.

Evenden, M.L., Judd, G.J.R. and Borden, J.H. (1999b) Pheromone-mediated mating disruption of *Choristoneura rosaceana*: is the most attractive blend really the most effective? *Entomologia Experimentalis et Applicata* 90, 37–47.

Frankenhuyzen, K. van (1990) Development and current status of *Bacillus thuringiensis* for control of defoliating forest insects. *Forestry Chronicle* 66, 498–507.

Hall, L.M., Huffman, J. and Topinka, K. (2000) Pollen flow between herbicide tolerant canola (*Brassica napus*) is the cause of multiple resistant canola volunteers. In: Wilcut, J.W. (ed.) *2000 Meeting of the Weed Science Society of America. 6–9 February 2000, Toronto, Ontario, Canada.* Weed Science Society of America Abstracts, p. 48.

Harker, K.N. and Hill, B.D. (1997) Herbicide leaching into shallow groundwater. In: Wood, C. (ed.) Agricultural Impacts on Water Quality in Alberta. Alberta Agriculture Food and Rural Development, Edmonton, Alberta, pp. 58–59.

Harris, C.R., Turnbull, S.A. and Whistlecraft, J.W. (1982) Multiple resistance shown by field strains of house fly, *Musca domestica* (Diptera: Muscidae), to organochlorine, organophosphorus, carbamate, and pyrethroid insecticides. *The Canadian Entomologist* 114, 447–454.

Hill, B.D., Inaba, D.J., Harker, K.N., Moyer, J.R. and Hasselback, P. (1999) Phenoxy herbicides in Alberta rainfall: cause for concern? http://res2.agr.ca/lethbridge/posters.htm (30 May 2000).

Johnson, D.L. (1997) Nosematidae and other Protozoa as agents for control of grasshoppers and locusts: current status and prospects. *Memoirs of the Entomological Society of Canada* 171, 375–389.

Lacey, L.A. and Undeen, A.H. (1986) Microbial control of black flies and mosquitoes. *Annual Review of Entomology* 31, 265–296.

Lenteren, J.C. van and Woets, J. van (1988) Biological and integrated control in greenhouses. *Annual Review of Entomology* 33, 239–269.

Marshall, G. (1998) Herbicide-tolerant crops – real farmer opportunity or potential environmental problem. *Pesticide Science* 52, 394–402.

Morris, O.N. (1986) Susceptibility of the bertha armyworm, *Mamestra configurata* (Lepidoptera: Noctuidae), to commercial formulations of *Bacillus thuringiensis* var. *kurstaki*. *The Canadian Entomologist* 118, 473–478.

Morris, O.N., Cunningham, J.C., Finney-Crawley, J.R., Jaques, R.P. and Kinoshita, G. (1986) Microbial insecticides in Canada: their registration and use in agriculture, forestry and public and animal health. *Bulletin of the Entomological Society of Canada, Supplement* 18(2), 43 pp.

Mwangala, F.S. and Galloway, T.D. (1993) Susceptibility of horn flies, *Haematobia irritans* (L.) (Diptera: Muscidae), to pyrethroids in Manitoba. *The Canadian Entomologist* 125, 47–53.

Pap, L. and Farkas, R. (1994) Monitoring of resistance of insecticides in house fly (*Musca domestica*) populations in Hungary. *Pesticide Science* 40, 245–258.

Shamoun, S.F. (2000) Application of biological control to vegetation management in forestry. In: Spencer, N.R. (ed.) *Proceedings of the X International Symposium on Biological Control of Weeds, 4–14 July 1999*. Montana State University, Bozeman, Montana, pp. 73–82.

Smirnoff, W.A. and Morris, O.N. (1984) Field development of *Bacillus thuringiensis* in Eastern Canada, 1970–80. In: Kelleher, J.S. and Hulme, M.A. (eds) *Biological Control Programmes Against Insects and Weeds in Canada 1969–1980*. Commonwealth Agriculture Bureaux, Slough, UK, pp. 238–247.

Surgeoner, G.A., Lindsay, L.R. and Heal, J.D. (1996) Assessment of resistance by horn flies to three insecticides impregnated into ear tags. *1996 Ontario Beef Research Update*, 85–88.

Turnock, W.J. (1991) Biological control of insect pests of field crops. Proceedings of the Workshop on Biological Control of Pests in Canada, Calgary, Alberta, Canada. *Alberta Environmental Centre Report* AECV91-P1, pp. 9–14.

Wagner, R.G. (1993) Research directions to advance forest vegetation management in North America. *Canadian Journal of Forest Research* 23, 2317–2327.

Watkinson, I. (1994) Global view of present and future markets for Bt products. *Agriculture, Ecosystems and Environment* 49, 3–7.

WCLP (1978) *Guide for Recommendations for the Control of Livestock Insects in Western Provinces*. Distributed by Crop Protection and Pest Control Branch, Alberta Department of Agriculture, Edmonton, Alberta.

WCLP (1999) Recommendations for the Control of Arthropod Pests of Livestock and Poultry in Western Canada. http://eru.usask.ca/livestok/wclp/ (13 May 1999).

3 Taxonomy and Biological Control

J.T. Huber, S. Darbyshire, J. Bissett and R.G. Foottit

Introduction

Many articles on the relationship of taxonomy to biological control exist, 36 being listed in Knutson and Murphy (1988), along with an additional 140 titles on systematics in relation to pest management, quarantine and regulatory activities, the environment, and biology and ecology in general. Danks and Ball (1993), Miller and Rossman (1995) and Eidt (1995) discussed the importance of systematics to entomology, agriculture and forestry, respectively. Although important to biological research in general, systematics historically has had a close relationship with classical biological control during the past 60 years, e.g. Clausen (1942), LaSalle (1993), Schauff and LaSalle (1998) and Gordh and Beardsley (1999).

Over the past three decades, a consistent worldwide decline in research in organism-based taxonomy and classical biological control has occurred, due mainly to a greatly reduced number of specialists working in these fields. In Canada, the number of taxonomists studying insects, arachnids, nematodes, vascular plants and fungi has declined steadily since its peak in the 1970s, e.g. at the Biosystematics Research Institute, Ottawa, there were 52 taxonomists (Hardwick, 1976), now there are 26, fewer than in 1951. The issues of biodiversity, sustainable agriculture and forestry, public concern for the environment, and increased introductions of foreign species have increased government awareness that more taxonomists are again needed to accurately identify species and carry out basic research. In the USA, a sharp increase in employment opportunities for plant taxonomists has occurred, such that the demand cannot be filled (Dalton, 1999) and in mycology so few trained taxonomists are graduating that it may be difficult to fill the available positions (e.g. Burdsall, 1993).

Taxonomy Defined

Wheeler (1995) reviewed the many definitions and concluded that taxonomy is the study of species, the phylogenetic relationships among them and, ultimately, the proposal of a predictive classification consistent with phylogeny. Biological systematics is a subset of taxonomy concerned specifically with analysing phylogenetic relationships, and is pursued so that classifications will summarize efficiently what we know about biological diversity and predict what we do not yet know. Ball and Danks (1993) discussed, among other things, the value of classifications, noting that they are the most widely used product of systematics. The science of taxonomy includes discovering, recognizing, identifying, describing and naming organisms (Gordh and Beardsley, 1999). Whereas diversity and variation hinder most other disciplines, they are the subject matter of taxonomy, and describing this diversity is taxonomy's backbone.

The essential taxonomic tools are reference collections of specimens and relevant literature. Taxonomists prepare comprehensive revisions containing illustrated species descriptions, identification keys, species catalogues, phylogenetic hypotheses and predictive classifications. Such research products may take many years to prepare, yet they permit the important ongoing and practical task of accurately and reliably identifying species. Recognizing undescribed species, as well as accurately naming those previously described, is an important part of a taxonomist's work. In poorly researched groups, far more undescribed than described species exist. LaSalle (1993) estimated that 75% of parasitic Hymenoptera have yet to be described and many of those described are not recognizable from their original descriptions alone. Specimens in such groups often cannot be correctly identified to species. Although an incomplete identification, e.g. to genus, does not help in accessing the literature on a particular species, it is still better than an incorrect species name, because misinformation is disseminated as a result of misidentifications. For example, in *Trichogramma* virtually all the research published before 1963 used only three species names, and now over 20 times that number of species are described (Pinto, 1998). Further, that research is invalidated because of a lack of voucher specimens to verify species' identities. Having the correct name for a species and voucher specimens deposited in a permanent collection, in contrast, permits access to published information about it and enables unambiguous communication about the species.

Problems Facing Taxonomists

The first problem, long recognized by taxonomists (e.g. Aldrich, 1927), is that the number of extant species is far greater than previously estimated and many of them are complexes of morphologically similar but biologically distinct species. Biological control specialists are well aware of this complexity and often the first to notice it while working out life histories and host specificity to determine which agents have biological control potential. For taxonomists, the decision as to how to treat the entities in such complexes rests on their concept of the nature of a species, a complex and refractory problem in itself (Unruh and Woolley, 1999). While only a small fraction of species are directly relevant to biological control programmes, taxonomists must be more inclusive and study a much wider range of taxa to understand the position of each species within the evolutionary history of the entire group. The respective agendas of taxonomists and biological control workers thus may have different goals and time frames. Biological control workers benefit from the large body of existing taxonomic work, although it often requires correcting and updating as new discoveries are made. However, many groups of organisms lack even the most preliminary and basic treatments, sometimes seriously impeding effective biological control work.

A second problem is scientific nomenclature that binds taxonomy to a history that is often obscure. The historical links are: (i) rules of priority for naming organisms; (ii) original descriptions that validate scientific names; and (iii) type specimens that objectively define those names. To avoid chaos in the naming of millions of species, international bodies of taxonomists have established rules that provide a workable structure for naming the seemingly endless number of species without restricting an individual's interpretation of a genus, species or other category. The resulting International Codes of Zoological, Botanical, Bacterial and Viral Nomenclature are thus relevant to biological control workers. Scientific names are often changed to conform to the rules. Taxonomic judgment, as exercised by different workers, may also lead to name changes, such as moving species from one genus to another as relationships become better understood or are re-interpreted. This may result in a taxon having several 'legal' names under one of the Codes, e.g. the weedy forage crop, tall wheatgrass, has been assigned to five different genera and two different species concepts (Darbyshire, 1997). The use of any particular name depends on the taxonomists' concept of a genus and a species. Although the Codes allow for a relatively stable system of scientific names, a taxonomic dilemma often arises with the immediate needs of biological control specialists for identifications and names. The dilemma is that while accurate and specific scientific names are needed, species names often cannot be correctly applied because of broken or missing type(s), incomplete or inadequate descriptions, and/or unfamiliarity of the taxonomist with the group in question, often due to lack of specimens. It may therefore be difficult or impossible to identify confidently and accurately specimens from a species complex, especially those whose differentiating features are biochemical, behavioural or discernible only by crossing experiments.

A third problem, more common to plants and fungi than to animals, is that of promiscuous sex or, conversely, a lack of sex. Self-fertilization, parthenogenesis, apomixis, hybridization and reticulate evolution, all sexual processes, cause no end of taxonomic difficulties. This is often the case with various plant complexes that arrived in North America from abroad and flourished as weeds. Plant populations that are relatively distinct morphologically and spatially separate in Eurasia may lose their geographic and reproductive barriers in North America, becoming a mixture of intergrading forms, e.g. leafy spurge, *Euphorbia esula* L. (see Bourchier *et al.*, Chapter 69 this volume), and knapweeds, *Centaurea* spp. (see Bourchier *et al.*, Chapter 63 this volume). Conversely, clonal evolution has produced intergrading strains and cryptic species among asexual fungi, which include most of the naturally occurring and commercial biological control agents of insects, weeds and soil-borne diseases. Identification and naming of these populations then becomes a matter of knowing the genotypic and phenotypic characteristics of the whole population, as well as the species concept employed. The problem of weed population differences between North America and Eurasia, regardless of the ability of taxonomists to name the weed, can be circumvented by doing the preliminary testing of a Eurasian biological control agent in Europe (or Asia) using North American target plants. Those species that feed readily on North American weed populations would be the ones to investigate more thoroughly. Eventually taxonomists will catch up with the biological control agents and fine-tune the target plant taxonomy. Of more critical concern initially is the non-target plant taxonomy, i.e. what related species should be tested for agent susceptibility (see Harris and McEvoy, 1992).

Finally, lack of sex is a major reason for nomenclatural instability in the fungi. At an early stage in developing fungal taxonomic principals, mycologists chose to maintain a dual nomenclature with separate names for sexual and asexual forms. An independent taxonomy and classification was established for asexually reproducing fungi (anamorphs), affecting classifications and nomenclatural stability. Currently, sexual states (teleomorphs) are not known for most asexually reproducing fungi and asexually reproducing lineages probably occur commonly in the fungi. A recent movement by taxonomists toward a unified classification and nomenclature, based on the integration of anamorphs into the teleomorph classification, should help eliminate the prevailing confusion (e.g. Seifert and Samuels, 2000).

Current Situation

Heraty (1998) entitled a paper 'Systematics: Science or Service?' The answer is both. All biological sciences sooner or later provide some service, even if only to support other basic research. Taxonomy has often been treated by non-taxonomists as a service – that of providing correct names of organisms. Production of robust phylogenies and the resulting new classifications are having a tremendous impact on biology, because they more accurately portray evolution and provide a better understanding of species relationships. Synthesized outputs in the form of comprehensive revisions and identification keys must be based on adequate collections of well-preserved specimens accessible to taxonomists who need to study them. Authoritatively identified specimen collections are a basic product of taxonomic research and are the fundamental source of information for taxonomy. Each specimen in a collection is a testable hypothesis – evidence for presence of a species at a particular time and place. If the basic scientific work and collection development is poorly supported, the service will suffer in the form of an increasing proportion of inaccurate or incomplete identifications. Accurate species names are needed for biological control, especially when introductions of organisms to new areas are being considered. The need for authoritative identifications, supported by voucher specimens (Huber, 1998; Gordh and Beardsley, 1999), is stipulated in international import standards (FAO, 1996).

Biological control research also provides a service. The obvious one is to control a pest while discovering new information about the biology of various species of biological control agents and their interactions with native species. For taxonomists, a particularly useful service is provision of fresh, well-preserved specimens from known hosts for study. Because different groups of organisms require different preservation methods, it is important that biological control workers contact taxonomists at the beginning of their investigations to learn the best methods for preserving the species under study for identification and future reference.

The past two decades have seen important changes in taxonomy. The greatly reduced numbers of taxonomists are spending an increasing proportion of their time seeking funding (usually available only for applied projects), sometimes to the detriment of doing basic research. Powerful new diagnostic tools, e.g. molecular techniques, are now part of the taxonomist's arsenal, not only to help in species identification but also to identify and characterize distinct populations (strains, races, biotypes) within species (Unruh and Woolley, 1999; Scott and Straus, 2000). 'Traditional', morphology-based taxonomy is often unable to differentiate organisms below the species level. Molecular identification is also increasingly important in: developing reliable diagnostic systems to monitor genetic variation both within and among strains of commercially important organisms; detecting genetic drift occurring during several generations of multiplication; certifying commercial lots of biological control agents for mass release (e.g. Landry *et al.*, 1993); monitoring and tracking genetically modified biological control agents; and, especially, developing taxonomic concepts and identification tools for microorganisms, which may lack useful morphological characters on which to base predictive phylogenies and reliable identification protocols.

The microbial communities in complex habitats such as soil and water are particularly difficult to monitor effectively, as shown by the recent appearance of potato wart fungus, *Synchytrium endobioticum* (Schilbersky) Percival, in Prince Edward Island (C.A. Lévesque, Ottawa, 2000, personal communication). Similarly, it is important to determine the fate and persistence of exotic organisms, including genetically modified organisms, employed as biological control agents. Analyses of clone libraries of 16S rDNA indicate that as many as 99% of procaryotes in nature cannot be isolated and are essentially 'invisible' to classical taxonomic methodologies. DNA sequencing has permitted the elucidation of phylogeny in many difficult taxonomic groups, e.g. bacteria (Fox *et al.*, 1980; Weisburg *et al.*, 1991; Pace, 1997) and fungi (Bruns *et al.*, 1991; Bowman *et al.*, 1992; Berbee *et al.*, 1995; Seifert *et al.*, 1995). However, the potential of DNA-based methods is far from being fully exploited for microorganism identification (Lévesque, 1997).

Automated identifications based on carbon substrate utilization patterns in microtitre plates are available for bacteria and are being developed for fungi, e.g. Seifert *et al.* (2000). These techniques may lack the comparative absolute reliability of sophisticated molecular techniques and require independent confirmation, but have the advantage of being much faster and more cost-effective than their macromolecular counterparts for microbial identification.

Examples

An example of the benefits that taxonomists and biological control workers obtain from close cooperation is the case of *Lygus* bugs and their parasitoids. Although Schwartz and Foottit (1998) provided a firm taxonomic foundation for accurate identification of *Lygus* spp., their nymphal parasitoids, *Peristenus* spp. and *Leiophron* spp., being studied for biological control (see Broadbent *et al.*, Chapter 32 this volume), present many problems despite revisions of the North American and European species (Loan, 1974a, b). These revisions were possible because of good rearing records and specimens supplied to Loan by biological control workers, permitting recognition of some biological species that otherwise would not have been formally named. Although *Lygus* nymphs have a high percentage of field parasitism, adult parasitoids are rarely collected and most of Loan's species are based on only a few individuals. Thus, morphological variation has not been assessed adequately and problems in obtaining accurate species identifications still exist. Although some introductions of European species into North America have been made since Loan's publications, the native parasitoid fauna was never adequately surveyed and consists of many more species than previously recognized (H. Goulet, Ottawa, 2000, personal communication). Detailed biological studies, better rearing techniques and intensive collecting of adults have resulted in a wealth of new material and host records. A new taxonomic revision, based substantially on reared specimens, will eventually permit accurate and reliable identifications and should be of long-lasting value.

A second example is the cabbage seed pod weevil, *Ceutorhynchus obstrictus* (Marsham) (see Kuhlmann *et al*., Chapter 11 this volume). Other *Ceutorhynchus* spp. have been introduced into North America to control weeds, and additional introductions are planned. To decide whether to introduce European parasitoids of *C. obstrictus*, the biological control worker must know how related it is to these other *Ceutorhynchus* spp. and the specificity of candidate parasitoids. Such information should help them decide if parasitoids of *C. obstrictus* are likely also to attack the beneficial *Ceutorhynchus* spp. used in weed biological control. A taxonomic revision and phylogeny of Holarctic *Ceutorhynchus* spp. and their parasitoids would help to determine the likelihood that the parasitoids would move from *C. obstrictus* to the beneficial species.

An increasingly important requirement, particularly in weed biological control programmes, is provision of a species list to test the host range of a potential biological control agent (Harris and McEvoy, 1992; Wan and Harris, 1997). Because it is impossible to test all potential host species, a sampling strategy must be adopted to provide a manageable list of likely candidates. Taxonomic information, in the form of robust phylogenetic hypotheses, is the only tool available to establish a non-random list. Host selection from this list is based on the theory that the likelihood of an agent attacking a non-target species is proportional to its genetic relatedness with the target, because related organisms are likely to be morphologically and physiologically more similar than unrelated ones. Wapshere (1974) proposed a centrifugal phylogenetic testing method in which taxa closely related to the target should be tested more thoroughly than distant taxa. He noted that it would only fail if an agent's host recognition systems were not phylogenetically distributed or if an agent utilized alternative, unrelated hosts – the latter a consideration for many parasitic wasps, rusts and aphids. Accurate phylogenies allow confidence in the derived list of candidate species chosen for testing, and reduce the risk of negative environmental impact. Figure 3.1 shows a series of concentric priorities for the main criteria that should be considered in developing a plant

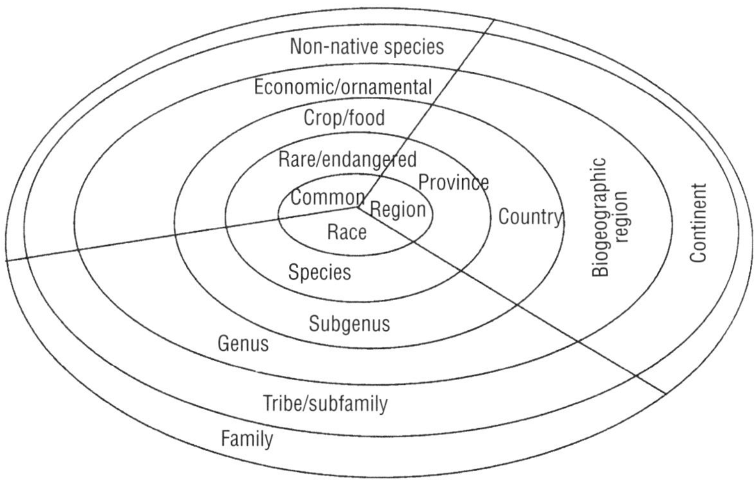

Fig. 3.1. Model for developing a list of non-target species for testing with potential biological control agents. The target species is at the centre of the model. Concentric rings of increasing radius indicate decreasing risk, and, therefore, testing priority. The three axes – taxonomy, geography and ecology/ethnobiology – must be considered together to optimize the predictive power of the phylogenetic hypothesis represented in the taxonomy axis.

test list. Increase in radius indicates a decreasing priority for testing, because it also predicts a decreasing risk to human interests and/or environmental integrity. Although three criteria axes are shown, for taxonomy, geography and ecology/ethnobiology, the latter two elements should also be considered with a taxonomic perspective. The phylogenetic aspects of systematics are thus not only useful for devising better classifications, but essential for developing reliable strategies for evaluating the safety of biological control agents.

Conclusions

Specimens in well-maintained biological collections, well-supported taxonomic libraries, and research based on these assets are the capital upon which applied taxonomy depends. To the extent that this basic work can be supported, taxonomists will be able to help biological control workers and others to solve pest problems using biological methods. This means not only providing accurate species identifications but also robust phylogenies that provide a framework for testing hypotheses, and durable classifications for cataloguing information. Other considerations, e.g. international trade, may depend on availability of taxonomic expertise for accurate identifications. Fair resolution of trade issues may be compromised at considerable expense, if a country depends on outside taxonomic help.

Biological control specialists can provide taxonomists with reared and properly preserved material from known hosts, often with detailed biological information. Information on the host range of a biological control agent can also supply useful data for taxonomic studies of the target species and its relatives. Such mutual help can only improve the sciences of taxonomy and biological control.

Acknowledgement

We thank John Heraty, University of California, Riverside, for reviewing the chapter and suggesting improvements.

References

Aldrich, J.M. (1927) The limitations of taxonomy. *Science* 65(1686), 381–385.

Ball, G.E. and Danks, H.V. (1993) Systematics and entomology: introduction. In: Ball, G.E. and Danks, H.V. (eds) *Systematics and Entomology: Diversity, Distribution, Adaptation, and Application.* Memoirs of the Entomological Society of Canada, 165, pp. 3–10.

Berbee, M.L., Yoshimura, A., Sugiyama, J. and Taylor, J.W. (1995) Is *Penicillium* monophyletic? An evaluation of phylogeny in the family Trichocomaceae from 18S, 5.8S and ITS ribosomal DNA sequence data. *Mycologia* 87, 210–222.

Bowman, B.H., Taylor, J.W., Brownlee, A.G., Lee, J., Lu, S.-D. and White, T.J. (1992) Molecular evolution of the fungi: relationship of the Basidiomycetes, Ascomycetes and Chytridiomycetes. *Molecular Biology and Evolution* 9, 285–296.

Bruns, T., White, T.J. and Taylor, J.W. (1991) Fungal molecular systematics. *Annual Review of Ecology and Systematics* 22, 525–564.

Burdsall, H.H. Jr (1993) Taxonomic mycology: the good, the bad, the optimistic. *Mushroom the Journal*, Fall 1993, 17–19.

Clausen, C.P. (1942) The relationship of taxonomy to biological control. *Journal of Economic Entomology* 35, 744–748.

Dalton, R. (1999) US universities find that demand for botanists exceeds supply. *Nature* 402, 109–110.

Danks, H.V. and Ball, G.E. (1993) Systematics and entomology: some major themes. In: Ball, G.E. and Danks, H.V. (eds) *Systematics and Entomology: Diversity, Distribution, Adaptation, and Application.* Memoirs of the Entomological Society of Canada, 165, pp. 257–272.

Darbyshire, S.J. (1997) Tall wheatgrass, *Elymus elongatus* subsp. *ponticus*, in Nova Scotia. *Rhodora* 99, 161–165.

Eidt, D.C. (1995) The importance of insect taxonomy and biosystematics to forestry. *The Forestry Chronicle* 71, 581–583.

FAO (1996) *International Standards for Phytosanitary Measures. Part 1 – Import Regulations. Code of Conduct for the Import and Release of Exotic Biological Control Agents.* Publication No. 3, Secretariat, International Plant Protection Convention, Food and Agriculture Organization of the United Nations, Rome.

Fox, G.E., Stackebrandt, E., Hespell, R.B., Gibson, J., Maniloff, J., Dyer, T.A., Wolfe, R.S., Balch, W.E., Tanner, R.S., Magrum, L.J., Zablen, L.B., Blakemore, R., Gupta, R., Bonen, L., Lewis, B.J., Stahl, D.A., Luehrsen, K.R., Chen, K.N. and Woese, C.R. (1980) The phylogeny of prokaryotes. *Science* 209, 457–463.

Gordh, G. and Beardsley, J.W. (1999) Taxonomy and biological control. In: Bellows, T.S. and Fisher, T.W. (eds) *Handbook of Biological Control: Principles and Applications of Biological Control.* Academic Press, San Diego, California, pp. 45–55.

Hardwick, D.F. (1976) The history and objectives of the Biosystematics Research Institute. *Bulletin of the Entomological Society of Canada* 8, 15–21.

Harris, P. and McEvoy, P. (1992) The predictability of insect host plant utilization from feeding tests and suggested improvements for screening weed biological control agents. In: *Proceedings of the 8th International Symposium on Biological Control of Weeds. Lincoln University, New Zealand. 2–7 February*, pp. 125–131.

Heraty, J. (1998) Systematics: science or service? In: Hoddle, M.S. (ed.) *Innovation in Biological Control Research.* California Conference on Biological Control, 10–11 June, University of California, Berkeley, California, pp. 187–190.

Huber, J.T. (1998) The importance of voucher specimens, with practical guidelines for preserving specimens of the major invertebrate phyla for identification. *Journal of Natural History* 32, 367–385.

Knutson, L. and Murphy, W.L. (1988) *Systematics: Relevance, Resources, Services, and Management. A Bibliography.* Association of Systematics Collections, Special Publication no. 1, Washington, DC.

Landry, B.S., Dextrase, L. and Boivin, G. (1993) Random amplified polymorphic DNA markers for DNA fingerprinting and genetic variability assessment of minute parasitic wasp species (Hymenoptera: Mymaridae and Trichogrammatidae) used in biological control programs of phytophagous insects. *Genome* 36, 580–587.

LaSalle, J. (1993) Parasitic Hymenoptera, biological control and biodiversity. In: LaSalle, J. and Gauld, I.D. (eds) *Hymenoptera and Biodiversity.* CAB International, Wallingford, pp. 197–215.

Lévesque, C.A. (1997) Molecular detection tools in integrated disease management: overcoming current limitations. *Phytoparasitica* 25, 3–7.

Loan, C.C. (1974a) The European species of *Leiophron* Nees and *Peristenus* Foerster (Hymenoptera: Braconidae, Euphorinae). *Transactions of the Royal Entomological Society of London* 126, 207–238.

Loan, C.C. (1974b) The North American species of *Leiophron* Nees, 1818 and *Peristenus Foerster,* 1862 (Hymenoptera: Braconidae, Euphorinae) including the description of 31 new species. *Le Naturaliste Canadien* 101, 821–860.

Miller, D.R. and Rossman, A.Y. (1995) Systematics, biodiversity, and agriculture. *Biosciences* 45, 680–686.

Pace, N.R. (1997) A molecular view of microbial diversity and the biosphere. *Science* 276, 734–740.

Pinto, J.D. (1998) The role of taxonomy in inundative release programs utilizing *Trichogramma.* In: Hoddle, M.S. (ed.) *Innovation in Biological Control Research.* California Conference on Biological Control, 10–11 June, University of California, Berkeley, California, pp. 45–49.

Schauff, M.E. and LaSalle, J. (1998) The relevance of systematics to biological control: protecting the investment in research. In: Zalucki, M.P., Drew, R.A.I. and White, G.G. (eds) *Pest Managment – Future Challenges,* Vol. 1. Proceedings of the 6th Australian Applied Entomological Conference, Brisbane, Australia, 29 September–2 October, pp. 425–436.

Schwartz, M.D. and Foottit, R.G. (1998) Revision of the Nearctic species of the genus *Lygus* Hahn, with a review of the Palaearctic species (Heteroptera: Miridae). *Memoirs on Entomology, International* 10, 428 pp.

Scott, J. and Straus, N. (2000) A review of current methods in DNA fingerprinting. In: Samson, R.A. and Pitt, J.I. (eds) *Integration of Modern Taxonomic Methods for* Penicillium *and* Aspergillus *Classification.* Harwood Academic Publishers, Amsterdam, The Netherlands, pp. 209–224.

Seifert, K.A. and Samuels, G.J. (2000) How should we look at anamorphs? *Studies in Mycology* 45, 5–18.
Seifert, K.A., Wingfield, B.D. and Wingfield, M.J. (1995) A critique of DNA sequence analysis in the taxonomy of filamentous Ascomycetes and ascomycetous anamorphs. *Canadian Journal of Botany* 73 (suppl. 1), 760–767.
Seifert, K.A., Bissett, J., Giuseppin, S. and Louis-Seize, G. (2000) Substrate utilization patterns as identification aids in *Penicillium*. In: Samson, R.A. and Pitt, J.J. (eds) *Integration of Modern Taxonomic Methods for* Penicillium *and* Aspergillus *Classification*. Harwood Academic Publishers, Amsterdam, The Netherlands, pp. 239–250.
Unruh, T.R. and Woolley, J.B. (1999) Molecular methods in classical biological control. In: Bellows, T.S. and Fisher, T.W. (eds) *Handbook of Biological Control. Principles and Applications of Biological Control*. Academic Press, New York, New York, pp. 57–85.
Wan, F.-H. and Harris, P. (1997) Use of risk analysis for screening weed biocontrol agents: *Altica carduorum* Guer. (Coleoptera: Chryomelidae) from China as a biocontrol agent of *Cirsium arvense* (L.) Scop. in North America. *Biocontrol Science and Technology* 7, 299–308.
Wapshere, A.J. (1974) A strategy for evaluating the safety of organisms for biological weed control. *Annals of Applied Biology* 77, 201–211.
Weisburg, W.G., Barns, S.M., Pelletier, D.A. and Lane, D.J. (1991) 16S ribosomal DNA amplification for phylogenetic study. *Journal of Bacteriology* 173, 697–703.
Wheeler, Q.D. (1995) The 'old systematics': classification and phylogeny. In: Pakaluk, J. and Slipinski, S.A. (eds) *Biology, Phylogeny, and Classification of Coleoptera: Papers Celebrating the 80th Birthday of Roy A. Crowson*. Museum i Instytut Zoologii PAN, Warsaw, Poland, pp. 31–62.

4 *Acantholyda erythrocephala* (L.), Pine False Webworm (Hymenoptera: Pamphiliidae)

D.B. Lyons, M. Kenis and R.S. Bourchier

Pest Status

The pine false webworm, *Acantholyda erythrocephala* (L.), distributed from Great Britain to Korea (Middlekauff, 1958), was introduced into eastern North America prior to 1925 (Wells, 1926). In the USA, it has spread as far west as Minnesota and Wisconsin (Middlekauff, 1958; Wilson, 1977). The first record of *A. erythrocephala* in Canada was from Scarborough township, Ontario, in 1961 (Eidt and McPhee, 1963). Syme (1981) reported the species as occurring south of a line joining Parry Sound and Ottawa, and in the Lake of the Woods area in northwestern Ontario. In North America, *A. erythrocephala* has been reported from red pine, *Pinus resinosa* Aiton, eastern white pine, *P. strobus* L., Scots pine, *P. sylvestris* L., mugho pine, *P. mugo* Turra, Austrian pine, *P. nigra* Arnold, Japanese red pine, *P. densiflora* Siebold, jack pine, *P. banksiana* Lambert, and western white pine, *P. monticola* Douglas (Howse, 2000).

In Ontario, *A. erythrocephala* was described as troublesome to pines grown as ornamentals or Christmas trees (Syme, 1981). Syme (1990) reported it as a serious defoliator of a number of *Pinus* spp. and it was the most destructive insect encountered in surveys of young *P. resinosa* plantations. In Ontario, throughout the 1980s and early 1990s, *A. erythrocephala* continued to be a chronic problem in young plantations. In 1993, the situation changed dramatically when a heavy infestation was discovered in 45–55-year-old *P. resinosa* in Simcoe county. Shortly thereafter, a similar situation was encountered in Ganaraska Forest, Northumberland county. This species has also been reported from Quebec, Edmonton, Alberta and St John's, Newfoundland (Howse, 2000). In New York, *A. erythrocephala* severely defoliated 185 ha of timber-size *P. sylvestris* in 1981 and has spread eastward and southward until, by 1995, about 5000 ha of pine plantations were annually experiencing moderate to severe defoliation (Asaro and Allen, 1999).

Lyons (1994, 1996) studied the phenology of the arboreal stages, and adult flight activity and oviposition of *A. erythrocephala*, respectively, and Lyons (1995) and Lyons and Jones (2000) summarized its biology. Overwintering larvae (pronymphs) pupate in earth cells in spring as soon as the soil begins to thaw under the host tree. As soil temperatures continue to warm, adults eclose and burrow up to the soil surface, emerge protandrously, and mate. Females begin to oviposit on host needles immediately after mating, by cutting a slit into the needle and inserting a crease of the egg chorion. Upon hatching, larvae crawl to the twig and begin to feed gregariously on the base of the needles. There, they form a web in which they feed. The webs, which consist of silk, uneaten needles, frass and exuviae, expand as the larvae develop until entire branches can be enclosed. Males pass through five instars and females six. When development is complete the larvae drop to the ground and burrow into the mineral soil where they construct an overwintering cell. The larvae, now referred to as eonymphs, undergo an aestival diapause, then transform into pronymphs characterized by a pupa-like eye. Some individuals may remain in the diapause stage for one or more years.

Background

Chemical control strategies have been developed for *A. erythrocephala,* using both conventional synthetic insecticides (Lyons *et al.*, 1993) and natural-product insecticides (Lyons *et al.*, 1996, 1998). Because of the desire to reduce dependence on chemical insecticides, biological controls were investigated.

No pathogens are known from North American populations of *A. erythrocephala*. A *Nucleopolyhedrovirus* (NPV) was reported from European populations (Jahn, 1967). Presumably, this is the *Acantholyda erythrocephala* NPV (Acer NPV) reported by Murphy *et al.* (1995). Wilson (1984) demonstrated in the laboratory that *A. erythrocephala* larvae were susceptible to infection by *Pleistophora schubergi* Zwolfer, but because of host-rearing problems, was unable to assess its potential impact. Asaro and Allen (1999) isolated *Steinernema* n. sp. near *kraussi* Steiner from a pronymph in New York. Related nematodes have been reported from conifer-feeding Pamphiliidae in Europe (Bednarek and Mracek, 1986; Mracek, 1986; Eichhorn, 1988).

A few parasitoids have been reared from *A. erythrocephala* in North America. Barron (1981) described *Ctenopelma erythrocephalae*, which oviposits in *A. erythrocephala* eggs. *Homaspis interruptus* (Provancher) was reported from *Acantholyda* sp. in Ontario (Barron 1990) and *A. erythrocephala* in New York (Asaro and Allen, 1999). *Sinophorus megalodontis* Sanborne, *Olesicampe* n. sp. (H. Townes, Gainsville, 1986, personal communication), and *Trichogramma minutum* Riley were reared from *A. erythrocephala* in Ontario (Lyons, 1995).

Biological Control Agents

Parasitoids

Lyons (1999) and Bourchier *et al.* (2000) described the biologies of *S. megalodontis* and *Olesicampe* sp. in *A. erythrocephala*. *S. megalodontis* emerges from the host prior to overwintering, whereas *Olesicampe* sp. overwinters in the host integument as a fully formed larva. Cocoons of the latter are only collected in spring. Both species are univoltine, and adults of both species emerged protandrously, beginning in late May. Emergence periods of *S. megalodontis* and *Olesicampe* sp. lasted for 17 and 16 days, respectively. The observed flight period of both species lasted 28 days. Unhatched eggs of *S. megalodontis* and *Olesicampe* sp. were found in all host instars. Annual variability in host stage attacked suggested that year-to-year variations occurred in synchronization with the host's phenology. Parasitoid larvae occurred in all host instars indicating that the eggs hatched soon after oviposition. Parasitoid larvae remained as first instars until some time after host larvae dropped to the ground to overwinter.

Eggs of *S. megalodontis* were found in final instar *A. erythrocephala* larvae collected in drop traps, suggesting that even late-instar larvae were being attacked. Parasitism by the two parasitoids increased throughout the drop period, perhaps due to a reduction in development rates of parasitized host larvae or increased parasitoid activity at the end of the larval period. For the entire drop period, the proportion of parasitized larvae was not significantly different between the host sexes. Total parasitism of *A. erythrocephala* by *S. megalodontis* and *Olesicampe* sp., for the period of larval drop, was 17.7% and 6.2%, respectively. Superparasitism and multiparasitism limited the effectiveness of both parasitoids. Encapsulation of parasitoid larvae, resulting in their death, was common, thus severely limiting the parasitoids' effectiveness in reducing host populations.

The transcontinental distribution of *S. megalodontis* (Sanborne, 1984) and the reports of unidentified *Sinophorus* spp. and *Olesicampe* spp. attacking *Cephalcia* spp. in Canada (Eidt, 1969) suggested that these species are endemic to North America. *S. megalodontis* and *Olesicampe* sp. are apparently native larval endoparasitoids that have adapted to attacking the introduced *A. erythrocephala*.

In Ontario, *T. minutum* Riley and *Trichogramma platneri* Nagarkatti were evaluated for inundative biological control of an infestation of *A. erythrocephala* in a *P. strobus* plantation near Owen Sound (Bourchier *et al.*, 2000). *T. minutum* used in the release were collected near Barrie from *A. erythrocephala* eggs. The parasitoid was selected from several *T. minutum* lines tested on *A. erythrocephala* eggs. The 'Barrie' line was mass-reared on Mediterranean flour moth, *Ephestia kuehniella* (Zeller), at Sault Ste Marie prior to the release. *T. platneri* (obtained from Beneficial Insectaries, Guelph, Ontario), normally used in apple orchards for codling moth, *Cydia pomonella* (L.), control, was included in the field test because it is arboreal and commercially available. Nominal release rates of *T. minutum* were 64,000, 16,000 and 8000 females per ten trees, while *T. platneri* was released at a rate of 64,000 females per ten trees. Actual release rates of female wasps were significantly lower than planned.

Parasitism of sentinel egg masses (*E. kuehniella* eggs pasted on cards) followed a similar pattern for both species, peaking 7 days after the beginning of parasitoid emergence and declining 6 days later, when the last sentinel egg masses were collected. The temporal pattern of parasitism of sentinel egg masses was similar for all *T. minutum* release rates and parasitism was positively correlated with release rates. Emergence of *T. minutum* was 65% and *T. platneri* almost 95% from parasitized eggs of the factitious host when the last sentinel egg masses were collected. Three days earlier, when branches containing *A. erythrocephala* were sampled, emergence was only 33% and 55% for *T. minutum* and *T. platneri*, respectively. The mean apparent parasitism of *A. erythrocephala* eggs by *T. platneri* was

10.9% with a maximum at one tree of 36.2%. The higher parasitism by *T. platneri* was matched with a lower rate of *A. erythrocephala* emergence. There was a non-significant trend towards increased *A. erythrocephala* mortality in all treated trees compared to controls. Parasitism by *T. minutum* was not significantly higher than on control trees and there were no effects of release rate on parasitism rates.

In Europe, natural enemies, especially parasitoids, are more numerous and outbreaks of *A. erythrocephala* are usually of lower density and of shorter duration than in North America (Kenis and Kloosterman, 2001). Eggs of European *A. erythrocephala* are attacked by several *Trichogramma* spp. The main larval parasitoids are *Myxexoristops hertingi* Mesnil, and several ichneumonids, the most common being *Xenochesis* sp. and *Sinophorus* sp. Investigations have focused mainly on *M. hertingi* and *Trichogramma acantholydae* Pintureau & Kenis (Pintureau *et al.*, 2001) from Poland, Switzerland and Italy.

T. acantholydae was collected from outbreak populations of *Acantholyda posticalis* Matsumura and low-density populations of *A. erythrocephala* in northern Italy. Unlike most other *Trichogramma* spp., *T. acantholydae* appears to be univoltine; mature larvae enter into an obligate diapause in *A. erythrocephala* eggs and, in spring, 3–12 individuals emerge per host egg. To assess host specificity of *T. acantholydae*, adults were screened against eggs of the *E. kuehniella*, black army cutworm, *Actebia fennica* (Tauscher), eastern spruce budworm, *Choristoneura fumiferana* (Clemens), hemlock looper, *Lambdina fiscellaria fiscellaria* (Guenée), *Diprion pini* L. and *Gilpinia frutetorum* F. (Bourchier *et al.*, 2000; Kenis and Kloosterman, 2001). Oviposition was observed only in *L. fiscellaria* eggs, but no parasitoids emerged. In contrast, successful parasitism of *A. erythrocephala* eggs was observed, confirming that *T. acantholydae* is more specific to *A. erythrocephala* than the *Trichogramma* spp. found attacking *A. erythrocephala* in North America. The latter species require alternate host eggs later in the season.

M. hertingi overwinters in soil as a mature larva within the dead host larval skin. In spring, the larva moves to the soil surface and forms a puparium. The adult emerges about a month later and mates. In the laboratory, mated females start to lay eggs less than 10 days after emergence. They deposit microtype eggs on the host plant foliage where they are consumed by host larvae. On average, 1500 eggs were found in gravid females. Most *M. hertingi* larval development occurs after the host larva leaves the foliage to enter the soil. The larva consumes the host before winter.

Releases and Recoveries

M. hertingi adults were released into two screen cages about 3 m tall × 1.8 m wide × 1.8 m long, each enclosing a single red pine infested with *A. erythrocephala*, in a mixed red and white pine plantation near Apto, Ontario (44°31.9′N, 79°46.7′W) (D.B. Lyons, unpublished). Adult *M. hertingi* were released when host larval development progressed to the third instar. In one cage 42 newly emerged adults (13 males and 29 females) and in the second cage 78 adults (12 males and 66 females) were released in the morning. None of the females was mated prior to being released. Collections of the overwintering larvae from within the two cages have been made, but no parasitoids have yet emerged.

Evaluation of Biological Control

Endemic parasitoids attacking *A. erythrocephala* in North America are ineffective in reducing host populations due to superparasitism, multiparasitism, encapsulation and variable synchronization with the host. Thus, the use of inundative and classical biological control strategies is warranted.

The release results were promising in that for *T. platneri* we were able to demonstrate a significant increase in parasitism of *A. erythrocephala* eggs. A key issue for both species was timing of the release. Observations of activity of *A. erythro-*

cephala adults indicated that an earlier release date might have better targeted the availability of host eggs. In addition, the emergence of both parasitoid species was slow and peaked after both our sampling of *A. erythrocephala* eggs and the start of *Trichogramma* spp. emergence from the host eggs. The impact of both species should be improved by synchronizing parasitoid emergence with the initiation of *A. erythrocephala* egg laying.

The cumulative emergence of 66% for *T. minutum* was lower than that observed in previous releases (Bourchier and Smith, 1998). Actual release rates of *T. minutum* females, on the date that *A. erythrocephala* eggs were sampled, were very low (900, 3600, 7200 actual females of 8000, 16,000 and 64,000 potential females, respectively) because of the delay in parasitoid emergence. Given the number of females available to attack the host on our sampling date, it is encouraging that there was any observable parasitism at all at the *T. minutum* trees. There is potential to make *T. minutum* more effective by better timing of emergence and improving the cumulative level of emergence to historical levels (about 85%).

T. acantholydae, with its single generation per year and restricted host specificity, is a promising classical biological control agent for *A. erythrocephala* in North America.

M. hertingi is considered the most promising candidate for introduction into North America because: it is the most frequently cited parasitoid *of A. erythrocephala* in Europe and the most important species in outbreak populations in Poland; it has a broad climatic distribution; it is apparently specific to *A. erythrocephala*, while closely related *Acantholyda* spp. and *Cephalcia* spp. are attacked by other *Myxexoristops* spp.; and there are no tachinids reported from *A. erythrocephala* in North America so *M. hertingi* would fill an empty ecological niche in the region of introduction.

Recommendations

Further work should include:

1. Improving the synchronization of *Trichogramma* emergence with host oviposition, and better release timing to coincide with *A. erythrocephala* emergence;
2. Developing mating, propagation and release strategies for *M. hertingi*;
3. Further assessing the host specificity of *T. acantholydae* to evaluate its potential interactions with native *Trichogramma* spp. used for inundative release.

Acknowledgements

We thank the following taxonomists for identification of the European parasitoids: K. Horstmann, J. LaSalle, L. Masner, B. Pintureau, A. Polaszek and H.-P. Tschorsnig.

References

Asaro, C. and Allen, D.C. (1999) Biology of pine false webworm (Hymenoptera: Pamphiliidae) during an outbreak. *The Canadian Entomologist* 131, 729–742.

Barron, J.R. (1981) The Nearctic species of *Ctenopelma* (Hymenoptera, Ichneumonidae, Ctenopelmatinae). *Le Naturaliste canadien* 108, 17–56.

Barron, J.R. (1990) The Nearctic species of *Homaspis* (Hymenoptera, Ichneumonidae, Ctenopelmatinae). *The Canadian Entomologist* 122, 191–216.

Bednarek, A. and Mracek, Z. (1986) The incidence of nematodes of the family Steinernematidae in *Cephalcia falleni* Dalm. (Hymenoptera: Pamphiliidae) habitat after an outbreak of the pest. *Journal of Applied Entomology* 102, 527–530.

Bourchier, R.S. and Smith, S.M. (1998) Interaction between large-scale inundative releases of *Trichogramma minutum* (Hymenoptera: Trichogrammatidae) and naturally occurring spruce budworm (Lepidoptera: Tortricidae) parasitoids. *Environmental Entomology* 27, 1273–1279.

Bourchier, R.S., Lyons, D.B. and Kenis, M. (2000) Biological control of the pine false webworm. In: Lyons, D.B., Jones, G.C. and Scarr, T.A. (eds) *Proceedings of a Workshop on the Pine False Webworm,* Acantholyda erythrocephala *(Hymenoptera: Pamphiliidae)*. Natural Resources Canada, Canadian Forest Service, Sault Ste Marie, Ontario, pp. 23–30.

Eichhorn, O. (1988) Untersuchungen über die fichtengespinstblattwespen *Cephalcia* spp. Panz. (Hym., Pamphiliidae) II. Die larven- und nymphenparasiten. *Journal of Applied Entomology* 105, 105–140.

Eidt, D.C. (1969) The life histories, distribution, and immature forms of the North American sawflies of the genus *Cephalcia* (Hymenoptera: Pamphiliidae). *Memoirs of the Entomological Society of Canada* No. 59.

Eidt, D.C. and McPhee, J.R. (1963) *Acantholyda erythrocephala* (L.) new in Canada. *Canada Department of Forestry, Forest Entomology and Pathology Branch, Bi-Monthly Progress Report* 19, 2.

Howse, G.M. (2000) The history, distribution and damage levels of the pine false webworm in Canada. In: Lyons, D.B., Jones, G.C. and Scarr, T.A. (eds) *Proceedings of a Workshop on the Pine False Webworm,* Acantholyda erythrocephala *(Hymenoptera: Pamphiliidae)*. Natural Resources Canada, Canadian Forest Service, Sault Ste Marie, Ontario, pp. 13–16.

Jahn, E. (1967) Population outbreak of the pine false webworm, *Acantholyda erythrocephala* Chr. in the Steinfeld, Lower Austria, in the years 1964–1967. *Anzeiger für Schädlingskunde* 39, 145–152.

Kenis, M. and Kloosterman, K. (2001) European parasitoids of the pine false webworm (*Acantholyda erythrocephala* (L.)) and their potential for biological control in North America. In: Liebhold, A.M. and McManus, M.L. (eds) *Proceedings: Population Dynamics, Impact, and Integrated Management of Forest Defoliating Insects 1999*, August 15–19, Victoria, British Columbia, United States Department of Agriculture, Forest Service General Technical Report NE-227, 65–73.

Lyons, D.B. (1994) Development of the arboreal stages of the pine false webworm (Hymenoptera: Pamphiliidae). *Environmental Entomology* 23, 846–854.

Lyons, D.B. (1995) Pine false webworm, *Acantholyda erythrocephala*. In: Armstrong, J.A. and Ives, W.G.H. (eds) *Forest Insect Pests in Canada*. Natural Resources Canada, Canadian Forest Service, Ottawa, Ontario, pp. 245–251.

Lyons, D.B. (1996) Oviposition and fecundity of pine false webworm (Hymenoptera: Pamphiliidae). *The Canadian Entomologist* 128, 779–790.

Lyons, D.B. (1999) Phenology of the native parasitoid, *Sinophorus megalodontis* (Hymenoptera: Ichneumonidae), relative to its host, the pine false webworm, in Ontario, Canada. *The Canadian Entomologist* 131, 787–800.

Lyons, D.B. and Jones, G.C. (2000) What do we know about the biology of the pine false webworm in Ontario? In: Lyons, D.B., Jones, G.C. and Scarr, T.A. (eds) *Proceedings of a Workshop on the Pine False Webworm,* Acantholyda erythrocephala *(Hymenoptera: Pamphiliidae)*. Natural Resources Canada, Canadian Forest Service, Sault Ste Marie, Ontario, pp. 3–12.

Lyons, D.B., Helson, B.V., Jones, G.C. and McFarlane, J.W. (1993) Development of a chemical control strategy for the pine false webworm, *Acantholyda erythrocephala* (Hymenoptera: Pamphiliidae). *The Canadian Entomologist* 125, 499–511.

Lyons, D.B., Helson, B.V., Jones, G.C., McFarlane, J.W. and Scarr, T. (1996) Systemic activity of neem seed extract containing azadirachtin in pine foliage for control of the pine false webworm *Acantholyda erythrocephala* (Hymenoptera: Pamphiliidae). *Proceedings of the Entomological Society of Ontario* 127, 45–55.

Lyons, D.B., Helson, B.V., Jones, G.C. and McFarlane, J.W. (1998) Effectiveness of neem- and diflubenzuron-based insecticides for control of the pine false webworm, *Acantholyda erythrocephala* (L.) (Hymenoptera: Pamphiliidae). *Proceedings of the Entomological Society of Ontario* 129, 115–126.

Middlekauff, W.W. (1958) The North American sawflies of the genera *Acantholyda, Cephalcia,* and *Neurotoma* (Hymenoptera: Pamphiliidae). *University of California Publications in Entomology* 14, 51–174.

Mracek, Z. (1986) Nematodes and other factors controlling *Cephalcia abietis* (Pamphiliidae: Hymenoptera), in Czechoslovakia. *Forest Ecology and Management* 15, 75–79.

Murphy, F.A., Fauquet, C.M., Bishop, D.H.L., Ghabrial, S.A., Jarvis, A.W., Martelli, G.P., Mayo, M.A.

and Summers, M.D. (eds) (1995) *Virus Taxonomy Classification and Nomenclature of Viruses. Sixth Report of the International Committee on Taxonomy of Viruses.* Springer-Verlag, Vienna.
Pintureau, B., Stefanescu, C. and Kenis, M. (2001) Two new species of *Trichogramma* (Hym.: Trichogrammatidae). *Annales de la Société Entomologique de France* 26: 417–422.
Sanborne, M. (1984) A revision of the world species of *Sinophorus* Foerster (Ichneumonidae). *Memoirs of the American Entomological Institute* No. 38.
Syme, P.D. (1981) Occurrence of the introduced sawfly, *Acantholyda erythrocephala* (L.) in Ontario. *Canadian Forest Service Research Notes* 1, 4–5.
Syme, P.D. (1990) Insect pest problems and monitoring in Ontario conifer plantations. *Revue d'Entomologie du Québec* 35, 25–30.
Wells, A.B. (1926) Notes on tree and shrub insects in southwestern Pennsylvania. *Entomological News* 37, 254–258.
Wilson, G.G. (1984) Infection of the pine false webworm by *Pleistophora schubergi* (Microsporida). *Canadian Forest Service Research Notes* 4, 7–8.
Wilson, L.F. (1977) A guide to the insect injury of conifers in the Lake Sates. *United States Department of Agriculture, Forest Service, Agricultural Handbook* 501.

5 *Acleris gloverana* (Walshingham), Western Blackheaded Budworm (Lepidoptera: Tortricidae)

I.S. Otvos, N. Conder and D.G. Heppner

Pest Status

The western blackheaded budworm, *Acleris gloverana* (Walshingham), a native defoliator in western North America, was recognized as a distinct species from its close relative the eastern blackheaded budworm, *Acleris variana* (Fernald), in 1962, but this status was not widely accepted until 1970 (Schmiege and Crosby, 1970). The preferred hosts for *A. gloverana* in British Columbia, Alaska and the northwestern USA are western hemlock, *Tsuga heterophylla* (Rafinesque-Schmaltz) Sargent, and, at higher elevations, mountain hemlock, *Tsuga mertensiana* (Bongard) Carrière (Anonymous, 1972). Other hosts include Sitka spruce, *Picea sitchensis* (Bongard) Carrière, Pacific silver fir, *Abies amabilis* (Douglas ex. Loudon) Douglas ex. J. Forbes, grand fir, *Abies grandis* (Douglas ex. D. Don) Lindley, alpine fir, *Abies lasiocarpa* (Hooker) Nuttall, and Douglas fir, *Pseudotsuga menziesii* (Mirbel) Franco (Keen, 1952). In British Columbia, severe infestations of *A. gloverana* tend to occur in mixed old-growth stands and young pure hemlock stands (Prebble and Graham, 1944). In Alaska, it was found to feed both on *T. heterophylla* and *P. sitchensis* in mixed stands. However, spruce stands suffered less severe defoliation than adjacent stands of pure hemlock (Schmiege and Hard, 1966).

Outbreaks of *A. gloverana* occur about 8–14 years apart. Populations build up over a 2–3-year period and generally remain high for another 2–3 years before collapsing. Occasionally an outbreak may last 4–5 years, in which case mortality in mature hemlock stands can be significant (Lejeune, 1975). Factors contributing to population collapse include parasitism, predation, competition, disease and weather (Prebble and Graham, 1945; Hard, 1974) but their exact roles are unknown. Larvae are wasteful feeders, causing defoliation, growth loss, top-kill, deformities and, in extreme cases, tree mortality (McCambridge, 1956; Lejeune, 1975; Eglitis, 1980). Trees surviving defoliation are weakened and susceptible to secondary insect attack (McCambridge and Downing, 1960).

In British Columbia, *A. gloverana* has one generation per year and overwinters as eggs. Larvae hatch from mid-May to early June (Brown and Silver, 1957) and mine into the expanding new growth. They have five instars; early instars feed on new shoots, whereas older instars can feed on old foliage. Pupation occurs on branches among the frass and dead needles from mid-July to late August. The pupal stage lasts about 2 weeks. Adults emerge and lay their eggs individually on the underside of needles from August to September (Shepherd and Gray, 1990).

Background

In British Columbia, several chemical insecticides were used to control *A. gloverana*, including calcium arsenate, DDT, fenitrothion and organophosphates (Lejeune, 1975; Heppner and Wood, 1986; Armstrong and Cook, 1993), until their use was banned in Canadian forests. Although about 50 parasitoid species have been reported to attack *A. gloverana*, causing about 30% parasitism, they are not generally considered to cause sufficient mortality to bring about the collapse of an outbreak (Allen and Silver, 1959; Gray and Shepherd, 1993). Parasitoids are generally considered to exert the greatest impact on populations that are already declining due to effects of weather and disease (Silver and Lejeune, 1956; Allen and Silver, 1959; Schmiege, 1966). For these reasons *Bacillus thuringiensis* serovar *kurstaki* (*B.t.k.*), a microbial insecticide already registered for other forest insects, was chosen for testing.

B.t.k. was first tried against *A. gloverana* on the Queen Charlotte Islands in 1960 (Kinghorn *et al.*, 1961), one of the first operational uses of *B.t.k.* for forest insect control in Canada. Heppner and Wood (1986) reviewed insecticide use, including *B.t.k.*, against *A. gloverana* and noted correctly that the early trials were generally applied too late in the insect's outbreak cycle (when populations were already declining) to allow for accurate assessment of the effects of *B.t.k.* They recommended that an experimental spray be conducted to properly evaluate *B.t.k.* efficacy against *A. gloverana*.

Biological Control Agents

Pathogens

Although *B.t.k.* is registered and used successfully to control several *Choristoneura* spp. and other forest Lepidoptera, it is not registered in Canada for either *A. variana* or *A. gloverana* (M. Furgiuele, Ottawa, 2000, personal communication).

An outbreak of *A. gloverana* on northern Vancouver Island from 1987 to 1991 provided an opportunity to test the efficacy of newer, high-potency *B.t.k.* products. During this outbreak, experimental trials were conducted in the Holberg area in 1989 and 1990 to collect field efficacy data to support registration of *B.t.k.* against *A. gloverana* (cooperative research by the British Columbia Ministry of Forests, the Canadian Forest Service and *B.t.k.* manufacturers).

Treatments were applied in both years by a fixed-wing aircraft equipped with four Micronair Atomizers (AU 4000). In 1989, Dipel® 176, an oil-based formulation of *B.t.k.*, was applied to three 50 ha plots (45 sample trees in each) at 30×10^9 International Units (IU) ha^{-1} at a rate of 1.8 $l\ ha^{-1}$. Controls were three untreated areas, similar in size. Population reduction was 90.1% and 74.8% in two of the plots by the third post-spray sample, but there was no detectable population reduction in the third plot. The inconsistent larval population reduction was attributed to the variable spray deposit in the plots caused by hilly terrain, especially in the third plot. The average population reduction for the treatment, using data from all three replicates, was 46%, but when the third plot was excluded, population reduction 3 weeks after application of Dipel® 176 was 88%. Based on these encouraging results, the experiment continued the following year.

In 1990, three products were tested: the oil-based Dipel® 176, and two water-based formulations, Foray® 48B and Futura XLV-HP. These were applied at 40×10^9 IU ha^{-1} in 2.4 l ha^{-1}, $40 \times 10_9$ IU ha^{-1} in 1.2 l ha^{-1}, and 50×10^9 IU ha^{-1} in 3.9 l ha^{-1}, respectively. Each product was applied to three separate plots, from 20 to 30 ha in size, and containing 45 sample trees in three separate sample lines of 15 trees each. Due to difficulties posed by the terrain, dense understory and closed tree canopy, sample trees were located along old logging roads and skid trails. Three separate plots of comparable size, 500–1500 m away from the treatment plots to minimize spray drift, were used as controls. Spray droplet analysis showed, as expected, a direct relationship between spray volume emitted and number of spray droplets per needle, averaging 0.30, 0.40 and 0.90 for Futura XLV-HP, Dipel® 176 and Foray® 48B treatments, respectively.

All three products provided good to excellent larval population reduction. Dipel® 176 caused 97.0% mortality, whereas Futura XLV-HP and Foray® 48B caused 83.2% and 69.4% mortality, respectively. The lower than expected population reductions caused by Foray® 48B were probably due to the poor spray deposit in one of the three replicates, where population reduction was only 55.2%. When this replicate was excluded from the analysis, Foray® 48B treatment was responsible for 95.0% larval mortality in the two remaining plots. Generally, most forest managers would gladly accept this level of protection because the goal is to reduce such impacts as top-kill and tree mortality and not necessarily to eliminate defoliation completely.

Evaluation of Biological Control

Application of all three products caused significant mortality of *A. gloverana* larvae in dense and young, 10–15 m tall, western hemlock stands. However, treating larval populations in all forest types, e.g. mountainous terrain with mature western hemlock stands, was a problem; not all larvae were exposed to *B.t.k.*

Recommendations

Further work should include:

1. Evaluating higher-potency *B.t.k.* products at somewhat higher doses in the 50 and 60×10^9 IU ha^{-1} range and higher volumes (about 3–5 l ha^{-1} range);
2. Confirming the promising results reported here in mature western hemlock stands.

References

Allen, S.J. and Silver, G.T. (1959) Brief history of the blackheaded budworm infestation on the Queen Charlotte Islands, 1952–1955. Canadian Department of Agriculture, Forest Biology Laboratory, Victoria, British Columbia, Unpublished Report 1959 (15).

Anonymous (1972) *Blackheaded Budworm: A Tree Killer?* Canadian Forest Service, Pacific Forestry Centre, Victoria, British Columbia, Pamphlet BC-P-4-72.

Armstrong, J.A. and Cook, C.A. (1993) *Aerial Spray Applications on Canadian Forests: 1945–1990.* Forestry Canada Information Report ST-X-2. Forestry Canada, Ottawa, Ontario.

Brown, G.S. and Silver, G.T. (1957) *Studies on the Blackheaded Budworm on Northern Vancouver Island.* Canadian Department of Agriculture, Forest Biology Laboratory, Victoria, British Columbia, Interim Report 1955–6.

Eglitis, A. (1980) *Western Black-headed Budworm on Heceta Island, Southeast Alaska – Tongass National Forest February 1980.* United States Department of Agriculture, Forest Service, Alaska Region, Forest Insect and Disease Management Biological Evaluation Report R10-80-2.

Gray, T.G. and Shepherd, R.F. (1993) Hymenopterous parasites of the blackheaded budworm, *Acleris gloverana,* on Vancouver Island, British Columbia. *Journal of the Entomological Society of British Columbia* 90, 11–13.

Hard, J.S. (1974) *The Forest Ecosystem of Southeast Alaska. 2. Forest Insects.* United States Department of Agriculture, Forest Service, Pacific Northwest Research Station, General Technical Report PNW-13.

Heppner, D.G. and Wood, P.M. (1986) *Blackheaded Budworm in the Vancouver Forest Region: Current Control Options.* Vancouver Forest Region, British Columbia Ministry of Forests, Burnaby, British Columbia, Internal Report PM-V-9.

Keen, F.P. (1952) *Insect Enemies of Western Forests.* United States Department of Agriculture, Miscellaneous Publication 273.

Kinghorn, J.M., Fisher, R.A., Angus, T.A. and Heimpel, A.M. (1961) Aerial spray trials against the blackheaded budworm in British Columbia. *Department of Forestry Bi-Monthly Progress Report* 17(3), 3–4.

Lejeune, R.R. (1975) Western black-headed budworm, *Acleris gloverana* (Wals.). In: Prebble, M.L. (ed.) *Aerial Control of Forest Insects in Canada.* Canadian Department of Environment, Ottawa, Ontario, pp. 159–166.

McCambridge, W.F. (1956) Effects of black-headed budworm feeding on second-growth western hemlock and Sitka spruce. *Proceedings of the Society of American Foresters* 1955/1956, pp. 171–172.

McCambridge, W.F. and Downing, G.L. (1960) *Black-headed Budworm.* United States Department of Agriculture, Forest Service Pest Leaflet No. 45.

Prebble, M.L. and Graham, K. (1944) *The Outbreak of Black-headed Budworm in the Coastal District of British Columbia. A Preliminary Report, 1940–1943.* Dominion Department of Agriculture, Forest Insect Investigations, Victoria, British Columbia, Unpublished Report.

Prebble, M.L. and Graham, K. (1945) The current outbreak of defoliating insects in coast hemlock forests of British Columbia. Part II. Factors of natural control. *British Columbia Lumberman* 29(3), 37–39, 88–92.

Schmiege, D.C. (1966) The relation of weather to two population declines of the blackheaded budworm, *Acleris variana* (Fernald) (Lepidoptera: Tortricidae), in coastal Alaska. *The Canadian Entomologist* 98, 1045–1050.

Schmiege, D.C. and Crosby, D. (1970) *Black-headed Budworm in Western United States.* United States Department of Agriculture, Forest Service, Forest Pest Leaflet No. 45.

Schmiege, D.C. and Hard, J.S. (1966) *Oviposition Preference of the Black-headed Budworm and Host Phenology.* United States Department of Agriculture, Forest Service, Northern Forest Experimental Station, Research Note NOR-16.

Shepherd, R.F. and Gray, T. (1990) Distribution of eggs of western blackheaded budworm, *Acleris gloverana* (Walshingham) (Lepidoptera: Tortricidae) and of foliage over the crowns of western hemlock, *Tsuga heterophylla* (Raf.) Sarg. *The Canadian Entomologist* 122, 547–554.

Silver, G.T. and Lejeune, R.R. (1956) Report on the black-headed budworm infestation on north Vancouver Island 1956. Canadian Department of Agriculture, Forest Biology Laboratory, Victoria, British Columbia, Unpublished Report 1956 (16).

6 *Aculops lycopersici* (Massee), Tomato Russet Mite (Acari: Eriophyidae)

J.L. Shipp, D.R. Gillespie and G.M. Ferguson

Pest Status

Tomato russet mite, *Aculops lycopersici* (Massee), native to North America, is a periodic pest of greenhouse tomato, *Lycopersicon esculentum* L., in British Columbia, Ontario and Quebec. In general, plant hosts are in the family *Solanaceae*. Nightshade, *Solanum* spp. and petunia, *Artemisia jussieana* Jussieu, are frequently sources of infestations. *A. lycopersici* can cause severe crop losses, but only a few such cases have occurred in Canada. Infestations cause the leaves to turn a yellowish-brown colour and the edges to curl. Infestations may also result in flower abortion and cause russetting cracks to form on infested fruit. Infested plants wilt and eventually die.

A. lycopersici females lay 10–50 eggs during their life span of 20–40 days. High reproductive rates and rapid development are favoured by moderate temperatures (21°C) and low humidities (30% RH). Under these conditions the life cycle can be completed in 6–7 days. The ability of *A. lycopersici* to survive winters in Canada is unknown.

Background

A. lycopersici infestations can be prevented by a strict greenhouse sanitation programme, especially thorough cleaning between crops. Humidities of 70–80% will help prevent infestations. Methods for early detection of *A. lycopersici* on greenhouse crops are needed.

Biological Control Agents

Predators

Various commercially available species, e.g. *Phytoseiulus persimilis* Athias-Henriot, *Amblyseius cucumeris* (Oudemans), *Amblyseius fallacis* Garman, *Metaseiulus occidentalis* (Nesbitt) and *Orius tristicolor* (White), will feed on *A. lycopersici* (Perring and Farrar, 1986; Brodeur *et al.*, 1997). Experimentally, *A. fallacis* and *M. occidentalis* were found to have the greatest potential as biological control agents for *A. lycopersici*.

Evaluation of Biological Control

One of the difficulties faced in biological control of *A. lycopersici* is that populations often increase to enormous numbers before being detected, making it difficult to introduce enough natural enemies to obtain effective control before economic damage has occurred.

Recommendations

Further work should include:

1. Continued evaluation of the natural enemy complex of *A. lycopersici* to find effective biological control agents.

References

Brodeur, J., Bouchard, A. and Turcotte, G. (1997) Potential of four species of predatory mites as biological control agents of the tomato russet mite, *Aculops lycopersici* (Massee) (Eriophyidae). *The Canadian Entomologist* 129, 1–6.

Perring, T.M. and Farrar, C.A. (1986) Historical perspective and current world status of the tomato russet mite (Acari: Eriophyidae). *Miscellaneous Publications of the Entomological Society of America* 63, 1–18.

7 *Adelphocoris lineolatus* (Goeze), Alfalfa Plant Bug (Hemiptera: Miridae)

J.J. Soroka and K. Carl

Pest Status

The alfalfa plant bug, *Adelphocoris lineolatus* (Goeze), native to Europe and western Asia, was introduced to North America in about 1917. The bugs are a major pest of seed alfalfa, *Medicago sativa* L., because they feed on buds, flowers and young pods, reducing the quantity and quality of seed produced. In severe infestations, *A. lineolatus* can totally destroy a alfalfa seed crop; the bugs are a chronic threat to the Can$50 million industry (Soroka and Murrell, 1993). Economic injury by *A. lineolatus* to sainfoin, *Onobrychis viciaefolia* Scopoli (Morrill *et al.*, 1984), and birdsfoot trefoil, *Lotus maizeiculatus* L. (Wipfli *et al.*, 1990; Peterson *et al.*, 1992), also occurs. *A. lineolatus* has become a pest on cotton, *Gossypium hirsutum* L. (Khamraev, 1993; Li *et al.*, 1994; Gao and Li, 1998), in Asia, and on such diverse crops as asparagus, *Asparagus officinalis* L., shoots (Wukasch and Sears, 1982) and blackberries and raspberries, *Rubus* spp. (Spangler *et al.*, 1993), in North America.

In North America, at latitudes below 51°N, two or more generations per year occur, and at latitudes above 53°N only one complete generation of *A. lineolatus* occurs (Craig, 1963). Eggs overwinter in stems of host plants, primarily legumes such as alfalfa, sainfoin, birdsfoot trefoil, red clover, *Trifolium pratense* L., and sweet clover, *Melilotus officinalis* Lamarck and *Melilotus alba* Desvaux. Nymphs emerge in spring; development proceeds through five nymphal instars, and first-generation adults appear about mid-June.

Background

Because *A. lineolatus* overwinters as eggs in crop residue, late autumn or early spring burning of alfalfa stubble is effective in controlling its populations. If burning is not feasible, *A. lineolatus* can be controlled by using a recommended insecticide when alfalfa is in early bud. The removal of biomass by ensiling, dehydrating, and pelleting or cubing alfalfa hay will usually limit the build-up of *A. lineolatus* populations in alfalfa hay fields. Removal of weeds near horticultural crops early in the season may help to control *A. lineolatus*.

Polynema pratensiphagum Walley parasitizes *A. lineolatus* eggs (Al-Ghamdi *et al.*, 1993). *Phasia robertsonii* (Townsend) reportedly parasitizes adult *A. lineolatus* at levels of 0.1% (Day, 1995). Wheeler (1972) found the fungus *Entomophthora erupta* (Dustan) infecting up to 33% of *A. lineolatus* nymphs in alfalfa near Ithaca, New York.

In North America, *Peristenus pallipes* (Curtis)[1] parasitizes first-generation *A. lineolatus* nymphs (Loan, 1965). Day (1987) found parasitism of *A. lineolatus* by *P. pallipes* in New Jersey to be 20%, considerably less than reported in Ontario (40–60%, [Loan, 1965]), but more than in Saskatchewan (0–4%, [Craig and Loan, 1987]), where it is primarily a parasitoid of *Lygus* spp. In areas where *A. lineolatus* is bi- or multivoltine, no parasitism of the second generation by *P. pallipes* has been found, although Day (1987) found 4% of third-generation *A. lineolatus* to be parasitized.

Biological Control Agents

Parasitoids

In western Europe, known parasitoids of *Adelphocoris* nymphs are *Peristenus adelphocoridis* Loan, *P. conradi* Marsh, *P. digoneutis* Loan, *P. pallipes*, *P. rubricollis* (Thomson) and *P. stygicus* Loan (Bilewicz-Pawinska, 1977; Loan, 1979; Day, 1987, 1997). This parasitoid complex is similar to that found on European tarnished plant bug, *Lygus rugulipennis* Poppius, except for *P. adelphocoridis*, which may be specific to *A. lineolatus*.

A. lineolatus is generally a rare species in European agroecosystems. Because *A. lineolatus* hibernates as eggs, in cultivated areas where females oviposit into the stalks of alfalfa or clovers, most of the eggs are removed from the field with the autumn harvest of the crop. Therefore, large collections of parasitized nymphs could only be made in the experimental fields that were strip-cut only twice in the season.

Releases and Recoveries

In Saskatchewan, eight separate releases of *P. adelphocoridis*, *P. digoneutis* and *P. rubricollis* were made in alfalfa fields in the early and mid-1980s (Table 7.1). The largest single release was of *P. digoneutis*, which, according to Day (1996), prefers to parasitize *Lygus lineolaris*. No recovery has been made of any of these introduced species. These parasitoid species are sympatric in their distribution, and *P. digoneutis*, released for control of *Lygus* spp. (see Broadbent *et al.*, Chapter 32, this volume), may become established on *A. lineolatus*.

Evaluation of Biological Control

Although not all of the introduced *Peristenus* spp. have established, their potential as biological control agents remains high. *P. conradi* is established in the USA (Day *et al.*, 1992). First discovered in 1989 near Newark, Delaware, it apparently was introduced accidentally along with an unsuccessful introduction of *P. rubricollis*. It has spread north-eastward along the eastern seaboard of the USA (Day *et al.*, 1992, 1998). This species has one generation a year, with moderate levels of parasitism of *A. lineolatus* (20–30%, [Day, 1997]). In Quebec, Broadbent *et al.* (1999)

[1]The status of *P. pallipes* and other *Peristenus* spp. is currently being reviewed. The North American *P. pallipes* is a new species (H. Goulet, Ottawa, 2000, personal communication).

Table 7.1. Introduction of *Peristenus* spp. into Saskatchewan (SK) for laboratory studies or field releases against *Adelphocoris lineolatus*, 1981–1999.

Year introduced	Site of introduction	Lab study (L) or field release (F)	Parasitoid species	Country of origin	Number introduced
1981a	Shellbrook, SK 53°13′N 106°24′W	F	*P. adelphocoridis* Loan	Austria	12
1981b	Yellow Creek, SK 52°45′N 105°15′W	F	*P. adelphocoridis*	Austria	16
1981c	Saskatoon, SK 52°07′N 106°38′W	F	*P. adelphocoridis*	Austria	23
1985a	Saskatoon, SK 52°07′N 106°38′W	L	*P. adelphocoridis*	Austria	14
1985b	Saskatoon, SK 52°07′N 106°38′W	L	*P. digoneutis* Loan	Austria	14
1985c	Saskatoon, SK 52°07′N 106°38′W	F	*P. digoneutis*	Austria	12
1985d	Saskatoon, SK 52°07′N 106°38′W	F	*P. rubricollis* (Thompson)	Austria	6
1986a	Saskatoon, SK 52°07′N 106°38′W	L	*P. adelphocoridis*	Austria, Germany	3
1986b	Saskatoon, SK 52°07′N 106°38′W	F	*P. digoneutis*	Austria, Germany	294
1986c	Saskatoon, SK 52°07′N 106°38′W	F (cage)	*P. digoneutis*	Austria, Germany	50
1986d	Saskatoon, SK 52°07′N 106°38′W	L	*P. digoneutis*	Austria, Germany	24
1986e	Saskatoon, SK 52°07′N 106°38′W	F (cage)	*P. rubricollis*	Austria, Germany	6

found *P. conradi* in 1998 on *L. lineolaris* nymphs.

P. digoneutis is established along the eastern seaboard of the USA on tarnished plant bug, *L. lineolaris* (Day *et al*., 1992; Day, 1996) and was recently found in Quebec (Broadbent *et al*., 1999). It also attacks *A. lineolatus* at low levels, especially if *Lygus* bug numbers are low (Day, 1996).

Because of the relatively recent introduction of *A. lineolatus* from Europe without its accompanying parasitoids, it is an excellent candidate for a biological control programme. The small numbers of parasitoids introduced into Canada in the past rendered their establishment improbable.

Recommendations

Future work should include:

1. Developing mass rearing for *P. adelphocoridis*, *P. conradi* and *P. rubricollis*, as is presently being done with *P. digoneutis*;
2. Release of *P. conradi* and *P. digoneutis* from established sites in North America into regions where they are needed;
3. Exploration of areas of eastern Europe and central Asia for additional biological control agents, particularly multivoltine species or those attacking the second generation of *A. lineolatus*;
4. Resolution of the taxonomy of *Peristenus* spp. in the Holarctic region.

References

Al-Ghamdi, K.M., Stewart, R.K. and Boivin, G. (1993) Note on overwintering of *Polynema pratensiphagum* (Walley) (Hymenoptera: Mymaridae) in southwestern Quebec. *The Canadian Entomologist* 125, 407–408.

Bilewicz-Pawinska, T. (1977) Parasitism of *Adelphocoris lineolatus* Popp. (Heteroptera) by braconids and their occurrence on alfalfa. *Ekologia Polska* 25, 539–550.

Broadbent, A.B., Goulet, H., Whistlecraft, J.W., Lachance, S. and Mason, P.G. (1999) First Canadian record of three parasitoid species (Hymenoptera: Braconidae: Euphoridae) of the tarnished plant bug *Lygus lineolaris* (Hemiptera: Miridae). *Proceedings of the Entomological Society of Ontario* 130, 1–3.

Craig, C.H. (1963) The alfalfa plant bug, *Adelphocoris lineolatus* (Goeze), in northern Saskatchewan. *The Canadian Entomologist* 95, 1–13.

Craig, C.H. and Loan, C.C. (1987) Biological control efforts on Miridae in Canada. In: Hedlund, R. and Graham, H.M. (eds) *Economic Importance and Biological Control of* Lygus *and* Adelphocoris *in North America.* United States Department of Agriculture, Agricultural Research Publication ARS 64, pp. 48–53.

Day, W.H. (1987) Biological control efforts against *Lygus* and *Adelphocoris* spp. infesting alfalfa in the United States, with notes on other associated species. In: Hedlund, R. and Graham, H.M. (eds) *Economic Importance and Biological Control of* Lygus *and* Adelphocoris *in North America.* United States Department of Agriculture, Agricultural Research Publication ARS 64, pp. 20–39.

Day, W.H. (1995) Biological observations on *Phasia robertsonii* (Townsend) (Diptera: Tachinidae), a native parasite of adult plant bugs (Hemiptera: Miridae) feeding on alfalfa and grasses. *Journal of the New York Entomological Society* 103, 100–106.

Day, W.H. (1996) Evaluation of biological control of the tarnished plant bug (Hemiptera: Miridae) in alfalfa by the introduced parasite *Peristenus digoneutis* (Hymenoptera: Braconidae). *Environmental Entomology* 25, 512–518.

Day, W.H. (1997) Biological control of mirids in northeastern alfalfa. In: Soroka, J. (ed.) *Proceedings of the Lygus Working Group Meeting, April 11–12, 1996, Winnipeg, MB.* Agriculture and Agri-Food Canada, Saskatoon Research Centre, Saskatoon, Saskatchewan, pp. 23–28.

Day, W.H., Marsh, P.M., Fuester, R.W., Hoyer, H. and Dysart, R.J. (1992) Biology, initial effect, and description of a new species of *Peristenus* (Hymenoptera: Braconidae), a parasite of the alfalfa plant bug (Hemiptera: Miridae), recently established in the United States. *Annals of the Entomological Society of America* 85, 482–488.

Day, W.H., Tropp, J.M., Eaton, A.T., Romig, R.F., van Driesche, R.G. and Chianese, R.J. (1998) Geographic distributions of *Peristenus conradi* and *P. digoneutis* (Hymenoptera: Braconidae), parasites of the alfalfa plant bug and the tarnished plant bug (Hemiptera: Miridae) in the northeastern United States. *Journal of the New York Entomological Society* 106, 69–75.

Gao, Z.R. and Li, Q.O. (1998) On the selectivity and dispersion of alfalfa plant bug among its host plants in eastern Henan cotton region. *Acta Phytophylacica Sinica* 25, 330–336.

Khamraev, A.S. (1993) Mirids as cotton pests. *Zaschita Rastenii 1993* No. 4, 25–26.

Li, Q.S., Liu, Q.X. and Deng, W.X. (1994) The effect of different host plants on the population dynamics of the alfalfa plant bug. *Acta Phytophylacica Sinica* 21, 351–355.

Loan, C.C. (1965) Life cycle and development of *Leophron pallipes* Curtis (Hymenoptera: Braconidae, Euphorinae) in five mirid hosts in the Belleville district. *Proceedings of the Entomological Society of Ontario* 100, 188–195.

Loan, C.C. (1979) Three new species of *Peristenus* Foerster from Canada and western Europe (Hymenoptera: Braconidae, Euphorinae). *Le Naturaliste Canadien* 106, 387–391.

Morrill, W.L., Ditterline, R.L. and Winstead, C. (1984) Effects of *Lygus borealis* Kelton (Hemiptera: Miridae) and *Adelphocoris lineolatus* (Goeze) (Hemiptera: Miridae) feeding on sainfoin seed production [*Onobrychis viciifolia*]. *Journal of Economic Entomology* 77, 966–968.

Peterson, S.S., Wedberg, J.L. and Hogg, D.B. (1992) Plant bug (Hemiptera: Miridae) damage to birdsfoot trefoil seed production. *Journal of Economic Entomology* 85, 250–255.

Soroka, J.J. and Murrell, D.C. (1993) The effects of alfalfa plant bug (Hemiptera: Miridae) feeding late in the season on alfalfa seed yield in northern Saskatchewan. *The Canadian Entomologist* 125, 815–824.

Spangler, S.M., Agnello, A.M. and Schwartz, M.D. (1993) Seasonal densities of tarnished plant bug, *Lygus lineolaris* (Palisot), and other phytophagous Heteroptera in brambles. *Journal of Economic Entomology* 86, 110–116.
Wheeler, A.G. (1972) Studies on the arthropod fauna of alfalfa. III Infection of the alfalfa plant bug, *Adelphocoris lineolatus* (Hemiptera: Miridae) by the fungus *Entomophthora erupta*. *The Canadian Entomologist* 104, 1763–1766.
Wipfli, M.S., Wedberg, J.L. and Hogg, D.B. (1990) Damage potentials of three plant bug (Hemiptera: Heteroptera: Miridae) species to birdsfoot trefoil grown for seed in Wisconsin. *Journal of Economic Entomology* 83, 580–584.
Wukasch, R.T. and Sears, M.K. (1982) Damage to asparagus by tarnished plant bugs, *Lygus lineolaris*, and alfalfa plant bugs, *Adelphocoris lineolatus* (Heteroptera: Miridae). *Proceedings of the Entomological Society of Ontario* 112, 49–51.

8 *Aedes, Culiseta* and *Culex* spp., Mosquitoes (Diptera: Culicidae)

T.D. Galloway, M.S. Goettel, M. Boisvert and J. Boisvert

Pest Status

Mosquitoes, particularly *Aedes* spp., *Anopheles* spp. and *Culex* spp. (Diptera: Culicidae), are important pests of humans and livestock in North America. Among the species known to bite humans or domestic animals and birds in Canada, *Culex tarsalis* Coquillett, *Mansonia perturbans* (Walker) and *Culex pipiens* L. are important vectors of arboviruses, e.g. western equine encephalitis, eastern equine encephalitis and St Louis encephalitis (Wood *et al.*, 1979) that endanger the health of domestic animals and humans in many parts of the country. Exotic pathogens may also be vectored by native mosquitoes, e.g. *Anopheles* spp., presenting ongoing disease threats. Floodwater and snowmelt *Aedes* spp. can be present in extraordinary numbers and constitute a major source of annoyance and stress to livestock, wildlife and humans (Laird *et al.*, 1982). Species that develop enormous populations, e.g. *Aedes vexans* (Meigen), particularly during wet summers (Wood *et al.*, 1979; Wood, 1985), have earned Canada a worldwide reputation for its mosquito pest populations.

Wood *et al.* (1979) and Wood (1985) summarized mosquito life cycles in Canada. Overwintering may occur in the egg, larval or adult stages. Females of pest species usually require a blood meal to produce large numbers of eggs. Eggs may be laid on permanent or semipermanent standing water, in tree holes, rock pools, man-made containers or on the soil at the margins of temporary pools. After the eggs hatch, the larvae pass through four instars, feeding on living or dead organic matter in water (except for a couple of uncommon, predacious species). Pupae are also aquatic, although they breathe surface air through thoracic trumpets. Fully developed adults eclose from floating pupae at the water surface. Males of many species form mating swarms to which females are attracted, and mating takes place in the air. Depending on the species and environmental factors, one or more generations occur each year.

Background

Significant annoyance and the potential for transmission of potentially lethal disease-causing organisms have made mosquito control programmes important requirements in many communities throughout Canada. The risks of contracting arboviruses and other diseases has led to development of detailed monitoring and control implementation procedures (Canada Biting Fly Centre, 1990). Although personal protection, in the form of repellents and protective clothing, is helpful, it has limited effectiveness. Mosquito control using chemical insecticides (adulticides or larvicides), applied by aircraft and by vehicle-mounted or backpack sprayers, is still widely practised in certain provinces, e.g. Manitoba. Because of the negative impacts of these chemicals on humans, wildlife and non-target aquatic invertebrates, alternative control strategies have been sought and implemented in some provinces, e.g. Quebec.

Biological Control Agents

George (1984) and Shemanchuk *et al.* (1984) reported on biological control of *Culex pipiens* L. and *Culiseta inornata* (Williston) using flatworms, *Dugesia tigrina* (Girard) and the fungus, *Coelomomyces psorophorae* Couch. Mermithid nematodes, e.g. *Hydromermis churchillensis*, associated with mosquitoes, e.g. *Aedes communis* (DeGeer), were reported in Canada almost 50 years ago (Beckel and Copps, 1955; Welch, 1960). Brust and Smith (1972) observed juvenile nematodes in adult *Aedes hexodontus* (Dyar) and *Aedes impiger* (Walker) near Baker Lake, Northwest Territories. Although Trpiš *et al.* (1968) and Trpiš (1971) examined the impact of mermithids, their potential as biological control agents in Canada was largely unexplored.

Pathogens

Nematodes

The potential of Mermithidae for mosquito biological control became apparent following development of mass-rearing procedures for *Romanomermis culicivorax* Ross and Smith from Louisiana (Petersen and Willis, 1972). This species showed a wide host range (Petersen and Chapman, 1979), could be easily applied to mosquito breeding sites, and was the first mermithid to be commercially available (Nickle, 1976). In Canada, work has focused on its morphology and physiological relationships with its host (Curran, 1981, 1982; Gordon *et al.*, 1981, 1982, 1989, 1990; Curran and Webster, 1983, 1984; Gordon and Burford, 1984; Galloway and Brust, 1985; Gordon, 1986, 1987; Gordon and Cornect, 1987; Jagdale and Gordon, 1994a, b). Because *R. culicivorax* is found naturally only in the southern USA, it was not surprising that its field use in Canada was restricted by low temperatures (Galloway and Brust, 1977), which caused low parasitism in field trials against spring *Aedes* spp. in Manitoba (Galloway and Brust, 1976). However, low temperatures (10°C and 15°C) favoured long-term storage of embryonated eggs (Thornton *et al.*, 1982). Unsuitable hosts may also limit the potential for *R. culicivorax* for biological control, e.g. even at very high application rates (10,000–100,000 preparasites m^{-2}), levels of infection in *Ae. vexans* larvae did not exceed 50% in artificial pools (Galloway and Brust, 1985).

Native Mermithidae besides *H. churchillensis* that parasitize mosquito larvae in northern Canada are *Romanomermis hermaphrodita* Ross and Smith, *R. kiktoreak* Ross and Smith, and *R. communensis* Galloway and Brust (Ross and Smith, 1976; Galloway and Brust, 1979). Thornton (1978) detailed the biology of *R. communen-*

sis and described the difficulties in stimulating synchronous hatch in embryonated eggs. Galloway and Brust (1982) discovered a limited capacity for cross-mating between *R. culicivorax* and *R. communensis*. Mermithids may also parasitize mosquito larvae but complete their development in the adult stage (e.g. Trpiš *et al.*, 1968). Galloway (1976), Thornton (1978) and Harlos *et al.* (1980) studied a *Culicimermis* sp. that emerged from adult *Ae. vexans* in Manitoba. Nearly 50% of field-collected larvae were parasitized by this species at one locality, and infected females never successfully laid eggs. This mermithid was reared through four successive generations in the laboratory (Harlos *et al.*, 1980) and, as a parasite of one of Canada's most important pest species, *Ae. vexans*, warrants further investigation for biological control.

Fungi

Culicinomyces clavisporus Couch, Romney and Rao and *Smittium* sp. were first recorded in Canada by Goettel (1987a). The Canadian isolate of *C. clavisporus* was compared with isolates from the USA and Australia with regard to growth rate, colonial morphology and pigmentation (Goettel *et al.*, 1984). The Canadian and Australian isolates were more similar to each other than to the American isolate.

Taylor *et al.* (1980) provided the first record of infection of *Aedes trivittatus* (Coquillett) by a *Coelomomyces* sp. Adult females were collected in 1978 from a scrub oak flood plain along the La Salle River near Winnipeg, Manitoba, and were provided with a blood meal in the laboratory. Within 5 or 6 days, about 50% of the females had died. Examination of the cadavers revealed mature *Coelomomyces* sporangia within the haemocoel. In subsequent studies in artificial pools in 1979, infections taking place during the fourth larval instar and/or during the pupal stage resulted in infected adults. In addition, sporangia were only found in blood-fed adults. *Aedes sticticus* (Meigen) was also found infected with *Coelomomyces* sp. at the same study site in 1977.

Laboratory host–pathogen studies between *Coelomomyces stegomyiae* Keilin and *Aedes aegypti* L. showed that production of infected females is affected by larval instar and inoculum at the time of infection, and by rearing temperature following infection (Shoulkamy *et al.*, 1997). After breaching the host cuticle, hyphae ramified throughout the fat body, leading to cell lysis and depletion of fat bodies (Shoulkamy and Lucarotti, 1998). Hyphae also invade muscle and gut tissues and the lumen of haemopoietic organs and imaginal discs.

Tolypocladium cylindrosporum Gams was evaluated as a potential biological control agent (Goettel, 1987b). This is a relatively slow-acting fungal pathogen with relatively low virulence to mosquitoes; large doses are required to elicit a response (Goettel, 1987c). LC_{50}s were about 10^4–10^5 conidia ml^{-1}; LT_{50}s were 3–14 days against larval *Ae. aegypti*, *Ae. vexans* and *Culiseta inornata*. No increased pathogenicity occurred after passage of the fungus 18 times through mosquito larvae (Goettel, 1987d). The fungus was easily propagated on a cellophane surface and wheat bran (Goettel, 1984). The half-life of conidia stored at −20°C was 12.8 months (Goettel, 1987e). Principal sites of invasion of *T. cylindrosporum* are through the base of the mandibles and maxillae and the anus of *Ae. aegypti* (Goettel, 1988a). Larvae were most susceptible immediately prior to moulting, although little fungal colonization of the haemocoel occurred at this time. Conidia ingested by larvae were still viable after excretion (Goettel, 1988b). In Alberta, mass applications of conidia in the field failed to induce an epizootic; however, infections were apparent in larvae transferred to laboratory conditions up to 29 days after application (Goettel, 1987f). In Quebec, *T. cylindrosporum* was active in laboratory bioassays against *Aedes triseriatus* Say (Nadeau and Boisvert, 1994). All larval instars of *Ae. triseriatus* were susceptible at temperatures of 18–25°C. Blastospores were more virulent than conidia. Use of blastospores and limiting exposure time were better methods for bioassay of *T. cylindrosporum* against mosquitoes, as compared to using conidia and continuous exposure. Laboratory challenge tests with conidia expanded the previous known host range of *T. cylindrosporum* to include species of Ceratopogonidae, Chaoboridae and Psychodidae (Lam *et al.*, 1988).

A Californian strain of *Lagenidium giganteum* Couch, recently registered in the USA, was evaluated in artificial pools in the southern coastal forest of British Columbia (Lux, 1995). First-instar mosquito larvae were added to each pool twice weekly, and the numbers of emerging adults were counted. In 1994, four pools were inoculated with zoospores and mycelia of *L. giganteum*. In 1995, three pools each received 5.4×10^6 zoospores. No significant reductions in the number of adults emerging from treated and untreated pools were noted in 1994. In contrast, significant reductions in adult emergence occurred for a period of 92 days in 1995. In field surveys, *L. giganteum* has not been found to occur naturally in the lower mainland of British Columbia. However, larvae of *Cs. inornata* infected with a *Lagenidium* sp. were collected near Lethbridge, Alberta, in 1973 (H.C. Whisler, Pullman, 1982, personal communication).

Bacteria

Bacillus thuringiensis Berliner serovar *israelensis* (*B.t.i.*), discovered in 1976, was registered in Canada for mosquito control shortly thereafter. At that time mosquito-borne viral encephalitides, especially western equine encephalitis, were a major concern in western Canada. Because of its high degree of specificity, *B.t.i.* was hailed as the solution to replace chemical insecticides.

In the early 1980s, research on *B.t.i.* was carried out in Manitoba, Ontario, Newfoundland and Quebec. In Manitoba, Sebastien and Brust (1981) first tested two formulations of *B.t.i.*, which gave good control of *Ae. vexans* and *Culex restuans* Theobald larvae in artificial, sod-lined pools, although residual activity was less than 24 h. Non-target, invertebrate predators (Odonata and Hemiptera) were not affected by the treatments over a 5-day period. This paper appears to be the only one published by Canadian researchers on the use of *B.t.i.* formulations to control mosquito larvae.

Dupont and Boisvert (1985) and Boisvert and Boisvert (1999) studied the persistence of *B.t.i.* activity in Canadian marshes. Diffusion chambers contained a *B.t.i.* formulation with and without natural substrates and were separated from the marsh water by a membrane. Contrary to findings in warmer climates, they showed that *B.t.i.* toxicity remained quite stable for nearly 3 weeks in chambers without natural substrates and then declined. *B.t.i.* toxicity against mosquito larvae persisted for up to 4–5 months in the presence of vegetation within these chambers. Recycling of *B.t.i.* spores could occur in the diffusion chambers but, under these conditions, the intensity of recycling would not be sufficient to maintain larvicidal activity.

In Canada, no studies have been conducted to determine the long-term effect of *B.t.i.* treatments on non-target organisms in mosquito control programmes (Lacoursière and Boisvert, 1994). Boisvert and Boisvert (2000) reviewed the effects of both unformulated and formulated *B.t.i.* on target and non-target species. Of the more than 300 articles studied, results from only one paper could be extrapolated to certain Canadian biotopes. In that study, intensive *B.t.i.* treatments over a 3-year period caused an important effect on insect diversity, richness and density in mosquito marshes. Municipalities in most provinces and the military currently use *B.t.i.* to control mosquitoes.

Evaluation of Biological Control

Nematodes have proved less than ideal for biological control of mosquitoes under Canadian conditions. Much of the work has been carried out on *R. culicivorax*, a species neither particularly well suited to survival in most parts of Canada nor very effective against some of our most important mosquito pests; however, endemic species require further investigation. Mass production of Mermithidae is difficult and expensive, being restricted, for the time being, to *in vivo* methods.

B.t.i. has been very successful and is generally used only in 'ecologically sensitive' areas. With public pressure to reduce or eliminate chemical pesticide use, especially within urban areas, it can be expected that use of *B.t.i.* will increase. In Quebec, *B.t.i.* has been used exclusively since 1984 to control nuisance mosquitoes in and around urban areas. In 2000, control programmes were undertaken in 25 municipalities to protect nearly 700,000 people (J.F. Bourque, Québec, 2000, personal communication). No resistance to *B.t.i.* has been observed (C. Black, Trois-Rivières, 2000, personal communication), most probably because of the small number of treatments per year. Although *B.t.i.* users are not required to report possible resistance problems to federal or provincial authorities, if use increases the possibility of selecting resistant populations will need to be considered in any long-term mosquito abatement programme.

Recommendations

Further work should include:

1. Additional surveys and taxonomic research to find mermithids that parasitize mosquitoes in Canada;
2. Determining the potential of *Culicimermis* sp. as a biological control agent for *Ae. vexans*;
3. Determining long-term, non-target effects and the possibility of resistance development, assuming that *B.t.i.* will be used exclusively in long-term mosquito abatement programmes;
4. Searching for, and selection of, pathogens adapted to the Canadian environment.

References

Beckel, W.E. and Copps, T.P. (1955) *Laboratory Rearing of the Adults of Northern* Aedes *Mosquitoes (Culicidae).* Report of the Defense Research Board of Canada, Ottawa, Ontario, DRNL 7/55.

Boisvert, M. and Boisvert, J. (1999) Persistence of toxic activity and recycling of *Bacillus thuringiensis* var. *israelensis* in cold water: field experiments using diffusion chambers in a pond. *Biocontrol Science and Technology* 9, 507–522.

Boisvert M. and Boisvert, J. (2000) Effects of *Bacillus thuringiensis* var. *israelensis* on target and non-target organisms: a review of laboratory and field experiments. *Biocontrol Science and Technology* 10, 517–561.

Brust, R.A. and Smith, S.M. (1972) Mosquito intersexes in the arctic of Canada (Diptera: Culicidae). *Proceedings of the XIII International Congress of Entomology, Moscow,* 3, pp. 135–136.

Canada Biting Fly Centre (1990) *A Manual on Guidelines for the Control of Arboviral Encephalitides in Canada.* Agriculture Canada, Research Branch, Ottawa, Ontario, Technical Bulletin 1990-5E.

Curran, J. (1981) Morphometrics of *Romanomermis culicivorax* Ross and Smith, 1976 (Nematoda: Mermithidae). *Canadian Journal of Zoology* 59, 2365–2374.

Curran, J. (1982) Morphological variation in *Romanomermis culicivorax* Ross and Smith, 1976 (Nematoda: Mermithidae). *Canadian Journal of Zoology* 60, 1007–1011.

Curran, J. and Webster, J.M. (1983) Post-embryonic growth of *Romanomermis culicivorax* Ross and Smith, 1976: an example of accretionary growth in Nematoda. *Canadian Journal of Zoology* 61, 1793–1796.

Curran, J. and Webster, J.M. (1984) Reproductive isolation and taxonomic differentiation of *Romanomermis culicivorax* Ross and Smith, 1976 and *R. communensis* Galloway and Brust, 1979. *Journal of Nematology* 16, 375–379.

Dupont, C. and Boisvert, J. (1985) Persistence of *Bacillus thuringiensis* serovar. *israelensis* toxic activity in the environment and interaction with natural substrates. *Water, Air, and Soil Pollution* 29, 425–438.

Galloway, T.D. (1976) Observations on mermithid parasites of mosquitoes in Manitoba. In: *Proceedings of the 1st International Symposium on Invertebrate Pathology,* Kingston, Ontario, pp. 227–231.

Galloway, T.D. and Brust, R.A. (1976) Field application of the mermithid nematode, *Romanomermis culicivorax* Ross and Smith, for the control of mosquitoes, *Aedes* spp., in spring in Manitoba. *Manitoba Entomologist* 10, 18–25.
Galloway, T.D. and Brust, R.A. (1977) Effects of temperature and photoperiod on the infection of two mosquito species by *Romanomermis culicivorax. Journal of Nematology* 9, 218–221.
Galloway, T.D. and Brust, R.A. (1979) Review of the genus *Romanomermis* (Nematoda: Mermithidae) with a description of *R. communensis* sp.n. from Canada. *Canadian Journal of Zoology* 57, 281–289.
Galloway, T.D. and Brust, R.A. (1982) Cross-mating of *Romanomermis culicivorax* and *R. communensis* (Nematoda: Mermithidae). *Journal of Nematology* 14, 274–276.
Galloway, T.D. and Brust, R.A. (1985) Results of field trials using *Romanomermis culicivorax* (Nematoda: Mermithidae) against *Aedes vexans* (Diptera: Culicidae), and the effects of parasitism on growth and development of larvae in laboratory and field tests. *Canadian Journal of Zoology* 63, 2437–2442.
George, J.A. (1984) *Culex pipiens* L., North House Mosquito (Diptera: Culicidae). In: Kelleher, J.S. and Hulme, M.A. (eds) *Biological Control Programmes against Insects and Weeds in Canada 1969–1980.* Commonwealth Agricultural Bureaux, Farnham Royal, Slough, UK, pp. 19–21.
Goettel, M.S. (1984) A simple method for mass culturing entomopathogenic hyphomycete fungi. *Journal of Microbiological Methods* 3, 15–20.
Goettel, M.S. (1987a) Field incidence of mosquito pathogens and parasites in central Alberta. *Journal of the American Mosquito Control Association* 3, 231–238.
Goettel, M.S. (1987b) Studies on microbial control of mosquitoes in central Alberta with emphasis on the hyphomycete *Tolypocladium cylindrosporum.* PhD thesis, University of Alberta, Edmonton, Alberta, Canada.
Goettel, M.S. (1987c) Studies on bioassay of the entomopathogenic hyphomycete fungus *Tolypocladium cylindrosporum* in mosquitoes. *Journal of the American Mosquito Control Association* 3, 561–567.
Goettel, M.S. (1987d) Serial *in vivo* passage of the entomopathogenic hyphomycete *Tolypocladium cylindrosporum* in mosquitoes. *The Canadian Entomologist* 119, 599–601.
Goettel, M.S. (1987e) Conidial viability of the mosquito pathogenic hyphomycete *Tolypocladium cylindrosporum* following prolonged storage at −20°C. *Journal of Invertebrate Pathology* 50, 327–329.
Goettel, M.S. (1987f) Preliminary field trials with the entomopathogenic hyphomycete *Tolypocladium cylindrosporum* in central Alberta. *Journal of the American Mosquito Control Association* 3, 239–245.
Goettel, M.S. (1988a) Pathogenesis of the hyphomycete *Tolypocladium cylindrosporum* in the mosquito *Aedes aegypti. Journal of Invertebrate Pathology* 51, 254–274.
Goettel, M.S. (1988b) Viability of *Tolypocladium cylindrosporum* (Hyphomycetes) conidia following ingestion and excretion by larval *Aedes aegypti. Journal of Invertebrate Pathology* 51, 275–277.
Goettel, M.S., Sigler, L. and Carmichael, J.W. (1984) Studies on the mosquito pathogenic hyphomycete *Culicinomyces clavisporus. Mycologia* 76, 614–625.
Gordon, R. (1986) Recent advances on the physiology of *Romanomermis culicivorax*, a mermithid parasite of mosquitoes. In: Samson, R.A., Vlak, J.M. and Peters, D. (eds) *Fundamental and Applied Aspects of Invertebrate Pathology.* Foundation of the Fourth International Colloquium on Invertebrate Pathology, Veldhoven, The Netherlands, pp. 292–295.
Gordon, R. (1987) Glyoxylate pathway in the free-living stages of the entomophilic nematode *Romanomermis culicivorax. Journal of Nematology* 19, 277–281.
Gordon, R. and Burford, I.R. (1984) Transport of palmitic acid across the tegument of the entomophilic nematode *Romanomermis culicivorax. Journal of Nematology* 16, 14–21.
Gordon, R. and Cornect, M. (1987) Nutrient composition of *Romanomermis culicivorax* in relation to egg production and metabolism. *Journal of Nematology* 19, 487–494.
Gordon, R., Squires, J.M., Babie, S.J. and Burford, I.R. (1981) Effects of host diet on *Romanomermis culicivorax*, a mermithid parasite of mosquitoes. *Journal of Nematology* 13, 285–290.
Gordon, R., Burford, I.R. and Young, T.L. (1982) Uptake of lipids by the entomophilic nematode *Romanomermis culicivorax. Journal of Nematology* 14, 492–495.
Gordon, R., Cornect, M., Walters, B.M., Hall, D.E. and Brosnan, M.E. (1989) Polyamine synthesis by the mermithid nematode *Romanomermis culicivorax. Journal of Nematology* 21, 81–86.

Gordon, R., Cornect, M., Young, T.L. and Kean, K.T. (1990) Empirical and physiological assessment of *in vitro* growth in the mermithid nematode *Romanomermis culicivorax*. *Canadian Journal of Zoology* 68, 511–516.

Harlos, J.A., Brust, R.A. and Galloway, T.D. (1980) Observations on a nematode parasite of *Aedes vexans* (Diptera: Culicidae) in Manitoba. *Canadian Journal of Zoology* 58, 215–220.

Jagdale, G.B. and Gordon, R. (1994a) Role of catecholamines in the reproduction of *Romanomermis culicivorax*. *Journal of Nematology* 26, 40–45.

Jagdale, G.B. and Gordon, R. (1994b) Caudal papillae in *Romanomermis culicivorax*. *Journal of Nematology* 26, 235–237.

Lacoursière J.O. and Boisvert, J. (1994) *Le* Bacillus thuringiensis *et le contrôle des insectes piqueurs au Québec*. Rapport présenté pour la Direction du Milieu Agricole et du Contrôle des Pesticides, Ministère de l'Environnement, Province de Québec, Quebec, QC, Canada.

Laird, M., Aubin, A., Belton, P., Chance, M.M., Fredeen, F.J.H., Haufe, W.O., Hynes, H.B.N., Lewis, D.J., Lindsay, I.S., McLean, D.M., Surgeoner, G.A. and Wood, D.M. (1982) *Biting Flies in Canada: Health Effects and Economic Consequences*. National Research Council of Canada, Ottawa, Ontario, No. 19248.

Lam, T.N.C., Soares, G.G., Jr and Goettel, M.S. (1988) Host records of the mosquito pathogenic hyphomycete *Tolypocladium cylindrosporum*. *Florida Entomologist* 71, 86–89.

Lux, D.K. (1995) Pathogenic efficacy of the Californian strain of *Lagenidium giganteum* (Oomycetes: Lagenidiales) on larval mosquitoes in the southern coastal forest of British Columbia with results of a field survey for native *Lagenidium* strains. MPM thesis, Simon Fraser University, Burnaby, British Columbia, Canada.

Nadeau, M.P. and Boisvert, J.L. (1994) Larvicidal activity of the entomopathogenic fungus *Tolypocladium cylindrosporum* (Deuteromycotina: Hyphomycetes) on the mosquito *Aedes triseriatus* and the black fly *Simulium vittatum* (Diptera: Simuliidae). *Journal of the American Mosquito Control Association* 10, 487–491.

Nickle, W.R. (1976) Toward commercialization of a mosquito mermithid. In: *Proceedings of the 1st International Symposium on Invertebrate Pathology*, Kingston, Ontario, Canada, pp. 241–244.

Petersen, J.J. and Chapman, H.C. (1979) Checklist of mosquito species tested against the nematode parasite *Romanomermis culicivorax*. *Journal of Medical Entomology* 15, 468–471.

Petersen, J.J. and Willis, O.R. (1972) Procedures for the mass rearing of a mermithid parasite of mosquitoes. *Mosquito News* 32, 226–230.

Ross, J.R. and Smith, S.M. (1976) A review of mermithid parasites (Nematoda: Mermithidae) described from North American mosquitoes (Diptera: Culicidae) with descriptions of three new species. *Canadian Journal of Zoology* 54, 1084–1102.

Sebastien, R.J. and Brust, R.A. (1981) An evaluation of two formulations of *Bacillus thuringiensis* var. *israelensis* for larval mosquito control in sod-lined simulated pools. *Mosquito News* 41, 508–512.

Shemanchuk, J.A., Whisler, H.C. and Zebold, S.L. (1984) *Culiseta inornata* (Williston), a mosquito (Diptera: Culicidae). In: Kelleher, J.S. and Hulme, M.A. (eds) *Biological Control Programmes against Insects and Weeds in Canada 1969–1980*, Commonwealth Agricultural Bureaux, Farnham Royal, Slough, UK, pp. 23–24.

Shoulkamy, M.A. and Lucarotti, C.J. (1998) Pathology of *Coelomomyces stegomyiae* in larval *Aedes aegypti*. *Mycologia* 90, 559–564.

Shoulkamy, M.A., Lucarotti, C.J., El-Ktatny, M.S.T. and Hassan, S.K.M. (1997) Factors affecting *Coelomomyces stegomyiae* infections in adult *Aedes aegypti*. *Mycologia* 89, 830–836.

Taylor, B.W., Harlos, J.A. and Brust, R.A. (1980) *Coelomomyces* infection of the adult mosquito *Aedes trivittatus* (Coquillett) in Manitoba. *Canadian Journal of Zoology* 58, 1215–1219.

Thornton, D.P. (1978) Studies on the biology of three mermithid parasites (Nematoda: Mermithidae) of mosquitoes. MSc thesis, University of Manitoba, Winnipeg, Manitoba, Canada.

Thornton, D.P., Brust, R.A. and Galloway, T.D. (1982) Effect of low temperatures on development and survival of postparasitic juveniles of *Romanomermis culicivorax* (Nematoda: Mermithidae). *Journal of Nematology* 14, 386–393.

Trpiš, M. (1971) Parasitical castration of mosquito females by mermithid nematodes. *Helminthologica* 10, 79–81.

Trpiš, M., Haufe, W.O. and Shemanchuk, J.A. (1968) Mermithid parasites of the mosquito *Aedes vexans* Meigen in British Columbia. *Canadian Journal of Zoology* 46, 1077–1079.

Welch, H.E. (1960) *Hydromermis churchillensis* n.sp. (Nematoda: Mermithidae) a parasite of *Aedes communis* (DeG.) from Churchill, Manitoba, with observations on its incidence and bionomics. *Canadian Journal of Zoology* 38, 465–474.
Wood, D.M. (1985) *Biting Flies Attacking Man and Livestock in Canada.* Agriculture Canada, Ottawa, Ontario, Publication 1781 E.
Wood, D.M., Dang, P.T. and Ellis, R.A. (1979) *The Insects and Arachnids of Canada. Part 6: The Mosquitoes of Canada Diptera: Culicidae.* Agriculture Canada, Ottawa, Ontario, Publication 1686.

9 *Aphis gossypii* Glover, Melon/Cotton Aphid, *Aulacorthum solani* (Kaltenbach), Foxglove Aphid, *Macrosiphum euphorbiae* (Thomas), Potato Aphid, and *Myzus persicae* (Sulzer), Green Peach Aphid (Homoptera: Aphididae)

D.R. Gillespie, J.L. Shipp, D.A. Raworth and R.G. Foottit

Pest Status

The melon/cotton aphid, *Aphis gossypii* Glover, the foxglove or glasshouse potato aphid, *Aulacorthum solani* (Kaltenbach), the potato aphid, *Macrosiphum euphorbiae* (Thomas), and the green peach aphid, *Myzus persicae* (Sulzer), are treated together here because of the common approaches to biological control applied against all four of these species in greenhouse vegetable crops. All are almost cosmopolitan pests of a wide range of crop plants (Blackman and Eastop, 1984) and occur on greenhouse crops across Canada. They cause damage through deposits of honeydew on fruit that encourage sooty moulds, retardation of plant growth, distortion of growing tips and fruit, and transmission of plant viruses. Crop losses result from a combination of plant defoliation, direct damage to fruit, costs of fruit washing, destruction of purchased biological control agents by pesticides applied against aphids, and subsequent damage by other pests as a result of their release from biological control.

Aphis gossypii in Canada is largely confined to greenhouses, and only anholocyclic (completely parthenogenetic) lines occur. In greenhouses, *A. gossypii* attacks cucumber, *Cucumis sativus* L., pepper, *Capsicum annuum* L., and a wide range of flower crops. Although populations have been recorded from tomato, *Lycopersicon esculentum* L., no damage has yet occurred. In British Columbia, damaging populations have been recorded from potato, *Solanum tuberosum* L. (Howard *et al.*, 1994).

Aulacorthum solani attacks potato outdoors, and pepper and tomato inside greenhouses (Howard *et al.*, 1994). Both anholocyclic and holocyclic (with sexual and parthenogenetic phases of the life cycle) races of this species are present. The foxglove aphid is unusual in that it can overwinter as eggs on various primary host species (Blackman and Eastop, 1984) including foxglove, *Digitalis purpurea* L., and buttercup, *Ranunculus* spp. The foxglove aphid vectors a wide range of viruses, and, on pepper, causes hypertoxic reactions that result in foliage and growing-point distortions, and abortion of flowers and fruit.

Macrosiphum euphorbiae is primarily a pest of *Solanaceae* that attacks potato outside greenhouses, and pepper and tomato inside greenhouses. Of the four aphid species, only *M. euphorbiae* is native to North America (Blackman and Eastop, 1984). It is holocyclic in north-eastern North America, and is mainly anholocyclic elsewhere (Blackman and Eastop, 1984). *Rosa* spp. are the overwintering (primary) hosts. In greenhouses, *M. euphorbiae* causes distortions of the growing points of pepper, and bud and flower abortion.

Myzus persicae overwinters on its primary hosts, *Prunus* spp., and during summer attacks secondary hosts, including many economically important crops species (Blackman and Eastop, 1984). In Canada, *M. persicae* is an important pest of asparagus, *Asparagus officinalis* L., spinach, *Spinacia oleracea* L., celery, *Apium graveolens* var. *dulce* (Miller) Persoon, crucifer crops, herb crops, potato, pepper, aubergine, *Solanum melongena* var. *esculentum* Nees, and tomato outdoors (Howard *et al.*, 1994). In greenhouses, *M. persicae* causes serious damage in sweet pepper but is rarely damaging on cucumber or tomato. In British Columbia, and probably elsewhere in Canada, damaging populations have occurred on greenhouse lettuce. Many flower crops are also attacked in greenhouses. In many parts of Canada, *M. persicae* survives in greenhouses, storage cellars and other protected environments as anholocyclic populations. It may occur infrequently as holocyclic populations where it overwinters as eggs on *Prunus* spp. (MacGillivray, 1972).

Background

Cultural approaches to control include screening, especially in greenhouses that are positively vented by fans. Weed control is an important adjunct to population management inside greenhouses, and in and around field crops. Use of oil- and soap-based pesticides is compatible to a degree with some natural enemies.

Given that all these aphids are important pests of greenhouse and horticultural crops, are widely distributed, and that three of the four species are of European or Asian origin, it is surprising that few classical biological control introductions have been made in Canada against these pests.

Four native parasitoid species were propagated at Belleville, Ontario, and shipped to greenhouse growers in Alberta, British Columbia, Ontario and Quebec, in 1938, 1939 and 1940 to control green peach aphid (McLeod, 1962), apparently successfully. Otherwise, there seem to have been no introductions or applications of biological control agents specifically against any of these pests until the mid-1980s.

In greenhouse vegetables, biological control of all four aphids by introductions of natural enemies has become the standard approach for their management. Factors that predispose greenhouse vegetable growers to use biological controls as the principal approach to IPM of aphids in greenhouses include: the negative effects of pesticide applications on natural enemies introduced to control other pest species and on bees used for pollination; pesticide resistance; withdrawal of specific aphicides or exclusion of their residues from exported produce (e.g. pirimicarb); and the periodic invasion of large numbers of winged aphids into greenhouses.

Biological Control Agents

Predators

Aphidoletes aphidimyza (Rondi), a virtually cosmopolitan aphid predator (Harris, 1973), is commonly introduced into greenhouses to control all four aphid pests. Gilkeson (1990) described release of this predator to control *M. persicae* on tomato and pepper, in combination with *Aphidius matricariae* Haliday. Adult *A. aphidimyza* lay eggs among colonies of aphids. Upon hatching, the larvae feed on all stages of *M. persicae*, eventually dropping to the soil to pupate. Adults feed on nectar only. The predator is shipped to growers from producers in Canada and Europe as pupae in bottles of vermiculite. The original purpose of introductions of *A. aphidimyza* was to establish populations that would persist throughout a growing season (Gilkeson, 1990). However, the shift of the greenhouse industry towards plastic floor coverings and soil-less culture has removed pupation sites from the greenhouse. Weekly introductions of pupae provide suppression of aphid populations, together with other natural enemies. Gilkeson *et al.* (1993) noted the presence of the parasitoid *Aphanogmus fulmeki* Ashmead in *A. aphidimyza* in Canada.

Hippodamia convergens Guerin, the convergent ladybird, collected in overwintering aggregations in California, is released inundatively in greenhouses to suppress *M. persicae* and *A. gossypii* outbreaks on pepper. However, *H. convergens* parasitized by either *Dinocampus* sp. or *Perilitus* sp. have inadvertently been imported in shipments. These parasitoids, which kill adult *H. convergens*, rapidly reduce the efficacy of beetle releases.

Harmonia axyridis (Pallas), the Asian ladybird, is reared commercially in insectaries in Canada and Europe. According to Gordon (1985) it was collected in Japan and the USSR, and introduced into North America several times between 1916 and 1981. However, Day *et al.* (1994) suggested establishment through accidental introductions at sea ports in eastern North America. The beetle is now distributed widely throughout North America, and is often the dominant species (H. Goulet, Ottawa, 2000, personal communication). Introductions of *H. axyridis* in greenhouse pepper establish breeding populations; its role in the control of aphids is still being evaluated.

When predator and parasitoid assemblages exist for a pest, emphasis must be placed on the effective use and management of the native species rather than the introduction of exotics. Exotic predators may simply displace native predators, with little gain in terms of pest control. *Coccinella septempunctata* L. was introduced into the USA after the 1950s. Coccinellid assemblages on alfalfa, *Medicago sativa* L., corn, *Zea mays* L., and small grains were monitored for 13 years before, and 5 years after, the establishment of *C. septempunctata* in South Dakota. Greatly reduced abundance of two species was observed, with no significant increase in total abundance of coccinellids in the crops (Elliott *et al.*, 1996). The rapid expansion of the range of another introduced ladybird, *H. axyridis* (e.g. Wheeler and Stoops, 1996), suggests that this introduced species may also affect species assemblages.

Parasitoids

Four parasitoid species are commonly released against aphid pests in greenhouse vegetable crops. These are *Aphidius matricariae, A. colemani* Viereck, *A. ervi* Haliday, and *Aphelinus abdominalis* (Dalman). *A. abdominalis* (Ferrière, 1965), *A. matricariae* and *A. ervi* are European in origin, and *A. colemani* originates from the Indian subcontinent (Mackauer and Starý, 1967). *A. matricariae* was originally introduced into North America in the 1950s (Clausen, 1978). Although establishment was not reported at that time, the species is now apparently widely distributed. All of these species are shipped as adults from producers in Canada and Europe to growers. Adults of all four species deposit eggs inside aphid nymphs. Larvae develop internally and eventually pupate inside a mummy formed from the exoskeleton of the dead aphid host. Gilkeson (1990) reported the successful use of inoculative releases of *A. matricariae* against *M. persicae* on greenhouse tomato and pepper. Since about 1990, *A. colemani* has been widely used in place of *A. matricariae* because the former has been shown to be superior for control of both *M. persicae* and *A. gossypii* (Van Steenis, 1993). *A. ervi* was introduced into the USA to control pea aphid, *Acrythosiphon pisum* (Harris) (Mackauer, 1971). Inoculative releases are made in greenhouses against *M. euphorbiae* in pepper and tomato. Similarly, *A. abdominalis* has been used preferentially against *A. solani* since about 1998. Hyperparasitoids of all four parasitoid species invade greenhouses in late spring and summer and can severely impair biological control, resulting in outbreaks. Although hyperparasitoid contamination of imported parasitoid shipments has not been demonstrated, it should be recognized as an important risk.

Pathogens

Presently, no pathogens (microbial pesticides) are registered for use against aphids on greenhouse crops in Canada. *Verticillium lecanii* (A. Zimmerman) Viegas has been shown to be effective for aphid control on pepper (Helyer, 1993). Fournier and Brodeur (1999) demonstrated effective control of *M. persicae*, *M. euphorbiae* and the lettuce aphid, *Nasonovia ribis-nigri* (Mosley), using *V. lecanii*. *Beauvaria bassiana* (Balsamo) Vuillemin is also an effective control agent for *M. persicae* and *A. gossypii*. This entomopathogen is currently being evaulated under commercial greenhouse vegetable production conditions (J.L. Shipp, unpublished).

Releases of Biological Control Agents

The approaches to release and release rates vary from crop to crop and among regions across Canada. However, there are some common approaches to application that are noteworthy.

Parasitoid species are increasingly being released in greenhouses using 'banker plant' approaches (e.g. Bennison and Corless, 1993). Potted grasses, usually wheat, *Triticum aestivum* L., oats, *Avena sativa* L., or barley, *Hordeum vulgare* L., inoculated with grass-feeding aphids, usually *Rhopalosiphum padi* (L.) or *Sitobion avenae* (Fabricius), are placed in the greenhouse. One of the parasitoid species is inoculated on to the aphids on the grass, which then serves as a source of parasitoids to attack aphids on the crop. Generally, the banker plants and parasitoids are placed in advance of the appearance of the pest species, which ensures that pest aphids are attacked by parasitoids before their numbers have increased to damaging levels. Fresh banker plants with unparasitized aphids are added periodically.

A. aphidimyza is generally applied after the first incidence of aphids on the crop, because otherwise it would attack and reduce the aphid populations on the banker plants. Routine inoculations (weekly or bi-weekly) are the usual approach.

H. convergens is used in inundative releases to reduce outbreaks of aphids when these occur on the crop though invasion of alates in the summer, or because of failure of banker plants. It is not yet clear what role *H. axyridis* will play in the biological control approaches in greenhouses, but currently this ladybird is too expensive to be considered for inundative releases and its status as a nuisance pest in homes in some jurisdictions may preclude its widespread use. Registration of microbial products such as *V. lecanii* and *B. bassiana* would replace the use of ladybirds for management of aphid outbreaks.

Evaluation of Biological Control

Application of natural enemies for biological control of pest aphids has become a standard approach in Canadian vegetable greenhouses. The use of banker plants has, in recent years, greatly improved the reliability of aphid biological control. The maintenance of parasitoid populations on alternate aphid species ensures that parasitism occurs at first presence of the pest.

Hyperparasitoid build-up during the summer months tends to reduce the efficacy of parasitoids and can result in outbreaks of pest aphids. Similary, parasitoids of the predator species reduce their efficacy.

Recommendations

Further work should include:

1. Testing augmentative introductions or modified conservation approaches, such as banker plants, for biological control of pest aphids on annual crops outside of greenhouses, because a sufficient diversity of natural enemies is readily available from commercial insectaries;
2. Registration and use of microbial pesticides to control aphid outbreaks, so as to reduce the frequency of inundative releases of exotic Coccinellidae that may have potentially negative environmental consequences;
3. Solving the problem of hyperparasitoids, which usually have an impact on parasitoids during the late summer months;
4. Developing cost-effective local mass-rearing techniques for native Coccinellidae to reduce or replace imports (i.e. *H. convergens*) that are collected out-of-doors, potentially resulting in overexploitation of these natural populations (the latter may also be heavily parasitized, which reduces their effectiveness in greenhouses);
5. Understanding predator–predator or predator–parasitoid interactions to develop optimal strategies for using the numerous aphidophagous and generalist biological control agents;
6. Linking (e.g. with molecular markers) invading populations to sources (e.g. invasions of alates from overwintering plants or from outbreaks on crop and non-crop plants) to facilitate prediction of invasion, thus allowing prophylactic introductions of natural enemies.

References

Bennison, J.A. and Corless, S.P. (1993) Biological control of aphids on cucumbers: further development of open rearing units or 'Banker plants' to aid establishment of aphid natural enemies. *International Organization for Biological Control/ West Palaearctic Regional Section, Bulletin* 16(2), 5–8.

Blackman, R.L. and Eastop, V.F. (1984) *Aphids on the World's Crops: An Identification and Information Guide.* John Wiley and Sons, Toronto, Ontario.

Clausen, C.P. (ed.) (1978) *Introduced Parasites and Predators of Arthropod Pests and Weeds: A World Review.* United States Department of Agriculture, Agriculture Research Service, Agriculture Handbook No. 480.

Day, W.H., Prokrym, D.R., Ellis, D.R. and Chianese, R.J. (1994) The known distribution of the predator *Propylea quatuordecimpunctata* (Coleoptera: Coccinellidae) in the United States, and thoughts on the origin of this species and five other exotic lady beetles in eastern North America. *Entomological News* 105, 244–256.

Elliott, N., Kieckhefer, R. and Kauffman, W. (1996) Effects of an invading coccinellid on native coccinellids in an agricultural landscape. *Oecologia* 105, 537–544.

Ferrière, C. (1965) *Hymenoptera: Aphelinidae d'Europe et du Bassin Méditerranéen.* Masson, Paris.

Fournier, V. and Brodeur, J. (1999) Biological control of lettuce aphids with the entomopathogenic fungus *Verticillium lecanii* in greenhouses. *International Organization for Biological Control/ West Palaearctic Regional Section, Bulletin* 22(1), 77–80.

Gilkeson, L.A. (1990) Biological control of aphids in greenhouse sweet peppers and tomatoes. *International Organization for Biological Control/ West Palaearctic Regional Section Bulletin* 13(5), 64–70.

Gilkeson, L.A., McLean, J.P. and Dessart, P. (1993) *Aphanogmus fulmeki* Ashmead (Hymenoptera: Ceraphronidae), a parasitoid of *Aphidoletes aphidimyza* Rondani (Diptera: Cecidomyiidae). *The Canadian Entomologist* 125, 161–162.

Gordon R.D. (1985) The Coccinellidae (Coleoptera) of America North of Mexico. *Journal of the New York Entomological Society* 93, 1–912.

Harris, K.M. (1973) Aphidophagous Cecidomyiidae (Diptera): taxonomy, biology and assessments of field populations. *Bulletin of Entomological Research* 63, 305–325.

Helyer, N. (1993) *Verticillium lecanii* for control of aphids and thrips on cucumber. *International Organization for Biological Control/ West Palaearctic Regional Section, Bulletin* 16(2), 63–66.

Howard, R.J., Garland, J.A. and Seaman, W.L. (eds) (1994) *Diseases and Pests of Vegetable Crops in Canada.* Canadian Phytopathology Society and Entomological Society of Canada, Ottawa, Ontario.

MacGillivray, M.E. (1972) The sexuality of *Myzus persicae* (Sulzer), the green peach aphid, in New Brunswick (Homoptera: Aphididae). *Canadian Journal of Zoology* 50, 469–471.

Mackauer, M. (1971) *Acrythosiphum pisum* (Harris), pea aphid (Homoptera: Aphididae). In: *Biological Control Programmes against Insects and Weeds in Canada, 1959–1968. Part I. Biological Control of Agricultural Insects in Canada, 1959–1968.* Technical Communication No. 4. Commonwealth Institute of Biological Control Trinidad. Commonwealth Agricultural Bureaux, Farnham Royal, UK, pp. 3–10.

Mackauer, M. and Starý, P. (1967) *Index of Entomophagous Insect: Hym. Ichneumonoidea; World Aphidiidae.* Le François, Paris, France.

McLeod, J.H. (1962) Part I. Biological control of pests of crops, fruit trees, ornamentals and weeds in Canada up to 1959. In: *A Review of the Biological Control Attempts Against Insects and Weeds in Canada.* Technical Communication No. 2, Commonwealth Institute of Biological Control, Trinidad. Commonwealth Agricultural Bureaux, Farnham Royal, UK, pp. 1–33.

Van Steenis, M.J. (1993) Suitability of *Aphis gossypii* Glov., *Macrosiphum euphorbiae* (Thom.), and *Myzus persicae* Sulz. (Hom.: Aphididae) as host for several aphid parasitoid species (Hym.: Braconidae). *International Organization for Biological Control/ West Palaearctic Regional Section, Bulletin* 16(2), 157–160.

Wheeler, A.G. Jr and Stoops, C.A. (1996) Status and spread of the Palearctic lady beetles *Hippodamia variegata* and *Propylea quatuordecimpunctata* (Coleoptera: Coccinellidae) in Pennsylvania, 1993–1995. *Entomological News* 107, 291–298.

10 *Bradysia* spp., Fungus Gnats (Diptera: Sciaridae)

D.R. Gillespie, V. Carney, C. Teerling and J.L. Shipp

Pest Status

Fungus gnats, *Bradysia* spp., attack a variety of crops in protected culture. Bedding plants, ornamentals, vegetables, and tree seedlings, in propagation, are attacked, as are greenhouse vegetable and flower crops in production (Howard *et al.*, 1994). Fungus gnats are also major pests of mushroom production (Harris *et al.*, 1996).

Fungus gnats in greenhouses used to be considered as symptomatic of overwatering and large numbers were tolerated because it was thought that damage caused by these pests was inconsequential. Many growers used an action threshold based strictly on the degree of annoyance caused by adult flies, which were often sufficiently numerous to be regularly inhaled. They have been shown to damage plants directly, through larvae feeding on roots and root hairs (reviewed in Harris *et al.*, 1996), and indirectly, through larval and adult transmission of disease (Gillespie and Menzies, 1993; Jarvis *et al.*, 1993).

The key species attacking greenhouse crops are most frequently identified as *Bradysia impatiens* (Johannsen) and *Bradysia coprophilia* Lintner, but other species are sometimes seen, e.g. *Corynoptera* sp. (Gillespie, 1986). Harris *et al.* (1996) reviewed the recent literature. Eggs are laid singly or in small groups in moist situations. Oviposition is encouraged by moisture and the presence of organic debris. Larvae, which are elongate and transparent, with a distinct, black head capsule, develop through five instars in the soil. Larvae pupate in the substrate and pupae wriggle to the surface at adult emergence. Males emerge slightly before females and there is a pre-oviposition period of about 24 h. Development from egg to adult at 20°C takes 16–20 days.

Background

Tolerance for fungus gnats decreased throughout the 1980s and early 1990s. However, no economic thresholds have been developed, partly due to the diversity of species and the lack of a useful field guide for identification of larvae and adults of economically important species. Yellow sticky traps were demonstrated to be an effective approach to measuring adult numbers, but these did not correlate with larval numbers, in media (Rutherford *et al.*, 1985).

Two factors combined to reduce tolerance for fungus gnats and prompted growers to seek biological control approaches in greenhouse vegetable production. First, vapours from applications of diazinon to the floor for fungus gnat control were found to interfere with the use of natural enemies such as *Encarsia formosa* Gahan, used for biological control of greenhouse whitefly, *Trialeurodes vaporariorum* (Westwood). Second, it was found that fungus gnat adults could spread plant root diseases such as *Pythium aphanidermatum* (Edson) and *Fusarium oxysporum* f. sp. *radicis-lycopersici* Jarvis and Shoemaker (Gillespie & Menzies, 1993; Jarvis *et al.*, 1993).

Biological Control Agents

Predators

The predatory mites *Hypoaspis aculeifer* (Canestrini) and *Hypoaspis miles* (Berlese), common in soils throughout the northern hemisphere, have been shown to control fungus gnats and western flower thrips, *Frankliniella occidentalis* (Pergande), in greenhouse cropping systems (Gillespie and Quiring, 1990; Wright and Chambers, 1994). Adults, protonymphs and deutonymphs are predatory and feed on fungus gnats, thrips pupae, and other small, soft-bodied organisms in greenhouse soils and substrates. Mites, in a bran substrate that usually contains mixed stages, are shipped to growers from insectaries in Canada and Europe. They are released in greenhouses by sprinkling the bran on to the substrate surface. *Hypoaspis* spp. have been used widely for fungus gnat control since the early 1990s. They are generally applied prophylactically to growing media, as a routine pest management measure, either at the beginning of each crop, or earlier, during plant propagation.

In Ontario, a cosmopolitan soil-dwelling rove beetle, *Atheta coriaria* (Kraatz), is currently being tested at Vineland as a potential biological control agent for fungus gnats and shore flies (Ephydridae). Miller (1981) and Miller and Williams (1983) studied the biology of *A. coriaria* and its functional response to prey densities of Nitidulidae and Muscidae. In Vineland, *A. coriaria* successfully reduced populations of fungus gnats, shore flies and *F. occidentalis*, in laboratory and greenhouse trials. All active beetle stages (the three larval instars and adult) readily consume fungus gnat and shore fly eggs, larvae and pupae. At high prey densities, a single adult *A. coriaria* has the potential to consume over 120 fungus gnat eggs in less than 24 hours. Comparative tests with thrips indicate that over 80 thrips pupae are eaten over a similar time period. Currently, efficacy of *A. coriaria* is being tested in greenhouses, monitoring techniques are being developed for both predator and prey, and mass rearing protocols on natural and artificial diet substrates are being perfected.

Pathogens

Bacteria

Bacillus thuringiensis (Berliner) serovar *israelensis* (*B.t.i.*) is effective for biological control of fungus gnats in ornamental crops (Osborne *et al.*, 1985). It is registered for use against fungus gnats in greenhouse ornamentals. Formulated products are applied in water to the growing media in response to outbreaks.

Nematodes

Steinernema carpocapsae (Weiser) and *Steinernema feltiae* (Filipjev) are useful for control of fungus gnats in greenhouses (Lindquist and Piatkowski, 1993). They are sometimes applied in water to greenhouse substrates for biological control of fungus gnats as required.

Evaluation of Biological Control

Applications of *Hypoaspis* spp. to substrates in advance of fungus gnat infestations provide reasonably good, long-term control of fungus gnats in greenhouse crops, provided that conditions favouring development of fungus gnat outbreaks, such as overwatered soils and accumulated organic debris, are avoided. Predator populations are sensitive to applications of pesticides for other pest problems. These applications can cause fungus gnat populations to be released from biological control, and will result in outbreaks. Applications of *B.t.i.* and nematodes aid in the supression of such outbreaks. The use of *A. coriaria* is still under investigation.

Recommendations

Further work should include:

1. Studies of intra-guild predation among predators, given the impending introduction of multiple, generalist predators into greenhouses;
2. Studies of parasitoids of fungus gnats to evaluate their potential as biological control agents in conjunction with generalist predators;
3. Studies of the potential of these natural enemies to provide biological control of fungus gnats in mushroom production;
4. Studies of the diversity of fungus gnats, leading to the production of a guide to common economic species, in order to facilitate the development of economic thresholds.

References

Gillespie, D.R. (1986) A simple rearing method for fungus gnats, *Corynoptera* sp. (Diptera: Sciaridae) with notes on life history. *Journal of the Entomological Society of British Columbia* 83, 45–48.

Gillespie, D.R. and Menzies, J.G. (1993) Fungus gnats vector *Fusarium oxysporum* f. sp. *radicis-lycopersici*. *Annals of Applied Biology* 123, 539–544.

Gillespie, D.R. and Quiring, D.M.J. (1990) Biological control of fungus gnats, *Bradysia* spp. (Diptera: Sciaridae), and western flower thrips, *Frankliniella occidentalis* (Pergande) (Thysanoptera: Thripidae), in greenhouses using a soil-dwelling predatory mite, *Geolaelaps* sp. nr. *aculeifer* (Canestrini) (Acari: Laelapidae). *The Canadian Entomologist* 122, 975–983.

Harris, M.A., Gardner, W.A. and Oetting, R.D. (1996) A review of the scientific literature on fungus gnats (Diptera: Sciaridae) in the genus *Bradysia*. *Journal of Entomological Science* 31, 252–276.

Howard, R.J., Garland, J.A. and Seaman, W.L. (eds) (1994) *Diseases and Pests of Vegetable Crops in*

Canada. Canadian Phytopathology Society and Entomological Society of Canada, Ottawa, Ontario.

Jarvis, W.R., Shipp, J.L. and Gardiner, R.B. (1993) Transmission of *Pythium aphanidermatum* to greenhouse cucumber by the fungus gnat *Bradysia impatiens* (Diptera: Sciaridae). *Annals of Applied Biology* 122, 23–29.

Lindquist, R. and Piatkowski, J. (1993) Evaluation of entomopathogenic nematodes for control of fungus gnat larvae. *International Organization for Biological Control/ West Palaearctic Regional Section, Bulletin* 16, 97–100.

Miller, K.V. (1981) The biology, host preference, and functional response of *Atheta coriaria* (Kraatz) (Coleoptera: Staphylinidae). MSc thesis, Ohio State University, Columbus, Ohio.

Miller, K.V. and Williams, R.N. (1983) Biology and host preference of *Atheta coriaria* (Coleoptera: Staphylinidae), an egg predator of Nitidulidae and Muscidae. *Annals of the Entomological Society of America* 76, 158–161.

Osborne, L.S., Boucias, D.G. and Lindquist, R.K. (1985) Activity of *Bacillus thuringiensis* var. *israelensis* on *Bradysia coprophilia* (Dipera: Sciaridae). *Journal of Economic Entomology* 78, 922–925.

Rutherford, T.A., Trotter, D.B. and Webster, J.M. (1985) Monitoring fungus gnats (Diptera: Sciaridae) in cucumber greenhouses. *The Canadian Entomologist* 117, 1387–1394.

Wright, E.M. and Chambers, R.J. (1994) The biology of the predatory mite *Hypoaspis miles* (Acari: Laelapidae), a potential biological control agent of *Bradysia paupera* (Dipt.:Sciaridae). *Entomophaga* 39, 225–235.

11 *Ceutorhynchus obstrictus* (Marsham), Cabbage Seedpod Weevil (Coleoptera: Curculionidae)

U. Kuhlmann, L.M. Dosdall and P.G. Mason

Pest Status

The cabbage seedpod weevil, *Ceutorhynchus obstrictus* (Marsham) [= *C. assimilis* (Paykull) Colonnelli (1990, 1993)], is native to Europe and a serious pest of canola, *Brassica napus* L. and *B. rapa* L., in North America. The weevil was recorded in Vancouver, British Columbia, in 1931 (McLeod, 1962), was first discovered in canola near Lethbridge, Alberta, in 1995 (Dosdall *et al.*, 1999), and in 2000 was reported in Quebec (J. Brodeur, Sainte-Foy, 2000, personal communication). Its discovery immediately raised concern among members of Canada's canola industry because *C. obstrictus* is the most significant insect pest of canola and rapeseed in Europe and the USA. In north-western USA, weevil infestations can reduce yields of winter (autumn-planted) canola by 15–35% in fields not treated with insecticides (McCaffrey *et al.*, 1986). Populations of *C. obstrictus* remained relatively low in southern Alberta from 1995 to 1998, but in 1999 outbreak densities occurred in about

100,000 ha of canola, resulting in crop losses estimated at Can$1 million (L.M. Dosdall, unpublished).

C. obstrictus completes a single generation in British Columbia, Washington, Idaho and Alberta (McLeod, 1962; L.M. Dosdall, unpublished). Kirk (1992) described its life cycle. Adult weevils overwinter in debris or soil, and in spring fly to flowering crucifers, where the females feed on pollen until ovarian development is completed. Eggs are laid in the pods through holes chewed by females. Each larva consumes about five seeds, to complete its development in about 4 weeks. The larva then bores through the pod wall and falls to the ground, where it pupates in a cocoon just below the soil surface. Adults emerge 2–4 weeks later to feed on green stems and canola pods. McLeod (1962) reported that *C. obstrictus* attacks wild *Brassicaceae*, e.g. wild mustard, *Brassica juncea* L., wild rape, *B. rapa* L., and wild radish, *Raphanus raphanistrum* L., as well as cultivated crucifers, and noted that wild host species provide a reservoir from which *C. obstrictus*, a strong flyer, can disperse over long distances.

Background

Control of *C. obstrictus* is only through prophylactic use of broad-spectrum chemical insecticides (McCaffrey *et al.*, 1986), but research is being conducted to develop canola germplasm resistant to *C. obstrictus* (McCaffrey *et al.*, 1999). No insecticides are yet registered in Canada to control this pest but, in 1999, applications of chemical insecticides (temporarily given emergency registration) were necessary in some fields in southern Alberta. Chemical insecticides can be toxic to pollinating insect species and, in Europe, Murchie *et al.* (1997) found that insecticides have a negative impact on the parasitoid *Trichomalus perfectus* (Walker). There is a critical need to develop alternatives to insecticides, including the more effective use of biological control.

In Europe, many parasitoids attack *C. obstrictus* (Dmoch, 1965; Herting, 1973; Dolinski, 1979; Kuhlmann and Mason, 1999), but the most important are *Microctonus melanopus* Ruthe, *Diospilus oleraceus* Haliday, *T. perfectus* and *Mesopolobus morys* L. (Kuhlmann and Mason, 1999).

Surveys in Washington (Doucette, 1948; Hanson *et al.*, 1948), Oregon (Doucette, 1948), California (Carlson *et al.*, 1951) and British Columbia (McLeod, 1952) determined that a maximum of 11 parasitoid species were associated with *C. obstrictus* in the USA and British Columbia, and that *M. morys* and *T. perfectus* were the most abundant and effective parasitoids of *C. obstrictus*. In northern Idaho, *T. perfectus* and *M. morys* were important parasitoids, but *Necremnus duplicatus* Gahan was also found to attack *C. obstrictus* in substantial numbers (Doucette, 1948; Walz, 1957). European parasitoids that already occur in some North American locations may have been introduced accidentally with *C. obstrictus*. Harmon and McCaffrey (1997) found that *M. melanopus* significantly reduced survival of overwintering adult weevils in Idaho and Washington, and parasitism levels were as high as 70%.

In Alberta, surveys in 1998 and 1999 determined that populations of *C. obstrictus* were almost free of parasitioids. Although one adult weevil specimen was parasitized, the parasitoid was an adult Chloropidae, not considered to be of importance in biological control because it attacks insects already wounded (T. Wheeler, Montreal, 1999, personal communication). Dissections of hundreds of canola pods collected in 1999 have not yielded evidence of larval parasitoids.

Given the potential importance of biological control agents in reducing populations of *C. obstrictus* in western Canada, a strategy for biological control of *C. obstrictus* involving both classical and inundative approaches is needed. Prior to importation, the host specificity of candidate European parasitoids must be determined in their native cultivated and non-cultivated habitats to evaluate potential non-target risks. This is especially important because several species of European Ceutorhynchinae have been introduced to North America to control weeds, and parasitoids of *C. obstrictus* that have a negative impact on these biological control agents must be avoided (Kuhlmann and Mason, 1999).

Biological Control Agents

Parasitoids

In Europe, the host specificity of parasitoids of *C. obstrictus* is being evaluated[1] for potential risks to non-target Ceutorhynchinae host species in North America. In 1999, target and non-target Ceutorhynchinae were sampled from April to July in cultivated and non-cultivated habitats in the canola-growing region of northern Germany (Kuhlmann *et al.*, 1999). Twelve Ceutorhynchinae species were found in the stems and seeds of canola and five weed species associated with canola plus one unassociated weed (Table 11.1). The target species, *C. obstrictus* and *C. pallidactylus* (Marsham), were found in canola seeds and stems, respectively. Primary European parasitoids of *C. obstrictus* are common but only four species have potential for selection as candidate biological control agents for introduction to Canada: *M. melanopus* and *D. oleraceus*, *T. perfectus* and *M. morys*.

M. melanopus is a solitary adult endoparasitoid parasitizing *C. obstrictus* adults. Jourdheuil (1960) described its biology. The parasitoid attacks the new generation of *C. obstrictus* and overwinters as a first instar larva within the adult weevil. The larva emerges from its host the following spring and pupates in the soil. The new generation of parasitoids attack the same overwintered generation of weevils, but the next generation of parasitoids attack the new overwintering weevil generation. Thus, there are two generations of the para-

Table 11.1. Ceutorhynchinae species collected, host plant species and feeding location during the 1999 survey in Northern Germany (Kuhlmann *et al.*, 1999).

Ceutorhynchinae species	Host plant	Feeding location
	Brassicaceae	
Ceutorhynchus obstrictus (Marsham) Syn.: *C. assimilis* Paykull	*Brassica napus* L.	Seed
C. pallidactylus (Marsham) Syn.: *C. quadridens* (Panzer)	*Brassica napus* L.	Stem
C. alliariae Brisout	*Alliaria petiolata* (M. Bieberstein) Cavara et Grande	Stem
C. roberti Gyllenhal	*Alliaria petiolata* (M. Bieberstein) Cavara et Grande	Stem
C. constrictus Marsh	*Alliaria petiolata* (M. Bieberstein) Cavara et Grande	Seed
C. floralis (Paykull)	*Capsella bursa-pastoris* (L.) Medicus	Seed
C. rapae Gyllenhal	*Sisymbrium officinale* (L.) Scopoli	Stem
	Asteraceae	
Microplontus rugulosus (Herbst)	*Tripleurospermum perforatum* Lainz	Stem
M. edentulus (Schultz)	*Tripleurospermum perforatum* Lainz	Stem
Hadroplontus litura (Fabricius)	*Cirsium arvense* (L.) Scopoli	Stem
	Boraginaceae	
Mogulones borraginis (Fabricius)	*Cynoglossum officinale* L.	Seed
M. trisignatus Gyllenhal	*Cynoglossum officinale* L.	Stem

[1]By AAFC and CABI *Bioscience* in collaboration with B. Klander, University of Kiel, Germany.

sitoid and one generation of *C. obstrictus* (Harmon and McCaffrey, 1997).

D. oleraceus is a primary solitary larval endoparasitoid. Jourdheuil (1960) determined that it is polyvoltine and probably two or three generations attack the same population of *Ceutorhynchus* spp. Maximum rates of parasitism were 34.7% from the first and 19.4% from the second generation of *Ceutorhynchus pleurostigma* (Marsham) in 1956, but low levels of parasitism, mostly 1–4%, were reported for *C. obstrictus* from 1952 to 1955. The parasitoid overwinters as a larva within the *Ceutorhynchus* larva in the soil. Important aspects of the biology and ecology of *D. oleraceus*, such as its dispersal behaviour and cold-hardiness, are unknown.

T. perfectus is a primary solitary larval ectoparasitoid of *C. obstrictus*. Its immigration into the crop occurs mainly 3–4 weeks after weevils infest the pods (Laborius, 1972; Dmoch, 1975). The parasitoid usually lays a single egg on the body surface, primarily of third-instar larvae of *C. obstrictus* (Nissen, 1997). Odour from the frass of final-instar larvae of *C. obstrictus* apparently enables female parasitoids to locate their hosts (Dmoch and Rutkowska-Ostrowska, 1978, in Lerin, 1987). The larva feeds externally and completes its development on one weevil larva. Pupation (without cocoon formation) occurs in the pod. The newly emerged parasitoid leaves the pod before the crop is harvested by boring an exit hole that is smaller than that made by the weevil larva. Complete development of one generation requires about 18 days: 3, 7 and 8 days for the egg, larva and pupa, respectively (Dmoch, 1975). Adult females can also kill some host weevils without laying eggs, apparently by feeding on *C. obstrictus* larvae. Parasitized *C. obstrictus* larvae stop feeding during the third instar and cause less damage than non-parasitized larvae. Szczepanski (1972) found *T. perfectus* in pine forests in central Poland in relatively large numbers. It was present from the beginning of the growing season until about mid-May and again from the end of July to the end of the season. It was concluded that adults of *T. perfectus* overwintered in the forest. In mid-May, parasitoid adults moved to flowering winter rapeseed and parasitized *C. obstrictus*; the first generation completed its development by the beginning of July. The reappearance of *T. perfectus* in the forest at the end of July suggests the possible presence there of an additional, as yet unidentified, host (Rosen, 1964; Szczepanski, 1972; Nissen, 1997).

M. morys is a primary solitary larval ectoparasitoid. It was found in rapeseed pods throughout the area of crop cultivation in Sweden, although few individuals were collected (Rosen, 1964). This species had two generations per year, at least in the south. It overwintered as adults, possibly on conifers (Rosen, 1964).

Evaluation of Biological Control

Biological control of *C. obstrictus* must be a 'safety-first approach' to ensure that European Ceutorhynchinae species introduced to North America to control weeds are not negatively affected by parasitoids introduced to control *C. obstrictus*. Although two braconids and two pteromalids are promising candidates, host specificity must be evaluated before considering introductions.

Previously established parasitoid populations, such as *T. perfectus*, may provide important North American sources for releases in regions of canola production and reduce the number of screenings before releases in Alberta, Saskatchewan and Manitoba.

A cautious approach is important in developing a biological control strategy for *C. obstrictus* in western Canada in view of the potential damage to existing weed biological control programmes.

Recommendations

Further work should include:

1. Surveying the parasitoid complex of *C. obstrictus* in the Creston Valley, British Columbia, where *C. obstrictus* has been

established for several years, in order to determine whether populations of effective biological control agents (e.g. *T. perfectus*, reported by McLeod as *T. fasciatus*) are already established in Canada;
2. Surveys in western Canada to determine the indigenous species of Ceutorhynchinae inhabiting regions of canola production, so as to assess possible risks involved with introducing exotic parasitoid species;
3. A retrospective summary of biological control work already undertaken in order to determine the origins of European populations of parasitoids already introduced to North America and the histories of releases of exotic biological control agents for *C. obstrictus* in the USA and Canada;
4. A summary of releases in Canada of Ceutorhynchinae species for biological control of weeds, to provide important information on successful and unsuccessful establishments and distributions;
5. Determining the ecological host ranges of candidate parasitoids for releases in Canada, to optimize their potential for successful establishment; and screening in Europe of these candidates, to ensure that Ceutorhynchinae species, e.g. *Microplontus edentulus* (Schultz), *Hadroplontus litura* (Fabricius) and *Mogulones cruciger* (Herbst), introduced for weed biological control are not significantly affected;
6. Clarifying the taxonomy and phylogeny of *Ceutorhynchus* in the Holarctic region by including taxonomists in the project to provide host (Ceutorhynchinae) and parasitoid identifications: this is crucial for assessing the potential impacts of introduced agents on non-target species and on the broader ecosystem, and to identify accurately the native species of Ceutorhynchinae collected in surveys;
7. Developing mass collection and mass rearing techniques of parasitoids of *C. obstrictus*, with emphasis on biotypes from Europe and Canada, to optimize the potential for the establishment and dispersal of promising candidate species, e.g. *T. perfectus*;
8. Screening of potential entomopathogens to evaluate the pathogenicity to *C. obstrictus* of the many known strains;
9. Once appropriate parasitoid or pathogen species are selected for release in Canada (and the USA), monitoring the establishment and dispersal of these species to determine their effectiveness for reducing populations of *C. obstrictus*;
10. Evaluating the effects of registered insecticides on biological control agents, e.g. Murchie *et al.* (1997) found that insecticide treatments with triazophos in Europe were detrimental to populations of *T. perfectus*, but treatments with alphacypermethrin were less harmful.

Acknowledgements

The Alberta Canola Producers Commission, the Saskatchewan Canola Development Commission and the Alberta Agricultural Research Institute funded investigations in Alberta.

References

Carlson, E.C., Lange, W.H. and Sciaroni, R.H. (1951) Distribution and control of the cabbage seedpod weevil in California. *Journal of Economic Entomology* 44, 958–966.

Colonnelli, E. (1990) Curculionidae Ceutorhynchinae from the Canaries and Macaronesia (Coleoptera) *Vieraea* 18, 317–337.

Colonnelli, E. (1993) The Ceutorhynchinae types of I.C. Fabricius and G. von Paykull (Coleoptera: Curculionidae). *Koleopterologische Rundschau* 63, 299–310.

Dmoch, J. (1965) The dynamics of a population of the cabbage seedpod weevil (*Ceutorhynchus assimilis* Payk.) and the development of winter rape. Part I. *Ekologia Polska Seria A* 13, 249–287.

Dmoch, J. (1975) Study on the parasites of the cabbage seed weevil (*Ceutorrhynchus assimilis* Payk.). I. Species composition and economic importance of the larval ectoparasites. *Roczniki Nauk Rolniczych* (E) 5, 99–112.

Dmoch, J. and Rutkowska-Ostrowska, Z. (1978) In: Lerin, J. (1987) A short bibliographical review of *Trichomalus perfectus* Walk., a parasite of seedpod weevil *Ceutorhynchus assimilis* Payk. *International Organization for Biological Control/Western Palaearctic Regional Section, Bulletin* 10(4), 74–78.

Dolinski, M.G. (1979) The cabbage seedpod weevil, *Ceutorhynchus assimilis* (Payk.) (Coleoptera: Curculionidae), as a potential pest of rape production in Canada. MSc thesis, Simon Fraser University, Vancouver, British Columbia.

Dosdall, L.M., McFarlane, M.A., Moisey, D., Dolinski, M.G. and Jones, J. (1999) The cabbage seedpod weevil, a new pest of canola in Alberta. *The Alberta Canola Grower,* March/April issue, pp. 8–9.

Doucette, C.F. (1948) Field parasitization and larval mortality of the cabbage seedpod weevil. *Journal of Economic Entomology* 41, 763–765.

Hanson, A.J., Carlson, E.C., Breakey, E.P. and Webster, R.L. (1948) Biology of the cabbage seedpod weevil in northwestern Washington. *Washington Agriculture Experimental Station, Bulletin* 498.

Harmon, B.L. and McCaffrey, J.P. (1997) Parasitism of adult *Ceutorhynchus assimilis* (Coleoptera: Curculionidae) by *Microctonus melanopus* (Hymenoptera: Braconidae) in northern Idaho and eastern Washington. *Journal of Agricultural Entomology* 14, 55–59.

Herting, B. (1973) *A Catalogue of Parasites and Predators of Terrestrial Arthropods. Section A. Host or Prey/enemy. Volume III. Coleoptera and Strepsiptera.* Commonwealth Agriculture Bureau, Wallingford, UK.

Jourdheuil, P. (1960) Influence de quelques facteurs écologiques sur les fluctuations de population d'une biocénose parasitaire: étude relative à quelques hyménoptères (Ophioninae, Diospilinae, Euphorinae) parasites de divers coléoptères inféodés aux crucifères. *Annales de Epiphytologie* 11, 445–658.

Kirk, W.D.J. (1992) Insects on cabbages and oilseed rape. *Naturalists' Handbooks 18.* Richmond Publishing, Slough, UK.

Kuhlmann, U. and Mason, P.G. (1999) *Natural Host Specificity Assessment of European Parasitoids for Classical Biological Control of the Cabbage Seedpod Weevil in North America: a Safety First Approach for Evaluating Non-target Risks.* Technical Report. CABI Bioscience, Delémont, Switzerland.

Kuhlmann, U., Bürki, H., White, H., Lauro, N., Klander, B., Reimer, L., Hunt, E., Rahn, J., Harris, S., Lachance, S. and Herrmann, D. (1999) *Summary Report, Progress in 1999. Agricultural Pest Research.* Technical Report. CABI Bioscience, Delémont, Switzerland.

Laborius, G.A. (1972) Untersuchungen über die Parasitierung des Kohlschotenrüsslers (*Ceuthorrhynchus assimilis* Payk.) und der Kohlschotengallmücke (*Dasyneura brassicae* Winn.) in Schleswig-Holstein. *Zeitschrift für angewandte Entomologie* 72, 14–31.

Lerin, J. (1987) A short bibliographical review of *Trichomalus perfectus* Walk., a parasite of seedpod weevil *Ceutorhynchus assimilis* Payk. *International Organization for Biological Control/Western Palaearctic Regional Section, Bulletin* 10(4), 74–78.

McCaffrey, J.P., O'Keeffe, L.E. and Homan, H.W. (1986) Cabbage seedpod weevil control in winter rapeseed. *University of Idaho, College of Agriculture, Current Information Series 782.*

McCaffrey, J.P., Harmon, B.L., Brown, J., Brown, A.P. and Davis, J.B. (1999) Assessment of *Sinapis alba, Brassica napus* and *S. alba* × *B. napus* hybrids for resistance to cabbage seedpod weevil, *Ceutorhynchus assimilis* (Coleoptera: Curculionidae). *Journal of Agricultural Science* 132, 289–295.

McLeod, J.H. (1952) Notes on the cabbage seedpod weevil, *Ceutorhynchus assimilis* (Payk.) (Coleoptera: Curculionidae), and its parasites. *Proceedings of the Entomological Society of British Columbia* 49, 11–18.

McLeod, J.H. (1962) Part I. Biological control of pests of crops, fruit trees, ornamentals and weeds in Canada up to 1959. In: *A Review of the Biological Control Attempts Against Insects and Weeds in Canada.* Technical Communication No. 2, Commonwealth Institute of Biological Control, Trinidad. Commonwealth Agricultural Bureaux, Farnham Royal, UK, pp. 1–33.

Murchie, A.K., Williams, I.H. and Alford, D.V. (1997) Effects of commercial insecticide treatments to winter oilseed rape on parasitism of *Ceutorhynchus assimilis* Paykull (Coleoptera: Curculionidae) by *Trichomalus perfectus* (Walker) (Hymenoptera: Pteromalidae). *Crop Protection* 16, 199–202.

Nissen, U. (1997) Oekologische Studien zum Auftreten von Schadinsekten und ihren Parasitoiden an Winterraps norddeutscher Anbaugebiete. Dissertation, Christian-Albrechts-Universität zu Kiel.

Rosen, H.V. (1964) Untersuchungen über die Verbreitung und Biologie von zwei Pteromaliden in Rapsschoten (Hymenoptera, Chalcidoidea). *Meddelanden Statens Växtskyddanstalt* 12, 449–465.
Szczepanski, H. (1972) The rape pteromalid *Trichomalus perfectus* (Walker) (Hymenoptera, Pteromalidae) in forest biocoenosis and the problem of the biological protection of rape. *Polskie Pismo Entomologiczne* 42, 865–871.
Walz, A.J. (1957) Observations on the biologies of some hymenopterous parasites of the cabbage seedpod weevil in northern Idaho. *Annals of the Entomological Society of America* 50, 219–220.

12 *Choristoneura fumiferana* (Clemens), Eastern Spruce Budworm (Tortricidae)

S.M. Smith, K. van Frankenhuyzen, V.G. Nealis and R.S. Bourchier

Pest Status

The eastern spruce budworm, *Choristoneura fumiferana* (Clemens), is a native defoliator of balsam fir, *Abies balsamea* (L.), white spruce, *Picea glauca* (Moench) Voss, red spruce, *P. rubens* Sargent, and black spruce, *P. mariana* (Miller) Britton, Sterns, and Poggenburg, throughout the spruce–fir forests of northern North America east of the Rocky Mountains. It is by far the most damaging forest pest in eastern Canada, with defoliation during any given epidemic year often exceeding 30 million ha (FIDS, 1987). From 1982 to 1987, *C. fumiferana* caused growth loss of 1.6 million m^3 and tree mortality of 7.2 million m^3 in Ontario alone (Gross *et al.*, 1992). Seven cyclical outbreaks, each lasting 25–30 years, are thought to have occurred in eastern Canada over the past 250 years (Royama, 1984); the most recent began in the late 1970s and lasted until the mid-1980s (Sanders, 1995).

C. fumiferana feeds preferentially on the current-year's shoots, but when populations are high or epidemics are extended, the larvae will also 'backfeed' on to needles of previous years' growth. Defoliation results in loss of radial increment and height growth in trees the year following defoliation (McLean, 1990). Trees may begin to die following as little as 3 years of severe defoliation and mortality may continue for 5–8 years after *C. fumiferana* populations collapse. Older balsam fir trees tend be more susceptible, followed by younger trees or white and red spruce (Blais, 1983).

C. fumiferana completes one generation per year and is subjected to substantial natural parasitism and disease (Régnière and Lysyk, 1995). Overwintering second-instar larvae emerge in spring and start feeding under the bud caps of expanding shoots. As they reach the fourth instar in early June, the larvae feed externally on new foliage until the end of the sixth instar and then pupate on the foliage. Adults emerge in early to mid-July and lay eggs in masses consisting of about 20 eggs. The eggs hatch and the first instars disperse, without feeding, to produce overwintering hibernacula on the tree branches where they moult to the second instar and enter winter diapause.

Background

By 2000, *B. thuringiensis* serovar *kurstaki* (*B.t.k.*) was established as the principal commercial alternative to chemical insecticides used against *C. fumiferana*. Development of this product was based on more than 20 years of collaborative research between the Canadian Forest Service, industry, and provincial forest protection agencies. During the early 1980s, *B.t.k.* had limited use (<5% of the total area sprayed) because of inconsistent results and high treatment costs relative to synthetic insecticides (Smirnoff and Morris, 1982). By the mid-1980s, however, operational use of *B.t.k.* for budworm control increased to 50–65% of the area sprayed, and by the early 1990s it was the only product applied aerially to forested crown land in Canada. This rapid escalation in use was the result of both a political agenda by various provinces to curb aerial applications of synthetic insecticides on public forests and extensive preliminary field trials that improved formulation and application, and therefore efficacy, of *B.t.k.* (van Frankenhuyzen, 1990, 1993).

While the commercial use of *B.t.k.* dominated the biological control efforts against *C. fumiferana* after the 1980s, the political shift that facilitated its development also promoted investigation into other microbial and macrobial biological control agents. A significant component of this research was based in Ontario, where aerial applications of chemical insecticides on public forests stopped after 1985 and support was provided to continue the development of viable alternatives for *C. fumiferana* management.

Since 1980, studies to improve the efficacy of viruses were continued, with emphasis on obtaining new, more virulent isolates to initiate epizootics for longer-term management of *C. fumiferana* rather than simply annual suppression of populations and foliage protection. Work on other microbials, e.g. microsporidia and fungi, declined and was not continued after the early 1980s, due to shifting institutional interests and the perceived lack of potential for microbials, other than viruses, to initiate large-scale epizootics in the host.

Efforts were continued to introduce European parasitoids from non-native hosts into Canada. The expectation was that *C. fumiferana*, a native species, would be poorly adapted to such 'new associations' and that this would increase the apparent 'virulence' of the introduced parasitoids (Mills, 1983). At the same time, as suggested by Hulme and Green (1984), augmentative and inundative releases of parasitoids were considered. In Ontario, a 12-year study was conducted jointly by university, government and industrial partners to determine the potential for augmenting populations of the native egg parasitoid, *Trichogramma minutum* Riley, against *C. fumiferana*. Earlier work to augment natural enemies by spraying attractants on to trees or introducing parasitoids from western populations, where they seemed more abundant, had been unsuccessful. However, when preliminary studies in Quebec (W. Quednau, Ste-Foy, 1985, personal communication; Varty, 1984) and Maine (Houseweart *et al.*, 1984) showed that an inundative approach using an egg parasitoid had potential, research focused throughout the 1980s and early 1990s to establish its commercial success.

Biological Control Agents

Royama (1984) speculated that high levels of late-larval mortality in *C. fumiferana* were due to an unknown complex of viral and protozoan diseases, including the microsporidian *Nosema fumiferanae* (Thompson) and the fungi *Entomophaga aulicae* (Reichardt in Bail) and *Erynia radicans* (Brefeld) Humber (Perry and Régnière, 1986). Studies conducted during the last outbreak in eastern Canada, however, suggest that late-instar parasitoids such as *Meteorus trachynotus* (Viereck), *Winthemia fumiferanae* Tothill, *Lypha setifacies* (West) and *Actia interrupta* Curran may be as important to natural population declines as this disease complex. While resource depletion and species of native natural enemies that commonly attack earlier stages of *C. fumiferana* across its distribution may also play a role in population decline, natural mortality during the late-larval stage appears to be particularly important to intergeneration rates of change in abundance of *C. fumiferana* (Royama, 1984).

Parasitoids

Collections of parasitoids from the European spruce budworm, *Choristoneura murinana* (Hübner), were evaluated in Canada during the 1980s. One parasitoid, *Apanteles murinanae* Čapek and Zwölfer, attacked and completed development in *C. fumiferana*. A rearing system developed for the closely related native *C. fumiferana* parasitoid *Apanteles fumiferanae Viereck* (Nealis and Fraser, 1988), was adapted for *A. murinanae*. This system permitted production of sufficient parasitoid material to permit studies to compare life-history traits of the two species and to make a field release of *A. murinanae*.

Laboratory investigations suggested that although *A. murinanae* could attack successfully and complete a generation in *C. fumiferanae*, it was unlikely to compete or hybridize with the native *A. fumiferanae* because of a relatively lower attack rate and fecundity. Given this low risk and the possibility that the European species might complement the native species, a single, free release of 200–300 female *A. murinanae* was made in an increasing population of *C. fumiferana* near Aylmer, western Quebec (45°26′N 75°52′W, elevation 135 m), in August 1990. Sentinel larvae of *C. fumiferana* in hibernacula (Nealis, 1988) were deployed in the stand at the time of release. When retrieved the following spring, laboratory rearing of these sentinel larvae showed no evidence of parasitism by *A. murinanae*. No follow-up monitoring was done.

Three parasitoid species reared from a European *Zeiraphera* sp. – *Tranosema carbonellum* (Thomson), *Phytodietus gelitorius* Thunberg (= *coryphaeus* Gravenhorst) and *Dolichogenidea lineipes* (Wesmael) – were all tested for their acceptance of *C. fumiferana*. *T. carbonellum* readily attacked diapausing *C. fumiferana* in the laboratory. However, dissections revealed that *T. carbonellum* eggs were melanized and did not develop successfully. *P. gelitorius* showed typical host-seeking behaviour in the presence of *C. fumiferana* larvae but no ovipositions were observed. *D. lineipes* showed no interest in *C. fumiferana*. No releases of any of these parasitoids were made.

Species of the *T. minutum* complex (Pinto, 1998) are the only known egg parasitoids of *C. fumiferana*. Although natural parasitism rates from 15 to 77% have been reported (Anderson, 1976), the paucity of overwintering host eggs normally limits the potential of *T. minutum* to increase in response to outbreak populations of *C. fumiferana*. Studies during the 1970s in Quebec suggested that a native species of this complex could be reared on a factitious host egg and be effective upon release against *C. fumiferana* (W.F. Quednau, Ste-Foy, 1985, personal communication). In Maine, Houseweart *et al.* (1984) reported measurable increases in egg parasitism following experimental field release in the late 1970s and early 1980s. In 1981, the Ontario Ministry of Natural Resources initiated research to examine the feasibility of using *T. minutum* in inundative releases against *C. fumiferana* to determine the potential for developing mass production and release technology, and to determine the operational parameters under which releases would reduce *C. fumiferana* populations below economic damage levels (Smith *et al.*, 1990a).

The Ontario Project developed a mass-rearing system for *T. minutum* based on the factitious moth host, *Sitotroga cerealella* (Olivier), and capable of producing 30 million *T. minutum* per week, programmed by emergence time (Smith *et al.*, 1990a). Broad applications of parasitized *S. cerealella* eggs were made in the field with both ground and aerial delivery systems. The latter used a Bell 47 helicopter modified with a seed planter/slinger to achieve minimal drift and a swath width of 10–15 m over varying rates of application (Hope *et al.*, 1990; Smith *et al.*, 1990b). Despite the fact that over 50% of the aerially released material fell to the ground, up to 79.5% of *C. fumiferana* eggs within the release area were parasitized. Reductions up to 82% in subsequent larval populations were observed in treated plots (Smith *et al.*, 1990c). The largest trial year (1985) saw the application of these rates on five 1 ha plots. In 1984, the effective integration of parasitoid releases with a spring application of *B.t.k.* resulted in a reduction in larval populations up to 93%. Three years were spent developing the best application rate on large (1 ha) and small (25 × 25 m) plots (Smith *et al.*, 1990b). A strategy for using catches of the first male moth in pheromone traps to initiate two releases of 12 million female parasitoids ha^{-1} per release, 1 week apart, was recommended. The project concluded in 1986 following a collapse of *C. fumiferana* populations.

In 1989, Ciba–Geigy Canada, jointly with the Ontario Ministry of Natural Resources and the Universities of Toronto and Guelph, was awarded 5 years of provincial funding (Premier's Technology Fund) to commercialize this prototype system. The objective was to produce high-quality *T. minutum* in large numbers, at low cost, on a regular and continuous basis for large-scale release (Wallace and Smith, 1995). By the end of the project in 1994, when the company's rearing facility was sold to Beneficial Insectary Inc. (California), it was capable of producing over 100 million female parasitoids per week on the factitious host *Ephestia kuehniella* Zeller. These parasitized eggs could be stored for up to 7 weeks so that very large numbers could be accumulated. By optimizing the field strategy for using *T. minutum*, the application rate was reduced to two aerial applications of 10 million females ha^{-1} per release, 1 week apart. In 1993, this rate was applied to 30 ha (three 10 ha plots), thereby demonstrating its operational potential. Resulting parasitism of *C. fumiferana* egg masses on the treated plots averaged 67.0% and led to a reduction in larval populations of 64.5% the following spring, consistent with the work in the 1980s (Bourchier and Smith, 1998).

Numerous studies associated with this large-scale project helped to improve understanding of the system's potential as well as its limitations. Mass production was improved through changes in parameters that affected host (diet, lighting, temperature and CO_2 levels) and parasitoid rearing (long-term storage and host diapause, emergence programming, sting ratio and runting) (Corrigan and Laing, 1994; Corrigan *et al.*, 1995). Work on parasitoid quality and in the field showed: (i) the usefulness of molecular rDNA markers in the 18S region to identify select species/strains; (ii) the relative merits (wing size, fecundity) and demerits (walking speed) of measures to predict parasitoid quality in the field (Bourchier *et al.*, 1993; van Hezewijk *et al.*, 2000; Liu and Smith, 2000); (iii) the high level of genetic variation from only a few founding individuals and its subsequent reduction during colonization (Bourchier *et al.*, 1994); (iv) the necessity of temperatures above 15°C for successful flight and parasitism (Forsse *et al.*, 1992; Bourchier and Smith, 1996), as well as the ability to select strains for cold tolerance (Tocheva, 1995); (v) the limited effect of a single release of parasitoids with staggered emergence (Smith and You, 1990) due primarily to high predation (Braybrooks, 1995); (vi) the potential to integrate releases with natural populations of other *C. fumiferana* parasitoids (Bourchier and Smith, 1998); and (vii) the potential for non-target effects on at least four of 39 lepidopteran species present at the time of release.

The *T. minutum* project also diversified to studies showing that releases of the appropriate species against other forest defoliators also had potential, e.g. *Zeiraphera canadensis* Mutuura and Freeman (Wang and Smith, 1996; see West *et al.*, Chapter 58 this volume), forest tent caterpillar, *Malacosoma disstria* Hübner (Smith and Strom, 1993), hemlock looper, *Lambdina fiscellaria fiscellaria* (Guenée), black army cutworm, *Actebia fennica* (Tauscher), and western spruce budworm, *Choristoneura occidentalis* Freeman, but not gypsy moth, *Lymantria dispar* (L.), or white-marked tussock moth, *Orgyia leucostigma* (J.E. Smith) (Bai *et al.*, 1994). Further work on the release system resulted in modifications to deliver parasitized eggs either 'neat' (eggs alone) or in a carrier of water, with or without sticker, by tractor, backpack sprayer (S.M. Smith, unpublished) or fixed-wing aircraft (N. Payne, Sault Ste Marie, 1993, personal communication). Despite these successes, which led to the commercial production and release system, the costs remained high (about Can\$400 ha^{-1}), and large-scale research was discontinued in 1994. The simultaneous decline of epidemic populations of *C. fumiferana* throughout its range contributed to the reduction in interest for alternative control methods.

Pathogens

Bacteria

From 1980 to 2000, *B.t.k.* was applied to about 4.7 million ha of forest infested by *C. fumiferana* in Canada, using a total of about 158×10^{15} international units (IU) (Tables 12.1 and 12.2). The pattern of use for the various provinces over the years reflects regional shifts in *C. fumiferana* populations as well as differences among provincial forest protection agencies with regard to spray policies. Quebec and New Brunswick launched aggressive, large-scale protection programmes against *C. fumiferana*, gradually shifting from fenitrothion to *B.t.k.* in the mid-1980s (Quebec) or early 1990s (New Brunswick) (Table 12.1). The collapse of the outbreak around 1993 suspended the need for further control operations. In western Canada, *C. fumiferana* populations reached epidemic levels in the early 1990s and aggressive control programmes were initiated in Alberta (1990) and Saskatchewan (1993). These control programmes have continued to the present, with more than 80,000 ha being sprayed in Alberta alone during 1999 (H. Ono, Edmonton, 2000, personal communication).

Aerial application of commercial high-potency *B.t.k.* formulations has become the most widespread method to protect coniferous forests from excessive defoliation by epidemic *C. fumiferana* populations. In eastern Canada, the primary objective is foliage protection, and spray application is timed for peak fourth-instar larvae (Carter, 1991) whereas in western Canada, Alberta in particular, spray applications are delayed to peak numbers of the fifth instar to maximize population suppression (H. Ono, Edmonton, 2000, personal communication). Currently, the recommended dosage rate of 30×10^9 IU ha^{-1} is applied undiluted, in volumes of 1.2–2.4 l ha^{-1} over one or two applications. In the field, this rate may need to be adjusted down to 15×10^9 IU ha^{-1} for populations of less than 15–20 larvae per 45 cm branch (Carter, 1991) or up to 50×10^9 IU ha^{-1} (which will require changes in registration) to achieve foliage protection at high larval densities (Régnière and Cooke, 1998).

Viruses

Prior to 1980, extensive field tests on a cumulative total of about 2500 ha were conducted with several viruses that naturally infect *C. fumiferana*, including a *Nucleopolyhedrovirus* (NPV), a *Granulovirus* (GV), a cytoplasmic polyhedrovirus (= *Cypovirus*) (CPV), and an *Entomopoxvirus* (EV) (see detailed review by Cunningham and Howse, 1984). Limited experimentation with NPV continued in 1980, 1981 and 1983 on a total of about 200 ha (Cunningham, 1985). The goal of initiating an epizootic to eventually regulate the population was never achieved in any of these trials.

New isolates of NPV and GV were investigated during the 1990s. Initial single-tree trials and ground applications in Quebec suggested that the NPV isolate (T3) was more efficacious than the wild-type virus (J. Valéro, Ste-Foy, 1998, personal communication). Higher virulence was later confirmed in laboratory bioassays (W. Kaupp,

Table 12.1. Operational use of *Bacillus thuringiensis* for control of *Choristoneura fumiferana* in eastern Canada.

	Ontario		Quebec		New Brunswick		Nova Scotia		Newfoundland	
Year	Area (ha)[a]	10^9 IU applied[b]	Area (ha)	10^9 IU applied	Area (ha)	10^9 IU applied	Area (ha)	10^9 IU applied	Area (ha)	10^9 IU applied
1980	4,305	96,825	20,971	629,130	0	0	25,645	572,720	7,537	150,740
1981	6,900	135,404	12,188	243,760	0	0	31,903	647,726	1,920	38,400
1982	3,093	59,975	22,971	582,870	0	0	15,304	306,080	4,724	94,480
1983	2,763	55,260	24,532	490,640	10,300	309,000	20,616	412,320	0	0
1984	3,145	62,900	296,568	7,792,320	34,300	1,119,000	20,365	414,340	3,110	62,200
1985	29,369	587,380	512,155	14,832,340	81,000	2,430,000	49,719	1,491,570	3,450	103,500
1986	155,466	3,684,790	32,789	837,380	111,500	3,345,000	56,155	1,684,650	0	0
1987	76,819	1,536,380	205,092	6,081,760	91,300	2,703,000	31,080	932,400	0	0
1988	14,023	420,690	208,073	6,082,190	211,100	6,333,000	0	0	0	0
1989	30,383	917,220	189,229	6,828,240	104,700	3,282,000	0	0	0	0
1990	49,627	1,948,590	479,896	20,753,760	170,600	4,372,500	0	0	0	0
1991	67,913	2,420,580	194,975	5,849,250	111,500	3,058,500	0	0	0	0
1992	0	0	5,670	170,100	89,295	2,543,850	0	0	7,757	480,210
1993	291	8,730	0	0	64,200	1,821,000	0	0	0	0
Total	444,097	11,934,724	2,205,109	71,173,740	1,082,795	31,316,850	250,787	5,461,806	28,498	929,530

[a] Number of hectares treated with one or more applications.

[b] Total dose (expressed in International Units) applied ha^{-1} (= number of ha treated × number of applications × 10^9 IU ha^{-1} per application).

Source: Forestry Insecticide Database, Canadian Forest Service, Great Lakes Forestry Centre, Sault Ste Marie, Ontario.

Table 12.2. Operational use of *Bacillus thuringiensis* for control of *Choristoneura fumiferana* in western Canada.

	Manitoba		Saskatchewan		Alberta		British Columbia	
Year	Area (ha)[a]	10^9 IU applied[b]	Area (ha)	10^9 IU applied	Area (ha)	10^9 IU applied	Area (ha)	10^9 IU applied
1981	365	7,300	0	0	0	0	0	0
1987	536	10,720	0	0	0	0	0	0
1988	1,182	35,460	0	0	0	0	150	4,500
1989	4,984	149,520	0	0	0	0	0	0
1990	4,362	143,980	0	0	10,000	300,000	0	0
1991	0	0	0	0	27,800	1,688,000	262	15,720
1992	0	0	7,734	232,020	35,100	1,755,000	570	34,200
1993	0	0	34,016	1,887,900	8,230	418,084	0	0
1994	0	0	10,500	630,000	14,253	673,275	0	0
1995	15,223	911,878	8,550	513,000	110,923	3,708,050	0	0
1996	6,933	416,022	0	0	110,247	5,070,426	0	0
1997	0	0	40,000	2,400,000	20,068	832,688	0	0
1998	0	0	93,646	5,618,760	7,098	360,578	0	0
1999	0	0	82,321	4,939,260	70,323	3,511,190	0	0
Total	33,585	1,674,880	276,767	16,220,940	414,042	18,337,292	982	54,420

[a] Number of hectares treated with one or more applications.
[b] Total dose (expressed in International Units) applied ha^{-1} (= number of ha treated × number of applications × 10^9 IU ha^{-1} per application).

Source: Forestry Insecticide Database, Canadian Forest Service, Great Lakes Forestry Centre, Sault Ste Marie, Ontario.

Sault Ste Marie, 1995, personal communication). Aerial applications of this isolate on three 10 ha plots in Quebec during 1997 produced inconclusive results (J. Valéro and N. Payne, Sault Ste Marie, 1998, personal communication). However, two applications of T3 at 5.0×10^{11} polyhedral inclusion bodies (PIB) ha^{-1} on three 10 ha plots in Manitoba during 1998 did not suppress *C. fumiferana* populations significantly (L. Cadogan, Sault Ste Marie, 1999, personal communication). Similarly, a new isolate of GV was obtained from *C. fumiferana* larvae collected in the Gaspé Peninsula, Quebec, during 1992. Results from preliminary field tests in 1997 were encouraging, but no details were provided (Forté *et al.*, 1999).

Evaluation of Biological Control

After decades of field experimentation, *B.t.k.* has become a commercial success for suppression of populations of *C. fumiferana*. Naturally occurring viral pathogens, however, do not appear to hold much promise for further development. For several reasons, viruses are unlikely to be effective and commercially attractive for controlling epidemic populations. First, they do not appear to play a key role in terminating outbreaks, as epizootics have never been observed in naturally collapsing *C. fumiferana* populations. Secondly, despite several attempts, introduction of a naturally occurring virus through aerial application has not yet been successful in initiating an epizootic. One key reason for this may be that viral sprays must be applied at bud flush on exposed larvae, and this is too late for secondary infection and subsequent horizontal transmission (Cunningham and Kaupp, 1995). Efforts to obtain a secondary infection by treating younger (second instar) larvae as they emerge from their hibernacula, but before they mine old needles, have met with limited success (Kaupp *et al.*, 1990) and are currently considered impractical. Finally, successful implementation of this approach will require expensive *in vivo* mass production, and this is unlikely because of the highly specific nature of viruses and the lack of monetary reward for their commercial production.

Since 1980, annual aerial releases of the egg parasitoid *T. minutum* were developed and shown to be highly effective at reducing epidemic populations of *C. fumiferana*, similar to applications of *B.t.k.* No carry-over effects were observed on the target host, with limited potential for parasitism on eggs of species from families such as Nymphalidae, Hesperiidae, Noctuidae and Geometridae. The cost of producing the large number of parasitoids required for forest applications (about 20 million female parasitoids ha^{-1} over 2 weeks) was not competitive with commercial *B.t.k.* products. The collapse of *C. fumiferana* populations limited further work.

In terms of introduced parasitoids, continued work on 'new associations' from Europe should be viewed cautiously for several reasons. First, the introduction of exotic species is a long-term tactic that raises concerns about their impact on native parasitoid and non-target host populations. Secondly, there are no obvious 'missing' parasitoids in Canada that could be filled by European introductions; *C. fumiferana* already has a rich native fauna (Huber *et al.*, 1996) ecologically similar to that of its European counterpart (Mills, 1983). Thirdly, recent work on the ecology of *C. fumiferana* suggests that if natural enemies are involved in its population fluctuations, they interact as a suite with other perturbations such as habitat structure, e.g. loss of host trees through defoliation, making it unlikely that a single introduced species will have much influence.

As always, shifting forest management priorities and conditions will affect significantly the way foresters perceive the *C. fumiferana* problem and the options for its control. Because the current spruce–fir forest is much younger and under a shorter rotation time than in the past, future damage by *C. fumiferana* will be less. Also, the industrial shift away from spruce to other tree species in many areas should concentrate pest management efforts against *C. fumiferana* even more towards annual foliage protection on selected sites. While *B.t.k.* has become the standard for management by suppression, foresters will continue to look for realistic biological alternatives because of persisting concerns about interventions with pesticides on public forests. Thus, it appears that future biological control efforts against *C. fumiferana* should focus on augmenting native natural enemies, especially those close to achieving commercial success, e.g. releases of *T. minutum*, because such approaches can be used as effectively as *B.t.k.*, only more selectively and with less ecological impact.

Recommendations

Further work should include:

1. Focusing on biological control agents that can be integrated with early intervention for the overall management of *C. fumiferana*;
2. Concentrating efforts on augmenting native parasitoids, especially *T. minutum*, by reducing production costs through the development of artificial host eggs, improved parasitoid quality during mass-rearing, and refinements to field applications.

References

Anderson, J.F. (1976) Egg parasitoids of forest defoliating Lepidoptera. In: Anderson, J.F. and Kaya, H.K. (eds) *Perspectives in Forest Entomology.* Academic Press, New York, New York, pp. 233–250.

Bai, B.B., Cobanoğlu, S. and Smith, S.M. (1994) Potential for using *Trichogramma* species for biological control of lepidopterous forest defoliators. *Entomologia Experimentalis et Applicata* 75, 135–144.

Blais, J.R. (1983) Trends in the frequency, extent, and severity of spruce budworm outbreaks in eastern Canada. *Canadian Journal of Forest Research* 13, 539–547.

Bourchier, R.S. and Smith, S.M. (1996) Influence of environmental conditions and parasiatoid quality on field performance of *Trichogramma minutum. Entomologia Experimentalis et Applicata* 80, 461–468.

Bourchier, R.S. and Smith, S.M. (1998) Interactions between large scale inundative releases of *Trichogramma minutum* and naturally occurring spruce budworm parasitoids. *Environmental Entomology* 27, 1273–1279.

Bourchier, R.S., Smith, S.M. and Song, S.J. (1993) Host acceptance and parasitoid size as predictors of parasitoid quality for mass-reared *Trichogramma minutum. Biological Control* 3, 135–139.

Bourchier, R.S., Smith, S.M., Corrigan, J. and Laing, J.E. (1994) Effect of host-switching on performance of mass-reared *Trichogramma minutum. Biocontrol Science and Technology* 4, 353–362.

Braybrooks, A. (1995) Impact of ant predation on eggs of the Mediterranean flour moth, *Ephestia kuehniella* Zeller, a factitious host of the inundatively released parasitoid, *Trichogramma minutum* Riley. MScF thesis, University of Toronto, Toronto, Ontario.

Carter, N.E. (1991) Efficacy of *Bacillus thuringiensis* in New Brunswick, 1988–1990. In: *Proceedings of the 72nd Annual Meeting, Woodlands Section, Canadian Pulp and Paper Association, Montreal.* Canadian Pulp and Paper Association, Montreal, Quebec, pp. 113–116.

Corrigan, J.E. and Laing, J.E. (1994) Effects of the rearing host species and the host species attacked on performance by *Trichogramma minutum* Riley (Hym.: Trichogrammatidae). *Environmental Entomology* 23, 755–760.

Corrigan, J.E., Laing, J.E. and Zubricky, J.S. (1995) Effects of parasitoid:host ratio and time of day of parasitism on development and emergence of *Trichogramma minutum* (Hym.: Trichogrammatidae) parasitizing eggs of *Ephestia kuehniella* (Lep.: Pyralidae). *Annals of the Entomological Society of America* 88, 773–780.

Cunningham, J.C. (1985) Status of viruses as biocontrol agents for spruce budworms. In: Grimble, D.G. and Lewis, F.B. (eds) *Microbial Control of Spruce Budworms and Gypsy Moth.* United States Department of Agriculture, Forest Service, Northeastern Forest Experiment Station, Broomall, Pennsylvania, pp. 61–67.

Cunningham, J.C. and Howse, G.M. (1984) Viruses: application and assessment. In: Kelleher, J.S. and Hulme, M.A. (eds) *Biological Control Programmes Against Insects and Weeds in Canada 1969–1980.* Commonwealth Agriculture Bureaux, Slough, UK, pp. 248–259.

Cunningham, J.C. and Kaupp, W.J. (1995) Insect viruses. In: Armstrong, J.A. and Ives, W.G.H. (eds) *Forest Insect Pests in Canada*. Natural Resources Canada, Canadian Forest Service, Ottawa, Ontario, pp. 328–340.

FIDS (1987) *Forest Insect and Disease Conditions in Canada 1987*. Canadian Forest Service, Ottawa, Ontario.

Forsse, E., Smith, S.M. and Bourchier, R.S. (1992) Flight initiation in the egg parasitoid *Trichogramma minutum*: Effect of temperature, mates, food, and host eggs. *Entomological Experimentalis et Applicata* 62, 147–154.

Forté, A.J., Guertin, C. and Cabana, J. (1999) Pathogenicity of a granulosis virus towards *Choristoneura fumiferana* (Lepidoptera: Tortricidae). *The Canadian Entomologist* 131, 725–727.

Frankenhuyzen, K. van (1990) Development and current status of *Bacillus thuringiensis* for control of defoliating forest insects. *Forestry Chronicle* 66, 498–507.

Frankenhuyzen, K. van (1993) The challenge of *Bacillus thuringiensis*. In: Entwistle, P.F., Cory, J.S., Bailey, M.J. and Higgs, S. (eds) Bacillus thuringiensis, *An Environmental Biopesticide: Theory and Practice*. John Wiley and Sons, New York, New York, pp. 1–35.

Gross, H.L., Roden, D.B., Churcher, J.J., Howse, G.M. and Gertridge, D. (1992) *Pest-Caused Depletions to the Forest Resource of Ontario, 1982–1987*. Forestry Canada Ontario Region–Great Lakes Forestry Centre Joint Report 17. Canadian Forest Service, Ontario Region, Sault Ste Marie, Ontario.

Hezewijk, B. van, Bourchier, R.S. and Smith, S.M. (2000) Searching speed of *Trichogramma minutum* and its potential as a measure of parasitoid quality. *Biological Control* 17, 139–146.

Hope, C.A., Nicholson, S.A. and Churchen, J.J. (1990) Aerial release system for *Trichogramma minutum* Riley in plantation forests. In: Smith, S.M., Carrow, J.R. and Laing, J.E. (eds) Inundative release of the egg parasitoid, *Trichogramma minutum* (Hymenoptera: Trichogrammatidae), against forest insect pests such as the spruce budworm, *Choristoneura fumiferana* (Lepidoptera: Tortricidae): The Ontario project 1982–1986. *Memoirs of the Entomological Society of Canada* 153, 38–44.

Houseweart, M.W., Jennings, D.T. and Lawrence, R.K. (1984) Field releases of *Trichogramma minutum* (Hym.: Trichogrammatidae) for suppression of epidemic spruce budworm, *Choristoneura fumiferana* (Lep.: Tortricidae), egg populations in Maine. *The Canadian Entomologist* 116, 1357–1366.

Huber, J.T., Eveleigh, E., Pollock, S. and McCarthy, P. (1996) The chalcidoid parasitoids and hyperparasitoids (Hymenoptera: Chalcidoidea) of *Choristoneura* species (Lepidoptera: Tortricidae) in America north of Mexico. *The Canadian Entomologist* 128, 1167–1220.

Hulme, M.A. and Green, G.W. (1984) Biological control of forest insect pests in Canada 1969–1980: Restrospect and prospect. In: Kelleher, J.S. and Hulme, M.A. (eds) *Biological Control Programmes Against Insects and Weeds in Canada 1969–1980*. Commonwealth Agriculture Bureaux, Slough, UK, pp. 215–227.

Kaupp, W.J., Cunningham, J.C. and Cadogan, B.L. (1990) Aerial application of high dosages of nuclear polyhedrosis virus to early instar spruce budworm, *Choristoneura fumiferana*. Information Report FPM-X-82, Forestry Canada, Forest Pest Management Institute, Sault Ste Marie, Ontario.

Liu, F.-H. and Smith, S.M. (2000) Measurement and selection of parasitoid quality for mass-reared *Trichogramma minutum* Riley used in inundative release. *Biocontrol Science and Technology* 10, 3–13.

McLean, D.A. (1990) Impact of forest pests and fire on stand growth and timber yield: Implications for forest management planning. *Canadian Journal of Forest Research* 20, 391–404.

Mills, N.J. (1983) Possibilities for the biological control of *Choristoneura fumiferana* (Clemens) using natural enemies from Europe. *Biocontrol News and Information* 4, 103–125.

Nealis, V.G. (1988) Weather and the ecology of *Apanteles fumiferanae* Vier. (Hymenoptera: Braconidae). *Memoirs of the Entomological Society of Canada* 146, 57–70.

Nealis, V.G. and Fraser, S. (1988) Rate of development, reproduction, and mass-rearing of *Apanteles fumiferanae* Vier. (Hymenoptera: Braconidae) under controlled conditions. *The Canadian Entomologist* 120, 197–204.

Perry, D. and Régnière, J. (1986) The role of fungal pathogens in spruce budworm population dynamics: frequency and temporal relationships. In: Samson, R.A., Vlak, J.M. and Peters, D. (eds) *Fundamental and Applied Aspects of Invertebrate Pathology*. Foundation of the Fourth International Colloquium of Invertebrate Pathology, Wageningen, The Netherlands, pp. 167–170.

Pinto, J. (1998) Systematics of the North American species of *Trichogramma* Westwood (Hym.: Trichogrammatidae). *Memoirs of the Entomological Society of Washington* 22, 287 pp.

Régnière, J. and Cooke, B.J. (1998) Validation of a process-oriented model of *Bacillus thuringiensis* variety *kurstaki* efficacy against spruce budworm (Lepidoptera: Tortricidae). *Environmental Entomology* 27, 801–811.

Régnière, J. and Lysyk, T.J. (1995) Population dynamics of the spruce budworm, *Choristoneura fumiferana*. In: Armstrong, J.A. and Ives, W.G.H. (eds) *Forest Insect Pests in Canada*. Natural Resources Canada, Canadian Forest Service, Ottawa, Ontario, pp. 95–105.

Royama, T. (1984) Population dynamics of the spruce budworm, *Choristoneura fumiferana*. *Ecological Monographs* 54, 429–462.

Sanders, C.J. (1995) Research on the spruce budworm, *Choristoneura fumiferana*. In: Armstrong, J.A. and Ives, W.G.H. (eds) *Forest Insect Pests in Canada*. Natural Resources Canada, Canadian Forest Service, Ottawa, Ontario, pp. 91–93.

Smirnoff, V. and Morris, O. (1982) Field development of *Bacillus thuringiensis* in eastern Canada, 1970–1980. In: Kelleher, J.S. and Hulme, M.A. (eds) *Biological Control Programmes against Insects and Weeds in Canada, 1969–1980*. Commonwealth Agricultural Bureaux, Slough, UK, pp. 238–247.

Smith, S.M. and Strom, K. (1993) Oviposition by the forest tent caterpillar (Lep.: Lasiocampidae) and acceptability of its eggs to parasitism by *Trichogramma minutum* (Hym.: Trichogrammatidae). *Environmental Entomology* 22, 1375–1382.

Smith, S.M. and You, M. (1990) A life system simulation model for improving inundative releases of the egg parasitoid, *Trichogramma minutum* (Hym.: Trichogrammatidae) against the spruce budworm (Lep.: Tortricidae). *Ecological Modelling* 51, 123–142.

Smith, S.M., Carrow, J.R. and Laing, J.E. (eds) (1990a) Inundative release of the egg parasitoid, *Trichogramma minutum* (Hymenoptera: Trichogrammatidae), against forest insect pests such as the spruce budworm, *Choristoneura fumiferana* (Lepidoptera: Tortricidae): The Ontario project 1982–1986. *Memoirs of the Entomological Society of Canada* 153, 87 pp.

Smith, S.M., Wallace, D.R., Laing, J.E., Eden, G.M. and Nicholson, S.A. (1990b) Deposit and distribution of *Trichogramma minutum* Riley following aerial release. In: Smith, S.M., Carrow, J.R. and Laing, J.E. (eds) Inundative release of the egg parasitoid, *Trichogramma minutum* (Hymenoptera: Trichogrammatidae), against forest insect pests such as the spruce budworm, *Choristoneura fumiferana* (Lepidoptera: Tortricidae): The Ontario project 1982–1986. *Memoirs of the Entomological Society of Canada* 153, 45–55.

Smith, S.M., Wallace, D.R., Howse, G. and Meating, J. (1990c) Suppression of spruce budworm populations by *Trichogramma minutum* Riley, 1982–1986. In: Smith, S.M., Carrow, J.R. and Laing, J.E. (eds) Inundative release of the egg parasitoid, *Trichogramma minutum* (Hymenoptera: Trichogrammatidae), against forest insect pests such as the spruce budworm, *Choristoneura fumiferana* (Lepidoptera: Tortricidae): The Ontario project 1982–1986. *Memoirs of the Entomological Society of Canada* 153, 56–81.

Tocheva, S. (1995) Host exploitation at low temperatures by *Trichogramma minutum* Riley (Hym.: Trichogrammatidae): heritability estimates, selection, and the effect of selection on associated biological characteristics. MScF thesis, University of Toronto, Toronto, Ontario.

Varty, I.W. (1984) Spruce budworm; D. Testing of parasitoids. In: Kelleher, J.S. and Hulme, M.A. (eds) *Biological Control Programmes Against Insects and Weeds in Canada 1969–1980*. Commonwealth Agriculture Bureaux, Slough, pp. 267–279.

Wallace, D.R. and Smith, S.M. (1995) Spruce bud moth, *Zeiraphera canadensis*. In: Armstrong, J.A. and Ives, W.G.H. (eds) *Forest Insect Pests in Canada*. Natural Resources Canada, Canadian Forest Service, Ottawa, Ontario, pp. 183–192.

Wang, Z. and Smith, S.M. (1996) Phenotypic differences between thelytokous and arrhenotokous members of the *Trichogramma minutum* (Hym.: Trichogrammatidae) complex from *Zeiraphera canadensis* (Lep.: Olethreutidae). *Entomologia Experimentalis et Applicata* 78, 315–323.

13 *Choristoneura occidentalis* Freeman, Western Spruce Budworm (Lepidoptera: Tortricidae)

I.S. Otvos, N. Conder and K. van Frankenhuyzen

Pest Status

The western spruce budworm, *Choristoneura occidentalis* Freeman, is a native defoliator of Douglas fir, *Pseudotsuga menziesii* (Mirbel) Franco, in western North America. Six outbreaks have occurred in British Columbia since 1909 (Harris *et al.*, 1985). The last outbreak started in 1967, and defoliation caused by *C. occidentalis* in the province was recorded every year in British Columbia until 1999, when the outbreak decreased to about 1100 ha. There were two peak periods of defoliation, the first in 1976, when about 258,000 ha were defoliated. The second period of peak defoliation built up gradually, from about 26,100 ha in 1982 to about 196,400 ha in 1985, and by 1987 the outbreak covered 838,000 ha of Douglas fir (Wood *et al.*, 1987; Parfett *et al.*, 1994). Some stands have been defoliated repeatedly, resulting in growth loss, top kill and some tree mortality (Alfaro *et al.*, 1982; Van Sickle *et al.*, 1983; Alfaro, 1986). Tree mortality occurs most frequently in the understorey below large, heavily infested trees, with serious consequences where multiple-age forest management is practised on dry sites. If the understorey is killed by *C. occidentalis*, the next crop of trees is lost and the mature trees cannot be removed until a new understory is well established, which may take a decade or more, resulting in considerable delays in harvesting.

C. occidentalis eggs are laid in late July in masses on the underside of needles of the host trees. The number of eggs in each egg mass can vary, averaging 35–45 eggs per mass in British Columbia (Silver, 1960; Harris and Dawson, 1982). Larvae hatch in August, move without feeding to protected sites, spin a silk shelter to hibernate, and emerge the following spring in late May or early June to mine old needles or swelling buds until bud flush occurs. Larvae prefer to feed on the tender new foliage and they web the needles together to form a feeding tunnel. They complete development (there are six instars) and pupate in mid-July among the residual foliage on the branches. Adults emerge about 2 weeks later and females lay an average of 117–216 eggs, depending on geographic location and host (Carolin, 1987).

Background

Chemical insecticides have been considered to control *C. occidentalis*, but such operations were never conducted in British Columbia due to public opinion and political pressure. Research was therefore directed towards developing biological control agents. Three experimental field trials were conducted using viruses (1976, 1978 and 1982), with unacceptably low population reduction, and four field trials were carried out using *Bacillus thuringiensis* serovar *kurstaki* (*B.t.k.*) (1978, 1986–1988) against *C. occidentalis*. Otvos *et al.* (1989), Shepherd *et al.* (1995) and DeBoo and Taylor (1995) summarized these results. Here, experimental field trials and operational use of *B.t.k.* since 1980 in British Columbia are reviewed.

Biological Control Agents

Pathogens

Viruses

C. occidentalis is infected by the same virus that attacks its close relative, *C. fumiferana* (Clemens) (Cunningham, 1985). Most of the work on *C. occidentalis* was done using *Nucleopolyhedrovirus* (NPV) and, to a lesser extent, a *Granulovirus* (GV) (Shepherd *et al.*, 1995). It was thought that these viruses would be self-propagating after application and would only need to be applied once during the outbreak cycle to achieve control, as for *Orgyia pseudotsugata* (McDunnough) (see Otvos *et al.*, Chapter 41, this volume) (Shepherd *et al.*, 1984; Otvos *et al.*, 1987a, b).

In 1981, in a small ground spray trial, NPV and GV were applied separately to individual, infested trees in a natural Douglas fir regeneration. No differences were observed in the population reduction caused by either NPV or GV at the two higher dosages, but at the lowest dosage GV appeared to cause much higher mortality (Cunningham *et al.*, 1983). Consequently, in 1982, both NPV and GV were applied aerially to two 172-ha plots. These plot sizes were selected to minimize invasion by *C. occidentalis* after application, as was thought to have occurred with the smaller plots used in previous experiments. In the year of application, population reductions caused by NPV and GV were 51.8 and 34.6%, respectively. Population reductions due to vertical transmission of NPV and GV were 33.7 and 14.7%, respectively, 1 year after treatment, and 14.4 and 25.6%, respectively, 2 years after treatment (Table 13.1).

Bacteria

Several operational trials were conducted from 1986 to 1988 to evaluate *B.t.k.* for foliage protection, and to gain operational experience in planning and implementing aerial spray programmes. Results were variable, possibly due to terrain, climate or inexperience (DeBoo and Taylor, 1995). In 1987, the British Columbia Ministry of Forests initiated a large-scale field study to determine whether foliage protection could be obtained at the registered dose of 30×10^9 international units (IU) ha^{-1} in 2.4 l ha^{-1} on overstorey Douglas fir trees and whether the understorey could be protected and mortality reduced. The area treated varied from year to year (Table 13.2). This operational study is ongoing and results of the *B.t.k.* applications after 1988 at 30×10^9 IU ha^{-1} remain highly variable.

The variability of protection afforded at the registered dose has been attributed to several factors, primarily: (i) difficulty in achieving uniform spray application and consistent and sufficient spray deposit in mountainous terrain; and (ii) variability of bud flush and insect development. Bud development varies from tree to tree and from area to area, due to differences in elevation, aspect, microclimate, etc. Successful coordination of aerial application with bud flush and larval development over large areas is extremely difficult. Unstable weather conditions and rugged terrain also make spray application difficult, resulting in uneven spray deposit. All these factors often lead to compromises in operational spray applications and less than desirable results (Shepherd *et al.*, 1995).

To clarify and solve the problems of variable population reductions obtained by a single application of *B.t.k.* at 30×10^9 IU ha^{-1}, co-operative experiments to determine the efficacy of several *B.t.k.* products at higher dose and volume application were initiated (by the Canadian Forest Service, the British Columbia Ministry of Forests, and some *B.t.k.* manufacturers). Ten experiments were conducted between 1989 and 1996, mostly in the Merritt Forest District, Kamloops Forest Region. These experiments were done concurrently with the above-mentioned, large-scale operational field study. Details of the 6-year field efficacy trials will be published separately.

Spray plots were established in areas containing increasing or stable populations of *C. occidentalis* on trees (6–10 m tall, 30–60 years old) suitable for sampling with pole pruners. Each product was applied to

Table 13.1. Experimental application of baculoviruses against *Choristoneura occidentalis* near Ashcroft, British Columbia, 1982. (Adapted from Otvos *et al.*, 1989.)

Virus	Aircraft and spray dispenser	Application rates		Droplets $cm^{-2} \pm$ SD	Pre-spray larval density per 100 buds	Population reduction (%)[b]		
		Dose[a]	Vol. ($l\ ha^{-1}$)			Year of application	1 year after application	2 years after application
NPV	Cessna Agwagon 42 Teejet flat-fan nozzles	5.4×10^{11} PIB ha^{-1}	9.4	10.0 ± 1.36	14.3 ± 0.82	51.8	33.7	14.4
GV	Cessna Agwagon 42 Teejet flat-fan nozzles	1.7×10^{14} Caps ha^{-1}	9.4	12.0 ± 1.26	10.8 ± 0.68	34.6	14.7	25.6

[a] PIB, polyhedral inclusion bodies; Caps, capsules.
[b] Corrected population reduction calculated using Abbott's formula (Abbott, 1925).

Table 13.2. Operational use of *Bacillus thuringiensis* serovar *kurstaki* against *Choristoneura occidentalis* in British Columbia, 1986–1999.

Year	Area treated (ha)	Application rates	
		Dose ($\times 10^9$ IU[a] ha^{-1})	Vol. ($l\ ha^{-1}$)
1986[b]	200	30	5.9
1987[b]	940	30	2.8, 3.1
1988[b]	1,550	30	2.0
1989[c]	500	30	–
1991[c]	3,000	30	–
1992[c]	35,585	30	–
1993[c]	34,245	30	–
1994[c]	21,025	30	–
1997[c]	16,150	30	–
1998[c]	20,597	30	–
1999[c]	21,725	30	–
Total	155,517		–

[a] International Units.

[b] Obtained from DeBoo and Taylor (1995).

[c] L. MacLauchlan, Kamloops, 2000, personal communication.

a 50 ha plot containing 45 sample trees in 1989 and 30 sample trees in the other years, replicated three times. Three untreated areas of comparable size were used as controls in each of the 6 years of field trials.

In 1989, Dipel® 264, a high potency product, was applied at 30 $\times$ 10^9 IU ha^{-1} in 1.2 l ha^{-1}, but reduced populations by only 51.5% (Table 13.3), considered unacceptable by forest managers. In 1992, when Foray® 48B was applied at 60 $\times$ 10^9 IU ha^{-1} in 4.8 l ha^{-1}, i.e. twice the dosage and four times the volume applied in 1989, population reduction was good, at 73.4% (Table 13.3).

In 1993, Foray® 48B at the same dose and volumes as in 1992, and Foray® 76B (a higher potency product) at 60 $\times$ 10^9 IU ha^{-1} in 3.0 l ha^{-1} were tested. Population reduction due to Foray® 48B was 84.0%, whereas Foray® 76B at the same dose but in a lower volume caused only 42% reduction.

In 1994, the same products, doses and volumes gave similar results, namely, 74.9% and 41.0% reductions in Foray® 48B- and Foray® 76B-treated plots, respectively. Mortality was therefore higher in the plots receiving the same dose in higher volume. Population reduction in the plots treated with Foray® 76B may have been low, in part, because of the high population levels before the spray (247 larvae m^{-2}), the highest average larval density observed during the study.

In 1995, Dipel® 48AF at 50 $\times$ 10^9 IU ha^{-1} in 3.9 l ha^{-1}, Dipel® 76AF at 60 $\times$ 10^9 IU ha^{-1} in 3.0 l ha^{-1}, and Foray® 48B at 60 $\times$ 10^9 IU ha^{-1} in 4.8 l ha^{-1} were tested. Population reduction was highest (about 95%) in plots receiving the higher dose in the highest volume (Foray® 48B), the second highest (about 80%) at the same dose but lower volume (Dipel® 76AF), and the lowest (about 73%) in plots treated at 50 $\times$ 10^9 IU ha^{-1} in 3.9 l ha^{-1} (Dipel® 48AF).

In 1996, only Foray® 48B was applied, at 60 $\times$ 10^9 IU ha^{-1} in 4.8 l ha^{-1}. Population reduction was only about 66%. This unexpectedly low reduction, following the earlier results, may have been due, in part, to the high initial (or pre-spray) larval density, 178 larvae m^{-2}, the second highest average larval density observed during the study. The 30–32 mm of precipitation that fell on the third day after treatment could also have contributed to this lower population reduction, by washing spray deposits from the foliage.

Evaluation of Biological Control

The epizootic initiated by NPV and GV applications did not control *C. occidentalis*

Table 13.3. Experimental applications of *Bacillus thuringiensis* serovar *kurstaki* against *Choristoneura occidentalis*, in British Columbia, 1986–1996.

Year	Location	Formulation	Area treated (ha)	Aircraft and spray dispenser	Application Rates		Droplets / needle ± SD	Pre-spray larval density (m^{-2})	Population reduction (%)[b]
					$\times 10^9$ IU[a] ha^{-1}	Vol. ($l\ ha^{-1}$)			
1989	Glen Rosa	Dipel® 264	150	Fixed-wing AU4000	30	1.2	0.74	95.9	51.5
1992	Kamloops	Foray® 48B	150	Cessna Agtruck, Micronair AU4000	2 × 30	2 × 2.4	0.33 ± 0.57	113.7	73.4
1993	Merritt	Foray® 48B	150	Hiller 12E Soloy	60	4.8	1.33 ± 2.04	68.4	84.0
		Foray® 76B	150	Beecomist AU4000	60	3.0	0.83 ± 1.17	167.9	42.1
1994	Merritt	Foray® 48B	150	Hiller 12E Soloy	60	4.8	0.36 ± 0.61	60.3	74.9
		Foray® 76B	150	Beecomist AU4000	60	3.0	0.48 ± 0.87	246.5	41.0
1995	Merritt	Dipel® 48AF	150	Hiller 12E Soloy	50	3.9	1.53 ± 1.36	115.3	72.7
		Dipel® 76AF	150	Beecomist AU4000	60	3.0	0.59 ± 0.59	71.8	79.2
		Foray® 48B	150		60	4.8	1.27 ± 1.83	96.4	94.4
1996	Merritt	Foray® 48B	150	Hiller 12E Soloy, Beecomist AU4000	60	4.8	2.22 ± 1.83	178.2	66.3

[a]International units.
[b]Corrected population reduction calculated using Abbott's formula (Abbott, 1925).

populations in the year of application. Although vertical transmission of the applied virus occurred for 2 years after treatment, viral infection decreased each year. Consequently, further field testing of viruses against *C. occidentalis* was not recommended until a more virulent strain is discovered and the ecology of the virus is better understood, i.e. why these viruses are not as efficacious in the field as they are in the laboratory (Otvos *et al.*, 1989).

The experimental applications of various *B.t.k.* formulations over 6 years showed that both a higher dose and volume are needed than the currently registered 30 $\times$ 10^9 IU ha^{-1} in 2.4 l ha^{-1} to achieve good and consistent population reduction. Of the products tested, Foray® 48B at 60 $\times$ 10^9 IU ha^{-1} in 4.8 l ha^{-1} gave the highest population reduction, followed by Dipel® 76AF and Dipel® 48AF (Table 13.3).

For consistently good and reproducible population reduction of *C. occidentalis* in British Columbia, with its mountainous terrain, *B.t.k.* should be used in the 50–60 $\times$ 10^9 IU ha^{-1} dose range and applied in the 3.0–4.8 l ha^{-1} volume range once *B.t.k.* products are registered for *C. occidentalis* control at these higher doses and volumes.

Recommendations

Further work should include:

1. Searching for and evaluating more virulent strains of *Nucleopolyhedrovirus* and *Granulovirus*;
2. Obtaining registration for *B.t.k.* products at the doses required for effective *C. occidentalis* control.

References

Abbott, W.S. (1925) A method of computing the effectiveness of an insecticide. *Journal of Economic Entomology* 18, 265–267.

Alfaro, R.I. (1986) Mortality and top-kill in Douglas-fir following defoliation by the western spruce budworm in British Columbia. *Journal of the Entomological Society of British Columbia* 83, 19–26.

Alfaro, R.I., Van Sickle, G.A., Thomson, A.J. and Wegwitz, E. (1982) Tree mortality and radial growth losses caused by the western spruce budworm in a Douglas-fir stand in British Columbia. *Canadian Journal of Forest Research* 12, 780–787.

Carolin, V.M. (1987) Life history and behavior. In: Brookes, M.H., Campbell, R.W., Colbert, J.J., Mitchell, R.G. and Stark, R.W. (technical coordinators) *Western Spruce Budworm*. United States Department of Agriculture, Forest Service, Cooperative State Research Service, Technical Bulletin No. 1694, pp. 30–42.

Cunningham, J.C. (1985) Status of viruses as biological control agents for spruce budworms. In: Grimble, D.G. and Lewis, F.B. (eds) *Proceedings, Symposium: Microbial Control of Spruce Budworms and Gypsy Moths, 10–12 April, 1984, Windsor Locks, Connecticut. Canada United States Spruce Budworms Program*. United States Department of Agriculture, Forest Service, General Technical Report NE-100, pp. 61–67.

Cunningham, J.C., Kaupp, W.J., McPhee, J.R. and Shepherd, R.F. (1983) Ground spray trials with two baculoviruses on western spruce budworm. *Canadian Forest Service Research Notes* 3, 10–11.

DeBoo, R.F. and Taylor, S.P. (1995) Insect Control in British Columbia, 1974–1988. In: Armstrong, J.A. and Ives, W.G.H. (eds) *Forest Insect Pests in Canada*. Natural Resources Canada, Canadian Forest Service, Ottawa, Ontario, pp. 709–716.

Harris, J.W.E. and Dawson, A.F. (1982) Estimating the number of western spruce budworm eggs from egg mass measurements in British Columbia. *The Canadian Entomologist* 114, 643–645.

Harris, J.W.E., Alfaro, R.I., Dawson, A.F. and Brown, R.G. (1985) The western spruce budworm in British Columbia, 1909–1983. Canadian Forest Service, Pacific Forestry Centre, *Information Report* BC-X-257.

Otvos, I.S., Cunningham, J.C. and Friskie, L.M. (1987a) Aerial application of nuclear polyhedrosis virus against Douglas-fir tussock moth, *Orgyia pseudotsugata* (McDunnough) (Lepidoptera: Lymantriidae): I. Impact in the year of application. *The Canadian Entomologist* 119, 697–706.

Otvos, I.S., Cunningham, J.C. and Alfaro, R.I. (1987b) Aerial application of nuclear polyhedrosis virus against Douglas-fir tussock moth, *Orgyia pseudotsugata* (McDunnough) (Lepidoptera: Lymantriidae): II. Impact 1 and 2 years after application. *The Canadian Entomologist* 119, 707–715.

Otvos, I.S., Cunningham, J.C. and Kaupp, W.J. (1989) Aerial application of two baculoviruses against the western spruce budworm (Lepidoptera: Tortricidae) in British Columbia. *The Canadian Entomologist* 121, 209–217.

Parfett, N., Clarke, D. and Van Sickle, A. (1994) Using a geographical information system for the input and analysis of historical western spruce budworm in British Columbia. *Canada-British Columbia Partnership Agreement on Forest Resources Development: Forest Resources Development Agency, Report 219.*

Shepherd, R.F., Otvos, I.S., Chorney, R.J. and Cunningham, J.C. (1984) Pest management of Douglas-fir tussock moth (Lepidoptera: Lymantriidae): prevention of an outbreak through early application with a nuclear polyhedrosis virus by ground and aerial applications. *The Canadian Entomologist* 116, 1533–1542.

Shepherd, R.F., Cunningham, J.C. and Otvos, I.S. (1995) Western spruce budworm, *Choristoneura occidentalis.* In: Armstrong, J.A. and Ives, W.G.H. (eds) *Forest Insect Pests in Canada.* Natural Resources Canada, Canadian Forest Service, Ottawa, Ontario, pp. 119–121.

Silver, G.T. (1960) Notes on the spruce budworm infestation in British Columbia. *Forestry Chronicle* 36, 362–374.

Van Sickle, G.A., Alfaro, R.I. and Thomson, A.J. (1983) Douglas-fir height growth affected by western spruce budworm. *Canadian Journal of Forest Research* 13, 445–450.

Wood, C.S., Van Sickle, G.A. and Humble, L.M. (1987) Forest insect and disease conditions, British Columbia and Yukon, 1987. *Canadian Forest Service, Pacific Forestry Centre Information Report BC-X-296.*

14 *Choristoneura pinus pinus* Freeman, Jack Pine Budworm (Lepidoptera: Tortricidae)

K. van Frankenhuyzen

Pest Status

The jack pine budworm, *Choristoneura pinus pinus* Freeman, is a native defoliator of jack pine, *Pinus banksiana* Lambert, in North America. Jack pine is the principal host species but other species of *Pinus* and *Picea* are attacked as well, especially when they occur as a minor component of jack pine stands. In Canada, outbreaks of *C. p. pinus* occur most commonly in the prairie provinces and Ontario at intervals of 6–10 years (Volney, 1988; Volney and McCullough, 1994). Outbreaks typically last 2–5 years, with near complete defoliation sustained for 2–3 years. Two major outbreaks have occurred since 1980. In 1982, population increases became apparent in Ontario and Manitoba. In 1983, moderate to severe defoliation was mapped on 67,000 ha in Ontario and 146,000 ha in Manitoba. The outbreak peaked in 1985 with 3.6 million ha of defoliation in Ontario, 2.0 million ha in Manitoba and a

130,000 ha spill over into Saskatchewan. Populations generally declined after 1986. In 1991, a resurgence was observed in the Sudbury district, Ontario. That outbreak eventually covered 419,000 ha of defoliation in 1994, before collapsing in 1997.

The life history of *C. p. pinus* closely resembles that of the eastern spruce budworm, *C. fumiferana* (Clemens). Moths emerge in July–August and females deposit clusters of eggs on the needles. The eggs hatch in about 10 days and larvae overwinter as second instars in hibernaculae. Larvae emerge in late May to early June and feed on flowers and new shoots, going through seven instars before pupation in July. During an outbreak, severe defoliation can occur locally and over widespread areas. Sustained defoliation can result in reduced tree growth, mortality of the terminal leader (top-kill) and tree mortality (Gross, 1992). Significant losses in merchantable volume of jack pine can result from a single outbreak episode (Gross and Meating, 1994).

Background

No attempts have been made to control *C. p. pinus* by manipulating its parasitoid fauna. Information on parasitoid prevalence can be used to better target the use of microbial pesticides (Nealis and Lysyk, 1988), which have been the main biological control agent used for operational control. Comparison of parasitism in populations of *C. p. pinus* and *C. fumiferana* revealed a great similarity in the parasitoid fauna attacking outbreak populations despite the marked differences in outbreak patterns (Nealis, 1991, and earlier studies). Not only are the species the same, but the patterns of parasitism are similar as well. In both *Choristoneura* spp., early larval instar specialists, e.g. *Apanteles fumiferanae* Viereck, are ubiquitous, and parasitoids attacking late larval instars, although diverse, are dominated by only a few species. Those species, e.g. *Meteorus trachynotus* Viereck and *Lypha setifacies* (Westwood), become more abundant during later outbreak stages and are often associated with rapid declines in population density. Nealis and Lomick (1994) suggested that a strong density-dependent relationship between mortality of early instars and production of pollen cones by the host tree reduces *C. p. pinus* populations more quickly to a level where relative rates of parasitism become very high, so that *C. p. pinus* populations collapse much sooner than do populations of *C. fumiferana*.

Biological Control Agents

Pathogens

Bacteria

The massive resurgence of *C. p. pinus* in the mid-1980s coincided with the waning popularity of aerial spraying using conventional chemical insecticides, leaving *Bacillus thuringiensis* Berliner serovar *kurstaki* (*B.t.k.*) as the only option. Laboratory bioassays confirmed larval susceptibility to the pathogen (van Frankenhuyzen and Fast, 1989), and resulted in the adoption of an ultra-low-volume application strategy for large-scale operational control programmes in Ontario from 1985 to 1987. A similar control strategy was used during the second, smaller outbreak in Ontario from 1994 to 1996. From 1981 to 1999, about 910,000 ha were treated with *B.t.k.*, using a total of about 20×10^{15} international units (IU) (Table 14.1). No control programmes were conducted against *C. p. pinus* in any other province.

Viruses

The 1985 outbreak in Ontario presented an opportunity to test the *Nucleopolyhedrovirus C. fumiferana* (ChfuNPV) of against *C. p. pinus*. Both species are equally susceptible to this virus in the laboratory. In 1985, a 50-ha plot near Gogama was aerially sprayed with 7.5×10^{11} polyhedral inclusion bodies (PIB) ha^{-1} in 9.5 litres when larvae were at peak fourth instar (Cunningham and Kaupp, 1995).

Table 14.1. Operational use of *Bacillus thuringiensis* against *Choristoneura pinus pinus* in Ontario. (Source: Forestry Insecticide Database, Canadian Forest Service, Great Lakes Forestry Centre, Sault Ste Marie, Ontario.)

Year	Province	No. ha treated[a]	Dose applied[b]
1985	Ontario	220,000	4,400,000
1986	Ontario	482,032	10,488,720
1987	Ontario	105,463	2,109,260
1989[c]	Ontario	4,763	285,780
1993	Ontario	122	3,660
1994	Ontario	21,449	644,970
1995	Ontario	51,015	1,530,450
1996	Ontario	25,636	769,080
Total		910,530	21,231,920

[a]Number of hectares treated with one or more applications.
[b]Total dose (expressed in 10^9 International Units) applied per ha (= number of ha treated × number of applications × 10^9 IU ha^{-1} per application).
[c]Isolated infestation in NW Ontario received three applications to prevent spreading.

Evaluation of Biological Control

The use of *B.t.k.* has been very successful in reducing defoliation by *C. p. pinus*. Operational foliage protection is usually achieved by one application of undiluted, high-potency product at 20–30 × 10^9 IU in 1.5–2.4 l ha^{-1}. Sprays are applied when jack pine needles are beginning to escape their fascicle sheaths, which usually coincides with the fourth larval instar.

Experimental application of the ChfuNPV resulted in a 65% population reduction, but did not provide any foliage protection. There was little carry-over of NPV the following year. Although the virus is virulent, larval feeding habits make it difficult to deliver the virus to early larval instars. Spraying of fourth instars is too late to initiate a viral epizootic, and no further work with this virus is recommended at this time.

Recommendations

Further work should include:

1. Integration of parasitoid population monitoring into the management programme to time application of *B.t.k.* sprays precisely for minimum impact on the parasitoids and maximum control of *C. p. pinus*.

References

Cunningham, J.C. and Kaupp, W.J. (1995) Insect viruses. In: Armstrong, J.A. and Ives, W.G.H. (eds) *Forest Insect Pests in Canada*. Canadian Forest Service, Natural Resources Canada, Ottawa, Ontario, pp. 328–340.

Frankenhuyzen, K. van and Fast, P.G. (1989) Susceptibility of three coniferophagous *Choristoneura* species (Lepidoptera: Tortricidae) to *Bacillus thuringiensis* var. *kurstaki*. *Journal of Economic Entomology* 82, 193–196.

Gross, H.L. (1992) Impact analysis for a jack pine budworm infestation in Ontario. *Canadian Journal of Forest Research* 22, 818–831.

Gross, H.L and Meating, J.H. (1994) Impact analysis for a jack pine budworm infestation in Ontario. *Great Lakes Forestry Centre, Canadian Forest Service, Sault Ste Marie, Ontario, Canada, Information Report O-X-431.*

Nealis, V.G. (1991) Parasitism in sustained and collapsing populations of the jack pine budworm, *Choristoneura pinus pinus* Free. (Lepidoptera: Tortricidae), in Ontario, 1985–1987. *The Canadian Entomologist* 123, 1065–1075.

Nealis, V.G and Lomick, P.V. (1994) Host-plant influence on the population ecology of the jack pine budworm, *Choristoneura pinus* (Lepidoptera: Tortricidae). *Ecological Entomology* 19, 367–373.
Nealis, V.G. and Lysyk, T. J. (1988) Sampling overwintering jack pine budworm, *Choristoneura pinus pinus* Free. (Lepidoptera: Tortricidae), and two of its parasitoids (Hymenoptera). *The Canadian Entomologist* 120, 1101–1111.
Volney, W.J.A. (1988) Analysis of historic jack pine budworm outbreaks in the Prairie provinces of Canada. *Canadian Journal of Forest Research* 18, 1152–1158.
Volney, W.J.A. and McCullough, D.G. (1994) Jack pine budworm population behaviour in northwestern Wisconsin. *Canadian Journal of Forest Research* 24, 502–510.

15 *Choristoneura rosaceana* (Harris), Obliquebanded Leafroller (Lepidoptera: Tortricidae)

S.Y. Li, S.M. Fitzpatrick, T. Hueppelsheuser, J.E. Cossentine and C. Vincent

Pest Status

The obliquebanded leafroller, *Choristoneura rosaceana* (Harris), is a native pest of raspberry, *Rubus* spp., apple, *Malus pumila* Miller (= *M. domestica* Borkhausen), pear, *Pyrus communis* L., cherry, *Prunus* spp., filbert, *Corylus avellana* L., and other deciduous trees and bushes in southern Canada (Schuh and Mote, 1948; Prentice, 1965; Madsen and Madsen, 1980; AliNiazee, 1986; Li and Fitzpatrick, 1997a). Early instar larvae cause bud and leaf damage. Superficial feeding damage on fruit occurs when the leaf is tied over the fruit. On apple, superficial damage is caused by summer larvae that are free-living or hidden in leafrolls. *C. rosaceana* larvae contaminate harvested raspberries when shaken off the plants by harvesting machine, which results in greater economic loss to growers than foliar damage. On cherry, larvae bore holes in the fruits and are difficult to remove from fruit clusters during the canning process (Madsen and Procter, 1982). *C. rosaceana* is increasing its pest status in tree fruit orchards where broad-spectrum insecticide control of key pests, e.g. codling moth, *Cydia pomonella* (L.), is being replaced by more specific non-chemical controls, although it is difficult to distinguish the damage caused by *C. rosaceana* to small apple fruitlets from that caused by other leafroller species (Vincent and Hanley, 1997). In apple-producing areas in Quebec and Ontario, the pest status of *C. rosaceana* has also increased where the populations of *C. rosaceana* have developed insecticide resistance.

There are one to two generations of *C. rosaceana* per year and females can lay up to about 600 eggs. The second-generation larvae occur between late summer and early fall. Early instar larvae overwinter in protected sites on or near host plants and resume activity the next spring.

Background

Control of *C. rosaceana* in fruit orchards has relied heavily on chemical insecticides. As a result, its populations in fruit-growing areas have developed resistance to several insecticides including cypermethrin, azinphosmethyl and phosmet (Bellerose *et al.*, 1992; Carrière *et al.*, 1994, 1996; Smirle *et al.*, 1998). Broad-spectrum insecticide control of *C. rosaceana* on raspberries also creates problems, because the occurrence of first-generation larvae usually coincides with berry harvesting time but the conventional insecticides cannot be used during harvest. Insecticide use also makes it difficult to integrate parasitoids into pest management programmes.

Pheromone traps (Vincent *et al.*, 1990; Thomson *et al.*, 1991; Delisle, 1992; Li and Fitzpatrick, 1997a) are used to monitor the adult flight period in fruit orchards, raspberry fields and mixed forests. Pheromones also have potential for use in mating disruption (Lawson *et al.*, 1996; Evenden *et al.*, 1999). Plant extracts such as neem (Lowery *et al.*, 1996; Smirle *et al.*, 1996) and tansy (Larocque *et al.*, 1999) have shown potential as alternatives to chemical insecticides. An integrated pest management programme with a strong biological control component is still needed, especially in regions where *C. rosaceana* has developed insecticide resistance.

Biological Control Agents

Parasitoids

In commercial raspberry fields in the Fraser Valley, British Columbia, Li *et al.* (1999) reared 14 species of primary endoparasitoids (six Braconidae, seven Ichneumonidae and one Tachinidae) from overwintered *C. rosaceana* larvae. Total parasitism ranged from 5 to 15% in managed fields, and was as high as 30% in abandoned fields. The polyembryonic *Macrocentrus nigridorsis* Viereck was the most abundant parasitoid found. In some fields as many as 25% of overwintered *C. rosaceana* larvae were parasitized by *M. nigridorsis* alone. On average, a single *C. rosaceana* larva produced 36 *M. nigridorsis* parasitoids (Li *et al.*, 1999).

In apple orchards in the southern interior of British Columbia, Vakenti *et al.* (2001) found 13 parasitoid species associated with *C. rosaceana*. The most common included *Glypta* sp., two *Diadegma* spp. and *Hemisturmia tortricis* (Coquillett). Parasitoids common to *C. rosaceana* in both the raspberry and fruit industries in British Columbia included *H. tortricis* (Coquillett), *Meteorus trachynotus* Viereck, *Apophua simplicipes* (Cresson) and *Diadegma interruptum pterophorae* (Ashmead). *Macrocentrus nigridorsis* was also found on alternative host plants of *C. rosaceana*.

In eastern Canada, Maltais *et al.* (1989) found several parasitoid species in both *C. rosaceana* and the eastern spruce budworm, *Choristoneura fumiferana* (Clemens), including *M. trachynotus*, *Itoplectis conquistor* (Say), *Phaeogenes maculicornis* (Cresson), *Ephialtes ontario* (Cresson), *Glypta fumiferana* (Viereck) and *Macrocentrus iridescens* French.

McGregor *et al.* (1998) showed that the egg parasitoid, *Trichogramma minutum* Riley, collected from *C. rosaceana* eggs on birch trees, parasitized more *C. rosaceana* eggs than *Trichogramma* sp. near *pretiosum* Riley, or *Trichogramma sibericum* Sorokina. Field trials confirmed that *T. minutum* is the most suitable of the three candidates for parasitization of *C. rosaceana* eggs on raspberry. *Trichogramma minutum* parasitized nearly 70% of *C. rosaceana* egg masses in plots treated with a weekly release rate of 25 *T. minutum* females m^{-2} for four consecutive weeks, using point-source release techniques (T. Hueppelsheuser, unpublished). There tended to be more eggs parasitized downwind of the release points. Air temperatures of 20°C or higher were most suitable for *Trichogramma* parasitism of *C. rosaceana* eggs in raspberry fields.

Lawson *et al.* (1997) found that *Trichogramma platneri* Nagarkatti parasitized more *C. rosaceana* eggs per egg mass in the laboratory and in an apple orchard than did *T. pretiosum* or *T. minutum*. *C.*

rosaceana parasitoids are listed on the web (O'Hara, 2000) and Huber *et al.* (1996) listed most of the parasitoids of Nearctic *Choristoneura* spp., including *C. rosaceana*. *Colpoclypeus florus* (Walker), introduced from Europe, was not included in their list of chalcidoids but has been reared from *C. rosaceana* from Quebec, Ontario and British Columbia (J.T. Huber, Ottawa, 2000, personal communication).

Predators

Demougeot *et al.* (1993) and Demougeot (1994) evaluated two predators, *Harmonia axyridis* Pallas and *Coccinella septempunctata* L., for their potential against *C. rosaceana* larvae. *H. axyridis* showed greater voracity and faster consumption of *C. rosaceana* larvae than *C. septempunctata*. *H. axyridis* is polyphagous and preys on aphids or phytophagous mites when *C. rosaceana* populations are low (Lucas *et al.*, 1997).

Pathogens

Nematodes

In the laboratory, all instars of *C. rosaceana* were susceptible to *Steinernema carpocapsae* (Weiser) All strain, with LD_{50} values of 13, 5, 3 and 2 infective juveniles for the third, fourth, fifth and sixth instars, respectively (Bélair *et al.*, 1999). *Steinernema riobrave* 335, *Steinernema feltiae* UK, *Steinernema carpocapsae* All and *Steinernema glaseri* 326 caused 85%, 55%, 45% and 8% mortality of third instars, respectively, when exposed to 25 infective juveniles per dish. A minimum of 8 h exposure was required for significant larval mortality. Under field conditions, foliar applications of *S. carpocapsae* All strain at the rate of 2×10^9 infective juveniles ha^{-1} resulted in 13–37% mortality of *C. rosaceana* larvae.

Bacteria

Laboratory experiments showed that third- and fourth-instar *C. rosaceana* are the stages most susceptible to Dipel® WP, a commercial formulation of *Bacillus thuringiensis* Berliner serovar *kurstaki* (*B.t.k.*) (Li *et al.*, 1995a), and that this is more effective against *C. rosaceana* larvae at 25°C than at 20°C or 12°C (Li *et al.*, 1995b). The addition of a feeding stimulant, Pheast®, to Dipel® WP or Foray® 48B increased larval mortality (Li and Fitzpatrick, 1997b). In raspberry field trials of these two microbial insecticides, larval mortality of *C. rosaceana* increased with application rate, and decreased with an increase of spray volume. The half-life of *B.t.k.* on raspberry leaves ranged from 2.5 to 6.7 days, depending on application rate and spray volume (Li and Fitzpatrick, 1996). With the addition of the feeding stimulant, larval mortality increased and insecticidal activity persisted about 1.5 times longer (Li and Fitzpatrick, 1999).

In the 1980s, *B.t.k.* was registered for use against *C. rosaceana* on tree fruits. Different formulations with long residual activity have been tested in an apple orchard in Quebec (Côté and Vincent, 1998), but none of these has yet been registered. Hardman and Gaul (1990) found that *C. rosaceana* damage to apple was lower in the treatment with mixture of Dipel® WP and pyrethroids, compared to those in pyrethroid-treated plots. The advantage of mixing of *B.t.k.* with pyrethroids is that effective management of lepidopteran pests is achieved, while minimizing negative effects of pyrethroids on predators of phytophagous mites.

Protozoa

In laboratory trials, *C. rosaceana* larvae were susceptible to *Nosema fumiferanae* (Thomson), originally isolated from *C. fumiferana* (Thomson, 1955). Cossentine and Gardiner (1991) found that larval mortality was age- and dose-dependent. *N. fumiferanae* spores were retained to the adult stage by hosts treated as fourth or fifth instars.

Viruses

A multinucleocapsid *Nucleopolyhedrovirus* (MNPV) was isolated from *C. rosaceana* populations collected from *Prunus* spp. (Lucarotti and Morin, 1997) and a single nucleocapsid NPV (SNPV) was also identified (Pronier *et al.*, 2000). At 24°C, larval mortality from SNPV infection was about 75% when third instars were subjected to a suspension of 1.7 × 10^8 polyhedral inclusion bodies ml^{-1}. The average time for larval mortality was 23 ± 3 days after treatment.

Evaluation of Biological Control

A monitoring programme for *C. rosaceana* larvae, combined with pheromone trapping for adults, can be used to determine when and where *C. rosaceana* larvae are likely to occur in raspberry fields (Li and Fitzpatrick, 1997a). *Trichogramma* can be released at the beginning of first-generation *C. rosaceana* adult flight, and continued successively for 4–5 weeks. *B.t.*-based insecticides can be applied to target summer generation larvae that hatch from any unparasitized eggs.

Although the introduction of *N. fumiferanae* into *C. rosaceana* populations may effectively suppress the host population below the economic threshold, host microsporidian infections can have a negative impact on the development of insect parasitoids (Cossentine and Lewis, 1986, 1987).

In British Columbia, the rich parasitoid complex associated with *C. rosaceana* populations on raspberry and apple has the potential to maintain host population densities below the economic threshold.

Recommendations

Further work should include:

1. Understanding the parasitoid biologies, particularly for the key species, and determining their potential for mass production and inundative release;
2. Better understanding the impact of *B.t.*-based insecticides and pheromone disruption on indigenous parasitism;
3. Determining the pathology and impact of *N. fumiferanae* on the host parasitoids before it is considered for biological control (as per Cossentine and Lewis, 1986, 1987).

References

AliNiazee, M.T. (1986) Seasonal history, adult flight activity, and damage of the obliquebanded leafroller, *Choristoneura rosaceana* (Lepidoptera: Tortricidae), in filbert orchards. *The Canadian Entomologist* 118, 353–361.

Bélair, G., Vincent, C., Lemaire, S. and Coderre, D. (1999) Laboratory and field assays of entomopathogenic nematodes for the management of the oblique banded leafroller, *Choristoneura rosaceana* (Harris) (Tortricidae). *Journal of Nematology* (*Supplement*) 31(4S), 684–689.

Bellerose, S., Vincent, C. and Pilon, J.-G. (1992) Résistance à trois insecticides synthétiques de la tordeuse à bandes obliques de la région de Deux-Montagnes. *Résumé des recherches de la Station d'Agriculture Canada, Saint-Jean-sur-Richelieu* 20, 5–6.

Carrière, Y., Deland, J.-P., Roff, D.A. and Vincent, C. (1994) Life history costs associated with the evolution of insecticide resistance. *Journal of the Royal Society of London* B258, 35–40.

Carrière, Y., Deland, J.P. and Roff, D.A. (1996) Obliquebanded leafroller (Lepidoptera: Tortricidae) resistance to insecticides: among-orchard variation and cross-resistance. *Journal of Economic Entomology* 89, 577–582.

Cossentine, J.E. and Gardiner, M. (1991) Susceptibility of *Choristoneura rosaceana* (Lepidoptera: Tortricidae) to the microsporidium *Nosema fumiferanae* (Thomson) (Microsporida: Nosematidae). *The Canadian Entomologist* 123, 265–270.

Cossentine, J.E. and Lewis, L.C. (1986) Impact of *Vairimorpha necatrix, Vairimorpha* sp. (Microsporida: Microsporida) on *Bonnetia comta* within *Agrotis ipsilon* (Lepidoptera: Noctuidae) hosts. *Journal of Invertebrate Pathology* 47, 303–309.

Cossentine, J.E. and Lewis, L.C. (1987) Development of *Macrocentrus grandii* Goidanich within microsporidian-infected *Ostrinia nubilalis* (Hübner) host larvae. *Canadian Journal of Zoology* 65, 2532–2535.

Côté, J.C. and Vincent, C. (1998) Trials with *Bacillus thuringiensis* var. *kurstaki* formulations in apple orchards. In: Vincent, C. and Smith, R. (eds) *Orchard Pest Management in Canada.* Technical Bulletin, Agriculture and Agri-Food Canada, Saint-Jean-sur-Richelieu, Québec, pp. 81–91.

Delisle, J. (1992) Monitoring the seasonal male flight activity of *Choristoneura rosaceana* (Lepidoptera: Tortricidae) in eastern Canada using virgin females and several different pheromone blends. *Environmental Entomology* 21, 1007–1012.

Demougeot, S. (1994) Efficacité de prédation des adultes de *Coccinella septempunctata* et de *Harmonia axyridis* (Coleoptera: Coccinellidae) contre *Choristoneura rosaceana* (Lepidoptera: Tortricidae) et *Aphis pomi* (Homoptera: Aphididae). Mémoire de MSc, Université du Québec à Montréal, Montreal, Quebec.

Demougeot, S., Vincent, C. and Coderre, D. (1993) Efficacité des coccinelles contre deux ravageurs dans les vergers québécois. *Résumé des recherches de la Station d'Agriculture Canada, Saint-Jean-sur-Richelieu* 22, 7–8.

Evenden, M.L., Judd, G.L.R. and Borden, J.H. (1999) Pheromone-mediated mating disruption of *Choristoneura rosaceana*: is the most attractive blend really the most effective? *Entomologia Experimentalis and Applicata* 90, 37–47.

Hardman, J.M. and Gaul, S.O. (1990) Mixtures of *Bacillus thuringiensis* and pyrethroids control winter moth (Lepidoptera: Geometridae) in orchards without outbreak of mites. *Journal of Economic Entomology* 83, 920–936.

Huber, J.T., Eveleigh, E., Pollock, E. and McCarthy, P. (1996) The chalcidoid parasitoids and hyperparasitoids (Hymenoptera: Chalcidoidea) of *Choristoneura* species (Lepidoptera: Tortricidae) in America north of Mexico. *The Canadian Entomologist* 128, 1167–1220.

Larocque, N., Vincent, C., Bélanger, A. and Bourassa, J.-P. (1999) Effects of tansy oil, *Tanacetum vulgare* L., on the biology of the obliquebanded leafroller, *Choristoneura rosaceana* (Harris) (Lepidoptera: Tortricidae). *Journal of Chemical Ecology* 25, 51–56.

Lawson, D.S., Reissig, W.H., Agnello, A.M., Nyrop, J.P. and Roelofs, W.L. (1996) Interference with the mate-finding communication system of the obliquebanded leafroller (Lepidoptera: Tortricidae) using sex pheromones. *Environmental Entomology* 25, 895–905.

Lawson, D.S., Nyrop, J.P. and Reissig, W.H. (1997) Assays with commercially produced *Trichogramma* (Hymenoptera: Trichogrammatidae) to determine suitability for obliquebanded leafroller (Lepidoptera: Tortricidae) control. *Environmental Entomology* 26, 684–693.

Li, S.Y. and Fitzpatrick, S.M. (1996) The effects of application rate and spray volume on efficacy of two formulations of *Bacillus thuringiensis* Berliner var. *kurstaki* against *Choristoneura rosaceana* (Harris) (Lepidoptera: Tortricidae) on raspberries. *The Canadian Entomologist* 128, 605–612.

Li, S.Y. and Fitzpatrick, S.M. (1997a) Monitoring obliquebanded leafroller (Lepidoptera: Tortricidae) larvae and adults on raspberries. *Environmental Entomology* 26, 170–177.

Li, S.Y. and Fitzpatrick, S.M. (1997b) Responses of larval *Choristoneura rosaceana* (Harris) (Lepidoptera: Tortricidae) to a feeding stimulant. *The Canadian Entomologist* 129, 363–369.

Li, S.Y. and Fitzpatrick, S.M. (1999) Feeding stimulant added to *Bacillus thuringiensis* based insecticides enhances activity against *Choristoneura rosaceana* (Lepidoptera: Tortricidae). *The Canadian Entomologist* 131, 451–453.

Li, S.Y., Fitzpatrick, S.M. and Isman, M.B. (1995a) Susceptibility of different instars of the obliquebanded leafroller (Lepidoptera: Tortricidae) to *Bacillus thuringiensis* var. *kurstaki. Journal of Economic Entomology* 88, 610–614.

Li, S.Y., Fitzpatrick, S.M. and Isman, M.B. (1995b) Effect of temperature on toxicity of *Bacillus thuringiensis* to the obliquebanded leafroller (Lepidoptera: Tortricidae). *The Canadian Entomologist* 127, 271–273.

Li, S.Y., Fitzpatrick, S.M., Troubridge, J.T., Sharkey, M.J., Barron, J.R. and O'Hara, J.E. (1999) Parasitoids reared from the obliquebanded leafroller (Lepidoptera: Tortricidae) infesting raspberries. *The Canadian Entomologist* 131, 399–404.

Lowery, D.T., Bellerose, S., Smirle, M.J., Vincent, C. and Pilon, J.-P. (1996) Effect of neem on growth and development of the obliquebanded leafroller, *Choristoneura rosaceana. Entomologia Experimentalis et Applicata* 79, 203–209.

Lucarotti, C.J. and Morin, B. (1997) A nuclear polyhedrosis virus from the obliquebanded leafroller,

Choristoneura rosaceana (Harris) (Lepidoptera: Tortricidae). *Journal of Invertebrate Pathology* 70, 121–126.

Lucas, E., Coderre, D. and Vincent, C. (1997) Voracity and feeding preferences of two aphidophagous coccinellids on *Aphis citricola* and *Tetranychus urticae*. *Entomologia Experimentalis et Applicata* 85, 151–159.

Madsen, H.F. and Madsen, B.J. (1980) Response of four leafroller species (Lepidoptera: Tortricidae) to sex attractants in British Columbia orchards. *The Canadian Entomologist* 112, 427–430.

Madsen, H.F. and Procter, P.J. (1982) *Insects and Mites of Tree Fruits in British Columbia*. Ministry of Agriculture and Food, Victoria, British Columbia.

Maltais, J., Régnière, J., Cloutier, C., Hébert, C. and Perry, D.F. (1989) Seasonal biology of *Meteorus trachynotus* Vier. (Hymenoptera: Braconidae) and of its overwintering host *Choristoneura rosaceana* (Harr.) (Lepidoptera: Tortricidae). *The Canadian Entomologist* 121, 745–756.

McGregor, R., Hueppelsheuser, T., Luczynski, A. and Henderson, D. (1998) Collection and evaluation of *Trichogramma* species (Hymenoptera: Trichogrammatidae) as biological controls of the oblique-banded leafroller *Choristoneura rosaceana* (Harris) (Lepidoptera: Tortricidae) in raspberries and blueberries. *Biological Control* 11, 38–42.

O'Hara, J. (2000) Insect Parasitoids of Obliquebanded leafroller. http://res.agr.ca/ecorc/isbi/pest/oblrpara.htm

Prentice, R.M. (1965) *Forest Lepidoptera of Canada Recorded by the Forest Insect Survey*, Vol. 4. Publication 1142, Canada Department of Forestry, Ottawa, Ontario.

Pronier, I., Paré, J., Wissocq, J.-C., Vincent, C. and Stewart, R.K. (2000) Étude préliminaire d'un virus agent de la polyédrose nucléaire dans les tissus de son hôte, la tordeuse à bandes obliques. *Bulletin de la Société Zoologique de France* 125, 174–176.

Schuh, J. and Mote, D.G. (1948) The obliquebanded leafroller on red raspberries. *Oregon Agriculture Experimental Station, Technical Bulletin* 13.

Smirle, M.J., Lowery, D.T., and Zurowski, C. (1996) Influence of neem oil on detoxification activity in the obliquebanded leafroller, *Choristoneura rosaceana*. *Pesticide Biochemistry and Physiology* 56, 220–230.

Smirle, M.J., Vincent, C., Zurowski, C. and Rancourt, C. (1998) Azinphos-methyl resistance in the obliquebanded leafroller, *Choristoneura rosaceana*: reversion in the absence of selection and relationship to detoxification enzyme activity. *Pesticide Biochemistry and Physiology* 61, 183–189.

Thomson, D.R., Angerilli, N.P.D., Vincent, C. and Gaunce, A.P. (1991) Evidence for regional differences in the response of obliquebanded leafroller, *Choristoneura rosaceana* (Lepidoptera: Tortricidae) to sex pheromone blends. *Environmental Entomology* 20, 935–938.

Thomson, H.M. (1955) *Perezia fumiferanae* n. sp., a new species of microsporidia from the spruce budworm *Choristoneura fumiferana* (Clem.). *Journal of Parasitology* 41, 416–511.

Vakenti, J., Cossentine, J.E., Cooper, B.E., Sharkey, M.J., Yoshimoto, C.M. and Jensen, L.B.M (2001) Host-plant range and parasitism of obliquebanded and three-lined leafrollers (Lepidoptera: Tortricidae) in the southern interior of British Columbia. *The Canadian Entomologist* 133, 139–146.

Vincent, C. and Hanley, J. (1997) Measure of agreement between experts on apple damage assessment. *Phytoprotection* 78, 11–16.

Vincent, C., Mailloux, M., Hagley, E.A.C., Reissig, W.H., Coli, W.M. and Hosmer, T.H. (1990) Monitoring the codling moth (Lepidoptera: Olethreutidae) and the obliquebanded leafroller (Lepidoptera: Tortricidae) with sticky and non-sticky traps. *Journal of Economic Entomology* 83, 434–440.

16 *Chrysops, Hybomitra* and *Tabanus* spp., Horse and Deer Flies (Diptera: Tabanidae)

M. Iranpour and T.D. Galloway

Pest Status

Horse and deer flies, particularly *Chrysops*, *Hybomitra* and *Tabanus* spp., are among the most important pests of humans and animals (Wood, 1985). Teskey (1990) reported that 11 genera and 144 species occur in Canada. Females of most species require vertebrate blood to mature their eggs. This makes tabanids extremely annoying to their hosts, especially when they occur near the larval habitats (Magnarelli *et al.*, 1979).

Adult tabanids vector several pathogens, including viruses, bacteria, rickettsia-like organisms, trypanosomes and filarial worms (Pechuman, 1981). Most disease-causing agents are transmitted mechanically (Krinsky, 1976). Because of the pain of tabanid bites, a host makes the effort to dislodge the flies. The dislodged flies return to complete their blood meals or may select a nearby host. The new host may receive pathogens if the first host was infected (Krinsky, 1976).

Livestock can be severely affected by tabanids. Unprotected animals may have reduced milk production and weight gains (Roberts and Pund, 1974). Not only do tabanids take a considerable quantity of blood (0.082–0.34 ml as an average single blood meal; Pechuman, 1981), but the annoyance and irritation caused by large numbers of these flies interrupt grazing and resting behaviour (Ralley *et al.*, 1992). Under tabanid attack, there are increases in head tosses, foot stomps, ear flicks and tail switches in individual animals, and herds form grazing lines or bunch up (Ralley *et al.*, 1992). Animals under prolonged stress become more susceptible to secondary infections, such as respiratory infections, foot rot and pinkeye.

In the USA, heifers exposed to attacks by an average of 90 horse flies per animal per day for 84 days gained 0.08 kg per animal per day less than protected heifers, and the potential total economic loss was estimated to be more than US$10 per head each year (Perich *et al.*, 1986). Beef cattle production losses due to tabanid attacks were estimated to be US$54 million in stocker cattle alone (Drummond, 1987).

Oviposition generally begins 4–8 days after a blood meal. Eggs in a compact mass with 1–5 layers are usually laid on vegetation overhanging water, but may be laid on any solid substrate. Embryonic development has been reported to be 4–6 days (Teskey, 1990) but we found development to take 10–12 days for *Hybomitra nitidifrons nuda* (McDunnough) in Manitoba. All eggs in a given mass hatch at the same time and larvae drop to the water or wet soil below. Larvae overwinter, undergoing 5–11 moults and taking 1–3 years to complete their development, depending on species and latitude (Pechuman, 1981). Fully grown larvae migrate to drier areas and pupate in a vertical position. Depending on temperature and species, adults emerge after 1–3 weeks (Teskey, 1990).

Background

Methods used to control Tabanidae can be categorized into three main groups: environmental modifications and physical control; chemical control; and biological control.

However, there are some problems in their applications (Anderson, 1985). Strong power of adult dispersal, prolonged emergence periods and extensive breeding sites have made it difficult to manage populations. Environmental and physical control are not practical methods on a large scale and do not seem to have an important impact on tabanid populations. The cost of insecticides, difficulty in applying them, potential environmental pollution and short-term effectiveness are some problems associated with chemical control (Anderson, 1985). However, according to investigators over the past 100 years, natural enemies exert considerable impact on tabanid populations.

All stages of Tabanidae are attacked by a large fauna and flora of predators, parasites and pathogens. Eggs are attacked by hymenopterous parasitoids, insect predators and fungi. Larvae and pupae are eaten by vertebrates and invertebrates, are parasitized by insects and nematodes, and may be infected by protozoa and fungi. Adults are eaten by vertebrate, insect and acarine predators and are infected by microbial pathogens (Anderson, 1985).

Many anecdotal reports exist on parasitism of tabanids in Canada, but no detailed studies have been conducted. In Saskatchewan and Ontario, up to 36% of deer fly eggs were parasitized by *Trichogramma minutum* Riley[1] (Cameron, 1926; James, 1963). James (1963) also reported 6% parasitism of horse fly egg masses by *T. minutum*. Cameron (1926) and James (1963) also reported up to 30% parasitism by *Telenomus emersoni* (Girault) of horse fly and deer fly eggs. In British Columbia, Hatton (1948) found 80% parasitism by *T. emersoni* of tabanid egg masses. Larvae and pupae of horse and deer flies in Ontario were parasitized by *Villa lateralis* (Say) and *Carinosillus tabanivorus* (Hall) (Teskey, 1969) and *Trichopria* sp. parasitized up to 4.6% of larvae and pupae (Magnarelli and Anderson, 1980). In Manitoba (Teskey, 1969), Saskatchewan (Burks, 1979), Alberta, British Columbia and Ontario (Burger *et al.*, 1981) up to 20.8% of larvae and pupae were parasitized by *Diglochis occidentalis* (Ashmead). In Saskatchewan, up to 50% of larval and pupal stages were parasitized by *Trichopria tabanivora* Fouts (Cameron, 1926).

In Alberta, Shamsuddin (1966) reported a *Bathymermis* sp. parasitizing 16–37% of tabanid larvae, and in Manitoba James (1963) reported 7.7% of tabanid larvae parasitized by this genus, as well as parasitism of *Tabanus* sp. larvae by a *Mermis* sp.

Biological Control Agents

Parasitoids

In southern Manitoba, surveys for egg parasitoids from 1996 to 1998 showed that 98.9% of 93 multilayered egg masses of *H. nitidifrons nuda* collected were parasitized by *Telenomus* spp., and a mean of 34.5% of eggs within individual egg masses were attacked. In addition, 36.3% of all unparasitized eggs failed to hatch. In another location, 121 (79.1%) of 153 single-layered egg masses of *Chrysops aestuans* Van der Wulp were parasitized by a *Telenomus* sp. and *Trichogramma semblidis* (Aurivillius). Of the other egg masses, 17 (11.1%) were attacked only by *Telenomus* sp., six (3.9%) only by *T. semblidis*, and nine (5.9%) were unparasitized. Within egg masses attacked by both species, the *Telenomus* sp. emerged from 44.1% and *T. semblidis* emerged from 9.9%. In egg masses where a single species of parasitoid attacked the eggs, 40.8% were killed by *Telenomus* sp. and 11.1% were killed by *T. semblidis*. Of the total eggs, 18.6% produced neither *C. aestuans* larvae nor parasitoids. There was a significant interaction between these two parasitoids in *C. aestuans* egg masses.

In host-finding studies, an attractant response of the *Telenomus* spp. to hexane extracts of fresh tabanid egg masses, the whole body of adult females, the tip of the abdomen of females, and the remainder

[1]This identification is incorrect; the species almost certainly is *T. semblidis* (Aurivillius) (Pinto, 1998).

of the body, was identified. There is a chemical component present on the surface of horse fly egg masses that causes host-seeking parasitoids to stop searching and investigate the egg mass. This is the first demonstration of such a chemical in horse fly–parasitoid interactions.

Evaluation of Biological Control

Egg parasitoids have potential as biological control agents against tabanids; however, their impact on populations, even of the most important pest species, is unknown.

Recommendations

Further work should include:

1. Determining the specific nature of the chemical attractant;
2. Identifying the *Telenomus* spp. and describing their behaviour;
3. Examining the relationships between the parasitoids and host egg masses.

References

Anderson, J.F. (1985) The control of horse flies and deer flies (Diptera: Tabanidae). *Myia* 3, 547–598.

Burger, J.F., Lake, D.J. and McKay, M.L. (1981) The larval habitats and rearing of some common *Chrysops* species (Diptera: Tabanidae) in New Hampshire. *Proceedings of the Entomological Society of Washington* 83, 373–389.

Burks, B.D. (1979) Family Pteromalidae. In: Krombein, K.V., Hurd, P.D. Jr, Smith, D.R. and Burks, B.D. (eds) *Catalog of Hymenoptera in America North of Mexico,* Vol. 1, *Symphyta and Apocrita (Parasitica).* Smithsonian Institution Press, Washington DC, pp. 769–835.

Cameron, A.E. (1926) Bionomics of the Tabanidae (Diptera) of the Canadian Prairie. *Bulletin of Entomological Research* 17, 1–42.

Drummond, R.O. (1987) Economic aspects of ectoparasites of cattle in North America. In: Leaning, W.D.H and Guerrero, J. (eds) *The Economic Impact of Parasitism in Cattle.* Twenty-third World Veterinary Congress, 19 August, Montreal, Québec, pp. 9–24.

Hatton, G.N. (1948) Notes on the life history of some tabanid larvae (Diptera). *Proceedings of the Entomological Society of British Columbia* 44, 15–17.

James, H.G. (1963) Larval habitats, development, and parasites of some Tabanidae (Diptera) in southern Ontario. *The Canadian Entomologist* 95, 1223–1232.

Krinsky, W.L. (1976) Animal disease agents transmitted by horse flies and deer flies. *Journal of Medical Entomology* 13, 225–275.

Magnarelli, L.A. and Anderson, J.F. (1980) Feeding behavior of Tabanidae (Diptera) on cattle and serologic analyses of partial blood meals. *Environmental Entomology* 9, 664–667.

Magnarelli, L.A., Anderson, J.F. and Thorne, J.H. (1979) Diurnal nectar-feeding of salt marsh Tabanidae (Diptera). *Environmental Entomology* 8, 544–548.

Pechuman, L.L. (1981) The horse flies and deer flies of New York (Diptera: Tabanidae), 2nd edn. *Cornell University Agricultural Experiment Station, Agriculture Bulletin* 18, 1–68.

Perich, M.J., Wright, R.E. and Lusby, K.S. (1986) Impact of horse flies (Diptera: Tabanidae) on beef cattle. *Journal of Economic Entomology* 79, 128–131.

Pinto, J.D. (1998) Systematics of the North American species of *Trichogramma* Westwood (Hymenoptera: Trichogrammatidae). *Memoirs of the Entomological Society of Washington* 22.

Ralley, W.E., Galloway T.D. and Crow, G.H. (1992) Individual and group behaviour of pastured cattle in response to attack by biting flies. *Canadian Journal of Zoology* 71, 725–734.

Roberts, R.H. and Pund, W.A. (1974) Control of biting flies on beef steers: effect on performance in pasture and feedlot. *Journal of Economic Entomology* 67, 232–234.

Shamsuddin, M. (1966) A *Bathymermis* species (Mermithidae: Nematoda) parasitic on larval tabanids. *Quaestiones Entomologicae* 2, 253–256.

Teskey, H.J. (1969) Larvae and pupae of some eastern North America Tabanidae (Diptera). *Memoirs of the Entomological Society of Canada* 63.

Teskey, H.J. (1990) *The Horse Flies and Deer Flies of Canada and Alaska (Diptera: Tabanidae).* Part 16. *The Insects and Arachnids of Canada.* Ministry of Supply and Services, Canada, Ottawa, Ontario.

Wood, D.M. (1985) *Biting Flies Attacking Man and Livestock in Canada.* Publication 1781E, Agriculture Canada, Ottawa, Ontario.

17 *Croesia curvalana* (Kearfott), Blueberry Leaftier (Lepidoptera: Tortricidae)

P.L. Dixon and K. Carl

Pest Status

The blueberry leaftier, *Croesia curvalana* (Kearfott),[1] is native to North America. It is one of the most destructive pests of lowbush blueberry, *Vaccinium angustifolium* Aiton, in the Atlantic provinces (Morris, 1981), and also occurs in British Columbia as one of a complex of leafrollers on highbush blueberry, *Vaccinium corymbosum* L. (Raine, 1984; Belton, 1988). Extensive crop loss can occur when emerging first-instar larvae bore into flower buds in early spring, destroying potential fruit (Ponder and Seabrook, 1988). Later-instar larvae exit the buds and move about freely, feeding on foliage, and webbing and rolling leaves. In Atlantic Canada, *C. curvalana* is most common on wild, unmanaged blueberry land, although outbreaks do occur on managed stands (Neilson and Crozier, 1989). The latter is thought to be due mainly to a change in pruning method from biennial burning to flail mowing, allowing more eggs to survive (Polavarapu and Seabrook, 1992). *C. curvalana* is univoltine and overwinters on surface litter in the egg stage.

Background

Current management practices include the application of insecticides against first-instar larvae or adults (Polavarapu and Seabrook, 1992). Limited records from Atlantic Canada indicate a low rate of parasitism of *C. curvalana* by local species. In 1984, two specimens of *Chorinaeus excessorius* Davies were reared from 102 *C. curvalana* larvae from Pouch Cove, Newfoundland, and 10% of 28 *C. curvalana* from Blackville, New Brunswick, were parasitized by an unidentified tachinid (Ponder and Seabrook, 1988). From 1982 to 1984 small numbers of several species, including *Itoplectis quadricingulata* (Provancher), *Pimpla aequalis* (Provancher), *Mesochorus* sp., *Glypta* sp. and *Orgilus* sp., were reared in Newfoundland.

A literature review and field collections for European blueberry-feeding tortricids revealed that the parasitoid complex of the closely related European species, *Acleris variegana* Denis and Schiffermüller, was the best prospect for biological control. *A. variegana* does not occur in North America (Hodges *et al.*, 1983) but in Europe it attacks a large range of host plants in several families, including *Vaccinium myrtillus* L., the mountain bilberry. In North America, *V. myrtillus* occupies just two areas, both in the Rocky Mountains (Vander Kloet, 1988).

About 15 parasitoid species were recovered from *A. variegana* during a survey in Switzerland (for details see IIBC European Station Annual Reports, 1988–1995). Two braconid parasitoids from the Swiss Alps,

[1]Razowski (1987) placed *Croesia* in *Acleris*, although the status of *Croesia* species in North America has not yet been clarified (P.T. Dang, Ottawa, 1999, personal communication).

Microgaster hospes Marshall and *Earinus gloriatorius*[2] (Panzer), were selected for further study as potential imports into Canada, based on the criteria that: they could not be strictly monophagous as they had to be able to attack a foreign host; they had to be known from blueberry in Europe; they had to be compatible with North American parasitoids; and they had to have an appreciable impact on the European host.

Biological Control Agents

Parasitoids

M. hospes and *E. gloriatorius* showed promise as biological control agents. Nixon (1968) revised the European Microgastrinae and suspected that *M. hospes* was a Holarctic species. He synonymized the North American *Microgaster comptanae* Viereck that occurs on *Ancylis comptana* (Frölich) under the European *M. hospes*. *C. curvalana* is not a known host of *M. comptanae*. If the two parasitoids are not conspecific, we must determine how to separate them in field collections after any release of European material. *M. hospes* is a univoltine endoparasitoid that prefers second-instar larvae of *A. variegana*, emerges from mature larvae and overwinters as a cocoon. It is strictly solitary, although superparasitism with up to five eggs or young larvae has been observed in the laboratory (Lewandowski, 1992). *E. gloriatorius* is a univoltine endoparasitoid of several tortricid species. It prefers third- and fourth-instar *A. variegana* larvae. The mature parasitoid larva emerges from the fifth-instar caterpillar and overwinters in a cocoon.

The biology and life history of *M. hospes* and *E. gloriatorius* were studied at Delémont, with *A. variegana* as the host. These were the dominant parasitoids in most years. Rates of parasitism of *A. variegana* by *M. hospes*, in particular, commonly exceeded 50%. Mating occurred more readily in *M. hospes* although both did mate in small screened cages (30 × 30 × 20 cm). Oviposition usually occurred within 5 minutes of presentation of host larvae, with 60–80% parasitism obtained for both parasitoids.

Confined host-suitability studies were undertaken in 1993 and 1994 with small numbers of both parasitoids and *C. curvalana* in laboratories at St John's, Newfoundland, and at Delémont. In 1993, mated female parasitoids were confined with the appropriate larval instar of *C. curvalana* individually in Petri dishes and in groups in screened cages. At St John's, parasitism was not successful, although both species, *E. gloriatorius* in particular, vigorously probed rolled leaves containing host larvae and stung exposed larvae as well as those in leafrolls. At Delémont, two cocoons of *E. gloriatorius* were obtained from a small number of *C. curvalana* larvae. No cocoons of *M. hospes* were obtained and, when host larvae exposed to *M. hospes* were dissected, no immature stages of the parasitoid could be found. Similar studies with *A. variegana* produced large numbers of cocoons of both parasitoid species.

In 1994, extensive laboratory studies at Delémont with *E. gloriatorius*, *C. curvalana* from Newfoundland, and *A. variegana* from Europe showed that females exhibited similar oviposition behaviour when exposed to either *C. curvalana* or *A. variegana*. Second- and third-instar larvae were accepted as hosts and stung for a few seconds. However, parasitism was not successful on either tortricid. The parasitoid females were dissected and all had fully developed ovaries but eggs were deformed and shrivelled. The reasons for this are unknown, but no disease was apparent. As there were several hundred parasitoids from several field populations, it is possible that some rearing condition was responsible, although they had been reared successfully in previous years.

Thus, although *C. curvalana* appears to

[2]*Earinus gloriatorius* was originally misidentified as *Microdus* (now *Bassus*) *clausthalianus* (Ratzeburg), and is discussed under this name in early IIBC project reports.

be acceptable to *E. gloriatorius* females for oviposition, the question of the suitability of this parasitoid and of *M. hospes* for introduction remains unresolved. It is not clear whether, in 1993, there was stinging without oviposition or oviposition without egg development, and the reasons for egg malformation in 1994 are not known.

Recommendations

Further work should include:

1. Continued host suitability studies in Europe and Canada, with emphasis on *E. gloriatorius*;

2. Resolving the taxonomy of *Microgaster*, especially *M. hospes* and *M. comptanae*.

References

Belton, E.M. (1988) *Lepidoptera on Fruit Crops in Canada.* Pest Management Paper 30, Simon Fraser University, Burnaby, British Columbia.

Hodges, R.W., Dominick, T., Davis, D.R., Ferguson, D.C., Franclemont, J.G., Munro, E.G. and Powell, J.A. (1983) *Check List of the Lepidoptera of America North of Mexico.* E.W. Classey, Oxford, UK.

IIBC (International Institute of Biological Control) (1988–1995) *Annual Reports.* International Institute of Biological Control, European Station, Delémont, Switzerland.

Lewandowski, C. (1992) Untersuchungen zur Biologie und Parasitierung ausgewählter Wicklerarten an Heidelbeeren. Diploma thesis, University of Kiel, Kiel, Germany.

Morris, R. (1981) *Fighting Blueberry Pests in Newfoundland.* Publication 1938, News and Features, Agriculture Canada, Ottawa, Ontario.

Neilson, W.T.A. and Crozier, L. (1989) Insects. In: Blatt, C.R., Hall, I.V., Jenson, K.I.N., Neilson, W.T.A., Hildebrand, P.D., Nickerson, N.L., Prange, R.K., Lidster, P.D., Crozier, L. and Sibley J.D. (eds) *Lowbush Blueberry Production.* Publication 1477/E, Agriculture Canada, Ottawa, Ontario, pp. 27–28.

Nixon, G.E.J. (1968) A revision of the genus *Microgaster* Latreille (Hymenoptera: Braconidae). *Bulletin of the British Museum of Natural History* 22(2), 31–72.

Polavarapu, S. and Seabrook, W.D. (1992) Evaluation of pheromone-baited traps and pheromone lure concentrations for monitoring blueberry leaftier (Lepidoptera: Tortricidae) populations. *The Canadian Entomologist* 124, 815–825.

Ponder, B.M. and Seabrook, W.D. (1988) Biology of the blueberry leaftier *Croesia curvalana* (Kearfott) (Tortricidae): a field and laboratory study. *Journal of the Lepidopterists' Society* 42, 120–131.

Raine, J. (1984) Leafrollers on blueberries in British Columbia. *Canada Agriculture* 30, 8–11.

Razowski, J. (1987) The genera of Tortricidae, Part I: Palaearctic Chlidanotinae and Tortricinae. *Acta Zoologica Cracowiensia* XXX 1–11, 181–185.

Vander Kloet, S.P. (1988) *The Genus* Vaccinium *in North America.* Publication 1828, Agriculture Canada, Ottawa, Ontario.

18 *Cydia pomonella* (L.), Codling Moth (Lepidoptera: Tortricidae)

J. Cossentine and C. Vincent

Pest Status

Codling moth, *Cydia pomonella* (L.), accidently introduced from Eurasia in the early 1800s, is a key pest wherever apple, *Malus pumila* Miller (= *M. domestica* Borkhausen), and pear, *Pyrus communis* L., are grown. Larval feeding in fruits renders them unsuitable for fresh consumption. In Quebec, which is typically colder and more humid than the fruit-growing regions of British Columbia, no sprays are specifically directed against *C. pomonella*. A mean of 17.5% *C. pomonella* damage was observed at harvest from 1977 to 1984 in an unsprayed orchard (Vincent and Bostanian, 1988). During the same years, *C. pomonella* damage at harvest ranged from 0.01 to 0.06% in commercial orchards, despite the fact that no sprays were targeted primarily against *C. pomonella*, but rather towards the plum curculio, *Conotrachelus nenuphar* Herbst, a key pest in Quebec (Vincent and Roy, 1992).

C. pomonella eggs are laid on apples and leaves, and larvae bore into the fruits. Two and a half generations occur annually in areas of commercial fruit production in British Columbia (Madsen and Procter, 1982) and one and a half generations in Ontario (Trimble, 1995). Mature larvae overwinter as cocoons under loose bark or in crevices.

Background

Until the 1990s, the only effective control methods used in commercial orchards were broad-spectrum chemical insecticides. Consequently, many non-target secondary and beneficial insects were affected, thus limiting the potential of biological control agents. Insecticides, e.g. azinphosmethyl, diflubenzuron, permethrin and methomyl, and the acaricide, cyhexatin, significantly reduced levels of *Trichogramma* spp. (Hagley and Laing, 1989). Fungicides, e.g. captan, dodine and polyram, did not affect parasitism levels. Because the insecticides in orchards degrade at different rates, they have a differential impact on parasitoids (Yu *et al.*, 1984a).

Effective management strategies control *C. pomonella* before larvae enter the fruit. Options to at least partially control *C. pomonella* increased in the 1990s, to include mating disruption (Trimble, 1995; Chouinard *et al.*, 1996; Judd *et al.*, 1997), the release of sterile adults in an area-wide eradication programme in British Columbia (Proverbs, 1982; Dyck and Gardiner, 1992), and the use of more specific insecticides, e.g. the insect growth regulator tebufenozide. Treatments can be timed accurately by monitoring adult male populations with sticky or non-sticky pheromone traps (Vincent *et al.*, 1990). The availability and use of control strategies specific to *C. pomonella*, e.g. mating disruption, has greatly increased the potential for biological control to be integrated into management programmes to increase the overall control of *C. pomonella* as well as other orchard pests. In Quebec and Ontario, where *C. pomonella* exerts less pressure, innovative approaches, such as the spraying of border rows (Trimble and Solymar, 1997; Vincent *et al.*, 1998), give *C. pomonella* control using minimum amounts of insecticide, while maintaining natural enemies. Other technologies, such as a combination of pheromone and small doses of insecticides, e.g. Attract and Kill®, have been tested in Nova Scotia (Smith *et al.*, 2000).

Biological Control Agents

Pathogens

Codling moth *Granulovirus* (CpGV) is highly virulent towards *C. pomonella* larvae and is used commercially in the USA and Europe but was not registered for use in Canada until 2000. Data from Canadian CpGV orchard and laboratory trials (Jaques *et al.*, 1981, 1987, 1994; Cossentine and Jensen, 1987; Hardman, 1987, 1988) indicated that virus applications are effective in significantly reducing deep-entry damage to apples by *C. pomonella* larvae.

Using a polymerase chain reaction technique, CpGV was found to be indigenous in an average of 23% of the wild *C. pomonella* populations in the interior of British Columbia (Eastwell *et al.*, 1999). It was questioned whether the CpGV found in wild *C. pomonella* were the result of large-scale releases of irradiated moths from a CpGV-infected colony in preliminary trials of the sterile *C. pomonella* release programme in British Columbia from 1976 to 1978 (Proverbs *et al.*, 1982). The virus however, was not only found naturally within the area where the moths were released in the 1970s, but throughout the Okanagan and Similkameen valleys and in the Kootenay valley, which is separated by mountain ranges and is over 200 km away (Eastwell *et al.*, 1999). Sequence analyses of portions of the *granulin* and *iap* genes suggest that the virus is identical, or very similar, to the CpGV-M1 genotype of the Mexican isolate. A granulovirus was also isolated from wild *C. pomonella* in commercial orchards of Deux-Montagnes, Quebec (C. Vincent *et al.*, unpublished). The role that indigenous CpGV plays in the biological control of wild *C. pomonella* populations is unknown.

Parasitoids

In southern Ontario, Hagley (1986) showed that naturally occurring parasitism by *Trichogramma pretiosum* Riley was highest in July and August, and *Trichogramma minutum* Riley migrated into the orchard from alternative hosts and occurred in low numbers at the beginning of the season. Yu *et al.* (1984a, b) studied the feasibility of using inundative releases of *Trichogramma* spp. to control *C. pomonella*. *T. minutum* parasitism depended on the age of the host eggs and the numbers of *T. minutum* released. After releases of *T. pretiosum* and *T. minutum*, distribution within the canopy and the influence of wind varied between parasitoid species. Rain and low temperatures reduced the overall rate of parasitism by *T. minutum.*

The potential of *Trichogramma* spp. to parasitize and control hosts depends partially on the density of host eggs (Parker *et al.*, 1971). In the sterile *C. pomonella* release programme in British Columbia, only male moths are needed for release. However, separation of the sexes is costly and therefore millions of sterile female moths are included in the orchard releases. All *C. pomonella* eggs resulting from at least one sterile partner are non-viable. *Trichogramma platneri* Nagarkatti, a species indigenous to *C. pomonella* in North America west of the Rockies (Pinto, 1998), developed successfully in non-viable *C. pomonella eggs.* The frequency of *T. platneri* parasitism, parasitoid size and emergence were significantly reduced in *C. pomonella* eggs from sterile female crosses (Cossentine *et al.*, 1996). Females of *T. platneri* reared on viable *C. pomonella* eggs parasitized significantly more viable than non-viable eggs (Zhang and Cossentine, 1995). Field releases of *T. platneri* were carried out to determine if it would use the non-viable eggs to increase parasitoid impact and thereby supplement the sterile

C. pomonella release programme. High parasitism was recorded in non-viable *C. pomonella* eggs. However, the number of non-viable eggs found in the tree canopies, particularly early in the season, was too low to maintain a high *T. platneri* population (Cossentine and Jensen, 2000).

Predators

Holliday and Hagley (1984) used pitfall traps to study the carabid fauna in Ontario in different sod types (natural, fescue and rye) and found several species known to be *C. pomonella* predators. The common carabids were *Pterostichus melanarius* Illiger, *Harpalus aeneus* Fabricius (= *H. affinis* Schrank), *Anisodactylus sanctaecrucis* Fabricius, *Amara* spp. and *Stenolopus comma* Fabricius. The abundance of carabid species was not affected by sod type, but was affected by soil type.

Using immunoelectro osmophoresis (Allen and Hagley, 1982), *P. melanarius*, the most abundant carabid found in pitfall traps deployed in blocks of apple trees at Jordan Station, Ontario, gave positive serological reactions to the antiserum against *C. pomonella* (Hagley and Allen, 1988). Other carabid species that also showed positive serological reactions included: *Amara aenea* DeGeer, *A. sanctaecrucis, Bembidion quadrimaculatum oppositum* Say, *Clivinia impressifrons* LeConte, *Diplochaeila impressicolis* (Dejean), *H. aeneus* and *Pterostichus chalcites* Say. The total number of all predator species caught was significantly related to the number of *C. pomonella* larvae present, but the proportion of larvae that pupated was not related to the number of predators. Hagley and Allen (1988) concluded that although the carabids feed on mature *C. pomonella* larvae, they did not significantly reduce their numbers.

Evaluation of Biological Control

Although it is unlikely that a single biological control technique could suppress *C. pomonella* populations below economically damaging thresholds, there are several, e.g. CpGV, indigenous and introduced parasitoids, and predators, that are potentially valuable supplements to reduced-pesticide and non-toxic control programmes.

Recommendations

Further work should include:

1. Studying the role of indigenous CpGV, parasitoids and predators in regulating wild *C. pomonella* populations, to better understand how they can best be manipulated;
2. Integrating biological control strategies, e.g. CpGV, with mating disruption and/or sterile *C. pomonella* release strategies when needed.

References

Allen, W.R. and Hagley, E.A.C. (1982) Evaluation of immunoelectroosmophoresis on cellulose polyacetate for assessing predation of Lepidoptera (Tortricidae) by Coleoptera (Carabidae) species. *The Canadian Entomologist* 114, 1047–1054.

Chouinard, G., Vincent, C., Roy, M. and Langlais, G. (1996) Régie des populations de *Cydia pomonella* (Lepidoptera: Olethreutidae), dans les vergers commerciaux du Québec avec des phéromones de synthèse. *Phytoprotection* 77, 57–64.

Cossentine, J.E. and Jensen, L.B. (1987) *Relative Effectiveness of Codling Moth Granulosis Virus and Impact of the Virus on Nontarget Apple Orchard Fauna.* Pesticide Research Report 1987, Expert Committee on Pesticide Use in Agriculture, Agriculture and Agri-Food Canada, Ottawa, Ontario, p. 5.

Cossentine, J.E. and Jensen, L.B.J. (2000) Releases of *Trichogramma platneri* (Hymenoptera: Trichogrammatidae) in apple orchards under a sterile codling moth release program. *Biological Control* 18, 179–186.

Cossentine, J.E., Lemieux, J. and Zhang, Y. (1996) Comparative host suitablility of viable and non-viable coding moth (Lepidoptera: Tortricidae) eggs for parasitism by *Trichogramma platneri* (Hymenoptera; Trichgorammatidae). *Environmental Entomology* 25, 1052–1057.

Dyck, V.A. and Gardiner, M.G.T. (1992) Sterile-insect release programme to control the codling moth *Cydia pomonella* (L.) (Lepidoptera; Olethreutidae) in British Columbia, Canada. *Acta Phytopathologica et Entomologica Hungarica* 27, 219–222.

Eastwell, K.C., Cossentine, J.E. and Bernardy, M.G. (1999) Characterisation of *Cydia pomonella* granulovirus from codling moths in a laboratory colony and in orchards of British Columbia. *Annals of Applied Biology* 134, 285–291.

Hagley, E.A.C. (1986) Occurrence of *Trichogramma* spp. (Hymenoptera: Trichogrammatidae) in apple orchards in southern Ontario. *Proceedings of the Entomological Society of Ontario* 117, 79–82.

Hagley, E.A.C. and Allen, W.R. (1988) Ground beetles (Coleoptera: Carabidae) as predators of the codling moth, *Cydia pomonella* (L.) (Lepidoptera: Tortricidae). *The Canadian Entomologist* 120, 917–925.

Hagley, E.A.C. and Laing, J.E. (1989) Effect of pesticides on parasitism of artificially distributed eggs of the codling moth, *Cydia pomonella* (Lepidoptera: Tortricidae) by *Trichogramma* spp. (Hymenoptera: Trichogrammatidae). *Proceedings of the Entomological Society of Ontario* 120, 25–33.

Hardman, J.M. (1987) *Evaluation of Granulosis Virus and Virus/mixture for Codling Moth Control.* Pesticide Research Report, Expert Committee on Pesticide Use in Agriculture, Agriculture and Agri-Food Canada, Ottawa, Ontario, p. 7.

Hardman, J.M. (1988) 1988 *Evaluation of Granulosis Virus and Virus/guthion Mixture for Codling Moth Control.* Pesticide Research Report, Expert Committee on Pesticide Use in Agriculture, Agriculture and Agri-Food Canada Ontario, p. 6.

Holliday, N.J. and Hagley, E.A.C. (1984) The effect of sod type on the occurrence of ground beetles (Coleoptera: Carabidae) in a pest management apple orchard. *The Canadian Entomologist* 116, 165–171.

Jaques, R.P., Laing, J.E., MacLellan, C.R., Proverbs, M.D., Sanford, K.H. and Trottier, R. (1981) Apple orchard tests on the efficacy of the granulosis virus of the codling moth, *Laspeyresia pomonella* (Lep.: Olethreutidae). *Entomophaga* 26, 111–118.

Jaques, R.P., Laing, J.E., Laing, D.R. and Yu, D.S.K. (1987) Effectiveness and persistence of the granulosis virus of the codling moth *Cydia pomonella* (L.) (Lepidoptera: Olethreutidae) on apple. *The Canadian Entomologist* 119, 1063–1067.

Jaques, R., Hardman, J., Laing, J., Smith, R. and Bent, E. (1994) Orchard trials in Canada on control of *Cydia pomonella* (Lep.: Tortricidae) by granulosis virus. *Entomophaga* 39, 281–292.

Judd, G.J.R, Gardiner, M.G.T. and Thomson, D.R. (1997) Control of codling moth in organically-managed apple orchards by combining pheromone-mediated mating disruption, post-harvest fruit removal and tree banding. *Entomologia Experimentalis et Applicata* 83, 137–146.

Madsen, H.F. and Procter, P.J. (1982) *Insects and Mites of Tree Fruits in British Columbia.* British Columbia Minstry of Agriculture and Food, Victoria, British Columbia.

Parker, F.D., Lawson, F.R. and Pinnell, R.E. (1971) Suppression of *Pieris rapae* using a new control system: mass release of both the pest and its parasites. *Journal of Economic Entomology* 64, 721–735.

Pinto, J.D. (1998) The systematics of the North American species of *Trichogramma* Westwood (Hymenoptera: Trichogrammatidae). *Memoirs of the Entomological Society of Washington* No. 22.

Proverbs, M.D. (1982) Sterile insect technique in codling moth control. In: *Sterile Insect Technique and Radiation in Insect Control.* Proceedings of the International Atomic Energy Agency, Vienna, Austria, 1981, AIEA-SM255/8, pp. 85–99.

Proverbs, M.D., Newton, J.R. and Campbell, C.J. (1982) Codling moth: a pilot program of control by sterile insect release in British Columbia. *The Canadian Entomologist* 114, 363–376.

Smith, R.F., Rigby, S., Mahar, A., Sheffield, C., O'Flaherty, C. and Trombley, M. (2000) *Evaluation of Last Call 'Bait and Kill' for Management of Codling Moth in Nova Scotia Apple Orchards.* Pest Management Research Report, 1999, Expert Committee on IPM, Agriculture and Agri-Food Canada, Ottawa, Ontario.

Trimble, R.M. (1995) Mating disruption for controlling the codling moth, *Cydia pomonella* (L.) (Lepidoptera: Tortricidae), in organic apple production in southwestern Ontario. *The Canadian Entomologist* 127, 493–505.

Trimble, R.M. and Solymar, B. (1997) Modified summer programme using border sprays for managing codling moth, *Cydia pomonella* (L.) and apple maggot, *Rhagoletis pomonella* (Walsh) in Ontario apple orchards. *Crop Protection* 16, 73–79.

Vincent, C. and Bostanian, N.J. (1988) La protection des vergers de pommiers au Québec: état de la question. *Le Naturaliste Canadien* 115, 261–276.

Vincent, C. and Roy, M. (1992) Entomological limits to the implementation of biological programs in Quebec apple orchards. *Acta Phytopathologica et Entomologica Hungarica* 27, 649–657.

Vincent, C., Mailloux, M., Hagley, E.A.C., Reissig, W.H.W., Coli, M. and Hosmer, T.H. (1990) Monitoring the codling moth (Lepidoptera: Olethreutidae) and the oblique-banded leafroller (Lepidoptera:Tortricidae) with sticky and non-sticky traps. *Journal of Economic Entomology* 83, 434–440.

Vincent, C., Chouinard, G., Bostanian, N.J. and Trimble, R.M. (1998) The concept of peripheral zone treatment and its application in commercial orchards. In: Vincent, C. and Smith, R. (eds) *Orchard Pest Management in Canada/La protection des vergers au Canada.* Bulletin Technique, Agriculture et agroalimentaire Canada, Saint-Jean-sur-Richelieu, Québec, pp. 93–103.

Yu, D.S.K., Hagley, E.A. and Laing, J.E. (1984a) Biology of *Trichogramma minutum* Riley collected from apples in southern Ontario. *Environmental Entomology* 13, 1324–1329.

Yu, D.S.K., Laing, J.E. and Hagley, E.A.C. (1984b) Dispersal of *Trichogramma* spp. (Hymenoptera: Trichogrammatidae) in an apple orchard after inundative releases. *Environmental Entomology* 13, 371–374.

Zhang, Y. and Cossentine, J.E. (1995) *Trichogramma platneri* (Hym.: Trichogrammatidae): Host choices between viable and nonviable coding moth, *Cydia pomonella*, and three-lined leafroller, *Pandemis limitata* (Lep.: Tortricidae). *Entomophaga* 40(3/4), 457–466.

19 *Cydia strobilella* (L.), Spruce Seed Moth (Lepidoptera: Tortricidae)

E.G. Brockerhoff, M. Kenis and J.J. Turgeon

Pest Status

The spruce seed moth, *Cydia strobilella* (L.), is an important Holarctic pest of spruce seed cones. In North America, where it was formerly known as *Cydia youngana* (Kearfott) (Brown and Miller, 1983), *C. strobilella* attacks mainly white spruce, *Picea glauca* (Moench) Voss, and Engelmann spruce, *Picea engelmannii* Parry ex Engelmann. It was also recorded from sitka spruce, *Picea sitchensis* (Bongard) Carrière, black spruce, *Picea mariana* (Miller) Britton, Sterns and Poggenburg, red spruce, *Picea rubens* Sargent, and blue spruce, *Picea pungens* Engelmann (Hedlin *et al.*, 1980; Miller and Ruth, 1989). In Europe and northern Asia, it attacks Norway spruce, *Picea abies* (L.) Karst, and many other spruces (e.g. Bakke, 1963; Stadnitzsky *et al.*, 1978; Da Ros *et al.*, 1993).

Because one larva can destroy about 40% of the seeds in a white spruce cone (Hedlin, 1973), *C. strobilella* can cause considerable damage. Seed cone infestation levels in natural stands vary considerably among years and regions, and range from 0 to 92% (Miller and Ruth, 1989; Fogal, 1990; Turgeon, 1990). Generally, damage by *C. strobilella* is negatively correlated with the cone crop (Annila, 1981; Fogal, 1990). Despite the current low level of damage caused by *C. strobilella* in seed orchards, where most seeds used in reforestation originate, control operations could become necessary again in the future. Marked changes in populations of *C. strobilella*, ranging over five orders of magnitude, have been documented (Annila, 1981).

Typically, female moths lay eggs between the scales of seed conelets, shortly after pollination. Larvae feed primarily on the developing seed, and overwinter in the cone axis (Tripp, 1954; Bakke, 1963).

Background

Research into the management of *C. strobilella* during the past two decades has focused on monitoring populations and damage, chemical control, and biological control using pathogens and parasitoids. Fogal (1989) and Sweeney *et al.* (1990) studied sampling methods and damage predictions by dissection of cones. A synthetic pheromone for monitoring male moth populations is available (Grant *et al.*, 1989). Fogal and Plowman (1989) and de Groot *et al.* (1994) reviewed chemical control trials against *C. strobilella* and other cone insects, as well as potential side-effects, such as insecticide resistance and phytotoxicity. Because *C. strobilella* spends most of its life cycle inside the cone, contact insecticides such as pyrethroids are usually not effective, but they can provide some control when applied during the oviposition period (Annila and Heliövaara, 1991). Systemic insecticides applied as foliar sprays, stem injections or implants may provide sufficient control.

Prior to the 1980s, information on the natural enemies of *C. strobilella* was limited to a few parasitoid records listed by Townes and Townes (1960) and Carlson (1979). Surveys for parasitoids of *C. strobilella* have been the subject of many studies, primarily in Europe, since the beginning of the century (Trägårdh, 1917). The first attempt at biological control was made in Ontario in 1947 when two braconids, the Holarctic *Ascogaster quadridentata* Wesmael and the North American *Macrocentrus ancylivorus* Rohwer, were released at a site where *C. strobilella* was common (McGugan and Coppel, 1962). The releases totalled 750 *A. quadridentata* and over 7000 *M. ancylivorus*, which originated from Canadian biological control programmes against codling moth, *Cydia pomonella* (L.), and *Cydia molesta* (Busck), respectively. It was assumed that these parasitoids might attack the closely related *C. strobilella*, but no evidence of this was found in later studies (McGugan and Coppel, 1962). Other attempts at biological control occurred in Latvia, where inundative releases of the egg parasitoid *Trichogramma cacoeciae* Marchal were considered promising against *C. strobilella* (Saksons *et al.*, 1973).

Since 1980, investigations on the biological control of *C. strobilella* in Canada have focused on the assessment of: (i) the effectiveness of microbial preparations (Timonin *et al.*, 1980; Fogal *et al.*, 1986a); (ii) surveys of the native parasitoid fauna of *C. strobilella* (Brockerhoff and Kenis, 1996); and (iii) the possibility of using European parasitoids of *C. strobilella* for its control in Canada (Brockerhoff and Kenis, 1996).

Biological Control Agents

Pathogens

Fungi

Beauveria bassiana (Balsamo) Vuillemin and *Metarhizium anisopliae* (Metschnikoff) Sorokin caused 100% mortality of *C. strobilella* in less than 48 h under laboratory conditions (Timonin *et al.*, 1980). In subsequent field studies (Fogal *et al.*, 1986a, b), white spruce conelets infested with *C. strobilella* and *Strobilomyia neanthracina* Michelsen were dusted with *B. bassiana* spore powder containing 7.3×10^7 viable spores mg^{-1}. Treated cones produced up to 55% more sound seed than untreated cones, but the results were inconsistent. Timing of the application appeared to be important.

Bacteria

Bacillus thuringiensis Berliner (*B.t.*) has not been used against spruce cone insects in Canada, but in Sweden applications to Norway spruce conelets did not reduce infestation levels of *C. strobilella* (Weslien, 1999).

Parasitoids

Substantial parasitoid records from British Columbia were obtained (G. Miller, Victoria, 1994, personal communication), but only recently published by Brockerhoff and Kenis (1996). At least six parasitoids are associated with *C. strobilella* in Canada, one larval endoparasitoid, and five larval ectoparasitoids (Table 19.1). The endoparasitoid *Phaedroctonus moderator* (L.) is the most common. Among the ectoparasitoids, *Exeristes comstockii* (Cresson) and two subspecies of *Scambus longicorpus* Walley were recorded most frequently. Two other *Scambus* spp. have been identified from *C. strobilella* (Brockerhoff and Kenis, 1996).

C. strobilella is known to support a rich parasitoid fauna in Europe. Eight parasitoid species of *C. strobilella* were reared and details of their life history and distribution recorded (Brockerhoff and Kenis, 1996). As in previous studies, e.g. Trägårdh (1917), Lovaszy (1941), Bakke (1963), and Stadnitzsky *et al.* (1978), *P. moderator*, *Bracon pineti* Thomson, *Liotryphon strobilellae* (L.) and *Elachertus geniculatus* (Zetterstedt), were the most common European larval parasitoids of *C. strobilella*, although some of these had previously been recorded under different names. An egg parasitoid, probably *T. cacoeciae*, and the larval ectoparasitoid *Scambus capitator* Aubert were reared for the first time from *C. strobilella*, and the pupal parasitoid *Tycherus fuscibucca* Berthoumieu is also likely to be associated with *C. strobilella* (Brockerhoff and Kenis, 1996). Although a literature review revealed 35 European parasitoids of *C. strobilella*, most of these probably represent misidentifications or incorrect host associations. A review of host records confirmed that the larval parasitoids we reared are host-specific specialists with a high degree of adaptation to the phenology of their host. All are thus theoretically suitable for biological control.

Our results indicate that the Canadian and European parasitoids of *C. strobilella* are more similar than previously thought, with several identical or closely related species in the two regions (Table 19.1). For example, the first candidate, *P. moderator*, a common and host-specific European parasitoid of *C. strobilella*, turned out to be a Holarctic species, a fact previously overlooked because the species was known under different names in the two continents. The other common larval parasitoids are not identical in Europe and North America, but they are closely related, have a similar biology and likely play a similar role in the population dynamics of this pest. Notable differences between the Canadian and European parasitoid faunas are the apparent absence of egg and pupal parasitoids of *C. strobilella* in North America (Table 19.1). Whether these differences represent a lack of study or an empty niche remains to be determined. Based on these results, the importation of European parasitoids showed only limited control prospects, and was not pursued further.

Evaluation of Biological Control

Because *C. strobilella* is currently not considered a major problem in seed orchards, research targeting biological control of this pest is not planned for the foreseeable future. However, populations of *C. strobilella* fluctuate widely, and it cannot be ruled out that control operations could again become necessary.

Promising control results were achieved with *B. bassiana*, although these varied

Table 19.1. North American and European parasitoids of *Cydia strobilella*. Closely related or identical species are shown on the same line. Of the European species, only those are listed that are either common or could be considered for biological control in North America. (After Brockerhoff and Kenis, 1996; and references therein.)

Parasitoid guild	North America	Europe
Egg parasitoids		
Trichogrammatidae	?[a]	*Trichogramma cacoeciae* Marchal
Larval endoparasitoids		
Ichneumonidae	*Phaedroctonus moderator* (L.)	*Phaedroctonus moderator* (L.)
Late larval ectoparasitoids		
Braconidae	*Bracon rhyacioniae* (Muesebeck)	*Bracon pineti* Thomson
Ichneumonidae	*Exeristes comstockii* (Cresson)	*Liotryphon strobilellae* (L.)
	Scambus spp.	*Scambus capitator* Aubert
Eulophidae	*Elachertus* sp.	*Elachertus geniculatus* (Zetterstedt)
Tachinidae	*Phytomyptera (Elfia)* sp.	?
Pupal parasitoids		
Ichneumonidae	?	*Tycherus fuscibucca* Berthoumieu

[a] Closely related species in this guild are not known from *C. strobilella* on this continent.

among applications. Compared with other biological control agents, this pathogen could control several cone pests, including *Strobilomyia* spp. (Sweeney *et al.*, Chapter 52 this volume) and *Choristoneura fumiferana* (Clemens) (Smith *et al.*, Chapter 12 this volume).

The biological control prospects of the use of parasitoids appear to be limited. *Trichogramma* sp. could be used for inundative releases, which have shown some control potential elsewhere. However, the logistics of rearing and supplying these at the right time would seem prohibitive unless they were used on a large scale. Furthermore, because no native egg parasitoid is known to attack *C. strobilella* in Canada; inundative releases would have to be made with an exotic, polyphagous species, such as *T. cacoeciae*, that might have non-target effects on the native fauna. Biological control of *C. strobilella* using larval parasitoids from Europe, e.g. *L. strobilellae* and *B. pineti*, does not appear promising because the ecological niches of the common species are already occupied by native parasitoids. This underpins the need for research aimed at elucidating gaps in our knowledge of the native parasitoid fauna of *C. strobilella*.

Recommendations

Further work should include:

1. Investigating whether sprayable, pathogen formulations could be commercialized as an alternative to chemical insecticides;
2. Examining the biology and impact of natural enemies, including potential egg and pupal parasitoids, on *C. strobilella* populations in Canada to determine whether strategies to conserve or enhance populations of native natural enemies (e.g. Brockerhoff and Kenis, 1998) could be sufficient or whether exotic parasitoids should be introduced;
3. Assessing whether the European pupal parasitoid *T. fuscibucca* is a suitable agent and whether it would fill an empty niche in Canada;
4. Investigating *Trichogramma* spp. (e.g. *T. cacoeciae*) as inundative agents.

Acknowledgements

We thank R.W. Carlson, E. Diller, K. Horstmann, D.R. Kasparyan, J. Papp, B. Pintureau and S. Vidal for the identification of specimens. Funding for this research was provided by the Canadian Forest Service (Green Plan).

References

Annila, E. (1981) Fluctuations in cone and seed insect populations in Norway spruce. *Communicationes Instituti Forestalis Fenniae* 101, 1–32.

Annila, E. and Heliövaara, K. (1991) Chemical control of cone pests in a Norway spruce seed orchard. *Silva Fennica* 25, 59–67.

Bakke, A. (1963) Studies on the spruce cone insects *Laspeyresia strobilella* (L.) (Lepidoptera: Tortricidae), *Kaltenbachiola strobi* (Winn.) (Diptera: Itonidae) and their parasites (Hymenoptera) in Norway. *Meddelelser fra det Norske Skogsförsöksvesen* 19, 1–151.

Brockerhoff, E.G. and Kenis, M. (1996) Parasitoids associated with *Cydia strobilella* (L.) (Lepidoptera: Tortricidae) in Europe, and considerations for their use for biological control in North America. *Biological Control* 6, 202–214.

Brockerhoff, E.G. and Kenis, M. (1998) Strategies for the biological control of insects infesting coniferous seed cones. In: Battisti, A. and Turgeon, J.J. (eds) *Proceedings, Cone and Seed Insect Working Party Conference (IUFRO S7.03–01). Sept. 1996, Monte Bondone, Italy.* Institute of Agricultural Entomology, University of Padova, Padova, Italy, pp. 49–56.

Brown, R.L. and Miller, W.E. (1983) Valid names of the spruce seed moth and a related *Cydia* species (Lepidoptera: Tortricidae). *Annals of the Entomological Society of America* 76, 110–111.

Carlson, R.W. (1979) Ichneumonidae. In: Krombein, K.V., Hurd, P.D. Jr, Smith, D.R. and Burke, B.D. (eds) *Catalog of Hymenoptera in America North of Mexico,* Vol. 1. Smithsonian Institute Press, Washington, DC, pp. 315–739.

Da Ros, N., Ostermeyer, R., Roques, A. and Raimbault, J.P. (1993) Insect damage to cones of exotic conifer species introduced in arboreta. I. Interspecific variations within the genus *Picea. Journal of Applied Entomology* 115, 113–133.

de Groot, P., Turgeon, J.J. and Miller, G.E. (1994) Status of cone and seed insect pest management in Canadian seed orchards. *Forestry Chronicle* 70, 745–761.

Fogal, W.H. (1989) Seed counts and cone insect foraging damage in relation to cone-collection date and stand type in white spruce. In: Miller, G.E. (ed.) *Proceedings of the 3rd Cone and Seed Insects Working Party Conference, Working Party S2.07–01, IUFRO, June 1988; Victoria, B.C.* Forestry Canada, Pacific Forestry Centre, Victoria, British Columbia, pp. 161–166.

Fogal, W.H. (1990) White spruce cone crops in relation to seed yields, cone insect damage, and seed moth populations. In: West, R.J. (ed.) *Proceedings, Cone and Seed Pest Workshop. 4 October 1989, St John's, Newfoundland.* Information Report N-X-274, Forestry Canada, Newfoundland and Labrador Region, St John's, Newfoundland, pp. 76–88.

Fogal, W.H. and Plowman, V.C. (1989) *Systemic Insecticides for Protecting Northern Spruce and Pine Seed Trees.* Information Report PI-X-92, Forestry Canada, Petawawa National Forestry Institute, St John's, Newfoundland.

Fogal, W.H., Thurston, G.S. and Chant, G.D. (1986a) Reducing seed losses to insects by treating white spruce conelets with conidiospores of *Beauveria bassiana. Proceeding of the Entomological Society of Ontario* 117, 95–98.

Fogal, W.H., Mittal, R.K. and Thurston, G.S. (1986b) *Production and Evaluation of* Beauveria bassiana *for Control of White Spruce Cone and Seed Insects.* Information Report PI-X-69, Canadian Forestry Service, Petawawa National Forestry Institute, St John's, Newfoundland.

Grant, G.G., Fogal, W.H., West, R.J., Slessor, K.N. and Miller, G.E. (1989) A sex attractant for the spruce seed moth, *Cydia strobilella* (L.), and the effect of lure dosage and trap height on capture of male moths. *The Canadian Entomologist* 121, 691–697.

Hedlin, A.F. (1973) Spruce cone insects in British Columbia and their control. *The Canadian Entomologist* 105, 113–122.

Hedlin, A.F., Yates, H.O. III, Tovar, D.C., Ebel, B.H., Koerber, T.W. and Merkel, E.P. (1980) *Cone and Seed Insects of North American Conifers.* Canadian Forestry Service; United States Department of Agriculture, Forest Service; Secretaria de Agricultura y Recursos Hidraulicos, Mexico.

Lovaszy, P. (1941) Beobachtungen über die Biologie und das Auftreten des Fichtenzapfenwicklers (*Laspeyresia strobilella* L.) und seiner Parasiten. *Annales entomologici Fennici* 7, 93–103.

McGugan, B.M. and Coppel, H.C. (1962) Part II. Biological Control of Forest Insects, 1910–1958. In: McLeod, J.H., McGugan, B.M. and Coppel, H.C. (eds) *A Review of the Biological Control Attempts Against Insects and Weeds in Canada.* Technical Communication No. 2, Commonwealth Institute of Biological Control, Trinidad. Commonwealth Agricultural Bureaux, Farnham Royal, UK, pp. 1–33.

Miller, G.E. and Ruth, D.S. (1989) The relative importance of cone and seed insect species on commercially important conifers in British Columbia. In: Miller, G.E. (ed.) *Proceedings of the 3rd Cone and Seed Insects Working Party Conference, Working Party S2.07–01, IUFRO, June 1988; Victoria, B.C.* Forestry Canada, Pacific Forestry Centre, Victoria, British Columbia, pp. 25–34.

Saksons, J., Saksons, Y.L. and Spalvins, Z. (1973) The entomofauna of the generative organs of Scots pine (*Pinus sylvestris* L.) and Norway spruce (*Picea abies* Karst.) in the Latvian SSR. *Zachita Lesa* 29–52.

Stadnitzsky, G.V., Lurchenko, G.I., Smetanin, A.N., Grebenshchikova, V.P. and Pribylova, M.V. (1978) *Vrediteli shishek i semian svoinykh porod.* Lesnaia promyshlennost, Moskow. [Translation: Yates, H.O. *Conifer Cone and Seed Pests.* Forestry Sciences Laboratory, Athens, Georgia]

Sweeney, J.D., Miller, G.E. and Ruth, D.S. (1990) Sampling seed and cone insects in spruce. In: West, R.J. (ed.) *Proceedings, Cone and Seed Pest Workshop. 4 October 1989, St John's, Newfoundland.* Information Report N-X-274, Forestry Canada, Newfoundland and Labrador Region, St John's, Newfoundland, pp. 63–75.

Timonin, M.I., Fogal, W.H. and Lopushanski, S.M. (1980) Possibility of using white and green muscardine fungi for control of cone and seed insects pests. *The Canadian Entomologist* 112, 849–854.

Townes, H. and Townes, M. (1960) Ichneumon-flies of America North of Mexico: 2. Subfamilies Ephialtinae, Xoridinae, Acaenitinae. *United States National Museum, Bulletin* 216, 3–11, 42–47, 60–63, 606–608.

Trägårdh, I. (1917) Undersökningar över gran- och tallkottarnas skadeinsekter. *Meddelelser Statens Skogsförsöksanstalt* 13–14, 1141–1204.

Tripp, H.A. (1954) Description and habits of the spruce seedworm *Laspeyresia youngana* (Kft.) (Lepidoptera: Olethreutidae). *The Canadian Entomologist* 86, 385–402.

Turgeon, J.J. (1990) Management of insect pests of cones in seed orchards in eastern Canada. In: West, R.J. (ed.) *Proceedings, Cone and Seed Pest Workshop. 4 October 1989, St John's, Newfoundland.* Information Report N-X-274, Forestry Canada, Newfoundland and Labrador Region, St John's, Newfoundland, pp. 89–99.

Weslien, J. (1999) Biological control of the spruce coneworm *Dioryctria abietella*: spraying with *Bacillus thuringiensis* reduced damage in a seed orchard. *Scandinavian Journal of Forest Research* 14, 127–130.

20 *Delia radicum* (L.), Cabbage Maggot (Diptera: Anthomyiidae)

J.J. Soroka, U. Kuhlmann, K.D. Floate, J. Whistlecraft, N.J. Holliday and G. Boivin

Pest Status

Cabbage maggot, *Delia radicum* (L.),[1] a pest of European origin, was introduced into eastern North America in the 19th century and is now common in cultivated regions across Canada (Griffiths, 1991). It attacks cruciferous crops, including canola, *Brassica napus* L. and *Brassica rapa oleifera* (De Candolle) Metzger, white mustard, *Sinapis*

[1]Prior to 1980, e.g. McLeod (1962) and Read (1971), *D. radicum* was called *Hylemyia brassicae* (Bouché).

alba L., rutabaga, *Brassica napus napobrassica* (L.) Hanelt, radish, *Raphanus sativus* L., turnip, *Brassica rapa rapa* L., and cole crops, e.g. broccoli, Brussels sprouts, cabbage and cauliflower (various varieties of *Brassica oleracea* L.). Damage is caused by larval feeding on and in roots of the host plant. Occasionally, larvae penetrate and damage the crucifer stem or head. High infestations can cause plant wilting, stunting, lodging, reduced flowering and seed set, and plant death. Secondary damage occurs when feeding sites provide entry points for bacterial and fungal pathogens that further stress the host plant. Levels of infestation and yield loss are most severe following cool, wet springs. The pest is becoming more of a problem in Alberta, Saskatchewan and Manitoba, where the incidence and severity of infestations have increased in canola crops in the past 15 years (Liu and Butts, 1982; Liu, 1984; Griffiths, 1986; Soroka *et al.*, 1999). In a year with heavy *D. radicum* infestations and poor canola growing conditions, yield losses have been estimated to be as high as Can$100 million (P. Thomas, Lacombe, 2000, personal communication).

Adult *D. radicum* overwinter in puparia located 5–20 cm below the soil surface. Oviposition begins shortly after spring emergence and continues for 5–6 weeks. Eggs are laid at or near the base of the host plant, usually in cracks or under a thin layer of soil. Upon hatching, maggots burrow deeper into the soil to feed on root hairs and on secondary roots. Late-instar maggots may tunnel into the tap root. Larvae feed for 3–4 weeks, then pupate in the soil near or in the tap root. Pupation lasts about 2 weeks. In Canada, 1–3 generations occur, depending on local climate.

Background

Some control of *D. radicum* in cruciferous vegetable crops can be achieved by timing planting operations to avoid peak fly emergence and egg-laying periods, by seeding resistant varieties (Mahr *et al.*, 1993), by use of row covers or other barriers, or by application of insecticidal drenches during the susceptible growing period. In cruciferous field crop production, tillage and plant densities affect *D. radicum* damage to plants (Dosdall *et al.*, 1995, 1996). Few insecticides are available for maggot control; when they are applied to control this pest they adversely affect its natural enemies and may lose efficacy with the development of pest resistance (Finlayson *et al.*, 1980, and references therein), hence the ongoing interest in developing biological controls.

Efforts to control *D. radicum* with biological agents were initiated in Atlantic Canada in 1949. At that time, the staphylinid *Aleochara bilineata* (Gyllenhal) and the eucoiline *Trybliographa rapae* (Westwood) were imported from Europe and released to control *D. radicum* pest populations in market-garden cole crops. However, subsequent studies revealed that these two species already were present in Canada, where *A. bilineata* had been misidentified as *Baryodma (Aleochara) ontarionis* Casey. In addition, *A. bilineata* and *T. rapae* were found to be widespread in populations of *D. radicum* infesting cole crops in eastern Canada, with rates of parasitism similar to those for cole crops in Europe. These results indicated that further releases of these species were unlikely to be of additional benefit and the importation programme was terminated in 1954. However, the expansion of this pest, particularly into canola, necessitates a re-evaluation of its biological control.

Biological Control Agents

Parasitoids

Many of the primary parasitoid species that attack *D. radicum* in Europe are already present in Canada, including *Aleochara bilineata* and *A. verna* Say [= *Aleochara bipustulata* (L.)], and *T. rapae*. Several thousand *A. bilineata*, and much smaller numbers *of A. verna*, *T. rapae* and the ichenumonid *Phygadeuon trichops* Thomson, were released from 1949 to 1954. Although the latter species did not establish, it is still a potential biological agent.

A. bilineata adults feed on host eggs and larvae. First-instar larvae actively search for host puparia, penetrate the puparial case, develop through three larval instars as ectoparasitoids, pupate within the host puparia and emerge as adults (Whistlecraft *et al.*, 1985).

A. verna is present in both Europe and Canada. A second, unnamed biotype of *A. verna*, reported to attack *Delia* spp. in Europe, has not yet been recorded from Canada (Klimaszewski, 1984). Clarification of the taxonomy and biology of the two biotypes would determine the suitability of the second biotype as a candidate for introduction.

T. rapae is a larval parasitoid. Eggs are laid in the first-, second- and third-instar larvae of *D. radicum*. Under laboratory conditions of 20°C, 60% RH, and L:D 16:8, larval development lasted 30–33 days, and pupal development about 25 days. Adults emerged from host pupae after about 61 days (Kacem *et al.*, 1996). Female longevity was 15 days at 20°C, and 11 days at 25°C, and females laid 46 and 35 eggs, respectively, under these conditions (Tamer, 1994). *T. rapae* is always outcompeted by *A. bilineata* when both are present in individual host puparia, but it maintains a substantial rate of parasitization in host populations even when the latter species is abundant.

P. trichops is a pupal parasitoid of several species of injurious Diptera, including *D. radicum* and the onion maggot, *Delia antiqua* (Meigen). When reared on *D. radicum*, adult male and female longevity averaged 57 days and 45 days, respectively (Plattner, 1974). Mating begins within 1 hour of adult emergence, with the onset of oviposition 2–4 days later. Eighty per cent of the eggs are laid during the first 20 days of oviposition, although Plattner (1974) reported that egg-laying may occur over 61 days. A host pupa may contain up to four parasitoid eggs, but only one larva completes its development.

Although *A. bilineata* is one of the most important natural enemies of *D. radicum* (references in Whistlecraft *et al.*, 1985), its effectiveness in eastern Canada is reduced because overwintering adults emerge several weeks after spring emergence of adult *D. radicum*. Thus, while early first-generation *D. radicum* are typically most injurious to crops, *A. bilineata* most effectively suppresses late first, second or third generations of the pest. Mass production of *A. bilineata* would permit releases of this beetle coincident with the emergence of first-generation *D. radicum*. Towards this end, a method to mass rear *A. bilineata* on *D. antiqua* was developed at London, Ontario, which permitted a weekly production of about 10,000 adult beetles with 5 h of labour per week (Whistlecraft *et al.*, 1985). In Ontario, releases of *A. bilineata* were made into home gardens over 2 years (Tomlin *et al.*, 1992), but to date no field releases have been made.

Floate *et al.* (1998) tested pupal parasitoids of house fly, *Musca domestica* L., as potential biological control agents of *D. radicum*. When puparia of the two pests were exposed simultaneously to parasitism in laboratory arenas, higher numbers of *Muscidifurax raptorellus* Kogan and Legner, *Muscidifurax zaraptor* Kogan and Legner, and *Trichomalopsis sarcophagae* Gahan emerged from puparia of *M. domestica* than from puparia of *D. radicum* (Table 20.1). Developmental times of the wasps either did not differ between the hosts or were longer on *D. radicum* (Table 20.2). Greenhouse studies suggested that the parasitoids were unable to locate *D. radicum* puparia under field conditions (J.J. Soroka and K.D. Floate, unpublished). *M. raptorellus* from 1728 *M. domestica* puparia exposed to parasitism were caged in pots containing a total of 840 *D. radicum* puparia placed at soil depths of 2.5 and 5.0 cm. Only one parasitoid was recovered subsequently from a *D. radicum* puparium.

Evaluation of Biological Control

Surveys during the mid-1990s characterized the parasitoid complex and levels of parasitism of *D. radicum* in cole crops. The

Table 20.1. Emergence of parasitoids from puparia of *Musca domestica* and *Delia radicum*.

Parasitoid	*Musca domestica* $\bar{X}$ ± SE (*n*[a])	*Delia radicum* $\bar{X}$ ± SE (*n*)
Muscidifurax raptorellus Kogan and Legner		
Pupae parasitized	18.2a ± 4.9 (10)	6.1b ± 0.9 (10)
Wasps/parasitized pupa	2.8a ± 0.3 (10)	2.2b ± 0.2 (10)
Muscidifurax zaraptor Kogan and Legner		
Pupae parasitized	9.9a ± 1.5 (22)	4.9b ± 0.8 (22)
Wasps/parasitized pupa	1.0 ± 0.0 (22)	1.0 ± 0.0 (22)
Trichomalopsis sarcophagae (Gahan)		
Pupae parasitized	5.3a ± 0.8 (10)	0.7b ± 0.3 (10)
Wasps/parasitized pupa	4.6a ± 0.3 (10)	1.4b ± 0.3 (10)

[a]Number of replications. Each replication contains 20 *M. domestica* and 20 *D. radicum* pupae simultaneously exposed to parasitism.
Means within a row that share a common letter do not differ ($P < 0.05$; *t*-test).

Table 20.2. Developmental time (days) of parasitoids at 25°C, when reared on puparia of *Musca domestica* and *Delia radicum*.

Sex/parasitoid	House fly $\bar{X}$ ± SE (*n*[a])	Cabbage maggot $\bar{X}$ ± SE (*n*)
Female		
Muscidifurax raptorellus Kogan and Legner	22a ± 0.1 (120)	23b ± 0.3 (33)
M. zaraptor Kogan and Legner	26a ± 0.1 (65)	26a ± 0.2 (27)
Trichomalopsis sarcophagae (Gahan)	21a ± 0.2 (79)	22a ± 1.0 (2)
Male		
Muscidifurax raptorellus	22a ± 0.1 (96)	22a ± 0.2 (29)
M. zaraptor	23a ± 0.2 (20)	25b ± 0.4 (11)
Trichomalopsis sarcophagae	21 ± 0.2 (56)	no data

[a]Number of replications. Each replication contains 20 *M. domestica* and 20 *D. radicum* pupae simultaneously exposed to parasitism.
Means within a row that share a common letter do not differ ($P < 0.05$; *t*-test).

most abundant parasitoids recovered from puparia collected in late fall at Winnipeg and Portage la Prairie (Manitoba), St.-Jean-sur-Richelieu (Quebec), St John's (Newfoundland) and London (Ontario) were *A. bilineata* and *T. rapae*. Except for puparia collected at London, parasitism by *A. bilineata* was high (up to 94%) and by *T. rapae* was low (<3%). At London, parasitism was 0–38% for *A. bilineata*, 2–6% for *T. rapae*, and 0–3% for the braconid *Aphaereta pallipes* Say. The latter species was not recovered from the other sites. At Winnipeg, host population density was related to parasitism by *A. bilineata* and to temperature and rainfall during June and July. Parasitism by *A. bilineata* may be related to cumulative degree-days over 5°C during June and July at Winnipeg and during June and September at London. Turnock *et al.* (1995) concluded that the existing complex of parasitoids was insufficient to stabilize host densities in cole crops. Additional natural enemy species need to be identified and evaluated as biological control agents to enhance biological control of *D. radicum*, particularly in canola-growing areas.

Recommendations

Further work should include:

1. Surveys for natural enemies in Europe and elsewhere in Canada to locate biotypes of *A. bilineata* that show better synchronization of spring emergence with that of *D. radicum*, and to find other potential biological control agents;
2. Investigating the effectiveness of releasing mass-reared *Aleochara* spp. at the time of fly oviposition for protection against early season damage;
3. Assessing the potential of the biotype of *A. verna* not present in Canada;
4. Assessing *P. trichops*;
5. Screening *Muscidifurax* and *Trichomalopsis* spp. for their potential against *D. radicum*, particularly if these agents can be mass-reared readily on *M. domestica*;
6. Continuing efforts to integrate use of chemical, cultural and biological methods of control.

References

Dosdall, L.M., Herbut, M.J., Crowle, N.T. and Micklich, T.M. (1995) The effect of plant density on root maggot (*Delia* spp.) (Diptera: Anthomyiidae) infestations in canola. *Proceedings of the Ninth International GCIRC Rapeseed Congress, Cambridge, UK,* Vol. 4, pp. 1306–1308.

Dosdall, L.M., Herbut, M.J., Crowle, N.T. and Micklich, T.M. (1996) The effect of tillage regime on the emergence of root maggots (*Delia* spp.) (Diptera: Anthomyiidae) from canola. *The Canadian Entomologist* 128, 1157–1165.

Finlayson, D.G., MacKenzie, J.R. and Campbell, C.J. (1980) Interactions of insecticides, a carabid predator, a staphylinid parasite, and cabbage maggots in cauliflower. *Environmental Entomology* 9, 789–794.

Floate, K.D., Soroka, J. and Spooner, R.W. (1998) Development of *Muscidifurax* and *Trichomalopsis* (Hymenoptera: Pteromalidae) on cabbage root maggot (Anthomyiidae: *Delia radicum*). 1998 PMR Report #58, In: *ECPM Pest Management Research Reports*, Agriculture and Agri-Food Canada, Research Branch, Ottawa, Ontario, pp. 168–170. http://res.agr.ca/lond/pmrc/download/pmrr_1998. pdf

Griffiths, G.C.D. (1986) Relative abundance of *Delia floralis* (Fallén) and *D. radicum* (L.) (Diptera: Anthomyiidae) in canola fields in Alberta. *Quaestiones Entomologicae* 22, 253–260.

Griffiths, G.C.D. (1991) Anthomyiidae. *Flies of the Nearctic Region* 8(2), No. 7, pp. 953–1040.

Kacem, N., Neveu, N. and Nénon, J.P. (1996) Development of *Trybliographa rapae*, a larval parasitoid of the cabbage root fly *Delia radicum*. *Bulletin OILB-SROP* 19, 156–161.

Klimaszewski, J. (1984) A revision of the genus *Aleochara* Gravenhorst of America North of Mexico (Coleoptera: Staphylinidae, Aleocharinae). *Memoirs of Entomological Society of Canada* 129, 1–211.

Liu, H.J. (1984) Surveys of root maggot damage to canola in Alberta and Northern British Columbia, 1981–1983. *Alberta Environmental Centre Report*, Vegreville, Alberta.

Liu, H.J. and Butts, R.A. (1982) *Delia* spp. (Diptera: Anthomyiidae) infesting canola in Alberta. *The Canadian Entomologist* 114, 651–653.

Mahr, S.E., Mahr, D.L. and Wyman, J.A. (1993) *Biological Control of Insect Pests of Crucifer Crops.* North Central Regional Publication 471. University of Wisconsin, Madison, Wisconsin.

McLeod, J.H. (1962) Part I. Biological control of pests of crops, fruit trees, ornamentals and weeds in Canada up to 1959. In: McLeod, J.H., McGugan, B.M. and Coppel, H.C. (eds) *A Review of the Biological Control Attempts Against Insects and Weeds in Canada.* Technical Communication No. 2, Commonwealth Institute of Biological Control, Trinidad. Commonwealth Agricultural Bureaux, Farnham Royal, UK, pp. 1–33.

Plattner, H.C. (1974) Contributions to the biology of *Phygadeuon trichops* Thomson (Hym., Ichneumonidae). *Anzeiger für Schadlingskunde, Pflanzenschutz, Umweltschutz* 48, 56–60.

Read, D.C. (1971) *Hylemya brassicae* (Bouché), Cabbage maggot (Diptera: Anthomyiidae). In: *Biological Control Programmes against Insects and Weeds in Canada 1959–1968.* Technical

Communication No. 4, Commonwealth Institute of Biological Control, Trinidad, Commonwealth Agricultural Bureaux, Farnham Royal, UK, pp. 20–22.
Soroka, J.J., Dosdall, L.M. and Olfert, O.O. (1999) Occurrence and damage potential of root maggots in canola. *Canola Council of Canada Project No. CA 96-16. Final Report.* Canola Council of Canada, Winnipeg, Manitoba.
Tamer, A. (1994) Laboratory investigations of the relationship between the cabbage root fly, *Delia radicum*, and its parasitoid, *Trybliographa rapae. Bulletin OILB-SROP* 17, 148–152.
Tomlin, A.D., McLeod, D.G.R., Moore, L.V., Whistlecraft, J.W., Miller, J.J. and Tolman, J.H. (1992) Dispersal of *Aleochara bilineata* (Col.: Staphylinidae) following inundative releases in urban gardens. *Entomophaga* 37, 55–63.
Turnock, W.J., Boivin, G. and Whistlecraft, J.W. (1995) Parasitism of overwintering puparia of the cabbage maggot, *Delia radicum* (L.) (Diptera: Anthomyiidae) in relation to host density and weather factors. *The Canadian Entomologist* 127, 535–542.
Whistlecraft, J.W., Harris, C.R., Tolman, J.H. and Tomlin, A.D. (1985) Mass-rearing technique for *Aleochara bilineata* (Coleoptera: Staphylinidae). *Journal of Economic Entomology* 78, 995–997.

21 *Dendroctonus ponderosae* Hopkins, Mountain Pine Beetle (Coleoptera: Scolytidae)

L. Safranyik, T.L. Shore, H.A. Moeck and H.S. Whitney

Pest Status

The mountain pine beetle, *Dendroctonus ponderosae* Hopkins, is the most destructive native insect pest of mature lodgepole pine, *Pinus contorta* var. *latifolia* Engelmann, in western North America (Wood, 1963). The blue stain fungi *Ophiostoma clavigerum* (Robinson, Jeffrey and David) and *Ophiostoma montium* (Rumbold) von Arx, commonly associated with *D. ponderosae*, kill resin-producing tissues in attacked trees (Reid *et al.*, 1967), resulting in reduced resinosis, tree death and successful beetle reproduction. Each year millions of trees are killed in this way. During an epidemic lasting 5–10 years, an infestation can spread to hundreds of thousands of hectares. Tree mortality of this magnitude seriously affects resource values and disrupts management plans. Well over 500 million trees have been killed by *D. ponderosae* in British Columbia during the past 80 years (Unger, 1993). From a forestry perspective, the problem is exacerbated by the beetles' preference for the largest trees.

D. ponderosae typically has a 1-year life cycle in British Columbia. Peak emergence, flight and attack by young adults on new host trees occur within a few days in mid to late July, but may continue until September. Eggs are laid in the phloem, and larvae overwinter and complete development the following year. The most common deviation from the 1-year life cycle is a partial or complete 2-year cycle, particularly at high elevations and near the northern edges of the beetle's range. Weather-driven changes in the life cycle may result in major shifts in the temporal distribution of the various brood developmental stages (McMullen *et al.*, 1986). Infestations tend to persist and expand in mature forests, usually ending when the host is exhausted or due to unseasonably cold weather before cold-hardiness has been established in early winter (Safranyik and Linton, 1991).

Background

Throughout the life cycle, beetles are exposed to many natural control factors (Amman and Cole, 1983). They are subject to predation from many families of insects and birds, as well as inter- and intraspecific competition, particularly when population levels are high. Low levels of mortality are caused by pathogens, e.g. *Beauvaria bassiana* (Balsamo) Vuillemin, again mostly during epidemics. Host resistance by resinosis and host drying also act to prevent infestation and reduce survival of broods. Because of the very short time of exposure outside the host, any potential control agent must be capable of affecting the beetles while they are protected by the bark.

Current direct control methods involve destruction of *D. ponderosae* in trees by harvesting and processing, felling and burning, debarking or using the systemic pesticide monosodium methanearsenate.

Field observations and experiments (Rankin, 1988) indicated the potential for using induced competition from secondary bark beetles to reduce *D. ponderosae* survival. This would be achieved by manipulating densities of competing species with behavioural chemicals. Competition for food and space within and among species is an important mortality factor for bark beetles (Berryman, 1974; Moeck and Safranyik, 1984; Poland, 1997). Trees killed by *D. ponderosae* are subsequently attacked by several normally less aggressive scolytid species (Safranyik *et al.*, 1999). These secondary species mainly breed in areas of the inner bark not colonized by the primary species, but overlapping attacks often occur. Two common scolytid associates of *D. ponderosae* are *Ips pini* (Say) (the pine engraver) and *Ips latidens* (LeConte).

A diverse complex of natural enemies, including insect and avian predators and disease organisms, is associated with *D. ponderosae* (Bushing, 1965; Edson, 1978; Dahlsten, 1982). Natural enemies of bark beetles contribute to keeping populations below the epidemic threshold (Moeck and Safranyik, 1984). In general, biological control of bark beetles as an alternative control strategy has received little attention (Stevens, 1981; Mills, 1983; Moeck and Safranyik, 1984), mainly because of insufficient knowledge of the nature and effects of natural enemies on the population dynamics of the target species.

Moeck and Safranyik (1984) reviewed the literature on insect predators, parasitoids and competitors of Scolytidae and recommended that inundative release of native clerid predators of *D. ponderosae* should be investigated. Hulme (1982) suggested that autoinfection with entomopathogens such as *B. bassiana* be investigated. Standard methods for rearing and handling *D. ponderosae* in the insectary and field (Linton *et al.*, 1987), modified as needed, were used for the study described below.

Biological Control Agents

Competitors

A series of experiments in south central British Columbia assessed the use of competing secondary bark beetle species to control *D. ponderosae*. Pheromones were used to manipulate the timing and intensity of secondary attacks by *Dryocoetes affaber* (Mannerheim). *I. pini* and *I. latidens* on lodgepole pine trees naturally attacked by *D. ponderosae*. However, results were inconclusive (Safranyik *et al.*, 1999).

In four experiments examining competitive interactions in mature lodgepole pine stands (Safranyik *et al.*, 1996, 1999), populations of *I. pini* as the main competitor species were manipulated using pheromone baits containing ipsdienol and lanierone. Young adults of *Ips pini* overwinter in duff at the base of their host trees, so the number emerging from the duff in spring was used as an index of brood production in analysing these experiments.

1. *I. pini* pheromones, lanierone and ipsdienol, applied in August and September to naturally infested trees, resulted in significantly lower numbers of *D. ponderosae* progeny emerging than from unbaited controls. In contrast, *I. pini* brood production was significantly increased.
2. *I. pini* baits were applied to *D. ponderosae* pheromone-baited trees 1 or 3 weeks after attack by *D. ponderosae*. Neither treatment significantly affected the number of attacks or brood produced by *D. ponderosae*, but did significantly increase *I. pini* brood production.
3. Dosages of one and six *I. pini* baits were placed on *D. ponderosae* pheromone-baited trees 1 and 3 weeks after *D. ponderosae* attack. The six-bait treatment did not significantly increase *I. pini* attack at breast height, but did result in significantly increased brood production over the whole bole. Early baiting, regardless of the dosage, also increased *I. pini* brood, but *D. ponderosae* was not affected. This was probably because of increased effectiveness of host resistance due to slow accumulation of *D. ponderosae* attacks, and to low *I. pini* attack density.
4. Lodgepole pine trees were baited for *D. ponderosae*; half were also simultaneously baited for *I. pini* and half had the *I. pini* bait applied 2 weeks later. At the time of the second bait application, half of the trees in both treatments were felled. Simultaneous baiting consistently, but not significantly, reduced *D. ponderosae* attack and egg-gallery length. However, these reductions were not reflected in reduced *D. ponderosae* brood. The highest *D. ponderosae* brood density was associated with the highest *I. pini* emergence in two trials. This suggested that *I. pini* densities were not high enough to affect *D. ponderosae* mortality through competitive exclusion. *I. pini* attack and brood production increased in trees baited following mass attack by *D. ponderosae*. In general, felling trees did not affect the attack or brood production by either species.

Predators

The feasibility of using inundative releases of the native clerids, *Enoclerus sphegeus* Fabricius, *Enoclerus lecontei* (Walcott), and *Thanasimus undatulus* Say, against spot infestations of *D. ponderosae* was tested. Native clerid adults collected in the field were laboratory reared to produce adequate numbers of beetles for test releases. Rearing clerids proved difficult (an average of 58% survival from egg to adult). Group rearing was not possible due to larval cannibalism; thus, individual, labour-intensive rearing was required. Also, larvae reared from eggs laid late in the season required 2 months' storage at 0°C to break prepupal diapause. No inundative releases were attempted.

Colour mutants in both *E. sphegeus* and *T. undatulus* were interesting. The larvae were pale green when young, as opposed to orange, and mature larvae were turquoise instead of purple. Mutant adults had a yellow abdomen instead of the normal reddish orange. This mutation was recessive. Such genetically marked clerids may prove to be useful in experimental field releases to test biological control efficacy because they could be distinguished from wild populations.

The European *Thanasimus formicarius* (L.) was imported for possible inoculative release against *D. ponderosae*. Because field-collected adults carried many mites and possibly internal parasites, they were unsuitable for direct release. Therefore they were laboratory reared to produce F_1 adults, which would have fewer parasites. Over 200 adults were reared for this purpose. However, further research into the biology of this species indicated that in Europe they feed on bark beetles other than *Dendroctonus* spp., including *Ips* spp. It was feared that in North America this could interfere with the competitive interaction between *I. pini* and *D. ponderosae* and thus hasten the development of *D. ponderosae* outbreaks. Also, crossbreeding experiments showed that *T. formicarius* and *T. undatulus* would interbreed, producing a low percentage of fertile hybrids.

For these two reasons, field releases of *T. formicarius* were not carried out.

Rhizophagus grandis Gyllenhal, used for biological control against the European *Dendroctonus micans* (Kugelmann), was imported into Canada to test its suitability against *D. ponderosae*. *R. grandis* adults and larvae fed on broods of *D. ponderosae* when presented one-on-one in dishes, with or without the presence of lodgepole pine bark. However, in slabs or bolts attacked by *D. ponderosae* the *R. grandis* adults produced very few offspring. This was probably because *D. ponderosae* larvae are solitary feeders, whereas *D. micans* larvae feed communally in large chambers under bark. No field releases of *R. grandis* were made.

Pathogens

Three fungi, *B. bassiana*, *Paecilomyces farinosus* (Holmskjold) A.H.S. Brown and G. Smith, and *Metarhizium anisopliae* (Metschnikoff) Sorokin, have been found on *D. ponderosae* throughout its range. *B. bassiana* was selected for inoculation experiments because of existing knowledge about its culture and mode of action, and because it has been used in previous attempts to control several species of insects, including beetles.

Attacking *D. ponderosae* were targeted for *B. bassiana* treatment with the goal of inducing mycosis early enough to prevent effective reproduction. Dry spore inoculum of the fungus was prepared from bran cultures, from spore masses harvested from agar plate cultures of fresh field isolates from *D. ponderosae*, and from well-known stock cultures originally from beetle hosts. *Beauveria* preparations from Abbott Laboratories and Mycotech Corporation were also evaluated.

New-generation adult *D. ponderosae* readily succumbed to virulent strains of *B. bassiana* applied topically by confining the beetles with an excess of spores and tumbling them for a few seconds just prior to bark penetration in laboratory, insectary and field tests on lodgepole pine bolts. Mortality of 90% or more developed in such treatments, with first symptoms appearing on about the third day, depending primarily on temperature. The reproduction potential of inoculated pairs of beetles in these conditions was essentially zero.

Autoinoculation of adult *D. ponderosae* was carried out by releasing them on the surface of lodgepole pine bark previously dusted with an excess of *B. bassiana* spores. Tests with autoinoculation on bolts in the insectary and in the forest showed reduced mortality compared to topical inoculation.

Neither topically inoculated nor autoinoculated *D. ponderosae* developed appreciable beauveriosis when caged on live lodgepole pine trees in the forest. It is believed that resinosis was involved by 'washing' *Beauveria* spores from the beetles or by inhibiting fungal growth.

Preliminary results from an experiment with autoinoculated and non-inoculated beetles caged on eight lodgepole pines successfully mass-attacked by *D. ponderosae* showed that the *Beauveria*-treated beetles produced only one-third the amount of brood compared to those that were not inoculated. Successfully attacked trees do not produce much resin, which may have enabled the mycosis to develop.

Evaluation of Biological Control

The competitive interactions revealed that the likely effect of *I. pini* on *D. ponderosae* survival would be greatest at moderate *D. ponderosae* densities. At high densities of *D. ponderosae*, intraspecific competition is probably more important than interspecific competition from *I. pini*. At low densities, host resistance reduces *I. pini* establishment and survival. *I. pini* and *D. ponderosae* attack densities are inversely related, with the engravers normally attacking bark areas not inhabited by *D. ponderosae*. The two species are repelled by each other's pheromones (Hunt and Borden, 1988). Placing *I. pini* bait on trees naturally infested by *D. ponderosae* resulted in increased engraver attack and brood production. *D. ponderosae* attack was not significantly affected because most of the attacks had already occurred. Such treatments showed promise in reducing *D. ponderosae* survival through larval competition from engraver broods. Simultaneous baiting results in extended attractancy for the engravers, causing them to attack earlier than they would naturally, thus coming into direct competition with the attacking *D. ponderosae*. When the *D. ponderosae* population is moderate or low, a lower *D. ponderosae* attack density and an elevated *I. pini* density result. When the overall population of *D. ponderosae* is high, the effects of intraspecific competition among *D. ponderosae* broods mask the effects of engraver competition. Individual rearing of clerids using live or frozen *D. ponderosae* as food is not practical.

Because one *Beauveria* isolate tested was pathogenic against some common beetle associates of *D. ponderosae* in *in vitro* tests, and other strains are known to have host ranges beyond beetles, these isolates cannot be used at this time.

Recommendations

Further work should include:

1. Determining the feasibility of increasing engraver populations locally by habitat manipulation, e.g. girdling of trees freshly killed by *D. ponderosae*;
2. Inundative field release of native clerids if a less labour-intensive method of rearing can be developed, using colour mutants as markers to evaluate experimental success;
3. Investigating host specificity of *Beauveria* isolates to develop highly specific strains;
4. Collaborating with industry partners that produce microbial biological agents, with emphasis on developing dust formulations, to ensure that the fungus inoculum becomes sufficiently attached to infection sites on *D. ponderosae* such that it cannot be washed away by resin or rubbed off by abrasion during bark penetration and gallery construction;
5. Trying to modify the symbiotic fungi of *D. ponderosae* to reduce their pathogenicity, thus making beetles with such modified fungi less successful in attacking trees.

Acknowledgements

N.J. Mills, Commonwealth Institute of Biological Control, supplied the European *T. formicarius* and, in cooperation with J.C. Grégoire of the Université Libre de Bruxelles, Belgium, supplied the European *R. grandis*.

References

Amman, G.D. and Cole, W.E. (1983) *Mountain Pine Beetle Dynamics in Lodgepole Pine Forests. Part II: Population dynamics.* General Technical Report INT-145, United States Department of Agriculture, Forest Service, Intermountain Forest and Range Experiment Station, Ogden, Utah.

Berryman, A.A. (1974) Dynamics of bark beetle populations: towards a general productivity model. *Environmental Entomology* 3, 579–585.

Bushing, R.W. (1965) A synoptic list of the parasites of Scolytidae (Coleoptera) in North America north of Mexico. *The Canadian Entomologist* 97, 449–492.

Dahlsten, D.L. (1982) Relationships between bark beetles and their natural enemies. In: Mitton, J.B. and Sturgeon, K.B. (eds) *Bark beetles in North American Conifers: A System for the Study of Evolutionary Biology.* University of Texas Press, Austin, Texas.

Edson, L.J. (1978) Host colonization and arrival sequence of the mountain pine beetle and its insectan associates. PhD thesis, University of California, Berkeley, California.

Hulme, M.A. (1982) *Biological Control in the Canadian Forestry Service.* Report DPC-X-11, Environment Canada, Canadian Forestry Service. Pacific Forestry Centre, Victoria, British Columbia.

Hunt, D.W.A. and Borden, J.H. (1988) Response of mountain pine beetle, *Dendroctonus ponderosae*

Hopkins, and pine engraver, *Ips pini* (Say), to ipsdienol in southwestern British Columbia. *Journal of Chemical Ecology* 14, 277–293.

Linton, D.A., Safranyik, L., McMullen, L.H. and Betts, R. (1987) Field techniques for rearing and marking mountain pine beetles for use in dispersal studies. *Journal of the Entomological Society of British Columbia* 84, 53–57.

McMullen, L.H., Safranyik, L. and Linton, D.A. (1986) *Suppression of Mountain Pine Beetle Infestations in Lodgepole Pine Forests.* Information Report BC-X-276, Canadian Forestry Service Pacific Forestry Centre, Victoria, British Columbia.

Mills, N.J. (1983) The natural enemies of bark beetles infesting conifer bark in Europe in relation to the biological control of *Dendroctonus* sp. in Canada. *Biocontrol News and Information* 4, 305–328.

Moeck, H.A. and Safranyik, L. (1984) *Assessment of Predator and Parasitoid Control of Bark Beetles.* Information Report BC-X-248, Environment Canada, Canadian Forestry Service, Pacific Forestry Centre, Victoria, British Columbia.

Poland, T.M. (1997) Competitive interactions between the spruce beetle, *Dendroctonus rufipennis* Kirby, and two secondary species, *Ips tridens* Mannerheim and *Dryocoetes affaber* Mannerheim (Coleoptera: Scolytidae). PhD thesis, Simon Fraser University, Burnaby, British Columbia.

Rankin, L.J. (1988) Competitive interactions between the mountain pine beetle and the pine engraver in lodgepole pine. Professional paper, Simon Fraser University, Burnaby, British Columbia.

Reid, R.W., Whitney, H.S. and Watson, J.A. (1967) Reactions of lodgepole pine to attack by *Dendroctonus ponderosae* Hopkins and blue stain fungi. *Canadian Journal of Botany* 45, 1115–1126.

Safanyik, L. and Linton, D.A. (1991) Unseasonably low fall and winter temperatures affecting mountain pine beetle and pine engraver populations and damage in the British Columbia Chilcotin Region. *Journal of the Entomological Society of British Columbia* 90, 17–21.

Safranyik, L., Shore, T.L. and Linton, D A. (1996) Ipsdienol and lanierone increase *Ips pini* Say (Coleoptera: Scolytidae) attack and brood density in lodgepole pine infested by mountain pine beetle. *The Canadian Entomologist* 128, 199–207.

Safranyik, L., Shore, T.L., Linton, D.A. and Rankin, L. (1999) *Effects of Induced Competitive Interactions with Secondary Bark Beetle Species on Establishment and Survival of Mountain Pine Beetle Broods in Lodgepole Pine.* Information Report BC-X- 384, Canadian Forest Service, Pacific Forestry Centre, Victoria, British Columbia.

Stevens, R.E. (1981) Natural enemies of bark beetles in the United States: potential for biological control. In: Coulson, J.R. (ed.) *Proceedings of the Joint American–Soviet Conference on the Use of Beneficial Organisms in the Control of Crop Pests, Washington, DC, 13–14 August, 1979.* College Park, Maryland, pp. 59–62.

Unger, L. (1993) Mountain pine beetle. *Forestry Canada. Forest Insect and Disease Survey.* Forest Pest Leaflet 76, Pacific Forestry Centre, Victoria, British Columbia.

Wood, S.L. (1963) A revision of bark beetle genus *Dendroctonus* Erichson (Coleoptera: Scolytidae). *Great Basin Naturalist Memoirs* 23, 1–117.

22 *Diuraphis noxia* (Kurdjumov), Russian Wheat Aphid (Homoptera: Aphididae)

O.O. Olfert, J.F. Doane, K. Carl, M.A. Erlandson and M.S. Goettel

Pest Status

The Russian wheat aphid, *Diuraphis noxia* (Kurdjumov), of Eurasian origin, attacks a wide range of plant species within the *Poaceae*. Wheat, *Triticum aestivum* L., barley, *Hordeum vulgare* L., and triticale, *Triticum* × *Triticosecale* Wittmack, are preferred hosts. Prior to 1935, it was not reported outside of the Ukraine and central Asia but it is now much more widespread, having colonized South Africa in 1978, Mexico in 1980 and South America in 1988. From Mexico, *D. noxia* spread rapidly northward, reaching Canada by 1988. By 1989, it was recorded in all states west of the 100th meridian and in the three most western provinces. On the Canadian prairies, infestations have been detected as far east as Parkbeg (50°26′N 106°17′W), as far north as Eatonia (51°13′N 109°23′W) and as far west as Fort McLeod (49°43′N 113°25′W) (Olfert *et al.*, 1990a, b).

Aphid feeding causes a variety of symptoms, including white or purple streaking and severe leaf rolling (Smith *et al.*, 1991). Under heavy attack, the crop appears to be under drought stress and tillers die off. Feeding also predisposes winter wheat to winter kill by reducing its freezing tolerance (Thomas and Butts, 1990). Because a large portion of the prairie region is seeded to cereal crops each year, the damage potential of *D. noxia* was considered to be high in the late 1980s (Butts *et al.*, 1997). Populations can overwinter in the Lethbridge area; however, in most years, western Canadian infestations result from migrations from the USA. In the 1990s, only light infestations occurred. Yield loss of small grains was reported by the US Great Plains Agricultural Council to be in excess of US$130 million in 1988 (Anonymous, 1989).

In Russia, both males and females of *D. noxia* exist (holocyclic) and the egg stage overwinters (Grossheim, 1914). However, no males have been reported in South Africa or North America. Both wingless and winged females occur in North America and they reproduce parthenogenetically. Generation time is about 10 days at temperatures of 19–21°C and the reproductive period lasts about 28 days (Webster and Starks, 1987). Limited overwintering of nymphs and adults occurs in southern Alberta but they have not been found to overwinter in Saskatchewan.

Background

In western Canada, insecticides are the first line of defence against *D. noxia* infestations (Hill *et al.*, 1993, 1995, 1996) but these are not integrated into a pest-management programme (Olfert and Johnson, 1998). A monitoring system was established to determine the timing and extent of migrations from the USA, and one suction trap continues to be in operation in the extreme south-western corner of Saskatchewan.

Natural enemies, including fungal diseases, are known to have a profound impact in regulation of aphid populations (Hagan and van den Bosch, 1968). In eastern Europe, cereal aphids in general, and *D. noxia* in particular, appear to be attacked by an impressive number of natural control agents (Berest, 1987).

Research began in 1989 to identify potential natural enemies of *D. noxia* (Doane *et al.*, 1991). In Europe, a study was initiated to evaluate parasitoids and predators from other regions in the world where *D. noxia* is indigenous for their potential as biological control agents should *D. noxia* become established in Canada (Krober and Carl, 1990). In the USA many species of parasitoids and predators of *D. noxia* were released with minimal study. In 1990, for example, 406,425 parasites (six species) and 646,817 predators (eight species) were released against *D. noxia* (Flanders and Burger, 1990).

Biological Control Agents

Parasitoids

In Saskatchewan, parasitoids associated with aphid populations were collected (Doane *et al.*, 1991) (Table 22.1). Yu (1992) studied and compared *Aphelinus varipes* (Förster) from Kazakhstan with the native *Aphelinus* sp. near *varipes* and concluded that the two species were ecologically compatible in that they occupied slightly different seasonal niches. This conclusion was based on the differences in the way the two responded to photoperiod and temperature (Yu, 1992).

Table 22.1. Parasitoids collected in sweep samples in Saskatchewan, 1989–1990.

Family	Species
Braconidae (Aphidiinae)	*Aphidius avenaphis* Fitch
	Aphidius matricariae Haliday
	Lysiphlebus testaceipes Cresson
Braconidae	*Opius* sp.
Ceraphronidae	*Ceraphron* sp.
Figitidae (Charipinae)	*Alloxysta* sp.
	Alloxysta victrix Westwood
Encyrtidae	*Syrphophagus* sp.
Eulophidae	*Aprostocetus* sp.
	Diglyphus sp.
Ichneumonidae	*Diplazon laetatorius* Fabricius
Megaspilidae	*Dendrocerus carpenteri* Kieffer
	Dendrocerus laticeps Hedicke
Platygastridae	*Platygaster* sp.
Pteromalidae	*Asaphes suspensus* Nees
	Asaphes vulgaris Walker
	Halticoptera triannulata Erdös
	Mesopolobus sp.
	Pachyneuron aphidis Bouché

Predators

In Saskatchewan, predators associated with aphid populations included eight insect families (Doane *et al.*, 1991) (Table 22.2). Of these, Coccinellidae (12 species) were the most important, particularly *Coccinella septempunctata* L., followed by Chrysopidae, Syrphidae, Nabidae and Anthocoridae. *Leucopis ninae* (Tanasijtshuk) and *Leucopis atritarsis* (Tanasijtshuk), the larvae of which feed on aphids, were imported from Yugoslavia and Kazakhstan, respectively.

Pathogens

In southern Alberta and Saskatchewan, cereal aphid populations were surveyed from 1989 to 1992 for promising pathogens (Goettel *et al.*, 1990; Doane *et al.*, 1991) (Table 22.3). All fungi found were new records for cereal aphids in Alberta and Saskatchewan and most were new Canadian records. In Saskatchewan, two fungi were recorded: *Conidiobolus obscurus* (Hall and Dunn) Remaudière and Keller (isolated from bird-cherry oat aphid, *Rhopalosiphum padi* L., and English grain aphid, *Sitobion avenae* Fabricius), and *Entomophthora chromoaphidis* Burger and Swain (isolated from several specimens of unidentified aphids). In each year, aphid populations were rarely high until late in summer and no major epizootics of either fungus occurred, although a minor outbreak of *C. obscurus* occurred in *R. padi* populations near Yorkton in 1989 (Doane *et al.*, 1991). In Alberta, *Pandora neoaphidis* (Remaudière and Hennebert) and *Baktoa apiculata* (Thaxter) Humber

Table 22.2. Predators collected in sweep samples in Saskatchewan 1989–1990.

Family	Species
Anthocoridae	*Orius tristicolor* White
Chrysopidae	*Chrysoperla carnea* Say
	Chrysopa oculata Stephan
Coccinellidae	*Adalia bipunctata* L.
	Anisosticta bitriangularis Say
	Brachiacantha albifrons Say
	Coccinella novemnotata Herbst
	Coccinella septempunctata L.
	Coccinella transversoguttata richardsoni Brown
	Coccinella trifasciata perplexa Mulsant
	Hippodamia convergens Guérin
	Hippodamia parenthesis Say
	Hippodamia sinuata crotchi Casey
	Hippodamia tredecimpunctata Say
	Hippodamia quinquesignata Kirby
	Hyperaspis inflexa Casey
	Hyperaspis lateris Mulsant
Nabidae	*Nabis alternatus* Parshley
	Nabis americoferus Carayon
	Nabis inscriptus Kirby
	Nabis subcoleoptrata Kirby
Syrphidae	*Eupeodes americanus* Wiedeman
	Helophilus latifrons Loew
	Paragus haemorrhous Meigen
	Scaeva pyrastri L.
	Sphaerophoria contigua Macquart
	Sphaerophoria philanthus Meigen
	Syritta pipiens L.
	Toxomerus marginatus Say

caused mortality in aphid populations (Goettel *et al*., 1990). In 1989, aphid populations were low until late July when they increased dramatically. In late August, fungal epizootics with infection rates of over 90% were common. The wet conditions in August (total precipitation of 78.4 mm versus the average of 41.4 mm) may have provided the conditions necessary for the epizootics. Although *D. noxia* was found late in the season, it was not infected with fungi. No epizootics were detected in 1990 or 1991 despite high late-season populations of aphids, including *D. noxia*. In 1992, *D. noxia* reached its highest numbers in many spring-seeded cereal fields and reached the economic threshold. Despite relatively high numbers, only a few individuals were found to be infected until late August when epizootics of *P. neoaphidis* were common (75% of 48 fields sampled contained significant numbers of diseased aphids) with infection rates reaching 50% in a number of fields (M.S. Goettel, unpublished).

Releases and Recoveries

Aphelinus varipes was released against *D. noxia* and other cereal aphids at Lethbridge, Alberta (51°45′N 106°49′W). A smaller release (6000 adults) was made near Shaunavon, south-western Saskatchewan (49°39′N 108°25′W), where the predominant aphid species were *R. padi*, *S. avenae* and corn leaf aphid, *Rhopalosiphum*

Table 22.3. Host records of entomopathogenic fungi parasitizing cereal aphids in southern Alberta and Saskatchewan during 1989.

Aphid	Fungus	Host plant
Alberta		
Diuraphis noxia (Mordvilko)	*Pandora neoaphidis* Remaudière and Hennebert	Barley
	Conidiobolus obscurus (Hall and Dunn) Remaudière and Hennebert	Barley
Rhopalosiphum maidis Fitch	*P. neoaphidis*	Barley
Schizaphis graminum Rondani	*Entomophthora chromaphidis* Burger and Swain	Barley
	P. neoaphidis	Barley
Sitobion avenae Fabricius	*Baktoa apiculata* (Thaxter) Humber	Barley
	P. neoaphidis	Barley
	C. obscurus	Barley
Saskatchewan		
Rhopalosiphum padi L.	*C. obscurus*	Wheat
S. avenae	*C. obscurus*	Wheat
Undetermined	*E. chromaphidis*	Bromegrass

maidis Fitch. Follow-up monitoring in 1995 at the Shaunavon release site failed to recover *A. varipes*.

Releases of 431 *L. atritarsis* were made near Shaunavon (50°26′N 106°17′W), in August 1991, into grain fields infested with *S. avenae* and *R. padi* because no infestations of *D. noxia* were available. *L. ninae* was released into barley infested with *D. noxia* near Lethbridge. Follow-up monitoring in 1995 failed to recover either *Leucopis* species.

Evaluation of Biological Control

In western Canada, a complex of native natural enemies controlling cereal aphids exists. These appear to be reasonably effective, particularly on late-seeded crops. It would be ideal to have natural enemies limit aphid numbers during their critical establishment phase (Edwards *et al.*, 1979).

In most years, *D. noxia* infestations in western Canada result from migrations from the USA and only periodically infest the southernmost portion of the prairies. As a result, research on integrated management of *D. noxia* has been suspended. However, in other regions of the world natural enemies can play a major role in controlling *D. noxia* and preventing outbreaks. Thus, natural enemies are likely to be important for controlling any future *D. noxia* outbreaks in Canada, particularly if climatic conditions become more favourable for its survival.

Recommendations

Future work should include:

1. Continued surveys of natural enemies to determine whether introduced agents have established and if native agents are controlling *D. noxia* effectively;
2. Assessing the potential to commercialize *D. noxia* pathogens, particularly to manage outbreaks if they occur in future.

References

Anonymous (1989) *Economic Impact of the Russian Wheat Aphid in the Western United States: 1987–1988.* Publication Number 129, Great Plains Agricultural Council, Fort Collins, Colorado.

Berest, Z.L. (1987) The trophic relations of grass aphid entomophages. *Journal of Zoology* 1, 45–48.

Butts, R.A., Thomas, J.B., Lukow, A. and Hill, B.D. (1997) Effect of fall infestations of Russian wheat aphid (Homoptera: Aphididae) on winter wheat yield and quality on the Canadian prairies. *Journal of Economic Entomology* 90, 1005–1009.

Doane, J.F., Matheson, M.M., Harris, J.L. and Erlandson, M.A. (1991) *Biological Control of the Russian Wheat Aphid.* Final Report to the Agricultural Development Fund.

Edwards, C.A., Sunderland, K.D. and George, K.S. (1979) Studies on polyphagus predation of cereal aphids. *Journal of Applied Ecology* 16, 811–823.

Flanders, R.V. and Burger, T.L. (1990) USDA-APHIS PPQ aphid biological control project summary of 1990 activities. In: *Proceedings of the Fourth Russian Wheat Aphid Workshop, 10–12 October, 1990.* Bozeman, Montana, pp. 177–183.

Goettel, M.S., Yu, D.S., Duke, G.M. and Erlandson, M.A. (1990) Potential of using biological control for the Russian wheat aphid in Canada. *Biocontrol News* 3, 32–39.

Grossheim, N.A. (1914) *Brachyolus noxius.* Memoirs of the Natural History Museum of the Zemstro of the Government of Taurida, Simferopol iii, 35–78 (*Review of Applied Entomology* (1915), 3, 307–308).

Hagen K.S. and van den Bosch, R. (1968) Impact of pathogens, parasites and predators on aphids. *Annual Review of Entomology* 13, 325–384.

Hill, B.D., Butts, R.A. and Schaalje, G.B. (1993) Reduced rates of foliar insecticides for control of Russian wheat aphid (Homoptera: Aphididae) in western Canada. *Journal of Economic Entomology* 86, 1259–1265.

Hill, B.D., Butts, R.A. and Schaalje, G.B. (1995) Mode of contact of chlorpyrifos with Russian wheat aphid (Homoptera: Aphididae) in wheat. *Journal of Economic Entomology* 88, 725–733.

Hill, B.D., Butts, R.A. and Schaalje, G.B. (1996) Factors affecting chlorpyrifos activity against Russian wheat aphid (Homoptera: Aphididae) in wheat. *Journal of Economic Entomology* 89, 1004–1009.

Krober, T. and Carl, K. (1990) *Cereal Aphids and Their Natural Enemies in Europe: a Literature Review.* International Institute of Biological Control Report, Delémont, Switzerland.

Olfert, O. and Johnson, D.L. (1998) Cereal crops and grain corn. *Western Committee on Crop Pests – 1998 Guide*, Lethbridge, Alberta, pp. 1–10.

Olfert, O., Doane, J.F. and Harris, J.L. (1990a) *Survey and Monitoring for Russian Wheat Aphid.* Biweekly Letter No. 235, Saskatoon Research Centre, Saskatoon, Saskatchewan.

Olfert, O., Doane, J.F. and Harris, J.L. (1990b) *Russian Wheat Aphid in Saskatchewan.* Research Centre, Saskatoon, Saskatchewan, Research Highlights – 1989. Saskatoon Research Centre, Saskatoon, Saskatchewan, p. 8.

Smith, C.M., Schotzko, D., Zemetra, R.S., Souza, E.J. and Schroeder-Teeter, S. (1991) Identification of Russian wheat aphid resistance in wheat. *Journal of Economic Entomology* 84, 328–332.

Thomas, J.B. and Butts, R.A. (1990) Effect of Russian wheat aphid on cold hardiness and winter kill of overwintering winter wheat. *Canadian Journal of Plant Science* 70, 1033–1041.

Webster, J.A. and Starks, K.J. (1987) Fecundity of *Schizaphis graminum* and *Diuraphis noxia* (Homoptera: Aphididae) at three temperature regimes. *Journal of the Kansas Entomological Society* 60, 580–582.

Yu, D.S. (1992) Effects of photoperiod and temperature on diapause of two *Aphelinus* spp. (Hymenoptera: Aphelinidae) parasitizing the Russian wheat aphid. *The Canadian Entomologist* 124, 853–860.

23 *Echinothrips americanus* (Morgan), *Frankliniella occidentalis* (Pergande), Western Flower Thrips, and *Thrips tabaci* Lindeman, Onion Thrips (Thysanoptera: Thripidae)

J.L. Shipp, D.R. Gillespie, K.M. Fry and G.M. Ferguson

Pest Status

Echinothrips americanus (Morgan), *Frankliniella occidentalis* (Pergande), western flower thrips, and *Thrips tabaci* Lindeman, onion thrips, are treated together here because of the common approaches to biological control applied against all three of these species on greenhouse crops.

E. americanus, native to eastern North America (Stanndard, 1968), is a relatively new pest of greenhouse crops in Canada. First reported in 1994 in British Columbia from commercial cucumber, *Cucumis sativus* L., in the Fraser Valley, *E. americanus* was also reported in 1995 from two greenhouse sweet pepper, *Capsicum annuum* L., crops (Opit *et al.*, 1997). To date, infestations have been found on poinsettia, *Euphorbia pulcherrima* Willdenow, cucumber and sweet pepper. In Ontario, *E. americanus* was first found on alstroemeria, *Alstroemeria* spp., and sweet pepper in 1999.

E. americanus causes extensive damage to foliage, such as silvering or streaking of the leaves. Damaged leaves may exhibit reduced photosynthesis (Buntin *et al.*, 1988) and severe infestations can result in plant death. Feeding damage has also been reported to sweet pepper fruit and is serious enough that the fruit must be culled.

F. occidentalis, a native species west of the Rocky Mountains, is a major pest of greenhouse vegetable and ornamental crops throughout Canada and a periodic pest of orchard crops in British Columbia. Its host plants include field and greenhouse vegetables, ornamentals, weeds, berries and tree fruits. Before the early 1980s, *F. occidentalis* was confined to western North America. During the late 1980s, it became a pest of greenhouse ornamentals and rapidly spread throughout Canada with plant shipments. Outside British Columbia, however, *F. occidentalis* is generally only a pest in greenhouses and does not overwinter outdoors (Broadbent and Hunt, 1991). In southern Ontario in 1989, *F. occidentalis* transmitted the tomato spotted wilt virus, *Tospovirus* (TSWV), to field tomato, *Lycopersicon esculentum* L., sweet pepper and potato, *Solanum tuberosum* L. (Pitblado *et al.*, 1990). It was speculated that infected thrips were brought into Ontario on infested tomato and sweet pepper transplants from Georgia, USA. TSWV is an important pathogen of ornamentals and occasionally tomato. This virus and *F. occidentalis* were not detected in Ontario the next year.

F. occidentalis causes damage by feeding on leaves, fruit or flower petals, and by ovipositing eggs in their tissues. Symptoms on greenhouse sweet pepper and cucumber are silvering striations, deformations and a dimpling or haloing effect from egg laying and hatching (Shipp *et al.*, 1998, 2000). Similar symptoms occur on other crops. Economic injury levels for fruit quality damage to cucumber, tomato and sweet pepper exist (Shipp, 1995; Shipp *et al.*, 1998, 2000). Only the larvae and adults feed and live on the plant. The pupal stage usually drops from the plant to complete development. Severe infestations of thrips can cause yield reduction through extensive feeding damage to the leaves and a resulting reduction in photosynthesis. *F. occidentalis* is also a major vector for TSWV. Its occurrence in greenhouses can result in complete crop destruction (Allen and Broadbent, 1986).

T. tabaci, a cosmopolitan species (Gentile and Bailey, 1968), was the predominant thrips pest of greenhouse crops across Canada before 1985. Since then, *F. occidentalis* has displaced *T. tabaci* as the major thrips pest. It is now difficult to find *T. tabaci* on greenhouse crops in many regions. *T. tabaci* differs from *F. occidentalis* in that it prefers foliage to flowers and fruit and is often restricted to the lower strata of the plants. Severe infestations can destroy leaves, eventually reducing the plant's photosynthetic capacity, but do not cause fruit curling or feeding damage to the fruit.

Background

E. americanus and *T. tabaci* are relatively easily controlled by insecticides because they prefer to feed on foliage. *F. occidentalis* is difficult to control with insecticides because nymphs and adults are sheltered in plant growing tips and unopened flower buds. It is also resistant to most currently registered insecticides (Hsu and Quarles, 1995). It can be controlled effectively using cultural and biological control methods, especially on greenhouse vegetables. Cultural strategies include strict sanitation programmes, e.g. maintaining a weed-free barrier around the outside of greenhouses, eliminating weeds inside greenhouses, screening vents, mass trapping adults using yellow or blue sticky tape or cards, removing and destroying crop residue, planting resistant cultivars, and manipulating greenhouse temperature and humidity to kill the thrips during crop clean-up or to optimize conditions for establishment and reproduction of natural enemies (Shipp *et al.*, 1991).

Biological Control Agents

Predators

Amblyseius barkeri (Hughes) and *Amblyseius cucumeris* (Oudemans) were evaluated for biological control of *F. occidentalis*. *A. cucumeris* was the more effective agent (Gillespie, 1989) but feeds only on first-instar thrips. During the 1990s, another predatory mite, *Amblyseius degenerans* Berlese, was shown to be effective in Europe (Ramakers and Voet, 1996), and was imported into Canada to control *F. occidentalis*. This species, which does not enter diapause under short day conditions, preys on both first- and second-instar thrips on leaves. A non-diapausing strain of *A. cucumeris* also became commercially available. The predatory soil mites *Hypoaspis miles* Berlese and *Hypoaspis aculeifer* (Canestrini) are used to control pupal stages on the ground (Gillespie and Quiring, 1990; Brødsgaard *et al.*, 1996).

A. cucumeris and *A. degenerans* were evaluated in field trials on sweet pepper plants twice infested artificially with 20 adult *E. americanus* per plant in biweekly applications. Infested plants were caged individually and 100 *A. cucumeris* or 20 *A. degenerans* per plant introduced. After 1 month, their effectiveness was evaluated. Neither mite species was observed feeding on immatures of *E. americanus* (Opit *et al.*, 1997).

Orius tristicolor (White) was also an effective biological control agent for *F. occidentalis* but was quickly replaced by the more widely distributed *O. insidiosus* (Say). Both species enter reproductive diapause under short-day conditions (<12 h day length) and prey on all stages of thrips on the plant, but do not feed on pupal stages on the ground. *O. insidiosus*, introduced at a rate of four adults per plant, significantly reduced population levels of *E. americanus* (Opit *et al.*, 1997). The generalist predator *Dicyphus hesperus* Knight is being evaluated against small insects, including thrips, on greenhouse tomato, sweet pepper and cucumber (Gillespie *et al.*, 1998, 1999, 2000; McGregor *et al.*, 1999).

Most arthropod natural enemies are shipped as reproductive adults in containers without carrier materials, and are released on the crop individually. *A. cucumeris*, however, is shipped either in a loose bran carrier that is sprinkled on to the crop, or in slow-release bags that are hung in the crop. The latter contain *A. cucumeris,* a food source (a stored-products mite) and a bran substrate as food for the prey. Predators reproduce in the bags and escape from an opening in the bags over 1–6 weeks with optimal release in weeks 3–4, ensuring continuous inoculation on the crop.

At low pest densities, the natural enemies are inoculated into the crop once, or up to three or four times, until populations are established. At moderate pest densities, inundative releases of either *A. cucumeris* or *O. insidiosus* may be necessary. Inundative releases of *A. cucumeris* are made by sprinkling large quantities of loose bran on the crop or by placing at least one slow-release bag per plant. Inundative releases of *O. insidiosus* are made by placing 2–4 adults per plant. Because of the relatively high cost of *O. insidiosus*, this is generally appropriate only for isolated groups of plants with high *F. occidentalis* densities.

In winter and early spring, vapour pressure deficit can be high and is detrimental to predatory mite survival (Shipp and van Houten, 1997). In the absence of thrips damage, slow-release bags of *A. cucumeris* may be used, but otherwise IPM-compatible pesticides are probably appropriate. *A. degenerans* requires pollen to establish, and therefore should not be applied on sweet pepper until open flowers are present. The absence of pollen in gynoecious cucumber varieties prevents their use on that crop. *O. insidiosus* enters reproductive diapause if the nymphs complete development at less than 12 hours light, so should only be introduced when it is certain that the offspring of introduced adults will complete development after the spring equinox.

Pathogens

Several species of entomopathogenic fungi have either been isolated from *F. occidentalis* or have shown potential as biological control agents in laboratory and/or greenhouse trials. The first was *Verticillium lecanii* (Zimmerman) Viegas (van der Schaaf *et al.*, 1991; Helyer *et al.*, 1992). It was commercialized initially in Europe and recently in the USA, but its success has been variable due to its requirement for high humidity (>90% RH). Subsequently, *Beauveria bassiana* (Balsamo) Vuillemin and *Metarhizium anisopliae* Sorokin have been found to infect *F. occidentalis* (Brownbridge, 1995; Vestergaard *et al.*, 1995). These fungi require lower humidity for germination and seem to show greater potential for use in greenhouses. In the USA, *B. bassiana* is presently registered for greenhouse crop pests.

Greenhouse efficacy trials with *B. bassiana* indicate that a relative humidity of 85–90% is required for 48 h for successful infection. Infection levels are greater in adults than larvae of *F. occidentalis*. Larval moulting negatively affects spore adhesion and penetration into the insect body. *B. bassiana* is compatible with predatory mites, but is not compatible with *Orius* spp.

Fry *et al.* (1999) tested three *B. bassiana* products from the USA against *F. occidentalis* on poinsettia. All three reduced thrips populations below those on controls when applied at the label rate or higher. Ten of 76 field isolates screened for pathogenicity against *F. occidentalis* showed higher mortality 6 days after application when compared to a commercially available strain (Fry *et al.*, 1999; K.M. Fry, unpublished).

Evaluation of Biological Control

Biological control of *F. occidentalis* is a critical component of crop-based IPM programmes in greenhouses. Factors such as pest density, time of year and presence of other pest species will modify the choice of natural enemies, introduction rates and method of introduction. Use of several natural enemy species and multiple introductions appear to give more stable control than single natural-enemy introductions. Biological control of *F. occidentalis* is generally successful on cucumber and sweet pepper, but less so on tomato. On flower crops, successful biological control of *F. occidentalis* depends to some degree on injury thresholds. In indoor plantings, e.g. conservatories, biological control against thrips can be quite good, and stable over a long time. A good potential to develop *B. bassiana*-based products suited for thrips control also exists. In Ontario, commercial growers tried unsuccessfully to use *O. insidiosus* to control *E. americanus* on alstroemeria and sweet pepper in the 2000 growing season.

Recommendations

Further work should include:

1. Evaluating new natural enemies, e.g. *D. hesperus* and *B. bassiana*;
2. Refining biological control of *F. occidentalis* on greenhouse tomato, ornamental plants and flower crops;
3. Evaluating the natural enemy complex of *E. americanus* in eastern North America to find specialist parasitoids that could be used in greenhouses.

References

Allen, W.R. and Broadbent, A.B. (1986) Transmission of tomato spotted wilt virus in Ontario greenhouses by *Frankliniella occidentalis*. *Canadian Journal of Plant Pathology* 8, 33–38.

Broadbent, A.B. and Hunt, D.W.A. (1991) Inability of western flower thrips, *Frankliniella occidentalis* (Pergande) (Thysanoptera: Thripidae), to overwinter in southern Ontario. *Proceedings of the Entomological Society of Ontario* 122, 47–49.

Brødsgaard, H.F., Sadar, M.A. and Enkegaard, A. (1996) Prey preference of *Hypoaspis miles* (Berlese) (Acarina: Hypoaspididae): Non-interference with other beneficials in glasshouse crops. *International Organization of Biological Control/West Palaearctic Regional Section, Bulletin* 19, 23–26.

Brownbridge, M. (1995) Prospects for mycopathogens in thrips management. In: Parker, B.L., Skinner, M. and Lewis, T. (eds) *Thrips Biology and Management.* Plenum Press, New York, New York, pp. 281–295.

Buntin, G.D., Harrison, R.D., Oetting, R.D. and Daniell, J.W. (1988) Response of leaf photosynthesis and water relations of impatiens and peach to thrips injury. *Journal of Agricultural Entomology* 5, 169–177.

Fry, K.M., Goettel, M.S. and Keddie, B.A. (1999) *Evaluation of the Fungus* Beauveria bassiana *for Management of Western Flower Thrips on Greenhouse Ornamentals.* Alberta Agricultural Research Institute Project 970766, Final Report.

Gentile, A.G. and Bailey, S.F. (1968) *A Revision of the Genus* Thrips *Linnaeus in the New World, with a Catalogue of the World Species.* University of California Press, Berkeley, California.

Gillespie, D.R. (1989) Biological control of thrips (Thysanoptera: Thripidae) on greenhouse cucumber by *Amblyseius cucumeris*. *Entomophaga* 34, 185–192.

Gillespie, D.R. and Quiring, D.M.J. (1990) Biological control of fungus gnats, *Bradysia* spp. (Diptera: Sciaridae), and western flower thrips, *Frankliniella occidentalis* (Pergande) (Thysanoptera: Thripidae), in greenhouses using a soil-dwelling mite, *Geolaelaps* sp. nr *aculeifer* (Canestrini) (Acari: Laelapidae). *The Canadian Entomologist* 122, 975–983.

Gillespie, D., McGregor, R., Quiring, D. and Foisy, M. (1998) Dicyphus hesperus – *This Bug's for You.* Agassiz Technical Report #148, Agriculture and Agri-Food Canada, Pacific Agri-Food Research Centre. http://res.agr.ca/summer/scrfrm2.htm

Gillespie, D., McGregor, R., Quiring, D. and Foisy, M. (1999) You are what you've eaten – prey versus plant feeding in *Dicyphus hesperus*. A second update on the development of an omnivorous predator for the British Columbia greenhouse industry. http://res.agr.ca/summer/scrfrm2.htm

Gillespie, D., McGregor, R., Quiring, D. and Foisy, M. (2000) *Biological Control of Greenhouse Whitefly with* Dicyphus hesperus – *an Update on the Development of an Omnivorous Predator for the British Columbia Greenhouse Industry.* Agassiz Technical Report # 157, Agriculture and Agri-Food Canada, Pacific Agri-Food Research Centre.

Helyer, N.L., Gill, G., Bywater, A. and Chambers, R. (1992) Elevated humidities for control of chrysanthemum pests with *Verticillium lecanii. Pesticide Science* 36, 373–378.

Hsu, C. and Quarles, W. (1995) Greenhouse IPM for western flower thrips. *The IPM Practitioner* 17, 1–11.

McGregor, R.R., Gillespie, D.R., Quiring, D.M.J. and Foisy, M.R.J. (1999) Potential use of *Dicyphus hesperus* Knight (Heteroptera: Miridae) for biological control of pests of greenhouse tomatoes. *Biological Control* 16, 104–110.

Opit, G.P., Peterson, B., Gillespie, D.R. and Costello, R.A. (1997) The life cycle and management of *Echinothrips americanus* (Thysanoptera: Thripidae). *Journal of Entomological Society of British Columbia* 94, 3–6.

Pitblado, R.E., Allen, W.R., Matteoni, J.A., Garton, R., Shipp, J.L. and Hunt, D.W.A. (1990) Introduction of the tomato spotted wilt virus and western flower thrips complex into field vegetables in Ontario, Canada. *Plant Disease* 74, 81.

Ramakers, P.M.J. and Voet, S.J.P. (1996) Introduction of *Amblyseius degenerans* for thrips control in sweet peppers with potted castor beans as banker plants. *International Organization of Biological Control/West Palaearctic Regional Section, Bulletin* 19, 127–130.

Schaaf, D.A. van der, Ravensberg, W.J. and Malais, M. (1991) *Verticillium lecanii* as a microbial insecticide against whitefly. In: Smits, P.H. (ed.) Proceedings, Third European Meeting on Microbial Control of Pests, IOBC Working Group on Insect Pathogens and insect Parasitic Nematodes, Wageningen, Netherlands. *International Organization of Biological Control/West Palaearctic Regional Section, Bulletin* 14, 120–123.

Shipp, J.L. (1995) Monitoring of western flower thrips on glasshouse and vegetable crops. In: Parker, B.L., Skinner, M. and Lewis, T. (eds) *Thrips Biology and Management.* Plenum Press, New York, New York, pp. 547–555.

Shipp, J.L. and van Houten, Y.M. (1997) Influence of temperature and vapor pressure deficit on survival of the predatory mite *Amblyseius cucumeris* (Acari: Phytoseiidae). *Environmental Entomology* 106, 106–113.

Shipp, J.L., Boland, G.J. and Shaw, L.A. (1991) Integrated pest management of disease and arthropod pests of greenhouse vegetable crops in Ontario: current status and future possibilities. *Canadian Journal of Plant Science* 71, 887–914.

Shipp, J.L., Binns, M.R., Hao, X. and Wang, K. (1998) Economic injury levels for western flower thrips (Thysanoptera: Thripidae) on greenhouse sweet pepper. *Journal of Economic Entomology* 91, 671–677.

Shipp, J.L., Wang, K. and Binns, M.R. (2000) Economic injury levels for western flower thrips (Thysanoptera: Thripidae) on greenhouse cucumber. *Journal of Economic Entomology* 93, 1732–1740.

Stanndard, L.J. (1968) The thrips, or Thysanoptera, of Illinois. *Illinois Natural History Survey, Bulletin* 29, 215–552.

Vestergaard, S., Butt, T.M., Gillespie, A.T., Schreiter, G. and Eilenberg, J. (1995) Pathogenicity of the hyphomycete fungi *Verticillium lecanii* and *Metarhizium anisopliae* to the western flower thrips, *Frankliniella occidentalis. Biocontrol Science and Technology* 5, 185–192.

24 *Eriosoma americanum* (Riley), Woolly Elm Aphid (Homoptera: Pemphigidae)

K.M. Fry

Pest Status

The woolly elm aphid, *Eriosoma americanum* (Riley), a native species, is one of two aphid species that attack the roots of saskatoon berry, *Amelanchier alnifolia* (Nuttall) Nuttall. *E. americanum* is considered by saskatoon growers to be the most serious insect pest attacking saskatoon seedlings. It feeds on the roots and is usually found attached to them under the soil surface. In severe infestations up to 85% of newly planted seedlings have been lost to *E. americanum*. On average, losses of Can\$120 ha^{-1}, or 5%, can be expected if the aphid is left untreated (Klein, 2000).

Woolly apple aphid, *Eriosoma lanigerum* (Hausmann), also native to North America, can attack *A. alnifolia* roots, although it is more commonly found on apple, *Malus pumila* Miller (= *M. domestica* Borkhausen) (Brown, 1986). The aphids typically occur above the soil surface and feed at the base of the stem. *E. lanigerum* is not considered to be a major pest of saskatoon bushes.

E. americanum exhibits dioecious heterocycly with the primary host being American elm, *Ulmus americana* L., and the secondary host being saskatoon berry (Patch, 1915; Neill *et al.*, 1994; Fry *et al.*, 1998). The life cycle begins on elm as overwintered eggs that hatch in spring to yield the stem mother or fundatrix. The fundatrix moves to an elm leaf bud and begins feeding on the underside of the leaf, maturing as the leaves flush from the bud, and gives birth parthenogenetically to young, the fundatrigenae. The action of feeding by the fundatrix and her offspring causes the leaves to curl under to form a protected habitat, a pseudogall. Within the pseudogall, the fundatrix and fundatrigenae feed and reproduce. The fundatrigenae develop either into apterous adults or alate adults, the spring migrants. These fly to saskatoon bushes and alight on the underside of leaves, where they deposit live young, the alienicolae, on to the lower surface. Alienicolae crawl down to the plant roots to suck plant juices from developing roots, causing a loss of plant vigour that commonly results in death. In late summer, a new generation of winged aphids arises, the fall migrants or gynoparae. These crawl up to the soil surface via earthworm tunnels or through soil disturbed by ants and fly to American elm. Once on the elm, they give birth to live young, the microsexuales, which are extremely small, non-feeding aphids that are either male or female. The microsexuales crawl around elm bark until they find a mate and copulate. Each female lays only one egg in a crack in the bark.

Background

The saskatoon berry industry has been expanding on the Canadian prairies. The area planted to saskatoon berry has increased from 500 ha in the early 1990s to nearly 2000 ha, with 1000 ha harvested mechanically. M. Steiner (Vegreville, 1993, personal communication) and Neill *et al.* (1994, 1995) evaluated chemical control, which resulted in a minor use registration for Orthene® on non-fruit-bearing plants.

Currently, root aphids are controlled most effectively with soil drench applications of systemic chemical insecticides. Particular care must be taken with these to allow for adequate time for residues to fall to acceptable levels prior to harvest. The timing of *E. americanum* on saskatoon berry is such that insecticides can be applied immediately following harvest thereby giving an effective pre-harvest interval of nearly 12 months. In Canada, no products are currently registered for use against *E. americanum* on fruit-bearing saskatoon plants. Carabidae and Anthocoridae, common soil-inhabiting predators, have been observed in the soil around aphid-infested plants. Although biological control organisms of *E. americanum* have been studied (Brown *et al.*, 1992; Mueller *et al.*, 1992), little work has been done on biological control.

Biological Control Agents

Pathogens

Fungi

Miranpuri and Khachatourians (1996) tested *Beauveria bassiana* (Balsamo) Vuillemin strains SG 8702 and SG 8601 and *Verticillium lecanii* (A.W. Zimmermann) Viegas on *E. americanum* infesting saskatoon berry seedlings in root trainers. Two applications of 1×10^8 spores ml^{-1} were effective at reducing aphid numbers. However, single applications to heavily infested plants did not provide significant control. Field trials of *B. bassiana* strains SG 8702, SG 8701, SG 8601, DN WU, DN 8803, DN 8806 and GK 2016 and *V. lecanii* ATCC478 reduced aphid numbers compared to controls.

Nematodes

Steinernema carpocapsae (Weiser) was evaluated for effectiveness in controlling *E. lanigerum* (Brown *et al.*, 1992), with foliar applications yielding significant control. Fry *et al.* (1998) evaluated four nematodes, *S. carpocapsae*, *Steinernema feltiae* (Filipjev), *Steinernema riobrave* Cabanillas, Poinar and Raulston, and *Heterorhabditis bacteriophora* Poinar, for management of edaphic populations of aphids infesting first-year Northline cultivar saskatoon seedlings. *H. bacteriophora* and, to a lesser extent, *S. feltiae* are cruisers and *S. carpocapsae* and *S. riobrave* are ambushers (Lewis *et al.*, 1992; Peters *et al.*, 1996).

In one experiment each nematode species was suspended in 100 ml of tap water and applied to each of 20 plants at a rate of 2000 nematodes per plant, or roughly 2×10^9 ha^{-1}, once a week for 3 or for 6 weeks. Only the 3-week treatment with *H. bacteriophora* reduced aphid numbers significantly.

In another experiment, the commercially available *Heterorhabditis megidis* Poinar, Jackson and Klein and *S. feltiae* were applied at rates of 2000, 20,000 and 200,000 nematodes per plant, on 21 July 1998 and 11 August 1999. Aphid infestation levels were rated on roots and at the crown 30 days after application. Infestation levels on roots were significantly lower at the highest treatment rate for both species tested (Fry and Pruski, 2001). No significant difference between the treatment rates for infestation levels at the crown occurred. *H. megidis* was more effective than *S. feltiae* for reducing aphids on both roots and crown.

In 1999, *E. americanum* infestation levels were substantially higher than in 1998. Neither species of nematode at any of the three rates tested reduced aphid infestations significantly at the crown or the roots. However, *H. megidis* was significantly more effective than *S. feltiae* at reducing *E. americanum* levels at the crown, similar to observations in 1998.

Evaluation of Biological Control

Given that new saskatoon plantings are regularly watered and *E. americanum* does not infest the plant until July, when soil temperatures are typically above 15°C on the prairies, *B. bassiana* is likely to perform well for managing *E. americanum*.

In 1998, aphid infestation levels at the

low and middle application rates of both nematode species were higher than on control plants. This may be explained, in part, if the rates were high enough to reduce the number of predators but not high enough to reduce *E. americanum* numbers. Neither nematode species was able to reduce aphid levels substantially at the rates tested when the plants were under severe aphid pressure.

Until nematodes, fungi or an alternative biological control agent can be proven to be as effective and of a comparable cost to current chemical applications, there will be little adoption of biological control of *E. americanum* by the saskatoon berry industry. However, organic producers can use nematodes or fungi to obtain limited control.

Recommendations

Further work should include:

1. Better understanding the interaction among the nematodes, predators and aphids;
2. Determining the optimal rate and timing of nematode and fungal applications.

Acknowledgements

H. bacteriophora was obtained from Natural Insect Control, Stevensville, Ontario, and *S. carpocapsae*, *S. feltiae* and *S. riobrave* were supplied by Biosys Inc., Palo Alto, California. Funding was received from Alberta Agriculture, Food and Rural Development (AAFRD), Alberta Agricultural Research Institute, Alberta Research Council, and the Fruit Grower's Society of Alberta. The author would like to acknowledge the contributions of Bruce Neill, Agriculture and Agri-Food Canada, and Kris Pruski, AAFRD.

References

Brown, M.W. (1986) Observations of woolly apple aphid, *Eriosoma lanigerum* (Hausmann) (Homoptera: Aphididae), root infestation in Eastern Virginia. *Melsheimer Entomological Series* 36, 5–8.

Brown, M.W., Jaeger, J.J., Pye, A.E. and Schmitt, J.J. (1992) Control of edaphic populations of woolly apple aphid using entomopathogenic nematodes and a systemic aphicide. *Journal of Entomological Science* 27, 224–232.

Fry, K.M. and Pruski, K. (2001) *Management of Woolly Elm Aphid using Entomopathogenic Nematodes.* Project Number 98M251, Final Report, Alberta Agricultural Research Institute Matching Grants Program.

Fry, K.M., Dosdall, L.M. and Steiner, M.Y. (1998) *Management of Woolly Elm Aphid in Saskatoons.* Project Number 940485 Final Report, Alberta Agricultural Research Institute, Farming For the Future.

Klein, K.K. (2000) Management of woolly elm aphids in saskatoons: economic impacts on Alberta's agri-food industry from this Alberta Research Council Project. Unpublished Report to the Alberta Research Council, Vegreville, Alberta.

Lewis, E.E., Gaugler, R. and Harrison, R. (1992) Entomopathogenic nematode host finding: Response to host cues by cruise and ambush foragers. *Parasitology* 105, 309–315.

Miranpuri, G.S. and Khachatourians, G.G. (1996) Bionomics and fungal control of woolly elm aphid, *Eriosoma americanum* (Riley) (Eriosomatidae: Homoptera) on saskatoon berry, *Amelanchier alnifolia*. *Journal of Insect Science* 9, 33–37.

Mueller, T.F., Blommers, L.H.M. and Mols, P.J.M. (1992) Woolly apple aphid (*Eriosoma lanigerum* Hausm., Hom., Aphidae) parasitism by *Aphelinus mali* Hal. (Hym., Aphelinidae) in relation to host stage and host colony size, shape and location. *Journal of Applied Entomology* 114, 143–154.

Neill, G.B., Reynard, D.A. McPherson, D.A. and Harris, J.L. (1994) *Woolly Elm Aphid Investigations – 1994.* Supplementary Report 94–1. Prairie Farm Rehabilitation Authority Shelterbelt Centre, Indian Head, Saskatchewan.

Neill, G.B., Reynard, D.A., McPherson, D.A. and Harris, J.L. (1995) *Woolly Elm Aphid Investigations – 1995.* Supplementary Report 95–3, Saskatchewan Agriculture and Food, Regina, Saskatchewan.

Patch, E. (1915) Woolly aphid of elm and juneberry. *Bulletin of the Maine Agricultural Experiment Station* 241, 197–204.
Peters, A., Huneke, K. and Ehlers, R.-U. (1996) Host finding by the entomopathogenic nematode *Steinernema feltiae*. *International Organization for Biological and Integrated Control of Noxious Animals and Plants/West Palaearctic Regional Section, Bulletin* 19, 99–102.

25 *Fenusa pusilla* (Lepeletier), Birch Leafminer, and *Profenusa thomsoni* (Konow), Ambermarked Birch Leafminer (Hymenoptera: Tenthredinidae)

D.W. Langor, S.C. Digweed and J.R. Spence

Pest Status

The birch leafminer, *Fenusa pusilla* (Lepeletier), and the ambermarked birch leafminer, *Profenusa thomsoni* (Konow), were inadvertently introduced from Europe into eastern North America in the early 1900s (Britton, 1924; Martin, 1960). These sawflies are common and often-abundant defoliators of native and exotic birches, *Betula* spp., and are now found from Newfoundland to British Columbia and north to the southern region of the Northwest Territories. Dispersal of these sawflies has probably been aided by shipment of infested ornamental birches. Populations of *F. pusilla* and *P. thomsoni* may remain high for many years in areas where no effective natural controls exist. Unsightly discoloration of foliage on ornamental birches caused by larval feeding in mines greatly reduces the tree's aesthetic value. Furthermore, repeated defoliation by these sawflies may weaken trees and predispose them to attack by other insects and fungi, resulting in die-back.

Typically, three overlapping generations of *F. pusilla* and one generation of *P. thomsoni* occur per year in most of Canada. The biology of *F. pusilla* has been studied in Alberta (Digweed, 1995; Digweed *et al.*, 1997), Quebec (Cheng and LeRoux, 1965, 1966, 1969, 1970) and Newfoundland (Jones and Raske, 1976); that of *P. thomsoni* has been studied in Alberta (Digweed, 1995; Digweed *et al.*, 1997) and northern Ontario (Martin, 1960).

Background

Prior to introduction of biological control agents, insecticide application was the most common method of control used in urban centres. Correct application of dimethoate by soil drench or by applying directly to the trunk of smaller trees (2.5–15.0 cm diameter) in late spring could control birch leafminers for a season. However, incorrect application was common, leading to incomplete control and damage to trees.

In Europe, *F. pusilla* has a parasitoid

complex of 17 species, causing as much as 38–47% parasitism of larvae (Eichorn and Pschorn-Walcher, 1973). The two most prevalent and specific Palaearctic parasitoids are *Lathrolestes nigricollis* (Thomson) and *Grypocentrus albipes* Ruthe. Detailed information on the biology of these species is available for Europe (Eichorn and Pschorn-Walcher, 1973) and Quebec (Quednau and Guèvremont, 1975; Guèvremont and Quednau, 1977a) and was summarized by Quednau (1984). Both species were introduced from Europe into Newfoundland (Raske and Jones, 1975) and Quebec (Guèvremont and Quednau, 1977b); *L. nigricollis* established in both provinces but *G. albipes* established only near Quebec City (Quednau, 1984). By the mid-1990s, neither parasitoid had spread to western Canada (S.C. Digweed, unpublished). Furthermore, a study in Alberta revealed that populations of *F. pusilla* suffered little mortality from native parasitoids (Digweed, 1998), suggesting that introduction of host-specific Palaearctic parasitoids may help achieve effective suppression of *F. pusilla* populations in western Canada.

P. thomsoni is rare in Europe, and only chalcidoid generalist parasitoids have been reared from it there, in small numbers (Eichorn and Pschorn-Walcher, 1973; Pschorn-Walcher and Altenhofer, 1989; Schönrogge and Altenhofer, 1992). In Alberta, where *P. thomsoni* has been abundant (Digweed *et al.*, 1997), similar chalcidoid generalists were occasionally found, but the most abundant parasitoid detected was the Holarctic ichneumonid *Lathrolestes luteolator* (Gravenhorst) (Digweed, 1998). The latter interaction was unexpected, as no association between these two species had been detected in Europe.

Biological Control Agents

Parasitoids

Since 1980, introductions of *L. nigricollis* and *G. albipes* have been made in western Canada following their successful establishment in eastern Canada. From 1994 to 1996, 348 *G. albipes* and 1167 *L. nigricollis*, collected from the Waldviertel region of Austria, were released at three locations in Edmonton, Alberta (Table 25.1).

The 1994 release site, Sunstar Nurseries (113°18′W 53°40′N), was monitored from 1995 to 1999 and *L. nigricollis* was seen to increase rapidly in abundance. No *G. albipes* were recovered from this site. The 1995 release site on the University of Alberta campus (113°32′W 53°32′N) was monitored from 1996 to 1998; *L. nigricollis* was recovered in 1996 and 1997 but not 1998, while

Table 25.1. Numbers of *Lathrolestes nigricollis* and *Grypocentrus albipes* shipped from Austria to Canada, emerged in quarantine in Ottawa, shipped to Edmonton as adults, and released from 1994 to 1996.

Year of release	Species	No. cocoons shipped to Canada	No. (%) adults emerged	No. adults shipped to Edmonton	No. adults (% females) released
1994	*Grypocentrus albipes* Ruthe	192	66 (34.4)	43	26 (57.7)
	Lathrolestes nigricollis (Thompson)	1490	158 (10.6)	139	103 (52.4)
1995	*G. albipes* (summer shipment)[a]	493	285 (57.8)	285	279 (51.6)
	L. nigricollis (spring shipment)[a]	2200	662 (30.1)	585	554 (50.0)
	L. nigricollis (summer shipment)[a]	972	195 (20.1)	106	98 (40.8)
1996	*G. albipes* (stock)[b]		66 (13.4)	43	43 (53.5)
	L. nigricollis (1995 summer stock)[b]		26 (2.7)		
	L. nigricollis	1435	471 (32.8)	431[c]	412 (49.8)

[a] Collected in Austria from first *F. pusilla* generation in 1995 and shipped on 1 August 1995.
[b] Individuals that experienced a prolonged diapause.
[c] Includes a small number of adults from 1995 stock.

G. albipes was recovered in all years. The 1996 release site, Howard Road (113°43′W 53°27′N) was flooded extensively in the spring and early summer of 1997, which may explain why the release there apparently failed and no parasitoids were recovered. Langor *et al.* (2000) provided additional details on releases and recoveries.

To date, only correlative *a posteriori* assessments have been made of the interaction between *L. luteolator* and *P. thomsoni*. Comparison of emergence trap catches in Edmonton from 1992 and 1995 showed a dramatic decrease in *P. thomsoni* captures (mean ± SE number of individuals per tree sampled per year: 1992, 47.7 ± 19.8 [n = 6]; 1995, 0.8 ± 0.8 [n = 8]) relative to those of *L. luteolator* (1992, 6.3 ± 4.0 [n = 6]; 1995, 1.3 ± 0.6 [n = 8]). Sticky-trap samples from 1994 and 1995 also reflect a high abundance of *L. luteolator* (1994, 598.8 ± 152.8 [n = 5]; 1995, 62.4 ± 24.5 [n = 8]) relative to that of *P. thomsoni* (1994, 139.8 ± 66.0 [n = 5]; 1995, 29.1 ± 4.6 [n = 8]).

Evaluation of Biological Control

In Edmonton, there is no doubt that *L. nigricollis* is well established, as two of the three releases were successful and individuals were recovered up to 5 years after release. A survey of parasitoid presence and abundance at 18 sites in and around Edmonton in 1999 revealed that *L. nigricollis* is spreading rapidly throughout the city and into the adjacent county of Strathcona, as individuals were recovered, often in relatively high abundance, at 13 of the sites, one of which was over 13 km from the nearest release site (Langor *et al.*, 2000). Thus, the failure of parasitoids to establish at the Howard Road site is inconsequential and is most likely explained by excessive and prolonged flooding of the release site.

Assessments of parasitism by *L. nigricollis* on first and second *F. pusilla* generations at Sunstar Nurseries in 1999 revealed that the percentage parasitism was 78% and 84%, respectively, and that 48% of eggs were encapsulated (Langor *et al.*, 2000). Thus, it is assumed that *L. nigricollis* is having a substantial impact on *F. pusilla* at that site. Given that Sunstar Nurseries is a commercial nursery that ships trees throughout Alberta and western Canada, it is likely that this has aided the spread of *L. nigricollis*, as parasitized cocoons are probably present in soil around the root mass.

G. albipes established at the university site and was recovered at, or within, a 400–500 m radius of the site for 3 years after release. However, the relatively small numbers of individuals captured may indicate that the population of this species, and hence its impact, is low, and that its establishment is tenuous. This species appears to be dispersing relatively slowly as no individuals were trapped during the city-wide survey in 1999.

Since the dramatic decreases in *P. thomsoni* during 1993 and 1994, larvae have been rare and difficult to find, with birch trees in Edmonton noticeably greener in late summer (S.C. Digweed, unpublished). This decrease in *P. thomsoni* is believed to be largely responsible for a 60–70% decrease in use of dimethoate in the Edmonton area from 1993 to 1998 (C. Saunders, Edmonton, 2000, personal communication). In addition, parasitism of *P. thomsoni* by *L. luteolator* may be widespread because these species co-occur at several sites across Alberta, as well as at Sudbury and Temagemi, northern Ontario (S.C. Digweed, unpublished).

Recommendations

Further work should include:

1. A wider survey in Alberta and neighboring provinces and territories to ascertain the ranges of *L. nigricollis* and *L. luteolator*, to determine whether there are climatic or other environmental barriers affecting these ranges, and to document the rate of spread of *L. nigricollis*;
2. Determining percentage parasitism of *F. pusilla* caused by *L. nigricollis*, and of *P. thomsoni* caused by *L. luteolator*, at a variety of sites;

3. Examining the effects of parasitism by *L. nigricollis* and *L. luteolator* on host abundance in the context of other mortality agents, to determine parasitoid impact and the success of the biological control programme;
4. Determining the fate of *G. albipes*;
5. More precise measurement of the reduced dimethoate use in urban areas and publicizing that this is attributable to biological control of *F. pusilla* and *P. thomsoni* populations.

Acknowledgements

We are thankful for the contributions of several collaborators: K. Carl, E. Altenhofer and M. Kenis (International Institute for Biological Control, Switzerland); M. Sarazin and J. Barron (Agriculture and Agri-Food Canada, Ottawa); D. Williams (Canadian Forest Service, Edmonton); and C. Saunders (Edmonton River Valley, Forestry, and Environmental Services). This study was funded by the Canadian Forest Service.

References

Britton, W.E. (1924) A European leafminer of birch. *Journal of Economic Entomology* 17, 601.

Cheng, H.H. and LeRoux, E.J. (1965) Life history and habits of the birch leaf miner, *Fenusa pusilla* (Lepeletier) (Hymenoptera: Tenthredinidae), on blue birch, *Betula caerulea grandis* Blanchard, Morgan Arboretum, Quebec, 1964. *Annals of the Entomological Society of Quebec* 1965, 173–188.

Cheng, H.H. and LeRoux, E.J. (1966) Preliminary life tables and notes on mortality factors of the birch leaf miner, *Fenusa pusilla* (Lepeletier) (Hymenoptera: Tenthredinidae), on blue birch, *Betula caerulea grandis* Blanchard, in Quebec. *Annals of the Entomological Society of Quebec* 1966, 81–104.

Cheng, H.H. and LeRoux, E.J. (1969) Parasites and predators of the birch leaf miner, *Fenusa pusilla* (Hymenoptera: Tenthredinidae), in Québec. *The Canadian Entomologist* 101, 839–846.

Cheng, H.H. and LeRoux, E.J. (1970) Major factors in survival of the immature stages of *Fenusa pusilla* in southwestern Quebec. *The Canadian Entomologist* 102, 995–1002.

Digweed, S.C. (1995) Effects of natural enemies, competition, and host plant quality on introduced birch leafminers (Hymenoptera: Tenthredinidae). MSc thesis, University of Alberta, Edmonton, Alberta, Canada.

Digweed, S.C. (1998) Mortality of birch leafmining sawflies (Hymenoptera: Tenthredinidae): impacts of natural enemies on introduced pests. *Environmental Entomology* 27, 1357–1367.

Digweed, S.C., Spence, J.R., and Langor, D.W. (1997) Exotic birch-leafmining sawflies (Hymenoptera: Tenthredinidae) in Alberta: distributions, seasonal activities, and the potential for competition. *The Canadian Entomologist* 129, 319–333.

Eichorn, O. and Pschorn-Walcher, H. (1973) The parasites of the birch leaf-mining sawfly (*Fenusa pusilla* [Lep.], Hymenoptera: Tenthredinidae) in central Europe. *Commonwealth Institute of Biological Control, Technical Bulletin* 16, 79–104.

Guèvremont, H.C. and Quednau, F.W. (1977a) Morphologie et biologie de *Grypocentrus albipes* (Hymenoptera: Ichneumonidae), parasite de la petite mineuse de bouleau, *Fenusa pusilla* (Hymenoptera: Tenthredinidae). *The Canadian Entomologist* 109, 1417–1424.

Guèvremont, H.C. and Quednau, F.W. (1977b) Introduction de parasites ichneumonides pour la lutte biologique contre *Fenusa pusilla* (Hymenoptera: Tenthredinidae) au Québec. *The Canadian Entomologist* 109, 1545–1548.

Jones, J.M. and Raske, A.G. (1976) Notes on the biology of the birch leafminer, *Fenusa pusilla* (Lep.), in Newfoundland (Hymenoptera: Tenthredinidae). *Phytoprotection* 57, 69–76.

Langor, D.W., Digweed, S.C., Williams, D.J.M., Spence, J.R. and Saunders, C. (2000) Establishment and spread of two introduced parasitoids (Ichneumonidae) of the birch leafminer, *Fenusa pusilla* (Lepeletier) (Tenthredinidae). *BioControl* 45, 415–423.

Martin, J. L. (1960) The bionomics of *Profenusa thomsoni* (Konow) (Hymenoptera: Tenthredinidae) a leaf-mining sawfly on *Betula* spp. *The Canadian Entomologist* 92, 376–384.

Pschorn-Walcher, H. and Altenhofer, E. (1989) The parasitoid community of leaf-mining sawflies (Fenusini and Heterarthrini): a comparative analysis. *Zoologischer Anzeiger* 222, 37–56.

Quednau, F.W (1984) *Fenusa pusilla* (Lepeletier), birch leaf-miner (Hymenoptera: Tenthredinidae). In: Kelleher, J.S. and Hulme, M.A. (eds) *Biological Control Programmes Against Insects and Weeds in Canada 1969–1980.* CAB International, Wallingford, UK, pp. 291–294.

Quednau, F.W. and Guèvremont, H. (1975) Observations on mating and oviposition of *Priopoda nigricollis* (Hymenoptera: Ichneumonidae), a parasite of the birch leaf-miner, *Fenusa pusilla* (Hymenoptera: Tenthredinidae). *The Canadian Entomologist* 107, 1199–1204.

Raske, A.G. and Jones, J.N. (1975) Introduction of parasitoids of the birch leaf-mining sawfly into Newfoundland. *Canadian Department of the Environment Bi-monthly, Research Notes* 31(2), 20–21.

Schönrogge, K. and Altenhofer, E. (1992) On the biology and larval parasitoids of the leaf-mining sawflies *Profenusa thomsoni* (Konow) and *P. pygmaea* (Konow) (Hymenoptera, Tenthredinidae). *Entomologist's Monthly Magazine* 128, 99–108.

26 *Forficula auricularia* L., European Earwig (Dermaptera: Forficulidae)

U. Kuhlmann, M.J. Sarazin and J.E. O'Hara

Pest Status

The European earwig, *Forficula auricularia* L., is native to Europe, western Asia and possibly North Africa (Clausen, 1978). In the early 1900s it was accidentally introduced into the Pacific coast of the USA, where it spread rapidly (Crumb *et al.*, 1941) and was involved in several outbreaks (Spencer, 1947). It was found on Rhode Island on the Atlantic coast in 1911 (Jones, 1917). In Canada, it was first recorded in 1916 from British Columbia (Treherne, 1923), in 1938 it was reported from Ontario (Smith, 1940), and was discovered on the east coast in the late 1940s. Although damage to vegetable and flower gardens was generally minor, when high population densities occur it is a major pest in gardens and a perpetual nuisance in households.

Adult *F. auricularia* overwinter in pairs in subterranean nests constructed in autumn. Females oviposit in the nest towards the end of winter. Soon after the eggs are laid females aggressively force males to leave the nest (Lamb, 1976). Hatching begins early in May (U. Kuhlmann, unpublished). After 1 week adult females and first-instar nymphs begin to leave the nests and appear on the soil surface. After five moults during summer, nymphal instars become adults, and the overwintered females die after rearing their brood (Lamb and Wellington, 1975).

Background

In North America, an integrated approach to control *F. auricularia* around houses involves the physical removal of insects by vacuuming, harbourage removal, perimeter spraying and baiting, modification of exterior lighting, pest proofing and trapping (Cooper, 1997). Preformulated cockroach baits were tested against *F. auricularia* in the form of bait stations, with and without bait, bark, water, leaves and/or cat food, but only produced useful levels of mortality after 3–10 weeks (Snell and Robinson, 1989). Lamb and Wellington (1975) reported that bait traps are frequently used by urban dwellers to control *F. auricularia* but almost all specimens found in traps in early spring are adult males; therefore this control is ineffective in reducing the spring population. Chemical sprays are ineffective against *F. auricularia* because of its widespread occurrence and great mobility (Santini and Caroli, 1992). Therefore, classical biological control of *F. auricularia* in Canada might offer an alternative solution by establishing natural enemies from Europe to reduce *F. auricularia* populations.

Two species of Tachinidae are important parasitoids of *F. auricularia* in central Europe. The most abundant, *Triarthria setipennis* (Fallén) (previously placed in *Digonochaeta* or *Bigonicheta*), is ovolarviparous and produces relatively few eggs, from which larvae emerge immediately after oviposition. The less abundant *Ocytata pallipes* (Fallén) (previously placed in *Rhacodineura*) is microoviparous and produces a large number of microtype eggs that are deposited on host food plants. After ingestion by the host, the eggs hatch in the gut and first-instar larvae penetrate the haemocoel.

During the 1930s, *O. pallipes* adults were released but only established temporarily in Oregon (Mote, 1931; Clausen, 1978). In 1924, *T. setipennis* from the Mediterranean region was introduced into Oregon, where it became established (Spencer, 1945). Since then, it has been reintroduced numerous times and is established in Oregon, Washington, California, Idaho, Utah, New Hampshire and Massachusetts (O'Hara, 1996). It was released in Connecticut and Rhode Island but has not been recovered there (O'Hara, 1994). No studies on the spread of *T. setipennis* in North America exist and consequently its current distribution is poorly known.

In Canada, releases of *T. setipennis* populations from Oregon were made in British Columbia (1934–1939), Ontario (1930–1941) and Newfoundland (1951–1953) (McLeod, 1962). It established in south-western British Columbia and Newfoundland but did not reach high population densities (Mote, 1931; Dimick and Mote, 1934; Spencer, 1947). It was assumed that this was partly due to poor adaptation of the parasitoid to local climatic conditions. Additional releases of *T. setipennis* collected from Switzerland, Germany and Sweden were made in the 1960s under the assumption that the climate at the sites of parasitoid collection in these countries is more similar to the climates at Canadian release sites. New introductions into Newfoundland were followed by an average increase in parasitism from 0.3% in 1955 to 2.1% in 1965, 12% in 1975 and 13.1% in 1985, and was coincident with reduced earwig numbers (Morris, 1971, 1984; Morry *et al.*, 1988).

In Nova Scotia, where earwigs are economically important because they infest cracks and crevices in leafy vegetables and fruit, no parasitoids have been reared. Therefore, in 1989 new surveys for natural enemies of *F. auricularia* in Switzerland, Austria, eastern France and northern Germany were carried out to clarify further the biology of *T. setipennis* and *O. pallipes* and identify new, effective biotypes for release. In addition, studies were initiated to develop methods to improve production of parasitoids for inoculative release.

Biological Control Agents

Parasitoids

Kuhlmann (1994, 1995) studied the biology of *T. setipennis*[1] and *O. pallipes* popula-

[1]Van Emden (1954) recognized two species and separated them on the basis of colour differences (*T. setipennis* being the dark form and *T. spinipennis* (Meigen) the light form). Mesnil (1973) also recognized two species, distinguished on the basis of structural differences. Herting (1984) synonymized the names under *T. setipennis*, indicating that there are intermediates and that male genitalia do not indicate a specific difference. Belshaw (1993) and Tschorsnig and Herting (1994) also placed *spinipennis* in synonymy with *setipennis*. Therefore, Kuhlmann (1995) concluded that cross-breeding experiments and electophoresis studies are needed to clarify whether there is one species or two.

tions in Germany, Austria, France and Switzerland. *T. setipennis* females laid, on average, 235 eggs. The oviposition period averaged 4–5 days. Larval development varied from 2 to 8 weeks during June and July. Overwintering occurred as puparia. In Germany and north-western Switzerland, one full and a partial second generation occurred per year. Emergence of the spring generation of *T. setipennis* in southern Austria was long and distinctly bimodal, with colour dimorphism between the first and second peak. This observation and cross-mating experiments cast doubt on a previous conclusion that these colour differences are seasonal dimorphisms, and lends support to the existence of two species.

The less abundant *O. pallipes* has a high reproductive potential with an average of 1040 eggs, although daily numbers and total number of eggs laid by females varied widely (Kuhlmann, 1994). The eggs must be eaten by the host to develop, and they hatch in the intestinal tract, where first-instar larvae subsequently penetrate the haemocoel. First-generation larvae take, on average, 300 days to develop, emerging the following spring. Larval development of the partial second generation takes, on average, 43 days. The parasitoid is partially bivoltine in all areas studied. In northern Germany, puparia of the first generation were formed between the end of May and the beginning of July, lasted on average 22 days, and adults emerged between the end of June and mid-July. For the partial second generation, puparia were formed from mid-August to the beginning of October, lasted on average 27 days, and adults emerged from September to the end of October. Second-instar larvae overwinter in the host and third instar larvae emerge in spring. Puparia of the first generation are formed from the end of May to early July. Thus, it is clear that most parasitized males of *F. auricularia* die before the parasitoid larvae have finished their development in the host.

Kuhlmann (1993) developed methods to parasitize hosts experimentally for studies on larval development and competition in super- and multiparasitized hosts, but also with the intention of improving laboratory production of both parasitoids for inoculative release into Canada. More than 84% parasitism was achieved by inoculating *F. auricularia* with larvae of *T. setipennis* when earwigs were immobilized. Lower rates of parasitism were obtained when earwigs were mobile and could fend off attacking parasitoid larvae before they could penetrate the host. For *O. pallipes*, incubation of starved earwigs with food items carrying 3–5 parasitoid eggs resulted in 60% parasitism. A problem still exists with successful hibernation of parasitized earwigs; if this cannot be overcome, then the summer generation of tachinids must be used for shipment to and release in Canada.

Releases and Recoveries

Five attempts were made in the 1980s and 1990s to establish *T. setipennis* in the Ottawa area. A total of 53 *T. setipennis* adults was released in 1986 (Sarazin, 1988a), 37 adults in 1987 (Sarazin, 1988b), 59 adults in 1988 (Sarazin, 1989) and 12 adults in 1991 (Sarazin, 1992). Each of those releases involved unmated adults of *T. setipennis*, as attempts to elicit mating among caged individuals in the laboratory were unsuccessful (O'Hara, 1994). In spring, 1992, an additional 1100 puparia of *T. setipennis* were sent from Europe for possible release. These were originally destined for Kentville, Nova Scotia, but were sent to Ottawa when the earwig biological control programme at the Kentville Research Centre was discontinued. A group comprising J. O'Hara, A. Schmidt (Agriculture and Agri-Food Canada), and B. Gill and D. Parker (Canadian Food Inspection Agency) released these *T. setipennis ad hoc* in Ottawa. The puparia and some recently emerged adults arrived in mid-June and were placed in cages (40 × 40 × 40 cm) for adult emergence and maintenance. Most of the puparia were dead on arrival or died later, but several hundred adults emerged over a 1-week period. Attempts were made to elicit mating indoors by placing cages in different lighting conditions (including near windows) and under varied humidity regimes, and by confining pairs of males and females in small vials, but no mating was observed. However, matings took place readily when cages were placed outdoors (O'Hara, 1994; see also Kuhlmann, 1995). Larval and adult earwigs were collected and artificially parasitized using the technique of Kuhlmann (1993). Maggots were applied singly to about 115 earwigs, and the earwigs and 20 adult flies were later released at one site in Ottawa. Although this *ad hoc* release programme was useful as a pilot study, the release of such a low number of potentially infected *F. auricularia* makes success of this introduction unlikely.

Evaluation of Biological Control Attempts

T. setipennis has established successfully in Newfoundland and British Columbia (Morris, 1984). Studies on the establishment of *T. setipennis* in Newfoundland indicated a considerable reduction in earwig numbers at St John's, which was most probably due to high levels of parasitism in the mid-1970s (Morris, 1984). Since 1978, no further evaluation of parasitoid impact has been undertaken.

Recommendations

Future work should include:

1. Assessing rates of parasitism by *T. setipennis* in Newfoundland and British Columbia;
2. Determining the impact of *T. setipennis* on *F. auricularia* populations in Newfoundland and British Columbia and clarifying the significance of the introduced European parasitoid;
3. Making introductions into Nova Scotia and Ontario;
4. Evaluating hibernation techniques for *F. auricularia* and its parasitoid *O. pallipes* to improve laboratory production of *O. pallipes* for inoculative releases;
5. Introducing *O. pallipes* as a second biological control agent potentially suitable for establishment in the Maritimes;
6. Clarifying the possible existence of two species, *T. setipennis* and *T. spinipennis*, based on the concept of van Emden (1954).

References

Belshaw, R. (1993) Tachinid flies. Diptera: Tachinidae. In: *Handbooks for the Identification of British Insects*, Vol. 10, Part 4a(I), Royal Entomological Society of London, London.

Clausen, C.P. (1978) Dermaptera – Forficulidae – European Earwig. In: Clausen, C.P. (ed.) *Introduced Parasites and Predators of Arthropod Pests and Weeds: A World Review*, Handbook No. 480, United States Department of Agriculture, Washington, DC, pp. 15–18.

Cooper, R. (1997) Handbook of Pest Control book excerpt. Earwigs An IPM approach to controlling this common household pest. *Pest Control Technology* 25, 1, 45, 48, 50.

Crumb, S.E., Eide, P.M. and Bonn, A.E. (1941) The European earwig. *United States Department of Agriculture, Technical Bulletin* 766, 1–76.

Dimick, R.E. and Mote, D.C. (1934) Progress report regarding the introduction in Oregon of *Digonocheata setipennis*, a tachinid parasite of the European earwig. *Journal of Economic Entomology* 27, 863–865.

Emden, F.I. van (1954) Diptera Cyclorrhapha. Calyptrata (I). Section (a). Tachinidae and Calliphoridae. In: *Handbooks for the Identification of British Insects*, Vol. 10, Part 4a. Royal Entomological Society of London, London.

Herting, B. (1984) Catalogue of Palaearctic Tachinidae (Diptera). *Stuttgarter Beiträge zur Naturkunde* (A) 369, 228pp.

Jones, D.W. (1917) The European earwig and its control. *United States Department of Agriculture Bulletin* 566, 1–12.

Kuhlmann, U. (1993) Techniques for rearing tachinid parasitoids of the European earwig *Forficula auricularia*. *Biocontrol Science and Technology* 3, 475–480.

Kuhlmann, U. (1994) *Ocytata pallipes* (Fallén) (Dipt., Tachinidae), a potential agent for the biological control of the European earwig. *Journal of Applied Entomology* 117, 262–267.

Kuhlmann, U. (1995) Biology of *Triarthria setipennis* (Fallén) (Diptera: Tachinidae), a native parasitoid of the European earwig, *Forficula auricularia* L. (Dermaptera: Forficulidae), in Europe. *The Canadian Entomologist* 127, 507–517.

Lamb, R.J. (1976) Parental behaviour in the Dermaptera with special reference to *Forficula auricularia* (Dermaptera: Forficulidae). *The Canadian Entomologist* 108, 609–619.

Lamb, R.J. and Wellington, W.G. (1975) Life history and population characteristics of the European earwig (*Forficula auricularia*) (Dermaptera: Forficulidae), at Vancouver, British Columbia. *The Canadian Entomologist* 107, 819–824.

McLeod, J.H. (1962) Part I. Biological control of pests of crops, fruit trees, ornamentals and weeds in Canada up to 1959. In: McLeod, J.H., McGugan, B.M. and Coppel, H.C. (eds) *A Review of the Biological Control Attempts Against Insects and Weeds in Canada.* Technical Communication No. 2, Commonwealth Institute of Biological Control, Trinidad. Commonwealth Agricultural Bureaux, Farnham Royal, UK, pp. 1–33.

Mesnil, L.P. (1973) Larvaevorinae (Tachininae). In: Lindner, E. (ed.) *Die Fliegen der palaearktischen Region 8*. E. Schweizerbart'sche Verlagsbuchhandlung, Stuttgart, Germany, pp. 1169–1232.

Morris, R.F. (1971) *Forficula auricularia*, European earwig (Dermaptera: Forficulidae). In: *Biological Control Programmes against Insects and Weeds in Canada 1959–1968.* Technical Communication No. 4, Commonwealth Institute of Biological Control Trinidad, Commonwealth Agricultural Bureaux, Farnham Royal, UK, pp. 18–20.

Morris, R.F. (1984) *Forficula auricularia*, European earwig (Dermaptera: Forficulidae). In: Kelleher, J.S. and Hulme, M.A. (eds) *Biological Control Programmes Against Insects and Weeds in Canada 1969–1980.* Commonwealth Agriculture Bureaux, Slough, UK, pp. 39–40.

Morry, H.G., Morris, R.F. and Proudfoot, K.G. (1988) An introduced parasite to control European earwig in Newfoundland. *Biocontrol News* 1, 32.

Mote, D.C. (1931) The introductions of the tachinid parasites of the European earwig in Oregon. *Journal of Economic Entomology* 24, 948–956.

O'Hara, J.E. (1994) Release of *Triarthria setipennis* in Ottawa and notes about the New World distribution of the genus. *The Tachinid Times* 7, 1–2.

O'Hara, J.E. (1996) Earwig parasitoids of the genus *Triarthria* Stephens (Diptera: Tachinidae) in the New World. *The Canadian Entomologist* 128, 15–26.

Santini, L. and Caroli, L. (1992) Damage to fruit crops by European earwig (*Forficula auricularia* L.). *Informatore-Fitopatologico* 42, 35–38.

Sarazin, M.J. (1988a) *Insect Liberations in Canada. Parasites and Predators 1986.* Agriculture Canada, Research Branch, Ottawa, Ontario.

Sarazin, M.J. (1988b) *Insect Liberations in Canada. Parasites and Predators 1987.* Agriculture Canada, Research Branch, Ottawa, Ontario.

Sarazin, M.J. (1989) *Insect Liberations in Canada. Parasites and Predators 1988.* Agriculture Canada, Research Branch, Ottawa, Ontario.

Sarazin, M.J. (1992) *Insect Liberations in Canada. For Classical Biological Control Purposes 1991.* Agriculture Canada, Research Branch, Ottawa, Ontario.

Smith, C.W. (1940) Successful hibernation of the earwig parasite, *Bigonicheta setipennis* Fall., in Ontario. *Report of the Entomological Society of Ontario* 71, 29–32.

Snell, E.J. and Robinson, W. (1989) An update on the European earwig. *Pest Control Technology* 17, 50–52.

Spencer, G.J. (1945) On the incidence, density and decline of certain insects in British Columbia. *Proceedings of the Entomological Society of British Columbia* 42, 19–23.

Spencer, G.J. (1947) The 1945 status of *Digonochaeta setipennis*, tachinid parasite of the European earwig in Vancouver. *Proceedings of the Entomological Society of British Colombia* 43, 8–9.

Treherne, R.C. (1923) The European earwig in British Columbia. *Proceedings of the Entomological Society of British Columbia, Economic Series* 17 and 19, 161–163.

Tschorsnig, H.P. and Herting, B. (1994) Die Raupenfliegen (Diptera: Tachinidae) Mitteleuropas: Bestimmungstabellen und Angaben zur Verbreitung und Oekologie der einzelnen Arten. *Stuttgarter Beiträge zur Naturkunde* (A) 506, 170pp.

27 *Haematobia irritans* (L.), Horn Fly (Diptera: Muscidae)

T.J. Lysyk and K.D. Floate

Pest Status

The horn fly, *Haematobia irritans* (L.), was introduced into North America from Europe in the late 1800s and became established in Canada by about 1900. It occurs from Canada to Argentina in the western hemisphere and from Europe to North Africa in the eastern hemisphere. Adult *H. irritans* typically occur on cattle, but may infest horses and game elk. *H. irritans* generally attacks larger animals, such as yearlings and cows, and tends to avoid calves. Feeding on yearlings can result in a reduction in weight gain of up to 18% (Haufe, 1982). *H. irritans* feeding on cows results in reduced milk production, which indirectly reduces weight gain in calves. Every 100 flies per cow (season average) can reduce calf weaning weights by 8% (Steelman *et al.*, 1991); reduction in calf weaning weight from 3 to 16% has been reported.

H. irritans is a pest only in pastures or rangeland. Adults become active early in spring and spend most of their lives on the backs of cattle. Both sexes feed on cattle blood by piercing the skin. They take numerous small meals from the host, and a single female can ingest from 11 to 21 mg of blood per day. The blood meal is required for egg production. Females feed for 2–3 days before laying their first batch of eggs, and continue laying eggs every 1–2 days until death, laying 8–13 eggs per day on average. The maximum female life span is 21 days, so a single female could potentially lay 100–200 eggs during her life. However, the average life span of *H. irritans* is less than 1 week, so each newly emerged female will, on average, contribute about 20 eggs to the next generation (Lysyk, 1991).

Females leave the host to lay eggs on cattle manure within 0.5 h of its deposition. Larvae hatch in 1 day, develop within the pat, pass through three instars in 1–2 weeks while feeding on bacteria, and pupate. Pupae develop for 1–2 weeks before adults emerge and reinfest cattle. The entire life cycle requires 2–4 weeks. Three or four generations are completed in southern Alberta during summer. In autumn, declining temperatures cause pupae to enter diapause and they overwinter under cow pats. Adults emerge the following spring as the temperature increases. In northern climates, adults emerge in May and populations increase steadily throughout spring and summer, reaching a single peak in early August. Populations decline from August to October as increasing numbers of pupae enter diapause.

Background

H. irritans management depends mainly on insecticides applied to cattle to control adult flies. These are applied using direct applications (sprays and pour-ons), self-applicating devices (oilers, dust bags) or sustained release devices. Direct applications are usually used in small pastures in which the producer can easily gather animals prior to application. These can provide up 2 weeks of satisfactory control. Self-treatment devices are placed in pastures where the animals visit them. They generally work best when placed near water or mineral blocks that animals visit frequently but must be serviced frequently, to ensure that insecticide is present and that the devices are working properly. Levels of control are variable, depending on the frequency of visits by cattle. Sustained release devices consist of insecticides placed within a matrix for slow release during the grazing season. These are either formulated as boluses or insecticidal ear tags. Boluses must be inserted into the animal's stomach before the animals are pastured. The insecticide acts either on the blood-feeding adults or kills larvae in the manure. Insecticidal ear tags are most commonly used. These consist of insecticides embedded in a plastic matrix. The insecticides diffuse over the surface of the animal and kill adult flies. Ear tags can be applied shortly before animals are turned out to pasture and can provide season-long control of *H. irritans*. The advantages of ear tags include ease of application, long-term efficacy, small amount of insecticide used and only on specific targets, and reduced risk of insecticide exposure to applicators. The major disadvantage of ear tags is that resistant populations of *H. irritans* developed quickly due to widespread tagging using a single chemical family. Ear tags were available in the late 1970s, and resistance had become widespread throughout the USA by the mid-1980s (Kunz and Schmidt, 1985), and throughout Canada by 1991 (Colwell *et al.*, 1992; Mwangala and Galloway, 1993).

Challenges to control of *H. irritans* in pasture and rangeland systems reflect the large areas over which cattle graze. Because it is difficult to round up cattle to apply insecticides, the tendency has been to develop persistent formulations to extend periods of control, resulting in widespread development of resistance. Because adult *H. irritans*, the usual target for insecticides, are closely tied to the host, selection for resistance has been intense. Immature *H. irritans* are protected within the manure pats that are well dispersed throughout the pasture. As a result, they are generally not the target of direct control measures. Some pour-on compounds applied to cattle are excreted in the faeces and will provide short-term control of *H. irritans* (Lysyk and Colwell, 1996). However, these compounds can reduce numbers of predacious beetles and parasitic wasps (Floate, 1998) that kill immature *H. irritans*. Development of biological control against *H. irritans* is attractive due to the difficulties inherent in applying conventional control methods.

Biological Control Agents

Parasitoids

In Alberta, Depner (1968) observed parasitism levels of up to 40% in all vegetative zones. The most common parasitoids were *Spalangia*, *Muscidifurax* and *Phygadeuon* spp. Peck (1974) updated these names and Gibson (2000) provided keys to *Spalangia* and *Muscidifurax* spp. associated with cattle manure. In an experimental study in British Columbia, *Spalangia haematobiae* Ashmead and *M. raptor* Girault and Sanders parasitized 0–19% of pupae (MacQueen and Beirne, 1974). Recent studies focused on the effects of microclimate on parasitism of pupae as part of a larger study on *H. irritans* diapause (T.J. Lysyk, unpublished). Pupal parasitism by *Spalangia*, *Muscidifurax* and *Phygadeuon* spp. averaged 21–23% during late June and late July, and declined to 5% in late August. More than 60% parasitism was observed in individual pats. Parasitism was affected by microclimate and was lower in pupae collected from pats shaded by tall grass. Temperature played an important role in determining parasitoid activity.

Predators

Early studies on predation in Canada indicated that predators reduced *H. irritans* emergence by 60–83%. The introduced staphylinid *Philonthus cruentatus* Gmelin was the most abundant and likely the most significant predator of immature *H. irritans* (MacQueen and Beirne, 1975). In southern Alberta, predation of *H. irritans* eggs was assessed using 'predator exclusion' cages without mesh (control) or with mesh sizes of 0.5 mm, 1.5 mm, or 3.0 mm. Percentage predation of eggs was 75 ± 10, 6 ± 6, 32 ± 11, and 30 ± 12, respectively. Differences in recovery were attributed to removal of eggs from the seeded sites by other arthropods. Based on their abundance in the exclusion cages, predators were most likely staphylinid species; at least 14 species were recovered from cattle dung (Floate, 1998). This indicates that predators can be a major mortality factor for *H. irritans* eggs.

Pathogens

The effects of various serovars of *Bacillus thuringiensis* Berliner on *H. irritans* larval survival are being examined (T.J. Lysyk, L.B. Selinger and D.D.S. Baines, unpublished). To date, we have examined toxicity of 85 isolates representing 57 serovars. Most isolates had relatively little effect on larval survival; however, five isolates were effective and are being studied for commercialization. The problem of application still remains a major hurdle.

Manipulation of microbial populations may be an effective means for reducing survival of *H. irritans* larvae. T.J. Lysyk (unpublished) identified 21 species of bacteria in 15 genera that occur in pats and larvae. The relative abundance of each species varied between the pat environment and the *H. irritans* larval gut. Twenty species were present in the manure, but only seven of these were detected in the gut. One species was present in the gut but not in the manure. Overall bacterial populations averaged 4.3×10^8 cells g^{-1} of manure and 5.6×10^5 cells per larval gut. *H. irritans* larval survival was highest in unsterilized manure, and extremely low in sterilized manure. Larval survival varied in sterilized manure that was augmented with individual species, and tended to be higher on species that were more abundant in manure. Survival was highest when reared on *Pseudomonas* and related species. This suggests that microbial manipulation of the manure may be a useful tactic for reducing *H. irritans* immature survival.

Evaluation of Biological Control

Manipulation of parasitoid, predator and microbial populations are promising areas for control.

Recommendations

Further work should include:

1. Resolving hurdles impeding the use of natural enemies, including dung pats (which protect immature stages), the dispersion of dung over large areas, and the relative inaccessibility of adult and immature *H. irritans*;
2. Developing biological control agents that are effective and can contact the target stage, which may require co-development of cultural methods that can be used to concentrate *H. irritans* populations into definable areas to allow natural enemies to exert their influence;
3. Faunal and impact studies over a wider geographic range to identify potential biological control agents.

References

Colwell, D.D., Lysyk, T.J., Whiting, A. and Philip, H. (1992) Horn fly resistance widespread. *Lethbridge Research Centre Weekly Letter* No. 3056.

Depner, K.R. (1968) Hymenopterous parasites of the horn fly, *Haematobia irritans* (Diptera: Muscidae) in Alberta. *The Canadian Entomologist* 100, 1057–1060.

Floate, K.D. (1998) Off-target effects of ivermectin on insects and on dung degradation in southern Alberta, Canada. *Bulletin of Entomological Research* 88, 25–35.

Gibson, G.A.P. (2000) Illustrated key to the native and introduced chalcidoid parasitoids of filth flies in America north of Mexico (Hymenoptera: Chalcidoidea). http://res2.agr.ca/ecorc/apss/chalkey/chalkey.htm
Haufe, W.O. (1982) Growth of range cattle protected from horn flies (*Haematobia irritans*) by ear tags impregnated with fenvalerate. *Canadian Journal of Animal Science* 62, 567–573.
Kunz, S.E. and Schmidt, C.D. (1985) The pyrethroid resistance problem in the horn fly. *Journal of Agricultural Entomology* 2, 358–363.
Lysyk, T.J. (1991) Use of life history parameters to improve a rearing method for horn fly, *Haematobia irritans irritans* (L.) (Diptera: Muscidae) on bovine hosts. *The Canadian Entomologist* 123, 1199–1207.
Lysyk, T.J. and Colwell, D. (1996) Duration of efficacy of diazinon ear tags and ivermectin pour-on for control of horn fly (Diptera: Muscidae). *Journal of Economic Entomology* 89, 1513–1520.
MacQueen, A. and Beirne, B.P. (1974) Insects and mites associated with fresh cattle dung in the southern interior of British Columbia. *Journal of the Entomological Society of British Columbia* 71, 5–9.
MacQueen, A. and Beirne, B.P. (1975) Influence of other insects on production of horn fly, *Haematobia irritans* (Diptera: Muscidae), from cattle dung in south-central British Columbia. *The Canadian Entomologist* 107, 1255–1264.
Mwangala, F.S. and Galloway, T.D. (1993) Susceptibility of horn flies, *Haematobia irritans* (L.) (Diptera: Muscidae), to pyrethroids in Manitoba. *The Canadian Entomologist* 125, 47–53.
Peck, O. (1974) Chalcidoid (Hymenoptera) parasites of the horn fly, *Haematobia irritans* (Diptera: Muscidae), in Alberta and elsewhere in Canada. *Canadian Entomologist* 106, 473–477.
Steelman, C.D., Brown, A.H., Gbur, E.E. and Tolley, G. (1991) Interactive response of the horn fly (Diptera: Muscidae) and selected breeds of beef cattle. *Journal of Economic Entomology* 84, 1275–1282.

28 *Hoplocampa testudinea* (Klug), European Apple Sawfly (Hymenoptera: Tenthredinidae)

C. Vincent, D. Babendreier and U. Kuhlmann

Pest status

The European apple sawfly, *Hoplocampa testudinea* Klug, is a host-specific, primary pest of apple, *Malus pumila* Miller (= *M. domestica* Borkhausen). In North America, it was first reported from Long Island, New York, in 1939 (Pyenson, 1943), and subsequently invaded the New England states (Anonymous, 1959, 1969). In Canada, *H. testudinea* was first discovered on Vancouver Island in 1940 (Downes and Andison, 1942) but apparently never established in continental western North America. In 1979, *H. testudinea* was discovered in southern Quebec (Huntingdon County) (Paradis, 1980), where it gradually spread throughout apple-growing areas of the province (Vincent and Mailloux, 1988). Since 1995, *H. testudinea* has been found in the Ottawa valley, Ontario (H. Goulet, Ottawa, 1999, personal communication) and apparently has spread westward.

Eggs of *H. testudinea* are laid singly in flower receptacles. First-instar larvae may feed on the epidermis of young fruits, leaving a ribbon-like scar called primary damage by Miles (1932), which often stay on the tree until harvest. Second- and third-instar larvae regularly enter another fruit, leaving a deep hole plugged by frass that Miles (1932) called secondary damage. Such fruit has a typical, strong odour and usually falls to the ground in June. Larvae then complete their development in the ground under trees at a depth of 10–25 cm and overwinter as eonymphs inside their cocoons.

In Quebec, primary damage caused by *H. testudinea* at harvest ranged from 0 to 14% in commercial orchards and from 0 to 4.1% in an unsprayed orchard (Vincent and Mailloux, 1988). In contrast to such key pests as the tarnished plant bug, *Lygus lineolaris* Palisot de Beauvois, the plum curculio, *Conotrachelus nenuphar* Herbst, and the apple maggot, *Rhagoletis pomonella* Walsh, *H. testudinea* was a secondary problem for Quebec apple growers (Vincent and Roy, 1992). However, by 1988, it became a serious concern because damage exceeded 5% of fruit at harvest in several commercial fields (Vincent, 1988).

Background

Although *H. testudinea* can be controlled with insecticides (Vincent and Rancourt, 1988), it proved difficult to harmonize treatments with those against *C. nenuphar* at petal fall. Effective larval control is now achieved through synthetic insecticides sprayed as a larvicide applied as first-instar larvae hatch. Adult *H. testudinea* can be trapped by using white sticky traps (Owens and Prokopy, 1978).

Biological control of the exotic *H. testudinea* might be an alternative to Canadian apple growers as there is real potential to reduce input costs through reduced insecticide use. As no known natural enemies occur in Canada, a study was initiated to introduce European parasitoids of *H. testudinea* into Quebec.

Biological Control Agents

Pathogens

Nematodes

Vincent and Bélair (1992) evaluated nematodes applied to the soil to control sawfly populations at Frelighsburg, Quebec. In laboratory assays, *Steinernema carpocapsae* (Weiser) strains DD 136 and All, *S. feltiae* (Filipjev) and *Heterorhabtidis bacteriophora* Poinar all caused 100% mortality 72 h after treatment. In field conditions, under dwarf apple trees, a single application of 40 or 80 *S. carpocapsae* All strain m^{-2} caused significant (>80%) larval mortality. Plots treated with nematodes had significantly less adult sawfly emergence the following year.

Bélair *et al.* (1998) also studied *H. testudinea* population control using nematodes as foliar applications. In some years, foliar applications of *S. carpocapsae* All strain reduced *H. testudinea* damage by 98–100%, but in other years these applications were ineffective.

Parasitoids

The European ichneumonid *Lathrolestes ensator* Brauns (Cakstynja, 1968; Jaworska, 1987; Zijp and Blommers, 1993; Babendreier, 1998; U. Kuhlmann, unpublished) was studied as a potential biological control agent. It is a univoltine, solitary, larval endoparasitoid specific to *H. testudinea* (Cross *et al.*, 1999) that is well synchronized with its host. Boevé *et al.* (1996) found that the spectrum of volatiles emitted by infested apples differs from that emitted by uninfested ones and hypothesized that female *L. ensator* may use chemical cues such as terpenoids during host location. Babendreier (1996) found that parasitoid females are able to detect infested fruitlets while hovering above trusses. When an infested fruitlet has been located, the female searches for a few seconds and probes her ovipositor into the apple near a sawfly mine, locates a suitable host, and oviposits. Only late first- and second-instar larvae can be parasitized successfully, even when there is a preponderance of later-instar larvae (Babendreier, 1996, 1998). Consequently, the parasitoid's searching period is limited to about 2 weeks (Babendreier, 1998), a possible limiting factor especially if environmental conditions are unfavourable during this narrow oviposition window.

The European cocoon ectoparasitoid, *Aptesis nigrocincta* Gravenhorst, obtained from *H. testudinea* cocoons exposed in an apple orchard, was also studied in Switzerland. Babendreier (1998, 1999) found that females often survive more than 6 weeks, thus being able to parasitize *H. testudinea* for a much longer period than *L. ensator*. The first generation emerges in June to coincide with the presence of *H. testudinea* cocoons. The nearly wingless females mate with winged males and then oviposit into the host cocoon. Eggs are laid singly on the surface of the eonymph and larvae of this idiobiont parasitoid hatch within 3–4 days before feeding externally on the host. A second generation occurs in close succession, with most individuals emerging during August. Oviposition by this generation probably continues until October. The phenology of *A. nigrocincta* is staggered, with fractions of the population emerging immediately, later the same year, and in the following year at different times (Babendreier, 1999). It was concluded that *A. nigrocincta* has one complete and one partial generation per year. Its main limitations as a biological control agent are that the number of mature eggs produced are small and females have to search for host cocoons in the soil. Furthermore, Babendreier (1998) showed that *A. nigrocincta* accepts, and is able to develop in, host cocoons already parasitized by *L. ensator*.

Apples damaged by late-instar *H. testudinea*, collected in central Europe, were maintained in Switzerland until larvae emerged. Parasitism was evaluated by examining mature host larvae for *L. ensator* eggs, visible through the host integument. Parasitized host larvae were overwintered and cocoons shipped to Canada. More than 2500 *H. testudinea* cocoons containing *L. ensator* were shipped from 1995 to 1999. A total of 604 *L. ensator* adults were released in an unsprayed orchard in Frelighsburg (45°03′N 75°50′W) (Vincent *et al.*, 2001).

Evaluation of Biological Control

To verify if *L. ensator* had established, fruit showing secondary damage by *H. testudinea* was collected and placed in experimental plots where the sawflies were allowed to pupate and overwinter. The following spring, two adult female *L. ensator* emerged, the first recovery of this parasitoid in North America. These females were released on 1 June and 8 June 1999 in the Frelighsburg orchard, when apples were about 1 cm in diameter. It is still too early to determine whether successful establishment has occurred.

Although effective, current nematode formulations have short residual activity on the foliage (about 40 h) and they are not likely to be used by apple growers.

Recommendations

Further work should include:

1. Ensuring that *H. testudinea* does not have an opportunity to establish outside the currently infested regions;
2. Continued monitoring and redistribution of *L. ensator* in infested areas as needed;
3. Testing the influence of insecticides used in IPM orchards on *L. ensator* to predict its potential for area-wide establishment;
4. Continued collections of *L. ensator* in different European regions to provide different biotypes of the biological control

agent, as well as to ensure its establishment;

5. Evaluating the potential influence of *A. nigrocincta* on the impact of *L. ensator* on *H. testudinea* should *A. nigrocincta* be released as a second biological control agent.

Acknowledgements

We thank Robert Trottier and Klaus Carl for facilitating this international cooperative project through the Agriculture and Agri-Food Canada/CABI Partnership Program.

References

Anonymous (1959) Status of some important insects in the United States–European apple sawfly (*Hoplocampa testudinea* (Klug)). *Cooperative Economic Insect Report* 18, 341–342.

Anonymous (1969) Summary of insect conditions in the United States – 1968. *Cooperative Economic Insect Report* 19, 196.

Babendreier, D. (1996) Studies on two ichneumonid parasitoids as potential biological control agents of the European apple sawfly, *Hoplocampa testudinea* Klug (Hymenoptera: Tenthredinidae). *International Organization for Biological Control/Western Palaearctic Regional Section, Bulletin* 19, 236–240.

Babendreier, D. (1998) Oekologie der Parasitoiden *Lathrolestes ensator* und *Aptesis nigrocincta* (Hymenoptera: Ichneumonidae) sowie deren Einfluss auf Populationen ihres gemeinsamen Wirtes, der Apfelsägewespe, *Hoplocampa testudinea* (Hymenoptera: Tenthredinidae). PhD thesis, University of Kiel, Kiel, Germany.

Babendreier, D. (1999) Observations on the biology and phenology of *Aptesis nigrocincta* (Hymenoptera: Ichneumonidae) parasitizing cocoons of the apple sawfly, *Hoplocampa testudinea* (Hymenoptera: Tenthredinidae). *International Organization for Biological Control/Western Palaearctic Regional Section, Bulletin* 22, 57–61.

Bélair, G., Vincent, C. and Chouinard. G. (1998) Foliar sprays with *Steinernema carpocapsae* against early season apple pests. *Journal of Nematology* 30, 599–606.

Boevé, J.L., Lengwiler, U., Tollsten, L., Dorn, S. and Turlings, T.C.J. (1996) Volatiles emitted by apple fruitlets infested by larvae of the European apple sawfly. *Phytochemistry* 4, 373–381.

Cakstynja, T. (1968) *Lathrolestes ensator* (Brauns) parazit jablonnogo pililscika (*Hoplocampa testudinea* Klug). [*Lathrolestes ensator* a parasitoid of the apple sawfly (*Hoplocampa testudinea*)]. *Biologicheskii Metod Bor'by s Vreditelyami Rastenii, Doklady Simpoziuma Riga*, 253–255.

Cross, J.V., Solomon, M.G., Babendreier, D., Blommers, L., Easterbrook, M.A., Jay, C.N., Jenser, G., Jolly, R.L., Kuhlmann, U., Lilley, R., Olivella, E., Töpfer, S. and Vidal, S. (1999) Biocontrol of pests of apples and pears in Northern and Central Europe: 2. Parasitoids. *Biocontrol Science and Technology* 9, 277–314.

Downes, W. and Andison, H. (1942) The apple sawfly *Hoplocampa testudinea* Klug on Vancouver Island, Bristish Columbia. *Proceedings of the Entomological Society of British Columbia* 39, 13–16.

Jaworska, M. (1987) Obserwacje nad *Lathrolestes marginatus* (Thompson), pasozytem owocnicy jablkowej – *Hoplocampa testudinea* (Klug) (Hymenoptera, Tenthredinidae). [Observations on *Lathrolestes marginatus* (Thompson), a parasite of apple sawfly, *Hoplocampa testudinea* (Klug) (Hymenoptera, Tenthredinidae)]. *Polskie Pismo Entomologiczne* 57, 553–567.

Miles, H.W. (1932) On the biology of the apple sawfly, *Hoplocampa testudinea* Klug. *Annals of Applied Biology* 39, 420–431.

Owens, E.D. and Prokopy, R.J. (1978) Visual monitoring trap for European apple sawfly. *Journal of Economic Entomology* 71, 576–578.

Paradis, R.O. (1980) L'hoplocampe des pommes, *Hoplocampa testudinea* (Klug) (Hymenoptera, Tenthredinidae) au Québec. *Phytoprotection* 61, 26–29.

Pyenson, L. (1943) A destructive apple sawfly new to North America. *Journal of Economic Entomology* 36, 218–221.

Vincent, C. (1988) The European Apple Sawfly: insect pest of apple orchards in Quebec. *Canadian Fruitgrower* 44(8), 8.

Vincent, C. and Bélair, G. (1992) Biocontrol of the apple sawfly, *Hoplocampa testudinea* Klug, with entomogenous nematodes. *Entomophaga* 37, 575–582.

Vincent, C. and Mailloux, M. (1988) Abondance, importance des dommages et distribution de l'hoplocampe des pommes au Québec de 1979 à 1986. *Annales de la Société Entomologique de France* 24, 39–46.
Vincent, C. and Rancourt, B. (1988) *Chemical control of the European apple sawfly with pre-bloom treatments.* Pesticide Research Report, Agriculture Canada, Ottawa, Ontario, p. 9.
Vincent, C. and Roy, M. (1992) Entomological limits to the implementation of biological programs in Quebec apple orchards. *Acta Entomologica et Phytopathologica Hungarica* 27, 649–657.
Vincent, C., Rancourt, B., Sarazin, M. and Kuhlmann, U. (2001) Releases and first recovery of *Lathrolestes ensator* Brauns (Hymenoptera: Ichneumonidae) in North America, a European parasitoid of the European apple sawfly *Hoplocampa testudinea* Klug (Hymenoptera: Tenthredinidae). *The Canadian Entomologist* 133, 147–149.
Zijp, J.P. and Blommers, L. (1993) *Lathrolestes ensator*, a parasitoid of the apple sawfly. *Proceedings of Experimental and Applied Entomology* 4, 237–242.

29 *Keiferia lycopersicella* (Walsingham), Tomato Pinworm (Lepidoptera: Gelechiidae)

J.L. Shipp, G.M. Ferguson and D.W.A. Hunt

Pest Status

Tomato pinworm, *Keiferia lycopersicella* (Walsingham), native to the southern USA, Mexico, the West Indies and South America (Zimmerman, 1978), was first introduced into Canada in 1946 on field and greenhouse crops in south-western Ontario, but it did not establish. The next occurrence in Ontario greenhouses was reported in 1991 from a single tomato grower (1.2 ha) in Essex County. Since 1991, *K. lycopersicella* has spread throughout the Leamington area, with 87 ha infested in 1999. It also occurred in greenhouses in British Columbia in 1970 and 1975 but failed to establish. High summer temperatures and increased immigration of moths from greenhouse to greenhouse are major factors resulting in the larger infestations during summer. Infestations are initially found close to doorways, along walkways and near wall vents.

K. lycopersicella damages both leaves and fruit of tomato, *Lycopersicon esculentum* L. Early-instar larvae mine through the leaves, whereas later instars are leaf rollers and may also burrow into the fruit (Trumble, 1994). Thus, the larvae can reduce photosynthesis and directly damage fruit. In 1994, one grower reported an estimated fruit loss of 32,000 kg due to direct fruit damage.

Background

Control of *K. lycopersicella* using pesticides is difficult because the larvae are concealed. In California, the use of sex pheromones for mating disruption has been successful for field tomato. In Ontario greenhouses, a synthetic microencapsulated formulation of *K.*

lycopersicella sex pheromone did not have any negative impact on *Encarsia formosa* Gahan or bumble bees, *Bombus* spp. (Ferguson *et al.*, 1999). Application of the sex pheromone at a rate of 150–200 ml ha^{-1} at 4-week intervals kept numbers of adult *K. lycopersicella* at a low level (<1–2 adults per trap). Use of the sex pheromone together with ultraviolet light traps and removal of infested leaves resulted in elimination or suppression of pest populations to non-economically damaging levels in winter. In summer, the sex pheromone was not as effective, especially at the lower rate, even at application intervals of 3 weeks. However, application at the higher rate, together with removal of infested leaves, usually resulted in successful control. In autumn, when outdoor temperatures do not exceed 10°C, the use of ultraviolet light traps is also recommended. Because pheromone use during summer is not completely successful, natural enemies may provide additional control.

Biological Control Agents

Parasitoids

Zimmerman (1978) reported three braconids, *Apanteles dignus* Muesebeck, *Horogenes blackburni* (Cameron) and *Panhormius pallidipes* (Ashmead), from *K. lycopersicella*.

Shipp *et al.* (1998) evaluated six commercially available *Trichogramma* spp. to control *K. lycopersicella. Trichogramma pretiosum* Riley and *T. brassicae* Bezdenko parasitized 40–50% of the eggs. In controlled environmental chamber trials *T. pretiosum* showed the greatest potential as a biological control agent. Adult females killed eggs by parasitizing or feeding on them. Temperatures of 28°C, compared to 20 and 25°C, significantly reduced parasitoid-induced mortality. Based on these trials, a parasitoid to host egg ratio of between 1:1 to 10:1 is recommended for inundative releases of *T. pretiosum* to control *K. lycopersicella*.

Recommendations

Further work should include:

1. Determining release rates of *T. pretiosum* under commercial production conditions;
2. Evaluating natural enemies that attack early larval stages of *K. lycopersicella* before economic damage occurs.

References

Ferguson, G.M., Shipp, J.L. and Hunt, D.W.A. (1999) Evaluation of pheromone concentrate for control of tomato pinworm in greenhouse tomatoes. *International Organization for Biological Control, West Palaearctic Regional Section, Bulletin* 22, 73–76.

Shipp, J.L., Wang, K. and Ferguson, G. (1998) Evaluation of commercially produced *Trichogramma* spp. (Hymenoptera: Trichogrammatidae) for control of tomato pinworm, *Keiferia lycopersicella* (Lepidoptera: Gelechiidae), on greenhouse tomatoes. *The Canadian Entomologist* 130, 721–731.

Trumble, J.T. (1994) Sampling arthropod pests in vegetables. In: Pedigo, L.P. and Buntin, G.D. (eds) *Handbook of Sampling Methods for Arthropods in Agriculture.* CRC Press, Boca Raton, Florida, pp. 604–621.

Zimmerman, E.C. (1978) *Insects of Hawaii.* Vol 9. *Microlepidoptera Part II Gelechioidea.* The University Press of Hawaii, Honolulu, pp. 883–1903.

30 *Lambdina fiscellaria fiscellaria* (Guenée), Hemlock Looper (Lepidoptera: Geometridae)

K. van Frankenhuyzen, R.J. West and M. Kenis

Pest Status

The eastern hemlock looper, *Lambdina fiscellaria fiscellaria* (Guenée), is a native species distributed in Canada from the Atlantic coast to Alberta. It feeds on a wide variety of coniferous and deciduous trees, but outbreaks occur predominantly in balsam fir, *Abies balsamea* (L.) Miller, stands. Although periodic outbreaks are found in all eastern provinces, Newfoundland has regular epidemics every 10–15 years that last from 3 to 6 years (Otvos *et al.*, 1979). Since larvae feed on both new and old foliage, severe defoliation can kill trees within 1 or 2 years. The largest outbreak recorded in Newfoundland's history lasted from 1966 to 1971 and killed 12×10^6 m^3 of merchantable wood (Otvos *et al.*, 1979). Since then, *L. f. fiscellaria* populations have reached epidemic levels there from 1984 to 1988, causing defoliation of about 330,000 ha during the peak in 1986, and again from 1994 to 1998, with peak defoliation of about 190,000 ha in 1996. Much smaller outbreaks were reported in New Brunswick (1989–1993), Nova Scotia (1991–1992, 1996), Ontario (1992–1994) and Quebec (1991–1992). Populations on the Gaspé Peninsula, Quebec, increased rapidly during 1996. A spray programme was planned to protect about half of the predicted 130,000 ha infestation in 1997. Because of high egg mortality in spring in most areas, high larval populations developed on only 13,000 ha. The Gaspé populations have returned to low levels since then. A new outbreak is developing on the north shore of the St Lawrence River, where the defoliation area increased dramatically from about 27,000 ha in 1998 to about 470,000 ha in 1999.

Adult *L. f. fiscellaria* lay eggs in cryptic locations on trees and stumps in late summer. They hatch the following June and larvae feed for 6 weeks in early summer before pupating in protected places on tree trunks or stumps (McLeod, 1962).

Background

Applying chemical pesticides was the main control strategy against *L. f. fiscellaria* but their use was eliminated until, for example, at the onset of the 1984 Newfoundland outbreak only fenitrothion was registered for use; from 1985 to 1988, 85% of the operational spray programmes against *L. f. fiscellaria* used it. The use of fenitrothion was discontinued in Newfoundland after 1988. In New Brunswick, fenitrothion was still used in 35% of the 1990–1993 *L. f. fiscellaria* control programme. The need to develop biological control alternatives was apparent.

Although few parasites and predators are closely and consistently associated with *L. f. fiscellaria*, Dupont (1998) reported that an egg parasitoid, *Telenomus* sp., caused the collapse of an outbreak in the Gaspé Peninsula. Egg parasitism averaged 64% in populations throughout the Gaspé in 1997, eliminating the need for aerial control operations in 92% of the area proposed for treatment (Dupont, 1998). Eggs of *L. f. fiscellaria* collected during the current outbreak on the north shore of the

St Lawrence River yielded three *Telenomus* spp. (G. Pelletier, Ste Foy, and L. Masner, Ottawa, 2000, personal communication). Hartling *et al.* (1999) showed that *Telenomus* sp. near *alsophilae* is abundant in New Brunswick. Otvos *et al.* (1973) also discovered a *Telenomus* sp. in Newfoundland. A tachinid fly, *Winthemia occidentalis* Reinhard introduced from British Columbia between 1949 and 1951 (ex. western hemlock looper, *Lambdina fiscellaria lugubrosa* Hulst, and oak looper, *L. fiscellaria somniaria* Hulst), is the most common parasitoid in Newfoundland and may infest more than 20% of larvae and pupae in declining populations (Otvos, 1973; Raske *et al.*, 1995).

The fungi *Entomophaga aulicae* (Reichardt in Bail) Humber (= *Entomophthora egressa* MacLeod and Tyrrell) and *Erynia radicans* (Brefeld) Humber (= *Entomophthora sphaerosperma* Freseneus) are often prevalent in declining *L. f. fiscellaria* populations (Otvos *et al.*, 1973). They can successfully infect both early and late larval instars. Fungal infections build up under favourable weather conditions in about 2 years from the time defoliation is first noticed, and are thought to contribute to, if not cause, subsequent collapse of outbreak populations (Otvos 1973). Other fungi, e.g. black yeast fungi (*Hormonema* spp. and *Aureobasidium* spp.), occur periodically at high levels in outbreak populations (West *et al.*, 1988).

Biological Control Agents

Parasitoids

Considering that the most important parasitoid of *L. f. fiscellaria* in Newfoundland is the introduced tachinid *W. occidentalis*, and following the suggestion by Mills and Räther (1990) that potential biological control agents may be found in Europe on other looper species, a survey was undertaken for parasitoids of conifer-feeding geometrids in central Europe. Three of 20 looper species studied, *Epirrita autumnata* (Borkhauser), *Agriopis aurantiaria* Hübner and *Poecilopsis isabellae* Harrison, were identified as potential sources of parasitoids to introduce into North America because their life cycles were similar to that of *L. f. fiscellaria* (M. Kenis, K. Herz and R. West, unpublished). Four larval parasitoids, *Dusona contumax* (Förster) from *A. aurantiaria*, *Dusona* sp. from *P. isabellae*, *Aleiodes* cf. *gastritor* (Thunberg) from *E. autumnata* and *Aleiodes* sp. from *P. isabellae*, were selected for further studies because of: (i) high incidence on their original host, especially at low host density; (ii) apparent specificity; (iii) good synchrony with *L. f. fiscellaria* phenology; and (iv) the likelihood they would fill poorly occupied ecological niches in the native parasitoid complex of *L. f. fiscellaria*. West and Kenis (1997) investigated the biology of these four parasitoids. After developing rearing methods and screening protocols in Europe on their natural hosts, small numbers were sent to Newfoundland in 1994 and 1995 and screened in the laboratory against *L. f. fiscellaria*. Although adult female *D. contumax*, *Dusona* sp. and *A.* cf. *gastritor* parasitized *L. f. fiscellaria* larvae, parasitoid development did not occur because all parasitoid eggs recovered from the host larvae were encapsulated. It was concluded that *L. f. fiscellaria* was not a suitable host for any of the four parasitoids and, in the absence of alternative parasitoids, the programme was discontinued.

Pathogens

At the onset of the 1984 Newfoundland outbreak the province received a minor-use registration for one *B. thuringiensis* Berliner serovar *kurstaki* (*B.t.k.*) product. About 15% of the operational spray programmes from 1985 to 1988 used *B.t.k.* Field trials from 1985 to 1996 contributed to the registration of various high-potency *B.t.k.* formulations (West *et al.*, 1987, 1989, 1997). In all of eastern Canada, a total of 265,000 ha were treated with *B.t.k.* to control *L. f. fiscellaria* from 1981 to 1999, using about 13×10^{15} international units (IU) (Table 30.1).

Table 30.1. Operational use of *Bacillus thuringiensis* against *Lambdina fiscellaria fiscellaria*.

Year	Province	No. ha treated[a]	Dose applied[b]
1985	Newfoundland	2,365	70,950
1986	Newfoundland	5,420	162,600
1987	Newfoundland	4,183	151,740
1988	Newfoundland	23,108	768,870
1989	Newfoundland	5,362	273,000
1990	Newfoundland	10,616	612,480
	Newfoundland	21,160	735,450
1991	Newfoundland	16,975	509,250
1992	Quebec	152	9,120
	Newfoundland	538	20,040
1993	Newfoundland	15,424	845,940
	New Brunswick	8,525	511,500
1994	Newfoundland	10,738	531,060
1995	Newfoundland	47,227	2,689,140
1996	Newfoundland	69,208	4,244,400
	Nova Scotia	2,000	12,000
1997	Newfoundland	4,316	169,950
	Nova Scotia	300	18,000
	Quebec	5,339	245,400
1998	Newfoundland	7,200	288,000
1999	Newfoundland	9,800	498,000
Total	All provinces	269,956	13,465,890

[a] Number of hectares treated with one or more applications.
[b] Total dose (expressed in 10^9 International Units) applied per ha (= number of ha treated × number of applications × 10^9 IU ha^{-1} per application).
Source: Forestry Insecticide Database, Canadian Forest Service, Great Lakes Forestry Centre, Sault Ste Marie, Ontario.

Evaluation of Control Attempts

West and Kenis (1997) concluded that *L. f. fiscellaria* was not a suitable target for classical biological control because parasitoids from closely related host species were either too specific to attack the target host or too polyphagous and, therefore, would likely show unwanted non-target effects.

Larvae of *L. f. fiscellaria* are highly susceptible to *B.t.k.* because all instars are exposed feeders. Sprays target early instar larvae, before extensive feeding damage has been done. Two applications are often needed due to extended egg hatch and prolonged development of early instars. The first application is timed for peak populations of first-instar larvae and the second is applied before larvae have completed the second instar. Foliage protection is usually achieved by two applications of undiluted high-potency products at 30 × 10^9 IU in 1.2–2.4 l ha^{-1}. The high efficacy of *B.t.k.* to control *L. f. fiscellaria* was clearly demonstrated in various experimental spray programmes (West *et al.*, 1987, 1989, 1997).

Recommendations

Future work should include:

1. Defining the exact role of native parasitoids and other natural enemies in the population dynamics of *L. f. fiscellaria* and integrating them into a pest management programme, e.g. Hartling *et al.* (1999) proposed monitoring egg parasitism to assess the need for spray programmes;
2. Determining the role of *Telenomus* spp. in controlling *L. f. fiscellaria*;

3. Determining if the *Telenomus* sp. present in Newfoundland is the same species as that occurring in continental North America and, if so, why its impact appears limited on the island;
4. Defining the role of *E. aulicae* as a causative agent in the decline of epidemic populations to improve targeting of spray programmes and accelerate population collapse through field release of infectious propagules early in the outbreak cycle;
5. Continuing to develop an inexpensive medium for mass production fermentation of hyphal bodies, initiated by Nolan (1993), and commercializing the technology.

References

Dupont, A. (1998) *Forest Protection Program Against Hemlock Looper in Eastern Quebec – 1997.* Société de protection des forêts contre les insects et maladies, Québec.

Hartling, L.K., Carter, N. and Proude, J. (1999) Spring parasitism of overwintered eggs of *Lambdina fiscellaria fiscellaria* (Guen.) (Lepidoptera: Geometridae) by *Telenomus* near *alsophilae* (Hymenoptera: Scelionidae). *The Canadian Entomologist* 131, 421–422.

McLeod, J.H. (1962) Part I. Biological Control of Pests of Crops, Fruit Trees, Ornamentals and Weeds in Canada up to 1959. In: McLeod, J.H., McGugan, B.M. and Coppel, H.C. (eds) *A Review of the Biological Control Attempts Against Insects and Weeds in Canada.* Technical Communication No. 2, Commonwealth Institute of Biological Control, Trinidad. Commonwealth Agricultural Bureaux, Farnham Royal, UK, pp. 1–33.

Mills, N.J. and Räther, M. (1990) Hemlock loopers in Canada; biology, pest status and potential for biological control. *Biocontrol News and Information* 11, 209–221.

Nolan, R.A. (1993) An inexpensive medium for mass fermentation production of *Entomophaga aulicae* hyphal bodies competent to form conidia. *Canadian Journal of Microbiology* 39, 588–593.

Otvos, I.S. (1973) *Biological Control Agents and their Role in the Population Fluctuation in the Eastern Hemlock Looper.* Information Report N-X-102, Canadian Forest Service, Newfoundland Forest Research Centre, St John's, Newfoundland.

Otvos, I.S., MacLeod, D.M. and Tyrrell, D. (1973) Two species of *Entomophtera* pathogenic to the eastern hemlock looper (Lepidoptera: Geometridae) in Newfoundland. *The Canadian Entomologist* 105, 1435–1441.

Otvos, I.S., Clarke, J.L. and Durling, D.S. (1979) *A History of Recorded Eastern Hemlock Looper Outbreaks in Newfoundland.* Information Report N-X-179. Canadian Forest Service, Newfoundland Forest Research Centre, St John's, Newfoundland.

Raske, A.G., West, R.J. and Retnakaran, A. (1995) Hemlock looper, *Lambdina fiscellaria.* In: Armstrong, J.A. and Ives, W.G.H. (eds) *Forest Insect Pests in Canada.* Canadian Forest Service, Natural Resources Canada, Ottawa, Ontario, pp. 141–147.

West, R.J. and Kenis, M. (1997) Screening four exotic parasitoids as potential controls for the eastern hemlock looper, *Lambdina fiscellaria fiscellaria* (Guenée) (Lepidoptera: Geometridae). *The Canadian Entomologist* 129, 831–841.

West, R.J., Raske, A.G., Retnakaran, A. and Lim, K.P. (1987) Efficacy of various *Bacillus thuringiensis* Berliner var. *kurstaki* formulations and dosages in the field against the hemlock looper, *Lambdina fiscellaria fiscellaria* (Guen.) (Lepidoptera: Geometridae), in Newfoundland. *The Canadian Entomologist* 119, 449–458.

West, R.J., Meades, J.P. and Dixon, P.L. (1988) *Efficacy of Single Applications of* Bacillus thuringiensis *and Diflubenzuron Formulations against the Hemlock Looper in Newfoundland in 1988.* Information Report N-X-284, Forestry Canada, Newfoundland and Labrador Region, St John's, Newfoundland.

West, R.J., Raske, A.G. and Sundaram, A. (1989) Efficacy of oil-based formulations of *Bacillus thuringiensis* Berliner var. *kurstaki* against the hemlock looper, *Lambdina fiscellaria fiscellaria* (Guen.) (Lepidoptera: Geometridae). *The Canadian Entomologist* 121, 55–63.

West, R.J., Thompson, D., Sundaram, K.M.S., Sundaram, A. Retnakaran, A. and Mickle, R. (1997) Efficacy of aerial applications of *Bacillus thuringiensis* Berliner and tefubenozide against the eastern hemlock looper (Lepidoptera: Geometridae). *The Canadian Entomologist* 129, 613–626.

31 *Leptinotarsa decemlineata* (Say), Colorado Potato Beetle (Coleoptera: Chrysomelidae)

C. Cloutier, G. Boiteau and M.S. Goettel

Pest status

The Colorado potato beetle, *Leptinotarsa decemlineata* (Say), originally from South America, moved north following the expansion of potato crops, reaching Canada in the late 19th century. It is the most important defoliator of potato, *Solanum tuberosum* L., in North America, and uses various other Solanaceae as host plants. With its invasive character and adaptability, its status as a potato pest continues to rise worldwide. In Canada, potential yield losses in potato due to *L. decemlineata* are estimated at 30–60% or more, with actual losses probably around 3%, or Can$15–18 million annually.

In Canada, *L. decemlineata* is mostly univoltine, overwintering as diapausing adults in soil within or near potato fields. Adults emerge in May and find new plants on which they feed and reproduce. Females deposit hundreds of eggs over several weeks, in masses of 2–3 dozen eggs that are glued to the foliage. The eggs hatch and larvae develop through four instars. Mature larvae drop to the soil to bury themselves and pupate. Complete development takes 4–6 weeks. New-generation adults emerge and start to feed and mate. Depending mainly on photoperiod, newly emerged females prepare either for a brief period of egg-laying that gives rise to a (generally incomplete) second generation or for diapause. Adults induced to diapause feed for 1–2 weeks before entering the soil.

Background

Control of *L. decemlineata* is mainly based on chemical insecticides. From 1980 to 1995, in Canada and the USA, attention was focused on the increasing threat of multiple insecticide resistance, paralleled by research to develop alternatives to chemical control (Boiteau *et al.*, 1987; Duchesne and Boiteau, 1995). In Canada, biological control of potato pests was evaluated by Boiteau (1987) for predators, by Sears (1987) for parasitoids, by Duchesne and Boiteau (1987) for pathogens and biotoxins, and generally by Cloutier *et al.* (1995). Because *L. decemlineata* is a relatively recent invader, it does not have a well-adapted, coevolved guild of natural enemies. *Myiopharus* spp. are the only known parasitoids. They attack *L. decemlineata* larvae but their impact is too limited or occurs too late in the season to prevent crop damage. In commercial potato fields, use of chemical insecticides is by far the main control method for *L. decemlineata*; thus parasitoids as well as potentially useful generalist predators are killed and have negligible impact on pest populations.

Biological Control Agents

Pathogens

Fungi

Beauveria bassiana (Balsamo) Vuillemin, formulated as Mycotrol®, was tested as a foliar spray in large field plots at

Fredericton in 1996 and 1997 (Boiteau and Osborn, 1997, 1998). It provided some protection in 1996 but was not as effective as the insecticide chlorfenapyr (Alert®). With two applications of Mycotrol, the defoliation index rose to over 3, which is above the critical level of 2 but, during the same period, untreated nearby plots went from below 2 to over 5. In 1997, a combination of Mycotrol and *Bacillus thuringiensis* Berliner serovar *tenebrionis* (Novodor®) provided a similar level of plant protection as chlorfenapyr. Two applications of these biological products kept defoliation under the index of 2, compared to nearby control plots that went from below 2 to above 6.

A strategy to contaminate adult *L. decemlineata* with *B. bassiana* as they leave potato fields in late summer in search of overwintering sites is being investigated, with the goal of contaminating these sites and providing long-term control and an overall reduction in beetle numbers (C. Noronha and M.S. Goettel, unpublished). In the laboratory, *B. bassiana* incites disease and causes mortality in soil-inhabiting stages of *L. decemlineata*. However, the digging behaviour of prediapausing adults dislodges spores; the deeper the adults dig, the fewer the number of spores that remain attached to their bodies. In a preliminary study to determine if adults overwintering in the field were susceptible, conidia at 1.6×10^8 and 1.6×10^7 cm^{-2} were applied to the soil surface. One hundred pre-diapausing adults were introduced into the plots and allowed to dig into the contaminated soil and overwinter. In spring, emergence cages were placed in the plots and, following adult emergence, the soil was removed in layers to a 20 cm depth and was examined for cadavers or surviving insects. No firm conclusions could be made because mortalities between treatment and control plots were equal and recovery of insects or insect parts were very low. Nevertheless, dead insects and insect parts homogenized and plated on to a selective medium from the treatment plots resulted in *B. bassiana* growth, whereas no *B. bassiana* growth was obtained from the cadavers or insect parts from the control plots.

Nematodes

Field and laboratory trials at Fredericton, New Brunswick, and Charlottetown, Prince Edward Island, from 1992 to 1993 documented the susceptibility of *L. decemlineata* larvae, pupae and adults to entomopathogenic nematodes, and determined their potential for reducing populations of late-season adults (Stewart *et al.*, 1998). In the laboratory, larvae, pupae and/or adults were exposed to *Steinernema carpocapsae* (All Strain) at 5×10^5 nematodes m^{-2}. Insect mortality was 100%; when larvae or pupae were treated, mortality was observed in subsequent growth stages. *S. carpocapsae* appears to persist through the larval-pupal or pupal-adult transitions. One application of *S. carpocapsae* was sufficient to control *L. decemlineata*.

In New Brunswick, a field trial showed that foliar application of *S. carpocapsae* was ineffective (Boiteau *et al.*, 1992). Soil applications were therefore used in further tests. Commercial preparations of *S. carpocapsae* (All Strain) were applied to coincide with the entry of mature larvae into the soil around potato plants (July in New Brunswick and August in Prince Edward Island). Additional water was applied to ensure that the nematodes penetrated the soil surface. Recommended rates of insecticides (fenvalerate or endosulfan in New Brunswick and phosmet in Prince Edward Island) were applied at the same time as the nematode treatments. In New Brunswick, late-season adults were reduced by 32% with *S. carpocapsae* and by 45% with insecticides, compared to the control. In Prince Edward Island, despite lower populations than in New Brunswick, *L. decemlineata* was reduced by 38% with the lower rate of *S. carpocapsae* and by 40% with insecticides.

Parasitoids

Edovum puttleri Grissell, from South America, released in potato plots during 1984 and 1985 in Ontario and New Brunswick, resulted in limited reduction of

L. decemlineata populations (Sears and Boiteau, 1989). Although parasitism of *L. decemlineata* egg masses was recorded in both years and at each location, the degree of parasitism and host feeding, and their effect on the population of *L. decemlineata*, was often minimal. Exceptions to this occurred at Cambridge during the second generation of 1984 and at Fredericton in 1985. In early summer, the effectiveness of *E. puttleri* was limited by temperatures below 20°C because oviposition and probing are reduced at temperatures below 19°C (as others have also noted), but Sears and Boiteau (1989) documented for the first time the degree to which this temperature limits effectiveness of *E. puttleri* as a mortality factor of *L. decemlineata* in the field.

Predators

Insect predators of *L. decemlineata* eggs and larvae have received considerable attention. *Coleomegilla maculata* DeGeer is a generalist predator in eastern North America. In early spring, adult activity has potential for destruction of *L. decemlineata* eggs and young larvae on potato, especially at the beginning of a rotation cycle in fields previously planted with corn, *Zea mays* L. In Quebec, releases of *C. maculata lengi* Timberlake in small plot tests produced short-term reductions in *L. decemlineata* larval densities, but significant potato foliage protection was not demonstrated (S. Giroux and D. Coderre, Quebec, 1996, personal communication).

Stinkbugs (Pentatomidae) are the most specialized insect predators of *L. decemlineata*. The spined soldier bug, *Podisus maculiventris* (Say), and the two-spotted stinkbug, *Perillus bioculatus* (F.), are the best known. *P. bioculatus* feeds on a variety of insect prey but appears to be more specialized than *P. maculiventris*, attacking eggs, all larval instars, and, infrequently, adults of *L. decemlineata* (Cloutier and Bauduin, 1995; Saint-Cyr, 1995; Saint-Cyr and Cloutier, 1996; Cloutier, 1997).

The influence of genetic versus maternal-diet factors on *P. bioculatus* prey preference was studied using *L. decemlineata* larvae versus yellow mealworms, *Tenebrio molitor* L., and house crickets, *Gryllus domesticus* L., as unusual prey (Saint-Cyr, 1995). During the course of a 50-day prey-alternation test, reproduction could be switched on and off up to 5 times for the longest-lived individuals. Even though *P. bioculatus* developed to maturity on *T. molitor*, they rarely reproduced on this prey, in contrast with *L. decemlineata* prey, which had a triggering effect on female reproduction.

Lachance and Cloutier (1997) investigated the effects of temperature, release density, nymphal stage and physiological age within stadium on movement potential of *P. bioculatus* in the laboratory and field. A temperature threshold of 19°C was necessary to initiate dispersal among small groups of second- and fourth-instar nymphs released on potato plants at 13–23°C in the laboratory. As expected, fourth instars were more dispersive than second instars. Field tests using second-instar *P. bioculatus* on potato plants confirmed laboratory findings regarding the effects of temperature, aggregation and feeding status on dispersal.

Lachance (1996) investigated field dispersal of *P. bioculatus* nymphs when released centrally and found that nymph densities dropped to 25–45% from release level in the first 2 days following release, but then tended to remain stable for weeks. Dispersal from the release point was 1.75 times faster (0.35 m day^{-1}) for fourth instars than second instars (0.2 m day^{-1}). The fourth instars dispersed equally fast both along and across rows but second instars moved along rows 1.2–1.4 times faster than across rows. Cloutier and Jean (1998) evaluated a transverse strip release pattern of small *P. bioculatus* nymphs and showed that egg predation decreased with distance from release plants. However, there was no difference between plot sections in the densities of larvae reaching pupation, indicating that predators eventually spread uniformly on all plants within plots. The release density of four nymphs per plant was necessary to control *L. decemlineata*. Compared to a central release, strip release allowed more uniform control.

In Quebec, Cloutier and Bauduin (1995) and Cloutier and Jean (1998) investigated augmentative biological control of *L. decemlineata* with *P. bioculatus*. Despite *P. maculiventris* also being a valuable candidate, the apparently *L. decemlineata*-specialized feeding by *P. bioculatus* was critical in its selection, as it implies less risk of non-target predation, should predators ever be used on a large scale. Cloutier and Bauduin (1995) and Cloutier and Jean (1998) designed tests to show efficacy of *P. bioculatus* under various conditions, including annual variation of *L. decemlineata* densities, experimental manipulation of predator density, timing of predator release, spatial release pattern, and use of *P. bioculatus* together with *Bacillus thuringiensis tenebrionis* (*B.t.t.*). During these studies, early season densities of *L. decemlineata* varied strongly. Cumulative egg recruitment over 4–6 weeks of oviposition by postdiapause beetles varied from a low average of 250 per potato plant in 1992 to a high of 700 in 1995. *L. decemlineata* egg recruitment tended to be highest on those plots that were best protected from defoliation. In contrast to defoliated potato plants or plants that are protected with chemical insecticides, those that are well protected with *P. bioculatus* remained suitable for oviposition by female *L. decemlineata* throughout the oviposition period.

Cloutier and Bauduin (1995) made field releases of *P. bioculatus* over three consecutive years. In 1992, two successive releases of three *P. bioculatus* nymphs per *L. decemlineata* egg mass were made uniformly over the test plots. The first batch was released at peak egg-laying and the second one near the end of egg-laying, resulting in a generation-wide egg destruction rate of 47–50%, 99% or better larval control, and excellent foliage protection in both a predator-alone and a predator + *B.t.t.* (Trident®) treatment. By comparison, 77% of hatched *L. decemlineata* larvae pupated in the control, versus 38% with *B.t.t.* alone. The relatively mild pressure from *L. decemlineata* in 1992 no doubt played a role in explaining this surprising success in the first year of field trials. In 1993, using a plot-central predator release pattern, a 3:1 ratio of *P. bioculatus* to *L. decemlineata* egg mass provided excellent foliage protection. Control was related more to larval rather than egg predation. This can be explained by the delay inherent in the central release, which resulted in predator dispersion and predation strongly interacting with release ratios, producing evident spatial patterns of control and foliage protection, including 'corner effects', in treated plots. *P. bioculatus* caused nearly complete destruction of the host egg masses found, and fourth-instar *L. decemlineata* larvae were susceptible to predation by *P. bioculatus* adults and fifth- and even fourth-instar nymphs. These observations also indicated that, contrary to earlier reports, *P. bioculatus* was active under cool conditions, including temperatures from 12 to 15°C.

In 1994, three release times of *P. bioculatus* relative to host oviposition were compared, i.e. early, normal and late (C. Cloutier *et al.*, unpublished). Early release coincided with the time when the first egg masses appeared on potato plants, normal release coincided with oviposition peak, and late release took place after the oviposition peak, when *L. decemlineata* larval emergence had already progressed to some extent. For the early treatment, predators were released at 2.5 nymphs per plant, an arbitrary rate because incoming *L. decemlineata* numbers could not be assessed precisely. For the normal and late release treatments, four *P. bioculatus* nymphs per plant were released, based on previous results.

Since *P. bioculatus* were uniformly released as small nymphs in all treatments, they reached the adult stage later and stayed 7 days longer in the late treatment plot. The maximum egg predation rate was observed in the normal treatment, but maximum larval mortality occurred in the late treatment (Table 31.1). The early treatment was the least efficacious because the

Table 31.1. *Leptinotarsa decemlineata* egg and prepupal recruitment for three different *Perillus bioculatus* release times, based on *L. decemlineata* oviposition.

Release period	n	Eggs laid, mean ± SE	Sucked eggs, mean ± SE	Predation (%)	Hatched eggs, mean ± SE	Hatching (%)	Prepupae dropping, mean ± SE	Survival L1-pupation (%)
Early	28	424 ± 36 a[b]	131 ± 13 a	30.8	129 ± 15 a	30.3	37 ± 5 b	28.9
Normal[a]	28	404 ± 27 a	161 ± 16 a	39.9	92 ± 10 b	22.8	28 ± 4 bc	30.7
Late	28	434 ± 28 a	129 ± 13 a	29.7	121 ± 12ab	28.0	19 ± 2 c	15.6
Control	28	304 ± 18 b	5 ± 2 b	1.6	123 ± 9 ab	40.5	88 ± 6 a	71.7

[a]At peak oviposition.
[b]In each column means followed by the same letter are not significantly different ($P > 0.05$).

release density (2.5 per plant) was too low with respect to *L. decemlineata* pressure. This study confirmed that releasing at an early stage before *L. decemlineata* egg mass density can be used to estimate pest density is a risky strategy, which would work only if cheap and easily available predators could be released as a preventive measure.

In New Brunswick, Boiteau *et al.* (1998) conducted large-scale field releases of *P. bioculatus* at Fredericton. In 1996, field release of 32,000 *P. bioculatus* (provided by the APHIS Mission Biological Control Laboratory, Texas) in discrete groups showed a rapid dispersal of the nymphs to adjacent plants and rows. However, it took about 10 days before the nymphs were recovered from the neighbouring control blocks. Multiple broadcast releases of *P. bioculatus* at the McCain Research Farm, Florenceville, New Brunswick, in 1997 (about 135,000) and 1998 (about 46,000) from the time of first egg-laying by *L. decemlineata* to the end of presence of *L. decemlineata* larvae gave excellent control. In 1997 in Fredericton, two *P. bioculatus* releases (about 96,000) at the time late-instar *L. decemlineata* larvae were present, gave lower efficacy than the earlier releases. The large-scale releases generated great interest among growers and confirmed observations made in Quebec and elsewhere that broadcast releases at the beginning of egg-laying provide the best level of *L. decemlineata* control. Broadcast *P. bioculatus* applied early against the hatching eggs should be considered in the development of an integrated pest-management strategy against *L. decemlineata*.

Cloutier and Bauduin (1995) showed that commercial *B.t.t.* formulations were compatible with *P. bioculatus* and suggested that sustained interactions can occur between them. Cloutier and Jean (1998) found that fourth-instar *L. decemlineata* larvae that ingested sublethal doses of *B.t.t.* (M-Trak®) experienced temporary anorexia and never recovered normal feeding rate, eventually delaying maturation by 2–4 days, 10–25% of the duration of the fourth instar. Affected larvae were 10 times more susceptible to predation by second-instar *P. bioculatus* nymphs than controls. The authors also showed that at low and high *L. decemlineata* densities, *P. bioculatus* heavily impacted host eggs, but larval survival varied. At low host egg densities, *P. bioculatus* dispersed relatively fast, reducing the potential for interaction with *B.t.t.* in killing larvae. At high host egg densities, *P. bioculatus* lingering after egg hatching was higher and resulted in a significant interaction between the predator and *B.t.t.* It was estimated that at high host egg densities, two *P. bioculatus* nymphs per potato plant with 2 l ha^{-1} of *B.t.t.* (Novodor®) produced 31% more larval mortality than expected from their simple additive effects at the doses used in 1995. Variable results at the two host egg mass densities may be due in part to different *B.t.t.* products.

Evaluation of Biological Control

Until recently, the most effective alternatives to chemical insecticides were commercial *B.t.t.* products, but unfortunately they are no longer available because of lack of demand. Their use greatly increased the possibility for bringing the predators and parasitoids back to commercial potato fields, as shown by Cloutier and Jean (1998). However, *B.t.t.* biopesticides were more expensive and more difficult to use than chemical insecticides. They did not kill as fast as the latter, and the large *L. decemlineata* larvae and adults were mainly resistant to commercial *B.t.t.* formulations. More recently, the introduction of highly resistant transgenic potato plants expressing *B.t.t.* proteins in foliage at levels that are highly toxic to all *L. decemlineata* stages has created even more uncertainty about the future of *B.t.t.* biopesticides in *L. decemlineata* control.

B. bassiana has some potential as an alternative control method. However, a better understanding of its timing in relation to the life stages of *L. decemlineata* and the mechanisms of infection will be required before it can be considered. Repeated applications of the fungus, higher rates, improved delivery systems or alternate targeting strategies are needed to provide a consistent level of control for *L. decemlineata* in potato-growing areas of Canada.

S. carpocapsae did not perform as well in the field as in the laboratory. In the field, higher rates of nematodes might have been more effective. However, the use of lower rates was an attempt to balance efficacy with a cost-effective control strategy. The foliar application of nematodes after sunset when ultraviolet radiation was not a problem and at relative humidities of near 75% did not provide economical control of *L. decemlineata*. Hardier, more virulent nematode formulations are needed.

E. puttleri is not well adapted to the temperate climate of eastern Canada. Its effective use will require application of compatible insecticides or selection of strains that can perform at temperatures as low as 15°C.

Commercial availability of generalist predators for wide application in biological control of *L. decemlineata* is critical if this approach is to be developed further. Currently, *P. maculiventris* may be a better prospect for future commercialization than *P. bioculatus*, but its less-specific prey selection characteristic may eventually be considered a disadvantage because of non-target impacts. Even if effective predators were widely available commercially, it is not clear that inundative release over large fields would be practical. For example, release rates ranging in the tens of thousands of *P. bioculatus* ha^{-1} would be required for spring eradication of *L. decemlineata* and good foliage protection in potato fields. Therefore, stinkbugs should be used for inoculative augmentation in integrated control that emphasizes non-chemical methods.

The success of stinkbug augmentation suggests that if generalist predators were not automatically excluded from potato crops by the frequent use of chemical insecticides, it might be possible to rely significantly on integrated pest management to control *L. decemlineata*. Considering that stinkbugs, *C. maculata*, *Myiopharus* spp. and other natural enemies are present at least at low population levels in many areas, control based on attracting, augmenting and facilitating their activity could become a viable alternative.

In Quebec in the late 1980s, Chagnon *et al.* (1990) estimated that 19% of all agro-insecticides sold in Quebec were applied strictly for *L. decemlineata* control. More than 10 years later, there is little reason to believe that this situation has changed. The swift introduction of imidacloprid by the chemical industry in 1995 has strongly reduced the general concern that existed 10 years ago about the *L. decemlineata* problem, and ensured that reliance on chemicals for *L. decemlineata* control will remain predominant in the near future. There are indications that the past 15 years or so of chemical control have been more damaging than ever before to the natural fauna in and around potato agroecosystems in Quebec. One indication is that in over 20 years of regular sampling for aphid parasitoids in commercial potato fields in the Quebec City region, it has become more and more difficult to find them at useful densities for field experimental research and observation

(Ashouri, 1999; J. Brodeur and J.N. McNeil, Québec, 1999, personal communication).

Recommendations

Further work should include:

1. Determining the role that *B. bassiana* could play in the reduction of overwintering *L. decemlineata* populations;
2. Documenting persistence, virulence and effects of *B. bassiana* contamination of overwintering sites on *L. decemlineata* for potential use in integrated management programmes;
3. Developing cost-effective mass-rearing techniques for stinkbug predators and genetically improving them;
4. Integrating use of predators, parasitoids and pathogens as an alternative to chemicals.

References

Ashouri, A. (1999) Interactions de la résistance aux ravageurs primaires avec les ravageurs secondaires et leurs ennemis naturels: le cas des pucerons (Homoptera: Aphididae) sur la pomme de terre (Solanaceae). PhD thesis, Université Laval, Québec.

Boiteau, G. (1987) The significance of predators and cultural methods. In: Boiteau, G., Singh, R. and Parry, R. (eds) *Potato Pest Management in Canada – Lutte Contre les Parasites de la Pomme de Terre au Canada.* Agriculture and Agri-Food Canada, Fredericton, New Brunswick, pp. 210–223.

Boiteau, G. and Osborn, W.P.L. (1997) Control of Colorado potato beetles with an experimental formulation of *Beauveria bassiana. 1997 Pest Management Research Report* 40, 91–93.

Boiteau, G. and Osborn, W.P.L. (1998) Control of Colorado potato beetles with an experimental formulation of *Beauveria bassiana. 1998 Pest Management Research Report* 41, 81–83.

Boiteau, G., Singh, R. and Parry, R. (1987) *Potato Pest Management in Canada–Lutte Contre les Parasites de la Pomme de Terre au Canada.* Agriculture and Agri-Food Canada, Fredericton, New Brunswick.

Boiteau, G., Eidt, E., Zervos, S., Drew, M.E. and Osborn. W.P.L. (1992) Biological control of the Colorado potato beetle. *1992 Pest Management Research Report* 35, 72–74.

Boiteau, G., Walsh, J.R. and Osborn, W.P.L. (1998) Utilisation of the two-spotted stinkbug to control Colorado potato beetles in New Brunswick. *1998 Pest Management Research Report* 59, 171–174.

Chagnon, M., Payette, A., Jean, C. and Cadieux, C. (1990) *Modes alternatifs de répression des insectes dans les agro-écosystèmes québécois, tome 2 : identification des insectes ravageurs et état de l'agriculture biologique au Québec.* Ministère de l'Environnement et Centre québécois de valorisation de la biomasse, Québec.

Cloutier, C. (1997) Facilitated predation through interaction between life stages in the stinkbug predator *Perillus bioculatus* (Hemiptera: Pentatomidae). *Journal of Insect Behavior* 10, 581–598.

Cloutier, C. and Bauduin, F. (1995) Biological control of the Colorado potato beetle, *Leptinotarsa decemlineata* (Coleoptera: Chrysomelidae) in Québec by augmentative releases of the two-spotted stinkbug *Perillus bioculatus* (Hemiptera: Pentatomidae). *The Canadian Entomologist* 127, 195–212.

Cloutier, C. and Jean, C. (1998) Synergism between natural enemies and biopesticides: a test case using the stinkbug *Perillus bioculatus* (Hemiptera: Pentatomidae) and *Bacillus thuringiensis tenebrionis* against Colorado potato beetle (Coleoptera: Chrysomelidae). *Journal of Economic Entomology* 91, 1096–1108.

Cloutier, C., Jean, C. and Bauduin, F. (1995) More biological control for a sustainable potato pest management strategy. In: Duchesne, R.-M. and Boiteau, G. (eds) *Proceedings, Symposium: Insect Pest Control on Potato: Development of a Sustainable Approach.* Ministère de l'Agriculture, des Pêcheries et de l'Alimentation du Québec, Québec.

Duchesne, R.-M. and Boiteau, G. (1987) Microbial control of insect pests of potatoes. In: Boiteau, G., Singh, R. and Parry, R. (eds) *Potato Pest Management in Canada–Lutte Contre les Parasites de la Pomme de Terre au Canada.* Agriculture and Agri-Food Canada, Fredericton, New Brunswick, pp. 112–132.

Duchesne, R.-M. and Boiteau, G. (1995) *Insect Pest Control on Potato: Development of a Sustainable Approach.* Ministère de l'Agriculture, des Pêcheries et de l'Alimentation du Québec, Québec.

Lachance, S. (1996) Lutte biologique contre le doryphore de la pomme de terre par des lâchers inondatifs de la punaise *Perillus bioculatus*: facteurs influençant la dispersion du prédateur. Mémoire de maîtrise, Université Laval, Québec.
Lachance, S. and Cloutier, C. (1997) Factors affecting dispersal of *Perillus bioculatus* (Hemiptera: Pentatomidae), a predator of the Colorado potato beetle (Coleoptera: Chrysomelidae). *Environmental Entomology* 26, 946–954.
Saint-Cyr, J.-F. (1995) Préférences alimentaires chez *Perillus bioculatus* (Hemiptera: Pentatomidae), un prédateur généraliste. Mémoire de maîtrise, Université Laval, Québec.
Saint-Cyr, J.-F. and Cloutier, C. (1996) Prey preference by stinkbug *Perillus bioculatus*, a predator of the Colorado potato beetle. *Biological Control* 7, 251–258.
Sears, M.K. (1987) Significance of parasitoids in control of insect pests of potatoes. In: Boiteau, G., Singh, R. and Parry, R. (eds) *Potato Pest Management in Canada–Lutte Contre les Parasites de la Pomme de Terre au Canada.* Agriculture and Agri-Food Canada, Fredericton, New Brunswick, pp. 193–200.
Sears, M.K. and Boiteau, G. (1989) Parasitism of Colorado potato beetle (Coleoptera: Chrysomelidae) eggs by *Edovum puttleri* (Hymenoptera: Eulophidae) on potato in Eastern Canada. *Journal of Economic Entomology* 82, 803–810.
Stewart, J.G., Boiteau, G. and Kimpinski, J. (1998) Management of late-season adults of the Colorado potato beetle (Coleoptera: Chrysomelidae) with entomopathogenic nematodes. *The Canadian Entomologist* 130, 509–514.

32 *Lygus* spp., Plant Bugs (Hemiptera: Miridae)

A.B. Broadbent, P.G. Mason, S. Lachance, J.W. Whistlecraft, J.J. Soroka and U. Kuhlmann

Pest Status

Native plant bugs, *Lygus* spp., cause economic damage to a wide variety of agricultural crops. The tarnished plant bug, *Lygus lineolaris* (Palisot de Beauvois), is the most widespread of the 29 Nearctic species (Schwartz and Foottit, 1998). It is polyphagous (Young, 1986) and damages vegetables, fruits, greenhouse crops, canola, *Brassica napus* L. and *B. rapa* L. and legume crops, primarily those grown for seed, e.g. alfalfa, *Medicago sativa* L. Other pest species, abundant in agricultural crops in western Canada, include *L. borealis* (Kelton), *L. elisus* (Van Duzee), *L. hesperus* Knight, *L. keltoni* Schwartz and *L. shulli* Knight.

Adult and immature *Lygus* spp. feed by piercing plant tissues, secreting digestive enzymes and pumping out the liquefied plant material (Tingey and Pillemer, 1977). Feeding damage on tomato results in malformed and dimpled fruit with 'cloud-spotting' on ripe fruit. Damage to celery, *Apium graveolens* var. *dulce* (Miller) Persoon, lettuce, *Lactuca sativa* L., spinach, *Spinacia oleraceae* L., and chinese cabbage, *Brassica chinensis* L., results in increased susceptibility to bacterial diseases (Chaput and

Uyenaka, 1998). In strawberry, *Fragaria* × *ananassa* Duschesne, feeding causes the fruit to develop abnormally and causes 'cat-facing' (Udayagiri and Welter, 2000). In canola, feeding injury consists of lesions on the surfaces of stems, buds, flowers and pods that cause buds and flowers to abscize and seeds to collapse, reducing the weight of healthy seeds (Butts and Lamb, 1990a, b, 1991). In alfalfa, *Lygus* spp. injure flowers and young seeds, causing premature dehiscence of flowers, and seeds to become distorted, shrunken and non-viable (Soroka, 1997). In greenhouse crops, nymphs and/or adults can cause flower buds and fruit abortion, side-shoot abscission, leaf tissue perforation and deformation or death of meristem tissue (Gillespie and Foottit, 1997). In all these cases, the crop's market value is reduced. In Ontario, *Lygus* damage to about 5% of both fruit and vegetable crops results in more than Can$12 million in annual losses; in Saskatchewan the Can$50 million alfalfa seed industry can be completely destroyed by *Lygus* and other plant bugs if not treated (Soroka, 1997); in southern Alberta, in 1997, although 200,000 ha of canola were sprayed (Mason and Soroka, 1998) an estimated 20% of the crop, valued at more than Can$70 million, was lost due to *Lygus* damage.

All *Lygus* spp. overwinter as adults in refuges such as shelter belts, which provide maximum winter protection (Cleveland, 1982; Craig and Loan, 1987; Gerber and Wise, 1995; Schwartz and Foottit, 1998). When they become active in spring they move to the first plants in flower, generally weeds and volunteer crop plants, e.g. canola, where the females lay eggs. The first generation develops on these plant hosts and new adults disperse to the next group of flowering plants, which includes many crops. Depending on climate, 1–5 generations of *Lygus* spp. occur in Canada (Craig and Loan, 1987; Gerber and Wise, 1995; Schwartz and Foottit, 1998).

Background

Cultural practices, e.g. crop rotation and weed management, do not successfully control *Lygus* spp. Control has been effected primarily through chemical insecticides, but few of them are registered for *Lygus* spp. in Canada and their toxicity to pollinating insects is a concern, particularly since spraying is often done during maximum pollination periods. Concerns about non-target safety of insecticides to humans and beneficial insects also makes biological control an important alternative management strategy. In greenhouses, biological control programmes are well developed for other pests so use of pesticides to control *Lygus* spp. would kill natural enemies, disrupting these programmes (Gillespie *et al.*, 2000).

Lygus spp. are susceptible to certain pathogens, including *Beauveria bassiana* (Balsamo) Vuillemin (Bidochka *et al.*, 1993). Formulations of this fungus have been tried against *L. lineolaris* in cotton fields in southern USA with some success, especially if combined with the insecticide imidacloprid (Steinkraus and Tugwell, 1997).

The most abundant predators on *Lygus* in an apple orchard were Hemiptera and spiders (Araneae) (Arnoldi *et al.*, 1991). All stages of *Lygus* spp. are attacked by the native generalist predators *Geocoris bullatus* (Say), *G. pallens* (Ståhl), *Nabis alternatus* (Parshley), *N. americoferus* Carayon, *Nabicula subcoleoptrata* Kirby, *Podisus maculiventris* Say, *Phymata pennsylvanica* Hanlisch, *Zelus renardii* Kolenati, *Z. socius* Ståhl, *Philodromus praelustris* Keyserling, and *Xysticus punctatus* Keyserling (Mason and Soroka, 1998).

Several native parasitoids attack *Lygus* eggs, nymphs and adults (Table 32.1). At least four egg parasitoids are known (Al-Ghamdi *et al.*, 1995) and one, *Anaphes iole* Girault, is commercially available for *Lygus* control in the USA. In Canada, two nymphal, univoltine *Peristenus* spp. and two multivoltine *Leiophron* spp. are considered relatively ineffective, with low field parasitism of *Lygus* spp. In Europe, several *Peristenus* spp. had significant impact on *Lygus* spp. (Bilewicz-Pawinska, 1982; Day, 1987). One of these, *P. digoneutis* Loan, was released in the early 1980s in New Jersey and has established successfully in the north-eastern USA on *L. lineolaris* (Day *et*

Table 32.1. Parasitoids reared from *Lygus* spp. in North America (modified from Mason and Soroka, 1998).

Parasitoid	Reference
Egg parasitoids	
Mymaridae	
Anaphes iole Girault	Jackson and Graham (1983), Graham *et al.* (1986), Sohati *et al.* (1992)
Erythmelus miridiphagus Dozier	Sohati *et al.* (1992)
Polynema pratensiphagum Walley	Sohati *et al.* (1992)
Scelionidae	
Telenomus sp.	Sohati *et al.* (1992)
Nymphal parasitoids	
Braconidae	
Leiophron lygivorus Loan (eastern)	Loan (1969, 1980)
L. uniformis (Gahan) (eastern)	Clancy and Pierce (1966), Graham *et al.* (1986), Arnoldi *et al.* (1991)
Peristenus howardi Shaw (western)	Day *et al.* (1999)
P. pallipes (Curtis)[a]	Clancy and Pierce (1966), Loan and Craig (1976), Loan (1980), Day (1999)
P. pseudopallipes Loan (eastern)	Loan (1969, 1980), Day (1999)
Adult parasitoids	
Tachinidae	
Phasia aeneoventris (Williston)	Arnaud (1978)
P. fumosa (Coquillett)	Arnaud (1978)
P. opaca (Coquillett)	Arnaud (1978)
P. pulverea (Coquillett)	Arnaud (1978)

[a]H. Goulet (Ottawa, 1999, personnal communication), presently reviewing the Nearctic *Peristenus* and *Leiophron*, has indicated that *P. pallipes* is a complex of at least five species.

al., 1990, 1992, 1998), resulting in decreased *Lygus* densities on alfalfa (Day, 1996). Broadbent *et al.* (1999) confirmed the presence of *P. digoneutis* in southern Quebec.

Increasing problems with *Lygus* spp., particularly in canola in western Canada; reluctance to spray insecticides during peak pollination periods; and success of *P. digoneutis* introductions have renewed interest in introducing European parasitoids into Canada.

Biological Control Agents

Parasitoids

In Europe, *P. digoneutis*, *Peristenus stygicus* Loan and *Peristenus rubricollis* (Thomson) attack the important *Lygus* pest *L. rugulipennis* Poppius (Bilewicz-Pawinska, 1982; Craig and Loan, 1984). *P. digoneutis* also attacks the alfalfa plant bug, *Adelphocoris lineolatus* (Goeze), (Coulson, 1987). Parasitism rates on alfalfa, rye, *Secale cereale* L., potato, *Solanum tuberosum* L., wheat, *Triticum aestivum* L., and barley, *Hordeum vulgare* L., are usually between 2 and 34% for *L. rugulipennis* (Bilewicz-Pawinska, 1973, 1977, 1982). *P. digoneutis* also has been found in clover, *Trifolium pratense* L., and corn, *Zea mays* L. (Bilewicz-Pawinska, 1982). In Poland, after a diapause of about 8 months, *P. digoneutis* emerges a few days earlier in spring than the other parasitoids (Bilewicz-Pawinska, 1969, 1974, 1976). After winter diapause, when moved to 21°C, *P. digoneutis* emerges in 2–22 days, whereas *P. stygicus* emerges in 11–32 days. Adults of second-generation *P. digoneutis* emerge in July and August (Bilewicz-

Pawinska, 1974, 1982) and attack second-generation *L. rugulipennis* nymphs on potato, alfalfa and goldenrod, *Solidago* spp. (Bilewicz-Pawinska, 1973).

Although widely distributed in Europe, *P. stygicus* is less abundant than *P. digoneutis* and *P. rubricollis* but is more common in the south (Coutinot, Montpellier, 1998, personal communication). It develops in *L. rugulipennis, Trigonotylus caelestialium* (Kirkaldy) (Bilewicz-Pawinska, 1982), *Polymerus unifasciatus* (Fabricius) (Drea *et al.*, 1973), and *Adelphocoris lineolatus* Goeze (Coulson, 1987). It has also been reared from *Lygus* sp. and *Adelphocoris* sp. in southern France, Turkey, Spain and Greece (Coulson, 1987). In Poland, parasitism rates recorded on *L. rugulipennis* in alfalfa, rye, wheat, barley and oats were never over 25% (Bilewicz-Pawinska, 1973, 1982). Mirid food plants where *P. stygicus* have been collected include alfalfa (Loan and Bilewicz-Pawinska, 1973; Van Steenwyck and Stern, 1976; Bilewicz-Pawinska, 1982), rye and potato (Loan and Bilewicz-Pawinska, 1973; Bilewicz-Pawinska, 1982), wheat, barley, oats, grasses near cereals, clover, corn (Bilewicz-Pawinska, 1982) and asparagus, *Asparagus officinalis* L. (Drea *et al.*, 1973). Second-generation *P. stygicus* emerge in July and August to parasitize second-generation *L. rugulipennis* (Bilewicz-Pawinska, 1982).

The univoltine *P. rubricollis* attacks first-generation *L. rugulipennis* and *A. lineolatus* (Bilewicz-Pawinska, 1982; Craig and Loan, 1987). In Poland, parasitism on *L. rugulipennis* in alfalfa, wheat, barley and oats can vary from 1 to 85% (Loan and Bilewicz-Pawinska, 1973; Bilewicz-Pawinska, 1977, 1982). On rye, average parasitism of nymphs was about 30% during June and July, maximum parasitism occurred during the first 10 days of July, and mean parasitism of adult *L. rugulipennis* was lower, varying from 0.5 to 11.0% (Bilewicz-Pawinska, 1969). From 1976 to 1979, *P. rubricollis* was the dominant species in cereals (rye, wheat, oats, barley), accounting for 45–85% of all *Peristenus* spp. observed (Bilewicz-Pawinska, 1982). *P. rubricollis* seems adapted to conditions with lower temperature (Bilewicz-Pawinska, 1977), as it was collected mostly in northern and central Poland. *P. rubricollis* emerges later than *P. digoneutis* (9–32 days and 2–22 days, respectively, after being moved from overwintering conditions to 21°C) (Bilewicz-Pawinska, 1982).

Releases and Recoveries

In 1978 and 1981, small numbers of *P. digoneutis* (total of 1419 in Saskatchewan and 544 in Alberta) and *P. stygicus* (total of 64 in both provinces) were released in batches in western Canada (Craig and Loan, 1984). No recoveries of either species were made. Therefore, several releases of *Peristenus* spp. were attempted again in western Canada from 1981 to 2000 (Table 32.2). Populations introduced were primarily from northern and southern Europe.

Five introductions of *P. digoneutis* adults were made near Saskatoon to suppress *A. lineolatus* in 1985 and 1986 (see Soroka and Carl, Chapter 7 this volume) and these parasitoids may also have attacked and developed in *Lygus* spp. present in the alfalfa, but so far no recoveries have been made. In western Canada and the USA, *P. stygicus* has also been released (Craig and Loan, 1984; Coulson, 1987) but is not established (VanSteenwyck and Stern, 1977; Craig and Loan, 1984). Despite the poor dispersal of *P. stygicus*, it seems to possess several other desirable qualities for release, e.g. facultative diapause, short developmental time, high level of parasitism and ease of mass-rearing (Broadbent, 1976). *P. rubricollis* was released in Arizona and Texas in the early 1970s (Coulson, 1987), in Delaware in the late 1970s (Day *et al.*, 1992) and in Saskatchewan in the 1980s (Craig and Loan, 1984, 1987), but apparently has not established.

Evaluation of Biological Control

Successful establishment of *P. digoneutis* has occurred in north-eastern USA and

Table 32.2. Introduction of *Peristenus* spp. into Canada for laboratory studies or field release against *Lygus* spp., 1981–1999.

Year	Site[a]	Latitude Longitude	Lab study (L) or field release (F)	Parasitoid species introduced	Country of origin	Number and stage introduced
1981	Saskatoon, SK	52°07′N 106°38′W	F	*P. digoneutis* Loan	Austria	240 adults
1990	Saskatoon, SK	52°07′N 106°38′W	F	*P. digoneutis*	USA	58 adults
1991	Saskatoon, SK	52°07′N 106°38′W	F	*P. digoneutis*	USA	19 adults
1992	St Jean-sur-Richelieu, QC	45°30′N 73°27′W	L	*P. digoneutis*	USA	22 adults
1992	Saskatoon, SK	52°07′N 106°38′W	F	*P. digoneutis*	USA	30 adults
1992	Guelph, ON	43°33′N 80°15′W	L	*P. digoneutis*	USA	13 adults
1994	Saskatoon, SK	52°07′N 106°38′W	L	*Peristenus* spp.	France	23 adults
1995	Saskatoon, SK	52°07′N 106°38′W	L	*Peristenus* spp.	France and Germany	492 cocoons
1996	London, ON	43°02′N 81°12′W	L	*P. digoneutis* *P. stygicus* Loan	Hungary and Switzerland	>1000 cocoons
1997	Reford, SK	52°14′N 108°18′W	F	*P. digoneutis*	Hungary	425 adults
1997	London, ON	43°02′N 81°12′W	L	*P. digoneutis* *P. stygicus*	Switzerland and Germany	485 cocoons
1998	London, ON	43°02′N 81°12′W	L	*P. digoneutis* *P. stygicus*	Germany and Italy	1238 cocoons
1999	London, ON	43°02′N 81°12′W	L	*P. digoneutis* *P. stygicus*	Germany, Italy and Switzerland	948 cocoons

[a] Ontario (ON), Quebec (QC), Saskatchewan (SK).

southern Quebec. Past introductions of *P. digoneutis*, *P. stygicus* and *P. rubricollis* into western Canada appear to have been unsuccessful, perhaps due to an inadequate number of adults released, poorly adapted populations or a male-biased sex ratio.

Recent concerns about the non-target host impact of biological control agents has led to more intensive study of the *Peristenus* candidates for introduction. Condit and Cate (1982) showed that in the laboratory *P. stygicus* will attack and complete development in *L. hesperus*, *L. lineolaris* and *Polymerus basalis* (Reuter), *Lindbergocapsus geminatus* (Johnston)[1] and *Pseudatomoscelis seriatus* (Reuter). Partial development was observed in the *Dicrooscytus* sp. (Mirinae); only attacks but no development on *Plagiognathus maculipennis* (Knight)[2] (Phylinae) and one species of Orthotylinae; and no attack on *Taedia johnstoni* (Knight) (Mirinae), two species of Bryocorinae and three species of Lygaeidae. These results need to be verified in the field in the area of origin (Kuhlmann *et al.*, 1998).

Recommendations

Future work should include:

1. Monitoring dispersal of the Quebec population of *P. digoneutis*;
2. Caged releases of mass-reared *P. digoneutis* to study establishment under Ontario conditions;
3. Post-release monitoring to determine percentage parasitism by *P. digoneutis* near release sites in various crops, particularly alfalfa;
4. Evaluating the impact of *P. digoneutis* on parasitism by native *Peristenus* and *Leiophron* spp.;

[1] *Labopidicola geminata* (Johnston) in Condit and Cate (1982).
[2] *Microphylellus maculipennis* (Knight) in Condit and Cate (1982).

5. Evaluating the biology and suitability of *P. stygicus* and *P. rubricollis* for future introductions into Canada, including European studies to assess the natural host range of *P. digoneutis*, *P. stygicus* and *P. rubricollis*, and host specificity testing in Europe and Canada to determine potential non-target impacts;

6. Evaluating pathogens for use as inundative microbial agents of *Lygus* spp.;

7. Evaluating the potential of *A. iole* as an inundative agent in high-value crops, e.g. strawberries;

8. Developing habitat management practices that enhance parasitism levels in non-crop and crop habitats.

References

Al-Ghamdi, K.M., Stewart, R.K. and Boivin, G. (1995) Synchrony between populations of the tarnished plant bug, *Lygus lineolaris* (Palisot de Beauvois) (Hemiptera: Miridae), and its egg parasitoids in southwestern Quebec. *The Canadian Entomologist* 127, 457–472.

Arnaud, P.H. (1978) *A Host–parasite Catalog of North American Diptera.* United States Department of Agriculture, Miscellaneous Publication No. 1319.

Arnoldi, D., Stewart, R.K. and Boivin, G. (1991) Field survey and laboratory evaluation of the predator complex of *Lygus lineolaris* and *Lygocoris communis* (Hemiptera: Miridae) in apple orchards. *Journal of Economic Entomology* 84, 830–836.

Bidochka, M.J., Miranpuri, G.S. and Khachatourians, G.G. (1993) Pathogenicity of *Beauveria bassiana* (Balsamo) Vuillemin toward lygus bug (Hemiptera: Miridae). *Journal of Applied Entomology* 115, 313–317.

Bilewicz-Pawinska, T. (1969) Natural limitation of *Lygus rugulipennis* Popp. by group of *Leiophron pallipes* Curtis on the rye crop fields. *Ekologia Polska* 16, 811–825.

Bilewicz-Pawinska, T. (1973) Uwagi o trzech gatunkach *Peristenus* Foerster (Hym., Braconidae) i ich pasozytach *Mesochorus* spp. (Hym., Ichneumonidae). *Polskie Pismo Entomologiczne* 44, 759–764.

Bilewicz-Pawinska, T. (1974) Emergence and longevity of two species of *Peristenus* Foerster (Braconidae) under laboratory conditions. *Ekologia Polska* 22, 213–222.

Bilewicz-Pawinska, T. (1976) Distribution of the insect parasites *Peristenus* Foerster and *Mesochorus* Gravenhorst in Poland. *Bulletin de l'Academie Polonaise des Sciences* 23, 823–827.

Bilewicz-Pawinska, T. (1977) Parasitism of *Adelphocoris lineolatus* Goeze and *Lygus rugulipennis* Popp. (Heteroptera) by braconids and their occurrence on lucerne. *Ekologia Polska* 25, 539–550.

Bilewicz-Pawinska, T. (1982) Plant bugs (Heteroptera: Miridae) and their parasitoids (Hymenoptera: Braconidae) on cereal crops. *Polish Ecological Studies* 8, 113–191.

Broadbent, A.B. (1976) Laboratory studies on the biology of *Peristenus stygicus* Loan (Hymenoptera: Braconidae), a parasitoid of *Lygus lineolaris* (P. de B.) (Hemiptera: Miridae). MSc Thesis, McGill University, Montreal, Quebec.

Broadbent, A.B., Goulet, H., Whistlecraft, J.W., Lachance, S. and Mason, P.G. (1999) First Canadian record of three parasitoid species (Hymenoptera: Braconidae: Euphorinae) of the tarnished plant bug, *Lygus lineolaris* (Hemiptera: Miridae). *Proceedings of the Entomological Society of Ontario* 130, 109–111.

Butts, R.A. and Lamb, R.J. (1990a) Injury to oilseed rape caused by mirid bugs (*Lygus*) (Heteroptera: Miridae) and its effect on seed production. *Annals of Applied Biology* 117, 253–266.

Butts, R.A. and Lamb, R.J. (1990b) Seasonal abundance of three *Lygus* species (Heteroptera: Miridae) in oilseed rape and lucerne in Alberta. *Journal of Economic Entomology* 84, 450–456.

Butts, R.A. and Lamb, R.J. (1991) Pest status of *Lygus* bugs (Hemiptera: Miridae) in oilseed *Brassica* crops. *Journal of Economic Entomology* 84, 1591–1596.

Chaput, J. and Uyenaka, J. (1998) *Tarnished Plant Bug Damage in Vegetable Crops in Ontario.* Fact Sheet No. 98–025. Ontario Ministry of Agriculture, Food and Rural Affairs.

Clancy, D.W. and Pierce, H.D. (1966) Natural enemies of some *Lygus* bugs. *Journal of Economic Entomology* 59, 853–858.

Cleveland, T.C. (1982) Hibernation and host plant sequence studies of tarnished plant bugs, *Lygus lineolaris*, in the Mississippi delta. *Environmental Entomology* 11, 1049–1052.

Condit, B.P. and Cate, J.R. (1982) Determination of host range in relation to systematics for *Peristenus stygicus* (Hymenoptera: Braconidae), a parasite of Miridae. *Entomophaga* 27, 203–210.
Coulson, J.R. (1987) Studies on the biological control of plants bugs (Heteroptera: Miridae): an introduction and history, 1961–83. In: Hedlund, R.C. and Graham, H.M. (eds) *Economic Importance and Biological Control of* Lygus *and* Adelphocoris *in North America.* ARS-64, United States Department of Agriculture, Agricultural Research Service, pp. 1–12.
Craig, C.H. and Loan, C.C. (1984) *Lygus* spp., plant bugs (Heteroptera: Miridae). In: Kelleher, J.S. and Hulme, M.A. (eds) *Biological Control Programmes against Insects and Weeds in Canada 1969–1980.* Commonwealth Agricultural Bureaux, Farnham Royal, UK, pp. 45–47.
Craig, C.H. and Loan, C.C. (1987) Biological control efforts on Miridae in Canada. In: Hedlund, R.C. and Graham, H.M. (eds) *Economic Importance and Biological Control of* Lygus *and* Adelphocorus *in North America.* ARS-64, United States Department of Agriculture, Agriculture Research Service, pp. 48–53.
Day, W.H. (1987) Biological control efforts against *Lygus* and *Adelphocoris* spp. infesting lucerne in the United States, with notes on other associated mirid species. In: Hedlund, R.C. and Graham, H.M. (eds) *Economic Importance and Biological Control of* Lygus *and* Adelphocorus *in North America.* ARS-64, United States Department of Agriculture, Agriculture Research Service, pp. 20–39.
Day, W.H. (1996) Evaluation of biological control of the tarnished plant bug (Hemiptera: Miridae) in lucerne by the introduced parasite *Peristenus digoneutis* (Hemiptera: Braconidae). *Environmental Entomology* 25, 512–518.
Day, W.H. (1999) Host preferences of introduced and native parasites (Hymenoptera: Braconidae) of phytophagous plant bugs (Hemiptera: Miridae) in alfalfa-grass fields in the northeastern USA. *Biocontrol* 44, 249–261.
Day, W.H., Hedlund, R.C., Saunders, L.B. and Coutinot, D. (1990) Establishment of *Peristenus digoneutis* (Hymenoptera: Braconidae), a parasite of the tarnished plant bug (Hemiptera: Miridae), in the United States. *Environmental Entomology* 19, 1528–1533.
Day, W.H., Marsh, P.M., Fuester, R.W., Hoyer, H. and Dysart, R.J. (1992) Biology, initial effect, and description of a new species of *Peristenus* (Hymenoptera: Braconidae), a parasite of the lucerne plant bug (Hemiptera: Miridae), recently established in the United States. *Annals of the Entomological Society of America* 85, 482–488.
Day, W.H., Tropp, J.M., Eaton, A.T., Romig, R.F., Driesche, R.G.V. and Chianese, R.J. (1998) Geographic distributions of *Peristenus conradi* and *P. digoneutis* (Hymenoptera: Braconidae), parasites of the lucerne plant bug and the tarnished plant bug (Hemiptera: Miridae) in the northeastern United States. *Journal of the New York Entomological Society* 106, 69–75.
Day, W.H., Baird, C.R. and Shaw, S.R. (1999) New native species of *Peristenus* (Hymenoptera: Braconidae) parasitizing *Lygus hesperus* (Hemiptera: Miridae) in Idaho: Biology, importance, and description. *Annals of the Entomological Society of America* 92, 370–375.
Drea, J.J., Dureseau, L. and Rivet, E. (1973) Biology of *Peristenus stygicus* from Turkey, a potential natural enemy of *Lygus* bugs in North America. *Environmental Entomology* 2, 278–280.
Gerber, G.H. and Wise, I.L. (1995) Seasonal occurrence and number of generations of *Lygus lineolaris* and *L. borealis* (Heteroptera: Miridae) in southern Manitoba. *The Canadian Entomologist* 127, 543–559.
Gillespie, D. and Foottit, R. (1997) *Lygus* bugs in vegetable greenhouses in B.C. In: Soroka, J.J. (ed.) *Proceedings of the* Lygus *Working Group Meeting, 11–12 April 1996, Winnipeg, Manitoba.* Agriculture and Agri-Food Canada, Research Branch, Saskatoon, Saskatchewan, pp. 7–9.
Gillespie, D., Foottit, R. and Shipp, J.L. (2000) Management of *Lygus* bugs on protected crops. In: Foottit, R. and Mason, P. (eds) *Proceedings of the* Lygus *Working Group Meeting, 26 September 1999, Saskatoon, Saskatchewan.* Agriculture and Agri-Food Canada, Research Branch, Ottawa, Ontario, pp. 1–8.
Graham, H.M., Jackson, C.G. and Debolt, J.W. (1986) *Lygus* spp. (Hemiptera: Miridae) and their parasites in agricultural areas of southern Arizona. *Environmental Entomology* 15, 132–142.
Jackson, C.G. and Graham, H.M. (1983) Parasitism of four species of *Lygus* (Hemiptera: Miridae) by *Anaphes ovijentatus* (Hymenoptera: Mymaridae) and an evaluation of other possible hosts. *Annals of the Entomological Society of America* 76, 772–775.
Kuhlmann, U., Mason, P.G. and Greathead, D.J. (1998) Assessment of potential risks for introducing European *Peristenus* species as biological control agents of native *Lygus* species in North America: a cooperative approach. *Biocontrol News and Information* 19, 83N-90N.

Loan, C.C. (1969) Two new parasites of the tarnished plant bug in Ontario: *Leiophron pseudopallipes* and *Euphoriana lygivora* (Hymenoptera: Braconidae, Euphorinae). *Proceedings of the Entomological Society of Ontario* 100, 188–194.
Loan, C.C. (1980) Plant bug hosts (Heteroptera: Miridae) of some Euphorine parasites (Hymenoptera: Braconidae) near Belleville, Ontario, Canada. *Le Naturaliste Canadien* 107, 87–93.
Loan, C.C. and Bilewicz-Pawinska, T. (1973) Systematics and biology of four Polish species of *Peristenus* Foerster (Hymenoptera: Braconidae, Euphorinae). *Environmental Entomology* 2, 271–278.
Loan, C.C. and Craig, C.H. (1976) Euphorine parasitism of *Lygus* spp. in lucerne in western Canada (Hymenoptera: Braconidae; Heteroptera: Miridae). *Le Naturaliste Canadien* 103, 497–500.
Mason, P.G. and Soroka, J.J. (1998) Plant bugs (*Lygus* spp.) An emerging problem in canola. In: *Soils and Crops '98.* Extension Division, University of Saskatchewan, Saskatoon, pp. 177–183.
Schwartz, M.D. and Foottit, R.G. (1998) *Revision of the Nearctic species of the genus* Lygus *Hahn, with a review of the Palaearctic species (Heteroptera: Miridae).* Associated Publishers, Gainesville, Florida.
Sohati, P.H., Boivin, G. and Stewart, R.K. (1992) Parasitism of *Lygus lineolaris* eggs on *Coronilla varia*, *Solanum tuberosum*, and three host weeds in southeastern Quebec. *Entomophaga* 37, 515–523.
Soroka, J.J. (1997) Plant bugs in lucerne. In: Soroka, J.J. (ed.) *Proceedings of the* Lygus *Working Group Meeting, 11–12 April 1996, Winnipeg, Manitoba.* Agriculture and Agri-Food Canada, Research Branch, Saskatoon, Saskatchewan, pp. 4–6.
Steinkraus, D.C. and Tugwell, N.P. (1997) *Beauveria bassiana* (Deuteromycotina: Moniliales) effects on *Lygus lineolaris* (Hemiptera: Miridae). *Journal of Entomological Science* 32, 79–90.
Tingey, W.M. and Pillemer, E.A. (1977) Lygus bugs: Crop resistance and physiological nature of feeding injury. *Bulletin of the Entomological Society of America* 23, 277–287.
Udayagiri, S. and Welter, S.C. (2000) Escape of *Lygus hesperus* (Heteroptera: Miridae) eggs from parasitism by *Anaphes iole* (Hymenoptera: Mymaridae) in strawberries: plant structure effects. *Biological Control* 17, 234–242.
VanSteenwyk, R.A. and Stern, V.M. (1976) The biology of *Peristenus stygicus* (Hymenoptera: Braconidae) a newly imported parasite of Lygus bugs. *Environmental Entomology* 5, 931–934.
VanSteenwyk, R.A. and Stern, V.M. (1977) Propagation, release, and evaluation of *Peristenus stygicus*, a newly imported parasite of Lygus bugs. *Journal of Economic Entomology* 70, 66–69.
Young, O.P. (1986) Host plants of the tarnished plant bug, *Lygus lineolaris* (Heteroptera: Miridae). *Annals of the Entomological Society of America* 79, 747–762.

33 *Lymantria dispar* (L.), Gypsy Moth (Lepidoptera: Lymantriidae)

V.G. Nealis, N. Carter, M. Kenis, F.W. Quednau and K. van Frankenhuyzen

Pest status

Gypsy moth, *Lymantria dispar* (L.), is one of the most notorious non-indigenous defoliators of broadleaf trees in Canada. Doane and McManus (1981) documented its spread in North America from an accidental introduction near Boston in 1869.

Griffiths and Quednau (1984) summarized its spread in Canada to 1979. Since 1980, *L. dispar* has greatly expanded its range in Canada. As of 2000, populations have become established throughout the St Lawrence–Great Lakes forests of Quebec and Ontario as far west and north as Lake Superior and eastward to Nova Scotia and New Brunswick (Nealis and Erb, 1993). Isolated infestations of the European strain have been found repeatedly in British Columbia since 1980, mostly associated with inadvertent movement of egg masses from eastern Canada. On several occasions, populations have persisted for more than one generation and eradication programmes have been undertaken. An introduction of the Asian strain to Vancouver from ships originating in Russian ports in 1991 also led to an eradication programme in 1992 (Humble and Stewart, 1994).

Establishment of *L. dispar* has frequently resulted in severe defoliation of primary host trees, especially oaks, *Quercus* spp. From 1981 to 1996, more than 1 million ha of moderate-to-severe defoliation were mapped in Ontario (Nealis *et al.*, 1999). Although the immediate impact of defoliation is obvious, our understanding of more long-term ecological impacts is fragmentary (Davidson *et al.*, 1999). Economic impacts are undeniable. For over 100 years, governments and private landowners have used various insecticides to control *L. dispar*. In addition to the cost of insecticides and their application, the ecological and social costs of spray programmes are increasingly debated. Costs associated with trade conditions imposed on regions infested by *L. dispar* also have become more sharply focused as global movement of commodities increases and uninfested jurisdictions attempt to maintain their gypsy moth-free status (Wallner, 1996).

Background

Griffiths and Quednau (1984) reported the presence in Canada of introduced natural enemies derived from extensive biological control programmes in the USA as well as their own releases of egg parasitoids. The earliest work with microbial insecticides had also begun at the time of their report. Despite these developments, Griffiths and Quednau (1984) questioned the value of further work on biological control because of the limited range and relatively low levels of *L. dispar*. Since then the established range of *L. dispar* has expanded greatly, resulting in large-scale annual suppression programmes in Ontario and eradication programmes in British Columbia and New Brunswick (see Table 33.1) (Jobin, 1995). Increasing public criticism of these programmes, especially in semi-urban habitats typical of *L. dispar* infestations, has led to a modest revival of interest in biological control alternatives. Most resources, however, have been directed at replacing chemical insecticides with microbial insecticides such as *Bacillus thuringiensis* Berliner serovar *kurstaki* (*B.t.k.*) and *Nucleopolyhedrovirus* isolated from *L. dispar* (LydiNPV). Hence, repeated and extensive aerial application of insecticides has remained the principal method used to suppress or eradicate *L. dispar* populations.

Griffiths (1976) listed 22 parasitoid species and 17 arthropod predators native to North America that attack *L. dispar*. Many of these also are native to Canada but relatively few have been recorded attacking *L. dispar* in Canadian forests (Griffiths and Quednau, 1984; Nealis *et al.*, 1999), probably reflecting the paucity of natural enemy surveys in Canada rather than an ecological situation greatly different from that in the USA.

Two of the most common and widespread introduced parasitoids of *L. dispar* are *Cotesia melanoscela* (Ratzeburg) and *Compsilura concinnata* (Meigen). They probably originated from a combination of successful releases in Canada against the related satin moth, *Leucoma salicis* (L.), and brown-tail moth, *Euproctis chrysorrhea* (L.) (McGugan and Coppel, 1962), and natural dispersal from successful releases in the USA. Other parasitoids released against these introduced lymantriids are recorded as parasitoids of *L. dispar*, but have not yet been observed attacking this host in Canada. These include *Meteorus versicolor* (Wesmael) and *Dolichogenidea lacteicolor* Viereck (McGugan and Coppel, 1962). The carabid *Calosoma sycophanta* L. was released into Canada in the early 1900s (McGugan and Coppel, 1962) but has not been recovered since (Griffiths and Quednau, 1984; D. Roden, Ontario, 1999, personal communication) despite its apparent success in some areas of the USA (Weseloh, 1985).

All other non-indigenous parasitoids reported attacking *L. dispar* (Griffiths and Quednau, 1984; Nealis *et al.*, 1999) except possibly the egg parasitoids *Ooencyrtus kuvanae* (Howard) and *Anastatus japonicus* Ashmead (formerly *Anastatus disparis* Ruschka), dispersed naturally from their established ranges in the USA. *O. kuvanae* was introduced to Canada near Kingston in 1976 and quickly became established and widespread throughout the expanding range of *L. dispar* (Griffiths and Quednau, 1984). In 1990, *O. kuvanae* was found in virtually every sampled population of *L. dispar* in southern Ontario (V. Nealis, unpublished). It was also recovered once from *L. dispar* in New Brunswick (Smith and Harrison, 1995) and has been reported to occur in Maine (Bradbury, 1991). Although established in Ontario, *A. japonicus* is uncommon and apparently has not moved beyond the release sites (Griffiths and Quednau, 1984; V. Nealis, unpublished). Nearly 6000 specimens were released in south-western New Brunswick in 1983 but no egg masses were found subsequently, so success of the release could not be measured (Magasi, 1984). It was not until 1996 that *A. japonicus* was encountered again in New Brunswick (Carter, 1996). Because *A. japonicus* also occurs in Maine (Bradbury, 1994), the population in New Brunswick may have dispersed from there.

Pathogens of *L. dispar* have not been surveyed extensively in Canada. Nealis *et al.* (1999) found the ubiquitous native fungi *Paecilomyces farinosus* (Holmskjold) A.H.S. Brown and G. Smith and *Beauveria bassiana* (Balsamo) Vuillemin present, but relatively uncommon, in Ontario *L. dispar* populations. An unidentified but common pathogen reported in that study raises the possibility that little-known native pathogens might play an important role in reducing *L. dispar* populations in some locations. In New Brunswick, Carter and Kettela (1993) reported *Paecilomyces* sp. and Lavigne and Carter (1996) added records of *B. bassiana* and *Verticillium* sp. to the list of native pathogens that infect *L. dispar.* The virus, LydiNPV, and the fungus, *Entomophaga maimaiga* Humber, Shimazu and Soper, have spread naturally throughout the range of *L. dispar* in Ontario (Nealis *et al.*, 1999), New Brunswick (Carter and Kettela, 1993) and, more recently, Nova Scotia (E. Georgeson, Halifax, 2000, personal communication).

Biological Control Agents

Pathogens

Bacteria

Commercial formulations of *B.t.k.* replaced the use of all synthetic insecticides in operational *L. dispar* control programmes in Canada after 1983. From 1985 to 1991, annual suppression programmes were carried out in Ontario (Table 33.1) (Jobin, 1995). Aerial spray programmes in other provinces were aimed mostly at eradicating or preventing the spread of small incipient infestations, e.g. the 1992 programme in the lower mainland of British Columbia to eliminate an infestation of the Asian and European strains of *L. dispar*, and the aerial spray programme on Vancouver Island in 1999 to eradicate the European strain. From 1981 to 1999, about 275,000 ha were treated with *B.t.k.* in Canada, with a total of about 20×10^{15} international units (IU) (Table 33.1).

Initial operational use of *B.t.k.* involved application of diluted product at 30×10^9 IU in 6.0 l ha^{-1}. When this was shown to be ineffective, application of undiluted high-potency products as two sprays of 30×10^9 IU ha^{-1} became the operational stan-

Table 33.1. Operational use of *Bacillus thuringiensis* against *Lymantria dispar* since 1980.

Year	Province	No. ha treated[a]	Dose applied[b]
1981	Quebec	29	870
1982	Ontario	270	13,120
1983	New Brunswick	182	16,380
1984	British Columbia	10	300
1985	Ontario	170	6,800
	British Columbia	160	14,400
1986	Ontario	103,094	6,488,220
	British Columbia	5	450
1987	Ontario	40,249	2,414,940
	British Columbia	25	2,250
1988	Ontario	13,784	827,040
	British Columbia	112	10,080
	New Brunswick	391	35,190
1989	Ontario	12,951	777,060
1990	Ontario	33,956	2,037,360
1991	Ontario	36,577	2,194,620
1992[c]	British Columbia	20,000	4,000,000
1993	British Columbia	730	131,000
1994	British Columbia	692	103,800
1995	British Columbia	352	52,800
1996	British Columbia	120	18,000
1999[d]	British Columbia	10,807	1,621,050
Total	All provinces	274,268	20,732,360

[a] Number of hectares treated with one or more applications.
[b] Total dose (expressed in 10^9 International Units) applied per ha (= number of ha treated × number of applications × 10^9 IU ha^{-1} per application).
[c] Asian gypsy moth eradication programme, Lower Mainland.
[d] European gypsy moth eradication programme, Vancouver Island.

dard for foliage protection (van Frankenhuyzen *et al.*, 1991). For eradication of incipient outbreaks, higher dosage rates (50×10^9 IU ha^{-1}) are used in 3–4 applications.

Viruses

A product containing LydiNPV was developed and registered as Disparvirus® in 1996 (Cunningham, 1998). Disparvirus® contains the same strain of LydiNPV registered in the USA under the name Gypchek® (Reardon *et al.*, 1996). A total of 784 ha was treated experimentally in Ontario with one or more applications of either Gypchek® or Disparvirus® from 1982 to 1994 (Table 33.2) (Cunningham and Kaupp, 1995; Cunningham *et al.*, 1996, 1997), demonstrating the effectiveness of application volumes as low as 2.5 l ha^{-1}.

Lack of Canadian registration before 1996 and of a commercial product were the main reasons for no operational use of Disparvirus® in Canada. From 1992 to 1995, the Canadian Forest Service and USDA Forest Service collaborated with American Cyanamid towards the development of a production facility, but the initiative was abandoned a year later. The Canadian Forest Service no longer produces Disparvirus® but the registration is still active and available for commercialization.

An alternative to aerial application of LydiNPV was tested in New Brunswick in 1995 in an effort to induce an epizootic in a recently established population of *L. dispar* (Lavigne and Carter, 1996). Five topical applications of virus were made directly to

Table 33.2. Experimental aerial applications of *Lymantria dispar Nucleopolyhedro virus*.

Year	No. ha treated	Dose (PIB ha^{-1})[a]	Tank mix (in water)	Volume (l ha^{-1})
1982	63	50×10^{11}	25% emulsifiable oil	18.8
1986	10	4.4×10^{12}	25% molasses, 6% Orzan LS	9.4
	88	5.4×10^{11}	25% molasses, 6% Orzan LS	9.4
1988	64	2.5×10^{12}	25% molasses, 6% Orzan LS	10.0
1989	90	1.0×10^{12}	25% molasses, 10% Orzan LS, Rhoplex	5.0, 10.0
1990	60	1.0×10^{12}	25% molasses, 10% Orzan LS, Rhoplex	2.5, 5.0
	30	1.0×10^{12}	25% emulsifiable oil	5.0
1992	48	1.0×10^{12}	25% molasses, 6% Orzan LS, 2% Bond	5.0
	43	1.0×10^{12}	American Cyanamid WP	5.0
	38	1.0×10^{11}	American Cyanamid WP, 1% Blankophor BBH	5.0
1993	100	1.0×10^{11}	American Cyanamid WP, 1% Blankophor BBH	5.0
	50	5.0×10^{10}	American Cyanamid WP, 1% Blankophor BBH	5.0
1994	100	1.0×10^{12}	Novo Carrier 244	2.5, 5.0

[a] Total dose (expressed in polyhedral inclusion bodies) applied per hectare in one or two applications.

the surface of 1570 egg masses on three private properties over a 3–4 day period. Rearing larvae collected biweekly between the beginning of June and the end of July, followed by use of a DNA probe and microscopic diagnosis of cadavers, revealed infection levels of 70–80% in later instars in two sites and about 40% in the third site. No viral infection was detected in larvae reared from egg masses collected before treatment, although DNA probing indicated that the virus was present at a low incidence (<10%). Associated reductions in egg mass densities ranged from 17% in the low-density and low-infection site to 65% in the high-density and high-infection site.

Parasitoids

Biological control efforts with parasitoids since 1980 have emphasized using more effective strains of established parasitoids and finding parasitoids associated with low-density levels of *L. dispar* in its native range.

Several species of parasitoids were imported from Europe or the USA and used in laboratory investigations but never released in Canada. These included *Glyptapanteles liparidis* (Bouché), *Glyptapanteles flavicoxis* (Marsh), *Hyposoter lymantriae* Cushman and *Parasetigena silvestris* (Robineau-Desvoidy). In 1987, release of about 1500 *G. flavicoxis* in southern Ontario by a private interest provided no evidence of successful parasitism.

In Ontario, high rates of hyperparasitism in *C. melanoscela* led to a programme to test the relative effectiveness of an Asian strain of this parasitoid, reputedly less vulnerable to hyperparasitism. Field trials showed that this Asian strain did indeed have higher levels of survival from native hyperparasitoids, because of the combination of its cocoon structure and its non-diapause characteristics (Nealis and Bourchier, 1995). This led to the recommendation that inundative releases of *C. melanoscela* should use non-diapause strains and should be made very early in the *L. dispar* larval period to avoid the period of high activity by hyperparasitoids and also to take advantage of a natural, second generation of the released parasitoids.

In 1980, a project began to survey low-density populations of *L. dispar* in Europe. Small trees were 'seeded' with larvae and these were subsequently collected and reared to recover parasitoids that had attacked them. The result was the discovery of the little-known parasitoid, *Aphantorhaphopsis samarensis* (Villeneuve) formerly reported as *Siphona samarensis* or *Ceranthia samarensis* (Mills, 1990). Since then the greatest emphasis in *L. dispar* biological control using parasitoids has been collection, rearing and release of this tachinid (Mills and Nealis, 1992; Nealis and Quednau, 1996). This programme was based on the premise that in relatively stable forest ecosystems a marked difference exists between natural enemy complexes attacking outbreak and non-outbreak phases of the same forest defoliators, and that the role of natural enemies may be different at these different phases (Pschorn-Walcher, 1977; Mills, 1990). Moreover, most of the parasitoids that dominate parasitism in outbreak populations of *L. dispar* in Europe had been established in North America during earlier biological control programmes and new opportunities from outbreak populations seemed limited.

From 1980 to 1998, except for 1996 when no *L. dispar* larvae were available, an annual programme of release, recollection and rearing of larvae was carried out at several sites in south-eastern France and neighbouring Germany and Switzerland. Although many species of *L. dispar* parasitoids were recovered, *A. samarensis* was found in almost every European location sampled although parasitism rates varied greatly. Spring weather and *L. dispar* density at the study site were important determinants of percentage parasitism by *A. samarensis*. Highest parasitism rates were observed in areas where natural *L. dispar* populations never reached outbreak densities or when the density of seeded *L. dispar* larvae was low because of poor survival. *A. samarensis* thus appears to be most effective in low-density *L. dispar* populations. One site, Plancher Bas in eastern France, consistently provided the highest parasitism, e.g. a mean of 53.6% and a maximum of 89.5% in 1997, and absolute number of *A. samarensis*. This species dominated the total parasitoid fauna at this site for 10 years.

Seasonal activity, stage of host attacked, environmental conditions inducing diapause, hyperparasitism, alternative hosts, etc. were investigated during the programme of collection, rearing, screening and release of *A. samarensis*. The adult parasitoid was found to be active in spring, attacking all stages of *L. dispar* larvae. Progeny exited from the third to sixth instars. Usually only one parasitoid is produced per host larva but occasionally as many as three maggots have been found exiting a late-instar host. Diapause in the pharate adult is facultative, although natural populations in Europe are mostly univoltine (Mills and Nealis, 1992). Continuous development in both the field and laboratory is promoted by warm conditions (>20°C) and diapause is induced if puparia are exposed to cooler, fluctuating temperatures (Quednau and Lamontagne, 1998). Once in diapause, *A. samarensis* requires at least 3 months of cold storage (2°C) but can survive as much as 10 months cold storage (Mills and Nealis, 1992). Survival in cold storage is greatly enhanced by covering the puparia in peat moss kept constantly moist (Quednau and Lamontagne, 1998).

Considerable effort has been expended in Europe to find alternate hosts of *A. samarensis*. More than 600 individual caterpillars belonging to 35 species in ten families have been screened. Only two species, both lymantriids, produced a few puparia that resembled *A. samarensis*. It is concluded that *A. samarensis* is host specific.

Releases and Recoveries

From 1984 to 1999, more than 10,000 *A. samarensis* were shipped to Canada as either active adults or diapausing insects in puparia, mostly from host exposures at Plancher Bas. The remainder came from other European localities or were produced in the laboratory from captive adults. After screening for hyperparasitoids, puparia were stored at 2°C. The first small-scale releases of insects imported directly from Europe were made in Ontario (Mills and Nealis, 1992). Eventually, critical features of the reproductive biology of *A. samarensis* were elucidated (Quednau, 1993) and an effective, albeit time-consuming, rearing programme was developed (Quednau and Lamontagne, 1998). The increased availability of insects resulting from the rearing programme made it possible to carry out additional investigations of the feasibility of establishing this parasitoid in Canada, and to share material with colleagues in the USA to try and improve rearing capabilities (Kauffman *et al.*, 1996). With an established colony annually infused with wild stock from Europe, a series of both caged and open releases of gravid female *A. samarensis* was carried out after 1990, first in Ontario (Nealis and Quednau 1996) and more recently in New Brunswick (D. Lavigne and N. Carter, Fredericton, 1997–1998, personal communication) and in Pennsylvania, USA (M. Blumenthal, Mifflin County, Pennsylvania, 1999, personal communication). Canadian releases were accompanied by observations on overwintering survival of *A. samarensis* and confirmed that it not only survives winters in southern Ontario and New Brunswick but that adult eclosion the following spring is well synchronized with the seasonal appearance of local *L. dispar* at their most vulnerable stages (Nealis and Quednau, 1996; D. Lavigne and N. Carter, Fredericton, 2000, personal communication).

Evaluation of Biological Control

B.t.k. has proven to be an effective control agent against *L. dispar*. Although rigorous evaluation of efficacy in suppression programmes is somewhat elusive, the demonstrable success of *B.t.k.* in eradication programmes ensures its continued use in most operational contexts.

LydiNPV has also proven to be a useful natural control against *L. dispar*. Availability of the product appears to be the single, greatest impediment to its more widespread use. None the less, public acceptance will most certainly be a challenge, especially if use is contemplated in urban settings.

Many *L. dispar* arthropod natural enemies in Canada are generalists commonly associated with outbreak populations both in Europe and the USA (Nealis *et al.*, 1999). The capability of these generalists to function effectively as biological control agents and to regulate *L. dispar* populations at lower densities is debatable. Without population studies of *L. dispar* in Canada, this issue will not be resolved. More specialized endoparasitoids such as *C. melanoscela* may be able to respond numerically to increases in *L. dispar* abundance, but suffer, in turn, high mortality from generalist, native hyperparasitoids (Bourchier and Nealis, 1992). Egg parasitoids, although widely distributed, appear to attack only a small proportion of available eggs within *L. dispar* egg masses and are therefore of limited benefit.

It is still too early to evaluate the releases of *A. samarensis* in terms of *L. dispar* control. This biological control programme has successfully completed its initial objectives, including discovery of a new, host-specific biological control agent that functions at a different phase of the outbreak cycle, development of a rearing system for multiplication of the stock, and elucidation of critical biological parameters to support releases in Canada. Following releases in Ontario (Nealis and Quednau, 1996) and New Brunswick (D. Lavigne, Fredericton, 1996, personal communication), there was evidence of successful parasitism by *A. samarensis* in the experimental populations in the same year (Nealis and Quednau, 1996). Follow-up studies at the Ontario sites (D. Ortiz, 1997, and D. Roden, Sault Ste Marie, 1998, personal communication), using laboratory-reared *L. dispar* larvae exposed at the release site and then re-collected and reared to determine parasitism, failed to recover *A. samarensis*. Whether this indicates failure to establish, or simply that parasitoids are too rare to be detected, cannot be determined at present. The Ontario release site may have been too open and dry to favour survival of adult *A. samarensis*. The release site in New Brunswick is thought to resemble more the habitat in Europe where *A. samarensis* has been collected. In New Brunswick the progeny of *A. samarensis* released in cages in 1998 successfully overwintered and attacked local, caged *L. dispar* larvae in 1999 (D. Lavigne, Fredericton, 1999, personal communication). Given the relatively small number of insects released at both locations and their low natural fecundity, permanent establishment by *A. samarensis* will take several years to confirm.

The impact of non-indigenous pathogens of *L. dispar* seems more significant, or at least more dramatic. LydiNPV is present throughout the established range of *L. dispar* in Canada but has a significant impact only at high host densities (Nealis *et al.*, 1999). The fungus *E. maimaiga* has spread rapidly from the USA into Canada and is reported as the single most important source of natural mortality, in both the established and leading-edge populations of *L. dispar* in eastern North America (Hajek *et al.*, 1996; Nealis *et al.*, 1999). While *E. maimaiga* appears to have reduced populations of *L. dispar*, it must be remembered that there have been population decreases before and the effectiveness of pathogens is notoriously dependent on ambient climatic conditions.

Despite expansion of the range of *L. dispar* in Canada since 1980, the severity of defoliation declined in the 1990s and research accomplishments did not receive as much attention as might otherwise have occurred. From 1997 to 1999, however, defoliation increased in Ontario and officials wondered whether this might be the start of a new outbreak (T. Scarr, Ontario, 1999, personal communication). Interest in biological control alternatives in other, more recently infested regions is very high, as resource managers attempt to forestall expansion of populations in their area and appease public distrust of widespread application of insecticides, including *B.t.k.*

One of the most important lessons from this programme has been the re-evaluation of the premises of classical biological control and recognition that pest populations are dynamic and may be more amenable to management by biological control at some stages than at others. A related notion is the idea that as *L. dispar* invades novel ecological habitats in Canada, different natural enemies could play important roles, and species or biotypes of parasitoids originally considered unsuitable for release, or relatively ineffective in eastern North America, might be more effective in these new environments. There also remain vast areas within the native range of *L. dispar* in Eurasia that have received limited exploration for natural enemies, notably China, eastern Russia and the Middle East (Kenis and Lopez-Vaamonde, 1998).

Perhaps the greatest challenge in managing *L. dispar* populations in Canada, however, is our lack of knowledge of the ecology of this exotic disturbance in the Canadian environment. Despite the expansion of *L. dispar* through eastern Canada and the severe defoliation and significant economic impact that has resulted, basic ecological work on these populations, except for the study by Nealis *et al.* (1999), has been practically non-existent since the report of Griffiths and Quednau (1984). Because of this, it is difficult to develop a truly scientific evaluation of the effectiveness of current biological control agents or to identify the best strategy for the future.

Recommendations

Future work should include:

1. Periodic surveillance of populations of *L. dispar* throughout its Canadian range, to identify and estimate the impact of both native and introduced natural enemies;
2. Continued search for potential natural enemies of *L. dispar* within its native range, especially in Asia;
3. Refining methods of harvesting, producing, storing and delivering LydiNPV at more modest levels for small-scale control of *L. dispar* in ecologically sensitive areas;
4. Maintaining stock colonies of *A. samarensis* in Canada and continuing releases in areas of promise and monitoring past release sites, using the collection methodologies developed in Europe;
5. Exchanging biological information on the international status of populations of *L. dispar* and its natural enemies that could be linked profitably to growing international trade.

References

Bourchier, R.S. and Nealis, V.G. (1992) Patterns of hyperparasitism of *Cotesia melanoscela* (Hymenoptera: Braconidae) in southern Ontario. *Environmental Entomology* 21, 907–912.

Bradbury, R. (1991) Gypsy moth in Maine – 1990. In: *Forest and Shade Tree Insect and Disease Conditions for Maine. A summary of the 1990 Situation.* Summary Report No. 5, Maine Forest Service, pp. 40–41.

Bradbury, R. (1994) Gypsy moth in Maine in 1993. In: *Forest and Shade Tree Insect and Disease Conditions for Maine. A Summary of the 1993 Situation.* Summary Report No. 8, Maine Forest Service, pp. 53–54.

Carter, N.E. (1996) *Status of Forest Pests in New Brunswick in 1996.* New Brunswick Department of Natural Resources and Energy, Fredericton, New Brunswick.

Carter, N.E. and Kettela, E.G. (1993) *A Preliminary Study of the Biology of Gypsy Moth in New Brunswick and Initial Examination of Selected Integrated Pest Management Techniques.* Final Report, Canada–New Brunswick Cooperative Agreement on Forest Development, Fredericton, New Brunswick.

Cunningham, J.C. (1998) North America. In: Hunter-Fujita, F.R., Entwistle, P.F., Evans, H.F. and Cook, N.E. (eds) *Insect Viruses and Pest Management.* John Wiley and Sons, Chichester, UK, pp. 313–331.

Cunningham, J.C. and Kaupp, W.J. (1995) Insect viruses. In: Armstrong, J.A. and Ives, W.G.H. (eds) *Forest Insect Pests in Canada.* Canadian Forest Service, Natural Resources Canada, Ottawa, Ontario, pp. 328–340.

Cunningham, J.C., Payne, N.J., Brown, K.W., Fleming, R.A., Burns, T., Mickle, R.E. and Scarr, T. (1996) Aerial spray trials with nuclear polyhedrosis virus and *Bacillus thuringiensis* on gypsy moth (Lepidoptera: Lymantriidae) in 1994. I. Impact in the year of application. *Proceedings of the Entomological Society of Ontario* 127, 21–35.

Cunningham, J.C., Brown, K.W., Payne, N.J., Mickle, R.E., Grant, G.C., Fleming, R.A., Robinson, A., Curry, R.D., Langevin, D. and Burns, T. (1997) Aerial spray trials in 1992 and 1993 against gypsy moth, *Lymantria dispar* (Lepidoptera: Lymantriidae), using nuclear polyhedrosis virus with and without an optical brightener compared to *Bacillus thuringiensis. Crop Protection* 16, 15–23.

Davidson, C.B., Gottschalk, K.W. and Johnson, J.E. (1999) Tree mortality following defoliation by the European gypsy moth (*Lymantria dispar* L.) in the United States: a review. *Forest Science* 45, 74–84.

Doane, C.C. and McManus, M.L. (1981) *The Gypsy Moth: Research Toward Integrated Pest Management.* Technical Bulletin 1584, United States Department of Agriculture.

Frankenhuyzen, K. van, Wiesner, C.J., Riley, C.M., Nystrom, C., Howard, C.A. and Howse, G.M. (1991) Distribution and activity of spray deposits in an oak canopy following aerial application of diluted and undiluted formulations of *Bacillus thuringiensis* Berliner against the gypsy moth, *Lymantria dispar* (Lepidoptera: Lymantriidae). *Pesticide Science* 33, 159–168.

Griffiths, K.J. (1976) *The Parasites and Predators of the Gypsy Moth: A Review of the World Literature with Special Application to Canada.* Report 0-X-243, Canadian Forestry Service, Sault Ste Marie, Ontario.

Griffiths, K.J. and Quednau, F.W. (1984) *Lymantria dispar* (L.) Gypsy Moth (Lepidoptera: Lymantriidae). In: Kelleher, J.S. and Hulme, M.A. (eds) *Biological Control Programmes Against Insects and Weeds in Canada 1969–1980.* Commonwealth Agriculture Bureaux, Slough, UK, pp. 303–310.

Hajek, A.E., Elkinton, J.S. and Witcosky, J.J. (1996) Introduction and spread of the fungal pathogen *Entomophaga maimaiga* (Zygomycetes: Entomophthorales) along the leading edge of the gypsy moth (Lepidoptera: Lymantriidae) spread. *Environmental Entomology* 25, 1233–1247.

Humble, L. and Stewart, A.J. (1994) *Gypsy Moth.* Forest Pest Leaflet 75, Cat. No. Fo29–6/75–1994E.

Jobin, L. (1995) Gypsy moth, *Lymantria dispar.* In: Armstrong, J.A. and Ives, W.G.H. (eds) *Forest Insects in Canada.* Canadian Forest Service, Natural Resources Canada, Ottawa, Ontario, Fo24-235/1995E, pp. 133–139.

Kauffman, W.C., Fuester, R.W. and Nealis, V.G. (1996) Non-target evaluation and release of two non-indigenous tachinids. *Proceedings of the United States Department of Agriculture, Interagency Gypsy Moth Research Forum 1996,* pp. 44–45.

Kenis, M. and Lopez-Vaamonde, C. (1998) Classical biological control of the gypsy moth, *Lymantria dispar* (L.), in North America: Prospects and new strategies. In: McManus, M.L. and Liebhold, A.M. (eds) *Population Dynamics, Impacts, and Integrated Management of Forest Defoliating Insects.* General Technical Report NE-247, United States Department of Agriculture, Forest Service, pp. 213–221.

Lavigne, D. and Carter, N.E. (1996) *Alternative Virus Application Strategy for Control of the Gypsy Moth.* New Brunswick Department of Natural Resources and Energy, Fredericton, New Brunswick.

Magasi, L.P. (1984) *Forest Pest Conditions in the Maritimes in 1983.* Information Report M-X-149, Canadian Forest Service, Fredericton, New Brunswick.

McGugan, B.M. and Coppel, H.C. (1962) II. Biological control of forest insects 1910–1958. In: McLeod, J.H., McGugan, B.M. and Coppel, H.C. (eds) *A Review of the Biological Control Attempts Against Insects and Weeds in Canada.* Technical Communication No. 2, Commonwealth Institute of Biological Control, Trinidad. Commonwealth Agricultural Bureaux, Farnham Royal, UK, pp. 35–216.

Mills, N.J. (1990) Are parasitoids of significance in endemic populations of forest defoliators? Some experimental observations from gypsy moth, *Lymantria dispar* (Lepidoptera: Lymantriidae). In: Watt, A.D., Leather, S.R., Hunter, M. and Kidd, N.A.C. (eds) *Population Dynamics of Forest Insects.* Intercept, Andover, UK, pp. 265–274.

Mills, N.J. and Nealis, V.G. (1992) European field collections and Canadian releases of *Ceranthia samarensis* (Dipt.: Tachinidae), a parasitoid of the gypsy moth. *Entomophaga* 37, 181–191.

Nealis, V.G. and Bourchier, R.S. (1995) Reduced vulnerability to hyperparasitism in nondiapause strains of *Cotesia melanoscela* (Ratzeburg) (Hymenoptera: Braconidae). *Proceedings of the Entomological Society of Ontario* 126, 29–35.

Nealis, V.G. and Erb, S. (1993) *A Sourcebook for Management of the Gypsy Moth.* Ministry of Supply and Services Canada, Ottawa, Ontario Fo42-193/1993E.

Nealis, V.G. and Quednau, F.W. (1996) Canadian field releases and overwinter survival of *Ceranthia samarensis* (Villeneuve) (Diptera: Tachinidae) for biological control of the gypsy moth, *Lymantria dispar* (L.) (Lepidoptera: Lymantriidae). *Proceedings of the Entomological Society of Ontario* 127, 11–20.

Nealis, V.G., Roden, P.M. and Ortiz, D.A. (1999) Natural mortality of the gypsy moth along a gradient of infestation. *The Canadian Entomologist* 131, 507–519.

Pschorn-Walcher, H. (1977) Biological control of forest insects. *Annual Review of Entomology* 22, 1–22.

Quednau, F.W. (1993) Reproductive biology and laboratory rearing of *Ceranthia samarensis* (Villeneuve) (Diptera: Tachinidae), a parasitoid of the gypsy moth, *Lymantria dispar* (L.). *The Canadian Entomologist* 125, 749–759.

Quednau, F.W. and Lamontagne, K. (1998) *Principles of mass culture of the gypsy moth parasitoid* Ceranthia samarensis (*Villeneuve*). Information Report LAU-X-121I, Canadian Forest Service, Sainte-Foy, Quebec.

Reardon, R.C., Podgwaite, J. and Zerillo, R. (1996) *Gypcheck – the Gypsy Moth Nucleopolyhedrosis Virus Product.* FHTET-96–16, United States Department of Agriculture, Forest Service.

Smith, G.A. and Harrison, K.J. (1995) New insect and fungus records in the Maritimes. In: Hurley, J.E. and Magasi, L.P. (eds) *Forest Pest Conditions in the Maritimes in 1994.* Information Report M-X-194E, Canadian Forest Service, Fredericton, New Brunswick, p. 36.

Wallner, W.E. (1996) Invasive pests ('biological pollutants') and US forests: whose problem, who pays? *OEPP/EPPO Bulletin* 26, 167–180.

Weseloh, R.M. (1985) Predation by *Calosoma sycophanta* L. (Coleoptera: Carabidae): evidence for a large impact on gypsy moth, *Lymantria dispar* L. (Lepidoptera: Lymantriidae), pupae. *The Canadian Entomologist* 117, 1117–1126.

34 *Mamestra configurata* Walker, Bertha Armyworm (Lepidoptera: Noctuidae)

P.G. Mason, W.J. Turnock, M.A. Erlandson, U. Kuhlmann and L. Braun

Pest Status

The bertha armyworm, *Mamestra configurata* Walker, native to North America, is one of several important insect pests of canola, *Brassica napus* L. and *B. rapa* L., although it ranges from British Columbia to Manitoba and south to Mexico (Crumb, 1956) and feeds on a wide variety of cultivated broadleaved plants (Turnock, 1985). King (1928) first reported *M. configurata* as a serious crop pest on flax, *Linum usitatissimum* L., sweet clover, *Melilotus* spp., and alfalfa, *Medicago sativa* L., in western Canada during the 1920s. Since 1971, it has been recognized as an important, although sporadic, pest in canola (Turnock, 1984a; Mason *et al.*, 1998b). The most recent widespread outbreak, in 1994–1996, resulted in yield losses of more than Can$50 million plus $40 million in control costs (Mason *et al.*, 1998b). Outbreaks generally last for 1–3 years, and those in one area do not coincide in time with outbreaks in other areas.

M. configurata overwinters as pupae in the soil, and adults emerge from early to mid-June until early August. Adults are attracted to flowering canola fields (Turnock, 1984b; Anonymous, 1995). Females lay eggs in masses of 20–200 on the underside of leaves. Larvae hatch about 1 week after the eggs are laid and immediately begin feeding. They are nocturnal and disperse from the oviposition sites. Second to fourth instars continue to feed on green foliage. Fifth to sixth instars cause the greatest damage because they feed on the developing seed pods (Bracken, 1984). Under field conditions, larvae mature in about 6 weeks, drop to the ground and seek cracks, where they shelter 5–16 cm under the surface. Here they pupate and most go into diapause and overwinter. In Canada, the species is univoltine, but in unusually warm summers some individuals continue to develop and emerge as adults in autumn.

Background

Crop protection usually involves insecticides, several being registered for use against *M. configurata*. The decision to apply insecticides is based on sampling to determine larval densities, estimated crop loss, crop value and cost of spraying (Anonymous, 1995). Commercial formulations of *Bacillus thuringiensis* Berliner serovar *kurstaki* (*B.t.k.*) give highly variable control (Morris, 1986). This and the decreasing number of new chemicals and undesirable environmental effects have stimulated interest in biological control alternatives.

Natural controls have an impact on *M. configurata* populations and may be important in determining the location and frequency of outbreaks. Winter mortality of pupae is an important population regulator; regions where snow cover is deepest (i.e. soil temperature is highest) tend to be where outbreaks occur (Lamb *et al.*, 1985). Tillage in either autumn or spring can increase pupal mortality by direct damage, predation or exposure to cold (King,

1929; Turnock and Bilodeau, 1984) but has not prevented outbreaks. Recent increases in use of minimum tillage techniques should increase overwintering survival of pupae.

Naturally occurring pathogens, including a fungus, *Entomophthora* sp., a *Granulovirus* (GV), a *Nucleopolyhedrovirus* (MacoNPV[1]), and a microsporidian, have been isolated from *M. configurata* populations across western Canada (Wylie and Bucher, 1977; Bucher and Turnock, 1983; Turnock, 1988; Erlandson, 1990; Li *et al.*, 1997). These have an impact only when host populations are high (Wylie and Bucher, 1977). The microsporidian appears to be a very minor mortality factor for *M. configurata* (Wylie and Bucher, 1977). Incidence of infection by *Entomophthora* sp. varies from 0 to 30% in area-wide surveys (Turnock, 1988) to more than 90% in some local *M. configurata* populations during fungal epizootics. Prevalence of the disease depends on environmental conditions, e.g. high humidity, which is required for good conidiospore germination and infection (M.A. Erlandson, unpublished). Isolates of MacoNPV from *M. configurata* are regularly found in field populations and some have been characterized biologically and biochemically (Bucher and Turnock, 1983; Erlandson, 1990; Li *et al.*, 1997). Impact varies but Erlandson (1990) recorded MacoNPV incidence of up to 95% in localized populations.

Insect parasitoids include *Trichogramma inyoense* Pinto and Oatman in eggs in Saskatchewan (Mason *et al.*, 1998a) and several larval parasitoids (Wylie and Bucher, 1977; Turnock, 1984a). Only the ichneumonid *Banchus flavescens* Cresson and the tachinid *Athrycia cinerea* (Coquillette) occur regularly in host populations (Wylie and Bucher, 1977; Turnock, 1988). Although they contribute to regulation of *M. configurata* populations, they do not prevent outbreaks from occurring (Turnock, 1988; Mason *et al.*, 1998b).

Biological Control Agents

Pathogens

Bacteria

Commercial formulations of *B.t.k.* and other *B.t.* varieties have not proved highly effective against *M. configurata* (Morris, 1986; Trottier *et al.*, 1988; Dosdall and Davies, 1993; Morris *et al.*, 1994, 1996; Masson *et al.*, 1998).

Viruses

Wylie and Bucher (1977) found that MacoNPV infections had a minor role in regulating *M. configurata* during periods of low population density but infected 1–12% of larvae at high population densities. In laboratory assays, MacoNPV isolates were shown to be as virulent as other NPVs being developed for other pests (Bucher and Turnock, 1983; Erlandson, 1990). The quantity of virus required to kill fifth- and sixth-instar larvae is significantly higher than for instars 1–3 (Bucher and Turnock, 1983), and the time between virus ingestion and mortality varies with dose, declining from 7.5 days at the LD_{50} to 4.5–5.0 days at 200 × the LD_{50} (Erlandson, 1990). More than 30 geographic isolates of the MacoNPV, distinguished by restriction endonuclease digestion patterns of genomic DNA, have been isolated and characterized from western Canadian *M. configurata* populations (M.A. Erlandson, unpublished). A single larva may contain multiple genotypes of MacoNPV (Erlandson, 1990; Li *et al.*, 1997).

Erlandson and Mason (1998) assessed several MacoNPV isolates in greenhouse and small cage trials to determine their efficacy to control second- to fourth-instar *M. configurata* larvae. Laboratory-reared larvae fed field-sprayed canola plants treated with MacoNPV at doses of 5.0×10^9 to 1.0×10^{12} polyhedral inclusion bodies (PIB) ha^{-1}

[1]International Committee on Taxonomy of Viruses nomenclature (Murphy *et al.*, 1995).

died 10–14 days after application with conventional spray equipment (8000067F nozzles delivering 35 l ha^{-1}). In field cages of treated canola significant MacoNPV infection was detected from 14 to 21 days after treatment. The trials also indicated that fourth-instar larvae can be more efficiently targeted that younger larvae.

Erlandson and Mason (1998) tested the compatibility of MacoNPV and the larval endoparasitoid *Microplitis mediator* Haliday in the laboratory. MacoNPV did not infect *M. mediator* larvae directly, and *M. mediator* could not vector MacoNPV to *M. configurata* larvae. However, if host *M. configurata* larvae were infected with MacoNPV prior to, or within, 48 h of parasitism, *M. mediator* died due to premature host death from virus infection. Although synergistic impact would be limited, these two biological control agents may be compatible because *M. mediator* emerges from the fourth-instar larvae of *M. configurata*, allowing MacoNPV to be used against later instars.

Parasitoids

In Canada, *B. flavescens* and *A. cinerea* are the only effective native parasitoids of *M. configurata*. *B flavescens* lays its eggs in first to third larval instars, and kills mature larvae after they enter the soil to pupate (Arthur and Mason, 1985). In the laboratory, *B. flavescens* consistently parasitized more *M. configurata* larvae in 24 h than did *M. mediator*, either when alone or when both were present (P.G. Mason, unpublished). Only three other hosts (all noctuids) of *B. flavescens* are known (Schaffner and Griswald, 1934; Wylie and Ayre, 1979) and this may explain why it is abundant only in the later years of *M. configurata* outbreaks.

A. cinerea is usually the second most important natural enemy (Turnock and Philip, 1977; Wylie and Bucher, 1977; Turnock, 1984a, 1988) but may be the dominant one in northern Alberta (L.M. Dosdall, Edmonton, 1999, personal communication). Eggs laid on third to sixth larval instars hatch within 10 min and the parasitoid larvae immediately burrow into the host (Wylie, 1977). Development from egg to pupa takes as little as 7 days, and several individuals may develop in the same host (Arthur and Powell, 1989). *A. cinerea* is univoltine and undergoes a facultative diapause (Wylie, 1977). Its host range includes three noctuids and a sawfly (O'Hara, 1999). Two more tachinid parasitoids, *Chetogena tachinomoides* (Townsend) and *Spallanzenia hebeus* (Fallén), were recently discovered; the latter overwinters in *M. configurata* pupae (O'Hara, 1999).

T. inyoense, recovered from *M. configurata* eggs in the field, was previously recorded only from arboreal host eggs (Pinto, 1998). It readily parasitizes *M. configurata* eggs, producing an average of 1.5 adults per host egg (P.G. Mason and L. Braun, unpublished). Its influence in the field is unknown because others (Turnock, 1984a; P.G. Mason, unpublished) failed to recover *T. inyoense* or any other egg parasitoid from wild or sentinel *M. configurata* eggs.

The parasitoid complex of *M. configurata* was compared to that of the closely related Eurasian pest, *Mamestra brassicae* L., on cabbage, *Brassica oleraceae* L., to determine if gaps existed that might be filled by introductions into Canada. Eggs from *M. brassicae* populations in Switzerland yielded two parasitoids, *Trichogramma buesi* Voegele (reported as *T. evanescens* Westwood by Turnock, 1984a) and *Telenomus* sp., with mortalities of up to 100% (Kählert and Carl, 1991; Carl *et al.*, 1995; Ziegler and Carl, 1996; Kuhlmann *et al.*, 1997; Lauro and Kuhlmann, 2000). Studies on their biology and competitive interactions suggest that both species are opportunists, and their relative abundance varies from year to year (U. Kuhlmann and P.G. Mason, unpublished). Johansen (1997) reported a third species, *Trichogramma semblidis* (Aurivillius), in low numbers from *M. brassicae* eggs.

Ernestia consobrina (Meigen) occupies a similar position in *M. brassicae* to that occupied by *Panzeria ampelus* (Walker) in

M. configurata. Because *P. ampelus* rarely occurs in *M. configurata* (Turnock, 1988), this niche was considered to be vacant and open to exploitation by an introduced species. *E. consobrina* was initially selected for pre-introduction study because of its abundance in cooler parts of the range of *M. brassicae*, e.g. in Russia, where both host and parasitoid are univoltine (Turnock, 1984a). Using reared material derived from collections in Germany (see Turnock and Bilodeau, 1999), Turnock and Carl (1995) showed that *E. consobrina* readily and successfully parasitized *M. configurata* and they then released *E. consobrina* in Canada.

Exetastes atrator (Förster)[2] has a similar life history to that of *B. flavescens* and is well synchronized with its host in both uni- and bivoltine Eurasian populations (Turnock, 1984a). It was not considered for introduction because its niche was effectively occupied by *B. flavescens*.

M. mediator is the most abundant larval parasitoid of all bivoltine and Norwegian univoltine *M. brassicae* populations (Carl and Sommer, 1975, 1976, 1977; Carl, 1978; Turnock, 1984a; Johansen, 1997) and could occupy a vacancy in the parasitoid guild of *M. configurata*. In the laboratory, parasitism of *M. brassicae* during 24 h increased with increasing host density up to 20, above which only seven cocoons were produced from up to 50 hosts (Lauro and Kuhlmann, 1999), suggesting that the number of eggs available for oviposition during a 24 h period is limited. Study of alternative hosts of *M. mediator* in Switzerland indicates that it can develop successfully on the noctuid *Autographa gamma* L. but not on *Pieris rapaa* L., *P. brassica* L. or *Plutella xylostella* L. in cabbage fields where *M. brassicae* is found (Lauro and Kuhlmann, 2000).

In Saskatchewan, overwintering survival of *M. mediator* was similar for populations reared in the laboratory for ten generations and those collected from the field in Switzerland (66.0% and 57.8%, respectively) (P.G. Mason, unpublished). In competitive interaction experiments between *M. mediator* and *B. flavescens*, larvae of both species are good competitors (P.G. Mason, unpublished). In Europe, *M. mediator* and *E. atrator* competition in the field is rare (Carl, 1978); thus, *M. mediator* and *B. flavescens* should coexist. Further, parasitism of *M. configurata* should be increased by adding *M. mediator* to the parasitoid complex, as shown in competitive oviposition experiments.

Releases and Recoveries

At three locations in Manitoba, *E. consobrina* was released in canola fields in which *M. configurata* was abundant (Turnock and Carl, 1995). Releases of 1455–2460 females and 559–666 males were timed to allow *E. consobrina* females to complete their prelarviposition period while host larvae were available. In 1986, releases were made at Kenville on 24 and 31 July, and 11, 12 and 21 August, when most host larvae were in the later instars. In 1987, the releases at Dauphin were made on 13 and 27 July and 4 August. The 1987 releases at Glenlea were not directed toward a specific host species. These adults emerged both before and particularly after larvae of *M. configurata* were present.

In Saskatchewan and Alberta, *M. mediator* populations collected in Switzerland from 1991 to 1999 were released at several locations (Mason and Youngs, 1994; Mason, 1999; J. Otani, Beaverlodge, 1999, personal communication). These included open-field liberations of adults, cocoons, and parasitized host larvae, and releases of adults into cages containing unparasitized larvae. Field liberations with 20–3000 individuals per release were timed to enable *M. mediator* females to attack early instar host larvae.

Evaluation of Biological Control

MacoNPV is a promising microbial control agent. The increased field activity demon-

[2]*E. atrator* (Förster) replaces *E. cinctipes* (Retzius) (CAB International, 1996).

strated by the numerous formulations evaluated suggests that improvements in MacoNPV virulence and virus formulation could make it commercially viable.

From 1986 to 1992 post-release sampling of *E. consobrina* showed that *M. configurata* larval density was below 2 larvae m^{-2} in the 10–16 fields sampled each year, and no *E. consobrina* were found (Turnock and Carl, 1995). The parasitoid was also absent from 39 larvae collected in an outbreak in 1987 of a potential alternative host, *Pseudaletia unipuncta* (Haworth), near the Kenville release site. The inability to recover *E. consobrina* was probably because host populations were low, so that host larval samples were too small to detect low levels of parasitism. The potential for success of these releases may have been reduced because the females were progeny (4–6 generations in culture) of a small number of individuals.

Since 1993, annual sampling has not recovered *M. mediator*, either as adults or larvae reared from field-collected *M. configurata*. Although releases were made in areas known to have outbreaks, subsequent host populations following releases have been too small for sampling to detect low parasitism levels.

Recommendations

Further work should include:

1. Field sampling to determine if populations of *E. consobrina* and *M. mediator* released in Manitoba, and Saskatchewan and Alberta, respectively, are established;
2. Investigating populations of *M. brassicae* in Russia and northern China as sources of *E. consobrina* and *M. mediator*;
3. Determining the impact of naturally occurring *T. inyoense* on eggs of *M. configurata* and determining the potential for using *Trichogramma* spp. as inundative biological control agents;
4. Additional screening and field testing of different classes of *B. thuringiensis*, particularly the Cry toxins active against *Spodoptera* spp.;
5. Determining the significance of pathogenesis and epizootiology of combinations of MacoNPV wild genotypes;
6. Examining the potential of novel formulations and inclusion of synergists on the field activity of MacoNPV;
7. Reviewing the taxonomy of *Microplitis* and *Telenomus* attacking *Mamestra* spp. to determine species limits and verify the names of *M. mediator* and its sister species *Microplitis tuberculata* Wesmael and provide a name for the *Telenomus* sp. parasitizing *M. brassicae* eggs.

Acknowledgements

The Saskatchewan Agriculture Development Fund (Projects R-89-05-0536 and 95000283) and the Canada–Saskatchewan Agriculture Green Plan Agreement (CPM 94-6) provided financial support for some aspects of the work.

References

Anonymous (1995) *Bertha armyworm*. Sustainable Agriculture Facts, Agriculture Canada, Alberta Agriculture, and British Columbia Ministry of Agriculture, Fisheries and Food.

Arthur, A.P. and Mason, P.G. (1985) Life history and immature stages of *Banchus flavescens* (Hymenoptera: Ichneumonidae), a parasitoid of the bertha armyworm, *Mamestra configurata* (Lepidoptera: Noctuidae) in western Canada. *The Canadian Entomologist* 117, 1249–1255.

Arthur, A.P. and Powell, Y.M. (1989) Descriptions of the immature stages and adult reproductive systems of *Athrycia cinerea* (Coq.) (Diptera: Tachinidae), a native parasitoid of *Mamestra configurata* (Walk.) (Lepidoptera: Noctuidae). *The Canadian Entomologist* 121, 1117–1123.

Bracken, G.E. (1984) Within plant preferences of larvae of *Mamestra configurata* (Lepidoptera: Noctuidae) feeding on oilseed rape. *The Canadian Entomologist* 116, 45–49.

Bucher, G.E. and Turnock, W.J. (1983) Dosage responses of the larval instars of the bertha armyworm, *Mamestra configurata* (Lepidoptera: Noctuidae) to a native nuclear polyhedrosis. *The Canadian Entomologist* 115, 341–349.

CAB International (ed.) (1996) *Arthropod Name Index*. CAB International, Wallingford, UK, CD-ROM.

Carl, K.P. (1978) Bertha armyworm (*Mamestra configurata* Walker). In: *Rapeseed Insects: Work in Europe in 1978 Annual Project Statement.* Commonwealth Institute for Biological Control, European Station, Delémont, Switzerland, p. 8.

Carl, K.P. and Sommer, G. (1975) Bertha armyworm (*Mamestra configurata* Walker). In: *Rapeseed Insects: Work in Europe in 1975.* Annual Project Statement, Commonwealth Institute for Biological Control, European Station, Delémont, Switzerland, p. 7.

Carl, K.P. and Sommer, G. (1976) Bertha armyworm (*Mamestra configurata* Walker). In: *Rapeseed Insects: Work in Europe in 1976.* Annual Project Statement, Commonwealth Institute for Biological Control, European Station, Delémont, Switzerland, p. 5.

Carl, K.P. and Sommer, G. (1977) Bertha armyworm (*Mamestra configurata* Walker). In: *Rapeseed Insects: Work in Europe in 1977.* Annual Project Statement, Commonwealth Institute for Biological Control, European Station, Delémont, Switzerland, p. 7.

Carl, K., Desmeules, H. and Nash, J. (1995) Bertha armyworm (*Mamestra configurata* Walker). In: *Annual Report 1995 for Agriculture and Agri-Food Canada.* International Institute of Biological Control, European Station, Delémont, Switzerland, pp. 25–30.

Crumb, S.E. (1956) *The larvae of the Phalaenidae.* Technical Bulletin 1135. United States Department of Agriculture.

Dosdall, L. and Davies, J.S. (1993) Production and evaluation of two strains of the bacterium, *Bacillus thuringiensis* Berliner, as biological insecticides for bertha armyworm, *Mamestra configurata* (Walker). *Alberta Environmental Centre, Vegreville, Alberta.* AECV 94-R2.

Erlandson, M.A. (1990) Biological and biochemical comparison of *Mamestra configurata* and *Mamestra brassicae* nuclear polyhedrosis virus isolates pathogenic for the bertha armyworm, *Mamestra configurata* (Lepidoptera: Noctuidae). *Journal of Invertebrate Pathology* 56, 47–56.

Erlandson, M.A. and Mason, P.G. (1998) *The Potential of Nucleopolyhedrovirus and an Imported Parasite as Biological Control Agents of Bertha Armyworm.* Final Report, Project CPM 94-6, Saskatchewan–Canada Green Plan Fund.

Johansen, N.S. (1997) Mortality of eggs, larvae and pupae, and larval dispersal of the cabbage moth, *Mamestra brassicae*, in white cabbage in south-eastern Norway. *Entomologia Experimentalis et Applicata* 83, 347–360.

Kählert, A. and Carl, K. (1991) Bertha armyworm (*Mamestra configurata*). In: *Agriculture Canada, Annual Project Reports 1991.* International Institute of Biological Control, European Station, Delémont, Switzerland, pp. 36–38.

King, K.M. (1928) *Barathra configurata* Walker, an armyworm with important potentialities on the northern prairies. *Journal of Economic Entomology* 21, 279–293.

King, K.M. (1929) *The bertha armyworm in the prairie provinces.* Dominion of Canada Department of Agriculture Pamphlet No. 103.

Kuhlmann, U., Babendreier, D., Hooper, L., Otten, N., Peddle, S. and Stahl, B. (1997) Bertha armyworm (*Mamestra configurata* Walker) In: *Summary Report, Progress in 1997 for Agriculture and Agri-Food Canada.* International Institute of Biological Control, European Station, Delémont, Switzerland, pp. 5–6.

Lamb, R.J., Turnock, W.J. and Hayhoe, H.N. (1985) Winter survival and outbreaks of bertha armyworm, *Mamestra configurata* (Lepidoptera: Noctuidae), on canola. *The Canadian Entomologist* 117, 727–736.

Lauro, N. and Kuhlmann, U. (1999) Bertha armyworm (*Mamestra configurata* Walker). In: *Annual Report 1998–1999 for Agriculture and Agri-Food Canada.* International Institute of Biological Control, European Station, Delémont, Switzerland, pp. 10–13.

Lauro, N. and Kuhlmann, U. (2000) Bertha armyworm (*Mamestra configurata* Walker). In: *Annual Report 1999–2000 for Agriculture and Agri-Food Canada.* International Institute of Biological Control, European Station, Delémont, Switzerland, pp. 13–16.

Li, S., Erlandson, M., Moody, D. and Gillott G. (1997) A physical map of the *Mamestra configurata* nucleopolyhedrovirus genome and sequence analysis of the polyhedrin gene. *Journal of General Virology* 78, 265–271.

Mason, P.G. (1999) *Release of the insect parasite* Microplitis mediator *for enhanced biological control of the bertha armyworm.* Final Report, Saskatchewan Agriculture Development Fund Project 95000283.

Mason, P.G. and Youngs, B.J. (1994) Biological control of a Canadian canola pest, the bertha armyworm (*Mamestra configurata*), with the European parasitoid *Microplitis mediator*. *Norwegian Journal of Agricultural Sciences, Supplement* 16, 405–406.

Mason, P.G., Pinto, J.D., Long, Z.L. and Harris, J.L. (1998a) First record of *Trichogramma inyoense* (Hymenoptera: Trichogrammatidae) attacking the bertha armyworm, *Mamestra configurata* (Lepidoptera: Noctuidae). *The Canadian Entomologist* 130, 105–106.

Mason, P.G., Arthur, A.P., Olfert, O.O. and Erlandson, M.A. (1998b) The bertha armyworm (*Mamestra configurata*) (Lepidoptera: Noctuidae) in western Canada. *The Canadian Entomologist* 130, 321–336.

Masson, L., Erlandson, M.A., Puzstai-Carey, M., Brousseau, R., Juárez-Pérez, V. and Frutos, R. (1998) A holistic approach for determining the entomopathogenic potential of *Bacillus thuringiensis* strains. *Applied and Environmental Microbiology* 64, 4782–4788.

Morris, O.N. (1986) Susceptibility of the bertha armyworm, *Mamestra configurata* (Lepidoptera: Noctuidae), to commercial formulations of *Bacillus thuringiensis* var. *kurstaki*. *The Canadian Entomologist* 118, 473–478.

Morris, O.N., Trottier, M., McLaughlin, N.B. and Converse, V. (1994) Interaction of caffeine and related compounds with *Bacillus thuringiensis* spp. *kurstaki* in bertha armyworm (Lepidoptera: Noctuidae). *Journal of Economic Entomology* 87, 610–617.

Morris, O.N., Trottier, M., Converse, V. and Kanagaratnam, P. (1996) Toxicity of *Bacillus thuringiensis* subsp. *aizawai* for *Mamestra configurata* (Lepidoptera: Noctuidae). *Journal of Economic Entomology* 89, 359–365.

Murphy, F.A., Fauquet, C.M., Bishop, D.H.L., Ghabrial, S.A., Jarvis, A.W., Martelli, G.P., Mayo, M.A. and Summers, M.D. (1995) *Virus Taxonomy, Classification and Nomenclature of Viruses*. Springer-Verlag, New York.

O'Hara, J.E. (1999) Tachinidae (Diptera) parasitoids of bertha armyworm (Lepidoptera: Noctuidae). *The Canadian Entomologist* 131, 11–28.

Pinto, J.D. (1998) Systematics of the North American species of *Trichogramma* Westwood (Hymenoptera: Trichogrammatidae). *Memoirs of the Entomological Society of Washington* 22, 1–287.

Schaffner, J.V. and Griswold, C.L. (1934) *Macrolepidoptera and their Parasites Reared from Field Collections in the Northeastern Part of the United States*. United States Department of Agriculture, Miscellaneous Publication 188.

Trottier, M.R., Morris, O.N. and Dulmage, H.T. (1988) Susceptibility of the bertha armyworm, *Mamestra configurata* (Lepidoptera: Noctuidae), to sixty-one strains from ten varieties of *Bacillus thuringiensis*. *Journal of Invertebrate Pathology* 51, 242–249.

Turnock, W.J. (1984a) *Mamestra configurata* Walker, bertha armyworm (Lepidoptera: Noctuidae). In: Kelleher, J.S. and Hulme, M.A. (eds) *Biological Control Programmes Against Insects and Weeds in Canada 1969–1980*. Commonwealth Agriculture Bureaux, Slough, pp. 49–55.

Turnock W.J. (1984b) Effects of the stage of development of canola (*Brassica napus*) on the capture of moths in sex attractant traps and on the larval density of *Mamestra configurata* (Lepidoptera: Noctuidae). *The Canadian Entomologist* 116, 579–590.

Turnock, W.J. (1985) Developmental, survival, and reproductive parameters of bertha armyworm, *Mamestra configurata* (Lepidoptera: Noctuidae) on four plant species. *The Canadian Entomologist* 117, 1267–1271.

Turnock, W.J. (1988) Density, parasitism, and disease incidence of larvae of the bertha armyworm, *Mamestra configurata* (Lepidoptera: Noctuidae), in Manitoba, 1973–1986. *The Canadian Entomologist* 120, 401–413.

Turnock, W.J. and Bilodeau, R.J. (1984) Survival of pupae of *Mamestra configurata* (Lepidoptera: Noctuidae) and two of its parasites in untilled and tilled soil. *The Canadian Entomologist* 116, 257–267.

Turnock, W.J. and Bilodeau, R.J. (1999) Rearing methods and developmental parameters for *Athrycia cinerea* (Coq.) and *Eurithia consobrina* Mg. (Diptera: Tachinidae). *Entomologist's Monthly Magazine* 135, 51–57.

Turnock, W.J. and Carl, K.P. (1995) Evaluation of the Palaearctic *Eurithia consobrina* (Diptera: Tachinidae) as a potential biocontrol agent for *Mamestra configurata* (Lepidoptera: Noctuidae) in Canada. *Biocontrol Science and Technology* 5, 55–67.

Turnock, W.J. and Philip, H.G. (1977) The outbreak of bertha armyworm, *Mamestra configurata* (Noctuidae:Lepidoptera), in Alberta, 1971 to 1975. *Manitoba Entomologist* 11, 10–21.

Wylie, H.G. (1977) Observations on *Athrycia cinerea* (Diptera: Tachinidae), a parasite of *Mamestra configurata* (Lepidoptera: Noctuidae). *The Canadian Entomologist* 109, 747–754.
Wylie, H.G. and Ayre, G.L. (1979) Hosts of *Banchus flavescens* (Hymenoptera: Ichneumonidae) and *Athrycia cinerea* (Diptera: Tachinidae) in Manitoba. *The Canadian Entomologist* 111, 747–748.
Wylie, H.G. and Bucher, G.E. (1977) The bertha armyworm, *Mamestra configurata* (Lepidoptera: Noctuidae): Mortality of immature stages on the rape crop 1972–1975. *The Canadian Entomologist* 109, 823–837.
Zeigler, C. and Carl, K. (1996) Cabbage Moth (*Mamestra brassicae* L.). In: *Annual Report 1996 for Agriculture and Agri-Food Canada.* International Institute of Biological Control, European Station, Delémont, Switzerland, pp. 16–21.

35 *Melanoplus* spp., *Camnula pellucida* (Scudder), Grasshoppers (Orthoptera: Acrididae)

M.A. Erlandson, M.S. Goettel and D.L. Johnson

Pest Status

North American grasshoppers, mainly *Melanoplus* spp. and *Camnula pellucida* (Scudder), are important pests of rangeland, forage, cereal and other crops. Of the 35 species typically found in grasslands, only a few species, including the migratory grasshopper, *Melanoplus sanguinipes* (Fabricius), *M. bivittatus* (Say), *M. packardii* Scudder and *C. pellucida*, cause significant damage in cereal crops (Madder and Stemeroff, 1988; Johnson, 1989a). They are also major pests of rangeland and forage crops; an additional 10–20 other species are common on rangeland (Hardman and Smoliak, 1982). Major grasshopper outbreaks have occurred at irregular intervals, and weather and natural enemies have been the primary factors responsible for reducing populations between outbreaks (Riegert, 1968; Smith and Smoliak, 1977).

A comprehensive grasshopper survey and predictive programme for the prairies (Johnson *et al.*, 1996a) provides extension agronomists and producers with area-specific advanced information about grasshopper population levels. Economic thresholds for infestations depend on crop species, weather conditions and stage of grasshopper development. In outbreak years large areas of cereal crops are treated with chemical insecticides, e.g. 7.7 million ha were sprayed in Alberta and Saskatchewan from 1985 to 1991 (Johnson *et al.*, 1996a). Despite such massive control programmes, crop losses and reduced yields can be enormous (Can$325 million for the above area and time frame). Between outbreak years, smaller areas are treated with chemical insecticides.

Most grasshopper species overwinter in the soil as clusters of eggs deposited in pods consisting of frothy glue and soil particles. Embryos develop to a specific stage and enter diapause. In spring to early summer, depending on species, hatchling grasshoppers crawl to the soil surface, shed their embryonic membrane, and begin feeding. Depending on species, grasshoppers undergo 4–6 moults during their nymphal life; development is dependent on weather conditions. After moulting to the adult stage, females typically have a 1–2 week preoviposition period during which feeding, growth and maturation of the first egg batch occur. Following mating, egg laying begins and continues throughout adult life. Most grasshopper species, including those of economic importance, complete their developmental cycle in 1 year; however, some species overwinter as nymphs, completing development the following summer.

Background

The negative environmental effects and costs of broad-spectrum insecticides have led to a renewed interest in biological control options. The potential for their successful use on rangeland and headland areas of crop production is higher than for direct crop protection because of the more stable nature of these systems and the higher levels of grasshopper feeding damage that can be tolerated.

Grasshoppers are hosts for various natural enemies, e.g. parasitoids, predators and pathogens, including viruses, rickettsia, bacteria, protozoa and fungi (Streett and McGuire, 1990; Mason and Erlandson, 1994; Goettel and Johnson, 1997). Many pathogens produce chronic infections and may have some potential as classical biological control agents for long-term grasshopper suppression. A few, more virulent, pathogens may have potential as microbial insecticides for short-term grasshopper suppression and crop protection.

Biological Control Agents

Pathogens

Viruses

Of the virus groups that infect Orthoptera, *Entomopoxvirus* (EV) has the best potential for development as microbial pesticides (Erlandson and Streett, 1997; Streett *et al.*, 1997). In laboratory assays, Olfert and Erlandson (1991) and Woods *et al.* (1992) showed that EV infection in *M. sanguinipes* produced mortality in two distinct phases: an early phase 2–10 days after virus ingestion but prior to completion of virus replication; and a late phase 15–40 days after ingestion, which produces large numbers of virus occlusion bodies or spheroids per host ($2–3 \times 10^8$ spheroids per grasshopper). Infected grasshoppers that die in the late phase typically become 'anaemic', with distended abdomens due to hypertrophy of the fat body, which is filled with EV spheroids. Aside from direct mortality, EV infections considerably delay grasshopper development (Olfert and Erlandson, 1991); infected insects moult only once following infection but often survive for extended periods in one nymphal stage. EV infection also leads to significantly reduced food consumption (less than 50% of normal consumption by 25 days post-infection) (Olfert and Erlandson, 1991). Thus, EV infection has at least three interrelated effects on grasshoppers that reduce potential economic damage.

In Saskatchewan, preliminary field-cage trials showed greater than 50% mortality in fourth-instar *M. sanguinipes* 3 weeks after exposure to EV on bran bait at a dose equivalent to 1.0×10^9 spheroids ha^{-1} (M.A. Erlandson, unpublished). In the USA from 1989 to 1991, Streett and Woods (1990) and McGuire *et al.* (1991) tested both wheat bran and encapsulated starch granule formulations of EV at 1.0×10^{10} spheroids ha^{-1} and found that prevalence of EV infection of grasshoppers in treated plots was 7.5–30% by 14 days after treatment, although little population suppression in EV-treated plots was observed compared to control plots.

Fungi

The *Entomophaga grylli* (Fresenius) Batko species complex occurs worldwide wherever grasshoppers are found, and commonly causes disease epizootics that significantly reduce outbreaks (Carruthers *et al.*, 1997). In 1962, an epizootic of *E. grylli* was a major factor in drastic reductions of *C. pellucida* populations in Saskatchewan (Pickford and Riegert, 1964). Erlandson *et al.* (1988) found *E. grylli* pathotypes 1 and 2 in Saskatchewan and Alberta. Pathotype 2 occurred in melanopline grasshoppers and was more prevalent and widespread in 1986 than in 1985 (44% versus 11% of the survey sites in Saskatchewan and 6% versus <1% of the sites surveyed in Alberta). Although sporadic in its distribution, the fungus was important as a natural control agent in localized areas, infecting up to 28% of a population. *M. bivittatus* was the predominant host for pathotype 2, with *M. sanguinipes* being affected to a lesser extent. Pathotype 1 was isolated only from *C. pellucida* from a few sites in both provinces. In Alberta, low infection levels were obtained in *M. bivittatus* after field application of pathotype 2 resting spores that had been held in cold storage and then exposed to light for 24 h prior to application in water (D.L. Johnson, unpublished).

Entomophaga grylli pathotype 3 was recently introduced into North America (North Dakota) from Australia in a classical biological control attempt (Carruthers *et al.*, 1997). Although it initially established and caused low infection levels, the frequency of infection has declined to the point that long-term survival of this pathotype in North America is questionable (Bidochka *et al.*, 1996). Nevertheless, the fungus may reappear during favourable environmental conditions and spread into Canada, so intermittent surveys for it should continue.

Aspergillus parasiticus Speare was found infecting *M. bivittatus* and *M. packardii* in Saskatchewan in 1986 and 1987 (Moore and Erlandson, 1988). Pathogenicity against *M. sanguinipes* in the laboratory showed that the fungus was most pathogenic after injection, mildly pathogenic after topical application, and not pathogenic after oral application.

Beauveria bassiana (Balsamo) Vuillemin is a well known and widespread insect pathogen that is common in grasshoppers (Goettel *et al.*, 1995; Jaronski and Goettel, 1997), with an LT_{50} in grasshoppers of 4–8 days, depending on the isolate (Khatchatourians, 1992). The fungus was isolated from *M. bivittatus* cadavers near Saskatoon (Moore and Erlandson, 1988). In the laboratory, the fungus was highly virulent when injected or applied topically, and only mildly virulent when applied orally. The biochemical, morphological and trophic mechanisms used by *B. bassiana* to adhere to, solubilize, penetrate and utilize the cuticle of *M. sanguinipes*, as well as to overcome resistance mechanisms in the host's haemocoel, have been elucidated (Bidochka and Khatchatourians, 1987, 1990, 1991, 1992, 1994a, b; Bidochka, 1989; Bidochka *et al.*, 1989; Gillespie *et al.*, 1991).

Mycotrol®, based on *B. bassiana* isolate GHA, originally from *Diabrotica undecimpunctata howardi* Barber, is commercially available in the USA (Bradley *et al.*, 1999). In Canada, Johnson *et al.* (1988) showed that a precursor to strain GHA infected grasshoppers fed wheat, *Triticum aestivum* L., leaves sprayed with an aqueous suspension of conidia. Goettel *et al.* (1995) demonstrated that the fungus was virulent against *M. sanguinipes* and *M. bivittatus* when conidia were applied directly to the integument or on food (leaf lettuce or wheat bran). Using an oil-bait inoculation method, Inglis *et al.* (1996a) showed that isolate GHA was the most virulent among four isolates tested, and Inglis *et al.* (1996b) showed that oil formulations were superior to aqueous ones. Ovipositing females and hatchling nymphs were susceptible to *B. bassiana* conidia incorporated into the oviposition substrate (Inglis *et al.*, 1995a); mortality was influenced by conidia dose and presence of microbial flora (Inglis *et al.*, 1998).

Inglis (1996) found that conidia of isolate GHA exposed to sunlight on alfalfa, *Medicago sativa* L., and crested wheatgrass, *Agropyron cristatum* (L.) Gaertner, were short-lived but conidia shaded in lower parts of the canopy were longer-lived (Inglis *et al.*, 1993). Persistence was highest in soil (Inglis *et al.*, 1997a). Formulation of conidia in sunscreens prolonged their survival (Inglis *et al.*, 1995b), substantial numbers of conidia were removed from leaf surfaces by simulated rain (Inglis *et al.*, 1995c), and oil formulations enhanced conidial retention on foliage (Inglis *et al.*, 2000).

Field applications of strain GHA gave mixed results. In 1992, an application during cloudy, cool weather reduced populations by 60% and 33%, by 9 and 15 days, respectively, after application (Johnson and Goettel, 1993). A subsequent application during hot, sunny weather gave no significant reduction in numbers, despite excellent targeting and up to 80% mortality in grasshoppers collected immediately and 2 days after conidial application and maintained in a greenhouse (Inglis *et al.*, 1997b). Mason and Erlandson (1994) also noted differences in *B. bassiana* virulence between field and cage environments.

Behavioural thermoregulation by grasshoppers may profoundly influence the ability of *B. bassiana* to overcome its host. In grasshoppers permitted to bask for 1 h day^{-1}, mycosis decreased by more than 46% (Inglis *et al.*, 1996c). They also exhibited a behavioural fever in response to mycosis. In a field experiment under hot, sunny conditions no population reductions were observed and field-collected grasshoppers kept in the greenhouse or in shaded field-cages succumbed, whereas those placed in full sunlight survived (Inglis *et al.*, 1997c).

Assay of combinations of *B. bassiana* and a thermotolerant fungus, *Metarhizium flavoviride* Gams and Rozsypal (= *Metarhizium anisopliae* var. *acridum* Driver and Milner), showed that *B. bassiana* was superior under simulated cool conditions, whereas *M. flavoviride* was superior under simulated hot conditions (Inglis *et al.*, 1997d). When both fungi were applied simultaneously, mortality was greater for the combination treatment than for *M. flavoviride* alone in a simulated hot environment and equal to *B. bassiana* alone in a simulated cool environment, suggesting that use of both fungi would overcome some thermal limitations due to cool weather or thermoregulation (Inglis *et al.*, 1999). Another strategy to overcome the detrimental effects of thermoregulation may be the use of sublethal quantities of chemicals that inhibit thermoregulatory behaviour (D.L. Johnson, unpublished).

A soil isolate of *Verticillium lecanii* (Zimmerman) Viegas, tested for activity against nine insect pests, showed virulence against *M. sanguinipes* (Harper and Huang, 1986). All grasshoppers dipped in a conidial suspension died, whereas 8–37% of grasshoppers kept in cages with plants that had been sprayed with spore suspensions succumbed within 10 days. Under laboratory conditions, significant mortality was obtained in *M. bivittatus*, *M. packardii* and *M. sanguinipes* after they were sprayed with a conidial suspension, or after they fed on wheat leaves that had been sprayed with conidia or on wheat bran that had been colonized by the fungus (Johnson *et al.*, 1988). No fungus transmission occurred when grasshoppers were confined with infected cadavers. In field-cage experiments, conidia sprayed on to grasshoppers resulted in 40% mortality, and application of wheat bran colonized by the fungus resulted in 48% mortality. Although moisture was a significant factor in the laboratory experiments, it was not an important factor in the field-cage experiments.

Protozoa

A diversity of protozoa, including amoebae, eugregarines, neogregarines and microsporidans, infect grasshoppers and locusts (Johnson, 1997). *Nosema locustae* Canning has a relatively wide host range among Acrididae and large quantities of infectious spores are relatively easily produced in laboratory-reared grasshoppers. Ewen and Mukerji (1984), Johnson and Henry (1987), Johnson (1989b) and Schaalje *et al.* (1992) obtained variable results in field evaluations despite the use of similar application rates and formulations (2×10^9–5×10^9 spores ha^{-1} on bran baits), methods of evaluating grasshopper densities (sweep net sampling or quadrat counts), and methods for determining infection levels (presence of spores by microscopic observation of hanging drop homogenate from randomly selected grasshoppers). Some field trials produced significant population reductions, e.g. 60% reduction in *M. sanguinipes* (Ewen and Mukerji, 1980). However, most showed more modest reductions in host grasshopper density, 30% reduction compared to control plots, with infection levels from 20 to 50% in survivors (Johnson, 1989b). In contrast, Johnson and Henry (1987) and Johnson and Dolinski (1997) noted population reductions of less than 10%. Some of the differences in efficacy are likely related to differences in timing (i.e. developmental stage of host species), host species composition, environmental conditions or differences in *N. locustae* spore viability. Greater population reduction in smaller-scale field trials also showed the importance of efficient host targetting of the bran bait formulation to obtain maximum impact (Johnson, 1997).

Johnson and Pavlikova (1986) showed that *N. locustae* infection had a dose-dependent negative impact on food consumption by *M. sanguinipes.* Infected grasshoppers appeared to be less active (Johnson, 1989b) and this may have important implications for differential dispersal of infected versus non-infected grasshoppers. *Locusta migratoria migratorioides* (Reich and Fairmarie) nymphs infected with *N. locustae* showed retarded development compared to controls, and impact on other grasshopper species may be similar (Schaalje *et al.*, 1992). The impact of *N. locustae* infection on fecundity and vertical transmission has varied, depending on host species tested and experimental design. However, Johnson (1997) cited unpublished data indicating that *N. locustae* reduced the mean number of eggs per pod and percentage hatch when *M. sanguinipes* were infected as nymphs. Ewen and Mukerji (1980) found large decreases in egg pod numbers in treated plots, but could not separate population reduction phenomena from *N. locustae* impact on reproduction. Field applications of *N. locustae* have led to the development of persistent enzootics, e.g. low infection levels (1.7–3.3%) were shown in populations from plots treated 6 years earlier with *N. locustae* (Johnson and Dolinski, 1997).

Two other *Nosema* spp., *N. acridophagus* Henry and *N. cuneatum* Henry, have been isolated from grasshoppers. Both are more virulent than *N. locustae* for grasshoppers (Erlandson *et al.*, 1985) and locusts (Shaalje *et al.*, 1992), producing mortality more quickly (LT_{50} values of 9.3 and 16.5 days after inoculation for *N. acridophagus* and *N. cuneatum*, respectively, compared to >25 days for *N. locustae*) and at lower doses. Early instars are much more susceptible to infection than mature nymphs and infection significantly delays nymphal development (Erlandson *et al.*, 1986). Although these species are more virulent than *N. locustae*, they produce fewer spores in infected grasshoppers and thus are difficult to produce in practical quantities unless *in vitro* production can be developed. Several other Microsporida have been isolated from grasshoppers and locusts and these may have potential as microbial insecticides (Johnson, 1997).

Parasitoids and Predators

In Canada, the few attempts to introduce exotic predators and parasitoids of grasshoppers have been largely unsuccessful. Mason and Erlandson (1994) reviewed the native parasitoids and predators of grasshoppers in various regions. Scelionid egg parasitoids are important population regulators. *Scelio calopteni* Riley, the primary species in western Canada, attacks *M. bivittatus* and *M. sanguinipes*, with maximum parasitism levels reaching 20% and 9%, respectively (Mukerji, 1987). Mason and Erlandson (1994) and Johnson *et al.* (1996b) reviewed the impact of fly parasitoids on grasshoppers. Danyk *et al.* (2000) studied the biology and ecology of *Blaespoxipha atlanis* (Aldrich). Sarcophagidae and Tachnidae are the predominant natural enemies of economically important grasshoppers, with parasitism rates of 0–26% depending on host species, location and date. These estimates, based on dissections and parasitoid rearing from grasshoppers samples, likely underestimate their overall impact (Johnson *et al.*, 1996b). Artificial parasitism of *M. sanguinipes* nymphs with *B. atlanis* led to mortality rates of 75–93% and had significant impacts on food consumption and reproduction in host grasshoppers (Danyk *et al.*, 2000). Pest-management systems that minimize the negative impact of chemical insecticides should be implemented.

Evaluation of Biological Control

Entomopoxvirus may not be virulent enough to be useful as a microbial insecticide for grasshopper control in crops such as cereals. In addition, efficient and economical *in vivo* or *in vitro* production has yet to be developed, limiting the utility of these viruses. However, grasshopper EVs are host specific, infectious at relatively low dose rates, produce significant reductions in developmental and food consumption rates, and thus have potential as biological control agents for longer-term suppression of grasshopper populations or for treatment of populations at the beginning of an outbreak phase.

Entomopathogenic fungi, in particular Deuteromycetes, including *B. bassiana* and *Metarhizium* spp., are attractive candidates as microbial control agents because of their rapid rate of kill in laboratory assays relative to other entomopathogens, the possibility for commerical scale production on artifical media, and their contact route of infection that allows for pest targetting using standard spray applications. Investigation into augmentative approaches with *E. grylli* are presently hampered by the lack of economical methods for mass producing the fungus.

In general, *N. locustae* has not met the requirements for a fast-acting biological control agent for grasshoppers. Assessment of its effects on grasshopper feeding, development and reproduction indicate that it may still have potential as an agent for long-term suppression and maintenance of low grasshopper densities (Henry, 1990; Johnson, 1997).

The generally recognized problems associated with assessing the effectiveness of microbial agents in field trials, particularly in estimating reductions in grasshopper populations in treatment versus control plots, stem from two factors: first, most potential microbial control agents produce mortality much more slowly than chemical insecticides; and secondly, the target population tends to be very mobile, leading to migration in and out of plots. Thus, many field trials have underestimated the population reductions by microbial control agents (Onsager, 1988) and in some cases this has had a negative impact on the overall view of the utility of some pathogens. In Canada, only *N. locustae* and *B. bassiana* have been field tested and both have produced mixed results. Mycotrol® produced adequate population reductions of 60% by 9 days after treatment in one test under cool field conditions but it failed to produce measurable reductions in several trials under hot, sunny conditions, despite evidence of good targetting. Evidence is mounting that control failures in the field with *B. bassiana* result from environmental factors combined with host behaviour. Field trials with *N. locustae* have also produced quite variable results, but it is clear that this pathogen produces much slower mortality results than the fungal agents, is more difficult to produce, and thus has limited potential as a biological insecticide against grasshoppers. However, it still may have a role as a long-term suppressor of rangeland grasshopper populations. Microbial pathogens, e.g. *B. bassiana*, may have less of a negative impact on native parasitoids than chemicals (Johnson *et al.*, 1996b).

Recommendations

Further work should include:

1. Quantitative estimation of the importance of natural enemies in regulating

grasshopper populations, and investigation of integrated pest management strategies that conserve and maximize the impact of parasites and predators;
2. Additional investigation of the impact of environmental conditions on the performance of microbial pathogens in field applications;
3. Continued assessment of formulation and application technology for microbial pathogens;
4. Additional research into the mechanisms of pathogenesis, e.g. mechanism of early mortality phase of EV infections, with the view to manipulate these for improved efficacy.

References

Bidochka, M.J. (1989) Interaction of the entomopathogenic fungus *Beauveria bassiana* with the migratory grasshopper, *Melanoplus sanguinipes*: a systematic study of pathogenesis. PhD Thesis, University of Saskatchewan, Saskatoon.

Bidochka, M.J. and Khatchatourians, G.G. (1987) Hemocytic defense response to the entomopathogenic fungus *Beauveria bassiana* in the migratory grasshopper *Melanoplus sanguinipes*. *Entomologia Experimentalis Applicata* 45, 151–156.

Bidochka, M.J. and Khatchatourians, G.G. (1990) Identification of *Beauveria bassiana* extracellular protease as a virulence factor in the pathogenicity toward the migratory grasshopper, *Melanoplus sanguinipes*. *Journal of Invertebrate Pathology* 56, 362–370.

Bidochka, M.J. and Khatchatourians, G.G. (1991) The implication of metabolic acids produced by *Beauveria bassiana* in pathogenesis of the migratory grasshopper, *Melanoplus sanguinipes*. *Journal of Invertebrate Pathology* 58, 106–117.

Bidochka, M.J. and Khatchatourians, G.G. (1992) Growth of the entomopathogenic fungus *Beauveria bassiana* on cuticular components from the migratory grasshopper, *Melanoplus sanguinipes*. *Journal of Invertebrate Pathology* 59, 165–173.

Bidochka, M.J. and Khatchatourians, G.G. (1994a) Hydrolysis of migratory grasshopper, *Melanoplus sanguinipes*, cuticle by proteolytic enzymes in culture supernatants of entomopathogenic fungi. *Journal of Invertebrate Pathology* 63, 7–13.

Bidochka, M.J. and Khatchatourians, G.G. (1994b) Basic proteases of entomopathogenic fungi differ in their adsorption properties to host cuticle substrate. *Journal of Invertebrate Pathology* 64, 26–32.

Bidochka, M.J., Gillespie, J.P. and Khatchatourians, G.G. (1989) Phenoloxidase activity of acridid grasshoppers from the subfamilies Melanoplinae and Oedipodinae. *Comparative Biochemistry and Physiology* 94B, 117–124.

Bidochka, M.J., Walsh, S.R.A., Ramos, M.E., St Leger, R.J., Silver, J.C. and Roberts, D.W. (1996) Fate of biological control introductions: monitoring an Australian fungal pathogen of grasshoppers in North America. *Proceedings of the National Academy of Sciences, USA* 93, 918–921.

Bradley, C.A., Wood, P. and Britton, J. (1999) Mycoinsecticide activity against grasshoppers produced by *Beauveria bassiana*. US patent 5,939,065, 17 August, 1999.

Carruthers, R.I., Ramos, M.E., Larkin, T.S., Hostetter, D.L. and Soper, R.S. (1997) The *Entomophaga grylli* (Fresenius) Batko species complex: its biology, ecology, and use for biological control of pest grasshoppers. In: Goettel, M.S. and Johnson, D.L. (eds) *Microbial Control of Grasshoppers and Locusts. Memoirs of the Entomological Society of Canada* 171, 329–353.

Danyk, T., Johnson, D.L. and Mackauer, M. (2000) Parasitism of the grasshopper *Melanoplus sanguinipes* by a sarcophagid fly, *Blaesoxipha atlanis*: influence of solitary and gregarious development on host and parasitoid. *Entomologia Expermentalis et Applicata* 94, 259–268.

Erlandson, M.A. and Streett, D.A. (1997) Entomopoxviruses associated with grasshoppers and locusts: Biochemical characterization. In: Goettel, M.S. and Johnson, D.L. (eds) *Microbial Control of Grasshoppers and Locusts. Memoirs of the Entomological Society of Canada* 171, 131–146.

Erlandson, M.A., Mukerji, M.K., Ewen, A.B. and Gillott, C. (1985) Comparative pathogenicity of *Nosema acridophagus* Henry and *Nosema cuneatum* Henry (Microsporida: Nosematidae) for *Melanoplus sanguinipes* (Fab.) (Orthoptera: Acrididae). *The Canadian Entomologist* 117, 1167–1175.

Erlandson, M.A., Ewen, A.B., Mukerji, M.K. and Gillott, C. (1986) Susceptibility of immature stages of *Melanoplus sanguinipes* (Fab.) (Orthoptera: Acrididae) to *Nosema cuneatum* Henry (Microsporida: Nosematidae) and its effect on host fecundity. *The Canadian Entomologist* 118, 29–35.

Erlandson, M.A., Johnson, D.L. and Olfert, O.O. (1988) *Entomophaga grylli* (Fresenius) infections in grasshopper (Orthoptera: Acrididae) populations in Saskatchewan and Alberta, 1985–1986. *The Canadian Entomologist* 120, 205–209.

Ewen, A.B. and Mukerji, M.K. (1980) Evaluation of *Nosema locustae* (Microsporida) as a control agent of grasshopper populations in Saskatchewan. *Journal of Invertebrate Pathology* 35, 295–303.

Ewen, A.B. and Mukerji, M.K. (1984) *Melanoplus* spp. *Camnula pellucida* (Scudder), and other grasshoppers (Orthoptera: Acrididae). In: Kelleher, J.S. and Hulme, M.A. (eds) *Biological Control Programmes Against Insects and Weeds in Canada 1969–1980.* Commonwealth Agriculture Bureaux, Slough, pp. 61–62.

Gillespie, J.P., Bidochka, M.J. and Khatchatourians, G.G. (1991) Separation and characterization of grasshopper hemolymph phenoloxidases by sodium dodecyl sulfate–polyacrylamide gel electrophoresis. *Comparative Biochemistry and Physiology* 93C, 351–358.

Goettel, M.S. and Johnson D.L. (1997) Microbial control of grasshoppers and locusts. *Memoirs of the Entomological Society of Canada* 171.

Goettel, M.S., Johnson, D.L. and Inglis, G.D. (1995) The role of fungi in the control of grasshoppers. *Canadian Journal of Botany* 73 (suppl. 1), S71–S75.

Hardman, J.M. and Smoliak, S. (1982) The relative impact of various grasshopper species on *Stipa–Agropyron* mixed prairie and fescue prairie in southern Alberta. *Journal of Range Management* 35, 171–176.

Harper, A.M. and Huang, H.C. (1986) Evaluation of the entomophagous fungus *Verticillium lecanii* (Moniliales: Moniliaceae) as a control agent for insects. *Environmental Entomology* 15, 281–284.

Henry, J.E. (1990) Control of insects by protozoa. In: Baker, R.E. and Dunn, P.E. (eds) *New Directions in Biological Control: Alternatives for Suppressing Agricultural Pests and Diseases.* A.R. Liss, New York, New York, pp. 161–176.

Inglis, G.D. (1996) Environmental factors influencing the efficacy of *Beauveria bassiana* against grasshoppers. PhD thesis, Department of Biological Sciences, Simon Fraser University, Burnaby, British Columbia.

Inglis, G.D., Goettel, M.S. and Johnson, D.L. (1993) Persistence of the entomopathogenic fungus, *Beauveria bassiana* on phylloplanes of crested wheatgrass and lucerne. *Biological Control* 3, 258–270.

Inglis, G.D., Feniuk, R.P., Goettel, M.S. and Johnson, D.L. (1995a) Mortality of grasshoppers exposed to soil-borne *Beauveria bassiana* during oviposition and nymphal emergence. *Journal of Invertebrate Pathology* 65, 139–146.

Inglis, G.D., Goettel, M.S. and Johnson, D.L. (1995b) Influence of ultraviolet light protectants on persistence of the entomopathogenic fungus, *Beauveria bassiana. Biological Control* 5, 581–590.

Inglis, G.D., Johnson, D.L. and Goettel, M.S. (1995c) Effects of simulated rain on the persistence of *Beauveria bassiana* conidia on leaves of lucerne and wheat. *Biocontrol Science and Technology* 5, 365–369.

Inglis, G.D., Johnson, D.L. and Goettel, M.S. (1996a) An oil-bait bioassay method used to test the efficacy of *Beauveria bassiana* against grasshoppers. *Journal of Invertebrate Pathology* 67, 312–315.

Inglis, G.D., Johnson, D.L. and Goettel, M.S. (1996b) Effect of bait substrate and formulation on infection of grasshopper nymphs by *Beauveria bassiana. Biocontrol Science and Technology* 6, 35–50.

Inglis, G.D., Johnson, D.L. and Goettel, M.S. (1996c) Effects of temperature and thermoregulation on mycosis by *Beauveria bassiana* in grasshoppers. *Biological Control* 7, 131–139.

Inglis, G.D., Duke, G.M., Kanagaratnam, P., Johnson, D.L. and Goettel, M.S. (1997a) Persistence of *Beauveria bassiana* in soil following application of conidia through crop canopies. In: Goettel, M.S. and Johnson, D.L. (eds) *Microbial Control of Grasshoppers and Locusts. Memoirs of the Entomological Society of Canada* 171, 253–263.

Inglis, G.D., Johnson, D.L. and Goettel, M.S. (1997b) Field and laboratory evaluation of two conidial batches of *Beauveria bassiana* against grasshoppers. *The Canadian Entomologist* 129, 171–186.

Inglis, G.D., Johnson, D.L. and Goettel, M.S. (1997c) Effects of temperature and sunlight on mycosis (*Beauveria bassiana*) (Hyphomycetes: Sympodulosporae) of grasshoppers under field conditions. *Environmental Entomology* 26, 400–409.

Inglis, G.D., Johnson, D.L., Cheng, K.-J. and Goettel, M.S. (1997d) Use of pathogen combinations to overcome the constraints of temperature on entomopathogenic Hyphomycetes against grasshoppers. *Biological Control* 8, 143–152.

Inglis, G.D., Johnson, D.L., Kawchuk, L.M. and Goettel, M.S. (1998) Effect of soil texture and soil sterilization on susceptibility of ovipositing grasshoppers to *Beauveria bassiana*. *Journal of Invertebrate Pathology* 71, 73–81.

Inglis, G.D., Duke, G.M., Kawchuk, L.M. and Goettel, M.S. (1999) Influence of oscillating temperatures on the competitive infection and colonization of the migratory grasshopper by *Beauveria bassiana* and *Metarhizium flavoviride*. *Biological Control* 14, 111–120.

Inglis, G.D., Ivie, T.J., Duke, G.M. and Goettel, M.S. (2000) Influence of rain and conidial formulation on persistence of *Beauveria bassiana* on potato leaves and Colorado potato beetle larvae. *Biological Control* 18, 55–64.

Jaronski, S.T. and Goettel, M.S. (1997) Development of *Beauveria bassiana* for control of grasshoppers and locusts. In: Goettel, M.S. and Johnson, D.L. (eds) *Microbial Control of Grasshoppers and Locusts. Memoirs of the Entomological Society of Canada* 171, 225–237.

Johnson, D.L. (1989a) Spatial autocorrelation, spatial modelling, and improvements in grasshopper survey methodology. *The Canadian Entomologist* 121, 579–588.

Johnson, D.L. (1989b) The effects of timing and frequency of application of *Nosema locustae* (Microspora: Microsporida) on the infection rate and activity of grasshoppers (Orthoptera: Acrididae). *Journal of Invertebrate Pathology* 54, 353–362.

Johnson, D.L. (1997) Nosematidae and other protozoa as agents for control of grasshoppers and locusts: Current status and prospects. In: Goettel, M.S. and Johnson, D.L. (eds) *Microbial Control of Grasshoppers and Locusts. Memoirs of the Entomological Society of Canada* 171, 375–389.

Johnson, D.L. and Dolinski, M.G. (1997) Attempts to increase the prevalence and severity of infection of grasshoppers with the entomopathogen *Nosema locustae* Canning (Microsporida: Nosematidae) by repeated field applications. In: Goettel, M.S. and Johnson, D.L. (eds) *Microbial Control of Grasshoppers and Locusts. Memoirs of the Entomological Society of Canada* 171, 391–400.

Johnson, D.L. and Goettel, M.S. (1993) Reduction of grasshopper populations following field application of the fungus *Beauveria bassiana*. *Biocontrol Science and Technology* 3, 165–175.

Johnson, D.L. and Henry, J.E. (1987) Low rates of insecticide and *Nosema locustae* (Microsporidia: Nosematidae) on baits applied to roadsides for grasshopper (Orthoptera: Acrididae) control. *Journal of Economic Entomology* 80, 685–689.

Johnson, D.L. and Pavlikova, E. (1986) Reduction of consumption by grasshoppers (Orthoptera: Acrididae) infected with *Nosema locustae* Canning (Microsporida: Nosematidae). *Journal of Invertebrate Pathology* 48, 232–238.

Johnson, D.L., Huang, H.C. and Harper, A.M. (1988) Mortality of grasshoppers (Orthoptera: Acrididae) inoculated with a Canadian isolate of the fungus *Verticillium lecanii*. *Journal of Invertebrate Pathology* 52, 335–342.

Johnson, D.L., Olfert, O., Dolinski, M. and Harris, L. (1996a) GIS-based forecasts for management of grasshopper populations in Western Canada. *Proceedings of the FAO International Symposium on Agricultural Pest Forecasting, Québec, 10–12 Oct, 1995*, pp. 109–112.

Johnson, D.L., Danyk, T.P., Goettel, M.S. and Rode, L.M. (1996b) *Use of Parasitic Flies, Pathogens and Insecticides for Sustainable Integrated Pest Management of Grasshoppers.* Final Report, Canada–Alberta Environmentally Sustainable Agriculture Agreement Project RES-082-94.

Khachatourians, G.G. (1992) Virulence of five *Beauveria* strains, *Paecilomyces farinosus*, and *Verticillium lecanii* against the migratory grasshopper, *Melanoplus sanguinipes*. *Journal of Invertebrate Pathology* 59, 212–214.

Madder, D.J. and Stemeroff, M. (1988) *The Economics of Insect Control on Wheat, Maize and Canola in Canada, 1980–1985. Insect Losses Committee Report, Part II.* Entomological Society of Canada, Ottawa, Ontario.

Mason, P.G. and Erlandson, M.A. (1994) The potential of biological control for management of grasshoppers (Orthoptera: Acrididae) in Canada. *The Canadian Entomologist* 126, 1459–1491.

McGuire, M.R., Streett, D.A. and Shasha, B.S. (1991) Evaluation of starch encapsulation for formulation of grasshopper (Orthoptera: Acrididae) entomopoxviruses. *Journal of Economic Entomology* 84, 1652–1656.

Moore, K.C. and Erlandson, M.A. (1988) Isolation of *Aspergillus parasiticus* Speare and *Beauveria bassiana* (Bals.) Vuillemin from melanopline grasshoppers (Orthoptera: Acrididae) and demonstration of their pathogenicity in *Melanoplus sanguinipes* (Fabricius). *The Canadian Entomologist* 120, 989–991.

Mukerji, M.K. (1987) Parasitism by *Scelio alopteni* Riley (Hymenoptera: Scelionidae) in eggs of the two dominant melanopline species (Orthoptera: Acrididae) in Saskatchewan. *The Canadian Entomologist* 119, 147–151.
Olfert, O.O. and Erlandson, M.A. (1991) Wheat foliage consumption by grasshoppers (Orthoptera: Acrididae) infected with *Melanoplus sanguinipes* entomopoxvirus. *Environmental Entomology* 20, 1720–1724.
Onsager, J.A. (1988) Assessing effectiveness of *Nosema locustae* for grasshopper control. *Montana AgResearch* 5, 12–16.
Pickford, R. and Riegert, P.W. (1964) The fungous disease caused by *Entomophaga grylli*, and its effects on grasshopper populations in Saskatchewan in 1963. *The Canadian Entomologist* 96, 1158–1166.
Riegert, P.W. (1968) A history of grasshopper abundance surveys and outbreaks in Saskatchewan. *Memoirs of the Entomological Society of Canada* 52.
Schaalje, G.B., Johnson, D.L. and van der Vaart, H.R. (1992) Application of competing risks theory to the analysis of effects of *Nosema locustae* and *N. cuneatum* on development and mortality of migratory locusts. *Environmental Entomology* 21, 939–948.
Smith, D.S. and Smoliak, S. (1977) The distribution and abundance of adult grasshoppers (Acrididae) in crops in Alberta, 1917–1975. *The Canadian Entomologist* 109, 575–592.
Streett, D.A. and McGuire, M.R. (1990) Pathogenic diseases of grasshoppers. In: Chapman, R.F. and Joern, A. (eds) *Biology of Grasshoppers*. Wiley Interscience, New York, New York, pp. 483–516.
Streett, D.A. and Woods, S.A. (1990) Grasshopper pathogen field evaluation: Virus. In: *Cooperative Grasshopper Integrated Pest Management Project 1990 Annual Report*. United States Department of Agriculture, Agriculture Plant Health Inspection Service, Washington, DC, pp. 210–217.
Street, D.A., Woods, S.A. and Erlandson, M.A. (1997) Entomopoxviruses of grasshoppers and locusts: Biology and biological control potential. In: Goettel, M.S. and Johnson, D.L. (eds) *Microbial Control of Grasshoppers and Locusts. Memoirs of the Entomological Society of Canada* 171, 115–130.
Woods, S.A., Streett, D.A. and Henry, J.E. (1992) Temporal patterns of mortality from an entomopoxvirus and strategies for control of the migratory grasshopper (*Melanoplus sanguinipes* (F.)). *Journal of Invertebrate Pathology* 60, 33–39.

36 *Mindarus abietinus* Koch, Balsam Twig Aphid (Hemiptera: Mindaridae)

C. Cloutier and C. Jean

Pest Status

The balsam twig aphid, *Mindarus abietinus* Koch, Holarctic in distribution, is a primary pest of balsam fir, *Abies balsamea* (L.) Miller, grown as Christmas trees in eastern North America. It is recorded from fir, *Abies* spp., spruce, *Picea* spp., pine, *Pinus* spp., and juniper, *Juniperus* spp. In eastern Canada, its main hosts are *A. balsamea* and white spruce, *Picea glauca* (Moench) Voss. In managed Christmas tree plantations in Quebec, *M. abietinus* is known to reduce the commercial value of trees at harvest when the incidence of aesthetic damage to current-year shoots exceeds 5%, which corresponds to 9% of opening buds carrying aphid colonies in spring (Deland *et al.*, 1998). In Quebec, Cloutier *et al.* (1998) documented the population cycle of *M. abietinus* in *A. balsamea* plantations over a 3-year period. At the peak in 1997, 95% of apical shoots were infested by an average of 6.5 fundatrices in spring. Deland *et al.* (1998) reported an average density of 154 aphids per shoot in June, shortly before dispersion, which was massive. Eggs laid as a result of incoming sexuparae in July in a plantation without a spring population were estimated at 28,000 per tree.

M. abietinus is monoecious and holocyclic, its cycle comprising three or four generations involving three different asexual viviparous female morphs, in addition to apterous sexuals. All morphs develop through four larval instars except males, which have only three. The apterous fundatrices hatch from overwintered eggs in late April. They feed initially on needles from the previous year's host shoots. Maturing fundatrices move on to opening buds in late May–early June where they initiate colonies by vivipositing up to 70 second-generation young. Heavily colonized *A. balsamea* shoots are stunted, causing the characteristic damage known as a pseudogall. Fundatrix progeny mostly develop into alate adults, the sexuparae, which eventually disperse by flight over several weeks, starting in early to mid-June. A minority of the fundatrix progeny remain apterous and reside on the host tree, where they produce third-generation sexuparae, hence the four-generation pathway of the aphid cycle. Following dispersal, sexuparae resettle on the host foliage and start vivipositing the sexuals, which feed little, are reduced in size and have a shortened development. Mated, reproductively mature oviparae lay one or two eggs, which are glued to the stem of maturing shoots in early to mid-July.

Background

Control of *M. abietinus* in managed plantations is mainly based on insecticide spraying against fundatrices, which prevents pseudogall formation. In the mid-1990s, it was estimated that 18,000 l of diazinon were applied annually to control *M. abietinus* in Quebec plantations (A. Pettigrew, Québec, 1995, personal communication). Insecticide spraying in plantations is known to disrupt bird nesting (Rondeau and Desgranges, 1991) and natural control of *M. abietinus* (Nettleton and Hain, 1982; Kleintjes, 1997).

Räther and Mills (1989) reviewed biological control agents of *M. abietinus*. In New Brunswick forests, Varty (1968, 1969) reported Araneae, Chrysopidae, Coccinellidae and Syrphidae as key predators. Starý (1975) described *Pseudopraon mindariphagum* Starý as a parasitoid of *M. abietinus* in Europe.

Biological Control Agents

Parasitoids

In Quebec, Cloutier *et al.* (unpublished) collected mummies of a primary parasitoid of *M. abietinus* fundatrices in *A. balsamea* plantations. The parasitoid may be an undescribed species closely resembling *P. mindariphagum* from Europe (M. Mackauer, Vancouver, 2000, personal communication). It occurred at low density, was limited to early season, and was frequently hyperparasitized.

Predators

In Quebec *A. balsamea* plantations, the indigenous coccinellid, *Anatis mali* Say, and the introduced *Harmonia axyridis* (Pallas), which recently invaded southern Canada from the USA (Coderre *et al.*, 1995), were consistent predators of *M. abietinus* in the absence of insecticide spraying (Cloutier *et al.*, 1998; Deland *et al.*, 1998; Berthiaume, 1998). Overwintered adults of both species hunted *M. abietinus* on *A. balsamea* in May, well before fundatrices invaded opening buds. They were by far the predominant species to oviposit and develop on *M. abietinus*-infested fir. *Mulsantina hudsonica* Casey, *Coccinella trifasciata* L. and *C. septempunctata* L. also reproduced on fir, but only the adults of *Adalia bipunctata* L., *Chilocorus stigma* Say and *Propylea quatuordecimpunctata* L. were observed foraging on *M. abietinus*-infested trees. Among syrphid predators, the larvae of *Syrphus ribesii* (L.) were also consistent predators, attacking *M. abietinus* within the pseudogalls.

Berthiaume *et al.* (2000) studied the impact of coccinellid predation on *M. abietinus* by experimentally excluding their egg masses in a 5–6-year-old *A. balsamea* plantation. They showed that coccinellid larvae, especially *A. mali*, were important natural mortality agents of *M. abietinus*, both during and after the phase of intense population growth resulting from fundatrix reproduction in early June. The density of *M. abietinus* eggs laid subsequently was reduced by 32% on trees with coccinellid larvae, and current-year shoots were 19% longer. However, the larvae acted too late to prevent aesthetic damage to developing shoots, which is critical in Christmas tree production.

In a parallel study (Berthiaume, 1998; Berthiaume *et al.*, 2001), adults of *Podabrus rugosulus* Leconte were commonly observed hunting on balsam fir at the time of *M. abietinus* dispersal. Adults fed on fourth instar and adult sexuparae, the largest and most mobile forms of the aphid. The beetle appeared to be an opportunist, benefiting from the sudden availability of aphid prey outside pseudogalls at the time of dispersal and sexuals production by *M. abietinus* sexuparae.

Cloutier and Jean (2000) experimented with predator augmentation to control *M. abietinus* in *A. balsamea* plantations. Initially, the potential of commercially available, generalist aphid predators was tested. *Chrysopa carnea* Stephens and *Aphidoletes aphidimyza* (Rondani) were released by hand on 5–6-year-old trees at 1000 first-instar *C. carnea* larvae per tree and 100 mature *A. aphidimyza* pupae per tree. Aphid density estimates up to 2–3 weeks after release indicated no impact of the predators, which apparently survived only briefly.

Cloutier and Jean (2000) tested *H. axyridis* larvae, from a lab colony established from adults collected during their autumn aggregation, as larval predators on *M. abietinus* fundatrices during the time that natural field populations of overwintered coccinellids were regenerating their reproductive potential or only starting to reproduce. Released *H. axyridis* larvae established well, with survival to the fourth larval instar averaging 10%. Contrary to expectations, peak second-generation *M. abietinus* density and aphid egg density at the end of the cycle reached higher levels on trees with released *H. axyridis* than on control trees exposed to natural mortality factors, including coccinellids. Aphid density on trees receiving 90 *H. axyridis* was similar to that on trees where coccinellid larvae were excluded. The coccinellids naturally present on trees had specific relationships to the *M. abietinus* system. *C. septempunctata*, followed by *A. mali*, *C. trifasciata* and *H. axyridis*, were initially most abundant as overwintered adults. However, *A. mali* exhibited by far the most egg-laying and larval survival on aphid-infested trees in the control blocks, compared to other species also hunting on the trees as adults. Interestingly, released *H. axyridis* larvae inhibited oviposition and directly affected immature survival of the naturally occurring species, which was most evident with *A. mali*; *A. mali* larval density was three times lower on trees treated with *H. axyridis* than on controls. Direct observations revealed that the released *H. axyridis* larvae, easily recognized by their development being ahead of the naturally occurring species, preyed both on each other and on egg masses, larvae and pupae of the other species.

Cloutier and Jean (2000) released adults of *H. axyridis* from a 'Flightless' strain obtained from Antibes, France (Ferran *et al.*, 1998). In spite of being unable to fly, 'Flightless' adults appear better adapted than larvae to hunt *M. abietinus* fundatrices, which occur mostly in isolation on the previous year's shoots, or small incipient colonies in opening buds and shoots. Because of exceptionally warm weather in late May, 1999, the *M. abietinus* cycle was advanced. Although the predators were available as planned, the first release occurred only shortly before initiation of the second aphid generation. Only the treatment involving two *H. axyridis* releases had an impact, reducing peak aphid density (10–20 aphids per shoot) by 30–50%. The 'Flightless' *H. axyridis* adults were much more mobile than expected, redistributing themselves by walking among trees during the experiment. Thus aphid egg densities were not different among treatments at the end of the test.

Natural predator activity was lower in this study than the previous one, probably because the experimental and nearby plantations had been sprayed with diazinon in previous years. *A. mali* was again the predominant natural, being active on *A. balsamea* both as overwintered adults and subsequently as larvae. Its reproductive success on control trees was estimated to be 42 times higher than that of 'Flightless' *H. axyridis* on trees where it was released. However, egg-laying by the 'Flightless' strain extended over 20 days compared with only 10 days for the natural *A. mali* population.

Evaluation of Biological Control

Coccinellids were the most prominent natural predators of *M. abietinus*. Among nearly ten species observed, *A. mali* and *H. axyridis* appeared to be the best adapted to impact on aphid populations. In addition to early predation on fundatrices in May, they reproduced on *A. balsamea* and their larvae hunted the aphid within pseudogalls during their second generation. Syrphid larvae clearly appeared to have a significant impact during the aphid's second generation but were observed attacking the fundatrices before they invaded new shoots (critical to aphid population regulation in this system) only during the peak aphid year.

Tests with *C. carnea* or *A. aphidimyza* showed little promise. In spite of the experimental conditions being favourable, with aphids being present at moderate to high density, and no extreme weather, there was no evidence that either species survived introduction. These commercially produced predators may not be adapted to hunt on *A. balsamea* or to kill or feed on *M. abietinus*, although we occasionally observed natural chrysopid or cecidomyiid predators.

In augmentation tests, early release of *H. axyridis* larvae interfered with natural predation, mostly by coccinellids. *H. axyridis* clearly is an aggressive competitor, capable of both intraguild predation and cannibalism. The experimental conditions evidently favoured the released *H. axyridis* in such interactions. It is possible that the release density was too high considering declining aphid population and relatively abundant naturally occurring species, but this could not be anticipated. Our study supports the contention that *H. axyridis* can act as a 'top predator' in aphid natural enemy guilds, which may partly explain its recent massive geographic expansion.

Although being limited to walking, 'Flightless' *H. axyridis* have not lost their tendency to fly. They were frequently observed climbing to exposed branch tips, spreading their elytra and attempting to take off, which often resulted in their falling to the ground. They thus redistributed themselves beyond treated trees following the initial release, which occurred before *M. abietinus* colonies had expanded significantly. That a release impact was evident only in the treatment involving two releases of *H. axyridis* likely resulted from the greater retention effect of the increase in aphid density that had occurred meanwhile. Although our single-year test was not entirely conclusive, 'Flightless' *H. axyridis* is the only predator whose augmentation had a measurable impact. They should be tested again at the onset of the next *M. abietinus* population rise, when the abundance of the naturally occurring coccinellids in plantations most likely would be low, implying less frequent negative intraguild interactions.

Recommendations

Further work should include:

1. Determining the ecological requirements of *A. mali*, especially its overwintering sites and minimum canopy characteristics for attraction to Christmas tree plantations;
2. Documenting syrphid biodiversity and requirements for early spring activity in plantations;
3. Documenting interactions between *M. abietinus* population cycling and the abundance of coccinellids and other aphid natural enemies in plantations;
4. Studying the potential of coccinellid releases, including 'Flightless' *H. axyridis*, with special attention to inherent risks to natural biological control agents;
5. Developing conservation programmes for coccinellids and other natural enemies of aphids and other pests in plantations.

References

Berthiaume, R. (1998) Les ennemis naturels du puceron des pousses du sapin, *Mindarus abietinus* Koch (Homoptera: Aphididae), avec une emphase particulière sur les coccinelles *Anatis mali* Say et *Harmonia axyridis* Pallas. Mémoire de maîtrise. Université Laval, Québec, Canada.

Berthiaume, R., Hébert, C. and Cloutier, C. (2000) Predation on *Mindarus abietinus* infesting balsam fir grown as Christmas trees: the impact of coccinellid larval predation with emphasis on *Anatis mali*. *Biocontrol* 45, 425–438.

Berthiaume, R., Hébert, C. and Cloutier, C. (2001) *Podabrus rugosulus* (Coleoptera: Cantharidae) as an opportunist predator of *Mindarus abietinus* (Hemiptera: Aphididae), in Christmas tree plantations. *The Canadian Entomologist*, 151–154.

Cloutier, C. and Jean, C. (2000) *Lutte biologique contre le puceron des pousses du sapin en plantations d'arbres de Noël.* Rapport final No. 4531, Conseil des Recherches en Pêches et Alimentation du Québec, Québec.

Cloutier, C., Deland, J.P., Berthiaume, R. and Hébert, C. (1998) *Programme alternatif de protection du sapin de Noël dans le contexte d'une saine gestion des ressources environnementales.* Rapport Synthèse, Environnement Québec, Québec.

Coderre, D., Lucas, E. and Gagné, I. (1995) The occurrence of *Harmonia axyridis* (Pallas) (Coleoptera: Coccinellidae) in Canada. *The Canadian Entomologist* 127, 609–611.

Deland, J.P., Berthiaume, R., Hébert, C. and Cloutier, C. (1998) *Programme alternatif de protection du sapin de Noël dans le contexte d'une saine gestion des ressources environnementales.* Rapport de Recherche, Environnement Québec.

Ferran, A., Giuge, L., Tourniaire, R., Gambier, J. and Fournier, D. (1998) An artificial non-flying mutation to improve efficiency of the ladybird *Harmonia axyridis* in biological control of aphids. *Entomophaga* 43, 53–64.

Kleintjes, P.K. (1997) Midseason insecticide treatment of balsam twig aphids (Homoptera: Aphididae) and their aphidophagous predators in a Wisconsin Christmas tree plantation. *Environmental Entomology* 26, 1393–1397.

Nettleton, W.A. and Hain, P. (1982) The life history, foliage damage, and control of the balsam twig aphid, *Mindarus abietinus* (Homoptera: Aphididae), in fraser fir Christmas tree plantations of western North Carolina. *The Canadian Entomologist* 114, 155–165.

Räther, M. and Mills, N.J. (1989) Possibilities for the biological control of the Christmas tree pests, the balsam gall midge, *Paradiplosis tumifex* Gagné (Diptera: Cecidomyiidae) and the balsam twig aphid, *Mindarus abietinus* Koch (Homoptera: Mindaridae), using exotic enemies from Europe. *Biocontrol News and Information* 10, 119–129.

Rondeau, G. and Desgranges, J.-L. (1991) *Effets des arrosages du diazinon (Basudin), du diméthoate (Cygon) et du savon insecticide (Safer) sur la faune avienne dans les plantations de sapins de Noël.* Série de rapports techniques no. 141, Service canadien de la faune, Région du Québec, Ste-Foy, Québec.

Starý, P. (1975) *Pseudopraon mindariphagum* gen. n., sp. n. (Hymenoptera, Aphidiidae) – Description and life history of a parasite of *Mindarus abietinus* (Homoptera, Mindaridae) in Central Europe. *Acta Entomologica Bohemoslovakia* 72, 249–258.

Varty, I.W. (1968) *The Biology of the Balsam Twig Aphid,* Mindarus abietinus *Koch, in New Brunswick: Polymorphism, Rates of Development, and Seasonal Distribution of Populations.* Internal report M-42, Department of Forestry and Environment Canada, Forest Research Laboratory, Fredericton, New Brunswick.

Varty, I.W. (1969) *Ecology of* Mulsantina hudsonica *Casey, a Ladybeetle in Fir–spruce Forest.* Internal report M-42, Department of Fisheries and Forestry Canada, Forest Research Laboratory, Fredericton, New Brunswick.

37 *Musca domestica* L., House Fly (Diptera: Muscidae)

K.D. Floate, T.J. Lysyk, G.A.P. Gibson and T.D. Galloway

Pest Status

The house fly, *Musca domestica* L., is a cosmopolitan pest that breeds in rotting organic matter associated with human habitation and livestock confinements, e.g. poultry houses, swine barns, cattle feedlots and dairies. It is a nuisance pest that annoys workers in livestock facilities. Adults disperse from breeding sites into surrounding areas where their presence can generate lawsuits by home owners against facility operators. Adult flies regurgitate while feeding, and leave unsightly vomit and faecal spots on resting areas. This latter behaviour, their association with manure and their tendency to enter homes makes them efficient disease vectors. *M. domestica* can mechanically transmit numerous pathogens, including the causative agents of amoebic dysentery, bacillary dysentery, cholera, mastitis, salmonellosis and tuberculosis (West, 1951). Their potential role in disease transmission is expected to lead to stricter regulations to control them at food production facilities that implement Hazard Analysis Critical Control Point standards. The life cycle of *M. domestica* is described by West (1951).

Background

M. domestica pupae are attacked by several species of parasitoids. In the USA, the parasitoid complexes are dominated by *Muscidifurax* spp. and *Spalangia* spp. (Legner, 1994), which became commercially available in the 1960s (McKay, 1997) and are now sold throughout North America. Commercialized species include *M. raptor* Girault and Saunders, *M. raptorellus* Kogan and Legner, *M. zaraptor* Kogan and Legner, *S. cameroni* Perkins, *S. endius* Walker, *S. nigroaenea* Curtis, and *Nasonia vitripennis* Walker (Cranshaw *et al.*, 1996). Parasitoids sold in Canada are shipped from insectaries in the USA, typically in units of 10,000 parasitized *M. domestica* pupae packaged in sawdust. Costs vary from Can$27.50–40.00 per unit, plus handling and shipping charges (K.D. Floate, unpublished). Units are advertised as producing from 30,000 to 80,000 parasitoids, depending upon the source. A common recommendation is the release of 500 parasitoids per large animal every 2–4 weeks during the fly season (Cranshaw *et al.*, 1996).

Maggots of the black dump fly, *Hydrotaea* (*Ophyra*) *aenescens* (Wiedemann), are predacious on maggots of other species. This species breeds in the same habitats as *M. domestica* and is sold as a biological control agent for this pest. Because *H. aenescens* is unknown from western Canada and has itself been viewed as a nuisance pest, sales are currently restricted to east of Manitoba.

Reasons for producer interest in biological agents for *M. domestica* control include failure of chemical products, concerns regarding non-target effects on animals and people, and the requirement for non-chemical alternatives in organic livestock operations.

Biological Control Agents

Pathogens

Fungi

Several pathogens, notably *Beauveria bassiana* (Balsamo) Vuillemin and *Entomophthora muscae* (Cohn) Fresenius, have been evaluated for *M. domestica* control in the USA (Watson *et al.*, 1995) but not in Canada. Epizootics of *E. muscae* typically occur at cooler times of the year, which may make this species attractive for use in colder climates. Although *E. muscae* occurs in Alberta (T.J. Lysyk, unpublished) and Manitoba (T.D. Galloway, unpublished), and *B. bassiana* occurs in Canada, their potential as biological control agents needs to be evaluated.

Nematodes

Belton *et al.* (1987) examined the use of *Heterorhabditis heliothidis* (Khan, Brooks and Hirschmann) against *M. domestica* in poultry barns. Applications of 2×10^6–4×10^6 nematodes m^{-2} reduced fly emergence by 80–86%. Fly populations in a treated barn increased at a slower rate than populations in an untreated barn. Predacious mites and beetles and *M. domestica* pupae were unaffected, so nematode application was not expected to affect parasitoid wasps within pupae. However, nematode infectivity in poultry manure was low in other studies, likely due to inactivation by ammonia, salts or other toxic products in the manure (Geden *et al.*, 1986; Mullens *et al.*, 1987). Microencapsulation may provide one solution to prevent inactivation and maintain nematode infectivity for longer periods (Renn, 1995).

Parasitoids

In Canada, parasitoid surveys have identified faunas distinct from those typically reported for the USA (Table 37.1). *Trichomalopsis sarcophagae* (Gahan) and *Phygadeuon fumator* Gravenhorst are common in Alberta and Manitoba, respectively, but are typically absent or rare in the USA. Conversely, *Spalangia* spp., virtually absent in Alberta and Manitoba, are abundant in the USA where they appear to be very important. *Spalangia* spp. and *Phygadeuon*, but not *Trichomalopsis* spp., appear to be important in Ontario (G.A.P. Gibson and K.D. Floate, unpublished). Recent surveys in Canada have led to systematic revisions of *M. domestica* parasitoids in the genera *Trichomalopsis* (Gibson and Floate, 2001) and *Urolepis* (Gibson, 2000).

In Alberta, parasitism of sentinel *M. domestica* pupae in dairies and feedlots has averaged less than 3% (Lysyk, 1995; Floate *et al.*, 1999, 2000). In Manitoba dairies, parasitism averaged 4–8% for sentinel *M. domestica* pupae and as high as 9% for naturally occurring pupae (McKay and Galloway, 1999).

Table 37.1. Species of parasitoids reported from sentinel *Musca domestica* pupae in Canada. Values are percentage of parasitized pupae.

Species	AB[a]	AB[b]	AB[c]	MB[d,e]
Braconidae				
Unidentified sp.	–	<1	–	–
Diapriidae				
Synacra sp.	–	<1	–	–
Eupelmidae				
Eupelmus (*Macroneura*) *vesicularis* (Retzius)	–	–	1	<1
Ichneumonidae				
Phygadeuon fumator Gravenhorst	–	–	–	67
Phygadeuon sp. (*P. ?fumator*)	3	3	12	–
Pteromalidae				
Dibrachys cavus (Walker)	<1	<1	<1	–
Muscidifurax raptor Girault and Saunders	54	37	18	6
Muscidifurax zaraptor Kogan and Legner	24	7	15	2
Nasonia vitripennis (Walker)	–	7	–	8[f]
Spalangia cameroni Perkins	2	–	–	<1
Spalangia nigra Latreille	–	–		<1
Spalangia subpunctata Förster	–	–	–	<1
Trichomalopsis americana (Gahan)	–	1[g]	2	–
Trichomalopsis dubia (Ashmead)	–	<1[g]	–	2[g]
Trichomalopsis sarcophagae (Gahan)	12	24	35	–
Trichomalopsis viridescens (Walsh)	–	2[g]	–	–
Urolepis rufipes (Ashmead)	5	19	16	14
Unidentified	–	–	–	1
Number of parasitized pupae examined	685	897	537	2999

Alberta (AB), Manitoba (MB).
[a]Lysyk (1995); [b]Floate *et al.* (1999); [c]Floate *et al.* (2000); [d]McKay (1997); [e]single specimens of Figitidae, *Aphaereta* sp. (Braconidae) and Staphylinidae were recovered from naturally occurring fly pupae; [f]inflated by mass-releases; [g]modified after Gibson and Floate (2001).

Predators

Ravinia spp., whose larvae are predacious and may have a minor role in reducing populations of pest flies, have been collected in Malaise traps in southern Alberta feedlots (O'Hara *et al.*, 2000), and have been shown to develop in feedlot cattle manure (Floate, 1998).

Eggs and larvae of *M. domestica* are undoubtedly consumed by many predacious mites (e.g. Macrochelidae, Uropodidae) and beetles (primarily Staphylinidae, but also Carabidae and Histeridae). *Carcinops pumilio* (Erichson) is an important predator of immature *M. domestica* in poultry houses in the USA (Geden and Axtell, 1988) and probably Canada (Davies, 1991).

Glofcheskie and Surgeoner (1990) tested Muscovy ducks, *Cairina moschata* L., as biological control agents of *M. domestica*. Ducks removed adult flies from enclosed 0.24 m^3 spaces at least 30-fold faster than did coiled fly paper rolls, fly traps, fly sheets, or bait cards. In calf pens, ducks consumed an average of 25 flies per 15 minute observation period when *M. domestica* populations were low to moderate. Ducks were identified as the most cost-effective of the five control methods considered. However, their potential to spread disease would prevent the use of ducks on commercial poultry operations (Glofcheskie and Surgeoner, 1990).

Releases and Recoveries

In Alberta, *M. raptorellus* was released at two sites near Lethbridge to establish local populations and reduce the need for releases in subsequent years (Floate *et al.*, 2000). Previous releases of this South American species have resulted in populations becoming established in Nebraska and Missouri. Following release of an estimated 1.3 million individuals in 1996 and 1997, 29 male *M. raptorellus* were collected from four sentinel *M. domestica* pupae parasitized at one site in spring, 1998. Hence, at least females can overwinter in southern Alberta, although the winter of 1997–1998 was unusually mild.

Weintraub (1985) reported the successful application of parasitoid wasps to reduce *M. domestica* in a swine barn, in conjunction with a sanitation programme and baited insecticide stations. Pre-release fly numbers declined 96–98% from an abundance index of near 60 following releases of 20,000–30,000 *M. raptor* every 2 weeks. Cessation of releases was followed by an increase in fly abundance, which declined with the resumption of *M. raptor* releases. Subsequent releases of *N. vitripennis* and *S. endius* provided an estimated 80–85% fly control. Similarly, 93–98% control of *M. domestica* was reported for April–November in a dairy barn following releases of *M. raptor* (Weintraub, 1985).

Bennett and Surgeoner (1996) reported that weekly releases of 5000 *M. raptor* in a weaning room (98 m^2) and two grower–finisher rooms (245 m^2) did not provide satisfactory control of *M. domestica* at a swine facility. Parasitism of *M. domestica* pupae averaged 27% and 39%, which was considered too low to justify the cost of the parasitoids (average Can$44.15 per 15,000). The difference between this result and those of Weintraub (1985) may reflect the use of sanitation and bait stations in the latter study.

McKay and Galloway (1999) reported that the release of 3.6 million parasitoids did not appreciably increase parasitism of *M. domestica* pupae in dairies. Failure was attributed, in part, to the poor quality of parasitoid shipments obtained from a commercial insectary. These produced an average of 43% fewer parasitoids than expected and contained only *N. vitripennis*, rather than two species as advertised. Although its gregarious habit lowers production costs, *N. vitripennis* is generally recognized as being ineffective in controlling *M. domestica* and stable fly, *Stomoxys calcitrans* (L.) (references in McKay and Galloway, 1999).

Poor quality control has been recognized previously as one reason for failures to control flies with parasitoid releases (Legner, 1981; Stage and Petersen, 1981), e.g. *M. domestica* pupae purchased in 1998 from a commercial insectary produced one species of parasitoid from 10% of pupae, rather than the advertised claim of four species with an estimated 90% of the pupae (H.G. Philip and K.D. Floate, unpublished). Shipments received from other insectaries have met advertised claims (K.D. Floate, unpublished). Insectaries in the USA are not legally required to ensure the quality of the parasitoids they ship.

In southern Alberta, the release of *M. raptorellus* can significantly enhance parasitism of *M. domestica* pupae in dairies and feedlots. Parasitism was 70% or less for sentinel pupae located up to 22.5 m away from the release points of about 690,000 parasitoids in an ungrazed paddock (Lysyk, 1996) and was up to 100% in samples of sentinel pupae (n = 30–50 pupae per sample) located up to 100 m distant (Floate *et al.*, 2000). In the latter study, release of 1.3 million parasitoids at a dairy and at the Lethbridge Research Centre averaged 34% parasitism for sentinel pupae located 1–100 m from the release site. In contrast, natural parasitism has averaged less than 3% (Lysyk, 1995; Floate *et al.*, 1999, 2000).

Releases of *Urolepis rufipes* (Ashmead) and *M. raptor* are currently being made to manage *M. domestica* populations at a swine facility near Edmonton (B.A. Khan, Edmonton, 2000, personal communication). Releases of *T. sarcophagae* are being made in feedlots near Lethbridge to assess their effect on parasitism rates of sentinel and naturally occurring *M. domestica* pupae (K.D. Floate, unpublished).

Bennett and Surgeoner (1996) did not recommend *H. aenescens* as a biological control agent of *M. domestica* because it displaced the *M. domestica* population in a swine facility and, therefore, could potentially become a pest itself. Other researchers suggest that this displacement is favourable because *H. aenescens* tend to stay at the breeding sites, e.g. in manure pits, whereas *M. domestica* are prone to disperse throughout livestock facilities and to nearby structures (Nolan and Kissam, 1987; Turner and Carter, 1990).

Evaluation of Biological Control

Early results of research on microbial agents for *M. domestica* control are promising, and should be continued, to develop commercializable products for the livestock industry. Efforts to educate livestock producers on the general biology of *M. domestica*, and their biological control agents, will be a critical step to enhance the successful use of biological control agents against *M. domestica*.

Regular innundative releases of parasitoid wasps can be used to manage populations of pest flies associated with confined livestock. However, use of parasitoids should be integrated with a programme of sanitation and judicious applications of insecticides.

Recommendations

Further work should include:

1. Assessing the potential of native species as inundative biological control agents;
2. Developing technologies that reduce the cost of parasitoids to livestock producers;
3. Identifying parasitoid species and rates of release (alone or in combination) best suited to different situations, e.g. swine barns versus feedlots.

References

Belton, P., Rutherford, T.A., Trottier, D.B. and Webster, J.M. (1987) *Heterorhabditis heliothidis*: a potential biological control agent of house flies in caged-layer poultry barns. *Journal of Nematology* 19, 263–266.

Bennett, W. and Surgeoner, G. (1996) Fly control in swine barns. *Agri-Food Research in Ontario* 19, 9–10.

Cranshaw, W., Sclar, D.C. and Cooper, D. (1996) A review of 1994 pricing and marketing by suppliers of organisms for biological control of arthropods in the United States. *Biological Control* 6, 291–296.

Davies, A. (1991) Histeridae. In: Bousquet, Y. (ed.) *Checklist of the Beetles of Canada and Alaska.* Publication 1861/E, Agriculture Canada, Ottawa, Ontario, pp. 135–141.

Floate, K.D. (1998) Off-target effects of ivermectin on insects and on dung degradation in southern Alberta, Canada. *Bulletin of Entomological Research* 88, 25–35.

Floate, K.D., Khan, B. and Gibson, G.A.P. (1999) Hymenopterous parasitoids of filth fly (Diptera: Muscidae) pupae in cattle feedlots of Alberta, Canada. *The Canadian Entomologist* 131, 347–362.

Floate, K.D., Coghlin, P. and Gibson, G.A.P. (2000) Dispersal distance and overwintering of the filth fly parasitoid *Muscidifurax raptorellus* (Hymenoptera: Pteromalidae) following mass-releases in cattle feedlots. *Biological Control* 18, 172–178.

Geden, C.J. and Axtell, R.C. (1988) Predation by *Carcinops pumilio* (Coleoptera: Histeridae) and *Macrocheles muscaedomesticae* (Acarina: Macrochelidae) on the house fly (Diptera: Muscidae): functional response, effects of temperature, and availability of alternative prey. *Environmental Entomology* 17, 739–744.

Geden, C.J., Axtell, R.C. and Brooks, W.M. (1986) Susceptibility of the house fly, *Musca domestica* (Diptera: Muscidae), to the entomogenous nematodes *Steinernema feltiae, S. glaseri* (Steinernematidae), and *Heterorhabditis heliothidis* (Heterorhabditidae). *Journal of Medical Entomology* 23, 326–332.

Gibson, G.A.P. (2000) Differentiation of the species of *Urolepis* (Hymenoptera: Chalcidoidea: Pteromalidae), potential biocontrol agents of filth flies (Diptera: Muscidae). *The Canadian Entomologist* 132, 391–410.

Gibson, G.A.P. and Floate, K. (2001) The species of *Trichomalopsis* (Hymenoptera: Pteromalidae) associated with filth flies (Diptera: Muscidae) in North America. *The Canadian Entomologist* 133, 49–85.

Glofcheskie, B.D. and Surgeoner, G.A. (1990) Muscovy ducks as an adjunct for the control of the house fly (Diptera: Muscidae). *Journal of Economic Entomology* 83, 788–791.

Legner, E.F. (1981) Improving commercial biological control of filth flies with parasites. In: *Status of Biological Control of Filth Flies.* Proceedings of a Workshop, 4–5 February 1981, Gainesville, Florida. Publication A106.2:F64, United States Department of Agriculture, Science and Education Administration, pp. 5–10.

Legner, E.F. (1994) Biological control of Diptera of medical and veterinary importance. *Journal of Vector Ecology* 20, 59–120.

Lysyk, T.J. (1995) Parasitoids (Hymenoptera: Pteromalidae, Ichneumonidae) of filth fly (Diptera: Muscidae) pupae at dairies in Alberta. *Journal of Economic Entomology* 88, 659–665.

Lysyk, T.J. (1996) *Development of Biological Control Methods for Stable Flies in Feedlots.* Farming for the Future Report 920100. Alberta Agricultural Research Institute.

McKay, T. (1997) Parasitoid wasps (Hymenoptera: Pteromalidae, Ichneumonidae) for control of house flies and stable flies (Diptera: Muscidae) in dairy operations in Manitoba. MSc thesis, The University of Manitoba, Winnipeg, Manitoba.

McKay, T. and Galloway, T.D. (1999) Survey and release of parasitoid wasps (Hymenoptera: Pteromalidae, Ichneumonidae) attacking house flies and stable flies (Diptera: Muscidae) in dairy operations in Manitoba. *The Canadian Entomologist* 131, 743–756.

Mullens, B.A., Meyer, J.A. and Georgis, R. (1987) Field tests of insect-parasitic nematodes (Rhabditida: Steinernematidae, Heterorhabditidae) against larvae of manure-breeding flies (Diptera: Muscidae) on caged-layer poultry facilities. *Journal of Economic Entomology* 80, 438–442.

Nolan, M.P. III and Kissam, J.B. (1987) Nuisance potential of a dump fly, *Ophyra aenescens* (Diptera: Muscidae), breeding at poultry farms. *Environmental Entomology* 16, 828–831.

O'Hara, J.E., Floate, K.D. and Cooper, B.E. (2000) The Sarcophagidae (Diptera) of cattle feedlots in southern Alberta. *Journal of the Kansas Entomological Society* 72, 167–176.

Renn, N. (1995) Mortality of immature houseflies (*Musca domestica* L.) in artificial diet and chicken manure after exposure to encapsulated entomopathogenic nematodes (Rhabditida: Steinernematidae, Heterorhabditidae). *Biocontrol Science and Technology* 5, 349–359.

Stage, D.A. and Petersen, J.J. (1981) Mass release of pupal parasitoids for control of stable flies and house flies in confined feedlots in Nebraska. In: *Status of Biological Control of Filth Flies.* Proceedings of a Workshop, 4–5 February 1981, Gainesville, Florida. Publication A106.2:F64, United States Department of Agriculture, Science and Education Administration, pp. 52–58.

Turner, E.C. Jr and Carter, L. (1990) Mass rearing and introduction of *Ophyra aenescens* (Weidemann) (Diptera: Muscidae) in high-rise caged layer poultry houses to reduce house fly populations. *Journal of Agricultural Entomology* 7, 247–257.

Watson D.W., Geden, C.J., Long, S.J. and Rutz, D.A. (1995) Efficacy of *Beauveria bassiana* for controlling the house fly and stable fly (Diptera: Muscidae). *Biological Control* 5, 405–411.

Weintraub, J. (1985) Biological control of house flies in confined livestock rearing. In: Sears, L.J.L. and Swierstra, E.E. (eds) *Research Highlights – 1984.* Agriculture Canada Research Station, Lethbridge, Alberta, pp. 26–28.

West, L.S. (1951) *The Housefly. Its Natural History, Medical Importance, and Control.* Comstock Publishing Company, Ithaca, New York.

38 *Neodiprion abietis* (Harris), Balsam Fir Sawfly (Hymenoptera: Diprionidae)

G.S. Thurston

Pest Status

The balsam fir sawfly, *Neodiprion abietis* (Harris), is a complex of closely related, little-studied species of defoliating sawflies native to North America (the taxonomy of this complex, currently under one name, *N. abietis*, still requires clarification). It occurs throughout Canada and the northern USA (Rose *et al.*, 1994; USDA Forest Service, 1985). According to Knerer and Atwood (1972), five 'races' occur throughout its range: a distinct western race primarily on white fir, *Abies concolor* (Gordon and Glendinning) Lindley, ex. Hildebrand, and four eastern ones, two on balsam fir, *A. balsamea* (L.) Miller, and one each on white spruce, *Picea glauca* (Moench) Voss, and black spruce, *P. mariana* (Miller) Briton, Sterns and Poggenburgg. In Atlantic Canada, it is found almost exclusively on balsam fir, with damage being noticed on black and white spruce in areas where larval numbers are high and feeding has reduced the amount of available balsam fir foliage. Larval feeding results in reduced tree growth and may cause mortality if feeding continues for more than 1 year. One year of severe feeding damage results in several years' growth reduction (D. Ostaff, Fredericton, 2000, personal communication). Open-grown trees (USDA Forest Service, 1985) and middle-aged forest stands of medium-to-thin density (Martineau, 1984) appear to be at greater risk. Outbreaks are usually of short duration, collapsing due to the actions of natural enemies, including a *Nucleopolyhedrovirus* (NPV) and parasitoids (Wallace and Cunningham, 1995). In Nova Scotia and Newfoundland, *N. abietis* is a serious concern; an outbreak in Newfoundland has persisted since 1991 and an outbreak in Cape Breton is expanding rapidly. In recent years more than 50,000 ha have been moderately to severely defoliated in Newfoundland and Nova Scotia. The potential loss to the forest industry, in terms of tree mortality and reduced growth, is significant.

Overwintering eggs hatch in spring and young larvae feed gregariously on 1-year-old and older foliage, rarely damaging the current year's growth. Later instars disperse from the colonies and become solitary feeders. Females lay eggs in current-year needles in autumn.

Background

Aerial protection programmes have been considered necessary to reduce damage by *N. abietis* even though no products are fully registered for aerial application. Several parasitoids have been reared from *N. abietis* (Martineau, 1984), with the most extensive list compiled by Raizenne (1957), who found 12 parasitoid species and two hyperparasitoid species. Although the individual impact of the parasitoid species is unknown, declines in *N. abietis* population numbers have been attributed to them (Carroll, 1962; Martineau, 1984). Steinhaus (1949) first recorded a virus from *N. abietis*, and isolates were collected in Canada in the 1950s (Wallace and Cunningham, 1995). Olofsson (1973) suggested that an NPV isolated from *N. abietis* could be used to prevent excessive feeding damage. The recent outbreak in

Atlantic Canada yielded NPV isolates from both Nova Scotia and Newfoundland.

Biological Control Agents

Parasitoids

The importance of *N. abietis* parasitoids is poorly understood. Hartling and O'Shea (New Brunswick Department of Natural Resources and Energy, Fredericton, 1999, personal communication) found that 26% of *N. abietis* eggs were parasitized by *Cirrospilus* sp., and 28% of cocoons were parasitized by one of several parasitoids, primarily *Mesopolobus verditer* (Norton). In Newfoundland, Nova Scotia and New Brunswick, 5–20% egg parasitism by *Cirrospilus* sp. occurred (D. Ostaff, Fredericton, 1999, personal communication). Larval parasitism was also significant in these studies, with indications that parasitoids may be important in maintaining low *N. abietis* populations.

Pathogens

Bacteria

No *Bacillus thuringiensis* Berliner (*B.t.*) strains have been been shown to be effective against Hymenoptera. However, *B. thuringiensis* serovar *israelensis* (*B.t.i.*) was tested against *N. abietis*. Laboratory spray-chamber trials with VectoBac 12AS (Valent Biosciences, Libertyville, Illinois, USA) produced a modest increase in larval mortality, sufficient to warrant a small-scale field trial. However, in a similar, leaf-dip trial, no increase in mortality was observed (S. Li, Corner Brook, 1999, personal communication). Vectobac 1200L (Valent Biosciences), applied at the rate of 72×10^9 international units (IU) ha^{-1} using a truck-mounted mistblower to 1 ha of thinned balsam fir infested with *N. abietis*, reduced larval survival by 20% and provided foliage protection of 41%. However, field mortality developed slowly (7+ days), suggesting that the *B.t.i.* crystal proteins might not be important in pathogenesis.

Viruses

Epizootics of NPV cause rapid decreases in *N. abietis* larval numbers under natural conditions. Because it is an open, communal feeder, *N. abietis* is particularly susceptible to disease spread. Olofsson (1973) showed that early instar larvae are more susceptible to NPV than later instars, and determined that small-scale ground sprays on individual trees using a backpack mistblower significantly reduced their numbers, especially when targeted against first-instar larvae. In western Newfoundland and the eastern shore area of Nova Scotia, NPV from *N. abietis* larvae collected in August, 1997 was isolated (C. Lucarotti, Fredericton, 1999, personal communication). By infecting laboratory-reared larvae with the isolated virus, sufficient material was obtained for small-scale field applications with both isolates. In summer, 1999, field trials were conducted against *N. abietis* in both provinces. In Nova Scotia, the virus was applied at the rate of 1×10^{10} polyhedral inclusion bodies (PIB) ha^{-1} to 3 ha of thinned balsam fir in the Cape Breton Highlands using a truck-mounted mistblower. Initial application was made later than planned (i.e. when most of the larvae were in first and second instars) with most of the larvae being in third and fourth instars (larval index = 3.67) on the pre-spray sample date. Even though many of the larvae were already in the fourth instar prior to virus application, survival was significantly reduced in the treated block. However, because of the amount of larval feeding that had occurred prior to virus application, no foliage protection was observed. Cadavers were recovered from the treated area and virus infection was confirmed. Larval collections from the block yielded considerable virus material for further work. In Newfoundland, the virus was applied aerially to 1.2 ha of thinned balsam fir, at a rate of 2.8×10^{11} PIB ha^{-1}. Larval mortality was observed but not quantified, as the main goals for this project were initiation of an epizootic and collection of large amounts of virus for future research.

Evaluation of Biological Control

Although *B.t.i.* may not be a good biological control candidate due to its uncertain efficacy, other *B.t.* strains active against sawflies may exist.

The use of NPVs for management of sawfly outbreaks is hampered by the fact that they are species-specific and cannot be produced *in vitro*. However, field trials indicated that they can be mass produced using field amplification techniques and that they are effective in reducing larval numbers when applied from the ground or the air.

Recommendations

Future work should include:

1. Monitoring sites in Nova Scotia and Newfoundland where NPV was applied to assess establishment;
2. Determining the minimum application rate of NPV that will give good *N. abietis* larval reduction and pathogen establishment;
3. Continued study of the *N. abietis* parasitoid complex;
4. Searching for *B.t.* strains effective against *N. abietis*;
5. Taxonomic elucidation of the *N. abietis* species complex.

References

Carroll, W.J. (1962) Some aspects of the *Neodiprion abietis* (Harr.) complex in Newfoundland. PhD thesis, State University of New York, Syracuse, New York.

Knerer, G. and Atwood, C.E. (1972) Evolutionary trends in the subsocial sawflies belonging to the *Neodiprion abietis* complex (Hymenoptera: Tenthredinoidea). *American Zoologist* 12, 407–418.

Martineau, R. (1984) *Insects Harmful to Forest Trees.* Forestry Technical Report #32, Environment Canada.

Olofsson, E. (1973) *Evaluation of a Nuclear Polyhedrosis Virus as an Agent for the Control of the Balsam Fir Sawfly,* Neodiprion abietis *Harr.* Canadian Forestry Service Information Report IP-X-2.

Raizenne, H. (1957) *Forest Sawflies of Southern Ontario and their Parasites.* Publication 1009, Canadian Department of Agriculture, Ottawa, Ontario.

Rose, A.H., Lindquist, O.H. and Syme, P. (1994) *Insects of Eastern Spruces, Fir and Hemlock.* Canadian Forest Service, Ottawa.

Steinhaus, E.A. (1949) *Principles of Insect Pathology.* McGraw-Hill, New York, New York.

USDA Forest Service (1985) *Insects of Eastern Forests.* United States Department of Agriculture, Forest Service, Miscellaneous Publication 1462.

Wallace, D.R. and Cunningham, J.C. (1995) Diprionid sawflies. In: Armstrong, J.A. and Ives, W.G.H. (eds) *Forest Insect Pests in Canada.* Publication FO24-235/1995E, Natural Resources Canada, Canadian Forest Service, Ottawa, Ontario, pp. 193–232.

39 *Neodiprion sertifer* (Geoffroy), European Pine Sawfly, and *N. lecontei* (Fitch), Redheaded Pine Sawfly (Hymenoptera: Diprionidae)

K. van Frankenhuyzen

Pest status

The European pine sawfly, *Neodiprion sertifer* (Geoffroy) was first recorded in Canada near Windsor, Ontario, in 1939. By 1978 it had spread as far east as Ottawa and as far north as Sault Ste Marie. It was recorded in Quebec and Newfoundland in 1974 and in Nova Scotia in 1980 (Griffiths *et al.*, 1984). Typically, outbreaks occur in young plantations of Scots pine, *Pinus sylvestris* L., red pine, *Pinus resinosa* Aiton, and jack pine, *Pinus banksiana* Lambert, a few years after establishment. Increasing population densities reach a peak in 4–7 years, followed by a gradual decline as natural control agents suppress the populations to a low density that generally does not resurge. *N. sertifer* is currently a minor pest in Canada, but local outbreaks are reported occasionally. It has one generation per year. Eggs laid in autumn, in slits cut in the needles of the host trees, overwinter and hatch the following April or May. The larvae feed gregariously on shoots and mature in 6–8 weeks, then drop to the ground where they spin cocoons in which they pupate. Adults emerge in September and October.

The redheaded pine sawfly, *Neodiprion lecontei* (Fitch), is native to North America. It is a serious defoliator of hard pine plantations in Ontario, Quebec, and, to a lesser extent, New Brunswick. Severe defoliation can cause extensive mortality of young trees, particularly when 2–3 years old. Infestations occur erratically and may persist for 3 years or longer (Prebble, 1975). Adult sawflies emerge in June and July from overwintered cocoons in the soil or litter. Females deposit about 120 eggs in slits in adjacent needles. Larvae feed from July until early autumn in colonies on foliage of all ages.

Background

Chemical insecticides are commonly used by plantation owners and Christmas tree growers to prevent serious sawfly damage, usually in ground-based applications. Aerial application of insecticides, primarily targeted against *N. sertifer*, involved the use of DDT, dieldrin and lindane until the mid-1960s, followed by phosphamidon and malathion in the late 1960s and 1970s (Prebble, 1975).

Natural control agents of *N. sertifer* include parasitoids introduced from Europe and a diverse complex of native parasitoids that moved on to it. Introductions of new parasitoids and relocation of established species continued until 1980 (Griffiths *et al.*, 1984). By that time, sawfly populations had generally declined to low levels, and field releases were stopped. Another natural control agent is a *Nucleopolyhedrovirus* (NPV) introduced from Europe in 1949 (Wallace *et al.*, 1975). The virus was applied to many infestations throughout southern Ontario during the 1950s and 1960s, which, together with the release of parasitoids, is credited for reducing *N. sertifer* to a minor pest (Cunningham and Kaupp, 1995).

In 1950, an NPV affecting *N. lecontei* (NeleNPV) was found in Ontario. During the 1960s and 1970s, various ground spray trials were conducted in Ontario and Quebec (Prebble, 1975). Aerial spray trials were conducted in Ontario from 1976 to 1980 on 20 plantations with a total area of 258 ha, and in Quebec a total area of 1015 ha was treated from the air or ground from 1978 to 1980 (Cunningham and de Groot, 1984). The viral product, Lecontvirus, produced by the Canadian Forest Service (CFS), received temporary registration in 1983 and full registration in 1987.

Biological Control Agents

Pathogens

Viruses

A petition to register European pine sawfly NPV (NeseNPV), called Sertifervirus, submitted by the CFS in 1985, has since been abandoned due to declining product demand. Only 152 ha were treated from 1975 to 1993 (Cunningham, 1998). Since 1983, small amounts of the virus have been used annually to treat ornamental trees in St John's, Newfoundland.

The redheaded pine sawfly NPV is the only viral product that is used routinely in Canada, albeit in small quantities (Cunningham, 1998). The virus is easily and inexpensively produced by treating a heavily infested plantation when larvae are in the fourth instar and removing dead and moribund larvae from the foliage. Cadavers are then lyophilized, ground to a fine powder and stored at 2°C (Cunningham and de Groot, 1984). From 1980 to 1990, the Ontario Ministry of Natural Resources treated 478 plantations totalling 3546 ha, using ground-spray equipment. An additional 1150 ha were treated in Ontario and Quebec from 1991 to 1995. No data are available on use after 1995.

Evaluation of Biological Control

The most effective and economical viral insecticides developed to date in Canada are the NPVs of *N. sertifer* and *N. lecontei*. These viruses are highly infectious because they replicate in midgut cells and are readily transferred to healthy larvae through defensive regurgitation and contaminated frass. Rapid spread of the virus through a colony (horizontal transmission) is aided by gregarious feeding habits of the larvae, while transmission from diseased to healthy colonies is thought to be aided by predators and parasitoids (Cunningham and Entwistle, 1981). About 50 diseased larvae produce enough inoculum to treat 1 ha (Cunningham, 1998). The recommended application rate is 5×10^9 polyhedral inclusion bodies (PIB) ha^{-1} in 5–10 l (aerial sprays) or 20 l (ground spray) of water, but the lowest effective application rate has not been determined experimentally. Multi-year surveys of treated plantations have shown that one application of virus can suppress population build-up for several years (Cunningham and de Groot, 1984), as a result of vertical transmission of the virus to subsequent sawfly generations (Cunningham and Entwistle, 1981). Although production technology has been developed and registration granted, Lecontvirus is not available as a commercial product. Production facilities and a small stockpile of the virus are available upon request from the Canadian Forest Service.

Recommendations

Further work should include:

1. Finding a commercial partner to produce and market Lecontvirus as a specialty product.

References

Cunningham, J.C. (1998) North America. In: Hunter-Fujita, F.R., Entwistle, P.F., Evans, H.F. and Cook, N.E. (eds) *Insect Viruses and Pest Management.* John Wiley and Sons, Chichester, UK, pp. 313–331.

Cunningham, J.C. and de Groot. P. (1984) *Neodiprion lecontei* (Fitch), Redheaded pine sawfly (Hymenoptera: Diprionidae). In: Kelleher, J.S. and Hulme, M.A. (eds) *Biological Control Programmes Against Insects and Weeds in Canada 1969–1980.* Commonwealth Agriculture Bureaux, Slough, UK, pp. 323–329.

Cunningham, J.C. and Entwistle, P.F. (1981) Control of sawflies by baculovirus. In: Burges, H.D. (ed.) *Microbial Control of Pests and Plant Diseases, 1970–1980.* Academic Press, London, UK, pp. 379–407.

Cunningham, J.C. and Kaupp, W.J. (1995) Insect viruses. In: Armstrong, J.A. and Ives, W.G.H. (eds) *Forest Insect Pests in Canada.* Publication FO24-235/1995E, Natural Resources Canada, Canadian Forest Service, Ottawa, Ontario, pp. 328–340.

Griffiths, K.J., Cunningham, J.C. and Otvos, I.S. (1984) *Neodiprion sertifer* (Geoffroy), European pine sawfly (Hymenoptera: Diprionidae). In: Kelleher, J.S. and Hulme, M.A. (eds) *Biological Control Programmes Against Insects and Weeds in Canada, 1969–1980.* Commonwealth Agricultural Bureaux, Slough, UK, pp. 331–340.

Prebble, M.L. (1975) Red-headed pine sawfly. In: Prebble, M.L. (ed.) *Aerial Control of Forest Insects in Canada.* Department of Environment, Ottawa, Ontario, pp. 220–223.

Wallace, D.R., Cameron, J.M. and Sullivan, C.R. (1975) European pine sawfly. In: Prebble, M.L. (ed.) *Aerial Control of Forest Insects in Canada.* Department of Environment, Ottawa, Ontario, pp. 224–230.

40 *Orgyia leucostigma* (J.E. Smith), Whitemarked Tussock Moth (Lepidoptera: Lymantriidae)

G.S. Thurston

Pest Status

The whitemarked tussock moth, *Orgyia leucostigma* (J.E. Smith), is a native defoliator of hardwood and softwood trees in eastern North America, with its range extending west into Alberta (Martineau, 1984). *O. leucostigma* is primarily a forest pest, although it does cause concern on ornamental trees and field crops such as blueberries, *Vaccinium* spp. It is capable of defoliating large areas of softwood forests, resulting in tree deformation and top-kill after 1 year and extensive tree mortality after 2 years of defoliation. *O. leucostigma* is particularly devastating to Christmas tree plantations of balsam fir, *Abies balsamea* (L.) Miller, where defoliation can lead to total crop loss in 1 year and the presence of egg masses results in unmarketable trees. Outbreaks of *O. leucostigma* have occurred in Atlantic Canada about every 9 years, with major outbreaks occurring every 20 years. The outbreaks tend to be of short duration, terminated by pathogens within about 3 years. Smaller outbreaks are localized in their effect but large ones can cause damage to large forest areas. In Nova Scotia, the 1998 outbreak caused defoliation to over 500,000 ha (E. Georgeson, Shubenacadie, 1999, personal communication).

O. leucostigma eggs overwinter in a frothy mass normally laid on foliage. Upon hatching in spring, larvae feed on the remains of the egg mass and then spin a silk thread with which they may balloon and be dispersed long distances. This is the primary method of dispersal, as adult females are wingless. Larvae feed for several weeks on the foliage and thin bark of host plants. During this time, and especially during severe outbreaks, the hairs from their bodies can cause severe allergic reactions, including rashes and anaphylaxis in sensitive people. After feeding is completed, the larvae pupate, usually on the underside of host tree branches. Adults emerge about 2 weeks later and females lay eggs on top of their empty cocoon. One full generation per year occurs in eastern Canada; some years may produce a partial second generation but the species cannot overwinter except in the egg stage.

Background

When foliage protection is needed, aerial application of *Bacillus thuringiensis* Berliner is the most commonly used treatment. Naturally occurring parasitoids have been identified from *O. leucostigma* populations (e.g. Martineau, 1984), but appear to be unimportant in eastern Canada. In 1998 only two of several hundred egg masses were found to have been parasitized and, in 1999, no parasitoids were reared from several hundred larvae of all instars collected from Nova Scotia.

Biological Control Agents

Pathogens

In Nova Scotia in 1999, a high-density population of *O. leucostigma* in the early stages of collapse was monitored for pathogens. Two were found to be the most important mortality factors. Larval mortality caused by the fungus *Entomophaga aulicae* (Reichardt in Bail) Humber and a *Nucleopolyhedrovirus* (OrleNPV) peaked at over 70% and 50%, respectively, on understory vegetation in the sample blocks. Infection rates on the balsam fir sample trees were considerably lower. Mortality caused by parasitoids was unimportant in this study (K. van Frankenhuyzen, Sault Ste Marie, 2000, personal communication).

Bacteria

Bacillus thuringiensis Berliner serovar *kurstaki* (*B.t.k.*) is an effective control agent when applied aerially at 30×10^9 international units (IU) ha^{-2}, giving high larval mortality and good foliage protection. In Nova Scotia in 1998, Foray® 76B (Valent Biosciences, Libertyville, Illinois, USA), Thuricide 48LV (Thermo Trilogy, Columbia, Maryland, USA) and Bioprotec (AEF Global, Sherbrooke, Quebec, Canada) were used in an experimental aerial spray programme. Larval mortality was as high as 82% and after-spray defoliation was near zero for all products, indicating good efficacy.

Viruses

The naturally occurring OrleNPV is largely responsible for the collapse of *O. leucostigma* outbreaks in eastern Canada. Another NPV, OrpsNPV, isolated from the Douglas-fir tussock moth, *O. pseudotsugata* (McDunnough), in western Canada, also infects *O. leucostigma*. Virtuss® and TM BioControl-1® are virus products registered for use in Canada against *O. pseudotsugata*. Both have been used successfully in trials against *O. leucostigma*. In Newfoundland, West *et al.* (1987) determined that Virtuss®, when applied by backpack mistblower, reduced *O. leucostigma* larval numbers and contributed to a population collapse. The same product applied aerially (West *et al.*, 1989) at 2.5×10^{11} polyhedral inclusion bodies (PIB) ha^{-1} to 25 ha of white birch infested with *O. leucostigma* significantly reduced larval numbers, and virus infection was confirmed in collected larvae. Egg mass numbers were low in the spray block in the autumn following the treatment, increasing in number with increased distance from the treatment area. In Nova Scotia, TM BioControl-1® was used in an experimental trial in 1998, causing over 85% larval mortality and apparently initiating an epizootic that resulted in no *O. leucostigma* being found in the spray blocks the following year.

Evaluation of Biological Control

Several *B.t.k.* products are now registered for use against *O. leucostigma* in Canada, and all provide good foliage protection when applied against early larval instars at the recommended rate of two applications of 30×10^9 IU ha^{-1} separated by 5 days. Use of *B.t.k.* is effective for short-term management of *O. leucostigma*, but does not contribute to long-term population suppression.

The use of NPV (both OrleNPV and OrpsNPV) for managing *O. leucostigma* outbreaks is effective if the virus can be obtained. Virus has been used successfully in an integrated management plan for *O. pseudotsugata* (Otvos *et al.*, 1998; see Otvos *et al.*, Chapter 41 this volume). A similar management plan could be implemented for *O. leucostigma*. At present, no virus is registered for use against *O. leucostigma*, but registration of TM BioControl-1® may soon be forthcoming.

Recommendations

Future work should include:

1. Determining the minimum application rate of NPV that will give good larval reduction and pathogen establishment;
2. Investigating epizootic development and spread after commercial virus application;
3. Determining conditions under which epizootics naturally cause population collapse;
4. Determining conditions that lead to *O. leucostigma* population outbreaks.

References

Martineau, R. (1984) *Insects Harmful to Forest Trees.* Forestry Technical Report #32, Environment Canada.

Otvos, I.S., Maclauchlan, L.E., Hall, P.M. and Conder, N. (1998) *A Management System to Control Douglas-Fir Tussock Moth,* Orgyia pseudotsugata, *using OpNPV.* Pacific Forestry Centre Technical Transfer Note #11, Natural Resources Canada.

West, R.J., Cunningham, J.C. and Kaupp, W.J. (1987) *Ground Spray Applications of Virtuss, A Nuclear Polyhedrosis Virus, Against White-marked Tussock Moth Larvae at Bottom Brook, Newfoundland in 1986.* Canadian Forestry Service Information Report N-X-257.

West, R.J., Kaupp, W.J. and Cunningham, J.C. (1989) *Aerial Application of Virtuss, A Nuclear Polyhedrosis Virus, Against Whitemarked Tussock Moth Larvae in Newfoundland in 1987.* Forestry Canada Information Report N-X-270.

41 *Orgyia pseudotsugata* (McDunnough), Douglas-fir Tussock Moth (Lepidoptera: Lymantriidae)

I.S. Otvos, R.F. Shepherd and J.C. Cunningham

Pest Status

The Douglas-fir tussock moth, *Orgyia pseudotsugata* (McDunnough), is a native defoliator of the interior dry-belt (semi-arid) forests in southern British Columbia and in the western USA (Beckwith, 1978). In Canada, the preferred host is the interior Douglas fir, *Pseudotsuga menziesii* var. *glauca* (Beissner) Franco, but later instars will also feed on Engelmann spruce, *Picea engelmannii* Parry ex Engelmann, and on ponderosa pine, *Pinus ponderosa* P. Lawson ex C. Lawson, when the preferred host is completely defoliated in mixed stands. In the western USA it also attacks grand fir, *Abies grandis* (Douglas ex D. Don) Lindley, and white fir, *Abies concolor* (Gordon and Glendinning) Lindley (Wellner, 1978). *O. pseudotsugata* also feeds on several ornamental conifer species, including Colorado blue spruce, *Picea pungens* Engelmann, in urban areas where population increases are often first noted. Flightless females lay a single egg mass on the cocoon from which she emerges. Eggs overwinter and hatch the following spring after bud flush of the host trees. Newly emerged larvae feed on the underside of new needles. Young larvae, the major dispersal stage, spin silk threads and can be blown a considerable distance by ballooning. Later-instar larvae will also feed on older needles if the new growth has been consumed. They are wasteful feeders, causing most of the damage by taking bite-sized chunks from the needles or clipping partially consumed needles off at the base. These dry out and change colour, giving severely defoliated stands a reddish-brown appearance.

O. pseudotsugata infestations can cause growth loss, top-kill, and considerable tree mortality at high population levels (Wickman, 1978; Alfaro *et al.*, 1987). Individual trees may die after 1 year of severe defoliation, but stand mortality is more common after 2 or more years of such defoliation (Johnson and Ross, 1967; Alfaro *et al.*, 1987). The defoliated trees are weakened and become more susceptible to attack by other insects, e.g. the Douglas-fir beetle, *Dendroctonus pseudotsugae* Hopkins.

O. pseudotsugata is a cyclic pest and outbreaks recur about every 7–11 years in western North America (Shepherd *et al.*, 1984b; Harris *et al.*, 1985; Otvos and Shepherd, 1991). In British Columbia, eight outbreaks have occurred since 1916, with the last from 1990 to 1993, controlled mainly by the operational use of the Douglas-fir tussock moth virus (Maclauchlan *et al.*, 2001). During the 1981–1984 outbreak in south-central British Columbia, about 26,000 ha were defoliated and up to 30% tree mortality occurred in severely affected stands (Ross and Taylor, 1985). Outbreaks begin at small epicentres, then expand and coalesce, generally collapsing after 2–4 years of defoliation (Mason and Luck, 1978) due to the combined action of several natural control agents, dominated by *Nucleopolyhedrovirus* (OrpsNPV). However, by the time the outbreak has collapsed on its own, extensive damage has occurred in the infested stand (Dahlsten and Thomas, 1969; Shepherd and Otvos, 1986).

Background

No attempts have been made to control *O. pseudotsugata* with introduced parasitoids or predators. Due to the effectiveness of the naturally occurring virus, earlier control emphasis was through manipulation of viral epizootics and aerial application of commercially available *Bacillus thuringiensis* Berliner (these trials were summarized by Cunningham and Shepherd, 1984). Following these extremely successful trials, the multicapsid isolate of the Douglas-fir tussock moth virus was registered in 1976 in the USA as TM BioControl-1®. This same virus, produced in the whitemarked tussock moth, *Orgyia leucostigma* (J.E. Smith) (which has a shorter feeding period and is easier to rear in the laboratory), received temporary registration in 1983 and full registration in 1987 in Canada under the name Virtuss®. TM BioControl-1® also received Canadian registration in 1987 (Otvos *et al.*, 1995). No further trials were done in British Columbia until the 1981–1984 outbreak. Concurrent with the virus research, a separate study focused on developing a sensitive and dependable sex pheromone monitoring system to provide early warning of impending outbreaks (Daterman *et al.*, 1976; Shepherd *et al.*, 1985a). The pheromone trapping study showed that when there are three consecutive years in which the number of male moths caught in baited traps increases to at least 25 males per trap, an outbreak will occur in 1 or 2 years (Shepherd *et al.*, 1985a). Pheromone trapping successfully detected the build-up of *O. pseudotsugata* populations in isolated stands in the Hedley area, southern British Columbia, in 1980 before any defoliation had occurred (Shepherd *et al.*, 1985a) and was confirmed by egg-mass surveys conducted in the autumn of 1980.

Biological Control Agents

Pathogens

Nucleopolyhedrovirus (NPV) replicates within the cell nuclei and the virus produces inclusion bodies. Two morphotypes of OrpsNPV have been isolated from *O. pseudotsugata* larvae, a unicapsid variety (OrpsSNPV) and a multicapsid variety (OrpsMNPV) (Hughes and Addison, 1970). Viruses of this group are generally slow acting and, like *Bacillus thuringiensis*, must be ingested to cause infection (Cunningham, 1982). OrpsMNPV used in research during the last two outbreaks (1981–1984 and 1990–1993) was propagated in *O. leucostigma* at the Great Lakes Forestry Centre.

The outbreak of 1981–1984 provided an opportunity to test: (i) whether viral epizootics could be initiated artificially by application of the virus during the early phase of the outbreak, before a natural epizootic might occur; (ii) whether it would result in reduced damage to the stands; and (iii) the effect of larval density on the development of viral epizootics. As outbreaks often begin close to ranches and residences in or near forested areas, ground applications were also tested to determine if epizootics could be initiated by this method of treatment.

Four plots, totalling about 20 ha, with light to moderate *O. pseudotsugata* population densities were aerially treated with OrpsMNPV at a rate of 2.2×10^{11} polyhedral inclusion bodies (PIB) in 11.3 l ha^{-1}, using a Bell 2006-B helicopter, equipped with a Simplex spray system with a boom 11 m long and nine flat fan nozzles (Teejet® 8010). The virus spray was an aqueous mixture containing powdered virus-killed larvae, 75% water, 25% molasses, 0.2% Chevron® sticker, and 0.04% Rhodamine B marker dye. Spray deposits were measured on Kromekote® cards, in both the aerially and ground-treated plots, and densities were found to be good in the four aerially treated plots, with 8.4, 8.1, 5.1 and 2.1 droplets cm^{-2}.

In the ground application trial, OrpsMNPV was applied to a line of 15 scattered trees extending through the centre of an infested stand, using a modified orchard-type hydraulic sprayer. An average of 4.5 l of aqueous spray mixture, containing 2.4×10^{10} PIB with 25% molasses and 0.2%

Chevron® sticker, was applied to each of the 7–15 m tall sample trees (Shepherd *et al.*, 1984b). Kromekote® cards were placed between the treated and intermediate trees, as well as between the line of treated trees and the three additional lines of sample trees that were parallel to and 50 m either side and 100 m to one side of the treated trees. To investigate virus spread, trees along these additional sample lines were sampled. A line of control trees was located 200 m from the treated trees.

Results showed that OrpsMNPV can be introduced at the early phases of an *O. pseudotsugata* outbreak and an epizootic can be caused by aerial or ground applications, even when populations are low (about 40 larvae m^{-2}). *O. pseudotsugata* populations were reduced and no egg masses were found in any of the treated plots in the autumn following treatment. The epizootic was directly proportional to the initial larval population densities. Initial levels of viral infection were higher in plots containing higher larval densities (Fig. 41.1). Levels of viral incidence increased slowly during the first 4 weeks, with the epizootic developing 5 weeks after treatment in the aerially treated plots. The epizootic occurred about 2 weeks earlier in plots with high larval densities. No spray drift was detected on the Kromekote® cards placed between the line of treated trees and the additional trees used to detect virus spread.

Virus incidence in larvae collected from sample trees followed the expected pattern of spread from an infection source; infection decreased in inverse proportion to the distance from the treated trees (Shepherd *et al.*, 1984b). Viral infection among larvae collected from the check trees located 200 m away was not detected until 5–8 weeks after spraying. However, the distance over which infection was observed suggests that flight lines located 200 m apart, instead of the usual 34 m, may be sufficient to initiate an epizootic in an infested stand (Otvos *et al.*, 1998). Feeding continued for most of the larval period and little foliage protection attributed to treatment could be detected in either the ground- or aerially treated plots (Otvos *et al.*, 1998).

In the second year of the outbreak, further experiments were conducted at a different location (Otvos *et al.*, 1987a, b). Population reduction in the plots treated

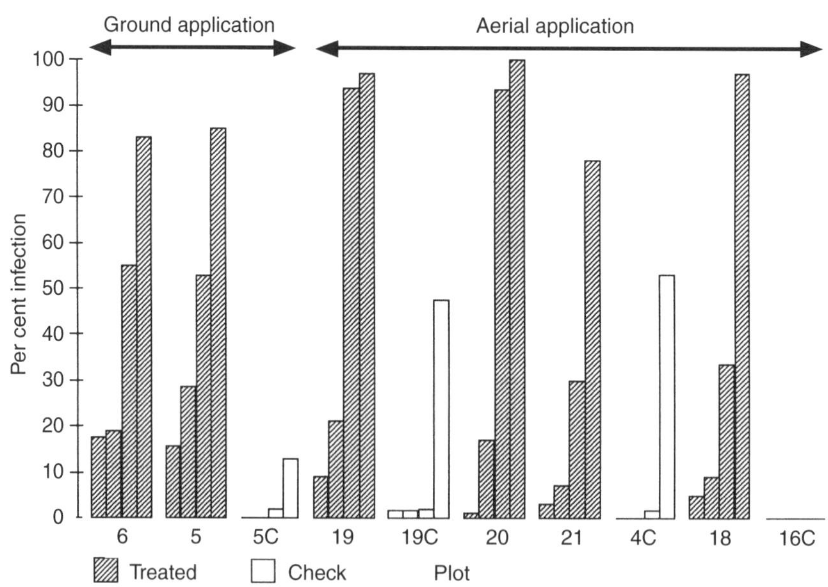

Fig. 41.1. Per cent infection of *Orgyia pseudotsugata* larvae with OrpsMNPV at 2, 4, 6 and 8 weeks after application for each study plot in British Columbia, 1981.

with the oil-based viral spray 6 weeks after treatment was in direct proportion to the dosage used. Larval mortality at full and 1/3 doses was similar (95% and 91%, respectively), and even the lowest dose (1/16 of full dose) gave about 65% population reduction; the label or full dose applied in aqueous mix with molasses gave about 87% larval mortality (Table 41.1). Following treatment, autumn egg-mass surveys showed that densities were reduced from outbreak levels to pre-treatment, endemic levels in all treatments compared to the control plots. The virus application prevented significant tree mortality up to 3 years after treatment (Table 41.1). One year after treatment less than 1% of the sample trees had died in the treated plots, but 38% were dead in the control plots; 2 years after treatment tree recovery was good and trees killed increased only slightly to about 3% in treated and 41% in control plots; in year 3 the trees continued to recover and no additional tree mortality attributable to *O. pseudotsugata* was observed (Otvos *et al.*, 1987b). Based on these results the recommended dose of OrpsMNPV can be reduced from 2.5×10^{11} to 8.3×10^{10} PIB and either aqueous or emulsifiable oil formulations are acceptable for application.

Shepherd *et al.* (1984a, b, 1985a) refined and calibrated these management tools, e.g. pheromone monitoring to predict impending outbreaks, egg-mass survey methods to predict expected defoliation levels, and the application of OrpsMNPV before significant damage occurs. These findings were integrated into a management system for *O. pseudotsugata* (Shepherd and Otvos, 1986; Otvos and Shepherd, 1991) that was operationally tested and proven during the next outbreak (1990–1993).

In the Kamloops Forest Region, British Columbia, pheromone trap catches of male moths followed by egg-mass surveys in the fall of 1990 indicated that population levels at 13 sites containing Douglas fir would reach population levels sufficient to cause defoliation (light defoliation at one site, moderate at seven sites, and severe at five sites) in 1991. In 1991, one of these sites was left untreated as a control and 12 sites were treated with stored virus products: four plots (about 40 ha in total) with 10-year-old TM BioControl-1®; four plots (about 100 ha) with 10-year-old Virtuss® at full dose of 2.5×10^{11} PIB ha^{-1}; and the remaining four areas (about 60 ha) with freshly produced Virtuss® at the same dose.

In 1991, both the stored and newly produced Virtuss® reduced *O. pseudotsugata* populations 8 weeks after treatment by 86% and 82%, respectively (Table 41.2). However, population reduction was only

Table 41.1. Population densities and reduction of *Orgyia pseudotsugata* larvae in four Virtuss®-treated experimental plots in the year of application, and cumulative proportion of trees killed 1 and 2 years after treatment in British Columbia (modified from Otvos *et al.*, 1987a).

Plot number	Treatments[a]	1982 pre-spray larvae m^{-2}	% Population reduction 1982[b]	% Sample trees killed	
				1983	1984
T1	1.6×10^{10}	182.8	64.7	0	0
T2	8.3×10^{10}	145.8	90.6	2	7
T3	2.5×10^{11}	302.0	95.1	0	4
T4	2.5×10^{11}	41.8	86.6	0	0
Mean				0.6	2.8
C1	Control	197.5		53	60
C2	Control	136.9		Logged	–
C3	Control	360.6		60	62
C4	Control	81.2		0	0
Mean				37.8	40.7

[a]Plots T1, T2 and T3 were treated with an oil-based virus formulation, Plot T4 with an aqueous virus formulation.
[b]Population reduction was calculated using a modified Abbott's formula (Abbott, 1925) in the year of application.

Table 41.2. Population reduction for experimental application of OrpsMNPV against *Orgyia pseudotsugata* in Kamloops, 1991.

Treatment	Plot	Number of *O. pseudotsugata* larvae m^{-2} at 0, 2, 4, 6 and 8 weeks after spray					Percentage population reduction at biweekly intervals after spray[a]		
		0[b]	2	4	6	8	4	6	8
Stored Virtuss®	3	52.34	64.96	22.34	14.31	4.91	42.6	57.6	79.0
	7	163.13	164.88	40.77	3.10	2.57	58.7	96.4	95.7
	7b	39.07	96.70	37.56	30.11	19.12	35.2	40.0	45.0
	9	130.51	121.92	67.43	28.18	2.84	7.7	55.5	93.5
	Mean	102.85	113.85	41.16	16.67	5.85	39.7	71.8	85.7
New Virtuss®[c]	1	7.92	39.76	14.28	7.47	3.09	40.1	63.8	78.4
	4	51.33	52.32	57.04	11.11	4.50	0	59.1	76.1
	5	86.57	179.70	64.62	15.80	10.36	40.0	83.1	84.0
	6	60.35	102.05	96.31	17.59	6.22	0	66.8	83.1
	Mean	53.07	90.13	64.69	13.58	5.89	0	71.0	81.8
Stored TM BioControl-1®	8	20.15	27.44	24.71	7.23	4.35	0	49.3	55.9
	10	30.98	23.35	19.04	14.25	8.85	0	0	0
	11	0.64	2.10	2.11	2.11	0.82	0	0	0
	12	7.25	8.85	9.48	7.10	5.49	0	0	0
	Mean	13.41	13.89	12.45	7.25	4.58	0	0	8.4
Treatment mean		61.75	75.16	36.93	12.83	5.44	18.0	67.1	79.9
Control		30.44	33.89	20.26	17.55	12.16			

[a]The calculations for Abbott's formula uses the mean value from collection time 2 weeks post-spray rather than pre-spray because there was an apparent increase in larval populations following the pre-spray sample.
[b]0 indicates pre-spray sample.
[c]Virus used was produced within 1 year prior to application.

about 8% in stands treated with stored TM BioControl-1®. This unacceptably low population reduction was probably due to the low larval densities in these plots prior to the spray, suggesting that it was unnecessary to treat stands containing such low densities. Egg-mass surveys taken the following autumn in the treated plots indicated that either trace or no defoliation would occur the following year. Defoliation surveys in 1992 confirmed this.

During autumn egg-mass surveys in 1991, two infested areas were located. One was used as a control in summer, 1991, and the infestation in this area had increased to about 260 ha. Egg-mass density in this area increased almost ninefold over the previous year, to 21 egg masses per branch. The second area was almost 460 ha in size, and average densities of over 0.6 egg masses per branch indicated that severe defoliation would occur (Shepherd *et al.*, 1984a, 1985b). Both infestations were treated in 1992 with an aqueous formulation of Virtuss® at full (label) dosage of 2.5×10^{11} PIB ha^{-1} in 9.4 l ha^{-1}. The second infestation (460 ha) was treated with the standard 'blanket' treatment, where flight lines were spaced 34 m apart. In the first infestation (260 ha), the flight lines were spaced 200 m apart to test the hypothesis that widely spaced swath applications of the recommended (label) dose of the virus might provide acceptable population reduction, even though it might not provide acceptable foliage protection. Thus, only about 10%, or 26 of the 260 ha, were sprayed directly. Virus infection, initially higher directly under the flight lines, gradually spread into the 'untreated' areas between flight lines. Population reduction 10 weeks after treatment along three of the four sample lines was more than 90%, while along the fourth sample line (line C) population reduction was about 76% (Table 41.3). This was similar to the population reduction of 94% and 96% seen 8 weeks after treatment in two of the plots (9, 6 and 7, respectively) receiving the complete or 'blanket' treatment the previous year (i.e. 1991, see Table 41.2) (Otvos *et al.*, 1998).

Ten weeks after treatment, it was difficult to collect larvae from directly under the flight lines even when mass collections were attempted by beating the branches of the sample trees. Dead, late-instar larvae were observed hanging from the branches on many trees within the sample plot. Practically all the living larvae, irrespective of where they were collected in the plot (i.e. from under the widely spaced flight lines or between them), were sluggish and showed signs of being infected with OrpsMNPV, and died either as larvae in partially completed cocoons or as pupae. The results of this control trial were encouraging, even though it was based on only one large, non-replicated treated area in 1 year with flight lines spaced 200 m apart. Because of these limitations it is imperative that this widely spaced flight line treatment be replicated, over more than 1 year and preferably in several locations with flight lines placed only 100 m apart, to confirm these promising results and provide more foliage protection.

In 1993, the British Columbia Ministry of Forests treated most infested stands (about 440 ha) with TM BioControl-1® or Virtuss® at full dose, so the widely spaced swath treatment could not be repeated. Population reduction averaged about 95% in these plots. The remaining stands received two-thirds the recommended dose and reductions of 68% for TM BioControl-1® and only 14% for Virtuss® occurred (Table 41.4).

Table 41.3. Population reduction of *Orgyia pseudotsugata* in plots treated experimentally with widely spaced swath applications of OrpsMNPV near Kamloops, 1992.

		O. pseudotsugata larvae m^{-2} at 2-week intervals after spray						
Treatment	Line	0[b]	2	4	6	8	10	% Population reduction[a]
Sample trees under flight lines[c]	A	34.55	1.56	15.44	13.58	5.14	1.7	95.1
	B	23.36	18.37	33.87	17.97	2.03	0.43	98.2
	C	50.33	18.76	32.38	48.23	7.27	12.30	75.6
	D	27.30	35.37	24.67	31.63	1.40	0.00	100.0
	Mean	32.64	20.28	26.77	27.62	3.59	3.06	90.6
Sample trees between flight lines	A	58.71	39.76	56.86	45.29	19.47	2.52	95.7
	B	32.32	52.32	21.49	50.27	1.48	1.07	96.7
	C	120.59	179.70	174.04	75.36	11.12	4.86	96.0
	D	60.71	102.05	47.66	27.72	7.47	0.29	99.5
	Mean	70.18	90.13	77.03	50.15	8.82	2.15	96.9
All sample trees (both under and between flight lines)	A	44.90	26.07	33.19	27.17	11.28	2.05	95.4
	B	27.61	29.14	27.68	34.12	1.76	0.75	97.3
	C	89.36	60.78	111.08	63.30	9.41	8.17	90.9
	D	42.48	54.52	35.12	29.77	4.16	0.13	99.7
	Mean	50.36	43.80	51.22	38.73	6.14	2.62	94.8

[a]No untreated check areas were available because all *O. pseudotsugata* infested stands were treated, therefore population reductions could not be corrected for natural mortality.
[b]0 indicates pre-spray sample.
[c]Flight lines were spaced 200 m apart.

Table 41.4. Results of operational treatment of *Orgyia pseudotsugata* infested stands, 1993.

Treatment	Hectares	Dosage[a]	% Population reduction[b]
Virtuss®	100	Full	89
	110	Full	98
	150	2/3	14
TM BioControl-1®	130	Full	90
	100	Full	98
	25	2/3	64

[a]Full dose = 2.5×10^{11} PIB in 11.3 l ha^{-1}.
[b]No untreated check areas were available because all infested stands were treated, therefore population reductions could not be corrected for natural mortality.

In 1991, a field experiment on 2 ha plots demonstrated successful mating disruption when high dosages (72 g ha^{-1}) of a synthetic form of the sex pheromone of the *O. pseudotsugata* were used (Hulme and Gray, 1994). In 1992, reduced dosages were tested in an attempt to make mating disruption more cost-effective. Results of the 1992 trials indicated that egg mass counts in the pheromone-treated plots were lower than in the untreated plots (Hulme and Gray, 1996). However, the high cost of the pheromone and the small size of the areas that could be treated prevented mating disruption from becoming an operational tool.

Evaluation of Biological Control

Development of *O. pseudotsugata* infestations can be prevented by a single application of oil- or water-based formulations of either Virtuss® or TM BioControl-1® at the beginning of an outbreak. Moreover, although foliage protection may be negligible or poor in the year of application, it is substantial the following year. Tree mortality is prevented when treatment is applied early in the outbreak cycle and to first- or second-instar larvae, permitting the development of two viral epizootics. If the treatment is applied correctly and early enough in the population cycle, the outbreak will not develop in treated stands. Use of the pheromone monitoring system (Shepherd *et al.*, 1985a) and early application of registered virus products was operationally proven during the 1990–1993 *O. pseudotsugata* cycle (Otvos *et al.*, 1998; Maclauchlan *et al.*, 2001). Tree mortality was minimal in the stands treated with OrpsMNPV because it prevented development of a full-blown outbreak. The integrated management system developed for *O. pseudotsugata* is the first and only one used operationally against a forest defoliator in Canada. Also, virus applications offer a cost-effective and practical alternative to chemical or other biological insecticides that would have to be applied annually.

Currently, about 320,000 acre-dose equivalents of TM BioControl-1® are held in cold storage by the United States Forest Service. An additional 80,000 acre-dose equivalents were used in Oregon in 2000 against *O. pseudotsugata* during the outbreak that probably started in 1999. In Canada, only the British Columbia Ministry of Forests has OrpsMNPV in storage, with enough TM BioControl-1® to treat about 1000 ha and enough Virtuss® to treat about 500 ha. The quantities of both of these products may be sufficient to treat the next Canadian outbreak, forecast to occur within 5 years.

Recommendations

Further work should include:

1. Determining the shelf-life and efficacy of the virus stocks currently held in storage;
2. Finding an industry partner to produce this virus – this can probably be done only

on contract because the virus has a limited market due to its host specificity to a few *Orgyia* spp. and relatively high production costs;

3. Investigating cost reduction of virus application, e.g. by applying lower doses or by spacing flight lines further apart.

References

Abbott, W.S. (1925) A method of computing the effectiveness of an insecticide. *Journal of Economic Entomology* 18, 265–267.

Alfaro, R.I., Taylor, S.P., Wegwitz, E. and Brown, R.G. (1987) Douglas-fir tussock moth damage in British Columbia. *Forestry Chronicle* 63, 351–355.

Beckwith, R.C. (1978) Biology of the insect. In: Brookes, M.H., Stark, R.W. and Campbell, R.W. (eds) *The Douglas-fir Tussock Moth: A Synthesis*. Technical Bulletin 1585, United States Department of Agriculture, Forest Service, Science and Education Agency, pp. 25–37.

Cunningham, J.C. (1982) Field trials with baculoviruses: control of forest insect pests. In: Kurstak, E. (ed.) *Microbial and Viral Pesticides*, Dekker, New York, pp. 335–386.

Cunningham, J.C. and Shepherd, R.F. (1984) *Orgyia pseudotsugata* (McDunnough), Douglas-fir tussock moth (Lepidoptera: Lymantriidae). In: Kelleher, J.S. and Hulme, M.A. (eds) *Biological Control Programmes Against Insects and Weeds in Canada 1969–1980*. Commonwealth Agriculture Bureaux, Slough, UK, pp. 363–367.

Dahlsten, D.L. and Thomas, G.M. (1969) A nucleopolyhedrosis virus in populations of Douglas-fir tussock moth, *Hemerocampa pseudotsugata*, in California. *Journal of Invertebrate Pathology* 13, 264–271.

Daterman, G.E., Peterson, L.J., Robbins, R.G., Sower, L.L., Daves, G.D. Jr and Smith, R.G. (1976) Laboratory and field bioassay of the Douglas-fir tussock moth pheromone, (Z)-6-heneicosen-11-one. *Environmental Entomology* 5, 1187–1190.

Harris, J.W.E., Dawson, A.F. and Brown, R.G. (1985) *The Douglas-fir Tussock Moth in British Columbia, 1916–1984*. Pacific Forestry Centre, Information Report BC-X-268, Canadian Forest Service.

Hughes, K.M. and Addison, R.B. (1970) Two nuclear polyhedrosis viruses of the Douglas-fir tussock moth. *Journal of Invertebrate Pathology* 16, 196–204.

Hulme, M.A. and Gray, T.G. (1994) Mating disruption of Douglas-fir tussock moth (Lepidoptera: Lymantriidae) using a sprayable bead formulation of Z-6-heneicosen-11-one. *Environmental Entomology* 23, 1097–1100.

Hulme, M.A. and Gray, T.G. (1996) Effect of pheromone dosage on the mating disruption of Douglas-fir tussock moth. *Journal of the Entomological Society of British Columbia* 93, 99–103.

Johnson, P.C. and Ross, D.A. (1967) Douglas-fir tussock moth, *Hemerocampa* (*Orgyia*) *pseudotsugata* McDunnough. In: Davidson, A.G. and Prentice, R.M. (eds) *Important Forest Insects and Diseases of Mutual Concern to Canada, the United States and Mexico*. Canada Department of Forestry and Rural Development, Ottawa, Ontario, pp. 105–107.

Maclauchlan, L.E., Hall, P.M. and Otvos, I.S. (2001) Successful implementation of research into an operational program to reduce damage caused by Douglas-fir tussock moth, *Orgyia pseudotsugata*, in British Columbia. *Forestry Chronicle* (in press).

Mason, R.R. and Luck, R.F. (1978) Population growth and regulation. In: Brookes, M.H., Stark, R.W. and Campbell, R.W. (eds) *The Douglas-fir Tussock Moth: a Synthesis*. Technical Bulletin 1585, United States Department of Agriculture, Forest Service, Science and Education Agency, pp. 41–46.

Otvos, I.S. and Shepherd, R.F. (1991) Integration of early virus treatment with a pheromone detection system to control Douglas-fir tussock moth, *Orgyia pseudotsugata* (Lepidoptera: Lymantriidae) populations at pre-outbreak levels. In: Raske, A.G. and Wickman, B. (eds) Proceedings of a symposium 'Towards Integrated Pest Management of Forest Defoliators'. *Forest Ecology and Management* 39, 143–151.

Otvos, I.S., Cunningham, J.C. and Friskie, L.M. (1987a) Aerial application of nuclear polyhedrosis virus against Douglas-fir tussock moth, *Orgyia pseudotsugata* (McDunnough) (Lepidoptera: Lymantriidae): I. Impact in the year of application. *The Canadian Entomologist* 119, 697–706.

Otvos, I.S., Cunningham, J.C. and Alfaro, R.I. (1987b) Aerial application of nuclear polyhedrosis virus against Douglas-fir tussock moth, *Orgyia pseudotsugata* (McDunnough) (Lepidoptera: Lymantriidae): II. Impact 1 and 2 years after application. *The Canadian Entomologist* 119, 707–715.

Otvos, I.S., Cunningham, J.C. and Shepherd, R.F. (1995) Douglas-fir tussock moth, *Orgyia pseudotsugata*. In: Armstrong, J.A. and Ives, W.G.H. (eds) *Forest Insect Pests in Canada.* Natural Resources Canada, Canadian Forest Service, Ottawa, Ontario, pp. 127–132.

Otvos, I.S., Cunningham, J.C., Maclauchlan, L., Hall, P. and Conder, N. (1998) The development and operational use of a management system for control of Douglas-fir tussock moth, *Orgyia pseudotsugata* (Lepidoptera: Lymantriidae), populations at pre-outbreak levels. In: McManus, M.L. and Liebhold, A.M. (eds) *Proceedings: Population Dynamics, Impacts and Integrated Management of Forest Defoliating Insects.* General Technical Report NE-247, United States Department of Agriculture, Forest Service, pp. 143–154.

Ross, D.W. and Taylor, S.P. (1985) Effects of the current (1980–1984) Douglas-fir tussock moth outbreaks on forest resources and other values. Internal Report PM-PB-9, British Columbia Ministry of Forests, Victoria, British Columbia.

Shepherd, R.F. and Otvos, I.S. (1986) Pest management of Douglas-fir tussock moth: procedures for insect monitoring problem evaluation and control actions. Pacific Forestry Centre, Information Report BC-X-270, Canadian Forest Service.

Shepherd, R.F., Otvos, I.S. and Chorney, R.J. (1984a) Pest management of Douglas-fir tussock moth (Lepidoptera: Lymantriidae): a sequential sampling method to determine egg mass density. *The Canadian Entomologist* 116, 1041–1049.

Shepherd, R.F., Otvos, I.S., Chorney, R.J. and Cunningham, J.C. (1984b) Pest management of Douglas-fir tussock moth (Lepidoptera: Lymantriidae): prevention of an outbreak through early treatment with a nuclear polyhedrosis virus by ground and aerial applications. *The Canadian Entomologist* 116, 1533–1542.

Shepherd, R.F., Gray, T.G., Chorney, R.J. and Daterman, G.E. (1985a) Pest management of Douglas-fir tussock moth: monitoring endemic populations with pheromone traps to detect incipient outbreaks. *The Canadian Entomologist* 117, 839–848.

Shepherd, R.F., Otvos, I.S. and Chorney, R.J. (1985b) *Sequential Sampling for Douglas-fir Tussock Moth Egg Masses in British Columbia.* Joint Report No. 15, Canadian Forestry Service, Pacific Forestry Research Centre, Victoria, BC, Canada and British Columbia Ministry of Forests, Kamloops, BC, Canada.

Wellner, C.A. (1978) Host stand biology and ecology. In: Brookes, M.H., Stark, R.W. and Campbell, R.W. (eds) *The Douglas-fir Tussock Moth: a Synthesis.* Technical Bulletin 1585, United States Department of Agriculture, Forest Service, Science and Education Agency, pp. 7–23.

Wickman, B.E. (1978) *Tree Mortality and Top-kill Related to Defoliation by the Douglas-fir Tussock Moth in the Blue Mountains Outbreak.* Research Paper PNW-233, United States Department of Agriculture, Forest Service.

42 *Panonychus ulmi* (Koch), European Red Mite (Acari: Tetranychidae)

J.M. Hardman and H.M.A. Thistlewood

Pest Status

The European red mite, *Panonychus ulmi* (Koch), is the most serious mite pest of apple, *Malus pumila* Miller (= *M. domestica* Borkhausen), and peach, *Prunus persica* (L.), in Canada and can be very damaging to a wide variety of other fruits, berry crops and even grape, *Vitis vinifera* L. (Lester *et al.*, 1998). The pest status of *P. ulmi* is due to its high reproductive rate, short generation time, multivoltinism and its ability to develop resistance rapidly to most pesticides used in orchards, berry plantations or vineyards. Mite damage to leaves reduces the rate of photosynthesis, thereby adversely affecting vegetative growth of trees, and yield and fruit quality, including size, firmness, flavour and storage life (Marini *et al.*, 1994).

P. ulmi has a generation time of 21 days at a mean temperature of 20°C, adult longevity is about 2.5 weeks, lifetime fecundity is 25 eggs per female and the maximum rate of oviposition is 1.9 eggs per female per day. There can be 4–8 generations per season, depending upon heat units above 10°C available in the growing season. The egg stage overwinters.

Background

In general, conservation of natural enemies will result in suppression of *P. ulmi* to subeconomic population levels. However, unrestrained outbreaks can follow the application of a wide variety of agricultural chemicals that are toxic to its predators, especially the phytoseiid mites (Bostanian *et al.*, 1985, 1998; Hardman *et al.*, 1988, 1991, 1995; Hardman and Gaul, 1990; Pree, 1990; Thistlewood and Elfving, 1992). Other factors that affect the relative importance of predatory species in different regions or on different crops include local climate and the use or avoidance of disruptive pesticides. Distinct regional differences exist in the degree to which biological control of mites has been practised since its widespread adoption in the late 1960s.

Most of the effort in the biological control of mites has centred on the conservation, augmentation or inundative release of indigenous predators or genetically improved strains. Phytoseiid mites are the most useful predators of *P. ulmi*, with an ability to develop resistance to toxic agricultural chemicals, a high searching capacity, multivoltinism and a high reproductive rate, which often exceeds that of their prey. Other important natural enemies are the stigmaeid and erythraeid mites, and coccinellid beetles (*Stethorus* spp.).

The phytoseiids can be grouped into specialist or generalist predators, each with distinctly different capabilities for biological control. The specialists, primarily *Typhlodromus* (= *Metaseiulus*) *occidentalis* Nesbitt in British Columbia, and *Amblyseius fallacis* (Garman) in British Columbia, Ontario, Quebec, New Brunswick and Prince Edward Island, are effective predators of web-building Tetranychidae mites but they also feed on *P. ulmi*, which is not a web builder. The specialists, which have greater reproductive rates, higher feeding rates and disperse more rapidly than the generalists, are better at colonizing new sites and can quickly suppress prey populations, even those that are increasing. Generalists, e.g. *Typhlodromus caudiglans* Schuster in British Columbia, Ontario and Quebec, and *Typhlodromus pyri* Scheuten in New Brunswick and Nova Scotia, are more strictly arboreal than the specialists and can more readily survive and reproduce on alternative foods, e.g. windborne pollen. Hence, the generalists often persist on trees where mite prey are scarce or absent (Walde *et al.*, 1992). Conversely, the specialists are more likely to starve or to disperse from trees where prey densities are too low (Bostanian and Hardman, 1998).

Biological Control Agents

Predators

In British Columbia, biological control of *P. ulmi* and other mite species occurs through active conservation of indigenous populations of predators, primarily *A. fallacis* on coastal berry crops (Henderson and Raworth, 1991; Elliott, 1997) and *T. occidentalis* or *T. caudiglans* on tree fruits in the dry interior (Angerilli and Brochu, 1987). On tree fruits, unrestrained spider mite populations are generally devastating due to a hot and dry climate, so researchers, extension agents and packing houses encourage the use of crop protection programmes that conserve mite predators. *T. occidentalis* is often transferred among local orchards and brought into new areas where the predator is absent. Predator establishment is particularly favoured in this arid region by the infrequency of use of fungicides, many of which are toxic to eggs and juvenile stages of phytoseiids.

In eastern Canada, it is more difficult to structure crop protection programmes around predator conservation, owing to intensive fungicide use (up to 13 sprays per year in Nova Scotia) and use of broad-spectrum insecticides against pest species that either do not occur in the west or are handled differently. Nevertheless, in Ontario and Quebec prior to the 1980s, spider mites were controlled mostly by indigenous populations of several species, primarily *A. fallacis*, that were resistant to organophosphate insecticides, supplemented by occasional use of miticides (Bostanian and Hardman, 1998; Pree, 1990; Thistlewood, 1991). In Quebec, there was also some success with augmentative release of *A. fallacis*. During the 1980s, growers across eastern Canada began using broad-spectrum pyrethroid insecticides to control key insects that had become resistant to organophosphates. Their use removed many predatory species, leading to serious outbreaks of *P. ulmi* and other Tetranychidae. In Ontario and Quebec, adoption in the 1990s of pest monitoring, treatment thresholds and a modified spray programme for the key pests is leading to conservation of populations of *A. fallacis* and *T. caudiglans*. However, because orchard repopulation could take 4 years without introductions of phytoseiids, releases of a commercially produced strain of *A. fallacis* resistant to organophosphate and pyrethroid insecticides (Thistlewood *et al.*, 1995) were made, with inconsistent results, for control of *P. ulmi* on apple and peach (Lester, 1998; Bostanian *et al.*, 1999; Lester *et al.*, 1999, 2000). In contrast, releases of this strain on to berries and other horticultural crops to control spider mites, e.g. *Tetranychus urticae* Koch, were highly effective in the Pacific Northwest (Elliott, 1997). Recent attention has shifted to conservation and release of other species, including the importation from Nova Scotia of a strain of *T. pyri*, originally from New Zealand, that also has resistance to organophosphates and pyrethroids (Hardman *et al.*, 1997). In one Ontario trial, this *T. pyri* strain has established and spread slowly, and has proven somewhat effective in biological control of mites on apple. Similar promising results have been observed with native strains of *T. pyri* to control *P. ulmi* on grapes.

In Quebec, attempts to establish populations of *T. pyri* from Nova Scotia failed, perhaps because Quebec winters are harsher than those in fruit-growing regions of Ontario and the Maritimes. The best results were obtained by adoption of a modified spray programme coupled with release of indigenous predators, including *A. fallacis* and *T. caudiglans* (Bostanian *et al.*, 1999). Sub-economic populations of *P. ulmi* in Ontario and Quebec are also associated with predation by indigenous populations of the erythraeid mite, *Balaustium* sp., and the stigmaeids, *Agistemus fleschneri* Summers and *Zetzellia mali* (Ewing).

In Nova Scotia, widespread use of pyrethroid insecticides and of the organophosphate dimethoate was associated with outbreaks of *P. ulmi* and the apple rust mite, *Aculus schlechtendali* (Nalepa), and more frequent use of miticides (Hardman *et al.*, 1988, 2000). Remedial actions included the importation and release of the pyrethroid-resistant strain of *T. pyri* (Hardman *et al.*, 1997), the development and use of a modified spray programme to conserve this strain, and use of selective miticides to supplement predator activity in the initial period after release. The modified spray programme employed *Bacillus thuringiensis* Berliner and a reduced concentration of pyrethroid to control winter moth, *Operophtera brumata* (L.), and avoided use of dimethoate and EBDC fungicides after bloom (Hardman and Rogers, 1998). Releases of *T. pyri* in commercial orchards began in Nova Scotia in 1993 and in New Brunswick in 1995 (Hardman *et al.*, 2000). Colonizations were nearly always successful in Nova Scotia and biological control was effective according to the degree to which growers followed the modified spray programme. In New Brunswick, the resistant strain has colonized well and provides effective biological control in orchards on the shore of the Bay of Fundy, but in regions on the Northumberland strait densities of *T. pyri* tend to increase later in the summer and control of *P. ulmi* is less effective. Recent releases have also been carried out in Prince Edward Island and *T. pyri* has been recovered from some of these sites.

Evaluation of Biological Control

Many growers and extension personnel realize the importance of conserving natural enemies for integrated mite control. None the less, use of agrochemicals of unknown toxicity to mite predators has often caused local annihilation of key natural enemies and subsequent mite outbreaks. Slow recolonization by the more useful, generalist species means mite problems often persist for several years after a toxin is no longer used. The longevity of perennial crops, such as tree fruits and grapes, carryover effects where pesticides applied in one season increase mite populations in subsequent seasons, and difficulties in determining which predators in a complex are most effective in mite suppression, have hampered implementation of biological control of mites.

Recommendations

Further work should include:

1. Elucidating interactions among mite species;
2. Assessing all new agricultural chemicals, particularly pesticides, for compatibility with predators, and replacement of harmful chemicals with less disruptive substitutes;
3. Studying the biology and dispersal of natural enemies, their genetic improvement, and methods of mass rearing;
4. Refining techniques for augmentative and inundative release of those predators that are best suited for different crops in different regions.

References

Angerilli, N.P.D. and Brochu, L. (1987) Some influences of area and pest management on apple mite populations in the Okanagan Valley of British Columbia. Journal of the Entomological Society of British Columbia 84, 3–9.

Bostanian, N.J. and Hardman, J.M. (1998) Phytophagous mite management in apple orchards in eastern Canada. In: Vincent, C. and Smith, R. (eds) *Orchard Pest Management in Canada*. Agriculture and Agri-Food Canada, Saint-Jean-sur-Richelieu, Quebec, pp. 53–69.

Bostanian, N.J., Belanger, A. and Rivard, I. (1985) Residues of four synthetic pyrethroids and azinphos-methyl on apple foliage and their toxicity to *Amblyseius fallacis* (Acari: Phytoseiidae). *The Canadian Entomologist* 117, 143–152.

Bostanian, N.J., Thistlewood, H.M.A. and Racette, G. (1998) Effects of five fungicides used in Quebec apple orchards on *Amblyseius fallacis* (Garman) (Phytoseiidae: Acari). *Journal of Horticultural Science* 73, 527–530.

Bostanian, N.J., Lasnier, J. and Racette, G. (1999) Biological control of mites in Quebec apple orchards. *Canadian Fruitgrower* Sept./Oct. 1999, 8–10.

Elliott, D. (1997) *Biological Control of Spider Mites on Fruit Crops.* National Agricultural Biotechnology Initiative Project BD 92 WD 071, Canada Department of Western Diversification, Ottawa, Ontario.

Hardman, J.M. and Gaul, S.O. (1990) Mixtures of *Bacillus thuringiensis* and pyrethroids control winter moth (Lepidoptera: Geometridae) in orchards without causing outbreaks of mites. *Journal of Economic Entomology* 83, 920–936.

Hardman, J.M. and Rogers, M.L. (1998) New opportunities for mite control in Nova Scotian apple orchards. *134th Annual Report of the Nova Scotia Fruitgrowers' Association 1997*, pp. 24–29.

Hardman, J.M., Rogers R.E.L. and MacLellan, C.R. (1988) Advantages and disadvantages of using pyrethroids in Nova Scotia apple orchards. *Journal of Economic Entomology* 81, 1737–1749.

Hardman, J.M., Rogers, R.E.L., Nyrop, J.P. and Frisch T. (1991) Effect of pesticide applications on abundance of European red mite (Acari: Tetranychidae) and *Typhlodromus pyri* (Acari: Phytoseiidae) in Nova Scotian apple orchards. *Journal of Economic Entomology* 84, 570–580.

Hardman, J.M., Smith, R.F. and Bent, E. (1995) Effects of different IPM programs on biological control of mites on apple by predatory mites (Acari) in Nova Scotia. *Environmental Entomology* 24, 125–142.

Hardman, J.M., Rogers, M.L., Gaul, S.O. and Bent, E.D. (1997) Insectary rearing and initial testing in Canada of an organophosphate/pyrethroid-resistant strain of the predator mite *Typhlodromus pyri* (Acari: Phytoseiidae) from New Zealand. *Environmental Entomology* 26, 1424–1436.

Hardman, J.M., Moreau, D.L., Snyder, M., Gaul, S.O. and Bent, E.D. (2000) Performance of a pyrethroid resistant strain of the predator mite *Typhlodromus pyri* (Acari: Phytoseiidae) under different insecticide regimes. *Journal of Economic Entomology* 93, 590–604.

Henderson, D.E. and Raworth, D.A. (1991) *Beneficial Insects and Common Pests on Strawberry and Raspberry Crops.* Publication 1863/E, Agriculture Canada, Ottawa, Ontario.

Lester P.J. (1998) An assessment of the predator *Amblyseius fallacis* for biological control of the European red mite. PhD thesis, Queen's University, Kingston, Ontario, Canada.

Lester, P.J., Thistlewood, H.M.A. and Ball, S. (1998) European red mite *Panonychus ulmi* (Koch): a new problem in Ontario vineyards. *Proceedings of the Entomological Society of Ontario* 128, 105–107.

Lester, P.J., Thistlewood, H.M.A., Marshall, D.B. and Harmsen, R. (1999) Assessment of *Amblyseius fallacis* (Garman) (Acari: Phytoseiidae) for biological control of tetranychid mites in an Ontario peach orchard. *Experimental and Applied Acarology* 23, 995–1009.

Lester P.J., Thistlewood, H.M.A. and Harmsen, R. (2000) Some effects of pre-release host plant on the biological control of *Panonychus ulmi* Koch by the predatory mite *Amblyseius fallacis* Garman. *Experimental and Applied Acarology* 24, 1–15.

Marini, R.P.D., Pfeiffer, D.G. and Sowers, D.S. (1994) Influence of European red mite (Acari: Tetranychidae) and crop density on fruit size and quality and on crop value of 'Delicious' apple trees. *Journal of Economic Entomology* 87, 1302–1311.

Pree, D.J. (1990) Resistance management in multiple-pest apple orchard ecosystems in Eastern North America. In: Roush, R.T. and Tabashnik, B.E. (eds) *Pesticide Resistance in Arthropods*, Chapman and Hall, New York, New York, pp. 261–276.

Thistlewood, H.M.A. (1991) Predatory mites in Ontario apple orchards with diverse pesticide programmes. *The Canadian Entomologist* 123, 1163–1174.

Thistlewood, H.M.A. and Elfving, D.C. (1992) Laboratory and field effects of chemical fruit thinners on tetranychid and predatory mites. *Journal of Economic Entomology* 85, 477–485.

Thistlewood, H.M.A, Pree, D.J. and Crawford, L.A. (1995) Selection and genetic analysis of permethrin resistance in *Amblyseius fallacis* (Garman) (Acari: Phytoseiidae) from Ontario apple orchards. *Experimental and Applied Acarology* 19, 707–721.

Walde, S.J., Nyrop, J.P. and Hardman, J.M. (1992) Dynamics of European red mite and *Typhlodromus pyri*: factors contributing to persistence. *Experimental and Applied Acarology* 14, 261–291.

43 *Phyllonorycter mespilella* (Hübner), a Tentiform Leafminer (Lepidoptera: Gracillariidae)

J.E. Cossentine

Pest Status

Phyllonorycter mespilella (Hübner) is a native North American tentiform leafminer that feeds within the leaves of apple, *Malus pumila* Miller (= *M. domestica* Borkhausen), cherry, *Prunus avium* L., and pear, *Pyrus communis* L. It was not found in commercial orchards in the interior of British Columbia until 1988. Presumably, it moved from Washington State into Canada (Cossentine and Jensen, 1992). Landry and Wagner (1995) suggested that *Phyllonorycter elmaella* Doğanlar and Mutuura, described as established in the western USA, was, in fact, *P. mespilella*. The leafminer had been a minor orchard pest in north-western USA, until it developed resistance to commonly used organophosphate and chlorinated-hydrocarbon insecticides (Hoyt, 1983). Although low *P. mespilella* infestations cause minimal fruit damage, severe infestations have resulted in premature ripening, leaf and fruit drop, reduction in apple firmness, size, colour and storage life, and reduced foliar absorption of growth regulators (Hoyt, 1983). Recently planted young trees are vulnerable to complete defoliation.

Three generations a year of *P. mespilella* occur in southern British Columbia. The first three larval instars are sap-feeders and create a blotch-shaped mine. The last two larval instars are tissue-feeders within a tent-shaped mine.

Background

Use of a carbamate insecticide to control *P. mespilella* in the USA resulted in disrupted integrated management programmes for mites, as the chemical is toxic to predacious mites (see Hardman and Thistlewood, Chapter 42 this volume). Consequently, to maintain a successful integrated mite management programme in apple orchards, producers in British Columbia were advised not to use a mite-toxic chemical insecticide. Alternative control methods, e.g. biological control were needed.

Biological Control Agents

Parasitoids

Surveys of *P. mespilella* were done in orchards in the Okanagan and Similkameen valleys from 1988 to 1990 to determine whether the eulophid parasitoids found in *Phyllonorycter* populations in Washington State would establish and suppress the leafminer populations in British Columbia.

Pnigalio flavipes (Ashmead), the primary tentiform leafminer parasitoid species in Washington State (Barrett, 1988) and the second most dominant *Phyllonorycter* parasitoid in Utah (Barrett and Jorgensen, 1986), was found in 87% of *P. mespilella* hosts in British Columbian orchards next to the American border in 1988 (Cossentine and Jensen, 1992). The parasitoid moved with the host as *P. mespilella* moved up the commercial fruit producing valley to more than 137 km north of the border in 1990. The ectoparasitoids *P. flavipes* and *Sympiesis marylandensis* Girault have been identified as the dominant parasitoids of *P. mespilella* in the interior of British Columbia

(Cossentine and Jensen, 1992). Adults can kill host larvae by stinging them while ovipositing and by feeding on them. Host feeding occurs predominantly on sap-feeding larvae and oviposition predominantly on tissue-feeding larvae (Barrett and Brunner, 1990). McGregor (1995) studied the possibility that the relative frequencies of sap-feeding and tissue-feeding larvae may alter the pattern of oviposition attack by parasitoids.

In 1991, the parasitoid complex was examined to determine if the relative roles of the two dominant parasitoid species influenced the level of host control. *P. flavipes* and *S. marylandensis* were found in 52.3% and 46.7% of the overwintering host mines, respectively. The parasitism by the two species did not account for differences in the numbers of overwintering or summer generation *P. mespilella* mines (Cossentine and Jensen, 1994). *P. flavipes* was the dominant parasitoid in orchard areas studied through three summer generations. A *Eulophus* sp., a *Cirrospilus* sp., and *Zagrammosoma multilineatum* (Ashmead) have been found occasionally parasitizing *P. mespilella* in British Columbia (Cossentine and Jensen, 1992, 1994).

Evaluation of Biological Control

The *P. flavipes* and *S. marylandensis* parasitoid complex provides efficient and effective biological control of *P. mespilella* in the interior of British Columbia. The parasitoid-induced mortality in the first *P. mespilella* generation was found to be negatively correlated with leafminer density in the second and third generations during a normal season, indicating that *P. flavipes* was able to successfully reduce intraseasonal leafminer populations (Cossentine and Jensen, 1992). However, in some years, unseasonably early warm temperatures appear to allow the leafminers to establish large populations before the parasitoids can have a significant controlling effect.

Recommendations

Future work should include:

1. Developing techniques for mass production of leafminer ectoparasitoids, to allow timely release in vulnerable orchards in problem years.

References

Barrett, B.A. (1988) The population dynamics of *Pnigalio flavipes* (Hymenoptera: Eulophidae), the major parasitoid of *Phyllonorycter elmaella* (Lepidoptera: Gracillariidae) in central Washington apple orchards. PhD thesis, Washington State University, Pullman, Washington.

Barrett, B.A and Brunner, J.F. (1990) Types of parasitoid-induced host preferences and sex ratios exhibited by *Pnigalio flavipes* (Hymenoptera: Eulophidae) using *Phyllonorycter elmaella* (Lepidoptera: Gracillariidae) as a host. *Environmental Entomology* 19, 803–807.

Barrett, B.A. and Jorgensen, C.D. (1986) Parasitoids of the western tentiform leafminer, *Phyllonorycter elmaella* (Lepidoptera: Gracillariidae) in Utah apple orchards. *Environmental Entomology* 15, 635–641.

Cossentine, J.E. and Jensen, L.B. (1992) Establishment of *Phyllonorycter mespilella* (Hübner) (Lepidoptera: Gracillariidae) and its parasitoid *Pnigalio flavipes* (Hymenoptera: Eulophidae) in fruit orchards in the Okanagan and Similkameen valleys of British Columbia. *Journal of the Entomological Society of British Columbia* 89, 18–24.

Cossentine, J.E. and Jensen, L.B. (1994) The role of two eulophid parasitoids in populations of the leafminer, *Phyllonorycter mespilella* (Lepidoptera: Gracillariidae) in British Columbia. *Journal of the Entomological Society of British Columbia* 91, 47–54.

Hoyt, S. (1983) Biology and control of the western tentiform leafminer. *Proceedings of the Washington Horticultural Association* 79, 115–118.

Landry, J.-F. and Wagner, D.L. (1995) Taxonomic review of apple-feeding species of *Phyllonorycter* (Hübner) (Lepidoptera, Gracillariidae) in North America. *Proceedings of the Entomological Society of Washington* 97, 603–625.

McGregor, R. (1995) The evolution of life history timing in a leafmining moth, *Phyllonorycter mespilella* (Hübner). PhD thesis, Simon Fraser University, Burnaby, British Columbia, Canada.

44 *Pikonema alaskensis* (Rohwer), Yellowheaded Spruce Sawfly (Hymenoptera: Tenthredinidae)

G.S. Thurston

Pest Status

The yellowheaded spruce sawfly, *Pikonema alaskensis* (Rohwer), is a native defoliator of spruce, *Picea* spp., trees throughout most of their range in Canada (de Groot, 1995). In Atlantic Canada, damage has most often been reported on black spruce, *Picea mariana* (Miller) Briton Sterns and Poggenberg, while other native spruces are at greater risk in other provinces (Martineau, 1984). *P. alaskensis* attacks young trees grown in open areas, e.g. plantations, hedgerows, windbreaks, ornamental plantings and nurseries (Martineau, 1984; Rose and Lindquist, 1994; de Groot, 1995) and has been reported to attack mature trees (Martineau, 1984). Trees in predominantly sunny locations (Martineau, 1984) and on hilltops, southern exposures, and in understocked plantations (Katovich *et al.*, 1995; Thurston, 1997) tend to be at greater risk. Problems associated with *P. alaskensis* appear to have increased with more spruce plantings in recent years.

P. alaskensis overwinters as larvae in cocoons in the soil. Pupation occurs in spring, and adults emerge in May or June. A small proportion of the population can have a prolonged diapause and emerges 1 year later (Duda, 1953). Females lay their eggs in slits in the needles on the newly expanding shoots of host trees. Unfertilized females lay viable eggs that produce only male offspring (Houseweart and Kulman, 1976). Initially larvae feed on new foliage, but once that is gone they move back to feed on older foliage. Feeding is wasteful and can result in defoliation of the entire tree. Moderately damaged trees can recover, but growth is reduced for several years. Moderate to severe feeding for a couple of years results in tree deformation (forked leaders, shrubby growth) and can contribute to mortality (Katovich *et al.*, 1995). Once the larvae finish feeding, they drop to the ground, burrow in a few centimetres and spin cocoons.

Background

A large suite of parasitoids attacks *P. alaskensis* (Thompson and Kulman, 1980), but chemical control may be required in many situations. Control has, to date, been restricted to insecticides, but the potential exists to use an entomopathogenic nematode in some circumstances.

Biological Control Agents

Parasitoids

Thompson and Kulman (1980) found that the parasitoid complex in Nova Scotia was similar to that in Maine, and that total apparent parasitism in the Nova Scotia population was 46.7%, somewhat higher than in Maine. All immature *P. alaskensis* life stages are parasitized, but few parasitoids affecting the egg or early larval stages have been identified (Katovich *et al.*, 1995). Houseweart and Kulman (1976) suggested that the most profitable targets for introduced biological control agents would be *P. alaskensis* eggs and early larval instars because of the high survival rate of these stages.

The egg parasitoid *Trichogramma minutum* Westwood was found in New Brunswick in 1995, but in very low numbers (Hartling and O'Shea, Fredericton, 1999, personal communication). Hartling *et al.* (1997) investigated inundative releases of *T. minutum* and *Trichogramma platneri* Nagarkatti to control *P. alaskensis*. The attempt proved unsuccessful because of poor weather and possible problems with synchrony, and they recommended further study.

Pathogens

Although diseases may cause some mortality in *P. alaskensis* populations (e.g. Smirnoff and Juneau, 1973), a survey in New Brunswick in 1995 and 1996 failed to find any pathogens.

Nematodes

Steinernema feltiae (Filipjev) was isolated from infected sawfly cocoons in New Brunswick in 1995, mass-reared, and used in field trials against *P. alaskensis* larvae. When applied by backpack mistblower, nematodes caused 50–75% reduction in larval numbers on treated trees (Thurston, 1997). When applied aerially, they were ineffective due to the need for more moisture than is normally supplied by an aerial spray. If sufficient water to allow for nematode movement could be provided at the time of spray, and the moisture level on the foliage remains high for several hours after spraying, then *S. feltiae* might be a useful management tool in isolated situations. This nematode is available commercially in Canada and could prove to be a viable alternative to chemicals to protect high-value ornamental trees and hedges.

Bacteria

A small-scale trial conducted in 1998 with a commercially available *Bacillus thuringiensis* Berliner serovar *israelensis* (*B.t.i.*) formulation suggested that it might be pathogenic to sawflies (E.G. Kettela, Fredericton, 1999, personal communication). The trial showed significantly higher larval mortality on treated trees compared to untreated trees. Based on these results, a small aerial trial using Vectobac 1200L (Valent Biosciences, Libertyville, Illinois, USA) was conducted in a black spruce plantation in 1999 in New Brunswick. Results suggested some effect (larval survival was 20% lower in the treated than the control block) but not enough to warrant further aerial trials with this product.

Evaluation of Biological Control

Although several insect parasitoids attack *P. alaskensis*, there appears to be room for more. The nematode *S. feltiae* provided good reduction of larval numbers in ground application when sufficient moisture was provided. Although the results of the trials with *B.t.i.* do not support its use against *P. alaskensis*, some effect was observed. This suggests that there may be other *Bacillus* strains with greater activity against foliage-feeding sawflies.

Recommendations

Future work should include:

1. Assessment of the parasitoid complex in a given area to determine if importation of parasitoids from other regions of Canada or elsewhere in the world would be useful;
2. Evaluation of *Trichogramma* spp. for inundative releases;
3. Enhancing survival of *S. feltiae* after application to improve its effectiveness, especially if aerial application is to be pursued;
4. Continuing to search for strains of *B.t.* that are pathogenic to sawflies.

References

de Groot, P. (1995) Yellowheaded spruce sawfly. In: Armstrong, J.A. and Ives, W.G.H. (eds) *Forest Insect Pests in Canada.* Natural Resources Canada, Ottawa, Ontario, pp. 241–244.

Duda, E.J. (1953) The yellow-headed spruce sawfly in Maine. MSc thesis, University of Massachusetts, Amherst, Massachusetts, USA.

Hartling, L., Bourchier, R. and Carter, N. (1997) *Assessment of Two Species of* Trichogramma *on Egg Parasitism of the Yellowheaded Spruce Sawfly* (Pikonema alaskensis (*Roh.*)). Spray Efficacy Research Group, Progress Report, April 1997.

Houseweart, M.W. and Kulman, H.M. (1976) Life tables of the yellowheaded spruce sawfly, *Pikonema alaskensis* (Rohwer) (Hymenoptera: Tenthredinidae) in Minnesota. *Environmental Entomology* 5, 859–867.

Katovich, S.A., McCullough, D.G. and Haack, R.A. (1995) *Yellowheaded Spruce Sawfly – Its Ecology and Management.* Technical Report NC-179, United States Department of Agriculture, Forest Service.

Martineau, R. (1984) *Insects Harmful to Forest Trees.* Forestry Technical Report #32, Environment Canada.

Rose, A.H. and Lindquist, O.H. (1994) *Insects of Eastern Spruces, Fir and Hemlock.* Natural Resources Canada, Ottawa, Ontario.

Smirnoff, W.A. and Juneau, A. (1973) Quinze années de recherches sur les micro-organismes des insectes forestiers de la province de Québec (1957–1972). *Annales de la Societé entomologique du Québec* 18, 147–181.

Thompson, L.C. and Kulman, H.M. (1980) Parasites of the yellowheaded spruce sawfly, *Pikonema alaskensis* (Hymenoptera: Tenthredinidae), in Maine and Nova Scotia. *The Canadian Entomologist* 112, 25–29.

Thurston, G.S. (1997) Control and management of the yellowheaded spruce sawfly in New Brunswick. In: Ostrofsky, W.D. and Krohn, W.B. (eds) *Our Forest's Place in the World: New England and Atlantic Canada's Forests.* Miscellaneous Publication 738, Maine Agricultural and Forestry Experiment Station, pp. 65–74.

45 *Pissodes strobi* (Peck), White Pine Weevil (Coleoptera: Curculionidae)

M.A. Hulme and M. Kenis

Pest Status

The white pine weevil, *Pissodes strobi* (Peck), native to North America, is a major plantation pest across most of Canada and the USA. It was earlier considered to comprise three species (Smith and Sugden, 1969). In eastern North America the main tree species damaged are eastern white pine, *Pinus strobus* L., jack pine, *Pinus banksiana* Lambert, and Norway spruce, *Picea abies* (L.) Karst. Damage to white pine alone can cause up to 25% of the timber value to be lost (Brace, 1972). In western North America the main tree species damaged are Engelmann spruce, *Picea engelmannii* Parry, white spruce, *Picea glauca* (Moench) Voss, and, in western coastal areas, Sitka spruce, *Picea sitchensis* (Bongard) Carrièrre. Damage to the latter species is so severe that Sitka spruce is not recommended for planting in most coastal areas or it should comprise, at most, only 20% of the planted stock.

P. strobi is generally univoltine. Adults overwinter in the duff and emerge in spring where, after feeding to reach sexual maturity, females lay eggs in 1-year-old terminal leaders near the apical bud. Larvae mine downwards, consuming the phloem and killing the leaders (Stevenson, 1967; Silver, 1968). Mature larvae usually excavate a cavity (often called a pupal cell or chip cocoon) in the xylem where they undergo pupation. New adults emerge in late summer and are sexually immature when they enter the duff for the winter. Adults may live for up to 4 years (McMullen and Condrashoff, 1973). In some cases, the living stem below the dead leader may be attacked the following year (Cozens, 1987) or, more usually, new replacement leaders produced from laterals are attacked, resulting in severe loss of height growth and the creation of stem deformities that may prevent any lumber being harvested from the tree.

Background

No satisfactory treatment has yet been found to control *P. strobi*. Several chemical insecticides have been tested by spraying them on tree leaders. The most effective was DDT, where damage was well controlled (Connola, 1961) but the environmental consequences of using DDT are unacceptable. Systemic insecticides have been tested with equivocal results (Fraser and Heppner, 1993). Cultural practices have been attempted to alter the microclimate of the tree leader, making conditions less suitable for tip weevils. For example, spruce or pine are less damaged when grown under a deciduous canopy (Wallace and Sullivan, 1985), but this shaded environment unacceptably reduces the growth of the crop trees. Attempts have been made to selectively destroy weevil broods by cutting and isolating infested leaders before new adult weevils emerge (Rankin and Lewis, 1994; Lavallée *et al.*, 1997). The method proved to be expensive because highly trained labour was required, damage was insufficiently controlled because mature adults outside leaders were not removed, and some weevil broods were inevitably missed through oversight or because leaders were not showing symptoms of attack at the time of leader removal (Rankin and Lewis, 1994).

One method so far little studied is biological control. Many native parasitoids and one dominant native predator are known to attack *P. strobi* broods. The most common parasitoids include *Dolichomitus terebrans nubilipennis* (Ratzeburg), *Eubazus strigitergum* (Cushman) (= *Allodorus crassigaster* (Provancher), see van Achterberg and Kenis, 2000), *Bracon pini* (Muesebeck), *Coeloides pissodis* (Ashmead), *Eurytoma pissodis* Girault, and *Rhopalicus pulchripennis* (Crawford) (Alfaro *et al.*, 1985). All but one of these parasitoids attack late larval stages of *P. strobi*, develop ectoparasitically and are polyphagous. The exception is *E. strigitergum*, an egg-larval endoparasitoid found only on *Pissodes* spp. While this braconid is found throughout the range of Nearctic *Pissodes* spp. other than *P. strobi*, on the latter it is only found throughout the range of Sitka spruce and is only abundant in western coastal regions (Hulme, 1994).

In addition to the parasitoids, *Lonchaea corticis* Taylor is a ubiquitous predator of *P. strobi* broods (Taylor, 1929; Hulme, 1989, 1990); however, its behaviour is atypical of a predator because females lay their eggs exclusively in the vicinity of *P. strobi* eggs, meaning that this fly has a highly refined searching capability and consumes only one prey species. The fly larvae mainly consume prepupal larvae or pupae of *P. strobi* and thus exert a large influence on *P. strobi* broods. Indeed, when the ratio of *L. corticis* larvae to *P. strobi* pupal cells exceeds a threshold value in natural conditions the brood totally fails (Hulme, 1990). However, *L. corticis* females do not always lay enough eggs to attain this threshold value. Hulme (1990) estimated that up to 10,000 additional *L. corticis* eggs ha^{-1}, i.e. perhaps just 20 additional gravid females, would be needed in a typical *P. strobi* infestation to ensure that the threshold was exceeded. Nealis (1998) later confirmed that predation by *L. corticis* larvae was the most important influence on the success of *P. strobi* broods. Biological control of *P. strobi* by augmentation of the numbers of *L. corticis* larvae thus seemed to be a promising approach.

Another approach to biological control of *P. strobi*, through importation of exotic insects, was first suggested by Taylor (1929). Some *Pissodes* spp. with similar ethology occur on other continents and could be used as a source of natural enemies to introduce into North America. A few specimens of a Palaearctic *Coeloides* sp. were released in 1950 in Quebec without subsequent monitoring (McGugan and Coppel, 1962), but no serious attempt to introduce exotic natural enemies was made until the 1980s. Studies of the parasitoid complex of European *Pissodes* spp. were begun to select promising parasitoid species for introduction into Canada. Predators were excluded because studies in Europe indicated that they are of minor importance in the natural control of European *Pissodes* spp. and, in Canada, *L. corticis* was a dominant natural enemy of *P. strobi* whereas several parasitoid guilds were poorly or not represented (Mills and Fisher, 1986).

Biological Control Agents

Predators

Augmentation of *L. corticis* larvae requires the release of adults to lay eggs in leaders infested with eggs of *P. strobi*. The fly overwinters in the tree leaders as mature larvae that readily pupate following an obligatory winter diapause. Adult eclosion follows shortly afterwards. Leaders infested by *P. strobi* were therefore collected in late autumn and held cold throughout the winter to fulfil the fly's diapause requirements. New *L. corticis* adults readily emerged the following spring and when the leaders were warmed about 5000 adults were released into a 5 ha plantation infested with *P. strobi*. The release had no impact on the success of the *P. strobi* brood and no increase in the number of *L. corticis* larvae was found. The female flies were neither gravid nor mated at the time of release, and evidently did not remain in the plantation to reach sexual maturity and lay eggs. Attempts were therefore made to rear gravid mated females in the laboratory before they were released outdoors. The female's metabolism proved to be anautogenous, and her maturation feeding required a diet containing protein in the form of peptides and amino acids typically found in autolysed yeast. Sources with higher molecular weight protein were unsuitable. With a suitable diet, fully gravid females were obtained in about 1 week, but dissection of the spermathecae showed that the females had not mated despite being caged with males. Adult eclosion of *L. corticis* is protandrous and copulation in such insects is often between a male that is several days old and a female that is freshly emerged. When these conditions were replicated in cage rearings mating still did not occur. Mating in many Tephritoidea requires elaborate mating rituals, including lecking and swarming. Several lonchaeids have been observed to swarm outdoors, possibly in behaviour associated with mating (McAlpine and Munroe, 1968), but attempts to mate *L. corticis* in the laboratory in various conditions that encouraged swarming were not successful. The difficulties of observing or obtaining mating of other Lonchaeidae are already recorded (e.g. Katsoyannos, 1983). It appears from our laboratory observations, and by drawing parallels with the behaviour of other members of the Tephritoidea (e.g. Drew, 1987), that *L. corticis* females disperse widely when they first emerge and complete their maturation feeding, perhaps partly on leaf bacteria, in areas that may be remote from *P. strobi* infestations (Drew *et al.*, 1983). At some point mating occurs and the gravid mated female then seeks out *P. strobi* infestations. This scenario would explain our repeated observation that *L. corticis* is rapidly able to find new infestations of *P. strobi*, no matter how remote they are from established infestations.

Parasitoids

To evaluate prospects for releasing exotic parasitoids, Mills and Fisher (1986) and Kenis and Mills (1994) surveyed in Europe to determine the parasitoid complex of the six most abundant European *Pissodes* spp. feeding on *Pinus*, *Abies* and *Picea* spp. Following these surveys, further work was focused on the egg–larval endoparasitoids *Eubazus* spp. and the larval ectoparasitoid *Coeloides sordidator* (Ratzeburg), because these species were the most common parasitoids of European *Pissodes* spp. and showed many charactersitics of successful biological control agents, e.g. broad geograghic range, a good capacity to locate small isolated host populations and a relative preference for *Pissodes* spp. (Kenis and Mills, 1994). Kenis (1996, 1997) studied the biology of *C. sordidator*, including developmental biology, competitive interactions, sex allocation and rearing techniques. He concluded that *C. sordidator* was less promising for biological control of *P. strobi* than the *Eubazus* spp., mainly because it is too habitat specific, less host specific than *Eubazus* spp., and it occupies an ecological niche (that of late-larval ectoparasitoids) that is already well represented by several native parasitoids of *P. strobi*.

Candidate exotic *Eubazus* spp. were first selected for testing based on the ecology and phenology of the host. No *Pissodes* spp. in Europe occupy the same ecological niche as *P. strobi*, and only the Palearctic *Pissodes validirostris* (Sahlberg) has a similar phenology to *P. strobi*. *Eubazus robustus* (Ratzeburg), reared from *P. validirostris*, was thus tested in British Columbia for its acceptance of *P. strobi*. About 850 male and female parasitoids were assessed in cages (Table 45.1).

Concurrent studies in Europe on several aspects of the biology, ecology and morphology of *Eubazus* spp. from a number of *Pissodes* spp. showed that several sibling species exist with different habitat preferences (Kenis *et al.*, 1996; Kenis and Mills, 1998; Achterberg and Kenis, 2000). Investigations on intra- and interspecific variation in developmental responses, and on host–parasitoid synchronization, were particularly important because all but one European *Pissodes* spp. have a different phenology to that of *P. strobi*. *Eubazus* spp. and their biotypes were reared in the laboratory and in field conditions on several Palaearctic *Pissodes* spp. and on *P. strobi*. Most species and biotypes developed without diapause, and adult parasitoids would thus emerge before or at the same time as *P. strobi*, i.e. when no host eggs are available.

Table 45.1. Number of *Eubazus* spp. shipped to Canada for cage testing against *Pissodes strobi*.

Year	Number of adults shipped	Species	*Pissodes* host
1986	183 males and females	*E. robustus* (Ratzeburg)	*P. validirostris* (Sahlberg)
1987	178 males and females	*E. robustus*	*P. validirostris*
1988	250 males and females	*E. robustus*	*P. validirostris*
1989	166 males and females	*E. robustus*	*P. validirostris*
1991	92 males and females	*E. robustus*	*P. validirostris*
1995	190 females	*E. semirugosus* (Nees)	*P. pini* (L.)
1996	300 females	*E. semirugosus*	*P. pini*
1997	310 females and 62 males	*E. semirugosus*	*P. pini*
1998	532 females and 71 males	*E. semirugosus*	*P. pini*
1999	246 females and 44 males	*E. semirugosus*	*P. pini*

However, a biotype of *E. semirugosus* (Nees), found on *Pissodes* spp. attacking *Pinus* spp. at high altitudes in the Alps, differed from other *Eubazus* spp. by having an obligatory diapause in normally non-diapausing host larvae (Kenis *et al.*, 1996), which allowed a significant part of the population to emerge the following spring, i.e. during the oviposition period of the target host. This biotype was chosen as the best candidate for introduction into North America. About 1500 females of the high-altitude biotype of *E. semirugosus*, collected from *Pissodes pini* (L.) in Europe, were tested in cages in British Columbia (Table 45.1). All female parasitoids were placed with males before or during shipment to ensure mating. Testing with the Nearctic *Eubazus crassigaster* (Provancher) indicated that maximum adult longevity in our cage conditions was near 1 month, and because the mean age of the Palaearctic adults was about 2 weeks when they arrived in Victoria, less than half of their adult life remained for our cage testing. Three test conditions were used. Cut leaders were placed in cages in the laboratory with the parasitoids. Similar cage tests were run outdoors. A final level of testing employed parasitoids enclosed in sleeve cages on plantation trees.

Initial laboratory tests with *P. strobi* employed *E. robustus*, which showed good acceptance of its new Nearctic host: about 80% of *P. strobi* in pupal cells in cut leaders produced parasitoids. However, the adult parasitoids emerged at about the same time as the adult weevils, and many months before new *P. strobi* eggs would be available, i.e. synchrony with the appropriate host stage needed for parasitism was so poor that the parasitoid generation could not be continued on this host. Dissection of leaders showed that no parasitized larvae remained to overwinter. This same parasitoid phenology was seen in repeated assays in all our testing conditions, and *E. robustus* was thus considered unsuitable for successful biological control. The focus of testing was therefore moved to biotypes of other *Eubazus* spp.

In outdoor cage tests, *E. semirugosus* from high-altitude *P. pini* parasitized about 80% of *P. strobi* in pupal cells. Many adult parasitoids emerged in late summer at about the same time as adult weevils emerged, and many months before new *P. strobi* eggs would be available. Dissection of leaders determined that parasitized *P. strobi* larvae continued to occupy up to half of the pupal cells at the onset of winter, although variability between leaders was high. Dissection of leaders the following spring showed that the proportion of pupal cells containing live parasitized *P. strobi* larvae was now below one-quarter: many cocoons contained shrivelled remains of *P. strobi* larvae where parasitism could not be determined. It appeared that many larvae desiccated during winter. Subsequent tests in sleeve cages on plantation trees with *E. semirugosus* from high-altitude *P. pini* gave similar results. The plantation was purposely established in a drier area, remote from other spruce trees, and thus remote from possible *P. strobi* infestations with their concomitant guild of parasitoids. While this site provided isolation, it still allowed *L. corticis* predation, and may not have provided sufficiently moist conditions needed for winter brood survival. More testing is required in more typical spruce habitats now that basic acceptance by the Palaearctic parasitoid of its new Nearctic host has been demonstrated.

Evaluation of Biological Control

Work to date shows promise for biological control of *P. strobi* using two different approaches: through augmentation of native insects or through introduction of exotic insects. Augmentation of *L. corticis* should successfully reduce weevil broods, based upon our observations of brood reduction by natural predation by *L. corticis*. Laboratory rearing is required for augmentation with mated gravid females, and we have already defined the conditions needed to rear gravid females. The remaining challenge is to find the mating trigger for these flies in captivity. Given the known specificity of this fly for just one prey, we anticipate that releases of mated gravid females will have a large impact on *P. strobi* broods. Introduction of Palaearctic *E. semirugosus* also shows promise. We have identified an ecotype that should thrive exclusively on *P. strobi* broods in natural conditions because its phenology is perfectly synchronized with that of *P. strobi* to ensure that the appropriate stage of the weevil is always present for exploitation by the parasitoid. The remaining challenge is to release the adult parasitoid when *P. strobi* is naturally laying eggs, and monitor the parasitoid's ability to replicate on *P. strobi* alone in a variety of natural conditions.

Recommendations

Further work should include:

1. Determining the conditions needed for the mating of *L. corticis* when reared in captivity;
2. Finding the natural mating conditions for *L. corticis* by observing the behaviour of newly emerging adults in the field;
3. Releasing mated gravid females of *L. corticis* into an infestation of *P. strobi*, monitoring the egg-laying behaviour of the released females, and monitoring the impact on *P. strobi* brood;
4. Determining how some adult *E. semirugosus* isolated from high-altitude *P. pini* delay their emergence, even when the parasitoid is transferred to other *Pissodes* spp. as hosts;
5. Making field releases in western Canada of *E. semirugosus* from high-altitude *P. pini*;
6. Monitoring for establishment of the parasitoid and its impact on *P. strobi* broods;
7. If western introduction is successful, making further introductions in eastern Canada, although it is possible that *E. semirugosus* may not survive extremely cold winters on *P. strobi* (Hulme *et al.*, 1986).

References

Achterberg, C. van and Kenis, M. (1999) The Holarctic species of the subgenus *Allodorus* Foerster s.s. of the genus *Eubazus* Nees (Hymenoptera: Braconidae). *Zoologische Mededelingen, Leiden* 73, 427–455.

Alfaro, R.I. (1985) Insects associated with the Sitka spruce weevil, *Pissodes strobi* (Coleoptera: Curculionidae), in Sitka spruce, *Picea sitchensis*, in British Columbia. *Entomophoga* 30, 415–418.

Brace, R.G. (1972) Weevil control could raise the value of white pine by 25%. *Canadian Forest Industries* 92, 42–45.

Connola, D.P. (1961) Portable mistblower spray tests against white pine weevil in New York. *Journal of Forestry* 59, 764–765.

Cozens, R.D. (1987) Second brood of *Pissodes strobi* (Coleoptera: Curculionidae) in previously attacked leaders of interior spruce. *Journal of the Entomological Society of British Columbia* 84, 46–49.

Drew, R.A.I. (1987) Behavioural strategies of fruit flies of the genus *Dacus* (Diptera: Tephritidae) significant in mating and host–plant relationships. *Bulletin of Entomological Research* 77, 73–81.

Drew, R.A.I., Courtice, A.C. and Teakle, D.S. (1983) Bacteria as a natural source of food for adult fruit flies (Diptera: Tephritidae). *Oecologia* 60, 279–284.

Fraser, R.G. and Heppner, D.G. (1993) Control of white pine weevil *Pissodes strobi* on Sitka spruce using implants containing insecticides. *Forestry Chronicle* 69, 600–603.

Hulme, M.A. (1989) Laboratory assessment of predation by *Lonchaea corticis* (Diptera: Lonchaeidae) on *Pissodes strobi* (Coleoptera: Curculionidae). *Environmental Entomology* 18, 1011–1014.

Hulme, M.A. (1990) Field assessment of predation by *Lonchaea corticis* (Diptera: Lonchaeidae) on *Pissodes strobi* in *Picea sitchensis*. *Environmental Entomology* 19, 54–58.

Hulme, M.A. (1994) The potential of *Allodorus crassigaster* for the biological control of *Pissodes strobi*. In: Alfaro, R.I., Kiss, G. and Fraser, R.G. (eds) *The White Pine Weevil: Biology, Damage and Management.* Symposium Proceedings, Canadian Forest Service and British Columbia Ministry of Forests, pp. 294–300.

Hulme, M.A., Dawson, A.F. and Harris, J. (1986) Exploiting cold hardiness to separate *Pissodes strobi* (Peck) (Coleoptera: Curculionidae) from associated insects in leaders of *Picea sitchensis* (Bong.) Carr. *The Canadian Entomologist* 118, 1115–1122.

Katsoyannos, B.I. (1983) Field observations on the biology and behavior of the black fig fly *Silba adipata* McAlpine (Diptera: Lonchaeidae), and trapping experiments. *Zeitschrift für Angewandte Entomologie* 95, 471–476.

Kenis, M. (1996) Factors affecting sex ratio in rearing of *Coeloides sordidator* (Hymenoptera: Braconidae). *Entomophaga* 41, 217–224.

Kenis, M. (1997) Biology of *Coeloides sordidator* (Hymenoptera: Braconidae), a possible candidate for introduction against *Pissodes strobi* (Coleoptera: Curculionidae) in North America. *Biocontrol Science and Technology* 7, 157–164.

Kenis, M. and Mills, N.J. (1994) Parasitoids of European species of the genus *Pissodes* (Coleoptera: Curculionidae) and their potential for biological control of *Pissodes strobi* (Peck) in Canada. *Biological Control* 4, 14–21.

Kenis, M. and Mills, N.J. (1998) Evidence for the occurrence of sibling species in *Eubazus* spp. (Hymenoptera: Braconidae), parasitoids of *Pissodes* weevils (Coleoptera: Curculionidae). *Bulletin of Entomological Research* 88, 149–163.

Kenis, M., Hulme, M.A. and Mills, N.J. (1996) Comparative developmental biology of populations of three European and one North American *Eubazus* spp. (Hymenoptera: Braconidae), parasitoids of *Pissodes* spp. weevils (Coleoptera: Curculionidae). *Bulletin of Entomological Research* 86, 143–153.

Lavallée, R., Bonneau, G. and Coulombe, C. (1997) *Mechanical and Biological Control of the White Pine Weevil.* Information Leaflet LFC 28, Laurentian Forestry Centre, Ste-Foy, Quebec.

McAlpine, J.F. and Munroe, D.D. (1968) Swarming of lonchaeid flies and other insects, with descriptions of four new species of Lonchaeidae (Diptera). *The Canadian Entomologist* 100, 1154–1178.

McGugan, B.M. and Coppel H.C. (1962) Biological control of forest insects 1910–1958. In: McLeod, J.H., McGugan, B.M. and Coppel, H.C. (eds) *A Review of the Biological Control Attempts Against Insects and Weeds in Canada.* Technical Communication No. 2, Commonwealth Institute of Biological Control, Trinidad. Commonwealth Agricultural Bureaux, Farnham Royal, UK, pp. 35–127.

McMullen, L.H. and Condrashoff, S.F. (1973) Notes on the dispersal, longevity and overwintering of adult *Pissodes strobi* (Peck) (Coleoptera: Curculionidae) on Vancouver Island. *Journal of the Entomological Society of British Columbia* 70, 22–26.

Mills, N.J. and Fischer, P. (1986) The entomophage complex of *Pissodes* weevils, with emphasis on the value of *P. validirostris* as a source of parasitoids for use in biological control. In: Roques, A. (ed.) *Proceedings of the second International Conference of the IUFRO Cone and Seed Insects Working Party, Briançon, Sept. 1986.* A. Olivet (France), INRA, pp. 297–305.

Nealis, V.G. (1998) Population dynamics of the white pine weevil, *Pissodes strobi*, infesting jack pine, *Pinus banksiana*, in Ontario, Canada. *Ecological Entomology* 23, 305–313.

Rankin, L.J. and Lewis, K. (1994) Effectiveness of leader clipping for control of the white pine weevil, *Pissodes strobi*, in the Cariboo forest region of British Columbia. In: Alfaro, R.I., Kiss, G. and Fraser, R.J. (eds) *The White Pine Weevil: Biology, Damage and Management.* Forest Research Development Agreement Report 226, British Columbia Ministry of Forests, Victoria, British Columbia, pp. 262–269.

Silver, G.T. (1968) Studies on the Sitka spruce weevil, *Pissodes sitchensis*, in British Columbia. *The Canadian Entomologist* 100, 93–110.

Smith, S.G. and Sugden, B.A. (1969) Host trees and breeding sites of native North American *Pissodes* bark weevils, with a note on synonymy. *Annals of the Entomological Society of America* 62, 146–148.

Stevenson, R.E. (1967) Notes on the biology of the Engelmann spruce weevil, *Pissodes engelmannii* (Curculionidae: Coleoptera) and its parasites and predators. *The Canadian Entomologist* 99, 201–213.

Taylor, R.L. (1929) The biology of the white pine weevil, *Pissodes strobi* (Peck), and a study of its insect parasites from an economic viewpoint. *Entomologica Americana* 9, 166–246; 10, 1–86.

Wallace, D.R. and Sullivan, C.R. (1985) The white pine weevil *Pissodes strobi* (Coleoptera: Curculionidae): a review emphasizing behaviour and development in relation to physical factors. *Proceedings of the Entomological Society of Ontario* 116 (Supplement), 39–62.

46 *Pristiphora geniculata* (Hartig), Mountain Ash Sawfly (Hymenoptera: Tenthredinidae)

R.J. West, P.L. Dixon, F.W. Quednau and K.P. Lim

Pest Status

The mountain ash sawfly, *Pristiphora geniculata* (Hartig), was accidentally introduced from Europe to Massachusetts in 1926 and was present in Canada by 1934 (Quednau, 1990). It is found throughout eastern Canada from south-western Ontario to eastern Newfoundland and, as a larva, defoliates species of mountain ash, *Sorbus americana* Marshall, a valued ornamental in urban settings. Trees can occasionally be completely defoliated, but seldom die as a result (Quednau, 1984).

P. geniculata normally has one generation a year, but may have a partial second generation (Quednau, 1984). Overwintering occurs in the cocoon stage, adults mate in late spring, and larvae feed throughout the summer (see also Forbes and Daviault, 1964).

Background

Organophospate insecticides easily kill *P. geniculata*; however, the desire to reduce pesticide use in urban areas led to research to find biological alternatives. A host-specific, solitary endoparasitoid, *Olesicampe geniculatae* Quednau and Lim, was successfully established in Quebec following introductions from its native Europe from 1976 to 1978 (Quednau and Lim, 1983; Quednau, 1984, 1990). Establishment of *O. geniculatae* in Newfoundland was expected to significantly reduce damage to mountain ash and thus reduce insecticide use in urban areas.

Biological Control Agents

Parasitoids

O. geniculatae generally attacks first- and second-instar larvae of its univoltine host, which is killed only just prior to parasitoid pupation in spring. The parasitoid overwinters inside the hibernating host cocoon in the soil. The availability of *O. geniculatae* from Quebec led to its introduction into Newfoundland from 1981 to 1984.

Releases and Recoveries

O. geniculatae, field-collected near Quebec City, was reared as described by Quednau (1990). Mated females were shipped to Newfoundland for release in a field cage at Oxen Pond Botanic Park, St John's, from

1981 to 1984. Mountain ash and *P. geniculata* were abundant at the release site and surrounding area. A total of 259 mated female *O. geniculatae* were released into the screened 2.4 × 2.4 × 2.4 m cage, constructed around a mountain ash tree. High numbers of *P. geniculata* larvae were released in the cage on mountain ash twigs prior to and following introduction of mated female parasitoids (West *et al.*, 1994).

An open release of 171 mated females of *O. geniculatae* shipped directly from Quebec was made near the cage from 27 July to 3 August 1984 (West *et al.*, 1994). The second open release was made on 22 September 1986, at the Canadian Forest Service Field Station, Pasadena, western Newfoundland. A total of 368, presumably parasitized, *P. geniculata* cocoons, collected from the cage as larvae on 1–15 August 1986, were buried 15 cm deep in a flower bed (West *et al.*, 1994).

Evaluation of Biological Control

Establishment of *O. geniculatae* was determined by dissections of hosts collected from mountain ash trees in 1 km^2 sample plots located within a 10 km^2 area with the release site at its centre (West *et al.*, 1994). An additional 36 sites in 1989 and 15 sites in 1990 were surveyed from late July to early August to determine spread of *O. geniculatae* across Newfoundland (West *et al.*, 1994).

O. geniculatae established successfully in the St John's area (West *et al.*, 1994). In 1985, 51% of 75 mountain ash sawfly collected just outside the cage and 42% of 75 larvae collected inside the cage were parasitized; however, no parasitoids were recovered from *P. geniculata* sampled from nearby plots. In 1986, the parasitism rate was 85% at the release site and 7% at one additional plot, 1 km away. In 1987, *P. geniculata* was not recovered from the release site but parasitism at the neighbouring plot rose to 50%. In 1988, *P. geniculata* was recovered from 78 of 92 plots surveyed. Parasitoids were present in 91% of the plots where *P. geniculata* was present and within-plot parasitism ranged from 2 to 97%. In 1989, the number of plots with *P. geniculata* continued to decline; parasitoids were present in all plots where it was present and within-plot parasitism ranged from 3 to 100%. In 1990, *P. geniculata* was recovered from only 6 of 82 plots surveyed and parasitism in these plots ranged from 37 to 77%.

In 1989–1990, *O. geniculatae* was recovered from 11 of 28 sites on the Avalon Peninsula, 2 of 5 sites on the Burin Peninsula, and 9 of 14 sites in western Newfoundland, but no parasitoids were recovered from two sites sampled in central Newfoundland (West *et al.*, 1994). Parasitism ranged from 6 to 85%. The establishment and spread of *O. geniculatae* was successful and has resulted in reduced *P. geniculata* populations.

Recommendations

Further work should include:

1. Surveying *P. geniculata* populations to determine if *O. geniculatae* has spread to central Newfoundland and determining its impact on *P. geniculata* populations throughout its range.

References

Forbes, R.S. and Daviault, L. (1964) The biology of the mountain-ash sawfly, *Pristiphora geniculata* (Htg.) (Hymenoptera: Tenthredinidae), in Eastern Canada. *The Canadian Entomologist* 96, 1117–1133.

Quednau, F.W. (1984) *Pristiphora geniculata* (Htg.), mountain ash sawfly (Hymenoptera: Tenthredinidae). In: Kelleher, J.S. and Hulme, M.A. (eds) *Biological Control Programmes Against Insects and Weeds in Canada 1969–1980.* Commonwealth Agriculture Bureaux, Slough, UK, pp. 381–385.

Quednau, F.W. (1990) Introduction, permanent establishment, and dispersal in eastern Canada of *Olesicampe geniculatae* Quednau and Lim (Hymenoptera: Ichneumonidae), an important biological control agent of the mountain ash sawfly, *Pristiphora geniculata* (Hartig) (Hymenoptera: Tenthredinidae). *The Canadian Entomologist* 122, 921–934.

Quednau, F.W. and Lim, K.P. (1983) *Olesicampe geniculatae*, a new palearctic ichneumonid parasite of *Pristiphora geniculata* (Hymenoptera: Tenthredinidae). *The Canadian Entomologist* 115, 109–113.

West, R.J., Dixon, P.L., Quednau, F.W., Lim, K.P. and Hiscock, K. (1994) Establishment of *Olesicampe geniculatae* (Hymenoptera: Ichneumonidae) to control the mountain ash sawfly, *Pristiphora geniculata* (Hymenoptera: Tenthredinidae), in Newfoundland. *The Canadian Entomologist* 126, 7–11.

47 *Prosimulium* and *Simulium* spp., Black Flies (Diptera: Simuliidae)

P.G. Mason, M. Boisvert, J. Boisvert and M.H. Colbo

Pest Status

Black flies, *Prosimulium* and *Simulium* spp., are important native pests of humans and livestock. Although not known as major disease vectors in Canada, black flies aggressively bite, causing itchy, painful wounds, and often elicit allergic reactions. Relatively few of the more than 100 Canadian species feed on mammals; most require blood from birds or no blood meal at all (Wood, 1985). The *Simulium venustum* Say and *S. verecundum* Stone and Jamnback species complexes, *S. arcticum* Malloch, *S. luggeri* Nicholson and Mickel, *S. decorum* Walker, *Prosimulium mixtum* Syme and Davies and *P. hirtipes* (Fries) are among the most bothersome to humans and livestock. Most pest species occur only from early May to mid-June. In addition to their annoyance, black flies cause economic losses through reduced beef and milk production, and losses in domestic birds from *Leucocytozoon* spp. infections (Cupp, 1987). In the Athabasca River region, Alberta, and the Saskatchewan River region around Prince Albert, Saskatchewan, high numbers of *S. arcticum* and *S. luggeri* limit cattle production and *S. arcticum* can kill cattle by anaphylaxis (Mason and Shemanchuk, 1990). Fredeen (1985) determined that summer-long outbreaks of *S. luggeri* in 1978 caused Can$1.4 million in losses to the livestock industry.

Wood (1985), Crosskey (1990) and Mason and Shemanchuk (1990) summarized black fly life cycles. Females of pest species require blood to produce at least one batch of viable eggs. Eggs, laid on various substrates (e.g. floating vegetation, small branches and rocks near or in running water) hatch within a few days or months, depending on species. Young larvae drift downstream until they find a suitable substrate to which they attach. The larvae filter food particles from the water. At maturity they pupate on the substrate and adults emerge 2–3 weeks later. Adults feed on nectar and honeydew (Burgin and Hunter, 1997) and females may then seek a blood meal. Black flies overwinter either as eggs in/on the substrate or as mature larvae in the water. Most species have only one generation per year, although *S. luggeri* can have up to five generations. Adult black flies disperse widely from larval habitats

(Mason and Kusters, 1990) and are influenced by meteorological factors, e.g. wind (McCreadie *et al.*, 1986; Shipp *et al.*, 1987, 1988; Fredeen and Mason, 1991).

Background

Many municipalities throughout Canada yearly fund black fly abatement programmes. Because the larvae are restricted to flowing water, control is most effective against this stage, for which monitoring and control programmes have been developed (e.g. Mason and Kusters, 1991). Although organochlorine insecticides, e.g. DDT and methoxychlor, were effective chemical agents for black fly control in cold-water streams, negative environmental impacts led to an urgent need to find alternative controls. This stimulated research on the use of natural enemies for biological control.

Several world reviews on the natural enemies of both immature and adult black flies exist, e.g. Davies (1981), Poinar (1981), Weiser and Undeen (1981), Molloy (1987) and Crosskey (1990). In Canada, natural enemies associated with black flies continue to be reported (e.g. Adler, 1986; Charpentier *et al.*, 1986; Erlandson and Mason, 1989; Adler and Mason, 1997). Although many species have been reported as predators of black flies, it is not possible to determine their impact because few or no quantitative data exist (Crosskey, 1990). Of the predator–black fly associations tabulated by Davies (1981) only trout, *Salmo* spp. and *Salvelinus fontinalis* (Mitchill), caddisflies, *Hydropsyche* spp. and *Cheumatopsyche* spp., and *Hydra* spp. are thought to be important in Canada.

Pathogens are prevalent in immature black flies (Laird *et al.*, 1980), although their significance is poorly understood. The microsporidians *Amblyospora bracteata* (Strickland), *Amblyospora fibrata* (Strickland), *Amblyospora varians* (Léger), *Caudospora pennsylvania* Beaudoin and Wills, *Caudospora polymorpha* (Strickland), *Caudospora simulii* Weiser, *Janacekia debaisieuxi* (Jírovec), *Nosema stricklandi* Jírovec and *Polydipyremia multispora* (Strickland) are specific to Simuliidae but their prevalence in black fly larvae is usually less than 1%, rarely more than 15%, and great variation exists within and between locations (Vávra and Undeen, 1981; Weiser and Undeen, 1981; Crosskey, 1990). Attempts to infect healthy larvae with spores from parasitized larvae have been unsuccessful, yet laboratory-reared larvae sometimes show microsporidial infections (Crosskey, 1990). Both iridescent viruses (IV) and cytoplasmic polyhedrosis viruses (CPV) are associated with black fly larvae, but they occur in less than 5% of populations (Weiser and Undeen, 1981; Crosskey, 1990). The most common and widely distributed pathogenic fungus known to infect black flies is *Coelomycidium simulii* Debaisieux (Weiser and Undeen, 1981). It has been associated with *P. mixtum*, *Stegopterna mutata* (Malloch), *Simulium aureum* Fries, *Simulium tuberosum* Lundström, the *Simulium vittatum* Zetterstedt complex, and the *S. venustum* complex in North America, and the pathogenesis described. Attempts to infect healthy *Prosimulium fuscum* Syme and Davies larvae with cysts of *Pythiopsis cymosa* de Bary obtained experimentally were unsuccessful (Crosskey, 1990). Several species of Trichomycetes are commonly present in the gut of black fly larvae and until recently were thought not to be pathogenic (Crosskey, 1990; Labeyrie *et al.*, 1996; Lichtwardt and Misra, 2000).

Mermithid nematodes are the most common parasites that attack black flies (Crosskey, 1990) and they kill larvae or sterilize adults (Poinar, 1981). *Gastromermis viridis* Welch, *Isomermis wisconsinensis* Welch and *Mesomermis flumenalis* Welch occur in the Nearctic region.

Adult black flies are attacked by a mite, *Sperchon ?jasperensis*, a protozoan *Tetrahymena rotunda* Lynn, Molloy and LeBrun, and the fungi *Entomophthora culicis*, *Erynia* spp. and *Harpella* sp. (Crosskey, 1990; Lichtwardt and Misra, 2000). The latter two species, which attack the ovaries and sterilize females, occurred in up to 40% of *P. mixtum* and *S. mutata* populations in Newfoundland (Yeboah *et al.*, 1984). Colbo (1982) noted that the size of Trichomycetes-infected adult flies was the same as that of uninfected flies. In Quebec, Nadeau *et al.* (1994, 1995, 1996a, b) studied the biology and pathogenesis of *Erynia conica* (Nowakowski) Remaudière and Hennebert and *Erynia curvispora* (Nowakowski) Nowakowski.

Biological Control Agents

Pathogens

Fungi

Tolypocladium cylindrosporum Gams, pathogenic to mosquitoes (Goettel, 1987), Psychodidae, Chaoboridae and Ceratopogonidae (Lam *et al.*, 1988), was evaluated against *S. vittatum*. Although pathogenic, *T. cylindrosporum* was less so to *S. vittatum* than to *Aedes triseriatus* (Say) (Nadeau and Boisvert, 1994).

Nematodes

Poinar (1981) summarized the biology of *G. viridis*, *I. wisconsinensis* and *M. flumenalis*. Black fly larvae are attacked and the nematodes usually emerge from this stage but will often remain in adults. Although commonly observed in various *Simulium* spp., nematode occurrence in a particular population varies greatly between and within streams. In Newfoundland, Colbo and Porter (1980) found mermithid prevalence ranging from 8 to 68% in an *S. venustum* complex population in a short stretch of stream in May–June. Also, Colbo (1982) found a population of mermithids in an *S. venustum/verecundum* complex where all mermithids were passed to the adult flies emerging at the oviposition site. All infected adults were morphologically female. He also noted that the greater the number of worms per fly, the smaller the adult fly.

Finney and Mokry (1980) and Finney (1981a, b) investigated *M. flumenalis*, *Isomermis* sp., *Romanomermis culicivorax* Ross and Smith (from mosquitoes) and determined that mass production and application were possible. However, important obstacles included the need to understand fully the host range and host–parasite dynamics, and to develop cost-effective *in vitro* production. In Newfoundland, Colbo (1990) studied the persistence of mermithid parasitism in *P. mixtum* and *S. venustum/verecundum* over a 10-year period, concluding that mermithids are host specific and that infection levels, whether high or low, are stable over several years. Further, survival of the free-living stages in the stream is the key to successful host infection; yet the biology of these stages is unknown.

Bacteria

In Canada, research on the use of *Bacillus thuringiensis* Berliner serovar *israelensis* (*B.t.i.*) against black flies was started soon after its discovery. Undeen and Nagel (1978) showed that *B.t.i.* as an unformulated bacterial powder had good activity against *S. verecundum*. Undeen and Colbo (1980), using an unformulated *B.t.i.* suspension, showed that stream discharge was important to determine the 'carry', i.e. the distance from an application point that a high level (usually >80%) of larval mortality is still present. Colbo and Undeen (1980) demonstrated that no significant decrease in numbers of major groups of non-target insects (other Diptera, Trichoptera, Coleoptera, Plecoptera, Ephemeroptera and Odonata) living on rocks occurred 3–7 days after treatment that resulted in more than 90% black fly larval mortality. This is one of the rare publications where non-target effects of unformulated *B.t.i.* were specifically tested.

Finney and Harding (1982) studied the efficacy of *B. thuringiensis* serovar *darmstadiensis* against *S. verecundum* larvae and concluded that it was 20 times less active than *B.t.i.* against black fly larvae.

In a pilot black fly control programme in

Newfoundland using the commercial *B.t.i.* formulation Teknar® WDC, Colbo and O'Brien (1984) found that *P. mixtum*, *S. vernum* Macquart, *S. tuberosum*, *S. venustum/verecundum*, *S. vittatum* and *S. mutata* were affected at a dose of 10 ppm for 1 min and, at temperatures near 0°C, longer application time was needed to obtain good control. The unpredictable relief pattern characteristic of northern streams in Newfoundland resulted in a highly variable downstream carry of the formulation. Colbo (1985, 1987) reported on the first successful large operational control programme in Canada using *B.t.i.*, carried out in Labrador City and Wabush, Labrador, in 1983 and 1984. Finney and Harding (as cited in Colbo and O'Brien, 1984) also reported that longer application times were needed in cold-temperature streams. Lacoursière and Charpentier (1988) also noted the important effect of temperature on the efficacy of *B.t.i.* against black flies in laboratory experiments. Nixon (1988) attributed reduced mortality in *S. vittatum* larvae fed *B.t.i.* at low temperatures to reduced feeding rates. Species with wintering larvae, e.g. *P. mixtum/fuscum*, were much less sensitive to *B.t.i.* than warmer-water species, e.g. *S. decorum*, when tested at the same temperature. Black fly mortality varied according to dose (concentration × duration of exposure) and temperature. Mortality is also reduced in response to increased concentrations of suspended solids (Nixon, 1988).

Although many *B.t.i.* formulations have been studied, e.g. Burton (1984) and Canada Biting Fly Centre studies, and several have been registered in Canada since 1980, Back *et al.* (1985) first tested a *B.t.i.* formulation (Teknar® WDC) at a high dose (86.6 mg l^{-1} for 1 min) in a stream with a discharge of 114 m^3 min^{-1}. They determined the effect of a single treatment on *Simulium* spp. and on non-target species, using drift nets, counting plates and artificial turf substrates over 1 km from the application point. Although the high-dose treatment created a substantial amount of larval black fly drift, no significant increase in drift was observed for Ephemeroptera, Plecoptera, Trichoptera and Chironomidae. However, larvae of *Blepharicera* sp. were severely affected by the treatment (a 50-fold increase in drift occurred) and on artificial substrates (not drift) some chironomids, e.g. *Eukieferiella* sp. and *Polypedilum* sp., were affected after the treatment. Blepharicidae and Chironomidae affected by the treatment were classified as periphyton-grazing species, indicating that possibly *B.t.i.* crystals could adsorb on to algae covering rocks (periphyton) and affect grazing Diptera. At that time it was assumed that only filtering Diptera could be affected by *B.t.i.* They used the currently recommended (label) dose for black fly treatment.

McCracken and Matthews (1997) reported that *B.t.i.* treatments at a high dose (25 mg l^{-1} for 1 min) in two different streams caused significant drift in both black fly and chironomid larvae, but were not correlated with changes in other non-target species (32 families in 12 orders). Many field trials indicated that, compared to chemical insecticides, *B.t.i.* formulations had a short carry. Many researchers had hypothesized that the loss of activity over distance from the application point was the result of *B.t.i.* toxic crystals sedimenting in small pools along a stream or river or adsorption on to moss and grass present in the treated streams. Tousignant *et al.* (1993) conducted a series of experiments in two streams in which hyporheic probes were driven at different depths into the streambed and at various distances from a *B.t.i.* application point. Water samples collected from the probes, vegetation, sediment and periphyton were analysed for *B.t.i.* toxic activity. *B.t.i.* activity was detected down to 65 cm under the streambed and up to 800 m from the application point, indicating that the hyporheic zone under the streambed and, to a lesser extent, adsorption to algae covering rocks, removed *B.t.i.* crystals from the open-channel water of a stream. Almost no activity was found in sediments from pools or from vegetation. In small streams, a major portion of the *B.t.i.* crystals flowed under the streambed, where sedimentation and adsorption on to the periphyton removed them quite readily because of the large surface areas present. In larger rivers, this hyporheic zone is much less important than in streams (Lacoursière and Vought, Lund, 1993, personal communication), which explained why the carry of *B.t.i.* is longer in rivers with greater discharge.

Most commercial formulations have remained relatively unchanged over the past 15 years, mainly because it is extremely difficult to compare a new or improved formulation to existing ones. Formulations are usually tested under very different field conditions (i.e. on different rivers, discharge, temperature, river profile, larval species, etc.) because once black fly larvae are eliminated after a single treatment, researchers would have to use another river. Therefore, some have claimed that possibly better formulations were overlooked because they were tested and eventually compared to other formulations but under very different field conditions.

Boisvert *et al.* (2001) tested *B.t.i.* formulations under the same biotic and abiotic conditions using a series of gutters along two streams. Two commercial formulations had the same toxic activity against black fly larvae when tested in the same streams in a cold temperature (about 15°C). But at a higher temperature (about 20°C), one of the formulations had a much longer carry. Both formulations kept at room temperature maintained their field activity for nearly 3 years (Boisvert and Boisvert, 2001) and the carry was mainly dependent on discharge, water temperature, physiological state of the larvae, and importance of the stream hyporheic zone.

Little recent work has been done in Canada on the effect of *B.t.i.* formulation additives on non-target fauna. Fortin *et al.* (1986) indicated that the presence of 2% xylene in a commercial formulation caused mortality in brook trout, *S. fontinalis*, but at a very high concentration of the product (6000 mg l^{-1}).

In Canada, no studies exist that determined the long-term effect of *B.t.i.* treatments on non-target organisms in black fly or mosquito control programmes (Lacoursière and Boisvert, 1994). Boisvert and Boisvert (2000) reviewed more than 300 articles for the effects of both unformulated and formulated *B.t.i.* on target and non-target species, and the only paper from which results could correlate with certain Canadian biotopes (the study was done in Minnesota) concluded that intensive *B.t.i.* treatments over a 3-year period could significantly affect insect diversity and density in mosquito marshes. They concluded that the additives present in the formulation used in this study could have been responsible for the observed effects.

Evaluation of Biological Control

Although many disease agents are commonly associated with black flies, their life cycles are often almost completely unknown, so knowledge of agent transmission among hosts is lacking (Crosskey, 1990). Mermithids are one of the most promising biological control agents but their commercial development is hindered by a lack of basic ecological information. The success of *B.t.i.* for black fly control has resulted in reduced research effort on other biological control agents.

B.t.i. successfully replaced chemical insecticides in some municipalities as early as 1984, and is used in most Canadian provinces for larval black fly control. In Saskatchewan, it is used annually in the Saskatchewan River to control *S. luggeri*. In Quebec, *B.t.i.* is used almost exclusively in nearly 30 towns, whereas other provinces use small amounts of *B.t.i.*, many preferring adulticiding with chemicals. Small to large municipalities and some military bases contract with private firms to successfully reduce the nuisance caused by black flies.

Recommendations

Future work should include:

1. Clarification of the life cycles of microsporidian, viral and fungal pathogens

to permit further assessment of their biological control potential;
2. Further study of the host range and parasite–host dynamics of mermithid species to facilitate mass production technology;
3. Studying the long-term effects of repeated *B.t.i.* treatments in streams and rivers;
4. Research on methodology to better evaluate use of *B.t.i.* formulations under Canadian conditions.

References

Adler, P.H. (1986) Ecology and cytology of some Alberta black flies (Diptera: Simuliidae). *Quaestiones Entomologica* 22, 1–18.

Adler, P.H. and Mason, P.G. (1997) Black flies (Diptera: Simuliidae) of east-central Saskatchewan, with a description of a new species and implications for pest management. *The Canadian Entomologist* 129, 81–91.

Back, C., Boisvert, J., Lacoursière, J.O. and Charpentier, G. (1985) High-dosage treatment of a Québec stream with *Bacillus thuringiensis* serovar *israelensis*: efficacy against black fly larvae (Diptera: Simuliidae) and impacts on non-target insects. *The Canadian Entomologist* 117, 1523–1534.

Boisvert, M. and Boisvert, J. (2000) Effects of *Bacillus thuringiensis* var. *israelensis* on target and non-target organisms: a review of laboratory and field experiments. *Biocontrol Science and Technology* 10, 517–561.

Boisvert, M. and Boisvert, J. (2001) Storage stability of two liquid formulations of *Bacillus thuringiensis* subsp. *israelensis* and effect of freezing over time. *Biocontrol Science and Technology* 11, 261–271.

Boisvert, M., Boisvert, J. and Aubin, A. (2001) A new field procedure and method of analysis to evaluate the performance of *Bacillus thuringiensis* subsp. *israelensis* liquid formulations in streams or rivers. *Biocontrol Science and Technology* 11 (in press).

Burgin, S.G. and Hunter, F.F. (1997) Nectar versus honeydew as sources of sugar for male and female black flies (Diptera: Simuliidae). *Journal of Medical Entomology* 34, 606–608.

Burton, D.K. (1984) Impact of *Bacillus thuringiensis* var. *israelensis* in dosages used for black fly (Simuliidae) control against target and non-target organisms in the Torch River, Saskatchewan. MSc thesis, Department of Entomology, University of Manitoba, Winnipeg, Manitoba.

Charpentier, G., Back, C., Garon, S. and Strykowski, H. (1986) Observations on a new intranuclear virus-like particle infecting larvae of the black fly *Simulium vittatum* (Diptera: Simuliidae). *Diseases of Aquatic Organisms* 1, 147–150.

Colbo, M.H. (1982) Size and fecundity of adult Simuliidae (Diptera) as a function of stream habitat, year and parasitism. *Canadian Journal of Zoology* 60, 2507–2513.

Colbo, M.H. (1985) Control of black flies (Simuliidae) using *Bacillus thuringiensis* var. *israelensis* (Bti) as a larvicide with emphasis on northern programs. *Proceedings of the Thirty-First Annual Meeting, Canadian Pest Management Society, Winnipeg, Manitoba, Canada, 20–22 August 1984*, pp. 98–109.

Colbo, M.H. (1987) Black fly control in northern Canada using *Bacillus thuringiensis* var. *israelensis. Proceedings of the Fortieth Annual Meeting of the Utah Mosquito Abatement Association, 27–29 September 1987, Park City Utah, USA*, pp. 54–55.

Colbo, M.H. (1990) Persistence of Mermithidae (Nematoda) infections in black fly (Diptera: Simuliidae) populations. *Journal of the American Mosquito Control Association* 6, 203–206.

Colbo, M.H. and O'Brien, H. (1984) A pilot black fly (Diptera: Simuliidae) control program using *Bacillus thuringiensis* var. *israelensis* in Newfoundland. *The Canadian Entomologist* 116, 1085–1096.

Colbo, M.H. and Porter, J.N. (1980) Distribution and specificity of Mermithidae (Nematoda) infecting Simuliidae (Diptera) in Newfoundland. *Canadian Journal of Zoology* 58, 1483–1490.

Colbo, M.H. and Undeen, A.H. (1980) Effects of *Bacillus thuringiensis* var. *israelensis* on non-target insects in stream trials for control of Simuliidae. *Mosquito News* 40, 368–371.

Crosskey, R.W. (1990) *The Natural History of Blackflies.* John Wiley & Sons, Toronto, Ontario.

Cupp, E.W. (1987) The epizootiology of livestock and poultry diseases associated with black flies. In: Kim, K.C. and Merritt, R.W. (eds) *Black Flies: Ecology, Population Management, and Annotated World List.* Pennsylvannia State University, University Park, Pennsylvania.

Davies, D.M. (1981) Predators upon blackflies. In: Laird, M. (ed.) *'Blackflies' The Future for Biological Control Methods in Integrated Control.* Academic Press, London, UK, pp. 139–158.

Erlandson, M.A. and Mason, P.G. (1989) An iridescent virus from *Simulium vittatum* (Diptera: Simuliidae) in Saskatchewan. *Journal of Invertebrate Pathology* 56, 8–14.

Finney, J.R. (1981a) Parasites – potential of mermithids for control and *in vitro* culture. In: Laird, M. (ed.) *'Blackflies' The Future for Biological Control Methods in Integrated Control.* Academic Press, London, UK, pp. 325–333.

Finney, J.R. (1981b) Mermithid nematodes: *in vitro* culture attempts. A review. *Journal of Nematology* 13, 275–280.

Finney, J.R. and Harding, J.B. (1982) The susceptibility of *Simulium verecundum* (Diptera: Simuliidae) to three isolates of *Bacillus thuringiensis* serotype 10 (*darmstadiensis*). *Mosquito News* 42, 434–435.

Finney, J.R. and Mokry, J.E. (1980) *Romanomermis culicivorax* and simuliids. *Journal of Invertebrate Pathology* 35, 211–213.

Fortin, C., Lapointe, D. and Charpentier, G. (1986) Susceptibility of brook trout (*Salvelinus fontinalis*) fry to a liquid formulation of *Bacillus thuringiensis* serovar *israelensis* (Teknar) used for black fly control. *Canadian Journal of Fisheries and Aquatic Sciences* 43, 1667–1670.

Fredeen, F.J.H. (1985) Some economic effects of outbreaks of black flies (*Simulium luggeri* Nicholson and Mickel) in Saskatchewan. *Quaestiones Entomologica* 21, 175–208.

Fredeen, F.J.H. and Mason, P.G. (1991) Meteorological factors influencing host-seeking activity of female *Simulium luggeri* (Diptera: Simuliidae). *Journal of Medical Entomology* 28, 831–840.

Goettel, M.S. (1987) Studies on microbial control of mosquitoes in central Alberta with emphasis on the hyphomycete *Tolypocladium cylindrosporum.* PhD Thesis, Department of Entomology, University of Alberta, Edmonton, Alberta.

Labeyrie, E.S., Molly, D.P. and Lichwardt, R.W. (1996) An investigation of Harpellales (Trichomycetes) in New York State black flies (Diptera: Simuliidae). *Journal of Invertebrate Pathology* 68, 293–298.

Lacoursière, J.O. and Boisvert, J. (1994) *Le* Bacillus thuringiensis *et le Contrôle des Insectes Piqueurs au Québec.* Rapport présenté pour la Direction du Milieu Agricole et du Contrôle des Pesticides, Ministère de l'Environnement, Province de Québec.

Lacoursière, J. and Charpentier, G. (1988) Laboratory study of the influence of water temperature and pH on *Bacillus thuringiensis* var. *israelensis* efficacy against black fly larvae (Diptera: Simuliidae). *Journal of the American Mosquito Control Association* 4, 104–116.

Laird, M., Colbo, M., Finney, J., Mokry, J. and Undeen, A. (1980) Pathogens of Simuliidae (Blackflies). In: Roberts, D.W. and Castillo, J.M. (eds) *Bibliography on Pathogens of Medically Important Arthropods 1980. Bulletin of the World Health Organization,* 58 (supplement), 105–124.

Lam, T.N.C., Soares, G.G. Jr and Goettel, M.S. (1988) Host records of the mosquito pathogenic hyphomycete *Tolypocladium cylindrosporum. Florida Entomologist* 71, 86–89.

Lichwardt, R.W. and Misra, J.K. (2000) *Illustrated Genera of Trichomycetes. Fungal Symbionts of Insects and other Arthropods.* Science Publishers, Enfield, New Hampshire.

Mason, P.G. and Kusters, P.M. (1990) Seasonal activity of female black flies (Diptera: Simuliidae) in pastures in northeastern Saskatchewan. *The Canadian Entomologist* 122, 825–835.

Mason, P.G. and Kusters, P.M. (1991) *Procedures Manual for the Saskatchewan Black Fly Program.* Miscellaneous Report, Research Branch Agriculture Canada, Saskatoon, Saskatchewan.

Mason, P.G. and Shemanchuk, J.A. (1990) *Black flies.* Publication 1499/E, Agriculture Canada, Ottawa, Ontario.

McCracken, I.R. and Matthews, S.L. (1997) Effects of *Bacillus thuringiensis* subsp. *israelensis* (B.t.i.) applications on invertebrates from two streams on Prince Edward Island. *Bulletin of Environmental Contamination and Toxicology* 58, 291–298.

McCreadie, J.W., Colbo, M.H. and Bennett, G.F. (1986) The influence of weather on host seeking and blood feeding of *Prosimulium mixtum* and *Simulium venustum/verecundum* complex (Diptera: Simuliidae). *Journal of Medical Entomology* 23, 289–297.

Molloy, D.P. (1987) The ecology of black fly parasites. In: Kim, K.C. and Merritt, R.W. (eds) *Black Flies: Ecology, Population Management, and Annotated World List.* Pennsylvannia State University, University Park, Pennsylvania.

Nadeau, M.P. and Boisvert, J.L. (1994) Larvicidal activity of the entomopathogenic fungus *Tolypocladium cylindrosporum* (Deuteromycotina: Hyphomycetes) on the mosquito *Aedes triseriatus* and the black fly *Simulium vittatum* (Diptera: Simuliidae). *Journal of the American Mosquito Control Association* 10, 487–491.

Nadeau, M.P., Dunphy, G.B. and Boisvert, J.L. (1994) Entomopathogenic fungi of the order Entomophthorales (Zygomytina) in adult black flies (Diptera: Simuliidae) in Quebec. *Canadian Journal of Microbiology* 40, 682–686.

Nadeau, M.P., Dunphy, G.B. and Boisvert, J.L. (1995) Effects of physical factors on the development of secondary conidia of *Erynia conica* (Zygomycetes: Entomophthorales), a pathogen of adult black flies (Diptera: Simuliidae). *Experimental Mycology* 19, 324–329.

Nadeau, M.P., Dunphy, G.B. and Boisvert, J.L. (1996a) Development of *Erynia conica* (Zygomycetes: Entomophthorales) on the cuticle of the adult black flies *Simulium rostratum* and *Simulium decorum* (Diptera: Simuliidae). *Journal of Invertebrate Pathology* 68, 50–58.

Nadeau, M.P., Dunphy, G.B. and Boisvert, J.L. (1996b) Replicative conidiospore formation and discharge by *Erynia conica* and *Erynia curvispora* (Zygomycetes: Entomophthorales). *Journal of Invertebrate Pathology* 68, 177–179.

Nixon, K.E. (1988) The effect of *Simulium vittatum* Zett. (Diptera: Simuliidae) larval feeding behaviour on the efficacy of *Bacillus thuringiensis* Serotype II-14 (De Barjac). MSc Thesis, Department of Entomology, University of Manitoba, Winnipeg, Manitoba.

Poinar, G.O. Jr (1981) Mermithid nematodes of blackflies. In: Laird, M. (ed.) *'Blackflies' The Future for Biological Control Methods in Integrated Control.* Academic Press, London, UK, pp. 159–170.

Shipp, J.L., Grace, B.W. and Schaalje, G.B. (1987) Effects of microclimate on daily flight activity of *Simulium arcticum* Malloch (Diptera: Simuliidae). *International Journal of Biometeorology* 31, 9–20.

Shipp, J.L., Grace, B.W. and Janzen, H.H. (1988) Influence of temperature and water vapour pressure on the flight activity of *Simulium arcticum* Malloch (Diptera: Simuliidae). *International Journal of Biometeorology* 32, 242–246.

Tousignant, M.E., Boisvert, J.L. and Chalifour, A. (1993) Loss of *Bacillus thuringiensis* var. *israelensis* larvicidal activity and its distribution in benthic substrates and hyporheic zone of streams. *Canadian Journal of Fisheries and Aquatic Sciences* 50, 443–451.

Undeen, A.H. and Colbo, M.H. (1980) The efficacy of *Bacillus thuringiensis* var. *israelensis* against blackfly larvae (Diptera: Simuliidae) in their natural habitats. *Mosquito News* 40, 181–184.

Undeen, A.H. and Nagel, W.L. (1978) The effect of *Bacillus thuringiensis* ONR-60A strain (Goldberg) on *Simulium* larvae in the laboratory. *Mosquito News* 38, 524–527.

Vávra, J. and Undeen, A.H. (1981) Microsporidia (Microspora: Microsporida) from Newfoundland blackflies (Diptera: Simuliidae). *Canadian Journal of Zoology* 59, 1431–1446.

Weiser, J. and Undeen, A.H. (1981) Diseases of blackflies. In: Laird, M. (ed.) *'Blackflies' The Future for Biological Control Methods in Integrated Control.* Academic Press, London, UK, pp. 181–196.

Wood, D.M. (1985) *Biting Flies Attacking Man and Livestock in Canada.* Publication 1781/E, Agriculture Canada, Ottawa, Ontario.

Yeboah, D.O., Undeen, A.H. and Colbo, M.H. (1984) Phycomycetes parasitizing ovaries of blackflies (Simuliidae). *Journal of Invertebrate Pathology* 43, 363–373.

48 *Rhagoletis pomonella* (Walsh), Apple Maggot (Diptera: Tephritidae)

T.S. Hoffmeister

Pest Status

The apple maggot, *Rhagoletis pomonella* (Walsh), is native to North America and occurs in certain areas of the USA and in southern Ontario and Quebec (Bush, 1966; Harris, 1989). The native hosts of *R. pomonella* are hawthorn, *Crataegus* spp., but it also attacks commercial apple, *Malus pumila* Miller (= *M. domestica* Borkhausen), and in some localities sour cherry, *Prunus cerasus* L., plum, *Prunus angustifolia* L., apricot, *Prunus armeniaca* L., peach, *Prunus persica* (L.), Siberian crabapple, *M. baccata* L., rose-hips, *Rosa rugosa* Thunberg and *R. carolina* L., and pear, *Pyrus communis* L. (Bush, 1966; Harris, 1989; White and Elson-Harris, 1992). In addition to fruit destruction, the stings (oviposition sites) made by *R. pomonella* females cause cosmetic damage that reduces the market value of apples (D. Pree, Vineland, 2000, personal communication). In Nova Scotia, *R. pomonella* is an important pest; about 40–50% of the 4000 ha of apple trees are treated annually (R. Smith, Kentville, 2000, personal communication). In southern Ontario, 80–100% of fruit in unmanaged orchards is damaged by *R. pomonella* in some years (D. Pree, Vineland, 2000, personal communication).

Adults of *R. pomonella* emerge from puparia in the soil under larval host plants. They are well synchronized with fruit ripening and, in Ontario, peak emergence (50% of the population) occurs around early August, i.e. after 809 degree-days above 8.7°C (Laing and Heraty, 1984). Mating occurs on the host plant. After a pre-oviposition period of about 2 weeks (103 degree-days above 8.7°C (Laing and Heraty, 1984), females start to lay single eggs into the host fruits, which are subsequently marked with a pheromone to avoid unrecognized multiple infestations (Prokopy, 1972). Eggs hatch within a few days and, after a larval period of about 2 weeks (depending on fruit-ripeness), mature larvae bore emergence holes through the fruit and drop to the ground where they pupate in the upper 10 cm of soil (Boller and Prokopy, 1976). Most flies undergo a winter diapause and emerge the following summer, but parts of the population may emerge within the same year or after diapausing for two winters (Boller and Prokopy, 1976).

Background

R. pomonella is controlled with 1–3 sprays of organophosphorus insecticides; border sprays are often successful (D. Pree, Vineland, 2000, personal communication). Baited, pesticide-treated spherical traps give effective control (Warner and Smith, 1989; Warner and Watson, 1991; Duan and Prokopy, 1995; Prokopy *et al.*, 1995; Reynolds *et al.*, 1998; Zhang *et al.*, 1999), but side-effects on non-target species may be a problem (Mondor, 1995). Attempts to use neem extract to control *R. pomonella* were unsuccessful (Prokopy and Powers, 1995). Examination of the impact of parasitoids and predators on *R. pomonella* populations in commercial orchards made it obvious that indigenous natural enemies cannot control this pest. Parasitism by indigenous parasitoids is low, varying from 2 to 7.4%

(Monteith, 1971b, 1977; E. Hagley, Vineland, 1992, personal communication). Similarly, epigeal arthropod predators have little impact on *R. pomonella* populations (Allen and Hagley, 1989). However, the egg parasitoid, *Anaphes conotracheli* (Gahan), which parasitizes both the plum curculio, *Conotrachelus nenuphar* Herbst, and sometimes also *R. pomonella*, may be worth investigating further (J. Huber, Ottawa, 2000, personal communication). Because indigenous pests can sometimes be controlled by introduced exotic natural enemies (Pimentel, 1963; Carl, 1982), a co-operative project between Agriculture Canada, Vineland Station, Ontario, and CABI *Bioscience* Centre, Switzerland (then the Commonwealth Institute of Biological Control) to study European parasitoids was initiated in the late 1960s (Monteith, 1971a). In Canada, experiments were carried out with two parasitoids of the European cherry fruit fly, *Rhagoletis cerasi* (L.) (Monteith, 1971a). Whereas the larval parasitoid *Opius rhagoleticola* Sachtleben failed to oviposit into *R. pomonella* larvae, three populations of *Phygadeuon wiesmanni* Sachtleben from Poland, Switzerland and Austria parasitized them. However, only the Austrian population, parasitizing the puparia instead of mature larvae, produced an F_1 generation (Monteith, 1971a).

Biological Control Agents

Parasitoids

In Europe, the parasitoid complexes of six fruit-attacking species of Tephritidae were studied in Germany, Switzerland, Austria and Hungary from 1984 to 1987 to find potential biological control agents against *R. pomonella*. A total of 17 parasitoid species was reared from populations of both host races of *R. cerasi* that develop in cherries, *Prunus* spp., and honeysuckle, *Lonicera xylosteum* L., respectively, *Myoleja lucida* Fallén, which also develops in honeysuckle fruits, *Rhagoletis berberidis* Jermy and *Rhagoletis meigenii* (Loew), which both develop in seeds of barberry, *Berberis vulgaris* L., *Rhagoletis alternata* (Fallén), which develops in rose-hips, and *Anomoia purmunda* (Harris), which develops in hawthorn fruits (Hoffmeister, 1992). Because *R. pomonella* larvae are well protected from parasitism while feeding in commercial apples, evaluation of pupal parasitoids attacking their fly hosts after formation of puparia in the ground was emphasized. Of the nine species of pupal parasitoids reared from the six host species, only two ichneumonids, *P. wiesmanni* Sachtleben and *Phygadeuon exiguus* Gravenhorst, reached more than 20% parasitism (up to 80% parasitism) in most of the samples and were thus considered for introduction into Canada (Hoffmeister, 1992). Whereas *P. wiesmanni* was found in all host species except *R. berberidis*, *P. exiguus* was only reared in larger numbers from *M. lucida* and *R. alternata*. Both *Phygadeuon* spp. develop as external parasitoids on fly pupae inside the puparium and produce two incomplete generations per year (Hoffmeister, 1988, 1990). The first generation seemed to depend on hosts such as *R. cerasi*, or *M. lucida* that develop in early summer. Because winter temperatures differ between Central Europe and Ontario, the ability of parasitoids to survive Canadian winters was studied. Cold-hardiness tests suggested that both *Phygadeuon* spp. should survive Ontario winters. Attempts to mass-rear the parasitoids in laboratory conditions failed. Therefore, parasitoids for shipment to Canada were produced by exposing fly puparia to parasitoid attack under natural host plants at sample sites in Austria, Switzerland and Germany. From 1984 to 1990, several thousand *P. wiesmanni* and *P. exiguus* specimens, as well as several hundred *O. rhagoleticola* specimens, were shipped to Canada.

In Vineland, laboratory and field-cage studies with *O. rhagoleticola*, *P. wiesmanni* and *P. exiguus* were conducted in 1985 and 1986 to determine if the parasitoids would parasitize immature stages of *R. pomonella*, and what levels of parasitism might be expected. Development rate, longevity and fecundity, and the successful survival of the parasitoids in southern Ontario were also investigated. Both species of *Phygadeuon* from all countries of Europe attacked and developed in puparia of *R. pomonella*. However, increasing puparial age negatively affected the success of parasitoid attack (Hagley *et al*., 1993). *P. wiesmanni* also developed successfully on puparia of the cherry fruit fly, *R. cingulata*. In 1986, *P. wiesmanni* that had hibernated under field conditions in Ontario were recovered, but only males emerged from the host puparia. In the laboratory, *O. rhagoleticola* successfully parasitized larvae of *R. pomonella*, but it could not be determined whether larvae were attacked inside apples or after they had emerged from the apples and were moving to their pupation sites. Because *O. rhagoleticola* was recovered in 1985 from outdoor cages placed over apples infested with *R. pomonella* larvae in 1984, this parasitoid obviously can also survive winters in southern Ontario.

Releases and Recoveries

Field releases of small numbers of *P. wiesmanni* and *O. rhagoleticola* were made from 1985 to 1991. Collections of pupae were made from all release areas, but neither parasitoid was ever recovered.

Evaluation of Biological Control

As a larval parasitoid, *O. rhagoleticola* is well synchronized with its European host, *R. cerasi*. Because *R. pomonella* occurs about 8 weeks later than European source populations of *O. rhagoleticola*, permanent establishment of *O. rhagoleticola* on *R. pomonella* is not likely. A similar problem might exist for the establishment of *Phygadeuon* spp. Because *R. pomonella* are available as puparia only in late summer, the developmental pattern of both *Phygadeuon* spp. could restrict parasitoid establishment to those areas where alternative hosts such as *R. cingulata* are available. However, the parasitoid even failed to establish in areas where these alternate hosts were present.

Both *Phygadeuon* spp. appear to be restricted to the puparia of Tephritidae developing in fruits or leaf-mines (Herting, 1978, 1982; Hoffmeister, 1992) but further testing to determine potential native non-target hosts have not been conducted.

Recommendations

Further work should include:

1. Determining the actual host range of *P. wiesmanni* and *P. exiguus* in their native European habitats and their potential host range in North America;
2. Study of the egg parasitoid, *A. conotracheli*, to evaluate impact and potential for mass-rearing.

References

Allen, W.R. and Hagley, E.A.C. (1989) Epigeal arthropods as predators of mature larvae and pupae of the apple maggot (Diptera: Tephritidae). *Environmental Entomology* 19, 309–312.

Boller, E.F. and Prokopy, R.J. (1976) Bionomics and management of *Rhagoletis*. *Annual Review of Entomology* 21, 223–246.

Bush, G.L. (1966) The taxonomy, cytology, and evolution of the genus *Rhagoletis* in North America. *Bulletin of the Museum of Comparative Zoology* 134, 431–562.

Carl, K.P. (1982) Biological control of native pests by introduced natural enemies. *Biocontrol News and Information* 3, 191–200.

Duan, J.J. and Prokopy, R.J. (1995) Control of apple maggot flies (Diptera: Tephritidae) with pesticide-treated red spheres. *Journal of Economic Entomology* 88, 700–707.

Hagley, E.A.C., Biggs, A.R., Timbers, G.E. and Coutu Sundy, J. (1993) Effect of age of the puparium of the apple maggot, *Rhagoletis pomonella* (Walsh) (Diptera: Tephritidae), on parasitism by *Phygadeuon wiesmanni* Sachtl. (Hymenoptera: Ichneumonidae). *The Canadian Entomologist* 125, 721–724.

Harris, E.J. (1989) Hawaiian islands and North America. In: Robinson, A.S. and Hooper, G. (eds) *Fruit Flies. Their Biology, Natural Enemies and Control.* Elsevier, Amsterdam, The Netherlands, pp. 73–81.

Herting, B. (1978) *A Catalogue of Parasites and Predators of Terrestrial Arthropods, Section A: Host or Prey/Enemy, Vol. V: Neuroptera, Diptera, Siphonaptera.* Commonwealth Agricultural Bureaux, Farnham Royal, UK.

Herting, B. (1982) *A Catalogue of Parasites and Predators of Terrestrial Arthropods, Section B: Enemy/Host or Prey, Vol. II: Hymenoptera Terebrantia.* Commonwealth Agricultural Bureaux, Farnham Royal, UK.

Hoffmeister, T. (1988) Biologie der Kirschfruchtfliege (*Rhagoletis cerasi* L.), verwandter Tephritiden und ihrer Parasiten. Diploma Thesis, Christian-Albrechts-University, Kiel, Germany.

Hoffmeister, T. (1990) Zur Struktur und Dynamik des Parasitoidenkomplexes der Kirschfruchtfliege *Rhagoletis cerasi* L. (Diptera: Tephritidae) auf Kirschen und Heckenkirschen. *Mitteilungen der Deutschen Gesellschaft für Allgemeine und Angewandte Entomologie* 7, 546–551.

Hoffmeister, T. (1992) Factors determining the structure and diversity of parasitoid complexes in tephritid fruit flies. *Oecologia* 89, 288–297.

Laing, J.E. and Heraty, J.M. (1984) The use of degree-days to predict emergence of the apple maggot, *Rhagoletis pomonella* (Diptera: Tephritidae), in Ontario. *The Canadian Entomologist* 116, 1123–1129.

Mondor, E.B. (1995) Syrphid captures on red sphere traps deployed for the apple maggot fly, *Rhagoletis pomonella* (Walsh). *Ecoscience* 2, 200–202.

Monteith, L.G. (1971a) *Rhagoletis pomonella* (Walsh), apple maggot (Diptera: Tephritidae). In: *Biological Control Programmes against Insects and Weeds in Canada 1959–1968.* Technical Communication No. 4, Commonwealth Institute of Biological Control, Trinidad, Commonwealth Agricultural Bureaux, Farnham Royal, UK, pp. 38–40.

Monteith, L.G. (1971b) The status of parasites of the apple maggot *Rhagoletis pomonella* (Diptera: Tephritidae) in Ontario. *The Canadian Entomologist* 103, 507–512.

Monteith, L.G. (1977) Additional records and the role of the parasites of the apple maggot *Rhagoletis pomonella* (Diptera: Tephritidae) in Ontario. *Proceedings of the Entomological Society of Ontario* 108, 3–6.

Pimentel, D. (1963) Introducing parasites and predators to control native pests. *The Canadian Entomologist* 95, 785–792.

Prokopy, R.J. (1972) Evidence for a marking pheromone deterring repeated oviposition in apple maggot flies. *Environmental Entomology* 1, 326–332.

Prokopy, R.J. and Powers, P.J. (1995) Influence of neem seed extract on oviposition and mortality of *Conotrachelus nenuphar* (Col., Curculionidae) and *Rhagoletis pomonella* (Dip., Tephritidae) adults. *Journal of Applied Entomology* 119, 63–65.

Prokopy, R.J., Duan, J.J. and Hu, X.P. (1995) Toxicant-treated red spheres for controlling apple maggot flies. *New England Fruit Meetings* 101, 71–77.

Reynolds, A.H., Kaknes, A.M. and Prokopy, R.J. (1998) Evaluation of two trap deployment methods to manage the apple maggot fly (Dipt., Tephritidae). *Journal of Applied Entomology* 122, 255–258.

Warner, J. and Smith, A. (1989) Apple maggot, *Rhagoletis pomonella* (Diptera: Tephritidae), response to traps, synthetic lures and adhesive in field tests in Ontario. *Proceedings of the Entomological Society of Ontario* 120, 55–64.

Warner, J. and Watson, A. (1991) Synthetic volatile lures improve the performance of apple maggot, *Rhagoletis pomonella* (Diptera: Tephritidae), traps in Ontario. *Proceedings of the Entomological Society of Ontario* 122, 9–13.

White, I.M. and Elson-Harris, M.M. (1992) *Fruits Flies of Economic Significance: Their Identification and Bionomics.* CAB International, Wallingford, UK.

Zhang, A.J., Linn, C. Jr, Wright, S., Prokopy, R., Reissig, W. and Roelofs, W. (1999) Identification of a new blend of apple volatiles attractive to the apple maggot, *Rhagoletis pomonella. Journal of Chemical Ecology* 25, 1221–1232.

49 *Rhopobota naevana* (Hübner), Blackheaded Fireworm (Lepidoptera: Tortricidae)

D.E. Henderson, S.Y. Li and R. Prasad

Pest Status

The blackheaded fireworm, *Rhopobota naevana* (Hübner), is an economically important, native pest of cranberry, *Vaccinium macrocarpon* (Aiton), in North America, especially in coastal British Columbia, Washington and Oregon (Cockfield *et al.*, 1994). Severe infestations of *R. naevana* can cause significant yield losses in cranberry, the third most valuable food crop in British Columbia (Baines, 1991).

This bivoltine species overwinters as flat, orange–yellow eggs on the underside of the previous season's evergreen cranberry leaves. In British Columbia, first generation larvae hatch in late April or early May to feed initially upon old leaves then later on developing buds or new foliage. Second-generation larvae hatch in June to feed on young foliage at the tips of cranberry runners and uprights (stems). They occasionally feed on flowers and developing fruit (Eck, 1990). Second-generation adults fly in August and occasionally September, and lay predominantly overwintering eggs. However, a small percentage of these eggs develop to produce a third cohort of adults (Fitzpatrick and Troubridge, 1993). Depending on the weather in autumn, this generation is capable of depositing a significant number of overwintering eggs. Adults have been caught in pheromone traps as late as mid-December in British Columbia (D.E. Henderson, unpublished).

Background

In British Columbia, management of *R. naevana* currently relies on insecticides. Although IPM-based monitoring has reduced pesticide use in cranberry from 5–6 sprays per season to 2–3, insecticides are still necessary to suppress fireworm populations below the economic threshold (Emery, 1994).

Alternatives to conventional insecticides for *R. naevana* control in cranberry include: essential-oil insecticides (Isman, 1999; M.B. Isman and D. MacArthur, Vancouver, 2000, personal communication); pheromone-based mating disruption, registered in 1999 in Canada (Fitzpatrick *et al.*, 1995); and use of parasitoids.

Biological Control Agents

Parasitoids

Following discovery of two indigenous egg parasitoids, *Trichogramma sibericum* (Sorokina) and *T. minutum* Riley, from natural *R. naevana* populations in British Columbia, efforts to develop a *Trichogramma*-based biological control programme for *R. naevana* were pursued (Li *et al.*, 1993). Both *Trichogramma* spp. display heavy female bias: 80% females for *T. minutum* and 95% females for *T. sibericum*. Parasitism of overwintering *R. naevana* eggs by *T. minutum* was found at one site on one farm to be over 90% but decreased to 46% by September because of insecticide applications. At an abandoned cranberry field with a high fireworm population, parasitism by *T. sibericum* increased steadily from 30% in overwintering eggs to 99% by the end of one growing season.

In the laboratory, *T. sibericum* parasitized *R. naevana* eggs to a higher degree than commercial *T. minutum* or *T. evanescens*, and *T. sibericum* showed a preference for parasitizing fresh overwintering *R. naevana* eggs (1–7 days old) over older ones (21 days old) (Li and Henderson, 1993; Li *et al.*, 1994). Unfed *T. sibericum* parasitized an average of 13 *R. naevana* eggs per female. Feeding *T. sibericum* females honey increased their longevity from 3 to 13 days and also their parasitizing capacity on Mediterranean flour moth, *Ephestia kuehniella* Zeller, from 17 to 59 eggs per female (Farrah, 1995). In the field, Henderson *et al.* (1996, 1997) found parasitism of 80.3% and 43.8% by native *T. sibericum* and commercial *T. minutum*, respectively. Parasitism was host density dependent in both laboratory (Li and Henderson, 1993) and field (Luczynski, 1993, 1994; Henderson *et al.*, 1997).

The optimal rearing temperature for *T. sibericum* is 22°C, given that daytime temperatures in cranberry fields from August to mid-September range from 15 to 24°C (Prasad, 1999). In particular, flight initiation was limited in insects reared at higher than ambient temperatures. Li *et al.* (1994) showed that *T. sibericum* maintained its acceptance of *R. naevana* eggs after mass-rearing. While parasitism on *R. naevana* eggs decreased somewhat after 17 successive generations on an alternative host, *T. sibericum* still displayed much higher parasitism than commercially available *Trichogramma* spp. To maintain this specificity to *R. naevana* eggs, *T. sibericum* are collected from cranberry fields annually. Each *Ephestia*-reared generation is tested for common quality control attributes such as fecundity, longevity, flight and sex ratio (100% female). Several generations, including those used for releases, are also tested for acceptance and successful parasitism of the natural host (R.P. Prasad, unpublished).

From 1993 to 1999, release rates of *T. sibericum* in field trials varied from 653,000 to 55.6 million acre^{-1}, and total area treated varied from less than 0.01 acre in 1993 to 66 acres in 1999 (Table 49.1). All field trials took place in commercial cranberry fields in Richmond, Pitt Meadows, Langely and Delta, British Columbia. To document egg parasitism after *Trichogramma* release, cranberry uprights were sampled and all leaves on uprights were examined visually for the presence of parasitized and unparasitized *R. naevana* eggs.

Two release methods were used for cranberry field trials: point source and broadcast

Table 49.1. Results of field release of *Trichogramma sibericum* against *Rhopobota naevana* eggs on cranberry in British Columbia.

Year	Release rate[a]	Total acres	Release method[b]	Host density[c]	Sample size[d]	Parasitism (%)	±SE
1993	55.6×10^6	0.009	P		624	93.4	1.6
	11.3×10^6	0.002	P			76.2	8.4
1994	914,760	0.15	P	0.53	43	24.8	37.6
	914,760	0.11	P	23.9	1915	78.2	12.9
1995	653,000	11.0	P	0.2		57.1	
	653,000	2.3	P	0.09		50.0	
1996	4.0×10^6	0.02	P		5952	45.7	26.5
	5.2×10^6	0.02	B		3732	36.8	13.1
1999	980,000	66.0	B		223	75.1	19.9

[a]Release rate = number of *T. sibericum* per acre.
[b]P = point source release; B = broadcast release.
[c]Host density = number of *R. naevana* eggs per 100 uprights.
[d]Sample size = total number of *R. naevana* eggs.

(Table 49.1). Point source releases involved placing *T. sibericum* pupae in containers before placing them in plots. Broadcast application involved mixing *T. sibericum* pupae in a carrier of moist vermiculite or perlite then spreading it evenly over the treated area (manually or with a seed spreader). Parasitism of *R. naevana* eggs in point source plots was highest around release points and decreased as distance from the release point increased, whereas in broadcast plots lower, but more even, parasitism resulted (Henderson *et al.*, 1997). Overall parasitism rates were similar for both release methods during the 1996 trial (Table 49.1) but numbers of flying females caught on sticky traps were up to 30% lower in broadcast plots, suggesting reduced survival of *T. sibericum*. The optimal distance between points in cranberry for *T. sibericum* was determined to be 6 m (Henderson *et al.*, 1997). Improvements to protocols and handling methods of *T. sibericum* in broadcast applications increased survival and parasitism in 1999 trials (Table 49.1).

Release rates of *T. sibericum* in 1993 were shown to be unnecessarily high after field trials with commercially available *T. minutum* had resulted in low parasitism. In 1996, extremely wet and cold weather accompanied releases and, despite the very high rates used, parasitism was low, reflecting poor survival of *T. sibericum*. Similar extremes in temperature and humidity have been associated with poor field performance of other *Trichogramma* spp. (Smith, 1996). In 1995, the lowest rate used (653,000 wasps acre^{-1}) resulted in 50–57% parasitism, even in low host densities. This rate approached adequate control for commercial purposes. Rates of 900,000–980,000 *Trichogramma* acre^{-1} released in 1994 and 1999 resulted in acceptable levels of parasitism and have been adopted for commercial purposes in cranberry.

Host density was studied as a factor affecting parasitism levels of *R. naevana* in 2 years of field trials. At 14 field sites in 1994, *T. sibericum* was applied at a rate of 914,760 wasps acre^{-1}. Intensive sampling revealed that parasitism was significantly and positively related to host density (r^2 = 0.87, $P < 0.0001$) (Luczynski, 1994). At a density of 5.28 host eggs in 100 cranberry uprights, parasitism reached 80%, and at higher host densities, approached 100%. Similar results were obtained in 1993 in a smaller study with eight plots ($r^2 = 0.57$, $P < 0.05$) (Luczynski, 1993).

Evaluation of Biological Control

Biological control of *R. naevana* with the egg parasitoid *T. sibericum* is successful and has evolved into a standardized protocol. Pheromone traps are used to monitor for the beginning of the second generation of *R. naevana* adults. *T. sibericum* applications are timed to occur for 2–3 weeks following peak trap catch and to correspond to at least a 3-day period of warm, dry weather. Growers refrain from irrigating with overhead sprinklers for 3 days following *Trichogramma* releases. *T. sibericum* is broadcast in a carrier of moist vermiculite or released in containers along edges or in hotspots. Two applications are made in each field, with the release rate split in half for each application. Applications are made in early morning and adult wasps emerge within 24 h of release.

This protocol for *T. sibericum* application is effective together with (but not dependent on) a season-long *R. naevana* monitoring programme. Biological control of *R. naevana* eggs in August or September is highly compatible with mating disruption of adults in spring and late summer. In 2000, *T. sibericum* was applied to 120 acres of cranberry. On one farm, it was used with mating disruption.

Application technology to address unique limitations to applying *T. sibericum* in cranberry is required. Cranberries grow as a matted vine and applications require walking on the crop. Not only is this damaging to the crop, but also time consuming and costly.

Recommendations

Further work should include:

1. Improving mass-rearing output, the biggest cost of which is *E. kuehniella* eggs;

2. Developing aerial application techniques combined with a biodegradable *Trichogramma* 'container' designed to protect pupae from weather and predators.

Acknowledgements

Many E.S. Cropconsult Ltd staff contributed to this project. The Industrial Research Assistance Programme, British Columbia Cranberry Growers Association, Ocean Spray Cranberries Inc., Cranberry Institute, Science Council of British Columbia, British Columbia Investment Agriculture Foundation, and British Columbia provincial Student Works and First Job in Science programmes provided financial assistance. Numerous British Columbia cranberry growers allowed us to work on their farms and more recently had the confidence to purchase *Trichogramma*.

References

Baines, P.S. (1991) *A Profile of the British Columbia Cranberry Industry.* Agriculture Canada and the British Columbia Ministry of Agriculture and Fisheries, Agri-food Regional Development Subsidiary Agreement (ARDSA 1985–90).

Cockfield, S.D., Fitzpatrick, S.M., Giles, K.V. and Mahr D.L. (1994) Hatch of blackheaded fireworm (Lepidoptera: Tortricidae) eggs and prediction with temperature-driven models. *Environmental Entomology* 23, 101–107.

Eck, P. (1990) *The American Cranberry.* Rutgers University Press, New Brunswick, New Jersey.

Emery, C. (1994) IPM programmes in British Columbia cranberries. *British Columbia Pest Monitor* 3(2), 1–2.

Farrah, G. (1995) Fecundity and longevity of *Trichogramma* sp. nr. *sibericum.* Internal Report, E.S. Cropconsult Ltd, Vancouver, British Columbia.

Fitzpatrick, S.M. and Troubridge, J.T. (1993) Fecundity, number of diapausing eggs, and egg size of successive generations of the blackheaded fireworm (Lepidoptera: Tortricidae) on cranberries. *Population Ecology* 22, 818–823.

Fitzpatrick, S.M., Troubridge, J.T., Maurice, C. and White, J. (1995) Initial studies of mating disruption of the blackheaded fireworm of cranberries (*Lepidoptera: Tortricidae*). *Journal of Economic Entomology* 88, 1017–1023.

Henderson, D.E., Luczynski, A. and Caddick, G. (1996) *The Use of* Trichogramma *sp. nr.* sibericum *to Control Blackheaded Fireworm in Commercial Cranberry Bogs.* Interim Report to Industrial Research Assistance Programme.

Henderson, D.E., Caddick, G. and Luczynski, A. (1997) *The Use of* Trichogramma sibericum *to Control Blackheaded Fireworm in Commercial Cranberry Bogs.* Final Report to Industrial Research Assistance Programme.

Isman, M.B. (1999) Pesticides based on plant essential oils. *Pesticide Outlook* 10, 68–72.

Li, S.Y. and Henderson, D.E. (1993) Response of *Trichogramma* sp. nr. *sibericum* (Hymenoptera: Trichogrammatidae) to age and density of its natural hosts, the eggs of *Rhopobota naevana* (Lepidoptera: Torticidae). *Journal of the Entomological Society of British Columbia* 90, 18–24.

Li, S.Y., Sirois, G.M., Luczynski, A. and Henderson, D.E. (1993) Indigenous *Trichogramma* (*Hymenoptera: Trichogrammatidae*) parasitizing eggs of *Rhopobota naevana* (*Lepidoptera: Tortricidae*) on cranberries in British Columbia. *Entomophaga* 38, 313–315.

Li, S.Y., Henderson, D.E. and Myers, J.H. (1994) Selection of suitable *Trichogramma* species for potential control of the blackheaded fireworm infesting cranberries. *Biological Control* 4, 244–248.

Luczynski, A. (1993) Comparison of the blackheaded fireworm parasitism between commercial and native species of *Trichogramma* in 1993, Field Releases. Internal Report, E.S. Cropconsult Ltd, Vancouver, British Columbia.

Luczynski, A. (1994) Field parasitism rates of blackheaded fireworm *Rhopobota naevana* by *Trichogramma* sp. nr. *sibericum*, 1994 trials. Internal Report, E.S. Cropconsult Ltd, Vancouver, British Columbia.

Prasad, R.P. (1999) The effect of rearing temperature on performance of *Trichogramma sibericum* at ambient temperature. MPM thesis, Simon Fraser University, Burnaby, British Columbia.
Smith, S.M. (1996) Biological control with *Trichogramma*: Advances, successes, and potential of their use. *Annual Review of Entomology* 41, 375–406.

50 *Sitodiplosis mosellana* (Géhin), Orange Wheat Blossom Midge (Diptera: Cecidomyiidae)

J.F. Doane, M.P. Braun, O.O. Olfert, F. Affolter and K. Carl

Pest Status

The orange wheat blossom midge, *Sitodiplosis mosellana* (Géhin), Palaearctic in origin, was apparently accidentally introduced into North America in the early 1800s (Felt, 1912). It is a major pest of spring wheat, *Triticum aestivum* L., in the northern Great Plains, including the Canadian prairies, and is widely distributed in many parts of the world where wheat production occurs, especially between the 42nd and 62nd parallels (Affolter, 1990). In western Canada, *S. mosellana* was first reported in Manitoba (Fletcher, 1902) but was not considered to be a pest until the 1950s (Allen, 1955). In 1983, *S. mosellana* emerged as an important pest in north-east Saskatchewan (Olfert *et al.*, 1985) and north-west Manitoba (Barker, 1984). The outbreak then spread throughout most of Manitoba, eastern Saskatchewan and North Dakota by the early 1990s (Barker *et al.*, 1995). The area of infestation in 2000 included much of the wheat-growing area of the northern Great Plains, including an incursion into Alberta (Hartley *et al.*, 2000). *S. mosellana* also occurs in wheat-growing areas of Nova Scotia, Ontario, Quebec and British Columbia.

Canadian varieties of hard red spring wheat, durum wheat, and soft spring wheat differ in their susceptibility to damage. Except for soft spring wheats, early maturing varieties suffer less damage than late-maturing varieties. The extent of crop damage due to *S. mosellana* depends on its population density, spatial distribution, and timing of oviposition relative to crop phenology (Wright and Doane, 1987; Elliott and Mann, 1996). Injury is caused by larvae feeding on the surface of developing kernels. Usually only some of the florets on a wheat head are infested and the infestation level can vary from one to eight or more larvae per floret. If three or more larvae develop within a floret, the kernel may abort or not fill properly. Mature kernels from infested florets are cracked, shrivelled or deformed. Small, lighter kernels are lost during harvesting operations, resulting in lower grain yield. If one larva develops on a kernel, the surface is scarred and slightly depressed, resembling drought or frost injury. Damaged kernels that are harvested lower grain quality, including milling and baking properties. In 1983, *S. mosellana* caused an estimated yield loss in spring wheat of Can$30 million in Saskatchewan (Olfert *et al.*, 1985).

On the northern Great Plains, adults emerge over a 6-week period, beginning in late June or early July. The highest populations usually occur during the second or third week of July. Adults are relatively poor fliers and may be distributed over long distances by thermal updrafts and wind. They are difficult to detect during the day because they remain within the crop canopy closer to ground level where it is more humid. Females become more active in the evening. Most egg-laying occurs at dusk when conditions are calm and temperatures above 10–11°C. Females live 3–7 days and lay an average of 80 eggs underneath the glumes or on grooves on the floret surface. Eggs are laid singly or in clusters of up to four eggs on the florets of emerging wheat heads. Larvae crawl into the floret and feed on the kernel surface for 2–3 weeks. Mature larvae remain within their cast skin in the wheat head when conditions are dry. Once moist conditions occur, larvae drop to the ground, burrow into the soil, spin a cocoon and overwinter. The following spring, further larval development depends on temperature and soil moisture; if conditions are dry during May and June, larvae remain dormant until the following year; if moist, larvae leave their cocoons and move to the soil surface to pupate.

Background

Insecticide treatments, applied at dusk, are recommended when there is at least one adult midge for every 4–5 wheat heads at several locations in the field (Elliott, 1988a, b).

Cultural practices are also an important management strategy (Elliott and Mann, 1996). Continuous wheat cropping should be avoided to discourage build-up of *S. mosellana* populations. In areas where populations exceed 1200 larvae m^{-2}, non-host crops, e.g. canola, *Brassica napus* L. and *B. rapa* L., flax, *Linum usitatissimum* L., and legumes should be grown instead. Other cereal crops such as barley, *Hordeum vulgare* L., oats, *Avena sativa* L., and annual canary grass, *Phalaris canariensis* L., can be grown with little or no risk of *S. mosellana* damage. For low to moderate infestations, damage can be reduced by selecting less susceptible varieties of spring wheat, planting early and at higher densities. These practices promote uniform, advanced heading to avoid high adult *S. mosellana* populations. Resistant varieties are currently being developed (Barker and MacKenzie, 1996).

In Europe, *S. mosellana* is a pest of secondary importance (Meier, 1985), probably because of the effect of natural enemies. On the northern Great Plains, the European parasitoid *Macroglenes penetrans* (Kirby), probably introduced with *S. mosellana*, is significant in reducing infestations. A parasitized *S. mosellana* larva completes development and overwinters in the soil. The next spring, the larval parasitoid consumes its host and emerges as an adult in July.

Biological Control Agents

Pathogens

Although the impact of disease organisms on *S. mosellana* mortality remains largely unquantified, diseases (e.g. the fungus, *Entomophthora brevinucleata* Keller and Wilding, and viruses) do not appear to be a significant mortality factor in most years (Affolter, 1990).

Predators

In Europe, several predators attack adults, eggs and larvae of *S. mosellana* (Affolter, 1990). Spiders are known to capture adults; eggs are preyed upon by thrips; larvae in the wheat head are eaten by Coccinellidae and Syrphidae; and larvae in or on the soil are eaten by Carabidae and Staphylinidae. In western Canada, Floate *et al.* (1990) documented Carabidae as predators of *S. mosellana*.

Parasitoids

A literature review suggested that 27 parasitoids attack *S. mosellana* and the related

Contarinia tritici (Kirby) in Europe. Most authors agree that the six most common species attack both hosts indiscriminately (Carl and Affolter, 1984).

In 1985, a study was begun to evaluate parasitoids that could be introduced to augment the biological control provided by *M. penetrans*. Affolter (1990) showed that the parasitoid complex of both *S. mosellana* and *C. tritici* comprises only eight of the 27 species recorded. Its composition does not change in different areas or farming systems. He provided information on the host specificity of individual parasitoids and showed that the parasitoid complex attacking *S. mosellana* is distinct from that associated with *C. tritici*. While it is not uncommon that literature records give an erroneous picture of parasitoid complexes, this is an extreme case of disagreement. Not only are there far too many parasitoid species on record (mainly due to misidentifications), also the hosts have apparently been mixed up.

M. penetrans had the highest constancy and frequency (present in 78% of the samples), followed by *Platygaster* sp. and *Euxestonotus error* Fitch. These parasitoids have a higher fecundity than their host, are well synchronized with their host, are widely distributed in Europe, and are tolerant to intensive farming practices. Moreover, they were shown to act as delayed, density-dependent regulating factors on their host. Because of their wide distribution, their potential for adaptation to a new environment appears to be good. They are host-specific. As a result, *Platygaster* sp. and *E. error* were recommended for introduction. Females of both species lay their eggs in *S. mosellana* eggs or early instar larvae, and the parasitoid adults emerge from the host's third larval instar.

Releases and Recoveries

In Saskatchewan, *Platygaster* sp. and *E. error* were released within the canopy of spring wheat fields heavily infested with *S. mosellana*. Releases were timed to coincide with occurrence of oviposition (July). The selected field sites were at Wakaw (52°39′N 105°44′W), Saltcoats (51°02′N 102°10′W), Langenburg (50°51′N 101°43′W) and Blaine Lake (52°50′N 106°54′W). In total, 2022 *Platygaster* sp. and 1397 *E. error* adults were released, the majority (1371 *Platygaster* sp. and 1094 *E. error*) at Langenburg.

From 1996 to 1998, about 20,000 wheat heads were collected annually from commercial fields in the Langenburg release area, spread out in an even layer, and left at room temperature to dry, after which they were threshed with a single-head thresher. Midge larvae were separated from the seeds and chaff with a seed cleaner. The harvested larvae were placed in a vermiculite and sphagnum mixture, and stored at 2°C for 5–6 months before incubating the mixture at 22°C until no more adults of *S. mosellana* or parasitoids emerged. Only three *E. error* were recovered in the year following the first release. Adult *Platygaster* sp. recovered in 1996, 1997 and 1998, were 7, 21 and 23, respectively.

Evaluation of Biological Control

M. penetrans continues to play a leading role in regulating *S. mosellana* infestations in western Canada. In many areas, 30–80% parasitism occurs. In addition, the introduction of *Platygaster* sp. to Saskatchewan was successful, but its impact on *S. mosellana* is still minimal. Since natural enemies are already playing a major role in regulating *S. mosellana* in western Canada, they should be preserved as much as possible.

Recommendations

Further work should include:

1. Monitoring to determine establishment of *E. error* and spread of *Platygaster* sp.;
2. Continued refinement of IPM practices to minimize insecticide impact on natural enemies.

References

Affolter, F. (1990) *Structure and Dynamics of the Parasitoid Complex of the Wheat Midges* Sitodiplosis mosellana *(Géhin) and* Contarinia tritici *(Kirby).* June 1990 Report, International Institute of Biological Control, Delémont, Switzerland.

Allen, W.R. (1955) The wheat midge, *Sitodiplosis mosellana* (Géhin). *Annual Conference of Manitoba Agronomists 1955*, pp. 28–29.

Barker, P.S. (1984) Distribution of wheat midge damage on wheat in Manitoba in 1984. *Proceedings of the Entomological Society of Manitoba* 40, 25–29.

Barker, P.S. and McKenzie, R.I.H. (1996) Possible sources of resistance to the wheat midge in wheat. *Canadian Journal of Plant Science* 76, 689–695.

Barker, P.S., McKenzie, R.I.H. and Czarnecki, E. (1995) Incidence of damage to spring wheat by the orange blossom wheat midge in Manitoba during 1993. *Proceedings of the Entomological Society of Manitoba* 51, 12–20.

Carl, K. and Affolter, F. (1984) *The Natural Enemies of the Wheat Blossom Midge,* Sitodiplosis mosellana *(Géhin) and a Proposal for Its Biological Control in Canada.* 1984 Report, International Institute of Biological Control, Delémont, Switzerland.

Elliott, R.H. (1988a) Evaluation of insecticides for protection of wheat against damage by the wheat midge, *Sitodiplosis mosellana* (Géhin) (Diptera: Cecidomyiidae). *The Canadian Entomologist* 120, 615–626.

Elliott, R.H. (1988b) Factors influencing the efficacy and economic returns of aerial sprays against the wheat midge, *Sitodiplosis mosellana* (Géhin) (Diptera: Cecidomyiidae). *The Canadian Entomologist* 120, 941–954.

Elliott, R.H., and Mann, L.W. (1996) Susceptibility of red spring wheat, *Triticum aestivum* L. cv. Katepwa, during heading and anthesis to damage by wheat midge, *Sitodiplosis mosellana* (Géhin) (Diptera: Cecidomyiidae). *The Canadian Entomologist* 128, 367–375.

Felt, E.P. (1912) Observations on the identity of the wheat midge. *Journal of Economic Entomology* 5, 286–289.

Fletcher, J. (1902) *Experimental Farms Reports for 1901, No. 16.* Government of Canada, Ottawa, Ontario, p. 212.

Floate, K.D., Doane, J.F. and Gillott, C. (1990) Carabid predators of the wheat midge, *Sitodiplosis mosellana* (Diptera: Cecidomyiidae), in Saskatchewan. *Environmental Entomology* 19, 1503–1511.

Hartley, S., Kaminski, L., Olfert, O. and Giffen, D. (2000) *Forecast of Wheat Midge in Saskatchewan for 2000.* Technical Bulletin No. 2000–01, Saskatoon Research Centre, pp. 21–23.

Meier, W. (1985) *Planzenschutz im Feldbau.* Tierische Schaedlinge und Pflanzenkrankheiten. Huber & Co., Frauenfeld, Germany.

Olfert, O., Mukerji, M.K. and Doane, J.F. (1985) Relationship between infestation levels and yield loss caused by wheat midge, *Sitodiplosis mosellana* (Géhin) (Diptera: Cecidomyiidae), in spring wheat in Saskatchewan. *The Canadian Entomologist* 117, 593–598.

Wright, A.T. and Doane, J.F. (1987) Wheat midge infestation of spring cereals in northeastern Saskatchewan. *Canadian Journal of Plant Science* 67, 117–120.

51 *Stomoxys calcitrans* (L.), Stable Fly (Diptera: Muscidae)

T.J. Lysyk

Pest Status

The stable fly, *Stomoxys calcitrans* (L.), is a worldwide pest, introduced to North America during the 1700s, and is widely distributed in Canada. It is a significant pest of cattle in confined rearing facilities and is becoming an increasing concern in pastures and rangeland. It is also a significant pest of humans in recreational areas. It has a broad host range, and will feed on a variety of larger mammals. Both sexes feed on blood and have painful and irritating bites. Livestock react by twitching, stamping their feet, and flicking their tails. The flies make numerous visits to the host and bite repeatedly to obtain a full blood meal. Attack by *S. calcitrans* reduces weight gains and feeding efficiency in feeder animals by up to 20% (Campbell *et al.*, 1987) with reductions occurring at densities as low as 1–2 flies per front leg. They reduce milk flow in dairy cattle by 0.7% per fly, with reductions as high as 40% (Bruce and Decker, 1958).

Female *S. calcitrans* lay about 60–120 eggs on moist, decaying organic matter. Preferred developmental sites include manure mounds, general lots and indoor accumulations of manure and feed (Lysyk, 1993b). Eggs hatch in less than 24 h, and larvae develop in 1–2 weeks before pupating. Larvae require microorganisms, e.g. bacteria, for growth and development. Adult flies emerge in 1–2 weeks. Adults feed and rest on protected surfaces along feedbunks, fences and the sides of buildings (Lysyk, 1993b). The life cycle from egg to egg-laying adult requires 3–5 weeks and several generations per year are produced. Seasonal activity varies with the regional climate. Peak activity usually occurs during warm periods following rainfall. In southern Alberta, flies are active from May to October with peak activity in August and September. The attack period ranges from 42 to 112 days and averages 73 days year^{-1} (Lysyk, 1993a).

Background

The main methods to control *S. calcitrans* are sanitation and insecticide applications. Sanitation reduces developmental sites for immatures, and consists of reducing manure volume through cleaning and moisture control. Sanitation can reduce adult populations by 36–51% if applied before adult populations peak (Thomas *et al.*, 1996). However, rigorous sanitation is costly and difficult to apply. It can be interrupted by weather and requires that sufficient land be available for manure incorporation or composting.

Insecticides are frequently used to reduce high adult populations. Aerial sprays can be used for immediate knockdown, but care must be taken to avoid contaminating feed and water. Additionally, spray deposition may interfere with populations of naturally occurring predators and parasitoids on the manure surface. Aerial sprays will not kill immatures within the developmental media, therefore repeated applications must be made to reduce populations of emerging flies. Residual sprays can be applied to walls and vertical surfaces for this purpose, but these are costly and difficult to apply thoroughly.

Insecticides are not effective against the immatures as they are protected within the developmental medium. These constraints indicate the need for biological control.

Biological Control Agents

Parasitoids

The parasitic wasp fauna of *S. calcitrans* in Canada is poorly known. Initial surveys focused on dairies in southern Alberta (Lysyk, 1995) and were expanded to include feedlots throughout the province (Floate *et al.*, 1999). Similar results were seen with both surveys, although Lysyk (1995) recovered seven species compared to ten reported by Floate *et al.* (1999). Both surveys indicated that *Muscidifurax raptor* Girault and Sanders, *Muscidifurax zaraptor* Kogan and Legner, *Trichomalopsis sarcophagae* Gahan, and *Urolepis rufipes* (Ashmead) were the most abundant species, accounting for 95% and 86% of the species recovered. The surveys showed that the parasitoid fauna in Alberta is distinct from that of the USA, where *Spalangia* spp. are common and *Trichomalopsis* spp. are rare.

The life histories of *S. calcitrans* and *M. raptor*, *M. zaraptor*, *T. sarcophagae* and an additional species, *Muscidifurax raptorellus* Kogan and Legner, imported from a colony initiated with material collected in Nebraska by J.J. Petersen, USDA, were compared (Lysyk, 1996, 1998a, b, 2000). The species' development ranking with respect to the lower temperature threshold (T_0) were *M. raptor* < *M. raptorellus* = *S. calcitrans* < *M. zaraptor* < *T. sarcophagae*. The upper threshold (T_L) for each species was ranked *T. sarcophagae* < *M. raptorellus* < *M. raptor* = *S. calcitrans* = *M. zaraptor*. The optimal development temperature (T_{OPT}) was ranked *T. sarcophagae* = *S. calcitrans* < *M. raptorellus* = *M. raptor* < *M. zaraptor*. These results suggest that *M. raptor* has the broadest thermal requirements and is capable of increasing populations at temperatures lower than for *S. calcitrans*. *M. raptorellus* would also increase populations at lower temperatures than for *S. calcitrans*, but had a lower T_L, suggesting that its efficacy would be limited at high upper temperatures. Even though *T. sarcophagae* had the highest T_0, it was still able to reproduce at temperatures lower than reported for *Spalangia* spp. (Lysyk, 1998b) and may have a role to play in early season control.

Comparison of the functional responses of *M. raptor*, *M. raptorellus*, *M. zaraptor* and *T. sarcophagae* on *S. calcitrans* pupae showed that ranking in terms of number of pupa killed per parasitoid was *M. zaraptor* = *M. raptor* > *M. raptorellus* >> *T. sarcophagae* (Lysyk, 1996). Parasitoid progeny/parasitoid were ranked *M. raptorellus* > *M. zaraptor* > *M. raptor* >> *T. sarcophagae*. *M. raptor* and *M. zaraptor* produce a single progeny per parasitized pupae whereas both *M. raptorellus* and *T. sarcophagae* are gregarious. However, *M. raptorellus* has a higher acceptance of *S. calcitrans* pupae compared to *T. sarcophagae*. Behavioural differences suggest that *Muscidifurax* spp. are more capable of accepting, killing and reproducing on *S. calcitrans* pupae than *T. sarcophagae*.

Pathogens

A 1981 review reported no studies on pathogens for *S. calcitrans* (Roberts *et al.*, 1983). Relatively little work has been done since then, although a mermithid nematode has been collected from engorged adults (Smith *et al.*, 1987) and a bacterium, *Serratia marcescens* Bizio, has been reported as a facultative pathogen of *S. calcitrans* (Watson and Peterson, 1991).

We have screened 85 isolates of *Bacillus thuringiensis* Berliner against larvae and adult *S. calcitrans* (T.J. Lysyk, L.B. Selinger and D.D.S. Baines, unpublished). Most isolates had relatively little influence on larval survival, but five were effective and are being studied to develop them as commercial products. A single isolate was toxic to adults, causing 50% mortality within a few days of application.

Larval survival and developmental time are influenced by the bacterial species composition of the diet (Lysyk *et al.*, 1999). Survival on single isolates was highest when larvae were fed *Acinetobacter* and *Flavobacterium*, compared with *Empedobacter* and *Escherichia coli* (Migula) Castellani and Chalmers. However, survival and developmental time were enhanced when larvae developed on mixed cultures of *Empedobacter* and *Flavobacterium*. These results suggest that *S. calcitrans* larvae are sensitive to the food composition, and suggest that microbial manipulation may be useful for reducing larval survival. Also isolated were strains of *Serratia* and *Aeromonas* that prevented larval development. Larval mortality was 100% when first instars were exposed to pure cultures of either species on egg-yolk media.

The effect of *Serratia*, *Aeromonas* and a *Pseudomonas* on survival of adult *S. calcitrans* showed that *Serratia* is pathogenic to them. Assays indicated that *Serratia* isolates grown on egg-yolk media were more virulent against adults compared with the same isolates grown on nutrient broth and a low-protein media. Most adult mortality occurred within 24 h after feeding on blood containing bacteria. *Aeromonas* cultured on egg-yolk media also caused significant mortality of adults, with most mortality occurring within 24 h following ingestion. *Aeromonas* cultured on low-protein media had minimal effect on adult survival. *Pseudomonas* appeared more virulent to adult *S. calcitrans* compared with either *Aeromonas* or *Serratia*, and caused greater mortality at lower ingested doses. Virulence was not influenced by the culture media. Mortality due to *Pseudomonas* ingestion peaked near 5 days after ingestion. In addition to live cell assays, it was determined that *Aeromonas* produces extracellular products active against *S. calcitrans* adults. Experiments with one of these products showed that flies treated with a relatively low dose of *Aeromonas* cells had an LT_{50} of 5 days, and those treated with spent broth had an LT_{50} of 7 days, compared with 10 days for an untreated control.

Evaluation of Biological Control

Because *S. calcitrans* occurs primarily in confined systems, it is an excellent candidate for developing biological control methods as its habitat is reasonably well-defined, accessible, and can be manipulated. A number of parasitoids and pathogens potentially useful for biological control have been identified. Life history information of *S. calcitrans* and many of the parasitoid species is being integrated into a biological control model.

Recommendations

Further work should include:

1. Integration of pest and parasitoid life history information into a management programme to define release strategies that make better use of parasitoids;
2. Microbial manipulation, either by replacing food bacteria with non-food antagonists or by introducing pathogenic bacteria;
3. Developing selective microbial larvicides because no selective larvicides are available.

References

Bruce, W.N. and Decker, G.C. (1958) The relationship of stable fly abundance to milk production in dairy cattle. *Journal of Economic Entomology* 51, 269–274.

Campbell, J.B., Berry, I.L., Boxler D.J., Davis, R.L., Clanton, D.C. and Deutscher, G.H. (1987) Effects of stable flies (Diptera: Muscidae) on weight gain and feed efficiency of feedlot cattle. *Journal of Economic Entomology* 80,117–119.

Floate, K.D., Khan, B. and Gibson, G.A.P. (1999) Hymenopterous parasitoids of filth fly (Diptera: Muscidae) pupae in cattle feedlots. *The Canadian Entomologist* 131, 347–362.

Lysyk, T.J. (1993a) Seasonal abundance of stable flies and house flies (Diptera: Muscidae) in Alberta dairies. *Journal of Medical Entomology* 30, 888–895.

Lysyk, T.J. (1993b) Adult resting and larval developmental sites of stable flies and house flies (Diptera: Muscidae) on dairies in Alberta. *Journal of Economic Entomology* 86, 1746–1753.
Lysyk, T.J. (1995) Parasitoids (Hymenoptera: Pteromalidae, Ichneumonidae) of filth fly pupae (Diptera: Muscidae) pupae on dairies in Alberta. *Journal of Economic Entomology* 88, 659–665.
Lysyk, T.J. (1996) *Development of Biological Control Methods for Stable Flies in Feedlots.* Farming for the Future, Direct Funding Program, Final Report, Alberta Agriculture Research Institute.
Lysyk, T.J. (1998a) Relationships between temperature and life-history parameters of *Stomoxys calcitrans*. *Journal of Medical Entomology* 35, 107–119.
Lysyk, T.J. (1998b) Relationships between temperature and life-history parameters of *Trichomalopsis sarcophagae*. *Environmental Entomology* 27, 488–498.
Lysyk, T.J. (2000) Relationships between temperature and life history parameters of *Muscidifurax raptor* (Hymenoptera: Pteromalidae). *Environmental Entomology* 29, 596–605.
Lysyk, T.J., Kalischuk-Tymensen, L., Selinger, L.B., Lancaster, R.C. and Cheng, K.-J. (1999) Rearing stable fly larvae on an egg yolk medium. *Journal of Medical Entomology* 36, 382–388.
Roberts, D.W., Daoust, R.A. and Wraight, S.P. (1983) *Bibliography on Pathogens of Medically Important Arthropods: 1981.* World Health Organization, Geneva, Switzerland.
Smith, J.P., Hall, R.D. and Thomas, G.D. (1987) Field parasitism of the stable fly (Diptera: Muscidae). *Annals of the Entomological Society of America* 80, 391–397.
Thomas, G.D., Skoda, S.R., Berkebile, D.R. and Campbell, J.B. (1996) Scheduled sanitation to reduce stable fly (Diptera: Muscidae) populations in beef cattle feedlots. *Journal of Economic Entomology* 89, 411–414.
Watson, D.W. and Peterson, J.J. (1991) Infectivity of *Serratia marcescens* (Eubacteriales: Enterobacteriaceae) in *Stomoxys calcitrans* (Diptera: Muscidae). *Journal of Medical Entomology* 28, 190–192.

52 *Strobilomyia neanthracina* Michelsen and *S. appalachensis* Michelsen, Spruce Cone Maggots (Diptera: Anthomyiidae)

J.D. Sweeney, E.G. Brockerhoff, M. Kenis and J.J. Turgeon

Pest Status

The white spruce cone maggot, *Strobilomyia neanthracina* Michelsen, and the black spruce cone maggot, *S. appalachensis* Michelsen, are native to North America. They are destructive pests of spruce, *Picea* spp., seeds in both natural stands and seed orchards (Turgeon and de Groot, 1992; Sweeney and Turgeon, 1994). Their pest status has grown in the past 20 years with the establishment of many spruce seed orchards and an increased reliance on genetically improved seed for accelerated regeneration of Canada's forests. For example, all of the seedlings used for artificial regeneration of New Brunswick crown lands are produced from seed orchard seed (K. Tosh, Fredericton, 1999, personal communication). Seed losses to *Strobilomyia* spp. range from less than 5% to more than 99%, with greater percentage losses usually occurring when cone abundance is low to moderate (Sweeney and Gesner, 1997).

All *Strobilomyia* spp. have a similar life cycle and biology (Michelsen, 1988). Typically, adult fly emergence is synchronized with bud burst of their respective host. Female *S. neanthracina* lay their eggs between the seed cone scales of *Picea glauca* (Moench) Voss, *P. sitchensis* (Bongard) Carrière and *P. engelmannii* Parry ex Engelmann, whereas those of *S. appalachensis* oviposit in *P. mariana* (Miller) Britton, Sterns and Poggenburg, and *P. rubens* Sargent (Michelsen, 1988; Turgeon and Sweeney, 1993). Upon hatching, larvae begin spiralling around the cone axis, feeding predominantly on seeds. Third-instar larvae drop to the ground, mostly during rainfall, pupate and enter either simple (1 year) or extended (2–5 years) diapause.

Background

When necessary, maggot damage is usually controlled with one application of a systemic insecticide, e.g. dimethoate, when cones are half to three-quarters pendant. No attempts at biological control of *Strobilomyia* spp. were made in Canada or elsewhere before 1979. Work since then has included: (i) studies on the diversity and impact of native parasitoids and predators (Fidgen *et al.*, 1999; J.D. Sweeney and G.N. Gesner, unpublished; E.G. Brockerhoff *et al.*, unpublished); (ii) foreign exploration to assess the diversity and impact of parasitoids of the Palaearctic *Strobilomyia anthracina* (Czerny), a pest of *Picea abies* (L.) Karsten in Europe and Asia (Brockerhoff and Kenis, 1997) and to compare it to that of Nearctic species; and (iii) laboratory and field assays on the use of entomopathogenic fungi (Fogal, 1986) and nematodes (Sweeney and Gesner, 1995; Sweeney *et al.*, 1998).

Biological Control Agents

Parasitoids

Surveys in Canada for parasitoids attacking eggs, larvae and pupae of *S. appalachensis* and *S. neanthracina* were conducted from 1992 to 1999 at numerous locations in the Maritimes, Ontario and British Columbia (Fidgen *et al.*, 1999; E.G. Brockerhoff *et al.*, unpublished). No egg parasitoids were found, but three species of endoparasitoids that attacked either eggs or larvae, at least one larval ectoparasitoid species, and two pupal parasitoids were recovered (Table 52.1). The most abundant endoparasitoid was an undescribed *Melanips* sp. (K. Schick, Sacramento, 1999, personal communication), which parasitized a mean (range) of 10% (0–36%) of *S. neanthracina* and 27% (0–79%) of *S. appalachensis* (E.G. Brockerhoff *et al.*, unpublished). An undescribed *Atractodes* sp. Gravenhorst (J. Luhman, Minneapolis, 1999, personal communication) was less common, with mean parasitism rates of 6% in both *S. neanthracina* (0–29%) and *S. appalachensis* (0–50%). Both species oviposit in host eggs and emerge from puparia (E.G. Brockerhoff *et al.*, unpublished). Mortality caused by *Melanips* sp. and *Atractodes* sp. may have been underestimated because some of the parasitized eggs probably had died as a result of the parasitoid ovipositing.

Parasitism rates by the Holarctic larval–pupal parasitoid, *Phaenocarpa seitneri* Fahringer, were about 5% (0–29%) in *S. neanthracina* and less than 1% (0–7%) in *S. appalachensis* (E.G. Brockerhoff *et al.*, unpublished). Parasitism by the ectoparasitoid(s) *Scambus longicorpus longicorpus* Walley (J. Luhman, 1999, Minneapolis, personal communication; E.G. Brockerhoff *et al.*, unpublished) or *Scambus* sp. (Fidgen *et al.*, 1999) varied considerably and averaged 6% (0–40%) in *S. neanthracina* and 7% (0–20%) in *S. appalachensis.* Occasionally, *Strobilomyia* sp. larvae that appeared like those paralysed by *Scambus* sp. were found but no parasitoid larva was observed. Possibly, this mortality was caused by probing females that did not

Table 52.1. Parasitoids of the European *Strobilomyia anthracina*, and North American *S. neanthracina* and *S. appalachensis*, compiled from the literature and Brockerhoff *et al.* (unpublished).

Parasitoid guild	North America	Europe
Egg parasitoids		
Trichogrammatidae	?	*Trichogramma* sp.[a]
Egg–pupal endoparasitoids		
Ichneumonidae	*Atractodes* sp. (not *scutellatus* Hellén) [b,c]	*Atractodes scutellatus* Hellén[d,e]
		Atractodes sp.[a]
Figitidae	*Melanips* sp.[b,c]	*Melanips* sp.[f], *Sarothrus* sp.[a,d]
Larval–pupal endoparasitoids		
Braconidae	*Phaenocarpa seitneri* Fahringer[b]	*Phaenocarpa seitneri* Fahringer[d]
Late larval ectoparasitoids		
Ichneumonidae	*Scambus longicorpus longicorpus* Walley[b]	*Scambus* sp.[a]
	Scambus sp.[c]	
Pupal parasitoids		
Pteromalidae	?	*Tritneptis* sp. near *lophyrorum* (Rushka)[a]
Megaspilidae	*Conostigmus* sp.[b]	?
Eulophidae	*Melittobia acasta* (Walker)[c]	?

[a]Brockerhoff and Kenis (1997); [b]Brockerhoff *et al.* (unpublished); [c]Fidgen *et al.* (1999); [d]Stadnitzskii *et al.* (1978); [e]Kangas and Leskinen (1943); [f]Annila (1981).

oviposit. Exposure of fresh *Strobilomyia* spp. puparia in mesh cages covered with soil and litter under spruce trees revealed the pupal parasitoids, *Conostigmus* sp. and *Melittobia acasta* (Walker) (Table 52.1).

In central Europe, at least six parasitoid species of *S. anthracina* occur: one egg parasitoid, one larval ectoparasitoid, three endoparasitoids that parasitize eggs or larvae and emerge from pupae, and one pupal parasitoid (Table 52.1) (Stadnitzskii *et al.*, 1978; Annila, 1981; Brockerhoff and Kenis, 1997). The parasitoid complex of *S. anthracina* in Europe is similar to that of *S. appalachensis* and *S. neanthracina* in North America and species of *Melanips*, *Atractodes*, *Phaenocarpa* and *Scambus* fill similar niches on both continents (Table 52.1). Apparent parasitism was generally low in a 3-year survey, especially when cones were abundant (Brockerhoff and Kenis, 1997). However, higher parasitism rates have been reported from Russia and Finland (Stadnitzskii *et al.*, 1978; Annila, 1981). The relative abundance of endoparasitoid species in Europe is similar to that found in Canada (E.G. Brockerhoff *et al.*, unpublished). However, *Sarothrus* spp. were not recovered in Canada whereas *Sarothrus austriacus* (Tavares) (M. Sporrong, Lund, 1999, personal communication) occurred in the Alps (Brockerhoff and Kenis, 1997), and *Sarothrus abietis* Belizin and *Sarothrus* sp. occur in Russia (Stadnitzskii *et al.*, 1978).

Only one of several hundred *S. anthracina* eggs examined was parasitized by a *Trichogramma* sp. (Brockerhoff and Kenis, 1997), possibly an accidental attack by *Trichogramma cacoeciae* Marchal, which parasitized eggs of *Cydia strobilella* (L.) in spruce cones at the same period (see Brockerhoff *et al.*, Chapter 19 this volume). No egg parasitoids are recorded from *S. neanthracina* or *S. appalachensis* in Canada and few pupal parasitoids were recovered in either Europe or Canada (Table 52.1). Parasitism of sentinel puparia in Europe was less than 5% by a gregarious wasp, *Tritneptis* sp. near *lophyrorum* (Rushka). The parasitism rate may be different under more natural conditions.

Predators

In New Brunswick, studies of stage-specific mortality factors affecting *S. appalachensis* (Fidgen *et al.*, 1999) and *S. neanthracina* (J.D. Sweeney and G.N. Gesner, unpublished) showed that soil-inhabiting predators caused higher mortality than parasitoids. For example, only 5–8% of *S. appalachensis* third-instar larvae died from parasitism (*Scambus* sp.) compared with 48–62% mortality of dispersing third-instar larvae (prepupae) due to predation on or in the soil; 25–33% of *S. appalachensis* pupae were killed by egg–pupal parasitoids (chiefly *Melanips* sp.) compared to 39–65% mortality due to predation and unknown factors (Fidgen *et al.*, 1999). Pitfall trap surveys for potential *Strobilomyia* spp. predators in seed orchards showed that Carabidae and Formicidae constituted more than 80% of the total invertebrate catch; six species of ants (but no ground beetles) preyed on *S. neanthracina* prepupae in field observations (J.D. Sweeney, unpublished).

Pathogens

Nematodes

Sweeney and Gesner (1995) and Sweeney *et al.* (1998) investigated the susceptibility of *S. appalachensis* and *S. neanthracina* to *Steinernema* spp. from 1990–1997 in New Brunswick. In the laboratory, *S. appalachensis* larvae were easily infected, although susceptibility varied somewhat among the nematode species and strains tested, whereas puparia were highly resistant to infection.

Attempts to infect larvae feeding within cones, in both the field and laboratory, were unsuccessful. Based on the life history of *Strobilomyia* spp. and the natural habitat of the nematodes, the strategy considered to have the greatest chance of success was to apply nematodes to soil to infect mature larvae after they had exited cones. Typically, a fly larva moults into a puparium within 2–5 days of leaving a cone (Sweeney and Turgeon, 1994), thus the window of opportunity for infection of a single larva by a nematode is narrow. Adequate population suppression would require the presence of sufficient concentrations of nematodes in the soil during the entire period of larval drop, which lasts 2–4 weeks, depending on frequency of rainfall (Sweeney and Gesner, 1995). In New Brunswick, field testing was conducted with *Steinernema feltiae* (Filipjev) (= *S. bibionis* (Bovien)), strains 27 and Umeå, and *Steinernema carpocapsae* (Weiser) All strain in *P. mariana* seed orchards at Bettsburg (1991–1992) and Sussex (1992–1994), and in a *P. glauca* seed orchard at Queensbury (1995–1997). Mean efficacy (percentage of maggots infected and killed) ranged from 20 to 95% on the day of application, but declined significantly 1 week after application. Nematode persistence and efficacy were not increased either by acclimatizing the infective juveniles to a fluctuating temperature regime that simulated field conditions, or by irrigating plots after application (J.D. Sweeney, unpublished). Applying a thin layer of peat or bark mulch to the soil following nematode application significantly reduced the rate of decline in nematode efficacy over time. However, the percentage of maggots infected immediately following nematode application was lower in mulched than in unmulched plots, so mean efficacy over a 3-week period was not increased by mulching (Sweeney *et al.*, 1998).

Fungi

Research on use of fungi to control *Strobilomyia* spp. and other seed and cone insects showed that larvae and pupae of *S. neanthracina* were susceptible to infection by *Beauveria bassiana* (Balsamo) Vuillemin and *Metarrhizium anisopliae* (Metchnikoff) Sorokin (Timonin *et al.*, 1980). Further experiments showed that *B. bassiana* efficacy was affected by soil moisture content (Fogal, 1986) and by timing of application relative to cone phenology (Fogal *et al.*, 1986a, b). Average mortality was 21% in field trials where third-instar larvae were dropped on soil treated with 3.0×10^5 conidia cm^{-2} (Fogal, 1986). Cones of *P. glauca* on which conidia of *B. bassiana* were applied directly (by hand using a camel-hair brush) shortly after closure of cone scales had significantly more filled seed (55%) than untreated cones. However, differences in filled seed between treated and untreated cones were not significant when conidia were applied on the same trees just 3 days earlier (Fogal *et al.*, 1986a, b).

Evaluation of Biological Control

Classical biological control of *Strobilomyia* spp. would have limited potential because the complex of natural enemies in Canada is similar to that observed in Europe. However, the action of native parasitoids could be enhanced, e.g. through conservation of natural enemies (Brockerhoff and Kenis, 1998). Unharvested cones are often removed from seed orchards and destroyed to reduce the populations of cone and seed insects overwintering in cones. This practice may also reduce natural enemy populations, e.g. *Scambus* sp., which also overwinter in the cones. Keeping cones in cages and selectively releasing parasitoids back into seed orchards may reduce pest problems (Brockerhoff and Kenis, 1998), but this has not been assessed.

Maggots of *Strobilomyia* spp. were susceptible to nematodes and fungi but field efficacy was variable and significantly affected by the timing of application relative to maggot phenology. Nematodes are currently exempt from registration under the Pest Control Products Act and could be used operationally in Canadian seed orchards. However, because of the rapid drop in efficacy following nematode application, 3–4 applications would be necessary to adequately target the population of dispersing larvae within an orchard. This would be costly. Also, the strategy of larval suppression would not reduce seed losses in the year of application nor necessarily reduce the risk of infestation in subsequent years because *Strobilomyia* spp. adults could emigrate from surrounding hosts or emerge from pupae in extended diapause within the orchard.

Recommendations

Future research should include:

1. Identifying the factors that affect the impact of natural enemies on *Strobilomyia* spp. populations and examining ways of conserving or enhancing their impact;
2. Reviewing the use of *Steinernema* spp. and *B. bassiana* as improved formulations with greater field persistence become available;
3. Clarifying the taxonomy of *Melanips* and *Sarothrus*.

Acknowledgements

Specimens were identified by A. Bennett, M. Fischer, N. Fergusson, K. Horstmann, J. Huber, D.R. Kasparyan, J. Luhman, L. Masner, B. Pintureau, J. Read, K. Schick, M. Sporrong, S. Vidal and R. Wharton. Funding for some of this research was provided by Canadian Forest Service's Green Plan.

References

Annila, E. (1981) Fluctuations in cone and seed insect populations in Norway spruce. *Communicationes Instituti Forestalis Fenniae* 101, 1–32.

Brockerhoff, E.G. and Kenis, M. (1997) Oviposition, life cycle, and parasitoids of the spruce cone maggot, *Strobilomyia anthracina* (Diptera: Anthomyiidae), in the Alps. *Bulletin of Entomological Research* 87, 555–562.

Brockerhoff, E.G. and Kenis, M. (1998) Strategies for the biological control of insects infesting coniferous seed cones. In: Battisti, A. and Turgeon, J.J. (eds) *Proceedings, Cone and Seed Insect Working Party Conference (IUFRO S7.03-01), September 1996, Monte Bondone, Italy.* Institute of Agricultural Entomology, University of Padova, Padova, Italy, pp. 49–56.

Fidgen, L.L., Sweeney, J.D. and Quiring, D.T. (1999) Stage-specific survival of *Strobilomyia appalachensis* (Diptera: Anthomyiidae). *The Canadian Entomologist* 131, 483–494.

Fogal, W.H. (1986) Applying *Beauveria bassiana* to soil for control of the spruce cone maggot. In: Roques, A. (ed.) *Proceedings of the Second Cone and Seed Insects Working Party Conference (IUFRO S7.03-01) 3–5 September 1986, Briançon, France.* Institut National de la Recherche Agronomique, Ardon, France, pp. 257–266.

Fogal, W.H., Thurston, G.S. and Chant, G.D. (1986a) Reducing seed losses to insects by treating white spruce conelets with conidiospores of *Beauveria bassiana. Proceedings of the Entomological Society of Ontario* 117, 95–98.

Fogal, W.H., Mittal, R.K. and Thurston, G.S. (1986b) *Production and Evaluation of* Beauveria bassiana *for Control of White Spruce Cone and Seed Insects.* Canadian Forestry Service Information Report PI-X-69.

Kangas, E. and Leskinen, K. (1943) *Pegohylemyia anthracina* Czerny (Dipt., Muscidae) als Zapfenschädling an der Fichte. *Annales Entomologici Fennici* 9, 195–212.

Michelsen, V. (1988) A world revision of *Strobilomyia* gen.n.: the anthomyiid seed pests of conifers (Diptera: Anthomyiidae). *Systematic Entomology* 13, 271–314.

Stadnitskii, G.V., Lurchenko, G.I., Smetanin, A.N., Grebenshchikova, V.P. and Pribylov, M.V. (1978) *Vrediteli shishek i semian svoinykh porod.* Lesnaia promyshlennost, Moskow, Russia. (Translation: Yates, H.O. *Conifer Cone and Seed Pests.* Forestry Sciences Laboratory, Athens, Georgia.)

Sweeney, J.D. and Gesner, G.N. (1995) Susceptibility of the black spruce cone maggot, *Strobilomyia appalachensis* Michelsen (Diptera: Anthomyiidae) to entomopathogenic nematodes (Nematoda: Steinernematidae). *The Canadian Entomologist* 127, 865–875.

Sweeney, J.D. and Gesner G.N. (1997) Effect of gibberellic acid$_{4/7}$ on cone crop of *Picea glauca* and prolonged diapause in *Strobilomyia neanthracina.* In: Battisti, A. and Turgeon, J.J. (eds) *Proceedings, Cone and Seed Insect Working Party Conference (IUFRO S7.03-01), September 1996, Monte Bondone, Italy.* Institute of Agricultural Entomology, University of Padova, Padova, Italy, pp. 141–148.

Sweeney, J.D. and Turgeon, J.J. (1994) Life cycle and phenology of a cone maggot, *Strobilomyia appalachensis* Michelsen (Diptera: Anthomyiidae), on black spruce, *Picea mariana* (Mill.) B.S.P. in eastern Canada. *The Canadian Entomologist* 126, 49–59.

Sweeney, J.D., Gesner, G.N., Bennett, R. and Vrain, T. (1998) Effect of mulches on persistence of entomopathogenic nematodes (*Steinernema* spp.) and infection of *Strobilomyia neanthracina* (Diptera: Anthomyiidae) in field trials. *Journal of Economic Entomology* 91, 1320–1330.

Timonin, M.I., Fogal, W.H. and Lopushanski, S.M. (1980) Possibility of using white and green muscardine fungi for control of cone and seed insect pests. *The Canadian Entomologist* 112, 849–854.

Turgeon, J.J. and de Groot, P. (1992) *Management of Insect Pests of Cones in Seed Orchards in Eastern Canada.* Ontario Ministry of Natural Resources and Forestry Canada, Sault Ste Marie, Ontario.

Turgeon, J.J. and Sweeney, J.D. (1993) Hosts and distribution of spruce cone maggots (*Strobilomyia* spp.) (Diptera: Anthomyiidae) and first record of *Strobilomyia appalachensis* Michelsen in Canada. *The Canadian Entomologist* 125, 637–642.

53 *Tetranychus urticae* Koch, Twospotted Spider Mite (Acari: Tetranychidae)

D.A. Raworth, D.R. Gillespie, M. Roy and H.M.A. Thistlewood

Pest Status

The twospotted spider mite, *Tetranychus urticae* Koch, originally from Eurasia, is now cosmopolitan and has a host range of more than 900 plant species (Navajas, 1998). Crops affected by this pest in Canada include: greenhouse tomato, *Lycopersicon esculentum* L., cucumber, *Cucumis sativus* L., pepper, *Capsicum annuum* L., chrysanthemum, *Chrysanthemum* spp., rose, *Rosa* spp., and other ornamentals, strawberry, *Fragaria* × *ananassa* Duchesne, raspberry, *Rubus idaeus* L., currant, *Ribes* spp., hop, *Humulus lupulus* L., apple, *Malus pumila* Miller (= *M. domestica* Borkhausen), pear, *Pyrus communis* L., peach, *Prunus persica* (L.), and grape, *Vitis* spp.

The impact of *T. urticae* on crop yields depends on host plant resilience. Raworth (1986) observed yield losses of more than 10% when mites increased to 65 per strawberry leaflet during spring, but no effect could be detected when mite densities were more than 100 mites per leaflet on raspberry fruiting canes (Raworth, 1989); only severe defoliation of primocanes appeared to have an effect on yield (Raworth and Clements, 1996). However, in Quebec, where *Tetranychus mcdanieli* McGregor is the most important spider mite on raspberry (Roy *et al.*, 1999a) and hot, dry conditions are frequently observed in July, high populations of spider mites on fruiting canes can affect yield and fruit quality. The greenhouse flower industry has a low tolerance for mite feeding damage. In general, the rapid rate of increase of *T. urticae* in field and greenhouse environments necessitates monitoring on a weekly basis and immediate remedial action to prevent defoliation and potential crop losses.

T. urticae readily forms host races (Gotoh *et al.*, 1993). The risk of accidental introduction of different races into Canada has increased with the global economy. The carmine mite, *Tetranychus cinnabarinus* (Boisduval), thought by some workers to be part of the *T. urticae* complex, was introduced into greenhouses in British Columbia, Alberta and Quebec during 1997–1999. On tomato, this form can cause chlorotic and necrotic lesions and premature leaf drop at very low mite densities (Scopes, 1985). Although *T. cinnabarinus* is typically found in countries with warm climates, and in greenhouses in Europe, a related form was found in 1990 on maize, *Zea mays* L., in Belgium (Hance *et al.*, 1998), suggesting that establishment of the strain outdoors in north temperate climates is possible. Research suggests that the strain introduced to Canada does diapause, so this form could pose a continuous threat to greenhouse tomato growers.

The life cycle of *T. urticae* includes: a translucent pearl-like egg; larva, protonymph and deutonymph with quiescent stages; and adult males and females. Spring and summer generations are typically light green in colour. Mated females overwinter as a diapausing red form. In warmer parts of Canada, e.g. south-western British Columbia and southern Ontario, they enter diapause in October–November and break diapause in February–March. However, in British Columbia greenhouses, timing mechanisms may be disturbed, because emerging overwintering red forms have been observed in April–May. Helle and Sabelis (1985) and Gillespie and Raworth (2001) provide further details on biology.

Background

The availability of chemical tools for spider mite control has declined during the past 20 years; companies have withdrawn products, resistance has occurred, and new products face strict registration requirements. At the same time, however, native arthropods with very different life histories have been developed and commercialized to control *T. urticae*, namely *Amblyseius fallacis* (Garman) (Elliott, 1997); *Feltiella acarisuga* (Vallot) (Gillespie *et al.*, 1998); *Stethorus punctillum* Weise (Raworth *et al.*, 1997); and a generalist predator, *Dicyphus hesperus* Knight (McGregor *et al.*, 1999).

Since the early 1980s, *Phytoseiulus persimilis* Athias-Henriot has been the principal predator used in greenhouses to control *T. urticae*, but growers increasingly rely on a community of predators rather than a single species. A predator community is, in principle, better able to regulate spider mites, given the variety of conditions found in greenhouses. The same principle is being applied to field crops; e.g. in Quebec an IPM programme, currently used on 30 raspberry farms, combines inoculative releases of *A. fallacis* and sound management practices to increase populations of *S. punctillum* (Roy *et al.*, 1999b).

Biological control programmes in the field have developed more slowly than in greenhouses. In British Columbia, introductions of *P. persimilis* (including a 'cold tolerant' strain from New Zealand) to control *T. urticae* on strawberry in the Fraser Valley during the early 1980s were unsuccessful and the predator was unable to overwinter. During the mid-1980s, *Typhlodromus* (= *Metaseiulus*) *occidentalis* Nesbitt was released on strawberry in the Okanagan. Although they established, and appeared to effect control (Elliott, 1987), the work ceased because predator production was targeted for apples and chemical controls interfered with that use. Biological control of tetranychid mites on fruit trees focused on *Panonychus ulmi* (Koch) (see Hardman and Thistlewood, Chapter 42 this volume) and secondarily on *T. urticae*. Considerable work was undertaken on small fruit crops across Canada in the mid-1990s, using the native predators *A. fallacis* and *S. punctillum*. Releases of these species are restricted by the cost relative to the crop value; natural populations are encouraged, and *A. fallacis* is released inoculatively (e.g. 17,000–25,000 ha^{-1} on strawberry) when native populations are absent.

Together with adoption of biological control in field and greenhouse crops, application of chemicals has been reduced through monitoring (e.g. Raworth and Strong, 1990). Miticides have been applied at half-rate to reduce *T. urticae* populations and maintain predator populations (Henderson and Matys, 1997). Caron *et al.* (2000) provided an IPM guide for spider mite control.

Biological Control Agents

Predators

The situation with respect to releases of biological control agents against spider mite has changed dramatically since 1980. In Canada, several insectaries produce agents for national and international distribution, and agents are also imported into Canada from insectaries in other countries. A logical source of information about releases of biological control agents is the sales records of the insectaries but confidentiality considerations restrict access to these data. We therefore provide some crude estimates based on area and average use.

P. persimilis originated from Chile (see Gillespie and Raworth, 2001). Egg-to-egg development is 91 degree-days above 11°C

(10 days at 20°C) (Sabelis, 1985). This predator does not have an overwintering form and survival depends on continuous warm conditions. It is released on greenhouse tomato, pepper, cucumber, rose and other ornamentals. A tomato-adapted strain was developed (Gillespie *et al.*, 1996) and produced commercially during the 1990s, but local production has been discontinued. A conservative estimate of the numbers necessary for annual spider mite control in greenhouses in the Fraser Valley is 12 million (230 ha × 0.5 proportion of affected ha × 2000 predators 0.1 ha^{-1} × 5 applications); similar estimates for Ontario and Quebec are 7 million (G.M. Fergeson, Guelph, 2000, personal communication) and 6 million (L. Lambert, Saint-Rémi, 2000, personal communication) annually. Although *P. persimilis* is not released for spider mite control outdoors in British Columbia, it has been found during summer on raspberry about 1 km from a greenhouse.

A. fallacis insectary stock originated from Vineland, Ontario, and has resistance to permethrin. Egg-to-egg development is 11 days at 20°C (McClanahan, 1968). Mated females overwinter in a short-day-induced facultative reproductive diapause and terminate diapause in early spring (Overmeer, 1985). *A. fallacis* is released on greenhouse pepper, cucumber, strawberry and rose, interior plantscapes, nurseries, and field raspberry and strawberry across Canada (D. Elliott, Sidney, 2000, personal communication).

F. acarisuga was originally Holarctic but is now cosmopolitan. Canadian insectary stock originated in the Fraser Valley. Egg-to-egg development at 20°C requires 11–15 days, and development time increases at RH below 50% (Gillespie *et al.*, 1999). There is no evidence for photoperiod-induced diapause in the British Columbia strain, but feeding on diapausing *T. urticae* seems to induce diapause (Gillespie *et al.*, 1998). *F. acarisuga* is released on greenhouse cucumber, pepper and tomato across Canada. Many releases are made as 'spot-treatments' in parts of crops where spider mite numbers are increasing rapidly, and not as a routine component of natural enemy releases. For this reason, it is difficult to estimate numbers released.

S. punctillum, an Old World species, was not recorded in North America until 1949 (Putman, 1955). Insectary stock originated from London, Ontario. Development from egg to adult is 21 days at 21°C (Putman, 1955). M. Roy (unpublished) demonstrated facultative, reproductive diapause for Quebec populations. *S. punctillum* is used on greenhouse pepper and cucumber across Canada (D. Elliot, Sidney, 2000, personal communication). Wild populations of *S. punctillum* are present in raspberry fields in Quebec (Roy *et al.*, 1999a) and British Columbia[1] (Raworth, 1989). In Quebec, these populations are enhanced through an IPM programme that reduces pesticide use.

D. hesperus insectary stock originated from the Okanagan valley but was replaced by material from California to help alleviate diapause problems. In the wild, 2–3 generations per year occur (McGregor *et al.*, 1999). All motile stages are omnivorous and must obtain water from plants to complete development and reproduce (Gillespie and McGregor, 2000). Although this agent may cause a few blemishes on tomato fruits that are allowed to ripen in greenhouses (McGregor *et al.*, 2000), it could become a major component in biological control programmes for *T. urticae* and Aleyrodidae. In 1999 and 2000, several thousand adults were released in greenhouses across Canada.

Given the number of species used for biological control, and the potential for contamination of insectary stocks, workers must take care to obtain valid identifications at regular intervals. Stocks of *A. fallacis*, for

[1]Identified as *Stethorus punctum picipes* Casey (I.D.J. McNamara, Ottawa, 1984), but now known to be *S. punctillum* (identified using genitalia by Y. Bousquet, Ottawa, 1997).

example, have been contaminated with *Amblyseius californicus* (McGregor).

Evaluation of Biological Control

Generalist and omnivorous natural enemies are increasingly being adopted as biological control agents, especially in greenhouses. Theoretical studies suggest that they may contribute to stability of predator–prey systems, although no empirical studies have yet confirmed this. Given the potential for unanticipated results with generalist and omnivorous natural enemies, regulatory agencies should continue to consider carefully proposals to introduce exotic species of this class of biological control agents.

P. persimilis has been studied extensively (e.g. Sabelis, 1981) and its efficacy demonstrated through years of experimental and commercial releases. However, growers consider that the efficacy of insectary stocks has decreased, many estimating that releases of twice as many *P. persimilis* are now required to control equivalent infestations of *T. urticae* as in previous years. Reproductive period of the predator and fecundity decreased from 1967 to 1999 (Raworth, 2000); and Bjørnson *et al.* (2000) documented a seasonal decline in fecundity.

A. fallacis was mass-reared (Applied Bio-nomics Ltd, Sidney, British Columbia; J. Whistlecraft, Agriculture and Agri-Food Canada, London, Ontario; and H.M.A. Thistlewood, AAFC, Vineland) and released on apple (H.M.A. Thistlewood, unpublished), peach (Lester *et al.*, 1999), strawberry, raspberry, currant, hops (Henderson and Matys, 1997) and greenhouse pepper (Luczynski and Matys, 1997) during experimental trials from 1992 to 1995. Totals of 18.8, 3.8 and 2.2 million were released in British Columbia, Ontario and Quebec, respectively (Elliott, 1997; H.M.A. Thistlewood, unpublished). That these predators established was verified on tree fruit crops in Ontario using genetic markers (Navajas and Thistlewood, 1996), and also by the detection of permethrin resistance in samples recovered from strawberry and hops in British Columbia. On fruit trees, *A. fallacis* failed to provide consistent control. Lack of adequate experimental controls made rigorous evaluation of the other field releases difficult, but previous work in experimental strawberry plots (Raworth, 1990) provided evidence of efficacy.

F. acarisuga was mass-reared (Natures Alternative International Inc., Nonoose Bay; and D. Gillespie, Agriculture and Agri-Food Canada, Agassiz) and released in commercial greenhouse tomato trials during 1994. Although poor establishment occurred in these trials, with a lack of spider mite control, concurrent trials in research greenhouses yielded good results. The commercial practice of removing leaves from the base of tomato plants as they grow appeared to be responsible for the poor results. In *ad hoc* trials in cucumber and pepper crops, where lower leaves were not all removed, *F. acarisuga* contributed to spider mite control.

S. punctillum was mass-reared (Applied Bio-nomics Ltd, Sidney, British Columbia; J. Whistlecraft, Agriculture and Agri-Food Canada, London, Ontario; and M. Roy, Sainte-Foy) and released on raspberry in Quebec field trials in 1994 and in commercial greenhouse trials on cucumber, pepper and tomato in British Columbia in 1996. The beetles moved throughout the greenhouses and established on pepper and cucumber, but not tomato (Raworth *et al.*, 1997). Cost prohibits their use in field crops.

D. hesperus has been mass reared (five commercial insectaries and D. Gillespie, Agriculture and Agri-Food Canada, Agassiz) and released in British Columbia, Ontario and Quebec in commercial greenhouse tomato trials since 1999. Gillespie *et al.* (2000) demonstrated control of greenhouse whitefly, *Trialeurodes vaporariorium* (Westwood), and McGregor *et al.* (1999) observed that the predator fed on spider mites in laboratory and greenhouse studies. However, biological control of spider mites by *D. hesperus* on greenhouse crops has not yet been demonstrated.

The increased movement of people and

products from other countries into Canada poses a serious risk to crops. A new strain of *T. urticae* has already been introduced into greenhouse crops and it is very possible that other mite species could be introduced. Many crops are at risk, but greenhouse crops are acutely so, given the warm conditions maintained therein, and the trend during the past 5 years towards continuous production. Growers, as well as federal and provincial agencies that deal with imports and plant protection, must be vigilant.

Recommendations

Further work should include:

1. Improving the efficacy of *P. persimilis*;
2. Studying predator–predator interactions to determine optimum control strategies;
3. Studies that clearly define the impact of spider mites on yield, particularly of greenhouse vegetables, so that growers can make informed decisions about application of chemical versus biological controls;
4. Improving techniques to evaluate efficacy of inoculative biological control in the field.

References

Caron, J., Laverdière, L. and Roy, M. (2000) *Guide de Lutte Intégrée Contre les Tétranyques dans la Production de la Framboise.* Hortiprotection, Breakeyville, Quebec.

Bjørnson, S., Raworth, D.A. and Bédard C. (2000) Abdominal discoloration and the predatory mite *Phytoseiulus persimilis* Athias-Henriot: prevalence of symptoms and their correlation with short-term performance. *Biological Control* 19, 17–27.

Elliott, D. (1987) *Mass Production of Pesticide Resistant Predatory Mites for use in Integrated Pest Management Programs in Fruit and Berry Crops.* Contribution CA910–3-0005/542, National Research Council of Canada, Ottawa, Ontario.

Elliott, D. (1997) Mass production. In: Elliott, D. (ed.) *Biological Control of Spider Mites on Fruit Crops.* Canada Department of Western Diversification, NABI Project # BC-92-WD-071, Section A, pp. 1–9.

Gillespie, D.R. and McGregor, R.R. (2000) The functions of plant feeding in the omnivorous predator *Dicyphus hesperus* (Heteroptera: Miridae): Water places limits on predation. *Ecological Entomology* 25, 380–386.

Gillespie, D.R. and Raworth, D.A. (2001) Biological control of twospotted spider mites on greenhouse vegetable crops. In: Heinz, K.M., van Driesche, R., and Parrella, M. (eds) *Biological Control of Arthropod Pests in Protected Culture* (in press).

Gillespie, D.R., Quiring, D.J.M., Foisy, M. and Contant, H. (1996) *An Evaluation of Characteristics of Tomato-adapted Strains of* Phytoseiulus persimilis. Technical Report 122, Agriculture and Agri-Food Canada, Pacific Agri-Food Research Centre, Agassiz, British Columbia.

Gillespie, D.R., Roitberg, B., Basalyga, E., Johnstone, M., Opit, G., Rodgers, J. and Sawyer, N. (1998) *Biology and Application of* Feltiella acarisuga *(Vallot) (Diptera: Cecidomyiidae) for Biological Control of Twospotted Spider Mites on Greenhouse Vegetable Crops.* Technical Report 145, Agriculture and Agri-Food Canada, Pacific Agri-Food Research Centre, Agassiz, British Columbia.

Gillespie, D.R., Opit, G. and Roitberg, B. (1999) Effects of temperature and relative humidity on development, reproduction and predation in *Feltiella acarisuga* (Vallot) (Diptera: Cecidomyiidae). *Biological Control* 17, 132–138.

Gillespie, D., McGregor, R., Quiring, D. and Foisy, M. (2000) *Biological Control of Greenhouse Whitefly with* Dicyphus hesperus. *An Update on the Development of an Omnivorous Predator for the British Columbia Greenhouse Industry.* Technical Report 157, Agriculture and Agri-Food Canada, Pacific Agri-Food Research Centre, Agassiz, British Columbia.

Gotoh, T., Bruin, J., Sabelis, M.W. and Menken, S.B.J. (1993) Host race formation in *Tetranychus urticae*: genetic differentiation, host plant preference, and mate choice in a tomato and a cucumber strain. *Entomologia Experimentalis et Applicata* 68, 171–178.

Hance, T., Neuberg, P. and Noèl-Lastelle, C. (1998) The use of fecundity, lobe biometry and the RAPD-PCR technique in order to compare strains of *Tetranychus* sp. *Experimental and Applied Acarology* 22, 649–666.

Helle, W. and Sabelis, M.W. (1985) *Spider Mites, Their Biology, Natural Enemies and Control*, Vol. 1A. Elsevier, Amsterdam, The Netherlands.
Henderson, D. and Matys, D. (1997) Field trials and monitoring. In: Elliott, D. (ed.) *Biological Control of Spider Mites on Fruit Crops*. Final Report for Canada Department of Western Diversification, NABI Project # BC-92-WD-071, Section B, pp. 1–79.
Lester, P.J., Thistlewood, H.M.A., Marshall, D.B. and Harmsen, R. (1999) Assessment of *Amblyseius fallacis* (Garmen) (Acari: Phytoseiidae) for biological control of tetranychid mites in an Ontario peach orchard. *Experimental and Applied Acarology* 23, 995–1009.
Luczynski, A. and Matys, D. (1997) Assessing the potential of *Amblyseius fallacis* as a biocontrol agent of the twospotted spider mite on indoor pepper during late fall and early spring. In: Elliott, D. (ed.) *Biological Control of Spider Mites on Fruit Crops*. Final Report for Canada Department of Western Diversification, NABI Project # BC-92-WD-071, Appendix G, pp. 1–10.
McClanahan, R.J. (1968) Influence of temperature on the reproductive potential of two mite predators of the two-spotted spider mite. *The Canadian Entomologist* 100, 549–556.
McGregor, R.R., Gillespie, D.R. Quiring, D.M.J. and Foisy, M.R.J. (1999) Potential use of *Dicyphus hesperus* Knight (Heteroptera: Miridae) for biological control of pests of greenhouse tomatoes. *Biological Control* 16, 104–110.
McGregor, R.R., Gillespie, D.R., Park, C.G., Quiring, D.M.J. and Foisy, M.R.J. (2000) Leaves or fruit? The potential for damage to tomato fruits by the omnivorous predator *Dicyphus hesperus* Knight (Heteroptera: Miridae). *Entomologia Experimentalis et Applicata* 95, 325–328.
Navajas, M. (1998) Host plant associations in the spider mite *Tetranychus urticae* (Acari: Tetranychidae): insights from molecular phylogeography. *Experimental and Applied Acarology* 22, 201–214.
Navajas, M. and Thistlewood, H.M.A. (1996) *Relevance of Genetic Markers in Evaluating Biological Control by Predatory Mites*. Bulletin 19, International Organization for Biological Control/Organization Internationale Lutte Biologique, p. 52.
Overmeer, W.P.J. (1985) Diapause. In: Helle, W. and Sabelis, M.W. (eds) *Spider Mites, Their Biology, Natural Enemies and Control*, Vol. 1B. Elsevier, Amsterdam, The Netherlands, pp. 95–101.
Putman, W.L. (1955) Bionomics of *Stethorus punctillum* Weise (Coleoptera: Coccinellidae) in Ontario. *The Canadian Entomologist* 87, 9–33.
Raworth, D.A. (1986) An economic threshold function for the twospotted spider mite, *Tetranychus urticae* (Acari: Tetranychidae) on strawberries. *The Canadian Entomologist* 118, 9–16.
Raworth, D.A. (1989) Towards the establishment of an economic threshold for the twospotted spider mite, *Tetranychus urticae* (Acari: Tetranychidae) on red raspberry, *Rubus idaeus*. *Acta Horticulturae* 262, 223–226.
Raworth, D.A. (1990) Predators associated with the twospotted spider mite, *Tetranychus urticae*, on strawberry at Abbotsford, BC, and development of non-chemical mite control. *Journal of the Entomological Society of British Columbia* 87, 59–67.
Raworth, D.A. (2000) Control of two-spotted spider mite by *Phytoseiulus persimilis*. In: *Proceedings of the International Symposium, Biological Control for Crop Protection, 24–25 February 2000*. Rural Development Administration, Suwon, Korea, pp. 171–186.
Raworth, D.A. and Clements S.J. (1996) Plant growth and yield of red raspberry following primocane defoliation. *HortScience* 31, 920–922.
Raworth, D.A. and Strong, W.B. (1990) Development of a management protocol for the twospotted spider mite, *Tetranychus urticae* Koch (Acari: Tetranychidae) on strawberries. In: Bostanian, N.J., Wilson, L.T. and Dennehy, T.J. (eds) *Monitoring and Integrated Management of Arthropod Pests of Small Fruit Crops*. Intercept Ltd, Andover, UK, pp. 103–116.
Raworth, D.A., Gillespie, D., Whistlecraft, J., Edmonds, R., Knott, M. and Davenport, A. (1997) *Biological Control of Twospotted Spider Mites on Pepper and Cucumber in Greenhouses*. Final Report for the British Columbia Western Greenhouse Growers' Society, Langley, British Columbia.
Roy, M., Brodeur, J. and Cloutier, C. (1999a) Seasonal abundance of spider mites and their predators on red raspberry in Québec. *Environmental Entomology* 28, 735–747.
Roy, M., Laverdière, L. and Caron, J. (1999b) Mise en place d'une stratégie de lutte intégrée contre les tétranyques dans les framboisières. Research Report. Programme 'Entente auxiliaire Canada-Québec pour le développement de l'agriculture'. Sainte-Foy, Québec.

Sabelis, M.W. (1981) *Biological Control of Two-spotted Spider Mites using Phytoseiid Predators. Part I. Modelling the Predator–Prey Interaction at the Individual Level.* Agriculture Research Reports No. 910. PUDOC (Centre for Agricultural Publishing and Documentation), Wageningen, The Netherlands.
Sabelis, M.W. (1985) Development. In: Helle, W. and Sabelis, M.W. (eds) *Spider Mites, Their Biology, Natural Enemies and Control,* Vol. 1B. Elsevier, Amsterdam, The Netherlands, pp. 43–52.
Scopes, N.E.A. (1985) Red spider mite and the predator *Phytoseiulus persimilis.* In: Hussey, N.W. and Scopes, N.E.A. (eds) *Biological Pest Control. The Glasshouse Experience.* Blanford Press, Poole, UK, pp. 43–52.

54 *Trialeurodes vaporariorum* (Westwood), Greenhouse Whitefly, and *Bemisia tabaci* (Gennadius), Sweetpotato Whitefly (Hemiptera: Aleyrodidae)

D.A. Raworth, D.R. Gillespie and J.L. Shipp

Pest Status

Greenhouse whitefly, *Trialeurodes vaporariorum* (Westwood), a cosmopolitan species (Russell, 1963), and sweetpotato whitefly, *Bemisia tabaci* (Gennadius), a tropical and subtropical species possibly originating in India or Pakistan (Brown *et al.*, 1995), are serious pests of greenhouse crops. *T. vaporariorum* occurs wherever greenhouse crops are grown, whereas *B. tabaci*, which was introduced into Canada in the late 1980s, is usually found associated with ornamental crops and occasionally with greenhouse vegetables (Howard *et al.*, 1994). Several biotypes of *B. tabaci* exist, including biotypes A and B; some workers regard biotype B as a separate species, *B. argentifolii* Bellows and Perring. Inadvertent transport on ornamental plants in 1985–86 established biotype B on nearly all continents, and subsequent dispersal has given it a worldwide distribution; biotype B displaced biotype A by 1991 in the south-western USA (Brown *et al.*, 1995). Broadleaf plants in over 100 genera are attacked by *T. vaporariorum* (Howard *et al.*, 1994), including tomato, *Lycopersicon esculentum* L., pepper, *Capsicum annuum* L., cucumber, *Cucumis sativus* L., lettuce, *Lactuca sativa* L., and flower crops. *B. tabaci*, biotype B in particular, also has a wide host range that includes these crops (Brown *et al.*, 1995).

Whiteflies reduce plant vigour and produce honeydew that coats the leaves and fruit. Honeydew provides a substrate for moulds; both substances must be washed from fruit before packing for the fresh market. *B. tabaci*, particularly biotype B, is known to induce phytotoxic disorders in some plants (Brown, 1994); blotchy ripening of tomato was associated with infestations in British Columbia greenhouses in 1988–89, resulting in significant crop losses (Howard *et al.*, 1994). In addition, whiteflies transmit several important geminiviruses, 'Subgroup III *Geminivirus*' and *Closterovirus*; among these, beet pseudo-yellows virus (BPYV), transmitted only by *T. vaporariorum* (Wisler *et al.*, 1998), can affect greenhouse cucumber production (Howard *et al.*, 1994) and has become an important disease in Ontario and Alberta. Given the global dissemination of biotype B, increases in whitefly populations in tropical and subtropical areas (Wisler *et al.*, 1998) – possibly due to the fact that biotype B is resistant to insecticides in several different classes of widely used chemicals (Brown, 1994) – and the global movement of people and goods, the potential exists for serious local, sporadic problems with whitefly-transmitted viruses in greenhouses and field crops in Canada.

Development time of *T. vaporariorum* from egg to adult on tomato is 381 degree-days above 8.3°C (22 days at 25.4°C) (Osborne, 1982), but estimates vary considerably with host plant, or whitefly strain. Development time of *B. tabaci* from egg to adult is 24 days at 25.4°C for both A and B biotypes, but biotype B lays more eggs, 68 versus 27 (Bethke *et al.*, 1991), and produces more honeydew (Byrne and Miller, 1990). *T. vaporariorum* tends to occupy higher leaf positions on tomato than *B. argentifolii* (Tsueda and Tsuchida, 1998). In Canada, *T. vaporariorum* may survive the winter outside on weeds near greenhouses, but *B. tabaci* cannot (Howard *et al.*, 1994). Girling (1990) provided further details about whitefly biology.

Background

T. vaporariorum was first controlled biologically through augmentative releases of *Encarsia formosa* Gahan on greenhouse tomatoes in the UK during the late 1920s. During the 1930s, the parasitoid was shipped to other European countries, Canada, Australia and New Zealand (van Lenteren and Woets, 1988). Chemical control prevailed from 1945 until the 1970s. In the 1970s, the establishment of a biological control programme for spider mites and the problems with whitefly resistance to chemical controls led to the re-adoption of biological control strategies for *T. vaporariorum*. *E. formosa* has been the predominant agent for whitefly control, but within the past 5–10 years other agents have been commercialized and growers are increasingly relying on a community of predators and parasitoids.

Biological Control Agents

Parasitoids

E. formosa was first described from parasitized whitefly specimens in the USA in 1924. Its native range is uncertain, but it is now cosmopolitan because of its use in greenhouses. It parasitizes at least 15 aleyrodid hosts in eight genera and is hyperparasitized by three species (Hoddle *et al.*, 1998). Thelytoky in this species is mediated by *Wolbachia* spp. (Zchori-Fein *et al.*, 1992). Adults kill the host by parasitism as well as by host-feeding (Hoddle *et al.*, 1998). Development from egg to adult requires 189 degree-days above 12.7°C (26 days at 20°C) (Osborne, 1982), but whitefly species and life stage, and the host plant, significantly affect development time and survival (Hoddle *et al.*, 1998). Juvenile mortality is higher and development time longer when *E. formosa* develops on *B. tabaci* rather than *T. vaporariorum* (Enkegaard, 1993). *E. formosa* is released weekly or biweekly, often before whitefly are detected, on all greenhouse vegetable crops. The introduction rate varies according to the crop, time of year, production practices, and grower experience with *Encarsia*. A conservative estimate of annual releases in British Columbia is 14 million (230 ha × 0.7 proportion affected ha × 450 parasitoids 0.1 ha^{-1} × 20 applications).

Eretmocerus eremicus Rose and Zolnerowich attacks *Bemisia* (*tabaci* complex) (Rose and Zolnerowich, 1997). Development from egg to adult requires 23 days at 28°C and net reproductive rate is profoundly affected by host plant: four on sweet potato, 12 on cotton, and 26 on cucumber (Headrick *et al.*, 1999). *E. eremicus* is released on greenhouse vegetable crops in Canada to control *T. vaporariorum* and *B. tabaci*, but numbers released are difficult to estimate.

Predators

Delphastus catalinae (Horn) insectary stocks originated from colonies at the University of Florida[1]. Development time from egg to adult is 29 days at 25°C (Gonzalez and Lopez, 1998). From 100 to 150 whitefly eggs are required per day to initiate and sustain oviposition, suggesting that populations can be maintained only in large prey populations (Hoelmer *et al.*, 1993). Numbers released in Canada are difficult to estimate.

Dicyphus hesperus Knight insectary stock was originally from the Okanagan Valley, British Columbia, but these were replaced by collections from California to help alleviate diapause problems. In the wild, 2–3 generations per year occur (McGregor *et al.*, 1999). All motile stages are omnivorous, and must obtain water from plants to complete development and reproduce (Gillespie and McGregor, 2001). Although this agent may cause few blemishes on tomato fruits that are allowed to ripen in greenhouses (McGregor *et al.*, 2000), it could become a major component in biological control programmes for whiteflies and *Tetranychus urticae* Koch. In 1999 and 2000, several thousand adults were released in greenhouses across Canada.

Macrolophus caliginosus Wagner, a generalist predator released for whitefly control in Europe and not known in North America, was found in a greenhouse in British Columbia in 1999. We do not know if this was an accidental or deliberate release, but no further specimens have been observed. Given its omnivorous feeding habits, *M. caliginosus* should not be released in Canada. The native *D. hesperus* serves the same function and is commercially mass-reared.

When introducing biological control agents, it is important to know when to introduce (i.e. time of year), pest density, final market for the product, past pesticide history for the greenhouse operation and production practices. All these variables can greatly affect the number and timing of releases, and the selection of biological control agent(s) to be introduced. Some biological control agents, e.g. *D. catalinae* and *D. hesperus*, are better suited for high population outbreaks, whereas *E. formosa* and *E. eremicus* are better introduced preventively or at low pest densities. *E. eremicus* is more effective at high temperatures and is more resistant to pesticides compared to *E. formosa*. Thus, growers must tailor their biological control programme to suit the needs of their greenhouse operation.

Evaluation of Biological Control

In the rush to find biological solutions to the outbreaks of *B. tabaci* biotype B, several new biological control agents were commercialized. These were selected for ecological and host-plant systems that are quite different from those in Canada. Further work must be done to evaluate the efficacy of these agents and develop protocols for their use in Canadian greenhouses.

E. formosa has been studied extensively (Hoddle *et al.*, 1998) and its efficacy demonstrated through years of experimental and commercial releases. There were concerns that reducing greenhouse temperatures to save energy would result in inadequate control of *T. vaporariorum* by *E. formosa*, but it was shown to be the best parasitoid for low-temperature programmes (van Lenteren and Woets, 1988). Laboratory and greenhouse trials, as well as growers' experiences, showed that residues from some chemical controls may seriously affect the behaviour and efficacy of *E. formosa* (Blümel *et al.*, 1999).

[1]Originally identified by R.D. Gordon as *D. pusillus* (LeConte) and sent to several insectaries including Applied Bio-nomics Ltd in 1990. In April, 1999, Gordon identified the stocks at Applied Bio-nomics as *D. catalinae* (D. Elliott, Sidney, 2000, personal communication).

E. eremicus has been evaluated to control *B. argentifolii* on poinsettia, *Euphorbia pulcherrima* Willdenow ex Klotzsch, in commercial greenhouses in the USA. Better control was achieved with inundative releases in which the insect acted as a predator rather than a parasitoid (Hoddle *et al.*, 1999).

D. catalinae was evaluated during the 1990s to control *B. argentifolii* outside and inside commercial greenhouses in the USA and elsewhere, but not Canada.

D. hesperus has been mass reared (five insectaries and D. Gillespie, Agriculture and Agri-Food Canada, Agassiz) and released in British Columbia, Ontario and Quebec, in commercial greenhouse tomato trials since 1999. Gillespie *et al.* (2000) demonstrated control of *T. vaporariorum* by *D. hesperus.*

Recommendations

Further work should include:

1. Evaluation of the efficacy of recently commercialized biological control agents in Canadian greenhouses;
2. Study of interactions among biological control agents;
3. Study of impacts of greenhouse biological control agents on neighbouring arthropod communities through emigration of generalist natural enemies.

References

Bethke, J.A., Paine, T.D. and Nuessly, G.S. (1991) Comparative biology, morphometrics, and development of two populations of *Bemisia tabaci* (Homoptera: Aleyrodidae) on cotton and poinsettia. *Annals of the Entomological Society of America* 84, 407–411.

Blümel, S., Matthews, G.A., Grinstein, A. and Elad, Y. (1999) Pesticides in IPM: Selectivity, side-effects, application and resistance problems. In: Albajes, R., Gullino, L.M., van Lenteren, J.C. and Elad, Y. (eds) *Integrated Pest and Disease Management in Greenhouse Crops.* Kluwer Academic Publishers, Boston, Massachusetts, pp. 150–167.

Brown, J.K. (1994) Current status of *Bemisia tabaci* as a plant pest and virus vector in agro-ecosystems worldwide. *FAO Plant Protection Bulletin* 42, 3–33.

Brown, J.K., Frohlich, D.R. and Rosell, R.C. (1995) The sweetpotato or silverleaf whiteflies: Biotypes of *Bemisia tabaci* or a species complex? *Annual Review of Entomology* 40, 511–534.

Byrne, D.N. and Miller, W.B. (1990) Carbohydrate and amino acid composition of phloem sap and honeydew produced by *Bemisia tabaci. Journal of Insect Physiology* 36, 433–439.

Enkegaard, A. (1993) *Encarsia formosa* parasitizing the Poinsettia-strain of the cotton whitefly, *Bemisia tabaci*, on Poinsettia: bionomics in relation to temperature. *Entomologia Experimentalis et Applicata* 69, 251–261.

Gillespie, D.R. and McGregor, R.R. (2001) The functions of plant feeding in the omnivorous predator *Dicyphus hesperus* (Heteroptera: Miridae): water places limits on predation. *Ecological Entomology* 25, 380–386.

Gillespie, D., McGregor, R., Quiring, D. and Foisy, M. (2000) *Biological Control of Greenhouse Whitefly with* Dicyphus hesperus. *An Update on the Development of an Omnivorous Predator for the British Columbia Greenhouse Industry.* Technical Report 157, Agriculture and Agri-Food Canada, Pacific Agri-Food Research Centre.

Girling, D. (1990) *Whiteflies: their Bionomics, Pest Status and Management.* Intercept, Andover, UK.

Gonzalez, J.G. and Lopez, A.A. (1998) Biology and feeding habits of *Delphastus pusillus* (Coleoptera: Coccinellidae) predator of whiteflies. *Revista Colombiana de Entomologia* 24, 95–102.

Headrick, D.H., Bellows, T.S. and Perring, T.M. (1999) Development and reproduction of a population of *Eretmocerus eremicus* (Hymenoptera: Aphelinidae) on *Bemisia argentifolii* (Homoptera: Aleyrodidae). *Environmental Entomology* 28, 300–306.

Hoddle, M.S., van Driesche, R.G. and Sanderson, J.P. (1998) Biology and use of the whitefly parasitoid *Encarsia formosa. Annual Review of Entomology* 43, 645–669.

Hoddle, M.S., Sanderson, J.P. and van Driesche, R.G. (1999) Biological control of *Bemisia argentifolii* (Hemiptera: Aleyrodidae) on poinsettia with inundative releases of *Eretmocerus eremicus* (Hymenoptera: Aphelinidae): does varying the weekly release rate affect control? *Bulletin of Entomological Research* 89, 41–51.

Hoelmer, K.A., Osborne, L.S. and Yokomi, R.K. (1993) Reproduction and feeding behavior of *Delphastus pusillus* (Coleoptera: Coccinellidae), a predator of *Bemisia tabaci* (Homoptera: Aleyrodidae). *Journal of Economic Entomology* 86, 322–329.
Howard, R.J., Garland, J.A. and Seaman, W.L. (eds) (1994) *Diseases and Pests of Vegetable Crops in Canada*. Canadian Phytopathology Society and Entomological Society of Canada, Ottawa, Ontario.
Lenteren, J.C. van and Woets, J. (1988) Biological and integrated pest control in greenhouses. *Annual Review of Entomology* 33, 239–269.
McGregor, R.R., Gillespie, D.R., Quiring, D.M.J. and Foisy, M.R.J. (1999) Potential use of *Dicyphus hesperus* Knight (Heteroptera: Miridae) for biological control of pests of greenhouse tomatoes. *Biological Control* 16, 104–110.
McGregor, R.R., Gillespie, D.R., Park, C.G., Quiring, D.M.J. and Foisy, M.R.J. (2000) Leaves or fruit? The potential for damage to tomato fruits by the omnivorous predator, *Dicyphus hesperus* Knight (Heteroptera: Miridae). *Entomologia Experimentalis et Applicata* 95, 325–328.
Osborne, L.S. (1982) Temperature-dependent development of greenhouse whitefly and its parasite *Encarsia formosa*. *Environmental Entomology* 11, 483–485.
Rose, M. and Zolnerowich, G. (1997) *Eretmocerus* Haldeman (Hymenoptera: Aphelinidae) in the United States with descriptions of new species attacking *Bemisia* (*tabaci* complex) (Homoptera: Aleyrodidae). *Proceedings of the Entomological Society of Washington* 99, 1–27.
Russell, L.M. (1963) Hosts and distribution of five species of *Trialeurodes* (Homoptera: Aleyrodidae). *Annals of the Entomological Society of America* 56, 149–153.
Tsueda, H. and Tsuchida, K. (1998) Differences in spatial distribution and life history parameters of two sympatric whiteflies, the greenhouse whitefly (*Trialeurodes vaporariorum* Westwood) and the silverleaf whitefly (*Bemisia argentifolii* Bellows & Perring), under greenhouse and laboratory conditions. *Applied Entomology and Zoology* 33, 379–383.
Wisler, G.C., Duffus, J.E., Liu, H.-Y. and Li, R.H. (1998) Ecology and epidemiology of whitefly-transmitted closteroviruses. *Plant Disease* 82, 270–280.
Zchori-Fein, E., Roush, R.T. and Hunter, M.S. (1992) Male production induced by antibiotic treatment in *Encarsia formosa* (Hymenoptera: Aphelinidae), an asexual species. *Experientia* 48, 102–105.

55 *Trichoplusia ni* Hübner, Cabbage Looper (Lepidoptera: Noctuidae)

D.R. Gillespie, D.A. Raworth and J.L. Shipp

Pest Status

The cabbage looper, *Trichoplusia ni* Hübner, is an important pest of greenhouse vegetable crops. Distribution is cosmopolitan, although it is only able to overwinter in warm-winter climates and in greenhouses. Canadian populations are established through annual migration of adult moths from the south (Lafontaine and Poole, 1991). *T. ni* became a chronic pest of greenhouse vegetable crops in British Columbia and Ontario in the early 1990s. Outside greenhouses, *T. ni* is an important pest of crucifers and many other crops in Ontario, but is less important in other vegetable-producing areas of Canada (Howard *et al.*, 1994). In greenhouse crops, caterpillars cause serious defoliation in cucumber, *Cucumis sativus* L., lettuce, *Lactuca sativa* L., pepper, *Capsicum annuum* L., and tomato, *Lycopersicon esculentum* Miller. Crop losses result from a combination of plant defoliation, direct damage to fruit, destruction of purchased biological control agents by pesticides applied against *T. ni*, and subsequent damage by other pests as a result of their release from biological control.

Eggs are laid singly, normally on the undersides of leaves. These hatch in 3–5 days. Caterpillars feed on leaves for 14–21 days, then pupate above ground in cocoons, either in loose leaf clusters or in small openings in the greenhouse structure. The pupal stage lasts for up to 2 weeks.

Background

In field crops, biological control programmes for *T. ni* based on a *Nucleopolyhedrovirus* (NPV) and on NPV plus *Bacillus thuringiensis* (Berliner) (*B.t.*) have been proposed (Jaques, 1971; Jaques and Laing, 1984) as a component of controls for other lepidopteran pests. *B.t.* is widely used but NPV has not yet been registered.

In greenhouses, although economic thresholds are not established, particularly for vegetable crops such as pepper, tomato and cucumber, on which the pest does not usually attack the fruit, control is necessary. Grower reluctance to apply chemical pesticides because of conflicts with biological control agents, and fears of development of resistance to the commonly used *B.t.*, prompted an increasing reliance on biological control using a community of natural enemies.

Biological Control Agents

Predators

Podisus maculiventris (Say) has been used to control several Noctuidae in greenhouses worldwide (DeClerq *et al.*, 1998). Its life cycle is up to 30+ days. Adult females live for several months, and lay hundreds of eggs in clusters. Of five nymphal instars, the first instar is phytophagous and the second to fifth instars are predacious, as are adults.

Other generalist predators commonly used in greenhouses, e.g. *Dicyphus hesperus* Knight and *Orius* spp., include Lepidoptera eggs and small caterpillars in their diets. The predators are usually introduced for other pests, but likely they exert some impact when *T. ni* populations are abundant relative to other prey.

Parasitoids

Trichogramma pretiosum Riley and *T. brassicae* Bezdenko are released inundatively against the freshly laid eggs of *T. ni*. Females lay eggs singly inside *T. ni* eggs. Developing larvae kill the host eggs, turning them dark-grey. Pupation is inside the host and adults emerge after 10–14 days at 20°C. Insectary stocks are reared on eggs of *Ephestia kuehniella* (Zeller), and are shipped to growers as pupae within host eggs, usually glued on cards for easy distribution.

Cotesia marginiventris (Cresson) parasitizes a wide range of Noctuidae, including *T. ni* (Marsh, 1979). Adult females lay single eggs inside first-instar caterpillars. A single larva completes development inside its host and emerges to spin a cocoon and pupate. At a temperature of 30°C, adults emerge about 9–12 days after oviposition (Boling and Pitre, 1970). Presently, parasitoid adults or pupae in cocoons are shipped to greenhouse growers from commercial insectaries, and are only introduced at inoculative rates because of cost.

Evaluation of Biological Control

Biological control of *T. ni* is most successful in greenhouses when a community of natural enemies is used in combination with other integrated pest management approaches. Otherwise, adult females invading from outside the greenhouse, and refugia provided by stages that are not susceptible to predation or parasitism, combine to produce an inevitable outbreak.

In pepper, *Trichogramma* spp. released at about 25 m^{-2} parasitize about 60–80% of eggs (R.A. Costello, Abbotsford, 2000, personal communication). Releases of *C. marginiventris* produced up to 100% parasitism of sentinel larvae (Gillespie *et al.*, 1997). In cucumber, *C. marginiventris* released at a rate of 0.25 females m^{-2} (with 1 male m^{-2}) parasitized up to 30% of first-instar caterpillars on plants and exhibited a type-II functional response to increasing caterpillar density (Gillespie *et al.*, 1999). Finally, in tomato at a predator–prey ratio of 1 : 3.3, *P. maculiventris* reduces caterpillar numbers and damage by tomato looper, *Chrysodeixis chalcites* (Esper) (DeClerq *et al.*, 1998). This agrees generally with experience with *P. maculiventris* released against *T. ni* in greenhouses in British Columbia. In commercial settings, *B.t.* is applied against outbreaks of *T. ni* and appears to be completely compatible with the other natural enemies. Biological control efforts presently rely largely on *B.t.* and egg parasitoids. The expense of rearing *C. marginiventris* precludes its use for inundative releases, but it may be effective if introduced before caterpillar populations have increased to damaging levels.

Recommendations

Further work should include:

1. Determining the biology of *T. ni* within greenhouses and the role of invasion by external populations (as an adjunct to this, care must be taken in all situations to correctly identify the pest; populations of *T. ni* may be confused with *Autographa californica* (Speyer) or other species);
2. Identifying predators that attack younger larval instars, and finding parasitoids that are less expensive to rear (perhaps polyembryonic parasitoids);
3. Quantifying the impact of *D. hesperus* in reducing *T. ni* populations, given different configurations of alternative prey availability;
4. Ensuring that new integrated pest-management techniques being developed for *T. ni* control, e.g. that utilize pheromones and other products with volatile components, are compatible with biological control efforts for *T. ni* and other pests in greenhouses.

References

Boling, J.C. and Pitre, H.N. (1970) Life history of *Apanteles marginiventris* with descriptions of immature stages. *Journal of the Kansas Entomological Society* 43, 465–470.

DeClerq, P., Merdlevede, F. , Mestdagh, I., Vandendurpel, K., Mohaghegh, J. and Degheele, D. (1998) Predation on the tomato looper, *Chrysodeixis chalcites* (Esper) (Lep., Noctuidae) by *Podisus maculiventris* (Say) and *Podisus nigrispinus* (Dallas) (Het., Pentatomidae). *Journal of Applied Entomology* 122, 93–98.

Gillespie, D., Opit, G., McGregor R., Johnston, M., Quiring, D. and Foisy, M. (1997) *Use of* Cotesia marginiventris *(Cresson) (Hymenoptera: Braconidae) for biological control of cabbage loopers,* Trichoplusia ni *(Lepidoptera: Noctuidae) in greenhouse vegetable crops in British Columbia.* Technical Report Number 141, 8 December 1997, Final report to the BC Greenhouse Vegetable Research Council. Projects 96–12 and 97–06. Agriculture and Agri-Food Canada, Pacific Agriculture Research Centre, Agassiz, British Columbia.

Gillespie, D.R., McGregor, R.R. and Opit, G. (1999) Evaluation of *Cotesia marginiventris* (Cresson) (Hymenoptera: Braconidae) for biological control of *Trichoplusia ni* (Hübner) (Lepidoptera: Noctuidae) in greenhouse vegetable crops in British Columbia. *International Organization for Biological Control/ West Palaearctic Regional Section, Bulletin* 22(1), 89–92.

Howard, R.J., Garland, J.A. and Seaman, W.L. (eds) (1994) *Diseases and Pests of Vegetable Crops in Canada.* Canadian Phytopathology Society and Entomological Society of Canada, Ottawa, Ontario.

Jaques, R.P. (1971) Miscellaneous agricultural insects. In: *Biological Control Programmes against Insects and Weeds in Canada 1959–1968.* Technical Communication No. 4, Commonwealth Institute of Biological Control, Trinidad. Commonwealth Agricultural Bureaux, Farnham Royal, UK, pp. 59–62.

Jaques, R.P. and Laing, J.E. (1984) *Artogeia rapae* (L.), imported cabbageworm (Lepidoptera: Pieridae), *Trichoplusia ni* (Hübner), cabbage looper (Lepidoptera: Noctuidae) and *Plutella xylostella* (L.) (Lepidoptera: Plutellidae). In: Kelleher, J.S. and Hulme, M.A. (eds) *Biological Control Programmes Against Insects and Weeds in Canada 1969–1980.* Commonwealth Agriculture Bureaux, Slough, UK, pp. 15–18.

Lafontaine, J.D. and Poole, R.W. (1991) Noctuoidea, Noctuidae (part) – Plusiinae. Fascicle 25.1. In: Hodges, R.W., Davis, D.R., Dominick, T., Ferguson, D.G., Franclemont, J.G., Munroe, E.G. and Powell, J.A. (eds) *The Moths of America North of Mexico.* The Wedge Entomological Research Foundation, Washington, DC.

Marsh, P.M. (1979) Family Braconidae. In: Krombein, K.V., Hurd, P.D., Smith, D.R. and Burks, B.D. (eds) *Catalogue of Hymenoptera in America North of Mexico*, Vol. 1. Smithsonian Institution Press, Washington, DC, pp. 142–195.

56 *Xanthogaleruca luteola* (Müller), Elm Leaf Beetle (Coleoptera: Chrysomelidae)

G.S. Thurston

Pest Status

The elm leaf beetle, *Xanthogaleruca luteola* (Müller), a native of Europe, is a defoliator of elms, *Ulmus* spp., in North America. Since its introduction into the eastern USA in the 1830s it has spread into Canada and across the continent. It is a pest throughout the entire range of elm in Canada and can damage all species, although European species are usually more susceptible (Martineau, 1984). Damaged leaves turn brown, dry up and drop off the tree, often resulting in complete defoliation of affected trees by midsummer during serious outbreaks. Heavily defoliated trees are stressed and may suffer considerable aesthetic damage, loss of vigour, and increased branch dieback and susceptibility to disease (Cranshaw *et al.*, 1989; Dmytrasz, 1998). Damage is often greater on trees near buildings, which provide a favourable overwintering habitat.

Adults overwinter in dry, sheltered locations. In urban areas this may create problems as large numbers of adults move into homes in autumn, causing considerable annoyance (USDA Forest Service, 1985). Adults emerge in spring and begin feeding on opening buds and newly flushed foliage. Severe leaf feeding results in a 'shotgun blast' appearance. After mating, females lay 400–800 eggs in clusters of 15–25 on the underside of leaves (Weber and Thompson, 1976; Dahlsten *et al.*, 1994). Larvae hatch about 1 week later and begin feeding, skeletonizing the lower leaf surface. After feeding is completed, the larvae crawl down the tree trunks or drop from the branches and pupate in cracks and crevices, usually with many clustered around the tree bases. Adults emerge in about 10 days and feed on elm leaves before seeking overwintering sites. This autumn feeding may cause heavy damage to the second flush of leaves of previously heavily defoliated trees (Thurston, 1998). There is one generation per year, with a possible second generation in southern Ontario (Rose and Lindquist, 1982). In years with prolonged warm seasons there may also be a partial second generation in other parts of Canada.

Background

Outbreaks of *X. luteola* do not occur frequently in any one location and do not last long, but can be intense when they do occur. In this situation, control measures may be required to prevent or limit the unsightly and stressful defoliation of ornamental and cityscape trees. Use of chemical insecticides is often restricted in urban areas, where this insect causes the most concern, so biological control must be employed.

Naturally occurring parasitoids and predators may be responsible for keeping populations low between outbreaks. However, because *X. luteola* is an introduced pest, it probably does not have its full complement of parasitoids and predators present in North America and the existing ones are incapable of containing the increases that lead to outbreaks. In Canada, the parasitoid complex, and the relative importance of individual parasitoids in regulating beetle populations, are not well studied.

Biological Control Agents

Pathogens

Bacteria

Bacillus thuringiensis Berliner serovar *tenebrionis* (*B.t.t.*) is effective against *X. luteola*. The host range of *B.t.t.* is limited to Coleoptera, making it suitable for use where non-target effects must be minimized. Laboratory assays indicated that first-instar larvae are tenfold more susceptible than third-instar larvae (Wells *et al.*, 1994). Cranshaw *et al.* (1989) evaluated its field activity against both larvae and adults and showed that the larvae are more susceptible. However, because *B.t.t.* is active on foliage for only a short time (Cranshaw *et al.*, 1989), and because early larval instars are more susceptible than later instars (Wells *et al.*, 1994), it should be applied as soon as possible after complete egg hatch. Moreover, to maximize product effectiveness, complete coverage is essential. Thurston (1998) found that mistblower application was more effective and less wasteful of product than high-pressure sprays.

Nematodes

Nematodes are effective biological control agents for many pest insects (Bedding *et al.*, 1993). *Steinernema carpocapsae* (Weiser) is pathogenic to *X. luteola* (Kaya *et al.*, 1984), but was ineffective when applied as a foliar spray (Kaya *et al.*, 1981). Thurston (1998) determined that *S. carpocapsae* killed a high proportion of migrating *X. luteola* larvae when added to tree bands containing cellulose mulch. The tree bands also enhanced the actions of generalist predators, e.g. ants (Formicidae) and predatory beetles, and the fungus *Beauveria bassiana* Balsamo (Vuillemin) (Thurston, 1998).

Predators

Several predators, including insectivorous birds, toads, *Bufo* spp., and many insect species, reduce *X. luteola* numbers (Martineau, 1984), but the importance of these natural enemies is unknown.

Parasitoids

Considerable work has been done in the USA with the egg parasitoid *Oomyzus gallerucae* Fonscolombe to control *X. luteola* (Dahlsten *et al.*, 1994). It has been released at many locations in California, has provided high levels of egg parasitism, but has failed to overwinter at most sites (Dahlsten *et al.*, 1993). In eastern Canada, sleeve-cage trials resulted in 15% average parasitism by introduced *O. gallerucae*, with an additional 35% of the exposed eggs damaged by host feeding, for a total impact of *Oomyzus* of 50% (D. Ostaff, Fredericton, 2000, personal communication). Although overwintering of *O. gallerucae* in Canada is unlikely, it may not be necessary for its successful use. Clair *et al.* (1987) devised a mass-rearing technique, and early spring inoculative or mass releases of the parasitoid may provide control in some situations, especially in areas with more than one *X. luteola* generation per year (Dahlsten *et al.*, 1994).

Evaluation of Biological Control

The importance of natural enemies in controlling population outbreaks is poorly known. Because *X. luteola* is an introduced pest in Canada, work to characterize its natural enemy complex, and especially to identify candidate agents for classical biological control, would contribute to lessening its damage.

O. gallerucae may be a useful tool for managing *X. luteola* outbreaks, based on the results of American research. However, a strain more tolerant of the cooler Canadian climates is needed, especially if long-term population suppression is to be achieved.

B.t.t. is effective at reducing larval damage and is now registered in Canada as Novodor® (Valent Biosciences, Libertyville, Illinois, USA) for mistblower application to urban trees. *S. carpocapsae* is effective in killing *X. luteola* larvae and is commercially available in Canada under several product names.

A population monitoring technique would be useful in areas where *X. luteola* is considered to be an ongoing problem. Dahlsten *et al.* (1993) developed a monitoring programme in California that is used to aid in determining when and where control is needed.

For effective *X. luteola* management, integration of all the available pest-management tools is needed. While it appears that several of the control strategies employed for *X. luteola* are compatible (Dahlsten *et al.*, 1994), an integrated pest management programme is not yet in place in Canada.

Recommendations

Further work should include:

1. Developing a strain of *O. gallerucae* more tolerant of Canadian climates to establish permanent populations for long-term *X. luteola* suppression;
2. Surveys of the natural enemy complex of *X. luteola* in Canada and in its native range, to identify potential biological control agents for introduction;
3. Integrating existing control methods into an effective management programme.

References

Bedding, R.A., Akhurst, R.J. and Kaya, H.K. (eds) (1993) *Nematodes and the Biological Control of Insect Pests.* CSIRO Publishing, Melbourne, Australia.

Clair, D.J., Dahlsten, D.L. and Hart, E.R. (1987) Rearing *Tetrastichus gallerucae* (Hymenoptera: Eulophidae) for biological control of the elm leaf beetle, *Xanthogaleruca luteola. Entomophaga* 32, 457–461.

Cranshaw, W.S., Day, S.J., Gritzmacher, T.J. and Zimmerman, R.J. (1989) Field and laboratory evaluations of *Bacillus thuringiensis* strains for control of elm leaf beetle. *Journal of Arboriculture* 15, 31–34.

Dahlsten, D.L., Tait, S.M., Rowney, D.L. and Gingg, B.J. (1993) A monitoring system and development of ecologically sound treatments for elm leaf beetle. *Journal of Arboriculture* 19, 181–186.

Dahlsten, D.L., Rowney, D.L. and Tait, S.M. (1994) Development of integrated pest management programs in urban forests: elm leaf beetle (*Xanthogaleruca luteola* (Müller)) in California, USA. *Forest Ecology and Management* 65, 31–44.

Dmytrasz, P. (1998) IPM for elm leaf beetle in Toronto. *The IPM Practitioner* 20(10), 1–7.
Kaya, H.K., Hara, A.H. and Reardon, R.C. (1981) Laboratory and field evaluation of *Neoaplectana carpocapsae* (Rhabditida: Steinernematidae) against the elm leaf beetle (Coleoptera: Chrysomelidae) and the western spruce budworm (Lepidoptera: Tortricidae). *The Canadian Entomologist* 113, 787–793.
Kaya, H.K., Joos, J.L., Falcon, L.A. and Berlowitz, A. (1984) Suppression of codling moth (Lepidoptera: Olethreutidae) with the entomogenous nematode, *Steinernema feltiae* (Rhabditida: Steinernematidae). *Journal of Economic Entomology* 77, 1240–1244.
Martineau, R. (1984) *Insects Harmful to Forest Trees.* Forestry Technical Report #32, Environment Canada.
Rose, A.H. and Lindquist, O.H. (1982) *Insects of Eastern Hardwood Trees.* Forestry Technical Report #29, Canadian Forestry Service.
Thurston, G.S. (1998) Biological control of elm leaf beetle. *Journal of Arboriculture* 24, 149–154.
USDA Forest Service (1985) *Insects of Eastern Forests.* Miscellaneous Publication 1462.
Weber, R.G. and Thompson, H.E. (1976) Oviposition site characteristics of the elm leaf beetle, *Pyrrhalta (Gallerucella) luteola* (Mueller) in north-central Kansas. *Journal of the Kansas Entomological Society* 49, 171–176.
Wells, A.J., Kwong, R.M. and Field, R. (1994) Elm leaf beetle control using the biological insecticide, Novodor® (*Bacillus thuringiensis* subsp. *tenebrionis*). *Plant Protection Quarterly* 9, 52–55.

57 *Yponomeuta malinellus* Zeller, Apple Ermine Moth (Lepidoptera: Yponomeutidae)

J. Cossentine and U. Kuhlmann

Pest Status

The apple ermine moth, *Yponomeuta malinellus* Zeller, an introduced species from Europe, is a defoliator of apples, *Malus pumila* Miller (= *M. domestica* Borkhausen). Climatic conditions in eastern Canada are ideal for *Y. malinellus* survival as the pest appears to prefer a humid temperate environment. Accidental introductions and subsequent eradications were reported for New Brunswick in 1917 (Hewitt, 1917) and Ontario in 1957 (Parker and Schmidt, 1985). The first *Y. malinellus*-infested tree in western Canada was found in Duncan, British Columbia, in 1981. By 1985, the pest was found throughout a wide area in the Fraser River Valley, and in 1989 it was detected in the fruit-producing region in the southern interior of the province. Unruh *et al.* (1993) summarized the distribution of *Y. malinellus* in Washington and reported on its spread by 1991 into north-western Oregon. Currently, *Y. malinellus* is considered to be a backyard and landscape problem on non-agricultural land in British Columbia. Frazer (1989) and Antonelli (1991) reported occasional trees completely defoliated by *Y. malinellus*.

Y. malinellus is univoltine. Eggs hatch in autumn and the neonate larvae overwinter under the egg mass remains. In spring, they form communal webs that extend as leaves are consumed (Menken *et al.*, 1992).

Pupation occurs within elongate cocoons, clustered within or adjacent to the feeding web. Adults emerge from late July to early September and egg masses are oviposited on the bark of smaller branches.

Background

By applying insecticides including *Bacillus thuringiensis* Berliner serovar *kurstaki* (*B.t.k.*) registered for use against lepidopterous pest of apples, *Y. malinellus* is easily kept below economically damaging levels. Indigenous biological control agents of *Y. malinellus* in British Columbia include the predators, *Balaustium* sp., *Atractotomas mali* Meyer-Dür, *Forficula auricularia* L., ants, spiders, birds, and occasionally the parasitoids *Itopletis quadricingulata* (Provancher), *Scabus deocorus* Walley, a Pimplini sp., *Hemisturmia tortricis* Coquillett, and *Compsilura concinnata* Meigen (Smith, 1990; J.E. Cossentine, unpublished). These are not abundant enough to provide significant control. An inventory of European natural enemies of *Y. malinellus* listed 35 primary parasitoids (Affolter and Carl, 1986). In 1987, a classical biological control project was begun to study and potentially introduce European parasitoids of *Y. malinellus* into British Columbia.

Biological Control Agents

Parasitoids

In Europe, life tables were developed to assess the significance of natural enemies on *Y. malinellus* poulations and to select candidates to introduce into Canada (Kuhlmann *et al.*, 1998a). The impact of egg predators accounted for 25–43% of total generational mortality, more than any other factor. Although parasitism varied from 18 to 30%, its impact was remarkably constant, averaging 11–14% of total generational mortality. *Y. malinellus* was attacked by five different obligate primary parasitoids, one obligate hyperparasitoid and three facultative hyperparasitoids. Of these, *Ageniaspis fuscicollis* Dalman and *Herpestomus brunnicornis* Gravenhorst were selected for introduction into Canada, due to a minimal degree of interspecific competition (Kuhlmann *et al.*, 1998a). In addition, the predatory fly *Agria mamillata* (Pandellé) was studied (Kuhlmann, 1995). The univoltine, solitary endoparasitoid, *Diadegma armillatum* Gravenhorst, was not considered suitable, despite its high impact, because it is a polyphagous parasitoid of microlepidoptera (Herting and Simmonds, 1982).

A. fuscicollis, an oligophagous egg-larval parasitoid, attacks *Y. malinellus* eggs, but the parasitoid eggs do not hatch until the host has reached the third instar. Each egg then develops polyembryonically and produces about 60–80 larvae (Junnikkala, 1960). Parasitized host larvae become swollen and are killed in the fifth instar by mummification as the parasitoids pupate. Based on a literature review by Blackman (1965), Affolter and Carl (1986) concluded that *A. fuscicollis* was well synchronized with its hosts, had a high capacity for rapid multiplication, and had few natural enemies. Kuhlmann *et al.* (1998a, b) studied the extent and distribution of parasitism by *A. fuscicollis* in the field and its oviposition behaviour in the laboratory. Parasitism was independent of host density at the whole-tree scale, but at the individual communal web scale, the probability of a communal web containing parasitized host larvae increased and percentage parasitism decreased with the number of host larvae per web (Kuhlmann *et al.*, 1998b). Observations on oviposition behaviour revealed that pre-patch experience affects the way *A. fuscicollis* females distribute their eggs within host egg batches (Kuhlmann *et al.*, 1998b; Hoffmeister *et al.*, 2000). Females of *A. fuscicollis* laid more eggs and self-superparasitized more often when kept under conditions that indicate competition for, and thus a limitation of, hosts (Hoffmeister *et al.*, 2000).

H. brunnicornis, a solitary, univoltine, oligophagous larval–pupal and pupal parasitoid, was also recognized as an important primary parasitoid of small ermine moths (Kuhlmann *et al.*, 1998a). It attacks fourth- and fifth-instar larvae and also pupae of *Y. malinellus*. Over a 4-year period, the biology of *H. brunnicornis*, its distribution of attack within the tree canopy and parasitism rate in relation to spatial variation in the host number per communal web were studied (Kuhlmann, 1996). Percentage parasitism among communal webs by *H. brunnicornis* was inversely related to the number of host larvae/pupae per web. Because *H. brunnicornis* is synovigenic, with females having a small maximum egg load, and handling time on host pupae is high, Kuhlmann (1996) concluded that these features adequately explained the inverse density-dependence in parasitism on a spatial scale.

A. mamillata is oligophagous and restricted to five *Yponomeuta* spp. in Europe. Kuhlmann (1995) studied its biology and predation to assess its potential. It is a univoltine pupal predator, well synchronized with its host. A predation rate per predator of five *Y. malinellus* pupae or larvae that had not yet spun their cocoons was observed. Based on this study, *A. mamillata* is probably one of the most important natural enemies destroying *Yponomeuta* spp. in Europe. Before considering it for introduction into Canada, further studies are needed on its prey range, impact on *Y. malinellus*, and interspecific competition with *A. fuscicollis* and *H. brunnicornis*, to avoid an adverse interaction.

Releases and Recoveries

In British Columbia, releases of *A. fuscicollis* to control *Y. malinellus* began in 1987. From 1987 to 1990 more than 15,700 *A. fuscicollis* collected in Switzerland and 3265 locally reared *A. fuscicollis* were released in the Fraser River Valley and on Galiano and Salt Spring Islands. These introductions resulted in initial low mean *A. fuscicollis* parasitism rates (0–6%) in the release areas (Smith, 1990). Parasitism was reassessed in 1995, and, from 1995 to 1998, 104,923 additional *A. fuscicollis* were imported and released on Vancouver and Salt Spring Islands, in the Fraser River Valley, and in infested areas in the interior (Cossentine and Kuhlmann, 1999).

H. brunnicornis individuals collected in Japan (288) and Europe (150) were released in 1990 in the Fraser River Valley and on Vancouver and Galiano islands. An additional 3225 *H. brunnicornis* females, collected in Switzerland, were released in *Y. malinellus*-infested orchards in 1996 and 1997, and 4010 female *H. brunnicornis* were released on Vancouver Island in 1998, as reported in the Liberation Bulletins (Sarazin, 1988, 1989, 1990, 1991, 1992; Sarazin and O'Hara, 1999).

Evaluation of Biological Control

A. fuscicollis appears to be well established in *Y. malinellus* populations in all the release areas (Cossentine and Kuhlmann, 2000). Mean parasitism by *A. fuscicollis* on Vancouver Island, where the host is common, was as high as 22.8 ± 12.6% in 1998. Host population densities have decreased in most *A. fuscicollis* release areas. However, it has not been confirmed that the parasitoid is wholly responsible for this effect. As of 2000, establishment of *H. brunnicornis* had not been confirmed.

Recommendations

Future work should include:

1. Monitoring and redistribution of *A. fuscicollis* and *H. brunnicornis* as needed.

References

Affolter, F. and Carl, K.P. (1986) *The Natural Enemies of the Apple Ermine Moth* Yponomeuta malinellus *in Europe. A Literature Review.* Report of the CAB International Institute of Biological Control, Delémont, Switzerland.

Antonelli, A.L. (1991) *Apple Ermine Moth.* Extension bulletin EB 1526, Cooperative Extension, College of Agriculture and Home Economics, Washington State University, Pullman, Washington.

Blackman, R.L. (1965) *A Review of the Literature on* Ageniaspis fuscicollis *(Dalm.).* Report of the Commonwealth Institute of Biological Control, Delémont, Switzerland.
Cossentine, J. and Kuhlmann, U. (1999) Successful establishment of European parasitoid in British Columbia. *Pest Management News* 10 (4), 1 p.
Cossentine, J.E. and Kuhlmann, U. (2000) Status of *Ageniaspis fuscicollis* (Hymenoptera: Encrytidae) in British Columbia: an introduced parasitoid of the apple ermine moth, *Yponomeuta malinellus* Zeller (Lepidoptera; Yponomeutidae). *The Canadian Entomologist* 132, 685–689.
Frazer, B.D. (1989) *Ageniaspis fuscicollis* (Dalman), a parasite of the apple ermine moth. *Biocontrol News* 2, 24.
Herting, B. and Simmonds, F.J. (1982) *A Catalogue of Parasites and Predators of Terrestrial Arthropods, Section B, Enemy/host or Prey, Volume II Hymenoptera Terebrantia.* Commonwealth Agriculture Bureaux, Farnham Royal, UK.
Hewitt, G. (1917) *The Discovery of European Ermine Moth* (Yponomeuta) *on Nursery Stock Imported into Canada.* Agricultural Gazette of Canada, Department of Agriculture, Ottawa, Ontario.
Hoffmeister, T.S., Thiel, A., Kock, B., Babendreier, D. and Kuhlmann, U. (2000) Pre-patch experience affects the egg distribution pattern in a polyembryonic parasitoid of moth egg batches. *Ethology* 106, 145–157.
Junnikkala, E. (1960) Life history and insect enemies of *Hyponomeuta malinellus* Zell. (Lep., Hponomeutidae) in Finland. *Annales Zoologici Societatis Zoologicae Botanicae Fennicae 'Vanamo'* 21, 3–44.
Kuhlmann, U. (1995) Biology and predation rate of the sarcophagid fly, *Agria mamillata*, a predator of European small ermine moths. *International Journal of Pest Management* 41, 67–73.
Kuhlmann, U. (1996) Biology and ecology of *Herpestomus brunnicornis* (Hymenoptera: Ichneumonidae), a biological control agent of the apple ermine moth (Lepidoptera: Yponomeutidae). *International Journal of Pest Management* 42, 131–138.
Kuhlmann, U., Carl, K.P. and Mills, N.J. (1998a) Quantifying the impact of insect predators and parasitoids on populations of the apple ermine moth, *Yponomeuta malinellus* (Lepidoptera: Yponomeutidae), in Europe. *Bulletin of Entomological Research* 88, 165–175.
Kuhlmann, U., Babendreier, D., Hoffmeister, T.S. and Mills, N. J. (1998b) Impact and oviposition behaviour of *Ageniaspis fuscicollis* (Hymenoptera: Encrytidae), a polyembryonic parasitoid of the apple ermine moth, *Yponomeuta malinellus* (Lepidoptera: Yponomeutidae). *Bulletin of Entomological Research* 88, 617–625.
Menken, S.B.J., Herrebout, W.M. and Wiebes, J.T. (1992) Small ermine moths (*Yponomeuta*): their host relations and evolution. *Annual Review of Entomology* 37, 41–66.
Parker, D.J. and Schmidt, A.C. (1985) *Apple Ermine Moth,* Yponomeuta malinellus. Report for Agriculture Agri-Food Canada. Plant Health Division, Ottawa, Ontario.
Sarazin, M.J. (1988) *Insect Liberations in Canada. Parasites and Predators 1987.* Liberation Bulletin 51, Agriculture Canada, Research Branch, Ottawa, Ontario.
Sarazin, M.J. (1989) *Insect Liberations in Canada. Parasites and Predators 1988.* Liberation Bulletin 52, Agriculture Canada, Research Branch, Ottawa, Ontario.
Sarazin, M.J. (1990) *Insect Liberations in Canada. Parasites and Predators 1989.* Liberation Bulletin 53, Agriculture Canada, Research Branch, Ottawa, Ontario.
Sarazin, M.J. (1991) *Insect Liberations in Canada. Parasites and Predators 1990.* Liberation Bulletin 54, Agriculture Canada, Research Branch, Ottawa, Ontario.
Sarazin, M.J. (1992) *Insect Liberations in Canada. For Classical Biological Control Purposes 1991.* Liberation Bulletin 55, Agriculture Canada, Research Branch, Ottawa, Ontario.
Sarazin, M.J. and O'Hara, J.E. (1999) Biocontrol liberations 1997–1998. http://res.agr.ca/ecorc/isbi/biocont/libhom.htm (Accessed 10 January 2000.)
Smith, R. (1990) *Biological Control of the Apple Ermine Moth in Southwestern British Columbia.* British Columbia Ministry of Agriculture, Fisheries and Food and Agriculture and Agri-Food Canada, Victoria, British Columbia.
Unruh, T.R., Congdon, B.D. and La Gasa, E. (1993) *Yponomeuta malinellus* Zeller (Lepidoptera: Yponomeutidae), a new immigrant pest of apples in the Northwest: phenology and distribution expansion, with notes on efficacy of natural enemies. *Pan-Pacific Entomologist* 69, 57–70.

58 *Zeiraphera canadensis* Mutuura and Freeman, Spruce Bud Moth (Lepidoptera: Tortricidae)

R.J. West, M. Kenis, R.S. Bourchier, S.M. Smith and G.W. Butt

Pest Status

The spruce bud moth, *Zeiraphera canadensis* Mutuura and Freeman, a native species found throughout Canada, feeds on white spruce, *Picea glauca* (Moench) Voss (Turgeon, 1992). *Z. canadensis* attack can result in crown deformation, multiple leaders and reduced growth (Carroll *et al.*, 1993). This damage may decrease lumber quality and value, as well as delay harvesting of infested stands (Turgeon, 1992). Populations of *Z. canadensis* decline following crown closure, thereby limiting economic damage to stands between 5 and 20 years of age. Most damage resulting from this pest has been reported from intensively managed white spruce plantations in New Brunswick where infestations during the early 1980s were severe on over 16,000 ha (Turgeon *et al.*, 1995).

Z. canadensis is univoltine. Eggs overwinter under bud scales on the upper side of the tree crown and hatch at bud burst. Larvae immediately bore into the current year's shoots, where they feed throughout their development. They pass through four instars, drop to the ground, and pupate in ground litter during June. Adults emerge, mate and lay eggs during mid- to late summer (Turgeon, 1992).

Background

Turgeon (1992) summarized the native natural enemy complex of *Z. canadensis* from records made in Quebec and New Brunswick. The egg parasitoids, *Trichogramma minutum* Riley and *Trichogramma* sp. have been reported at levels above 50% (Ostaff and Quiring, 1994; Ostaff, 1995). Larval and pupal parasitoids include nine species of Braconidae, nine of Ichneumonidae, three of Pteromalidae, and three of Eulophidae. In Quebec, Pilon (1965) recorded larval parasitism of less than 13%; however, in New Brunswick, *Earinus zeirapherae* (Walley) occurred in over 50% of the larvae collected (Turgeon, 1992).

Despite the number of native parasitoids occurring on *Z. canadensis*, natural control does not appear to keep populations down during early establishment of young white spruce plantations. Similarly, while several chemical control options exist (Turgeon, 1992), none has been found to be effective against *Z. canadensis*, because of its cryptic habits. Thus, biological control was investigated during the 1980s and early 1990s to find potentially more effective species for introduction into Canada, as well as ways to improve the effects of the native parasitoids.

Biological Control Agents

Parasitoids

In Europe, several conifer-feeding *Zeiraphera* spp. closely related to *Z. canadensis* exist, the most similar being the European spruce bud moth, *Z. ratzeburgiana* (Saxxasen) (Mutuura and Freeman, 1966). Other species include *Z. rufimitrana* (Herrich-Schäffer) on fir, *Abies* spp., and *Z. diniana* (Guenée) on larch, *Larix decidua* Miller, spruce, *Picea* spp., and pine, *Pinus* spp. In 1983, studies began in Europe on the biology of parasitoids from these *Zeiraphera* spp. and other closely related conifer tortricids, to assess their potential as biological control agents against *Z. canadensis* in Canada. Mills (1993) and Schönberg (1993) identified the pupal parasitoid *Tycherus* (= *Phaeogenes*) *osculator* (Thünberg) and the larval parasitoid *Tranosema carbonellum* (Thomson) as the most promising agents, based on their impact on *Z. ratzeburgiana*, their apparent specificity to *Zeiraphera* spp. and their synchrony with *Z. canadensis*. Other species studied were *Phytodietus griseanae* Kerrich, *Chorinaeus christator* (Gravenhorst), *Triclistus* spp., *Dolichogenidea lineipes* (Wesmael) and *Trichogramma cacoeciae* Marchal.

From 1995 to 1997, *T. osculator* was evaluated for its suitability against *Z. canadensis* and screening protocols were developed. In Europe, studies on its biology on Palaearctic *Zeiraphera* spp. showed that females overwintered and ovarian maturation did not occur until after several months of exposure to near-freezing temperatures. *T. osculator* successfully parasitized prepupae and pupae of *Z. diniana* of all ages but, in the laboratory, appeared to prefer pupae. Host-feeding by *T. osculator* was common but not necessary for ovarian maturation. In Newfoundland, West *et al.* (1999) showed that *T. osculator* parasitized and developed in *Z. canadensis* as well as in its natural hosts. In the laboratory, females attacked *Z. canadensis* and their offspring developed successfully. Specimens reared from *Z. canadensis*, however, were smaller than those reared from *Z. diniana*. Despite promising results, the project was discontinued in 1997 because a naturally occurring pupal parasitoid, identified as *T. osculator*, was found on *Z. canadensis* in Newfoundland. This was surprising because neither *T. osculator* nor any similar pupal parasitoid had been found during earlier, extensive surveys in Quebec and New Brunswick.

In eastern Newfoundland, subsequent surveys from 1994 to 1996 showed that endemic parasitism by *T. osculator* could be as high as 50% on larvae and pupae. Similarly, natural parasitism by *E. zeirapherae* was as high as 15%, that by *Ascogaster* sp. and *Clinocentrus* sp. under 3%, and that by *Lamachus* sp. and *Triclistus* sp. under 1% (West *et al.*, 1999). The records for *E. zeirapherae*, *Ascogaster* sp. and *Clinocentrus* sp. represent range extensions into Newfoundland, and, for *T. osculator*, into the Nearctic region.

T. minutum was studied from 1992 to 1994 as a potential inundative biological control agent against *Z. canadensis*, concurrent with a 5-year project in Ontario to assess this parasitoid's feasibility against *Choristoneura fumiferana* (Clemens) (see Smith *et al.*, Chapter 12 this volume). *Z. canadensis* was an optimal candidate for such a strategy because: (i) it was a natural overwintering host for native *T. minutum* with relatively high levels of egg parasitism; (ii) its eggs remain available for parasitism in a relatively undifferentiated state for several months during summer and early autumn; and (iii) the cryptic feeding pattern of *Z. canadensis* meant that insecticide applications (chemical or biological) were unlikely to be effective and a biological control agent that could search out host eggs might be more effective.

Two *T. minutum* strains were collected from eggs of *Z. canadensis* in northern New Brunswick, one was an arrenotokous (male/female) population, and the other a thelytokous (female-only) population that dominated the collections (>80%). Both strains were identified as *T. minutum* complex (J.D. Pinto, Riverside, 1993, personal communication; Pinto, 1998), but laboratory studies showed them to differ significantly in biological, biochemical and behavioural characteristics (Wang and Smith, 1996; Van Hezewijk *et al.*, 2000).

Laboratory studies showed that, unlike other hosts examined, *T. minutum* diapaused successfully in *Z. canadensis* eggs, possibly due to the undifferentiated state of their naturally overwintering eggs. Parasitized eggs held either outside or with a 12 : 12 photoperiod and 15°C yielded a significant number of viable adult parasitoids (Table 58.1). In the laboratory, a shoot assay showed that the native thelytokous strain was better than the arrenotokous strain in locating *Z. canadensis* eggs, either exposed or hidden (in nature they are hidden under budscales), suggesting that the thelytokous strain should be used in inundative releases.

In New Brunswick, arrenotokous *T. minutum* originally collected from *C. fumiferana* were released against *Z. canadensis* during 1993. Parasitoids were released from the ground using mistblowers on six 20 × 20 m plots. Half received an early application of 12 million parasitoids ha^{-1} (31 July) and half received the same rate in a late application (3 weeks later, on 21 August). Despite considerable predation by ants and low field temperatures (<20°C), parasitoids were able to complete at least one generation in the field on *Z. canadensis* eggs, and this resulted in a significant increase in overall mean parasitism from 35% in the control plots to 42% in the release plots (P = 0.04). No differences were observed between the two timings. Although the final parasitism level was considerably less than that observed following inundative releases against *C. fumiferana*, the field trial demonstrated that *T. minutum* could be used against *Z. canadensis*; in particular, it pointed to the need for using the local thelytokous strain present in the control plots. Based on these results, research the following year focused on determining quality attributes of the thelytokous strain in order to provide better material for future releases. The completion of the Ontario Project on *C. fumiferana* in 1994 ended work on *Z. canadensis*.

Pathogens

Bacteria

Aerial applications of *Bacillus thuringiensis* serovar *kurstaki* Berliner (*B.t.k.*) against *Z. canadensis* were ineffective in reducing larval populations and leader damage during 1980. It appeared that the larvae were not directly exposed to the spray droplets while feeding under the needles (Turgeon, 1992).

Nematodes

When the nematode *Steinernema carpocapsae* (Weiser) was applied as a foliar spray with the carrier used in the *B.t.k.* formulation Futura XLV, mortality of *Z. canadensis* larvae was increased by 82% in field trials during 1989 in New Brunswick (Eidt and Dunphy, 1991). Applications of *S. carpocapsae* alone at doses of 28–55 million infective juveniles per square metre reduced moth emergence by 68–78%.

Evaluation of Biological Control

T. osculator obtained from either Newfoundland or Europe may have potential as a biological control agent of *Z. canadensis* in mainland Canada, where it is presently absent. It is not clear whether

Table 58.1. Number of adult *Trichogramma minutum* successfully emerging from *Zeiraphera canadensis* eggs kept under different overwintering conditions during 1993.

Treatment	No. parasitized eggs	Total no. adults emerging	% emergence after 6 Jan 1994
16:8 L:D and 25°C	154	36	5
12:12 L:D and 15°C	183	129	67
Outdoors in Sault Ste Marie, Ontario	90	78	100

L:D, light:dark.

the *T. osculator* strain found in Newfoundland belongs to the same species as its European counterpart. *T. carbonellum* may also have potential as a biological control agent in Newfoundland and mainland Canada (see Schönberg, 1993). In Europe, this species is one of the main larval parasitoids of *Z. ratzeburgiana* and *Z. rufimitrana*, which are its only known hosts. Specimens identified as *T. carbonellum* at the Canadian National Collection of Insects, Ottawa, were collected from other hosts. Further work should continue to examine European parasitoids for their potential to be introduced, although this should proceed with caution given the lack of information about the natural parasitoid guild on *Z. canadensis* in Canada.

Native parasitoids in the *T. minutum* complex have potential for use to increase natural egg parasitism and provide annual suppression of *Z. canadensis* populations. In this strategy, it is important that the local thelytokous strain be used rather than the *Trichogramma* strains provided by commercial rearing facilities.

Recommendations

Further work should include:

1. Morphological, behavioural and genetic studies to compare the Newfoundland and European *T. osculator* strains to clarify their taxonomic status;
2. Introduction of *T. osculator* from Newfoundland to mainland Canada;
3. Further study of *T. carbonellum* host specificity and compatibility with *Z. canadensis*;
4. Continued examination of the potential of *T. minutum* for inundative releases with emphasis on studying biological parameters of the native *T. minutum* thelytokous strain to determine whether it can be mass-produced on a factitious host and used in a manner similar to the strain commercially available for *C. fumiferana*.

References

Carroll A.L., Lawlor, M.F. and Quiring, D.T. (1993) Influence of feeding by *Zeiraphera canadensis*, the spruce bud moth, on stem-wood growth of young white spruce. *Forest Ecology and Management* 58, 41–49.

Eidt, D.C. and Dunphy, G.B. (1991) Control of spruce bud moth, *Zeiraphera canadensis* Mut. and Free., in white spruce plantations with entomopathogenic nematodes, *Steinernema* spp. *The Canadian Entomologist* 123, 379–385.

Mills, N.J. (1993) Observations on the parasitoid complexes of budmoths (Lepidoptera: Tortricoidea) on larch in Europe. *Bulletin of Entomological Research* 83, 103–112.

Mutuura, A. and Freeman, T.N. (1966) The North American species of the genus *Zeiraphera*, Treit. (Olethreutidae). *Journal of Research in Lepidoptera* 5, 153–176.

Ostaff, D.P. (1995) Population dynamics of a specialist herbivore, *Zeiraphera canadensis*, on young white spruce. PhD thesis, University of New Brunswick, Fredericton, New Brunswick, Canada.

Ostaff, D.P. and Quiring, D.T. (1994) Seasonal distribution of adult eclosion, oviposition, and parasitism and predation of eggs of the spruce bud moth, *Zeiraphera canadensis* (Lepidoptera: Tortricidae). *The Canadian Entomologist* 126, 995–1006.

Pilon, J.G. (1965) Bionomics of the spruce bud moth, *Zeiraphera ratzeburgiana* (Ratz.) Lepidoptera: (Olethreutidae). *Phytoprotection* 46, 5–13.

Pinto, J.T. (1998) Systematics of the North American species of *Trichogramma* Westwood (Hymenoptera: Trichogrammatidae). *Memoirs of the Entomological Society of Washington* 22, 1–287.

Schönberg, F. (1993) Congeneric European *Zeiraphera* species (Lepidoptera: Tortricidae) and their parasitoid complexes – with implications for the biological control of *Zeiraphera canadensis* Mut. and Free. in Canada. PhD thesis, Christian-Albrechts-Universität, Kiel, Germany.

Turgeon, J.J. (1992) Status of research on the development of management tactics and strategies for the spruce bud moth in white spruce plantations. *Forestry Chronicle* 68, 614–622.

Turgeon, J.J., Kettela, E.G. and Jobin, L. (1995) Spruce bud moth, *Zeiraphera canadensis*. In: Armstrong, J.A. and Ives, W.G.H. (eds) *Forest Insect Pests in Canada*. Natural Resources Canada, Canadian Forest Service, Ottawa, Ontario, pp.183–192.

Van Hezewijk, B., Bourchier, R.S. and Smith, S.M. (2000) Searching speed of *Trichogramma minutum* and its potential as a measure of parsitoid quality. *Biological Control* 17, 139–146.
Wang, Z. and Smith, S.M. (1996) Phenotypic differences between thelytokous and arrhenotokous members of the *Trichogramma minutum* (Hym.: Trichogrammatidae) complex from *Zeiraphera canadensis* (Lep.: Olethreutidae). *Entomologica Experimentalis et Applicata* 78, 315–323.
West, R.J., Kenis, M., Butt, G.W. and Bennet, S.M. (1999) Parasitoid complex of *Zeiraphera canadensis* (Lepidoptera: Tortricidae) and evaluation of *Tycherus osculator* (Hymenoptera: Ichneumonidae) as a biological control agent. *The Canadian Entomologist* 131, 465–474.

59 *Acer, Alnus, Betula, Populus* and *Prunus* spp., Weedy Hardwood Trees (*Aceraceae, Betulaceae, Salicaceae, Rosaceae*)

S.F. Shamoun, D.E. Macey, R. Prasad and R.S. Winder

Pest Status

Competition from fast-growing hardwood trees, e.g. alders, *Alnus* spp., maples, *Acer* spp., birches, *Betula* spp., poplars, *Populus* spp., cherry, *Prunus* spp., and other species is a major problem endemic to conifer regeneration sites following harvest in plantations. This competition results in conifer mortality, reduced growth, delays in harvesting time, increased costs related to forest management and decreases in annual allowable cut (Wall *et al.*, 1992). In addition, an estimated 4 million ha of power-line rights-of-way occur, where control of hardwood species is an essential practice to maintain uninterrupted power supply and avoid fire hazard (Gosselin, 1996; Shamoun and Hintz, 1998b).

Of the approximately 417.6 million ha of productive forest land, about 1 million ha (0.4%) are harvested annually (Natural Resources Canada – Canadian Forest Service, 1999). A significant component consists of hardwood trees belonging to 65 genera, native to several forest ecosystems (Farrar, 1995). Hardwood trees are increasing in economic value but require sound management on forest lands dedicated to production of commercially valuable softwoods as well as on utility rights-of-way and thinned hardwood or mixed wood stands.

Background

Control of weedy hardwoods includes application of chemical herbicides, e.g. glyphosate, triclopyr, hexazinone, and imazapyr (Campbell, 1990; Prasad and Cadogan, 1992). Manual brushing of hardwoods often results in more vigorous regrowth from basal stump sprouts and denser cover than was originally removed. Recent public concerns about herbicides and interest in developing integrated management strategies have resulted in increased demand for alternative control methods for competing hardwood vegetation in conifer regeneration sites and utility rights-of-way (Dorworth, 1990; Wall *et al.*, 1992; Wagner, 1993; Watson and Wall, 1995). Because most of the hardwood species are native, many are ecologically useful, and few also have commercial value in certain situations, classical biological control is not a suitable option for most hardwood tree weeds. New approaches to their management are therefore urgently needed.

Mycoherbicides provide an attractive, relatively new weed control method that involves applying fungal propagules, often in a manner similar to chemical herbicides, and is based on epidemiological principles. Plant disease is often suppressed by host resistance, low pathogen inoculum levels and weakly virulent strains, and unfavourable moisture and/or temperature conditions. Periodically applying high levels of a formulated inoculum of a virulent pathogen on to target weed populations may bypass many of these constraints on disease development. The inundative mycoherbicide strategy for hardwood weed tree management, conducted by the Federal government and several universities[1], is reviewed here.

Biological Control Agents

Pathogens

Fungi

In a survey of endophytic fungi of aerial tissues of *Acer macrophyllum* Pursh, Sieber *et al.* (1990a) and Sieber and Dorworth (1994) identified *Cryptodiaporthe hysterix* (Tode) Petrak (teleomorph of *Diplodina acerina* (Passerini) Sutton) and *Glomerella cingulata* (Stoneman) Spaulding and H. Schrenk (teleomorph of *Colletotrichum gloeosporioides* (Penzig) Penzing and Saccardo) as potential candidates for biological control. Stem wound inoculations with endophytic *C. hysterix* induced circumferential cankers in 6-month-old *A. macrophyllum* seedlings and growth of host callus was inhibited in dual culture (Sieber *et al.*, 1990b). In a similar survey of *Alnus rubra* Bongard endophytes, *Melanconis alni* E. and C. Tulasne (teleomorph of *Melanconium sphaeroideum* Link: Fries) and *Nectria* spp. were identified as the most promising biological control candidates, although the previously reported pathogens *Diaporthe eres* Nitschke [teleomorph of *Phomopsis oblonga* (Desmazières) Hoehm], *Gnomonia setacea* (Persoon: Fries) Cesati and DeNotaris, *Gnomoniella tubaeformis* (Fries) Saccardo, and *Mycosphaerella punctiformis* Persoon: Fries (teleomorph of *Septoria alni* Saccardo) were considered worthy of further investigation (Sieber *et al.*, 1991a). Other *A. rubra* fungi suggested as biological control agents were *Didymosphaeria oregonis* Gooding, *Entoleuca mammata* (Wahlenberg: Fries) J.D. Rogers and Y.-M. Ju [= *Hypoxylon mammatum* (Wahlenberg) P. Karsten] and *Melanconis marginalis* (Peck) Wehmeyer (Sieber *et al.*, 1990b). Host plant manipulation to induce conversion from mutualistic saprophyte to pathogen may be exploited to expand usefulness of endophyte-base mycoherbicides (Sieber and Dorworth, 1994). Biochemical and cultural studies (Seiber *et al.*, 1991b; Shamoun and Sieber, 1993) of symptomless and disease-associated *Melanconium* spp. (anamorphs of *Melanconis* spp.) demonstrated that endophyte pathogenicity may be controlled by external factors and not exclusively by genotype. Furthermore, development of formulation and application technologies may be used to enhance the biological control potential of endophytic fungi (Dorworth and Callan, 1996). In field tests, *Melanconis* spp., *Nectria* spp. and other sap-rot pathogens produced significant cankers and growth reductions when formulated and inserted into *A. rubra* stems (Dorworth, 1995; Dorworth *et al.*, 1996).

Trichaptum biforme (Fries) Ryvarden [=

[1]The Canadian Forest Service and Laval, McGill and Simon Fraser Universities, the Nova Scotia Agricultural College, and University of Victoria.

Polyporous pargamenus (Fries) Klotzsch], *Schizophyllum commune* (Fries) Fries and *Cerrena unicolor* (Bulliard: Fries) Murrill, have been used to inoculate stumps of aspen, *Populus* spp. The efficacy of *C. unicolor* and *S. commune* appears to be similar to that of *Chondrostereum purpureum* (Persoon ex Fries) Pouzar (see below) in the first year, but *C. unicolor* appears to have better efficacy thereafter – fewer sprouts are produced and the inoculated stumps are thoroughly decayed. *T. biforme* does not seem to be very effective (M. Dumas, Sault Ste Marie, 1998, personal communication).

In laboratory studies of several pathogenic fungi with potential as endemic defoliators or stem pathogens, Winder *et al.* (1988–1991) inoculated *Populus tremuloides* Michaux, with spores from *Pollaccia* sp. (anamorph of *Venturia* sp.), *Ciborinia whetzelii* (Seaver) Seaver, and two unknown species. *Pollaccia* sp. demonstrated the highest potential for causing foliar damage (about 50%), the others caused about 25% leaf area damage. On *Acer spicatum* Lambert, *Colletotrichum gloeosporioides* (Penzig) Penzig and Saccardo [anamorph of *Glomerella cingulata* (Stoneman) Spaulding and Schrenck] caused about 50% leaf area damage and *Phyllosticta* sp., *Candida* sp. and *Ascochyta* sp. caused about 25% leaf area damage.

In Quebec, *Nectria* sp. and black knot disease, *Aspiosporina morbosa* (Schweinitz: Fries) von Arx [= *Dibotryon morbosum* (Schweinitz: Fries) Theissen & Sydow], were collected as potential agents from *Prunus pennsylvanica* L., but they were difficult to culture.

In the Maritimes, epiphytotics of *D. morbosum*, were started by introducing mature, sporulating ascostromata from local sources into four young *P. pennsylvanica* stands prior to bud break (Wall, 1985). Severe disease symptoms attributable to these pathogens were observed 1–2 years after their introduction, with a subsequent decline in *P. pennsylvanica* growth and an increase in mortality recorded within a 50 m radius of the introductions (Wall, 1986). It was concluded that augmentative introductions of *D. morbosum* could be used to suppress *P. pennsylvanica* in 1–2-year-old stands. Because the fungus usually attacks the current year's shoots, later introductions would likely not have much effect. *D. morbosum* invades meristematic or recently differentiated tissues (Wainwright and Lewis, 1970). Therefore, introduction of ascostromata that will sporulate during bud break and early shoot elongation is essential. In New Brunswick and Nova Scotia, ascostromata were collected and used to inoculate stands during the dormant season (late autumn–early spring). In some years, sound, potentially fertile ascostromata could not be found, suggesting a biennial cycle.

Several Ascomycotina isolated from symptomless and disease-associated *A. rubra* were field tested for biological control potential on southern Vancouver Island, British Columbia (Dorworth, 1995; Dorworth *et al.*, 1996). Stems of *A. rubra* in different size classes were inoculated with 11 isolates of five species, including *Melanconis marginalis*, *M. alni*, *Nectria distissima* Tulasne, *Phomopsis* sp. (anamorph of *Diaporthe* sp.), *E. mammata*, and *Xylaria hypoxylon* (L.: Fries) Greville. *Nectria ditissima*, isolate PFC-082, was found to be sufficiently virulent when formulated and inserted into stem punctures, resulting in nearly 100% colonization, significant tissue damage, reduced health and subsequent mortality of inoculated trees. Unwanted dispersion of the pathogen beyond the treated area was non-existent as no reproductive structures were observed on the inoculated stems, nor were natural infections of *N. ditissima* found in *A. rubra* trees surrounding the experimental plots.

Chondrostereum purpureum, a common pathogen found in temperate regions in orchards, urban areas and forests (Setliff and Wade, 1973; Ginns and Lefebvre, 1993), is the causal agent of silverleaf disease and mortality in various hardwood shrubs and trees, invading xylem vessels in the wood via wounds less than 1 month old (Brooks and Moore, 1926; Spiers and Hopcroft, 1988; Wall, 1990). It is an early colonizer of wounds on many hardwood species, often being displaced by other fungi over time (Mercer and Kirk, 1984). The fungus exists as mycelia in infected trees and is spread by airborne basidiospores released by fruiting bodies (basidiocarps) found on tree wounds, cut stumps and slash during humid, cool weather (Spiers, 1985). The forest industry and provincial forestry services are interested in *C. purpureum* as a selective biological control agent. The fungus has been shown to be an effective mycoherbicide to control stump sprout in *Prunus serotina* Ehrhart, *P. pennsylvanica*, *Populus* spp., *Alnus* spp. and *Betula papyrifera* Marsham in conifer regeneration sites (DeJong *et al.*, 1990; Wall, 1994; Dumas *et al.*, 1997; Jobidon, 1998; Harper *et al.*, 1999; Pitt *et al.*, 1999) and in utility rights-of-way, where it could widen the treatment window available for manual brushing of *A. rubra* in western Canada, and *B. papyrifera*, *P. tremuloides*, *P. pennsylvanica* and *Acer saccharum* Marshall in eastern Canada (Gosselin, 1996; Shamoun and Hintz, 1998b). Additional work has shown that *C. purpureum* has potential use on girdled weed trees (Shamoun and Wall, 1992; Wall, 1994; R. Prasad, unpublished).

In Quebec, a study was begun in 1992 to test the efficacy of two isolates of *C. purpureum*, CQP1 and IB, on *P. pennsylvanica* and *P. tremuloides* (Ste-Agathe site), and *B. papyrifera* and *A. saccharum* (St-Michel site). Two other sites containing *P. pennsylvanica* (Hunterstown site) and *P. tremuloides* (Hervey-Jonctione site) were added in 1993. The four sites were located on Hydro-Quebec 700 or 350 kV powerline corridors. The four target hardwood species were cut mechanically and stumps were treated with one of the two isolates in June and August, 1992 and 1993. Stump sprouting was greatly reduced by both isolates in the first year of evaluation and the treatment was even more successful in ensuing years. Three years after treatment, control levels varied from 76 to 100% using either test isolate on either species (Gosselin, 1996).

In British Columbia, Shamoun and Hintz (1998b) studied use of *C. purpureum* against *A. rubra* in hydro rights-of-way. In 1994, two formulated *C. purpureum* isolates (PFC 2139 and PFC 2140), a control formulation treatment, two chemical herbicide treatments (12% Vision® spray and carbopaste formulation of Vision®), and manual cutting (slash) were compared. Although re-sprouting of cut *A. rubra* stumps occurred throughout the six treatments after 18 months (spring of 1995), by mid-summer re-sprout mortality of 65–100% occurred on many stumps. *A. rubra* stumps treated with *C. purpureum* and with herbicides showed significantly fewer living sprouts than other treatments, with a mean of less than one living re-sprout per stump. *C. purpureum* and chemical herbicide treatments resulted in similar levels of stump mortality and re-sprouting of *A. rubra*, and were significantly different from the formulation control and slash treatments. Treatments with either fungal isolate gave similar results. Two years after treatment (1996), more than 95% mortality occurred on stumps with fungal and herbicide treatments, and up to 100% mortality with isolate PFC 2139 and Vision®. Compared to the 1995 results, all treatment plots had less re-sprouting and higher stump mortality. Fruiting bodies of *Coriolus versicolor* Fries, *Schizophyllum commune* Fries and other basidiomycetes were also observed on many stumps in all treatment plots (Shamoun and Hintz, 1998b). *C. purpureum* treatments were largely ineffective at controlling and reducing sprout vigour in a similar trial against *A. macrophyllum* in the lower mainland (S.F. Shamoun, unpublished; Comeau *et al.*, 1994).

In 1995, a nationwide field trial of *C. purpureum* evaluated its efficacy in conifer regeneration sites against major competitive weeds, including *P. tremuloides*, *Alnus viridis sinuata* (Regel) A. Love and D. Love, *Alnus rugosa* Dupon/Sprengel, and *Acer rubrum* L. Two formulations (one developed at the Pacific Forestry Centre, British Columbia, and the other at Great Lakes Forestry Centre, Ontario) combined with two fungal isolates (PFC 2139 and JAM6), control (blank formulation), cutting only, triclopyr herbicide applications and an uncut control were compared. Two growing seasons after treatment, results in eastern Canada showed that triclopyr herbicide provided greater control of *P. tremuloides*, *A. rugosa* and *A. rubrum* than the *C. purpureum* formulations (Pitt *et al.*, 1999). The fungus was most effective on *A. rugosa*, resulting in 72% reduction in volume index and 19% clump mortality. On *A. rubrum*, isolate PFC 2139 reduced volume of stem sprouts by only 32%. On *P. tremuloides*, both isolates caused 35% reduction in volume of stump resprouts and isolate PFC 2139 provided 88% reduction. Efficacy appeared to vary among fungal isolates and target species, while formulation was less important. Analysis of the British Columbia trial (Harper *et al.*, 1999) revealed that *A. viridis sinuata* clump mortality caused by both isolates was high (90% and 88%, respectively). The control treatment induced the lowest clump mortality and appeared to promote sprouting and growth of *A. viridis sinuata* when compared with culturing alone. However, efficacy of both formulations was different on *P. tremuloides* in British Columbia; only the British Columbia formulation with isolate PFC 2139 was an effective fungal treatment, resulting in 84% mortality. Results suggested that *C. purpureum* efficacy was dependent upon isolate virulence and formulation.

Genetic characterization, epidemiology and environmental fate studies were completed as essential components to register *C. purpureum* (DeJong *et al.*, 1996; Ramsfield *et al.*, 1996; Shamoun and Wall, 1996; Shamoun and Hintz, 1998a; Becker *et al.*, 1999; Ramsfield *et al.*, 1999; Hintz *et al.*, 2000). Gosselin *et al.* (1996, 1999) conducted extensive genetic variability, population structure and environmental fate investigations of *C. purpureum* in Quebec.

Competitive Interactions

In British Columbia, the influence of *C. purpureum* treatments on release of lodgepole pine, *Pinus contorta* var. *latifolia* Englemann, was tested. As part of the national trials to measure efficacy of *C. purpureum* formulations on *P. tremuloides*, *P. contorta* var. *latifolia* seedlings were planted in treated and untreated plots to study the effects of release from competition. Data collected from 1995 to 1998 provided strong evidence that *C. purpureum* formulations PFC 2139 and JAM6 not only suppressed *P. tremuloides* resprouting by 80% but also (indirectly) enhanced growth and development (height and volume) of *P. contorta* var. *latifolia* seedlings by 250–300%, suggesting that *C. purpureum* applications are as effective as manual cutting and triclopyr applications and that this biological control agent might be a preferred option in environmentally sensitive areas (Prasad, 2000). This increased growth was largely due to release from competition for light. No additive effects of the two types of formulation, manual cutting or triclopyr application were found.

Evaluation of Biological Control

In 1999, MycoLogic Inc., University of Victoria, submitted a registration package to the Pest Management Regulatory Agency, Canada, and the Environmental Protection Agency, USA, for registration of *C. purpureum* as Chontrol™. Despite the efficacy of the formulation, an improved application technology (spray) is needed for operational effectiveness.

Recommendations

Further work should include:

1. Exploitation of plurivorous wood-rot fungi, Ascomycotina, endophytic fungi and other necrotrophic fungi that have shown promise as biological control agents;
2. Searching for and evaluating more effective biological control agents to target other tolerant hardwood species, including some larger shrubs;
3. Integrating mycoherbicides that cannot give adequate control alone, with adju-

vants, synergists, disease vectors or silvicultural practices;
4. Developing better formulations to improve inoculum viability, efficacy, affordable mass production systems, and application technology;
5. Better understanding the molecular and cellular basis for virulence and host specificity of biological control agents;
6. Educating end-users and scientists unfamiliar with mycoherbicides to improve technology transfer.

References

Becker, E.M., Ball, L.A. and Hintz, W.E. (1999) PCR-based genetic markers for detection and infestion frequency analysis of the biocontrol fungus *Chondrostereum purpureum* on sitka alder and trembling aspen. *Biological Control* 15, 71–80.

Brooks, F.T. and Moore, W.C. (1926) Silver-leaf disease. V. *Journal of Pomology and Horticulture Science* 5, 11–97.

Campbell, R.A. (1990) Herbicide use for forest management in Canada: Where we are and where we are going. *Forestry Chronicle* 66, 355–360.

Comeau, P.E., Wall, R.E. and Prasad, R. (1994) *Control of Bigleaf Maple Using Cut Stump and Basal Treatments.* Research Report, Expert Committee on Weeds, Canada Department of Agriculture, Saskatoon, Dec 1–2, p. 1015.

DeJong, M.D., Scheepens, P.C. and Zadocks, J.C. (1990) Risk analysis for biological control: A Dutch case study in biocontrol of *Prunus serotina* by the fungus *Chondrostereum purpureum. Plant Disease* 74, 189–194.

DeJong, M.D., Sela, E., Shamoun, S.F. and Wall, R.E. (1996) Natural occurrence of *Chondrostereum purpureum* in relation to its use as a biological control agent in Canadian forests. *Biological Control* 6, 347–352.

Dorworth, C.E. (1990) Mycoherbicides for forest weed biocontrol – the P.F.C enhancement process. In: Bassett, C., Whitehouse, L.J. and Zabliewicz, J.A. (eds) *Alternatives to the Chemical Control of Weeds.* Bulletin 155, Forest Research Institute, Rotorua, New Zealand, pp. 116–119.

Dorworth, C.E. (1995) Biological control of red alder (*Alnus rubra*) with the fungus *Nectria ditissima. Weed Technology* 9, 243–248.

Dorworth, C.E. and Callan, B.E. (1996) Manipulation of endophytic fungi to promote their utility as vegetation biocontrol agents. In: Redlin, S. (ed.) *Systematics, Ecology and Evolution of Endophytic Fungi in Grasses and Woody Plants.* APS Press, Minneapolis, Minnesota, pp. 209–219.

Dorworth, C.E., Macey D.E., Sieber, T.N. and Woods, T.A.D. (1996) Biological control of red alder (*Alnus rubra*) with indigenous pathogenic Ascomycotina. *Canadian Journal of Plant Pathology* 18, 315–324.

Dumas, M.T., Wood, J.E., Mitchell, E.G. and Boyonoski, N.W. (1997) Control of stump sprouting of *Populus tremuloides* and *P. grandidentata* by inoculation with *Chondrostereum purpureum. Biological Control* 10, 37–41.

Farrar, J.L. (1995) *Trees in Canada.* Fitzhenry and Whiteside Limited and the Canadian Forest Service, Natural Resources Canada, in cooperation with the Canada Communication Group – Publishing, Supply and Services Canada, Ottawa, Ontario.

Ginns, J. and Lefebvre, M.N.L. (1993) *Lignocolous Corticoid Fungi (Basidiomycota) of North America, Systematics, Distribution, and Ecology.* Mycologia Memoir No. 19, The Mycological Society of America.

Gosselin, L. (1996) Biological control of stump sprouting of broad-leaf species in rights-of-way with *Chondrostereum purpureum*: I. Virulence of tested strains and susceptibility of target hosts. PhD thesis, Laval University, Quebec, Canada.

Gosselin, L., Jobidon, R. and Bernier, L. (1996) Assessment of genetic variation within *Chondrostereum purpureum* from Quebec by random amplified polymorphic DNA analysis. *Mycological Research* 100, 151–158.

Gosselin, L., Jobidon, R. and Bernier, L. (1999) Genetic variability and structure of Canadian populations of *Chondrostereum purpureum*, a potential biophytocide. *Molecular Ecology* 8, 113–122.

Harper, G., Comeau, P.G., Hintz, W., Wall, R.E., Prasad, R. and Becker, E.M. (1999) *Chondrostereum purpureum* as a biological control agent in forest management. 2. Efficacy on Sitka alder and aspen in Western Canada. *Canadian Journal of Forestry Research* 29, 852–858.

Hintz, W.E., Becker, E.M. and Shamoun, S.F. (2000) Development of genetic markers for risk assessment of biological control agents. *Canadian Journal of Plant Pathology* 23, 13–18.

Jobidon, R. (1998) Comparative efficacy of biological and chemical control of the vegetative reproduction in *Betula papyrifera* and *Prunus pensylvanica*. *Biological Control* 11, 22–28.

Mercer, P.C. and Kirk, S.A. (1984) Biological treatments for the control of decay in tree wounds. I. Laboratory tests. *Annals of Applied Biology* 104, 211–219.

Natural Resources Canada – Canadian Forest Service (1999) *The State of Canada's Forests*. Natural Resources Canada, Canadian Forest Service, Ottawa, Ontario.

Pitt, D.G., Dumas, M.T., Wall, R.E., Thompson, D.G., Lanteigne, L., Hintz, W., Sampson, G. and Wagner, R.G. (1999) *Chondrostereum purpureum* as a biological control agent in forest management. 1. Efficacy on speckled alder, red maple and aspen in Eastern Canada. *Canadian Journal of Forestry Research* 29, 841–851.

Prasad, R. (2000) Influence of a bioherbicide agent (*Chondrostereum purpureum*) on conifer release of lodgepole pine in British Columbia. *Canadian Journal of Plant Pathology* 22, 190.

Prasad, R. and Cadogan, B.L. (1992) Influence of droplet size and density on phytotoxicity of three herbicides. *Weed Technology* 6, 415–423.

Ramsfield, T., Becker, E., Rathlef, S., Tang, Y., Vrain, T., Shamoun, S.F. and Hintz, W.E. (1996) Geographic variation of *Chondrostereum purpureum* detected by polymorphisms in the ribosomal DNA. *Canadian Journal of Botany* 74, 1919–1929.

Ramsfield, T., Shamoun, S.F., Punja, Z. and Hintz, W.E. (1999) Variation in the mitochondrial DNA of the potential biological control agent *Chondrostereum purpureum*. *Canadian Journal of Botany* 77, 1490–1498.

Setliff, E.C. and Wade, E.K. (1973) *Stereum purpureum* associated with sudden decline and death of apple trees in Wisconsin. *Plant Disease Reporter* 57, 473–474.

Shamoun, S.F. and Hintz, W.E. (1998a) Development and registration of *Chondrostereum purpureum* as a mycoherbicide for hardwood weeds in conifer reforestation sites and utility rights-of-way. In: Burge, M. (ed.) *Proceedings of the IV International Bioherbicide Workshop Programme and Abstracts, 6–7 August 1998*. University of Strathclyde, Glasgow, UK, p. 14.

Shamoun, S.F. and Hintz, W.E. (1998b) Development of *Chondrostereum purpureum* as a biological control agent for red alder in utility rights-of-way. In: Wagner, R.G. and Thompson, D.G. (Compilers) *Third International Conference on Forest Vegetation Management*. Forestry Research Information Paper No. 141, Ontario Ministry of Natural Resources Institute, Ontario Forestry Research Institute, pp. 308–310.

Shamoun, S.F. and Sieber, T.N. (1993) Isozyme and protein patterns of endophytic and disease syndrome associated isolates of *Melanconium apiocarpum* and *Melanconium marginale* collected from alder. *Mycotaxon* 49, 151–166.

Shamoun, S.F. and Wall, R.E. (1992) *Chondrostereum purpureum*, a potential mycoherbicide for red alder in British Columbia. *Phytopathology* 82, 1154.

Shamoun, S.F. and Wall, R.E. (1996) Characterization of Canadian isolates of *Chondrostereum purpureum* by protein content, API ZYM and isozyme analyses. *European Journal of Forest Pathology* 26, 333–342.

Sieber, T.N. and Dorworth, C.E. (1994) An ecological study about assemblages of endophytic fungi in *Acer macrophyllum* in British Columbia: in search of candidate mycoherbicides. *Canadian Journal of Botany* 72, 1397–1402.

Sieber, T.N., Sieber-Canavesi, F. and Dorworth, C.E. (1990a) *Identification of Key Pathogens of Major Coastal Forest Weeds*. FRDA Report No. 113, Forestry Canada, Pacific Forestry Centre, Victoria, British Columbia.

Sieber, T.N., Sieber-Canavesi, F. and Dorworth, C.E. (1990b) Simultaneous stimulation of endophytic *Crytodiaporthe hystrix* and inhibition of *Acer macrophyllum* callus in dual culture. *Mycologia* 82, 569–575.

Sieber, T.N., Sieber-Canavesi, F. and Dorworth, C.E. (1991a) Endophytic fungi of red alder (*Alnus rubra*) leaves and twigs in British Columbia. *Canadian Journal of Botany* 69, 407–411.

Sieber, T.N., Sieber-Canavesi, F., Petrini, O., Edramoddoullah, A.K.M. and Dorworth, C.E. (1991b) Characterization of Canadian and European *Melanconium* from some *Alnus* species by morphological, cultural and biochemical studies. *Canadian Journal of Botany* 69, 2170–2176.

Spiers, A.G. (1985) Factors affecting basidiospore release by *Chondrostereum purpureum* in New Zealand. *European Journal of Forest Pathology* 15, 111–126.

Spiers, A.G. and Hopcroft, D.H. 1988. Ultrastructural studies of basidial and basidiospore development and basidiospore release in *Chondrostereum purpureum*. *European Journal of Forest Pathology* 18, 367–381.

Wagner, R.G. (1993) Research directions to advance forest vegetation management in North America. *Canadian Journal of Forestry Research* 23, 2317–2327.

Wainwright, S.H. and Lewis, F.H. (1970) Developmental morphology of the black knot pathogen on plum. *Phytopathology* 60, 1238–1244.

Wall, R.E. (1985) The role of disease in removal of weed species from developing forest stands. In: Delfosse, E.S. (ed.) *Proceedings of the VI International Symposium on Biological Control of weeds, August 1984, Vancouver*. Agriculture Canada, Ottawa, Ontario, pp. 673–676.

Wall, R.E. (1986) Effects of black knot disease on pin cherry. *Canadian Journal of Plant Pathology* 8, 71–77.

Wall, R.E. (1990) The fungus *Chondrostereum purpureum* as a silvicide to control stump sprouting in hardwoods. *Northern Journal of Applied Forestry* 7, 17–19.

Wall, R.E. (1994) Biological control of red alder using stem treatments with the fungus *Chondrostereum purpureum*. *Canadian Journal of Forestry Research* 24, 1527–1530.

Wall, R.E., Prasad, R. and Shamoun, S.F. (1992) The development and potential role of mycoherbicides for forestry. *Forestry Chronicle* 68, 736–741.

Watson, A.K. and Wall, R.E. (1995) Mycoherbicides: their role in vegetation management in Canadian forests. In: Charest, P.J. and Duchesne, L.C. (eds) *Recent Progress in Forest Biotechnology in Canada*. Information Report PI-X-120, Canadian Forest Service, pp. 74–82.

Winder, R.S., Cartier, J., Ciatola, M., Roy, G., Tourigny, G. and Watson, A.K. (1988–1991) Programme de recherche et de developpment sur les bioherbicides pour les secteurs urbain et forestier. Rapports d'avancement. (Unpublished internal reports to Quebec Ministry of the Environment 1988–1991.)

60 *Ambrosia artemisiifolia* L., Common Ragweed (*Asteraceae*)

M.P. Teshler, A. DiTommaso, J.A. Gagnon and A.K. Watson

Pest Status

Common or short ragweed, *Ambrosia artemisiifolia* L., a native North American species, has been collected from all Canadian provinces and the Northwest Territories but is far more abundant in eastern Canada, particularly southern Ontario and Quebec (Bassett and Crompton, 1975). As European settlers cleared land and intensified agriculture, *A. artemisiifolia* spread widely and became a serious pest in eastern Canada. It is listed as a 'secondary noxious weed' under the Federal Seeds Act and a 'noxious weed' in many provincial statutes. The most abundant of the four Canadian *Ambrosia* spp., *A. artemisiifolia*, is a monoecious, wind-pollinated plant with numerous staminate flowers containing prodigious numbers of pollen grains. *A. artemisiifolia* pollen is considered to be a biological pollutant that is the primary cause of allergenic hay fever, asthma and eczema. The complex mixture of 22 proteins that are released from ragweed pollen grains have been shown to be among the most powerful antigens/allergens known (Bagarozzi and Travis, 1998). Susceptible individuals have a histamine reaction to *A. artemisiifolia* pollen, mainly in August and September.

As a pioneer species, *A. artemisiifolia* flourishes in disturbed habitats, e.g. along rights-of-way and in vacant lots. In southwestern Quebec and Ontario, it has become a serious agricultural weed. Seeds germinate in spring, plants are in the vegetative phase from May to August, begin flowering in early August, and produce 3000–62,000 seeds per plant that can remain viable for 39 years or more in soil (Bassett and Crompton, 1975). *A. artemisiifolia* plants vary greatly in size and shape and are very competitive, with a high level of allelopathic activity.

Background

In most soils, *A. artemisiifolia* can easily be uprooted, but it readily adapts to mowing by quickly developing new stems below cutting height (Vincent and Ahmin, 1985). It is susceptible to the herbicides 2,4-D (2,4-dichlorophenoxyacetic acid), MCPA (4-chloro-2-methylphenoxyacetic acid), 2,4-DB (4-(2,4-dichlorophenoxy) butyric acid), MCPB (4-(chloro-2-methylphenoxy)-butyric acid), mecoprop and dicamba. Bentazon and imazethapyr provide *A. artemisiifolia* control in soybean, *Glycine max* (L.) Merrill, and various herbicides and herbicide mixtures provide control in corn, *Zea mays* L. Populations of *A. artemisiifolia* have developed resistance to atrazine and linuron (Heap, 1997; St-Louis *et al.*, 2000), thus restricting control options in vegetable crops.

Until recently, herbicides such as 2,4-D and dicamba have been the mainstay of *A. artemisiifolia* control strategies in urban areas. However, widescale herbicide use has declined in recent years, especially along highways and rights-of-way, because of increasing public concern about health and environmental effects. These reductions have resulted in increased *A. artemisiifolia* infestations, and associated increases in the incidence of allergenic reactions. Effective non-chemical strategies to control *A. artemisiifolia* in urban and suburban areas, as well as in agricultural fields, are required.

A. artemisiifolia is amenable to biological control. Mountainous regions of Mexico and South America are potential sources for biotic agents adapted to a cold climate to control *A. artemisiifolia* in Canada (Harris and Piper, 1970). Faunistic surveys in Canada, southern California and Mexico list 894 insect species (86 monophagous and 31 oligophagous) known to attack the 15 most common plant species representing all of the genera of North American Ambrosiinae (Goeden and Palmer, 1995). In Canada, some native insects and fungi of *A. artemisiifolia* are being studied as inundative biological control agents. Phytocenotic plant competition is also being pursued.

Biological Control Agents

Pathogens

Fungi

The white rust fungus, *Albugo tragopogi* Persoon ex S.F. Gray, an obligate parasite isolated from *A. artemisiifolia*, has a restricted host range (Hartmann and Watson, 1980b). When inoculated on to *A. artemisiifolia* seedlings at the two-leaf stage it reduced pollen production by 99%, seed production by 98%, and top weight by 79% for plants eventually developing systemic disease symptoms. However, only 14% of inoculated plants developed systemic symptoms (Hartman and Watson, 1980a). Because mature staminate flowers shed pollen over a relatively long time period, Hartman and Watson (1980a) suggested that multicyclic applications of *A. tragopogi* suspensions in field environments would increase infection level. Difficulties in mass producing *A. tragopogi* have limited its potential use.

In Quebec, a *Phoma* sp. was isolated from *A. artemisiifolia*. Inoculated plants frequently exhibited systemic infections in leaf petioles and stems, which eventually die back substantially. In many plants, the growing points and developing staminate flowers were colonized by the fungus, resulting in little or no pollen production (Brière *et al.*, 1995). In laboratory trials, feeding by the native chrysomelid, *Ophraella communa* LeSage, predisposed *A. artemisiifolia* plants to attack by *Phoma* sp. When applied alone, *Phoma* sp. caused systemic infection but rarely killed the whole plant. Combinations of *O. communa* and *Phoma* sp. were synergistic, resulting in high plant mortality (Teshler *et al.*, 1996). Unfortunately, the culture of *Phoma* sp. lost its virulence and attempts to revive or re-isolate it have failed.

Insects

Zygogramma suturalis (Fabricius) and *O. communa* are natural enemies of *A. artemisiifolia* being studied as inundative biological control agents. Teshler *et al.* (1996) determined the feeding potential of different life stages of *Z. suturalis* and *O. communa.* A high intrinsic reproductive rate, absence of an obligatory diapause and pupation directly on *A. artemisiifolia* plants has greatly facilitated *O. communa* mass-rearing on potted plants in the greenhouse. In contrast, the reduction or cessation of oviposition by *Z. suturalis* on extensively damaged plants, as well as pupation in soil, are important limitations for mass-rearing this beetle (Teshler *et al.*, 1998).

In Quebec, inundative cage releases of *O. communa* were made in fields of carrot, *Daucus carota sativus* Hoffman, and cabbage, *Brassica oleraceae* L., in 1998 and 1999 in Sherrington and St-Jacques le Mineur. Four to five *O. communa* adults per plant in the 4–6-leaf stage caused complete defoliation and death within 14 days (Teshler *et al.*, 2000). By the end of summer, the generalist Pentatomidae predators *Podisus maculiventris* (Say), *Picromerus bidens* (L.), *Perillus bioculatus* (Fabricius) and *Apateticus cynicus* (Say), various Coccinellidae, and the gregarious pupal parasitoid, *Asecodes mento* (Walker), significantly reduced *O. communa* density, but early season releases of *O. communa* were not affected (Teshler *et al.*, 2000).

Competitive interactions

In Quebec, Massicotte *et al.* (1998) investigated the effect of establishing a competitive vegetative cover to control *A. artemisiifolia* along highways. Previous research had demonstrated that in the presence of perennial grasses capable of forming dense canopies, *A. artemisiifolia* is a less effective competitor and therefore less abundant than in sparsely vegetated areas (Maryushkina, 1991). The ability of relatively low-growing herbaceous grass and broad-leaved species to become established on roadsides and to effectively suppress *A. artemisiifolia* was assessed. Among the potential competitor species evaluated were three commercially available perennial grasses: *Puccinellia distans* L., *Festuca rubra* L. and *Lolium perenne* L.; and three legumes: *Trifolium repens* L., *Medicago lupulina* L. and *Lotus corniculatus* L. Several test species, e.g. *T. repens*, *L. perenne*, had poor germination and low overwintering survival rates in sites with relatively high soil salinity concentrations (>100 mmol). *P. distans* and *F. rubra* showed the greatest potential for use as competitor species against *A. artemisiifolia* along roadways (Massicotte *et al.*, 1998). Seeds of *A. artemisiifolia* from roadside populations had a significantly greater salinity tolerance than seeds from the six potential competitor plant species used in field trials (DiTommaso *et al.*, 2000).

Evaluation of Biological Control

The pathogens *A. tragopogi* and *Phoma* sp. can cause considerable damage to *A. artemisiifolia* populations. Although *O. communa* is a promising candidate for use in inundative biological control, its efficacy may be reduced by the presence of indigenous predators and parasitoids. This negative impact can be diminished by early-season releases of adults, which are less vulnerable to predator or parasitoid attack. Moreover, early-season insect releases are more practical because host plants emerging in mid-May produce about 10 times more seeds than plants emerging in early July (Bassett and Crompton, 1975).

The use of *P. distans* and *F. rubra* as competitors of *A. artemisiifolia* appears to be promising.

In North America, phytophagous insects are important natural enemies of *A. artemisiifolia* that have been used successfully for classical biological control in other countries (Goeden and Teerink, 1993), e.g. Kovalev (1989) reported that *Z. suturalis* spread rapidly throughout ragweed-infested areas of southern Russia, attaining population densities as high as 5000 insects per m^2 and eliminating *A. artemisiifolia* within localized areas. Unfavourable climatic conditions and intense predation prevented the establishment of *Z. suturalis* in China and Australia (Wan *et al.*, 1995; Julien and Griffiths, 1998). Overwintering mortality also prevented population build-up of *Z. suturalis* in former Yugoslavia (Igrč *et al.*, 1995). In Australia, the widespread *Epiblema strenuana* (Walker) and the localized *Zygogramma bicolorata* Pallister provide effective control of *A. artemisiifolia* (McFadyen, 1992).

O. communa was evaluated for introduction into Australia but was rejected because it was found that sunflower, *Helianthus annuus* L., sustained some feeding by the beetle (Palmer and Goeden, 1991). Similarly, in no-choice tests conducted at Macdonald Campus, McGill University, *O. communa* adults and larvae fed on *H. annuus*. However, *O. communa* caused significantly more damage to *A. artemisiifolia* than to *H. annuus*; beetle fecundity was severely inhibited and mortality of neonate larvae increased significantly when *O. communa* fed on *H. annuus*.

The current trend in biological control research is not to automatically exclude oligophagous insects as potential control agents for *A. artemisiifolia* (Goeden and Palmer, 1995; McFadyen and Weggler-Beaton, 2000).

Recommendations

Further work should include:

1. Testing new isolates of *Phoma* sp. and other pathogenic fungi for their biological control potential;
2. Developing and evaluating semi-artificial diets for *O. communa* mass-rearing to significantly reduce contamination problems and labour costs;
3. Encouraging the commercial seed industry to select species for seeding along rights-of-ways that are well adapted to severe winters and the relatively high saline conditions typically found along roadsides following spring snowmelt;
4. Evaluating seeds harvested from desirable, competing species that occur naturally along roadsides for use in seeding operations to suppress *A. artemisiifolia*.

References

Bagarozzi, D.A. and Travis, J. (1998) Ragweed pollen proteolytic enzymes: possible roles in allergies and asthma. *Phytochemistry* 47, 593–598.

Bassett, I.J. and Crompton, C.W. (1975) The biology of Canadian weeds. 11. *Ambrosia artemisiifolia* F. and *A. psilostachya* DC. *Canadian Journal of Plant Science* 55, 463–476.

Brière, S.C., Watson, A.K., Paulitz, T.C. and Hallett, S.G. (1995) First report of a *Phoma* sp. on common ragweed in North America. *Plant Disease* 79, 968.

DiTommaso, A., Choy, J. and Watson, A.K. (2000) Seed germination of common ragweed (*Ambrosia artemisiifolia* L.) roadside populations and of potential competitor species under saline conditions. *Weed Science Society of America Abstracts* (6–10 February 2000, Toronto, Ontario, Canada) 40, 17–18.

Goeden, R.D. and Palmer, W.A. (1995) Lessons learned from studies of the insects associated with Ambrosiinae in North America in relation to the biological control of weedy members of this group. In: Delfosse, E.S. and Scott, R.R. (eds) *Proceedings of the VIII International Symposium of Biological Control of Weeds, 2–7 February 1992, Canterbury, New Zealand*, pp. 565–573.

Goeden, R.D. and Teerink, J.A. (1993) Phytophagous insect faunas of *Dicoria canescens* and *Iva axillaris*, native relatives of ragweeds, *Ambrosia* spp., in Southern California, with analyses of insect associates of Ambrosiinae. *Annals of the Entomological Society of America* 86, 38–50.

Harris, P. and Piper, G.L. (1970) *Common Ragweed (*Ambrosia *spp.: Compositae): its North American Insects and Possibilities for its Biological Control.* Commonwealth Institute of Biological Control Technical Bulletin 13, 117–140.

Hartmann, H. and Watson, A.K. (1980a) Damage to common ragweed (*Ambrosia artemisiifolia*) caused by the white rust fungus (*Albugo tragopogi*). *Journal of Weed Science* 28, 632–635.

Hartmann, H. and Watson, A.K. (1980b) Host range of *Albugo tragopogi* from common ragweed. *Canadian Journal of Plant Pathology* 2, 173–175.

Heap, I.M. (1997) The occurrence of herbicide resistant weeds, worldwide. *Pesticide Science* 51, 235–243.

Igrč, J., DeLoach, C.J. and Zlof, V. (1995) Release and establishment of *Zygogramma suturalis* F. (Coleoptera: Chrysomelidae) in Croatia for control of common ragweed (*Ambrosia artemisiifolia* L.). *Biological Control* 5, 203–208.

Julien, M.H. and Griffiths, M.W. (eds) (1998) Biological control of weeds. *A World Catalogue of Agents and their Target Weeds*, 4th edn. CAB International, Wallingford, UK.

Kovalev, O.V. (1989) New factors of efficiency of phytophages: a solitary population wave and succession process. In: Delfosse, E.S. (ed.) *Proceedings of the VII International Symposium of Biological Control of Weeds. 6–11 March 1988*, MAF, Rome, Italy, pp. 51–53.

Maryushkina, V.Y. (1991) Peculiarities of common ragweed (*Ambrosia artemisiifolia* L.) strategy. *Agriculture, Ecosystems and Environment* 36, 207–216.

Massicotte, R., DiTommaso, A., Beaumont, J.-P. and Watson, A.K. (1998) Establishment of competitive vegetation cover to reduce common ragweed (*Ambrosia artemisiifolia*) along roadsides. *Proceedings of the Expert Committee on Weeds (ECW), 7–9 December*, Winnipeg, Manitoba, p. 85.

McFadyen, R.E. (1992) Biological control against parthenium weed in Australia. *Crop Protection* 11, 400–407.

McFadyen, R.E. and Weggler-Beaton, K. (2000) The biology and host specificity of *Liothrips* sp. (Thysanoptera: Phlaeothripidae), an agent rejected for biocontrol of annual ragweed. *Biological Control* 19, 105–111.

Palmer, W.A. and Goeden, R.D. (1991) The host range of *Ophraella communa* (Coleoptera, Chrysomelidae). *Coleopterists' Bulletin* 45,115–120.

St-Louis, S., DiTommaso, A. and Watson, A.K. (2000) Resistance of common ragweed (*Ambrosia artemisiifolia* L.) to the herbicide linuron in carrot fields of southwestern Québec. *Weed Science Society of America Abstracts* (6–10 February 2000, Toronto, Ontario, Canada) 40, 92.

Teshler, M.P., Brière, S.G., Stewart, R.K., Watson, A.K. and Hallett, S.G. (1996) Life tables and feeding ability of *Ophraella communa* LeSage (Coleoptera: Chrysomelidae), a potential biocontrol agent of *Ambrosia artemisiifolia* L. In: Morin, V.C. and Hoffman, J.H. (eds). *Proceedings of the IX International Symposium of Biological Control of Weeds, 21–26 January 1996, Stellenbosch, University of Cape Town, South Africa*, p. 420.

Teshler, M.P., Teshler, I.B., DiTommaso, A., Gagnon, J.A. and Watson, A.K. (1998) Evaluation of two herbivorous insects (Coleoptera: Chrysomelidae) for biocontrol of common ragweed (*Ambrosia artemisiifolia* L.). *Proceedings of the Expert Committee on Weeds (ECW), 7–9 December 1998, Winnipeg, Manitoba*, p. 74.

Teshler, M.P., Teshler, I.B., DiTommaso, A. and Watson, A.K. (2000) Inundative biological control of common ragweed (*Ambrosia artemisiifolia*) using *Opraella communa* (Coleoptera: Chrysomelidae). *Weed Science Society of America Abstracts* (6–10 February 2000,Toronto, Ontario, Canada) 40, 29.

Vincent, G. and Ahmim, M. (1985) Note sur le comportement de *l'Ambrosia artemisiifolia* après fauchage. *Phytoprotection* 66,165–168.

Wan, F., Wang, R. and Ding, J. (1995) Biological control of *Ambrosia artemisiifolia* with introduced insect agents, *Zygogramma suturalis* and *Epiblema strenuana*, in China. In: Delfosse, E.S. and Scott, R.R. (eds) *Proceedings of the VIII International Symposium of Biological Control of Weeds, 2–7 February 1992, Canterbury, New Zealand.* DSIR/Commonwealth Scientific and Industrial Research Organization, Melbourne, Australia, pp. 193–200.

61 *Avena fatua* L., Wild Oat (*Poaceae*)

S.M. Boyetchko

Pest Status

Wild oat, *Avena fatua* L., native to Europe and Asia (Baum, 1977), is one of the most economically important weeds in cultivated crops in temperate and north-temperate areas (Sharma and Vanden Born, 1978; O'Donovan *et al.*, 1985). It is believed to have originated in south-west Asia and was introduced into other countries, e.g. Argentina, Australia, Canada, South Africa and the USA, as a contaminant in seed and feed transported by early settlers (Thurston and Phillipson, 1976). Recent surveys rank *A. fatua* as the second most abundant weed in the northern Great Plains, where it was found in 64% of fields surveyed (Thomas *et al.*, 1996, 1998a, b). In Canada, yield losses can vary from Can$120 million to Can$500 million and the amount of yield loss in cereals such as wheat, *Triticum aestivum* L., and barley, *Hordeum vulgare* L., is increased the earlier *A. fatua* emerges relative to the crop (Friesen, 1973; O'Donovan *et al.*, 1985). Yield loss decreases as the weed emerges later in the growing season. *A. fatua* is responsible for lower grade and quality of grain, dockage losses, and increased costs associated with chemical and cultural control (Sharma and Vanden Born, 1978). It is also a problem in canola, *Brassica napus* L. and *B. rapa* L., in the prairie provinces.

Cool, moist soils, prevalent in the spring and early autumn, favour germination and emergence of *A. fatua*, and seeds buried as deep as 20 cm can emerge (Sharma *et al.*, 1976). Seeds of *A. fatua* exhibit both primary and secondary dormancy, allowing them to persist in soil for up to 3–6 years, depending on environmental factors, particularly moisture and temperature (Banting, 1962; Hsiao, 1987). Secondary dormancy can occur when seeds are exposed to high moisture conditions.

Background

Pre- and post-emergent chemical herbicides are available to control *A. fatua* (Sharma and Vanden Born, 1978; Anonymous, 2000); for example triallate, trifluralin and diallate are soil-applied herbicides whereas chemicals such as difenzoquat, fenoxaprop, imazamethabenz and sethoxydim are foliar-applied. Frequent use of Group 1 herbicides (ACCase (acetyl coenzyme A carboxylase) inhibitors) has led to an increase in incidence of Group 1 herbicide-resistant *A. fatua* populations (Beckie *et al.*, 1999). More than 50% of the fields in Alberta, Saskatchewan and Manitoba were found to have herbicide-resistant *A. fatua*. In addition, 18% of the Group 1 herbicide-resistant *A. fatua* in Saskatchewan were found to have resistance to acetolactate synthase inhibitors or Group 2 herbicides, while 27% of Manitoba fields contained herbicide-resistant *A. fatua* populations with resistance to more than one herbicide group. The use of herbicides is therefore compromised by resistant weed populations, so a biological control alternative is worth pursuing.

Several fungal pathogens causing disease on cultivated oats, *Avena sativa* L., have also been reported on *A. fatua* (Conners, 1967; Ginns, 1986), including leaf blotch or stripe caused by *Drechslera avenacea* (M.A. Curtis ex Cooke) Shoemaker; stem rust and crown rust caused by *Puccinia graminis* Persoon: Persoon f. sp. *avenae* Eriksson and E.

Hennings and *P. coronata* Corda f. sp. *avenae* W.P. Fraser and Ledingham, respectively; and loose smut and covered smut caused by *Ustilago avenae* (Persoon) Roussel and *Ustilago kolleri* Wille, respectively.

More concerted efforts to investigate the utility of microbial agents for inundative biological control have identified the potential of several fungal and bacterial pathogens (Charudattan, 1991; Kremer and Kennedy, 1996; Boyetchko, 1999).

Biological Control Agents

Pathogens

Fungi

D. avenacea, a seed-borne pathogen of *A. sativa* and *A. fatua* that causes seedling blight and leaf blotch, was evaluated as a potential bioherbicide for *A. fatua* control (Prusinkiewicz and Mortensen, 1989). Although three isolates of *D. avenacea* reduced *A. fatua* biomass by 50–74% when applied as a granular inoculant to soil, the weed was able to outgrow the disease. Moreover, the fact that *D. avenacea* causes significant disease on *A. sativa* and that it has been reported on wheat makes this pathogen a risky candidate to pursue for biological control of *A. fatua*. It was concluded that *D. avenacea* was not a good candidate as a bioherbicide.

Bacteria

Bacteria isolated from the roots and rhizosphere of *A. fatua* were screened and evaluated as control agents (Boyetchko, 1997, 1998). In laboratory bioassays, a wide range of activities was exhibited; some isolates caused significant inhibitory activity to germination and root growth while others caused plant growth promotion. To develop effective weed biological control agents, only those isolates with greater than 80% weed suppressive activity were selected from an extensive screening programme and evaluated in the field. One bacterial strain has undergone 4 years of field testing. A dose–response field study comparing the efficacy of this bacterial strain using two formulations indicated that the bacterial agent in a pesta formulation reduced weed emergence by up to 35% and above-ground biomass by up to 23%. When using a peat-based formulation, emergence of *A. fatua* was reduced by as much as 57% and above-ground biomass by 64%. These results are extremely encouraging, considering that crop competition has not been factored in, and further selection and development of appropriate formulations are being conducted. In addition, laboratory bioassays evaluating the effect of the bacterial strain on herbicide-resistant *A. fatua* clearly demonstrated that it can significantly inhibit the growth of Group 1 herbicide-resistant *A. fatua* (Boyetchko, 1999).

Surveys for soil-borne and foliar fungal pathogens were conducted from 1994 to 1996 and their potential against *A. fatua* evaluated (Boyetchko *et al.*, 1998). Nine of 70 fungal agents isolated from *A. fatua* roots and evaluated in growth-pouch bioassays reduced *A. fatua* germination by more than 90%. Many of these pathogens have not been identified to species. From the foliar pathogens surveyed, a total of 73 fungal isolates in 12 genera showed pathogenicity; the most commonly isolated fungi from *A. fatua* were found to be *D. avenacea* and *Cephalosporium* spp. Other fungi identified, *Colletotrichum* spp., *Fusarium* spp. and *Verticillium* spp., are being evaluated further.

Recommendations

Further work should include:

1. Continued evaluation and development of soil bacteria, including development of suitable formulations for the pathogenic bacterial strains and determination of the optimum rate of application, and size and placement of granules in relation to the crop and weed;

2. Evaluation of bacterial strains capable of controlling herbicide-resistant *A. fatua* populations, including the discovery of microbes with novel modes of action differing from those of existing chemical herbicides;

3. Conducting surveys, particularly in Europe and Asia where *A. fatua* originated, or in geographic areas (i.e. South America) of greater microbial diversity, to find foliar and soil-borne fungal pathogens that can be developed as biological control agents of *A. fatua*.

References

Anonymous (2000) *Guide to Crop Protection 2000. Weeds, Plant Disease, Insects.* Bi-Provincial Publication. Regina, SK: Saskatchewan Agriculture and Food; Winnipeg, MB: Manitoba Agriculture.

Banting, J.D. (1962) The dormancy behavior of *Avena fatua* L. in cultivated soil. *Canadian Journal of Plant Science* 42, 22–39.

Baum, B.R. (1977) *Oats: Wild and Cultivated. A Monograph of the Genus* Avena *L. (Poaceae).* Monograph No. 14, Biosystematics Research Institute, Canada Department of Agriculture, Research Branch, Ottawa, Ontario.

Beckie, H.J., Thomas, A.G., Legere, A., Kelner, D.J., Van Acker, R.C. and Meers, S. (1999) Nature, occurrence, and cost of herbicide-resistant wild oat (*Avena fatua*) in small-grain production areas. *Weed Technology* 13, 612–625.

Boyetchko, S.M. (1997) Efficacy of rhizobacteria as biological control agents of grassy weeds. In: *Proceedings, Soils and Crops Workshop '97.* Extension Division, University of Saskatchewan, Saskatoon, Saskatchewan, pp. 460–465.

Boyetchko, S.M. (1998) Evaluation of deleterious rhizobacteria for biological control of grassy weeds. In: Burge, M. (ed.) *Proceedings of the IV International Bioherbicide Workshop, 6–7 August 1998.* University of Strathclyde, Glasgow, UK, p. 16.

Boyetchko, S.M. (1999) Innovative applications of microbial agents for biological weed control. In: Mukerji, K.G., Chamola, B.P. and Upadhyay, K. (eds) *Biotechnological Approaches in Biocontrol of Plant Pathogens.* Kluwer Academic/Plenum Publishers, London, UK, pp. 73–97.

Boyetchko, S.M., Wolf, T.M., Bailey, K.L., Mortensen, K. and Zhang, W.M. (1998) Survey and evaluation of fungal pathogens for biological control of grass weeds. In: *Proceedings, Soils and Crops Workshop '98.* Extension Division, University of Saskatchewan, Saskatoon, Saskatchewan, pp. 424–429.

Charudattan, R. (1991) The mycoherbicide approach with plant pathogens. In: TeBeest, D.O. (ed.) *Microbial Control of Weeds.* Chapman and Hall, New York, New York, pp. 24–57.

Conners, I.L. (1967) *An Annotated Index of Plant Diseases in Canada.* Publication 1251, Canada Department of Agriculture, Ottawa, Ontario.

Friesen, H.A. (1973) Identifying wild oats yield losses and assessing cultural control methods. In: *Let's Clean Up Wild Oats.* Agriculture Canada and United Grain Growers Ltd, Saskatoon, Saskatchewan, pp. 20–25.

Ginns, J.H. (1986) *Compendium of Plant Disease and Decay Fungi in Canada 1960–1980.* Publication 1813. Biosystematics Research Centre, Ottawa, Ontario, Research Branch Agriculture Canada.

Hsiao, A.I. (1987) Mechanisms of dormancy in wild oats (*Avena fatua*). In: Mares, D.J. (ed.) *Fourth International Symposium on Pre-Harvest Sprouting in Cereals.* Westview Press, Boulder, Colorado, pp. 425–440.

Kremer, R.J. and Kennedy, A.C. (1996) Rhizobacteria as biocontrol agents of weeds. *Weed Technology* 10, 601–609.

O'Donovan, J.T., de St Remy, E.A., O'Sullivan, P.A., Dew, D.A. and Sharma, A.K. (1985) Influence of the relative time of emergence of wild oat (*Avena fatua*) on yield loss of barley (*Hordeum vulgare*) and wheat (*Triticum aestivum*). *Weed Science* 33, 498–503.

Prusinkiewicz, E. and Mortensen, K. (1989) *Potential of Granular Formulation of* D. avenacea *as a Bioherbicide for Wild Oat Control.* Report, Agriculture Canada Regina Research Station, Regina, Saskatchewan.

Sharma, M.P. and Vanden Born, W.H. (1978) The biology of Canadian weeds. 27. *Avena fatua* L. *Canadian Journal of Plant Science* 58, 141–157.

Sharma, M.P., McBeath, D.K. and Vanden Born, W.H. (1976) Studies on the biology of wild oats. I. Dormancy, germination and emergence. *Canadian Journal of Plant Science* 56, 611–618.

Thomas, A.G., Wise, R.F., Frick, B.L. and Juras, L.T. (1996) *Saskatchewan Weed Survey: Cereal, Oilseed and Pulse Crops 1995*. Weed Survey Series Publication 96–1, Agriculture and Agri-Food Canada, Saskatoon, Saskatchewan.
Thomas, A.G., Frick, B.L. and Hall, L.M. (1998a) *Alberta Weed Survey: Cereal and Oilseed Crops 1997*. Weed Survey Series Publication 98–2, Agriculture and Agri-Food Canada, Saskatoon, Saskatchewan.
Thomas, A.G., Frick, B.L., Van Acker, R.C., Knezevic, S.Z. and Joosse, D. (1998b) *Manitoba Weed Survey: Cereal and Oilseed Crops 1997*. Weed Survey Series Publication 98–1, Agriculture and Agri-Food Canada, Saskatoon, Saskatchewan.
Thurston, J.M. and Phillipson, A. (1976) Distribution. In: Jones, D.P. (ed.) *Wild Oats in World Agriculture, an Interpretive Review of World Literature*. Agricultural Research Council, London, UK, pp. 19–64.

62 *Calamagrostis canadensis* (Michaux) Palisot de Beauvois, Marsh Reed Grass (*Poaceae*)

K.I. Mallett, D.E. Macey and R.S. Winder

Pest Status

Marsh reed grass, *Calamagrostis canadensis* (Michaux) Palisot de Beauvois, is a perennial, tussock-forming, rhizomatous grass with a circumpolar distribution. In North America it is found from Newfoundland to Alaska in a wide variety of habitats, but is particularly abundant in mesic-to-wet sites with high nutrient content. The grass has become a weed in western Canada especially in white spruce, *Picea glauca* (Moench) Voss, plantations, but can cause problems in regenerating lodgepole pine, *Pinus contorta* Douglas ex Loudon var. *latifolia* Engelmann, and trembling aspen, *Populus tremuloides* Michaux.

Lieffers *et al.* (1993) reviewed the ecology of *C. canadensis*. It becomes established in newly disturbed sites after fire or harvest, via growth from rhizomes and/or wind-dispersed seed. Warm soils, abundant moisture, soil disturbance or compaction, and high light levels allow it to grow profusely in a short time. Clonal expansion occurs and full occupation of a site can take 1–3 years (Lieffers *et al.*, 1993). *C. canadensis* forms thick sods with tall shoots (60–120 cm). Because of this, it can form a large biomass that can reduce tree growth through competition and inhibition (Blackmore and Corns, 1979; Eis, 1981; McDonald, 1986; John and Lieffers, 1991). With time (about 10–20 years or more) and without further disturbance from fire or grazing, *C. canadensis* loses dominance of the site. A woody shrub and tree canopy causes the grass to die back and become almost inconspicuous. It is not uncommon for plantations in western Canada to be retreated and replanted several times because of *C. canadensis*.

Background

Control of *C. canadensis* has been achieved chemically using glyphosate (Blackmore and Corns, 1979) and hexazinone (Otchere-Boateng and Herring, 1990); however, there is growing public resistance to herbicide use in forests. Mechanical site-preparation techniques, e.g. mounding, scalping, mixing, inverting and burial, have been used with some success (Lieffers *et al.*, 1993). The use of grazing has been limited, due to the marginal nutritive value of the grass and its regenerative capacity (Corns and Schraa, 1962). Prescribed fire has been successful but requires that the burn be deep enough to kill rhizomes, and this is often difficult to achieve or is harmful to regeneration. The forestry industry has been looking towards a biological control method because of public concern. Research since 1990 has focused on identifying pathogens effective for control because little is known about herbivores that attack *C. canadensis.*

Biological Control Agents

Insects

Sap-sucking insects have been reported to cause chlorosis and dwarfing of *C. canadensis* in Alaska rangeland. *Irbisia sericans* Stål has been identified with the damage, and preliminary results suggested that there might be a positive correlation between its damage and desirable forage-qualities of the grass (McKendrick and Bleicher, 1980). In Europe, several herbivores feed on *Calamagrostis epigeios* (L.) Roth (Dubbert *et al.*, 1998) but their presence in North America is unknown.

Pathogens

Fungi

Conners (1967) listed 37 species of fungi found on *C. canadensis* in North America and Greenland. From collections made in British Columbia, Winder (1992) found more than 30 endemic fungi capable of causing disease in *C. canadensis.* Laboratory pathogenicity tests revealed that *Colletotrichum* sp. (= *Vermicularia affinis* var. *calamagrostidis* Karsten, affin. *C. graminicola* [Cesati] G.W. Wilson, anamorph of *Glomerella* sp.), *Fusarium* spp. (anamorphs of *Giberella* spp.) and *Dilophorspora alopecuri* (Fries: Fries) Fries (anamorph of *Lidophia graminis* (Saccardo) Walker and Sutton) were pathogenic to the grass (Winder, 1999a). Winder (1999a) showed that, of the fungi tested, *Colletotrichum* sp. and *Fusarium* spp., particularly *Fusarium avenaceum* (Fries) Saccardo, provided the greatest opportunities for biological control. While capable of causing foliar damage on various hosts in the Poaceae, they did not cause symptoms on black spruce, *Picea mariana* (Miller) Britten, Sterns, and Poggenburg, or white spruce, *P. glauca* (Moench) Voss. Winder (1999b) tested formulations and did experiments on the influence of substrate and temperature on sporulation of *F. avenaceum* and their effect on *C. canadensis. F. avenaceum* and the *Colletotrichum* sp. isolate have also been applied in field tests, alone and in combination. Inoculations were performed in winter, spring or summer, with or without straw mulch from *C. canadensis*, using a powder made from water, flour and inoculum. Both fungi caused foliar symptoms, but the plants were able to recover. The only significant growth reduction occurred in plots with mulch, particularly in the winter application where snow was compacted on the plots (R.S. Winder, unpublished).

Snow moulds (*Typhula incarnata* Lasch ex Fries, *Microdochium nivale* (Fries) Samuels and Hallet, and a low-temperature basidiomycete) have been listed as pathogens of *C. canadensis* (Conners, 1967). In Alaska, Lebeau and Logsdon (1958) first reported a low-temperature basidiomycete infecting *C. canadensis.* This species is endemic to the boreal forest, growing at low temperatures under snow cover. Schreiner *et al.* (1995) reported that it was pathogenic to *C. canadensis* but not to white spruce. Mallett *et al.* (2000) showed that the low-temperature basidiomycete can cause mortality and up to 50% loss in biomass in low-temperature growth chamber and greenhouse experiments.

Bacteria

Deleterious rhizobacteria effectively suppress growth of weed grasses (Kremer and Kennedy, 1996). Growth-suppressive activity was recorded in 20% of the rhizobacteria collected from *C. canadensis* in British Columbia (D.E. Macey, unpublished). These reduced root growth by 32–54%, shoot growth by 16–61% and germination by 26–70% in laboratory assays (Macey and Winder, 1996). In greenhouse tests, rhizobacteria caused various responses, ranging from slight stimulation to 30% reduction in seedling biomass. However, selected rhizobacteria applied in combination with *F. avenaceum* resulted in biomass reduction greater than 75%, with no adverse effects on white spruce, lodgepole pine or trembling aspen (Winder and Macey, 1997). Efficacy of the co-inoculated root and shoot pathogens could be improved or constrained by environmental and nutritional factors (Winder and Macey, 1998).

Allelopathy

Winder (1997) reported that leachate from *C. canadensis* straw inhibited root growth and caused foliar necrosis in grass seedlings. Straw leachate coupled with certain endophytic fungi could enhance the effect by increasing the virulence of *Colletotrichum* sp. when this fungus was used as a biological control agent.

Evaluation of Biological Control

The field trials of *F. avenaceum* and *Colletotrichum* sp. demonstrated that foliar pathogens, while capable of causing short-term effects, will not be effective as the only control method. Some component attacking the below-ground portions of the plant will be necessary. The powdering process seemed to reduce inoculum viability, suggesting that further improvements in formulation and delivery are also necessary. With seedlings being the most susceptible host stage in trials with rhizobacteria and *F. avenaceum*, the most practical use of such organisms would probably involve application to mature panicles to prevent dissemination of viable seed. For *in situ* control, the low-temperature basidiomycete could provide the necessary level of suppression. Initial field trial results in Alberta suggest that *C. canadensis* biomass is reduced by up to 50% over that of control plots. This reduction occurs for up to 3 years after the application, suggesting that the low-temperature basidiomycete remains active after the initial application (K.I. Mallett, unpublished). Although chemical control will probably be its preferred method as long as it is available, the forest industry is interested in developing biological control agents for *C. canadensis*.

Recommendations

Future work should include:

1. Using rhizobacteria and various fungi as biological control agents in an integrated pest management programme;
2. Investigating European phytophagous insects reported from *C. epigeios* for potential introduction as biological control agents.

References

Blackmore, D.G. and Corns, W.G. (1979) Lodgepole pine and white spruce establishment after glyphosate and fertilizer treatments of grassy cutover forest land. *Forestry Chronicle* 55, 102–105.

Conners, I.L. (1967) *An Annotated Index of Plant Diseases in Canada.* Publication 125, Canada Department of Agriculture, Ottawa, Ontario.

Corns, W.G. and Schraa, R.J. (1962) Seasonal productivity and chemical composition of marsh reed grass (*Calamagrostis canadensis*) harvested periodically from fertilized and unfertilized native sod. *Canadian Journal of Plant Science* 42, 651–659.

Dubbert, M., Tscharntke, T. and Vidal, S. (1998) Stem-boring insects of fragmented *Calamagrostis* habitats: herbivore–parasitoid community structure and the unpredictability of grass shoot abundance. *Ecological Entomology* 23, 271–280.

Eis, S. (1981) Effects of vegetation competition on revegetation of white spruce. *Canadian Journal of Forest Research* 11, 1–8.

John, S.E.T. and Lieffers, V.J. (1991) *Analysis of Monitor Plot Data for Alberta.* Alberta Reforestation Branch, Alberta Forest Service, Edmonton, Alberta.

Kremer, R.J. and Kennedy, A.C. (1996) Rhizobacteria as biocontrol agents of weeds. *Weed Technology* 10, 601–609.

Lebeau, J.B. and Logsdon, C.E. (1958) Snow mold of forage crops in Alaska and Yukon. *Phytopathology* 48, 148–150.

Lieffers, V.J., MacDonald, S.E. and Hogg, E.H. (1993) Ecology of and control strategies for *Calamagrostis canadensis* in boreal forest sites. *Canadian Journal of Forest Research* 23, 2070–2077.

Macey, D.E. and Winder, R.S. (1996) Development of a co-inoculation strategy for biological control of marsh reed grass (*Calamagrostis canadensis*). In: Comeau, P. and Harper, G. (eds) *Proceedings of Expert Committee on Weeds 1996 National Meeting, 9–12 Dec. 1996, Victoria, BC.* British Columbia Ministry of Forests, Research Branch, Victoria, British Columbia, pp. 161–162.

Mallett, K.I., Schreiner, K.A. and Gaudet, D.A. (2000) Effect of cottony snow mould on mortality and biomass of *Calamagrostis canadensis* under controlled-environment conditions. *Biological Control* 18, 193–198.

McDonald, P.M. (1986) Grasses in young conifer plantations – hindrance and help. *Northwest Science* 60, 271–277.

McKendrick, J.D. and Bleicher, D.P. (1980) Observations of a grass bug on bluejoint ranges. *Agroborealis* 12, 15–18.

Otchere-Boateng, J. and Herring, L.T. (1990) Site preparation: chemical. In: Laveneder, D.P., Parish, R., Johnson, C.M., Montgomery, G., Vyse, A., Willis, R.A. and Winston, D. (eds) *Regenerating British Columbia's Forests.* University of British Columbia Press, Vancouver, British Columbia, pp. 164–178.

Schreiner, K., Mallett, K.I., Leiffers, V.J. and Gaudet, D. (1995) Biocontrol of bluejoint grass (*Calamagrostis canadensis*) using low-temperature basidiomycete. *Canadian Journal of Plant Pathology* 17, 362.

Winder, R.S. (1992) The potential for biological control of bluejoint (*Calamagrostis canadensis* [Michx.] Beauv.) in reforestation areas in British Columbia. In: Dorworth, C.E. (ed.) *Biocontrol of Forest Weeds. Proceedings of the Biocontrol of Forest Weeds Workshop. Western International Forest Disease Work Conference, Vernon, BC, 9 August 1991.* Canadian Forest Service, Victoria, British Columbia, pp. 30–36.

Winder, R.S. (1997) The in vitro effect of allelopathy and various fungi on marsh reed grass (*Calamagrostis canadensis*). *Canadian Journal of Botany* 75, 236–241.

Winder, R.S. (1999a) Evaluation of *Colletotrichum* sp. and *Fusarium* spp. as potential biological control agents for marsh reed grass (*Calamagrostis canadensis*). *Canadian Journal of Plant Pathology* 21, 8–15.

Winder, R.S. (1999b) The influence of substrate and temperature on the sporulation of *Fusarium avenaceum* and its virulence on marsh reed grass. *Mycological Research* 103, 1145–1151.

Winder, R.S. and Macey, D.E. (1997) Co-inoculation of marsh reed grass (*Calamagrostis canadensis* [Michx.] Beauv.) with a fungal shoot pathogen (*Fusarium avenaceum* [Fr.] Sacc.) and rhizobacteria. In: Murray D.S. (ed.) *Proceedings of the 1997 Meeting of the Weed Science Society of America, 3–6 February, 1997, Orlando FL.* Weed Science Society of America, Champaign, Illinois, p. 150.

Winder, R.S. and Macey, D.E. (1998) Biological control of grasses in reforestation areas: Problems and prospects. In: Wagner, R.G. and Thompson, D.G. (eds) *Third International Conference on Forest Vegetation Management: Popular Summaries.* Forest Research Information Paper No. 141, Ontario Forest Research Institute, Sault Ste Marie, Ontario, pp. 360–362.

63 *Centaurea diffusa* Lamarck, Diffuse Knapweed, and *Centaurea maculosa* Lamarck, Spotted Knapweed (*Asteraceae*)

R.S. Bourchier, K. Mortensen and M. Crowe

Pest Status

Spotted and diffuse knapweeds, *Centaurea maculosa* Lamarck and *C. diffusa* Lamarck, were introduced into British Columbia from Europe and Asia about 50 and 90 years ago, respectively. At least 40,000 ha of rangeland were infested by 1989 (Muller-Scharer and Schroeder, 1993) and the infestation continues to spread, with new areas being reported each year (S. Turner, Kamloops, 2000, personal communication). In British Columbia, the potential area of invasion is estimated at 1.1 million ha of grassland (Harris and Cranston, 1979). In western North America, over 3 million ha in 14 states and two provinces are affected (Story *et al.*, 2000). In Alberta, knapweeds are present but the eradication programme begun in 1974 (Ali, 1989) has been relatively successful at containing them.

Background

Because of the scale of the problem in western Canada and the cost and difficulty of conventional treatments, *C. diffusa* and *C. maculosa* were among the first weeds targeted for biological control. The early focus was on limiting seed production because of the high reproductive potential of *C. diffusa* and *C. maculosa*: 36,000 and 25,000 seeds m^{-2}, respectively (Harris, 1980). Of the first agents released, *Urophora affinis* Frauenfeld, *Urophora quadrifasciata* Meigen and *Metzneria paucipunctella* Zeller attack the seed head, and *Sphenoptera jugoslavica* Obenberger attacks the roots. While establishment of these agents has drastically reduced seed production, there has been limited progress in controlling the weeds (Harris and Myers, 1984).

Biological Control Agents

Since 1980, eight more European insect species have been released, four that attack part of the seed head, *Chaetorellia acrolophi* White and Marquardt, *Larinus minutus* Gyllenhal, *Larinus obtusus* Gyllenhal and *Terellia virens* Loew, and four that attack the root, *Agapeta zoegana* L., *Cyphocleonus achates* Fahraeus, *Pelochrista medullana* Staudinger and *Pterolonche inspersa* Staudinger. Additional agents were released to address the need for control in a variety of habitats, and to increase the stress on the weeds to achieve the reductions in seed production required for population declines of both *Centaurea* spp. (Myers, 1995).

In British Columbia, nine insects are now established for biological control. Since 1980, over 3200 releases have been made, with almost half occurring in the Nelson region (Table 63.1; V. Miller, Nelson, 2000, personal communication). Two fungi, *Sclerotinia sclerotiorum* (Libert) de Bary and *Puccinia jaceae* Otth, have also been studied as potential stress factors on knapweeds.

Table 63.1. Number of releases (number of insects released) in British Columbia from 1981 to 1999 to control *Centaurea diffusa* and *Centaurea maculosa*.

Year	Agent										
	Agapeta zoegana	*Chaetorellia acrolophi*	*Cyphocleonus achates*	*Larinus minutus*	*Larinus obtusus*	*Metzneria paucipunctella*	*Pelochrista medullana*	*Pterolonche inspersa*	*Sphenoptera jugoslavica*	*Terellia virens*	*Urophora* spp.
1981						2 (unknown)			2 (800)		
1982	2 (300)					1 (50)	1 (75)		3 (210)		
1983	1 (199)						1 (53)		2 (unknown)		7 (31,535)
1984							2 (69)		2 (48)		
1985	2 (45)					19 (3800)	2 (278)		24 (4838)		
1986	20 (480)					71 (34,700)	2 (27)	4 (128)	86 (17,020)		34 (19,500)
1987	23 (1121)		3 (51)			77 (21,800)		6 (12,267)	77 (16,700)		1 (unknown)
1988	72 (5021)		6 (134)			41 (9300)	1 (16)		66 (14,430)		
1989	16 (592)		11 (113)					4 (309)	91 (22,942)		
1990	30 (4133)		6 (360)			75 (14,300)		1 (25)	124 (24,635)		2 (21,000)
1991	69 (9370)	6 (164)	9 (947)	6 (4466)		70 (16,600)		3 (371)	103 (19,010)	5 (312)	
1992	79 (8405)	4 (1408)	20 (1898)			8 (2430)	2 (67)		30 (5020)	9 (3056)	1 (60)
1993	117 (11,767)		26 (2227)	1 (104)	9 (2805)		1 (133)		58 (9300)		
1994	197 (20,462)		89 (9242)	7 (873)	9 (3000)	1 (1000)	2 (134)		71 (7030)		2 (2000)
1995	180 (21,588)	8 (1439)	83 (8803)	20 (2872)					31 (6265)	9 (3231)	7 (unknown)
1996	92 (10,970)		161 (14,615)	30 (5800)			2 (865)				
1997	75 (8588)		114 (12,682)[a]	66 (19,300)[a]			1 (108)				
1998	79 (12,547)		114 (23,365)[a]	23 (4900)[a]							
1999	16 (4588)[a]		9 (3157)[a]	21 (8600)[a]	39 (14,800)						
Total	1070 (120,176)	18 (3011)	651 (77,594)	174 (46,915)	57 (20,605)	363 (103,980)	17 (1825)	18 (13,100)	770 (148,248)	23 (6599)	54(74,095)

[a]Nelson Region only.

Pathogens

Fungi

S. sclerotiorum (see Huang *et al.*, Chapter 99 this volume) was first recorded on *C. diffusa* near Vernon, British Columbia, in 1971 (Watson *et al.*, 1974). Normally the fungus does not seriously affect knapweed populations; about 10% of *C. diffusa* plants were observed wilting near Summerland (Mortensen and Hogue, 1995). Under dry conditions, infections occur mostly below ground from soil-borne sclerotia, so the disease spreads slowly in rangeland, e.g. in interior British Columbia. In the early 1980s, Mortensen and Hogue (1995) investigated *S. sclerotiorum* as a control for *C. diffusa*. In late autumn 1981, the fungus, applied as a granular inoculum to plots severely infested with *C. diffusa* at Summerland, resulted in a population reduction of about 25% in summer, 1982. By the following summer, however, knapweed populations in treated plots rebounded to levels in control plots. To obtain even this impact, at least 15 g m^{-2} of inoculum was required, i.e. 150 kg ha^{-1}. To be cost effective on rangeland, lower concentrations that provide longer-lasting control are required.

P. jaceae, reported on *C. diffusa* in Europe (Gaümann, 1959), was not found in North America until 1988 (Mortensen *et al.*, 1989). Watson *et al.* (1981) collected several isolates from *C. diffusa* in eastern Europe. The rust was not released in North America because host-range tests of these isolates showed that safflower, *Carthamus tinctorius* L., seedlings were susceptible to European *P. jaceae* (Watson and Alkhoury, 1981; Mortensen, 1985; Hasan *et al.*, 1990). In 1988, however, *P. jaceae* was discovered on *C. diffusa* at Oliver, British Columbia, (Mortensen *et al.*, 1989, 1991) and in 7 years the rust had spread more than 1400 km. By 1989, it had spread to most populations in interior British Columbia, by 1991, to Washington (Dugan and Carris, 1992; Palm *et al.*, 1992), and since then to Oregon, Idaho, Montana and South Dakota (Richard *et al.*, 1996).

Although *P. jaceae* was initially found on *C. diffusa*, it later became increasingly common on *C. maculosa*, suggesting that either the rust is changing in virulence or resistance in *C. maculosa* is changing. To test this, growth chamber inoculation tests on *C. maculosa* plants were conducted using rust isolates from the Kamloops area (one from 1993 from *C. maculosa* and two from 1998, one each from *C. maculosa* and *C. diffusa*). These isolates were compared with the isolate collected at Oliver in 1988, and two Romanian isolates, R11 and R13h2 (Mortensen *et al.*, 1989, 1991). Plants originating from *C. maculosa* and *C. diffusa* seeds, collected in 1992 in the Lillooet area, were grown individually in 10 cm^2 pots in a mixture of soil–peat–vermiculite (3:2:1, v/v), at 24°C. At the 4–6-leaf stage, plants were inoculated with an airbrush sprayer until runoff with a urediospore suspension at about 0.10×10^6 spores ml^{-1}. After a 24 h dew-period at 17°C in dark, inoculated plants were placed in a greenhouse for 5 weeks. Rust development ratings (0–9, Mortensen, 1985) were done at 3 and 5 weeks. A total of 72 plants per isolate were inoculated in two separate trials. Both susceptible and resistant *C. maculosa* plants were found when inoculated with the Oliver 1988 isolate as well as with isolates collected either from *C. maculosa* or *C. diffusa* plants 5 and 10 years later (Table 63.2). Romanian isolates showed virulence on some *C. maculosa* plants. It is unlikely that the differences among the Canadian isolates are significant. The data indicate that rust resistance in *C. maculosa* populations is segregating. As only one source of *C. maculosa* seed (Lillooet – 1992) was tested, it cannot be confirmed whether the ratio of susceptible and resistant plants has changed since the rust was discovered in Canada. Watson and Renney (1974) showed that hybridization can occur between *C. maculosa* and *C. diffusa*. Thus, it is not surprising that some *C. maculosa* plants are susceptible to the rust and that resistant plants existed in the *C. diffusa* population tested (Table 63.2). The impact of the rust on knapweed populations is unknown. A biomass reduction caused by *P. jaceae* occurred on young *C. diffusa* under controlled conditions (Mortensen *et al.*, 1991), so its presence is an additional stress to knapweeds.

Table 63.2. Pathogenicity of *Puccinia jaceae* inoculated on *Centaurea diffusa* and *C. maculosa* in greenhouse trials.

P. jaceae host/isolates[b]	Per cent of inoculated plants rated for rust attack[a] Resistant (%)	Moderately resistant (%)	Moderately susceptible (%)	Susceptible (%)
Centaurea maculosa				
R11	44.4	15.7	13.9	25
R13h2	81.9	15.3	2.8	0
Oliv.88	37.5	15	21.3	26.3
Kaml.I93	38	19.7	14.1	28.2
Kaml.I98	27.8	44.4	23.6	4.2
Kaml.II98	50.7	26.8	5.6	16.7
Centaurea diffusa				
Oliv.88	37.7	26.1	36.2	0
Kaml.I98	11.1	26.4	40.3	22.2
Kaml.II98	12.5	31.9	43.1	12.5

[a]Based on a rating scale (0–9, Mortensen, 1985). A total of 72 plants per isolate were inoculated with a urediospore suspension (0.10×10^6 spores ml^{-1}) in two separate trials.
[b]R11 and R13h2, collected from *C. diffusa* in Romania in 1978; Oliv.88, collected from *C. diffusa* at Oliver, 1988; Kaml.I93 and Kaml.I98, collected from *C. diffusa* in the Kamloops area in 1998.

Insects

A. zoegana biology has been studied extensively by Muller *et al.* (1988), Muller (1989a, b), Muller-Scharer (1991) and Powell *et al.* (2000). Larvae mine the roots of *C. diffusa* and *C. maculosa*; early instars damage epidermal tissues of the root crown and later instars cause serious damage, particularly to smaller plants or those containing more than one larva (Muller *et al.*, 1988). The insect can reduce plant survivorship, plant height and seed production, delay flowering time and decrease rosette survival (Muller *et al.*, 1988; Muller, 1989b; Muller and Schroeder, 1989). Rosette survival has been identified as a key factor for knapweed population dynamics (Myers and Risley, 2000). New root growth above a larval feeding site can offset the impact of limited water and nutrient uptake resulting from larval feeding (Steinger and Muller-Scharer, 1992). Larval infestations in Austria and parts of Hungary averaged 23.6% (Muller *et al.*, 1988). *A. zoegana* is compatible with the root feeders *C. achates*, *P. medullana* and *S. jugoslavica*.

C. acrolophi attacks *C. maculosa* capitula (Groppe and Marquardt, 1989a). Larvae feed on the seed head: early instars in the immature florets in the centre of the bud and later instars on developing seeds and florets. One larva can destroy the entire contents of a seed head. Larvae can develop on *C. diffusa*, but oviposition has only been observed on *C. maculosa* (Lang, 1997a). *C. acrolophi* prefers dry, south-facing slopes with scattered plants, rather than dense *C. maculosa* stands (Powell *et al.*, 1994). In Europe, *C. acrolophi* occurs on sparse and remote knapweed plants; Harris (1990) suggested that it would fill an unoccupied seed-feeding niche in North America because *Urophora* spp. density declines as knapweed plant density decreases and individual plants or isolated patches are often missed.

C. achates, from eastern and southern Europe and Asia Minor, attacks both *C. maculosa* and *C. diffusa* (Wikeem and Powell, 1999) and causes considerable damage, particularly to the taproot interior. Most damage occurs during bolting and early shoot development (Steinger and Muller-Scharer, 1992). Larval feeding within the root crown impedes nutrient flow into the stems, resulting in fewer flowers, lower seed production and stunting (Muller and Schroeder, 1989; Stinson *et al.*, 1994). In the weevil's native range, larval infestation near

60% may occur in *C. maculosa* (Volovnik, 1989). Adults may completely defoliate rosettes and even kill them if the central buds are attacked (Stinson *et al.*, 1994). The major limiting factor for *C. achates* is availability of plants large enough to sustain an individual larva. It prefers *C. maculosa* because its roots tend to be much larger and can support multiple larvae. On larger plants, *A. zoegana* and *P. medullana* can live on the same root as *C. achates* because their larvae feed on the outer root layers whereas *C. achates* larvae mine the centre (Stinson *et al.*, 1994). *C. achates* does best on sunny, south-facing slopes with light soils because high soil temperatures are necessary for complete development (Lang, 1997b). Habitats include disturbed hillsides, overgrazed range, and recent fallow (Stinson *et al.*, 1994).

L. minutus will attack both *C. diffusa* and *C. maculosa* (Groppe, 1990). Larvae begin feeding on pappus hairs and mine through to the capitulum to consume the seeds. A single larva may completely destroy all the seeds of a small capitulum (Kashefi and Sobhian, 1998). In larger heads, multiple larvae can destroy all the seeds (Groppe, 1990). Adults feed on rosettes in spring and later in flowers (Jordan, 1995). Although *L. minutus* used together with other seed-attacking insects may result in competition among agents, it does coexist with *U. affinis*, which initiates gall formation before weevil attack occurs (Groppe, 1990). In its native range *L. minutus* is particularly adapted to very dry sites (Jordan, 1995) and has the greatest impact on patch edges where knapweed densities are lower (Lang, 1997c). These preferences may limit some competitive interactions with other seed feeders.

L. obtusus occupies the same niche as *L. minutus*, attacking both *C. diffusa* and *C. maculosa*. It prefers moist sites, whereas *L. minutus* prefers drier sites (Groppe, 1992). In Europe, *L. obtusus* attacks 37–76% of capitula. More than one larva is common in a single flower head. Developing larvae destroy most, and often all, of the seeds and each larva uses additional seeds when constructing its cocoon.

M. paucipunctella is a seed feeder that was released to control *C. maculosa*. Its biology was reviewed in Harris and Myers (1984); interactions with *Urophora* spp. were examined by Story *et al.* (1991).

P. medullana prefers *C. diffusa* and only attacks *C. maculosa* in its absence (Muir, 1986). Larvae develop only on rosettes (Powell *et al.*, 1994); early instars mine the root cortex and later instars mine deeper. As many as four larvae per plant may occur but one is most common (Muir, 1986). Larval damage, similar to that of *A. zoegana*, results in reduced root storage capacity, limited nutrient uptake, fewer flowering heads, smaller plant size and exposure to pathogens (Gassmann *et al.*, 1982). Plants with a root diameter of less than 5 mm are usually completely destroyed. *P. medullana* can coexist with *A. zoegana* (Smith, 2000) and prefers sites with high knapweed densities and moderate moisture (Powell *et al.*, 1994).

P. inspersa attacks *C. diffusa* and *C. maculosa* but strongly prefers *C. diffusa* (Dunn *et al.*, 1989; Powell *et al.*, 1994). Attacked plants are recognized by silk tubes around the rosette crown. One or two larvae can cause significant root damage, resulting in stunted growth, reduced flower-head production, and swollen, spongy roots with reduced storage and limited nutrient transport capacity (Dunn *et al.*, 1989; Campobasso *et al.*, 1994). In Europe, *P. inspersa* infestations range from 20 to 30% but may reach 75% at sites with lower plant densities (Campobasso *et al.*, 1994). *P. inspersa* competes with *S. jugoslavica* (Muller, 1989b).

S. jugoslavica strongly prefers *C. diffusa*, but can be found attacking *C. maculosa* on dry summer sites (Julien and Griffiths, 1998). Although adults feed on leaves of seedlings, rosettes, and flowering plants, the most significant damage is caused by larval mining and occurs in the roots (Zwoelfer, 1976; Powell and Myers, 1988). Rosettes are often killed; mortality depends on the root being large enough to support early instar development, but not large enough to sustain complete development. Surviving plants are usually stunted and produce fewer flowers as a result of depleted root stores (Powell and Myers, 1988). These impacts diminish the aggressiveness of *C. diffusa*. After 5 or 6 years, once beetle numbers are high enough, knapweed populations collapse (Lang, 1998a). The beetle prefers dry environments with summer drought periods.

T. virens prefers *C. maculosa* but will also attack *C. diffusa*. Larval feeding reduces seed production; germination viability of *C. maculosa* seeds decreased from 77.6% to 6.6% as a result (Groppe and Marquardt, 1989b). *T. virens* coexists with *U. affinis* and *U. quadrifasciata* but has reduced survival if *L. minutus* is present at the same site (Groppe and Marquardt, 1989b). *T. virens* prefers dry, south facing slopes.

U. affinis attacks *C. diffusa* and *C. maculosa*. Harris and Shorthouse (1996) reported on its effectiveness, together with that of other gall inducers. The galls act as metabolic sinks by draining nutrients from other plant parts, thus extending plant damage well beyond that incurred in attacked seed heads (Harris, 1990). As many as 8 (average 1.2–1.6) galls per attacked head are formed. Densities above 1000 galls m^{-2} are common and may exceed 3000 galls m^{-2} (Harris and Shorthouse, 1996). In galled plants, new flower buds tend to abort due to lack of nutrients, and viable seed production is reduced. Harris (1980) showed that one *U. affinis* gall per seed head decreased flower head number by 9.2 per plant, reducing seed production in *C. diffusa* by 2.4 seeds per head and in *C. maculosa* by two seeds per head.

U. quadrifasciata attacks *C. diffusa* and *C. maculosa*. Gall production reduces seed and flower production (Powell *et al.*, 1994). The floret occupied by the larva is destroyed and adjacent florets tend to abort (Lang, 1998b). Each gall displaces 1.9 seeds in a *C. diffusa* seed-head (Harris, 1980). The two *Urophora* spp. combine to reduce *C. diffusa* seed production from 30,000–40,000 m^{-2} to about 1500 m^{-2} (Schroeder, 1985).

Releases and Recoveries

Most of the potential knapweed biological control agents from Europe have been successfully established in Canada or the western USA. *A. zoegana* was first released in 1982 (Table 63.1) but did not establish (Schroeder, 1985). A second release was made in 1983 and adults emerged successfully in 1984 (Muir, 1986). Adults can disperse up to 5 km and, with extensive release efforts, the moth has become widely distributed throughout knapweed-infested areas. In 2000, establishment was confirmed at 105 of 124 sites in the Nelson district (V. Miller, Nelson, 2000, personal communication). A 1989 release in Alberta did not establish (Table 63.3).

C. acrolophi, released in 1991, 1992 and 1995 (Table 63.1), has not been confirmed as established. Attempts to rear it in propagation tents at Kamloops were unsuccessful (S. Turner, Kamloops, 2000, personal communication). In Alberta, two releases were made in 1995 outside Waterton Lakes National Park (Table 63.3) but no establishment occurred.

C. achates was first released in 1987 (Table 63.1). Extensive redistribution (more than 600 releases from 1988 to 1999) resulted in establishment in much of British Columbia, as confirmed in 2000 at 131 of 173 sites (V. Miller, Nelson, 2000, personal communication). In Alberta, a release was made in 1996 outside Waterton Park but *C. achates* did not establish due to flooding. Additional releases were made in 2000 (Table 63.3). In Ontario, a release was made in 1993 but establishment is unknown.

L. minutus, first released in 1991, has been continuously released since 1993 (Table 63.1). In 1997, 66 releases (almost 20,000 insects) were made in the Nelson region alone. In 2000, establishment was confirmed at 31 of 36 sites (V. Miller, Nelson, 2000, personal communication) and numbers are now sufficient to allow for redistribution.

L. obtusus was released from 1992 to 1994 and in 1999 about 15,000 were redistributed in the Nelson region (Table 63.1). Establishment has been confirmed in 29 of

Table 63.3. Number of releases (number of insects released), in provinces other than British Columbia, from 1981 to 2000.

Insect	Year	Province			
		Alberta	Saskatchewan	Manitoba	Ontario
Agapeta zoegana L.	1989	1 (25)			
Cyphocleonus achates Fahraeus	1993				1 (175)
	1996	1 (529)			
	2000	3 (1150)			
Chaetorellia acrolophi White and Marquardt	1995	2 (284)			
Metzneria paucipunctella Zeller	1985	2 (400)			
	1986		1 (100)		
	1993				1 (1089)
Terellia virens Loew	1995	1 (736)			
Urophora spp.	1993				2 (16236)
	1994			Unknown	
Total		10 (3124)	1 (100)		4 (17500)

33 release sites (V. Miller, Nelson, 2000, personal communication).

M. paucipunctella was released in 1973 (Harris and Myers, 1984) but establishment was unsuccessful because it is susceptible to winter cold (Good *et al.*, 1997). Thus, little effort was made prior to 1985 to redistribute it in British Columbia (Muir, 1986). It was released more extensively from 1985 to 1994 (Table 63.1) and is now widely distributed in British Columbia. It has failed, however, to establish in Ontario.

P. medullana was released from 1982 to 1986 (Table 63.1). Establishment has not been confirmed and problems with overwintering survival exist. It overwintered for 2 years in field cages at Kamloops but did not persist (S. Turner, Kamloops, 2000, personal communication). Successful establishment requires 3–4 weeks with mean summer temperatures above 18°C (Muir, 1986).

P. inspersa was first released on *C. diffusa* at four locations in 1986 and on *C. maculosa* in 1991 (Table 63.1). Larval chimneys were discovered in spring, 2000, on both *Centaurea* spp. (V. Miller, Nelson, 2000, personal communication).

S. jugoslavica was released on *C. diffusa* in 1976 and extensively redistributed from 1985 to 1995 (Table 63.1). It now occurs throughout the driest range of *C. diffusa* (Julien and Griffiths, 1998).

T. virens was released against *C. maculosa* at three locations in 1991. Additional releases were made in 1992 and 1995 (Table 63.1). In Alberta, a release was made in 1995 (Table 63.3). There are no confirmed field establishments. As with *C. acrolophi*, attempts to rear *T. virens* in propagation tents at Kamloops have been unsuccessful (S. Turner, Kamloops, 2000, personal communication).

U. affinis is already established in British Columbia, Alberta and Quebec (Harris and Myers, 1984). New releases were made in Manitoba but establishment has not been confirmed. It failed to establish in Ontario.

U. quadrifasciata was released prior to 1980 against *C. diffusa* and *C. maculosa* (Harris and Myers, 1984). It is established in British Columbia, Ontario and Quebec but did not survive Saskatchewan winters. It is a strong flier (Lang, 1998b) and is establishing throughout much of the Canadian knapweed range.

Evaluation of Biological Control

S. sclerotiorum did not show potential for *C. diffusa* control under dry rangeland conditions. The impact of *P. jaceae* is unknown, although it stresses knapweeds.

The number and type of insects required for successful knapweed control has been debated (Myers, 1985; Harris, 1991; Muller-Scharer and Schroeder, 1993; Harris, 1998; Myers and Risley, 2000). Impact studies that assess multiple agents have only considered the first four agents released, *Urophora* spp., *M. paucipunctella* and *S. jugoslavica*, because they were the only insects established well enough to study (Powell and Myers, 1988; Powell, 1989; Story *et al.*, 1991; Myers, 1995; Myers and Risley, 2000). Myers (1995) reported that reduction in seed production in *C. diffusa* would have to be more than 99.7% to reduce plant densities, and that the combined impact of the two *Urophora* spp. and *S. jugoslavica* was not at this level; maximum combined seed mortality was 95% in one location. At high densities, *S. jugoslavica* reduced the densities of knapweed seedlings and rosettes; however, fluctuations in beetle populations resulted in only isolated impacts (Powell and Myers, 1988). *U. affinis* reduces above-ground biomass of *C. diffusa* by up to 71% (Harris, 1980). Myers (1985) noted that attack levels remained unchanged when it was used in conjunction with *S. jugoslavica*, suggesting that the two are compatible, whereas *U. quadrifasciata* occurs in significantly lower numbers on plants attacked by *S. jugoslavica*. *U. quadrifasciata* densities of 1.9 larvae per *C. maculosa* flower head reduced seed production by 67% (Harris, 1980). *U. quadrifasciata* together with *M. paucipunctella* resulted in only a 40% reduction in seed numbers, suggesting that the combined attack of the two agents is not significantly greater than on sites where *U. quadrifasciata* is well established by itself (Myers *et al.*, 1989).

Although some impact studies of the new biological control agents have been published, e.g. Callaway *et al.* (1999) and Story *et al.* (2000), studies that assess the interactions and cumulative impact of all biological control agents in established populations are just beginning. *A. zoegana*, *C. achates* and *Larinus* spp. have all been reported, in anecdotal cases, to have significant impact on knapweed populations in both Canada and the USA. There is, however, a lack of quantitative data. *A. zoegana* has been reported to cause reductions of knapweed at a site of up to 90%, with the fastest declines at sites where *C. achates* is also present (Julien and Griffiths, 1998). Results from Story *et al.* (2000) are less conclusive and emphasize the complexity of the interactions between *A. zoegana* and its habitat. Preliminary studies of *C. achates* populations have found 50% knapweed mortality within 3 months if populations reach 6 adults per plant (Story *et al.*, 1996). The early population studies (Powell and Myers 1988; Powell, 1989; Myers, 1995) need to be extended to consider these new agents. Such data are essential to make the best use of available agents and integrate biological control with other strategies for knapweed management.

Quantitative data become even more important as the merits and deficiencies of biological control are debated in the scientific literature, e.g. Cory and Myers (2000), Strong and Pemberton (2000). Two studies, Callaway *et al.* (1999) and Pearson *et al.* (2000), reporting non-target interactions that specifically concerned knapweed biological control agents have also fuelled the debate. Story *et al.* (2000) raised several concerns about the methods used by Callaway *et al.* (1999), primarily arguing in favour of trials under natural field situations. Unfortunately, the data required to compare methods (pot/plot studies versus field studies) are very limited. Pearson *et al.* (2000) demonstrated the potential cascade effects that can occur from the release of a biological control agent, in this case *U. affinis* and *U. quadrifasciata*. With eight biological control agents now well established, it becomes even more imperative to understand their combined impact and interactions on the plant.

Recommendations

Further work should include:

1. Extending population studies of knapweeds to include combinations of more

recently established agents using long-term monitoring plots that cover the range of release conditions and agent combinations;
2. Conducting controlled experiments to assess the effect of recently established agents in combination on knapweeds, because most release sites in British Columbia have at least three agents;
3. Investigating potential non-target impacts of *A. zoegana* (e.g. Callaway *et al.*, 1999) under natural field conditions and together with other agents;
4. Assessing the impact of knapweed outbreaks on native flora and fauna, to determine quantitatively both the environmental and economic rationale for control of these invasive species.

Acknowledgements

We thank D. Brooke, V. Miller and S. Turner, British Columbia Ministry of Forests, for extensive efforts in the propagation, release and record-keeping of knapweed biological control agents, and V. Miller for taking colour-coded filing to the next level. S. Turner supplied *C. diffusa* and *C. maculosa* seeds and rust-infected plants. P. Harris collected the rust-infected knapweed plants from Kamloops area. Funds for the ongoing insect research programme on knapweed were provided by the British Columbia Ministry of Forests, Canadian Pacific Railway, and the Agriculture and Agri-Food Canada Matching Investments Initiative.

References

Ali, S. (1989) Eradication program in Alberta. In: Fay, P. and Lacey, J. (eds) *Proceedings of the 1989 Knapweed Symposium*. Montana State University, Bozeman, Montana, pp. 105–106.

Callaway, R.M., Deluca, T.H. and Belliveau, W.M. (1999) Biological-control herbivores may increase competitive ability of the noxious weed *Centaurea maculosa*. *Ecology* 80, 1196–1201.

Campobasso, G., Sobhian, R., Knutson L., Pastorino, A.C. and Dunn, P.H. (1994) Biology of *Pterolonche inspersa* (Lep.: Pterolonchidae), a biological control agent for *Centaurea diffusa* and *C. maculosa* in the United States. *Entomophaga* 39, 377–384.

Cory, J.S. and Myers, J.H. (2000) Direct and indirect ecological effects of biological control. *Trends in Ecology and Evolution* 15, 137–139.

Dugan, F.M. and Carris, L.M. (1992) *Puccinia jaceae* var. *diffusa* and *P. acroptili* on knapweeds in Washington. *Plant Disease* 76, 972.

Dunn, P., Rosenthal, S.S., Campobasso, G. and Tait, S.M. (1989) Host specificity of *Pterolonche inspersa* (Lep.: Pterolonchidae) and its potential as a biological control agent for *Centaurea diffusa*, diffuse knapweed and *C. maculosa*, spotted knapweed. *Entomophaga* 34, 435–446.

Gassmann, A., Schroeder, D. and Muller, H. (1982) *Investigations on* Pelochrista medullana *(Stgr.) (Lep.: Tortricidae), a Possible Biocontrol Agent of Diffuse and Spotted Knapweed,* Centaurea diffusa *Lam., and* C. maculosa *Lam. (Compositae) in North America.* Final Report, Commonwealth Agriculture Bureaux, Delémont, Switzerland.

Gaümann, E. (1959) Beiträge zur Kryptogamenflora der Schweiz, Bd XII. *Die rostpilze Mitteleuropas.* Büchler and Co., Bern, Switzerland.

Good, W.R., Story, J.M. and Callan N.W. (1997) Winter cold hardiness and supercooling of *Metzneria paucipunctella* Zeller (Lepidoptera: Gelechiidae). *Environmental Entomology* 26, 1131–1135.

Groppe, K. (1990) *Screening Report.* Larinus minutus *Gyll. (Coleoptera: Curculionidae), a Suitable Candidate for the Biological Control of Diffuse and Spotted Knapweed in North America.* International Institute of Biological Control, European Station, Delémont, Switzerland.

Groppe, K. (1992) *Final Report.* Larinus obtusus *Gyll. (Coleoptera: Curculionidae), a Candidate for Biological Control of Diffuse and Spotted Knapweed.* International Institute of Biological Control, European Station, Delémont, Switzerland.

Groppe, K. and Marquardt, K. (1989a) *Screening Report.* Chaetorellia acrolophi *White and Marquardt (Diptera: Tephritidae), a Suitable Candidate for the Biological Control of Diffuse and Spotted Knapweed in North America.* International Institute of Biological Control, European Station, Delémont, Switzerland.

Groppe, K. and Marquardt, K. (1989b) *Screening Report.* Terellia virens *(Loew) (Diptera: Tephritidae), a Suitable Candidate for the Biological Control of Diffuse and Spotted Knapweed in North America.* International Institute of Biological Control, European Station, Delémont, Switzerland.

Harris, P. (1980) Effects of *Urophora affinis* Frfld, and *U. quadrifasciata* (Meig.) (Diptera: Tephritidae) on *Centaurea diffusa* Lam. and *C. maculosa* Lam. (Compositae). *Zeitschrift für Angewandte Entomologie* 90, 190–201.

Harris, P. (1990) The Canadian biocontrol of weeds program. In: Roche, B.F. and Roche C.T. (eds) *Range Weeds Revisited, Symposium Proceedings, 24–26 January 1989, Spokane, Washington.* Washington State University, Pullman, Washington, pp. 61–68.

Harris, P. (1991) Classical biocontrol of weeds: its definition, selection of effective agents, and administrative-political problems. *The Canadian Entomologist* 123, 827–849.

Harris, P. (1998) Evolution of classical weed biocontrol: meeting survival challenges. *Bulletin of the Entomological Society of Canada* 30, 134–143.

Harris, P. and Cranston, R. (1979) An economic evaluation of control methods for diffuse and spotted knapweed in Western Canada. *Canadian Journal of Plant Science* 59, 375–382.

Harris, P. and Myers, J.H. (1984) *Centaurea diffusa* Lam. and *C. maculosa* Lam. s. lat. diffuse and spotted knapweed (Compositae). In: Kelleher, J.S. and Hulme, M.A. (eds) *Biological Control Programmes Against Insects and Weeds in Canada 1969–1980.* Commonwealth Agriculture Bureaux, Slough, UK, pp. 127–137.

Harris, P. and Shorthouse, J.D. (1996) Effectiveness of gall inducers in weed biological control. *The Canadian Entomologist* 128, 1021–1055.

Hasan, S., Chaboudez, P. and Mortensen, K. (1990) Field experiment with the European rust *Puccinia jaceae* on safflower, sweet sultan, and bachelor's button. In: Delfosse, E.S. (ed.) *Proceedings of the VII International Symposium on Biological Control of Weeds, 6–11 March, 1988, Rome, Italy.* Istituto Sperimentale per la Patologia Vegetale Ministero dell'Agricoltura e delle Foreste, Rome, Italy, pp. 499–509.

Jordan, K. (1995) Host specificity of *Larinus minutus* Gyll. (Col., Curculionidae), an agent introduced for the biological control of diffuse and spotted knapweed in North America. *Journal of Applied Entomology* 119, 689–693.

Julien, M.H. and Griffiths M.W. (eds) (1998) *Biological Control of Weeds: a World Catalogue of Agents and Their Target Weeds,* 4th edn. CABI Publishing and the Australian Centre for International Agricultural Research, Antony Rowe, Chippenham, UK.

Kashefi, J.M. and Sobhian, R. (1998) Notes on the biology of *Larinus minutus* Gyllenhal (Col., Curculionidae), an agent for biological control of diffuse and spotted knapweeds. *Journal of Applied Entomology* 122, 547–549.

Lang, R.F. (1997a) *Chaetorellia acrolophi* Diptera: Tephritidae. http://www.nysaes.cornell.edu/ent/biocontrol/weedfeeders/chaetorellia_acrolophi.html (18 April 1997)

Lang, R.F. (1997b) *Chaetorellia acrolophi* Diptera: Tephritidae. http://www.nysaes.cornell.edu/ent/biocontrol/weedfeeders/cyphocleonus.html (7 March 1997)

Lang, R.F. (1997c) *Larinus minutus* Coleoptera: Curculionidae. http://www.nysaes.cornell.edu/ent/biocontrol/weedfeeders/larinus_minutus.html (18 April 1997)

Lang, R.F. (1998a) *Sphenoptera jugoslavica* Coleoptera: Burprestidae. http://www.nysaes.cornell.edu/ent/biocontrol/weedfeeders/sphenoptera.html (20 March 1998)

Lang, R.F. (1998b) *Urophora quadrifasciata* Diptera: Tephritidae. http://www.nysaes.cornell.edu/ent/biocontrol/weedfeeders/urophora_quad.html (28 August 1998)

Mortensen, K. (1985) Reaction of safflower cultivars to *Puccinia jaceae*, a potential biocontrol agent for diffuse knapweed. In: Delfosse, E.S. (ed.) *Proceedings of the VI International Symposium on Biological Control of Weeds (1985), Vancouver.* Agriculture Canada, Ottawa, Ontario, pp. 447–452.

Mortensen, K. and Hogue, E.J. (1995) *Sclerotinia sclerotiorum* as a potential biological control agent for diffuse knapweed on dry rangeland in interior British Columbia. *Proceedings of the VIII International Symposium on Biological Control of Weeds (1992), Lincoln, New Zealand.* Commonwealth Scientific and Industrial Research Organization, Melbourne, Australia, pp. 397–406.

Mortensen, K., Harris, P. and Makowski R.M.D. (1989) First occurrence of *Puccinia jaceae* var. *diffusae* in North America on diffuse knapweed (*Centaurea diffusa*). *Canadian Journal of Plant Pathology* 11, 322–324.

Mortensen, K., Harris, P. and Kim, W.K. (1991) Host ranges of *Puccinia jaceae, P. centaureae, P. acroptili,* and *P. carthami,* and the potential value of *P. jaceae* as a biological control agent for diffuse knapweed (*Centaurea diffusa*) in North America. *Canadian Journal of Plant Pathology* 13, 71–80.

Muir, D.M. (1986) *Knapweed in British Columbia: A Problem Analysis.* Province of British Columbia, Range Management Branch, Research Branch, British Columbia Ministry of Forests and Lands, Victoria, BC, Canada.

Muller, H. (1989a) Structural analysis of the phytophagous insect guilds associated with the roots of *Centaurea maculosa* Lam., *C. diffusa* Lam., and *C. vallesiaca* Jordan in Europe. *Oecologia* 78, 41–52.

Muller, H. (1989b) Growth pattern of diploid and tetraploid spotted knapweed, *Centaurea maculosa* Lam. (Compositae), and effects of the root-mining moth *Agapeta zoegana* (L.) (Lep.: Cochylidae). *Weed Research* 29, 103–111.

Muller, H. and Schroeder, D. (1989) The biological control of diffuse and spotted knapweed in North America: What did we learn? In: Fay, P.K. and Lacey, J.R. (eds) *Proceedings of the 1989 Knapweed Symposium, 4–5 April 1989, Bozeman, MT.* Plant and Soil Department and Extension Service, Montana State University, Bozeman, Montana, pp. 151–169.

Muller, H., Schroeder, D. and Gassmann, A. (1988) *Agapeta zoegana* (L.) (Lepidoptera: Cochylidae), a suitable prospect for biological control of spotted and diffuse knapweed, *Centaurea maculosa* Monnet de la Marck and *C. diffusa* Monnet de la Marck (Compositae) in North America. *The Canadian Entomologist* 120, 109–124.

Muller-Scharer, H. (1991) The impact of root herbivory as a function of plant density and competition: survival, growth and fecundity of *Centaurea maculosa* in field plots. *Journal of Applied Ecology* 28, 759–776.

Muller-Scharer, H. and Schroeder, D. (1993) The biological control of *Centaurea* spp. in North America: do insects solve the problem? *Pesticide Science* 37, 343–353.

Myers, J.H. (1985) How many insect species are necessary for successful biocontrol of weeds. In: Delfosse, E.S. (ed.) *Proceedings of the VI International Symposium on Biological Control of Weeds (1985), Vancouver.* Agriculture Canada, Ottawa, Ontario, pp. 77–82.

Myers, J.H. (1995) Long term studies and predictive models in the biological control of knapweed. In: Delfosse, E.S. and Scott, R.R. (eds) *Proceedings of the VIII International Symposium on Biological Control of Weeds (1992),* Lincoln University, Canterbury, New Zealand, pp. 221–224.

Myers, J.H. and Risley, C. (2000) Why reduced seed production is not necessarily translated into successful biological weed control. In: Spencer, N.R. (ed.) *Proceedings of the X International Symposium on Biological Control of Weeds, 4–14 July 1999.* Montana State University, Bozeman, Montana, pp 569–581.

Myers, J.H., Risley, C. and Eng, R. (1989) The ability of plants to compensate for insect attack: why biological control of weeds is so difficult. In: Delfosse, E.S. (ed.) *Proceedings of the VII International Symposium on Biological Control of Weeds, 6–11 March 1988, Rome, Italy.* Istituto Sperimentale per la Patologia Vegetale Ministero dell'Agricoltura e delle Foreste, Rome, Italy, pp. 67–73.

Palm, M.E., Richard, R.D. and Parker, P. (1992) First report of *Puccinia jaceae* var. *diffusae* on diffuse knapweed in the United States. *Plant Disease* 76, 972.

Pearson, D.E., Mckelvey, K.S. and Ruggiero, L.F. (2000) Non-target effects of an introduced biological control agent on deer mouse ecology. *Oecologia* 122, 121–128.

Powell, G.W., Sturko, A.,Wikeem, B.M. and Harris, P. (1994) *A Field Guide to the Biological Control of Weeds in British Columbia.* Ministry of Forests, Victoria, British Columbia.

Powell, G.W., Wikeem, B.M. and Sturko, A. (2000) Biology of *Agapeta zoegana* (Lepidoptera: Cochylidae), propagated for the biological control of knapweeds (Asteraceae). *The Canadian Entomologist* 132, 223–230.

Powell, R.D. (1989) The functional forms of density-dependent birth and death rates in diffuses knapweed (*Centaurea diffusa*) explain why it has not been controlled by *Urophora affinis, U. quadrifasciata* and *Sphenoptera jugoslavica.* In: Delfosse, E.S. (ed.) *Proceedings of VII International Symposium on Biological Control of Weeds, 6–11 March 1988, Rome, Italy,* pp. 195–202.

Powell, R.D. and Myers, J.H. (1988) The effect of *Sphenoptera jugoslavica* Obenb. (Col., Buprestidae) on its host plant *Centaurea diffusa* Lam. (Compositae). *Journal of Applied Entomology* 106, 25–45.

Richard, R.D., Parker, P.E., Palm, M.E. and Coombs, E. (1996) Spread of *Puccinia jaceae* var. *diffusae. Phytopathology* 86 (Suppl.), 81.

Schroeder, D. (1985) The search for effective biological control agents in Europe. 1. Diffuse and spotted knapweed. In: Delfosse, E.S. (ed.) *Proceedings of the VI International Symposium on*

Biological Control of Weeds (1985), Vancouver. Agriculture Canada, Ottawa, Ontario, pp. 103–119.

Smith, L. (2000) *Pelochrista medullana* Lepidoptera: Tortricidae. http://nysaes.cornell.edu/ent/biocontrol/weedfeeders/pelochrista_medullana.html (30 January 2000)

Steinger, T. and Muller-Scharer, H. (1992) Physiological and growth responses of *Centaurea maculosa* (Asteraceae) to root herbivory under varying levels of interspecific plant competition and soil nitrogen availability. *Oecologia* 91, 141–149.

Stinson, C.S.A., Schroeder, D. and Marquardt, K. (1994) Investigations on *Cyphocleonus achates* (Fahr.) (Col., Curculionidae), a potential biological control agent of spotted knapweed (*Centaurea maculosa* Lam.) and diffuse knapweed (*Centaurea diffusa* Lam.) (Compositae) in North America. *Journal of Applied Entomology* 117, 35–50.

Story, J.M., Boggs, K.W., Good, W.R., Harris, P. and Nowierski, R.M. (1991) *Metzneria paucipunctella* Zeller (Lepidoptera: Gelechiidae), a moth introduced against spotted knapweed: its feeding strategy and impact on two introduced *Urophora* spp. (Diptera: Tephritidae). *The Canadian Entomologist* 123, 1001–1007.

Story, J.M., White, L.J. and Good, W.R. (1996) Propagation of *Cyphocleonus achates* (Fahraeus) (Coleoptera: Curculionidae) for biological control of spotted knapweed: procedures and cost. *Biological Control* 7, 167–171.

Story, J.M., Good, W.R., White, L.J. and Smith, L. (2000) Effects of interaction of the biocontrol agent *Agapeta zoegana* L. (Lepidoptera: Cochylidae) and grass competition on spotted knapweed. *Biological Control* 17, 182–190.

Strong, D.R. and Pemberton, R.W. (2000) Biological control of invading species: risk and reform. *Science* 288, 1969.

Volovnik, S.V. (1989) On distribution and ecology of some species of Cleonine weevils (Coleoptera: Curculionidae) 1. Tribe Cleonini. *Entomological Review* 68, 138–144.

Watson, A.K. and Alkhoury, I. (1981) Response of safflower cultivars to *Puccinia jaceae* collected from diffuse knapweed in eastern Europe. In: Delfosse, E.S. (ed.) *Proceedings of the V International Symposium on Biological Control of Weeds (1980), Brisbane, Australia*. Commonwealth Scientific and Industrial Research Organization, Melbourne, Australia, pp. 301–305.

Watson, A.K. and Renney, R.J. (1974) The biology of Canadian weeds. *Centaurea diffusa* and *C. maculosa*. *Canadian Journal of Plant Science* 54, 687–701.

Watson, A.K., Copeman, R.J. and Renney, A.J. (1974) A first record of *Sclerotinia sclerotiorum* and *Microsphaeropsis centaureae* on *Centaurea diffusa*. *Canadian Journal of Botany* 52, 2639–2640.

Watson, A.K., Schroeder, D. and Alkhoury, I. (1981) Collection of *Puccinia* species from diffuse knapweed in Eastern Europe. *Canadian Journal of Plant Pathology* 3, 6–8.

Wikeem, B.M. and Powell, G.W. (1999) Biology of *Cyphocleonus achates* (Coleoptera: Curculionidae), propagated for the biological control of knapweeds (Asteraceae). *The Canadian Entomologist* 131, 243–250.

Zwoelfer, H. (1976) Investigations on *Sphenoptera (Chilostetha) jugoslavica* Obenb. (Col.: Buprestidae), a possible biocontrol agent of the weed *Centaurea diffusa* Lam. (Compositae) in North America. *Zeitschrift für Angewandte Entomologie* 80, 170–190.

64 *Chamerion angustifolium* (L.) Holub, Fireweed (*Onagraceae*)

R.S. Winder

Pest Status

Fireweed, *Chamerion angustifolium* (L.) Holub, also known as willowherb, great willowherb or rosebay willowherb (Mitich, 1999), is an important species native to boreal forest ecosystems throughout the Northern Hemisphere (Broderick, 1990). North American researchers often refer to it as *Epilobium angustifolium* L., but taxonomic and molecular analyses have shown that it belongs to *Chamerion* (Husband and Schemske, 1998). *C. angustifolium* is a honey-producing plant with showy pink to purple, or sometimes white, inflorescences, and it is admired enough to be the official flower of Yukon Territory. The plant is prevalent in areas exposed to fire or other disturbances, e.g. logging. Although moderate populations are actually beneficial for conifer regeneration, dense populations suppress conifer seedlings through competition and snow press. *C. angustifolium* may also act as a reservoir for root-rotting *Armillaria* spp., a serious problem for lodgepole pine, *Pinus contorta* (Douglas ex Loudon) (Klein-Gebbinck *et al.*, 1993). Among the forb species that inhabit Canadian forests, *C. angustifolium* is probably the most frequent cause of regeneration failures, although quantitative information on the exact extent of the problem is lacking. The plant is both annual and perennial, and it rapidly seeds-in to disturbed areas (Solbreck and Andersson, 1987).

Background

Glyphosate and hexazinone are the principal chemical herbicides used to control *C. angustifolium*, with poor to excellent results depending on site and type of application (Hauessler *et al.*, 1990). Other chemical herbicides have also been used or studied (Etherington, 1983; Bailey and Hoogland, 1984; Haeussler *et al.*, 1990; Winder and Watson, 1994; Siipilehto and Lyly, 1995). However, use of herbicides in Canadian forests has come under increasing restrictions, leading to development of other control methods. Moreover, tree seedlings can be damaged by herbicides, and the regenerative capacity of the rhizomes makes it difficult to control established perennial populations. Changing harvesting practices may also limit the practicality of aerial spraying, because small patch cuts and variable retention schemes are gaining in favour over clear-cutting.

Fire can be used to control *C. angustifolium* under certain conditions (Myerscough, 1980), but it usually encourages the plant to proliferate. Also, dense populations of live plants can actually suppress fire (Sylvester and Wein, 1981) and forest fires with sufficient intensity to kill *C. angustifolium* rhizomes can harm young conifers.

Biodegradable plastic Brush Blankets® have been used for *C. angustifolium* suppression. Their practicality depends on site, type of tree planted and labour costs. Similar approaches using mulches (Siipilehto and Lyly, 1995) and allelopathic conifer litter (Jobidon, 1986) have also been studied. Management schemes that shade out *C. angustifolium* and other competing vegetation, while permitting the growth and harvest of shade-tolerant conifers, have been suggested (Lieffers and Stadt, 1994).

Biological control is also being studied. Because *C. angustifolium* and its natural enemies are native throughout their northern circumpolar distribution, classical biological control methods are unlikely to be productive. Research has therefore focused on grazing by livestock, development of endemic biological herbicides and study of defoliator population dynamics.

Biological Control Agents

Vertebrates

In British Columbia, two grazing methods have been employed to control *C. angustifolium*. The first involves use of fences to manage cattle on cut blocks, and is largely an informal practice (Kerr, 1998). The second, sheep grazing, has been encouraged by the provincial Ministry of Forests as a management tool for fireweed and other competing vegetation (Cayford, 1993). In 1992, sheep were used to control about 6600 ha of vegetation. Grazing was conducted as a contractual arrangement in which the livestock were monitored by a veterinarian and provided access to browse in exchange for their use on cut-blocks, where needed. Rather than using fences, shepherds move the flock through an area with the assistance of border collies, while larger dogs patrol the surrounding area to ward off threats from wildlife. This method can be a very effective control in some situations, but not all. The sheep may trample conifer seedlings if left in an area too long, and they cannot browse in difficult terrain. Grizzly bears can become a nuisance to the flocks, and deploying sheep to an area can be a relatively expensive proposition.

Pathogens

Fungi

At least 40 species of fungi have been reported on *C. angustifolium* (Barr, 1953; Corlett, 1991; Winder and Watson, 1994; Fernando *et al.*, 1999). The fir-fireweed rust, *Pucciniastrum epilobii* Otth, is perhaps the most widespread disease of *C. angustifolium* (Broderick, 1990). This obligate pathogen may have locally severe effects on its alternative hosts, including alpine fir, *Abies lasiocarpa* (Hooker) Nutall, balsam fir, *A. balsamea* (L.) Miller, grand fir, *A. grandis* (Douglas ex David Don) Lindley, noble fir, *A. procera* Rehder, Pacific silver fir, *A. amabilis* Douglas ex Forbes, and white fir, *A. concolor* (Gordon and Glendinning) Lindley ex Hildebrand. Although transitory in nature, the severe foliar browning caused on firs near heavy infestations of *C. angustifolium* (Sinclair *et al.*, 1987) probably rules out its development as a biological control agent. Winder and Watson (1994) reported other naturally occurring diseases that appear to be widespread, including *Alternaria alternata* (Fries: Fries) Von Kiesler (anamorph of *Lewia* sp.), *Diploceras* (= *Seimatosporium*) *kriegerianum* (Bresadola) Nag Raj (anamorph of *Discostromopsis callistemonis* Swart), and *Colletotrichum dematium* (Persoon ex Fries) Grove (anamorph of *Glomerella* sp.). The extent to which any of these diseases suppresses *C. angustifolium* populations, or competition under natural conditions, is largely unknown.

Winder and Watson (1994) and Léger (1997) studied *C. dematium* as a potential bioherbicide. It produced up to 97% leaf area damage when seedlings in a growth chamber were treated with 10^9 conidia m^{-2} and an 18 h dew period. Similar results were obtained in field trials, although larger plants were not controlled. Host range tests indicated that *C. dematium* from fireweed is host-specific, and the isolate was later named *C. dematium* f. sp. *epilobii* (Abou-Zaid *et al.*, 1997). Various factors affected the virulence of this form-species, including inoculum density (optimum = 10^9 conidia m^{-2}), inoculum age (optimum < 20 days), dew period (optimum > 18 h), and seedling stage (optimum < 10 weeks) (Winder and Watson, 1994). Although virulence of the isolate was attenuated in subsequent testing, the fungus has been reported to produce potent phytotoxic compounds in culture filtrates (Abou-Zaid *et al.*, 1997).

D. kriegerianum, tested as a potential mycoherbicide, produces lesions on inoculated seedlings, but it grows slowly in culture and only affects a portion of the host (Winder and Watson, 1994).

Bacteria

Crude extracts from cultures of *Pseudomonas syringae* van Hall have been shown to control *C. angustifolium* seedlings at a rate of 10 ppm (extract : sand) (Norman *et al.*, 1994).

Insects

A wide variety of insects have been reported as defoliators of *C. angustifolium* (Macgarvin, 1982; Lempke and Stolk, 1986; Broderick, 1990; Pashchenko, 1993). In North America, there has been extensive investigation of the population dynamics of Aphididae, their predators and their tenders (Formicidae) (Robinson, 1979; Antolin and Addicott, 1991; Bretton and Addicott, 1992; Morris, 1992; Ives *et al.*, 1993; Pike *et al.*, 1996). Larvae of bedstraw hawk moth, *Hyles gallii* Rottemburg, form occasional epiphytotics in North America, as occurred in peak infestations in British Columbia (Costello, 1997). Bronze flea beetle, *Altica tombacina* Mannerheim, prevalent throughout the northern hemisphere, can also create serious epiphytotics on *C. angustifolium* (Michaud, 1990). *A. tombacina* and *Bromius obscurus* L., while occupying relatively few plants, may damage a considerably greater proportion of the host population (S. Hicks, R. Russel and J. Myers,Vancouver, 1999, personal communication). In Europe, *Craspedolepta nebulosa* Zetterstedt and *Craspedolepta subpunctata* Förster are differentially distributed along latitudinal and altitudinal gradients (Bird and Hodkinson, 1999). In British Columbia, *C. subpunctata* occurred on over 10% of *C. angustifolium* surveyed near Williams Lake. Among 17 other insects observed on *C. angustifolium* at Williams Lake, *Mompha nodicolella* Fuchs (= M. sturnipennella (Treitschlee)) and the larva of an unidentified lepidopteran were also prevalent. In caged experiments, *Mompha albapalpella* Chambers significantly reduced plant height and flowering after larvae fed on leaf tips and apical meristems (S. Hicks, R. Russel and J. Myers, Vancouver, 1999, personal communication).

Evaluation of Biological Control

Much of the biological control research mentioned above is at a developmental, rather than practical, stage. Because *C. angustifolium* is native to Canada, further development of biological control agents should focus on use of endemic natural enemies.

Recommendations

Further work should include:

1. Evaluating and eventually registering fungal pathogens as potential biopesticides;
2. Enhancing insect epiphytotics through study of the attractive effects of smoke, small fires or pheromones on *C. angustifolium* defoliators, e.g. *Actebia fennica* Tauscher, especially if conifer planting is delayed until after attack;
3. Understanding the timing and population dynamics of fire-following insects to control *C. angustifolium* in areas previously cleared by fire, as part of integrated management that includes selective logging and shading.

References

Abou-Zaid, M., Dumas, M., Charuet, D., Watson, A. and Thompson, D. (1997) C-Methyl flavonols from the fungus *Colletotrichum dematium* f. sp. *epilobii*. *Phytochemistry* 45, 957–961.

Antolin, M. and Addicott, J. (1991) Colonization, among shoot movement, and local population neighborhoods of two aphid species. *Oikos* 61, 45–53.

Bailey, J. and Hoogland, D. (1984) The response of *Epilobium* species to a range of soil and foliar acting herbicides. *Aspects of Applied Biology* 8, 43–52.

Barr, M. (1953) Pyrenomycetes of British Columbia. *Canadian Journal of Botany* 31, 810–831.

Bird, J. and Hodkinson, I. (1999) Species at the edge of their range: the significance of the thermal environment for the distribution of congeneric *Craspedolepta* species (Sternorrhyncha: Psylloidea) living on *Chamerion angustifolium* (Onagraceae). *European Journal of Entomology* 96, 103–109.

Bretton, L. and Addicott, J. (1992) Density-dependent mutualism in aphid–ant interaction. *Ecology* 73, 2175–2180.

Broderick, D. (1990) The biology of Canadian weeds. 93. *Epilobium angustifolium* L. (Onagraceae). *Canadian Journal of Plant Science* 70, 247–259.

Cayford, J. (1993) Sheep for vegetation management. *Forestry Chronicles* 69(1), 27.

Corlett, M. (1991) *An Annotated List of the Published Names in* Mycosphaerella *and* Sphaerella*: Mycologia memoir no. 18.* J. Cramer, Berlin, Germany.

Costello, B. (1997) Hornworms galore. *British Columbia Ministry of Agriculture and Food, Crop Protection Newsletter* 19(2), 1.

Etherington, J. (1983) Control of germination and seedling morphology by ethene: differential responses related to habitat of *Epilobium hirsutum* and *Chamerion angustifolium*. *Annals of Botany* 52, 653–658.

Fernando, A., Ring, F., Lowe, D. and Callan, B. (1999) *Index of Plant Pathogens, Plant-associated Microorganisms, and Forest Fungi of British Columbia.* Information Report BC-X-385, Natural Resources Canada, Canadian Forest Service, Victoria, British Columbia.

Hauessler, S., Coates, D. and Mather, J. (1990) *Autecology of Common Plants in British Columbia: A Literature Review.* Forest Resource Development Agreement, Report no. 158, Forestry Canada and British Columbia Ministry of Forests, Victoria, British Columbia.

Husband, B. and Schemske, D. (1998) Cytotype distribution at a diploid–tetraploid contact zone in *Chamerion (Epilobium) angustifolium* (Onagraceae). *American Journal of Botany* 85, 1688–1694.

Ives, A., Kareiva, P. and Perry, R. (1993) Response of a predator to variation in prey density at three hierarchical scales: lady beetles feeding on aphids. *Ecology* 74, 1929–1938.

Jobidon, R. (1986) Allelopathic potential of coniferous species to old-field weeds in eastern Quebec (Canada). *Forest Science* 32, 112–118.

Kerr, S. (1998) Northwood Pulp and Timber uses cattle for vegetation management. *Beef in British Columbia* 13, 73–74.

Klein-Gebbinck, H., Blenis, P. and Hiratsuka, Y. (1993) Fireweed as a possible inoculum resevoir for root-rotting *Armillaria* species. *Plant Pathology* 42, 132–136.

Léger, C. (1997) Development of a *Colletotrichum dematium* as a bioherbicide for the control of fireweed. MSc thesis, Macdonald Campus, McGill University, Montreal, Quebec.

Lempke, B. and Stolk, J. (1986) An interesting new form of *Deilephila elpenor* (Linnaeus) (Lepidoptera: Sphingidae). *Entomologische Berichten* 46, 157–158.

Lieffers, V. and Stadt, K. (1994) Growth of understory *Picea glauca*, *Calamagrostis canadensis*, and *Epilobium angustifolium* in relation to overstory light transmission. *Canadian Journal of Forest Research* 24, 1193–1198.

Macgarvin, M. (1982) Species–area relationships of insects on host plants: herbivores on rosebay willowherb (*Chamerion angustifolium*). *Journal of Animal Ecology* 51, 207–224.

Michaud, J. (1990) Observations on the biology of the bronze flea beetle *Altica tombacina* (Coleoptera: Chrysomelidae) in British Columbia (Canada). *Journal of the Entomological Society of British Columbia* 87, 41–49.

Mitich, L. (1999) Fireweed, *Epilobium angustifolium*. *Weed Technology* 13, 191–194.

Morris, W. (1992) The effects of natural enemies, competition, and host plant water availability on an aphid population. *Oecologia* 90, 359–365.

Myerscough, P.J. (1980) Biological flora of the British Isles: *Epilobium angustifolium* L. *Journal of Ecology* 68, 1047–1074.

Norman, M., Patten, K. and Gurusiddaiah, S. (1994) Evaluation of a phytotoxin from *Pseudomonas syringae* for weed control in cranberries. *Hortscience* 29, 1475–1477.

Pashchenko, G. (1993) Aphids of the genus *Aphis* (Homoptera, Aphidinea, Aphididae) living on plants of the families Lamiaceae, Lioniaceae, Onagraceae, Polemoniaceae, Primulaceae, and Santalaceae in the Russian Far East. *Zoologicheskii Zhurnal* 72, 41–53.

Pike, K., Starý, P., Miller, R., Allison, D., Boydton, L., Graf, G. and Miller, T. (1996) New species and host records of aphid parasitoids (Hymenoptera: Braconidae: Aphidiinae) from the Pacific Northwest, USA. *Proceedings of the Entomological Society of Washington* 98, 570–591.
Robinson, A. (1979) Annotated list of aphids (Homoptera: Aphididae), collected at Churchill, Manitoba, Canada, with descriptions of new species. *The Canadian Entomologist* 111, 447–458.
Siipilehto, J. and Lyly, O. (1995) Weed control trials with fibre mulch, glyphosate, and terbuthylazine in Scots pine plantations. *Silva Fennica* 29, 41–48.
Sinclair, W., Lyon, H., and Johnson, W. (1987) *Diseases of Trees and Shrubs*. Cornell University Press, Ithaca, New York.
Solbreck, C. and Andersson, D. (1987) Vertical distribution of fireweed, *Epilobium angustifolium*, seeds in the air. *Canadian Journal of Botany* 65, 2177–2178.
Sylvester, T.W. and Wein, R.W. (1981) Fuel characteristics of arctic plant species and simulated plant community flammability by Rothermel's model. *Canadian Journal of Botany* 59, 898–907.
Winder, R.S. and Watson, A.K. (1994) A potential microbial control for fireweed (*Epilobium angustifolium*). *Phytoprotection* 75, 19–33.

65 *Cirsium arvense* (L.) Scopoli, Canada Thistle (*Asteraceae*)

A.S. McClay, R.S. Bourchier, R.A. Butts and D.P. Peschken

Pest Status

Canada thistle, *Cirsium arvense* (L.) Scopoli, is one of the most widespread and competitive European weeds. It is probably originally native to south-eastern Europe and the eastern Mediterranean but now occurs throughout Europe, parts of North Africa, and Asia south to Afghanistan, Iran and Pakistan, and east to Japan (Moore, 1975). In North America, *C. arvense* occurs in all Canadian provinces and is listed as a noxious weed in 35 US states (Skinner *et al.*, 2000).

C. arvense causes extensive crop losses. At 20 shoots m^{-2} estimated yield losses are 34% in barley, *Hordeum vulgare* L. (O'Sullivan *et al.*, 1982), 26% in canola, *Brassica napus* L. and *B. rapa* L. (O'Sullivan *et al.*, 1985), 36% in winter wheat, *Triticum aestivum* L. (McLennan *et al.*, 1991) and 48% in seed corn, *Medicago sativa* L. (Moyer *et al.*, 1991). Actual shoot densities of *C. arvense* in field infestations can be up to 173 shoots m^{-2} (Donald and Khan, 1996). *C. arvense* was rated as a moderately invasive species of natural areas in Canada but it is mainly a problem in disturbed sites (White *et al.*, 1993).

Donald (1994) reviewed the biology of *C. arvense*. It is a dioecious, perennial herb with an extensive, creeping, deep root system. New stems arise each spring from old stem bases or from adventitious buds on the roots. Existing infestations spread mainly by horizontal root growth. Tillage can disperse *C. arvense* root fragments throughout cultivated fields. Seed dispersal has not generally been considered important, although *C. arvense* can produce abundant, fertile seeds. Heimann and

Cussans (1996) suggested that the importance of seed dispersal has been underestimated. At Vegreville, Alberta, patches of *C. arvense* produced a mean of 2840 seeds m^{-2} (A.S. McClay, unpublished).

Background

Chemical control of *C. arvense* is difficult, due to regrowth from roots. Many herbicides are registered for use in cereals, although most give only top growth control (Donald, 1990). Fewer herbicides are available for use in oilseeds, with clopyralid being the most effective (Ali, 1999). Summer cultivation followed by treatment of regrowing rosettes with glyphosate reduced *C. arvense* density by 98% after 2 years (Hunter, 1996).

Peschken (1984a) summarized work on the biological control of *C. arvense* in Canada up to 1980. Piper and Andres (1995) and McClay (2001) reviewed the status of biological control in western and eastern USA, respectively. The arthropods and pathogens attacking *C. arvense* have been surveyed extensively in Europe and some parts of Asia (Zwölfer, 1965, 1988; Schroeder, 1980; Winiarska, 1986; Freese, 1994; Berestetsky, 1997), and further surveys in southern Russia and central Asia are currently under way (Gassmann, Delémont, 2000, personal communication). Larvae of *Phtheochroa inopiana* (Haworth) were found mining *C. arvense* roots at Vegreville (A.S. McClay, unpublished). Its host specificity has not been studied but it is recorded in Europe from *Pulicaria dysenterica* (L.) Bernhardi and *Artemisia campestris* L. (Bradley *et al.*, 1973). Unidentified eriophyid mites have been found on *C. arvense* at Vegreville but cause little damage (A.S. McClay, unpublished); it is not known if this mite is *Aceria anthocoptes* (Nalepa), found on *C. arvense* in Serbia by Petanović *et al.* (1997).

There are 92 native *Cirsium* spp. in North America (USDA Natural Resources Conservation Service, 1999), including three endangered and two threatened taxa in the USA. In Canada, 11 native *Cirsium* spp. occur (Scoggan, 1979). One of these, *C. pitcheri* (Torrey) Torrey and Gray, which occurs in sand dunes along the shores of Lakes Michigan, Huron and Superior, is also listed as endangered in Canada (Promaine, 1999). Because many thistle-feeding insects in Europe are specific only to genus or subtribe of host plant, the perceived risk of damage to non-target native species is a major limiting factor in the biological control of *C. arvense* and other introduced *Cirsium* spp. in North America.

Biological Control Agents

Pathogens

Bacteria

Bailey *et al.* (2000) isolated *Pseudomonas syringae* pv. *tagetis* (Hellmers) Young, Dye and Wilkie from *C. arvense* in the prairies.

Fungi

Alternaria cirsinoxia Simmons and Mortensen, causing severe foliar necrosis, was isolated from diseased *C. arvense* plants in Saskatchewan (Green and Bailey, 2000a, b). *Puccinia punctiformis* (Strauss) Röhling is a widespread rust on *C. arvense* in Canada, although more frequent in the east and in moister sites. Systemic infestations resulting from teliospore infection can cause severe damage (Thomas *et al.*, 1994) and infected shoots rarely survive the season (Forsyth and Watson, 1985). The conditions required to induce such infections in the field are not yet well understood (French *et al.*, 1994).

Bailey *et al.* (2000) isolated 287 pathogenic fungi, including species of *Phoma*, *Phomopsis*, *Colletotrichum* and *Fusarium*, from *C. arvense* in the prairies.

Insects

Altica carduorum Guérin-Méneville, originating from Switzerland and France, was released in 1969 and 1970 but did not establish (Peschken, 1984a). Its life history is similar to that of *Lema cyanella* (L.). A biotype of *A. carduorum* from Xinjiang, north-western China, may be better adapted to the climate of the Canadian prairies than the European biotype (Wan *et al.*, 1996a) and was screened. Lactin *et al.* (1997) predicted that the Chinese biotype should complete development throughout most of the range of *C. arvense* on the prairies, if adults can thermoregulate. Wan *et al.* (1996b) found that in no-choice tests it would complete development on 18 *Cirsium* spp. and *Silybum marianum* (L.) Gaertner. A risk analysis approach, however, predicted that North American *Cirsium* spp. would be safe from attack in the field because host selection requires a series of sequential steps, with the native species being less preferred than *C. arvense* at each stage (Wan and Harris, 1997). Wan and Harris (1996) suggested that in the field *A. carduorum* is monophagous because host finding is dependent on aggregation to wound and frass substances specific to *C. arvense*. However, the Chinese biotype of *A. carduorum* was not approved for release in Canada.

Cassida rubiginosa Müller adults and larvae feed on foliage of *C. arvense* and many other Cardueae (Zwölfer, 1969). In Virginia, adults appear in late winter and oviposit, mainly on the underside of thistle leaves, from mid-March to early July. About five eggs are laid in oothecae. Development from egg to adult requires 435 degree-days above 10.4°C. New-generation adults begin to appear in late spring and can be found on plants up to November. Females produce an average of 815 eggs under laboratory conditions (Ward and Pienkowski, 1978). Bousquet (1991) recorded *C. rubiginosa* from Alberta, Saskatchewan, Manitoba, Ontario, Quebec and New Brunswick but we have not observed this species in the prairies.

In China, a *Cassida* sp. was observed defoliating *C. arvense* at Yining, Xinjiang. Slight feeding damage but no beetles were found on adjacent stands of *Cirsium alberti* Regel and Schmalhausen (P. Harris, Lethbridge, 2000, personal communication). Quarantine studies in Lethbridge in 1996 showed that significant feeding and oviposition occurred on *Carduus* and *Arctium* spp., and adult feeding on safflower, *Carthamus tinctorius* L., occurred (Table 65.1) so work on this insect was suspended. These results were similar to those for *C. rubiginosa* from Europe in no-choice tests (Zwölfer and Eichhorn, 1966).

Cleonis pigra (Scopoli), a univoltine European weevil, was first found in New York in 1929, and in Quebec in 1933 (Brown, 1940), from where it spread to Ontario. Females lay eggs in *C. arvense* stem bases, and larvae mine the root crown and form a spindle-shaped gall. *C. pigra* attacks *Cirsium*, *Carduus*, *Cynara*, *Onopordum*, *Arctium* and *Silybum* spp. (La Ferla, 1939; Scherf, 1964; Zwölfer, 1965).

Hadroplontus litura (Fabricius) (previously *Ceutorhynchus litura*), a stem- and root-mining weevil, oviposits into the midveins of *C. arvense* rosette leaves in early

Table 65.1. Host-plant testing for *Cassida* sp. from China on *Cirsium* spp. and related genera in choice tests on leaf disks, Lethbridge, 1996.

Species	Replicates	Adult feeding damage
Carthamus tinctorius L.	4	Yes
Cichorium sp.	4	No
Cirsium arvense (L.) Scopoli	4	Yes
Cirsium flodmanii (Rydberg) Arthur	4	Yes
Echinops sphaerocephalus L.	4	Yes
Helianthus sp.	4	No
Silybum marianum (L.) Gaertner	4	Yes
Arctium minus (Hill) Bernhardi	4	Yes

spring. Larvae mine down through the vein into the stem base and upper part of the tap root. Mature larvae emerge from the stem and pupate in soil and adults emerge to feed on *C. arvense* foliage in late summer (Peschken, 1984a).

Larinus planus (Fabricius), native to Europe, was accidentally introduced and has established in the eastern USA (Wheeler and Whitehead, 1985) and southern British Columbia. It oviposits into unopened flower buds of *C. arvense*. Larvae feed on developing achenes and receptacle tissue and pupate in a cocoon formed of chewed host plant tissue. A single larva can complete development in each head. Adults emerge in late summer. McClay (1989) found that *L. planus* would not feed on ornamental or economic species in the tribe *Cardueae* and that *C. arvense* was preferred over other *Cirsium* spp. for feeding and oviposition.

Lema cyanella oviposits on leaf undersurfaces and stems of *C. arvense*. Larvae feed on leaf undersurfaces, leaving the upper epidermis to form a characteristic feeding window. Mature larvae drop to the soil or leaf litter in mid-summer, where they secrete a foam cocoon in which they pupate. Adults emerge in late summer and feed on *C. arvense* foliage before overwintering in the soil (Zwölfer and Pattullo, 1970). In 1983, *L. cyanella* was approved for release in Canada. Approval was based on field records from the native range suggesting that it was specific to *C. arvense*, on choice and no-choice feeding tests in Petri dishes, and on field-cage tests (Peschken and Johnson, 1979; Peschken, 1984b). In these tests, feeding, oviposition and development occurred on some native North American *Cirsium* spp. However, it was argued that, according to the resource concentration hypothesis (Root, 1973), rare or scattered non-target *Cirsium* spp. would be less susceptible to attack by *L. cyanella* than the abundant target. Open-field and large-cage tests in Alberta, however, have shown that some native *Cirsium* spp. are readily attacked by *L. cyanella* even when adjacent to much more abundant *C. arvense* (A.S. McClay, unpublished).

Lixus sp., from populations attacking *C. arvense* in Yining, China, was studied in 1997. Screening was discontinued when it was found that oviposition and larval development occurred on several other *Cirsium* spp. and *Silybum marianum* (L.) Gaertner (Table 65.2).

Rhinocyllus conicus (Frölich) has a similar life cycle to that of *L. planus*, except that eggs are laid externally on flower buds and are covered with a cap of chewed host plant tissue. This species was originally released to control introduced *Carduus* spp. but has also colonized *C. arvense* and other *Cirsium* spp. (Rees, 1977; Youssef and Evans, 1994; Louda *et al.*, 1997). *R. conicus* has spread gradually northwards in Alberta since the mid-1980s on *C. arvense* and the native *Cirsium undulatum* (Nuttall) Sprengel and *Cirsium flodmanii* (Rydberg) Arthur (A.S. McClay and R.S. Bourchier, unpublished).

Terellia ruficauda (Fabricius) (previously *Orellia ruficauda*), unintentionally introduced from Europe, oviposits into female *C. arvense* flower heads 1 day before blooming. Larvae feed on developing achenes and overwinter in the seed head in cocoons of pappus hairs; pupation and emergence take place the following spring (Lalonde and Roitberg, 1992). In Europe, *T. ruficauda* attacks six *Cirsium* spp. (Zwölfer, 1965).

Urophora cardui (L.), a stem-galling fly, oviposits in axillary and terminal buds of *C. arvense*. Larvae induce development of a multi-chambered stem-gall up to 23 mm in diameter (Lalonde and Shorthouse, 1985). Pupation and overwintering occur in the gall and adults emerge in early summer.

Releases and Recoveries

Biological control agent releases and recoveries against *C. arvense* are listed in Table 65.3.

H. litura established at nine release sites in British Columbia, Alberta, Saskatchewan and Ontario (Peschken and Wilkinson, 1981) but did not establish in New

Table 65.2. Host-plant testing for *Lixus* sp. on *Cirsium* spp. and related genera in no-choice tests, Lethbridge, 1997–1998.

Species	Number of replicates	Adult feeding damage	Oviposition attempts	Successful development
Cirsium arvense (L.) Scopoli	4	Yes	Yes	Eggs and larvae
Cirsium undulatum (Nuttal) Sprengel	2	Yes	No	No
Cirsium flodmanii (Rydberg) Arthur	2	Yes	No	No
Cirsium hookerianum Nuttall	2	Yes	Yes	No
Cirsium japonicum De Candolle	3	Yes	Yes	Larvae
Cirsium ochrocentrum A. Gray	2	Yes	Yes	No
Cirsium discolor (Mühlenberg ex Willdenow) Sprengel	2	Yes	No	No
Cirsium edule Nuttall	2	Yes	Yes	Larvae
Cirsium scariosum Nuttall	1	Yes	Yes	Larvae
Silybum marianum Gaertner	3	Yes	Yes	Larvae
Sonchus sp.	1	Yes	No	No
Carthamus tinctorius L.	6	Yes	No	No
Centaurea maculosa Lamarck	1	Yes	No	No
Centaurea macrocephala Puschkarew ex Willdenow	3	Yes	No	No
Onopordum acanthium L.	1	–	No	No
Helianthus sp.	3	Yes	No	No
Echinops sphaerocephalus L.	1	Yes	No	No

Brunswick (Maund *et al.*, 1993). The population at Ladner, British Columbia, survived from 1975 (Peschken and Wilkinson, 1981) until at least 1994, when it was used as the source for a release at Kamloops (S. Turner, Kamloops, personal communication, 2001). In Alberta, one colony from a 1978 release near Busby persisted until at least 1991 (A.S. McClay, unpublished) but another, from a 1975 release at Lacombe, was destroyed when the field was cultivated after 1980 (D.P. Peschken, unpublished). *L. planus* was released on at least 85 occasions in five provinces from 1989 to 1996. Most releases were in British Columbia, with over 71 redistribution releases by 2000. The weevil established and spread readily. It now occurs widely in the Kamloops and Nelson Forest Regions. Its establishment status further north in the Cariboo, Prince Rupert and Prince George Regions is unknown (S. Turner, Kamloops, personal communication, 2001). In Alberta, releases were made using material from Maryland. Adults bred well near Tofield and Grande Prairie, Alberta, and many adults emerged. However, at all sites numbers declined annually, suggesting that adult overwinter survival was poor. The longest survival was 3 years at the Tofield site. There was also heavy attack by two native parasitoid species, *Itoplectis viduata* (Gravenhorst) and *Scambus tecumseh* (Viereck) (A.S. McClay, unpublished). *L. planus* did not establish in New Brunswick (Maund *et al.*, 1995).

Because of rearing problems, only a few small releases of *L. cyanella* were initially made after its approval for release in 1983. In 1992, a healthy colony, derived from material originally collected in Switzerland and France, was obtained from New Zealand. Four releases were made from 1993 to 1997 in Alberta using material from this colony. Some overwinter survival and breeding occurred at all these sites but only one population, at Vegreville, persisted for more than 2 years. This population remained at a low density, and efforts are now under way to eradicate it because of concerns about potential effects on native *Cirsium* spp. A field experiment suggested that *L. cyanella* had no significant impact on the growth or reproduction of *C. arvense*

Table 65.3. Open releases and recoveries of biological control agents against *Cirsium arvense* in Canada, 1981–2000.

Province and species	Location	Year	Number	Stage	Recoveries
British Columbia					
Hadroplontus litura (Fabricius)	Brentwood Bay	1987	117	Adult	Unknown
	Duncan	1987	117	Adult	Unknown
	Kamloops	1994	?	Adult	Not established 1999
Larinus planus (Fabricius)	Kamloops Forest Region (21 releases at 14 sites)	1991–1997	*c.* 4000	Adult	Established at 13 sites in 1999–2000
	Cariboo Forest Region (1 release)	1994	100	Adult	Unknown
	Prince Rupert Forest Region (10 releases)	1990–1998	1700	Adult	Unknown
	Prince George Forest Region (13 releases)	1990–1996	*c.* 2300	Adult	Unknown
	Vancouver Forest Region (3 releases)	1990–1996	450	Adult	Unknown
	Nelson Forest Region (12 releases)	1989–1998	*c.* 2400	Adult	Established at 7 sites by 2000
Urophora cardui (L.)	Brentwood Bay	1987	367	Adult	Unknown
	Duncan	1987	959	Adult	Unknown
	Nelson Region Kootenay Lake	1989	40	Adult	Unknown
	Vancouver	1990	200	Gall	1990–1994
	Fort St John district	1991	87	Adult	Unknown
	Paul Lake	1991	202	Gall	None
	Chilliwack	1991	400	Larva	Unknown
	Chilliwack	1991	320	Adult	Unknown
	Cariboo Region (2)	1995	?	Larva	Unknown
	Kamloops Region (3)	1994/95	900	Larva	Unknown
	Nelson Region (1)	1996	300	Larva	Unknown
	Pr. George Region (10)	1994–1996	*c.* 2500	Larva	Unknown
	Pr. Rupert (6)	1994–1996	*c.* 1620	Larva	Unknown
	Vancouver Region (3)	1996	800	Larva	Unknown
Alberta					
Hadroplontus litura	Eaglesham	1983	278	Adult	Unknown
	Kleskun Hill	1988	223	Adult	Unknown
Larinus planus	Hay Lakes	1990	126	Adult	None
	Grande Prairie	1991	140	Adult	1992
	Tofield	1991	107	Adult	1994
	Vegreville	1991	50	Adult	1992
	Edmonton	1994	120	Adult	None
	Vegreville	1994	73	Adult	None
Lema cyanella (L.)	Vegreville	1993	222	Adult	1995
	Edmonton	1994	150	Adult	None
	Vegreville	1994	183	Adult	1995–2000
	Edmonton	1997	100	Adult	1999
Urophora cardui	Edmonton	1996	149	Gall	1997
	Lethbridge	1996	400	Pupa	None
	Nanton	1996	800	Pupa	None
Saskatchewan					
Hadroplontus litura	Regina Research Station	1985	55	Adult	1986
	Echo Valley Provincial Park	1989	29	Adult	1990
Larinus planus	Ridgedale	1990	150	Adult	None
Lema cyanella	Regina	1982	31	Adult	None
	Indian Head	1983	48	Adult	None
Urophora cardui	Echo Valley Provincial Park	1984	3052	Adult	1985–2000
	Regina	1984	420	Adult	None
	Regina	1984	104	Adult	None

Continued

Table 65.3. *Continued.*

Province and species	Location	Year	Number	Stage	Recoveries
	Regina Research Station	1985	180	Adult	None
	Echo Valley Provincial Park	1986	287	Adult	1987–2000
	Regina Research Station	1986	261	Adult	None
	Regina Research Station	1986	31	Adult	Died out 1987
	Regina Research Station	1986	124	Adult	None
	Regina Research Station	1986	26	Adult	None
	Regina Research Station	1991	85	Adult	1992–1994
Manitoba					
Hadroplontus litura	Winnipeg	1989	285	Adult	Site destroyed 1990
Larinus planus	Grosse Isle	1996	100	Adult	Unknown
	Morris	1996	200	Adult	Unknown
	Stonewall	1996	100	Adult	Unknown
	Tyndall	1996	100	Adult	Unknown
New Brunswick					
Hadroplontus litura	Sussex	1984	300	Adult	None
	Sussex Corner	1985	300	Adult	None
	Sussex Corner	1986	51	Adult	None
	Sussex	1991	197	Adult	None
Larinus planus	Sussex	1990	82	Adult	None
	Sussex	1991	300	Adult	None
	Bear Island	1993	200	Adult	None
Lema cyanella	Sussex	1983	55	Adult	None
	Sussex	1984	367	Adult	None
	Sussex Corner	1986	23	Adult	None
	Sussex Corner	1986	24	Larva	None
	Sussex Corner	1986	30	Pupa	None
Urophora cardui	Multiple sites	1990	1063	Adult	
	Multiple sites	1991	1100	Adult	Established
	Multiple sites	1992	2856	Adult	At most sites
	Multiple sites	1993	4205	Adult	At most sites
	Multiple sites	1994	5071	Adult	At most sites
	Multiple sites	1995	7809	Adult	At most sites
Nova Scotia					
Hadroplontus litura	Eastville	1984	301	Adult	Unknown
	Eastville	1985	474	Adult	Unknown
	Eastville	1988	285	Adult	Unknown
	Old St Croix	1989	200	Adult	Unknown
	St Croix	1990	111	Adult	Unknown
Rhinocyllus conicus (Frölich)	Old St Croix	1989	110	Adult	Unknown
Urophora cardui	Antigonish	1991	600	Gall	Unknown
	Bridgewater	1991	1011	Gall	Unknown
	Inverness	1991	600	Gall	Unknown
	Merigomish	1991	250	Gall	Unknown
	New Glasgow	1991	600	Gall	Unknown
	Port Hawksberry	1991	600	Gall	Unknown
	Shelburne	1991	1500	Gall	Unknown
	Stewiacke	1991	600	Gall	Unknown
	Truro	1991	1351	Gall	Unknown
	10 sites	1996	7212	Both	Unknown
Prince Edward Island					
Hadroplontus litura	Charlottetown	1992	108	?	Unknown

(A.S. McClay, unpublished). No further releases of *L. cyanella* are planned.

Earlier releases of *U. cardui* resulted in establishment in Ontario, Quebec and New Brunswick (Peschken, 1984a; Peschken and Derby, 1997) but not in western Canada; a small colony surviving at Camrose, Alberta, from a 1977 release had died out by 1984 (A.S. McClay, unpublished). By 1984, galls were found over an area of 3000 ha from releases in Sussex, New Brunswick (Finnamore, 1984). In 1990–1995 there was extensive redistribution in New Brunswick (86 releases of 22,104 adults) using galls collected at previous release sites, resulting in establishment at most sites (Maund *et al.*, 1992, 1993, 1994, 1995). Redistribution was also done in Nova Scotia in 1991 and 1996. There have been no further releases in Quebec but *U. cardui* is well established and widespread from releases in the 1970s (A. Watson, Ste-Anne-de-Bellevue, 2000, personal communication). It is probably also widespread in Ontario; a large colony was found in High Park, Toronto (D.P. Peschken, unpublished). Releases in Echo Valley Provincial Park, Saskatchewan, using populations from Finland and New Brunswick, resulted in establishment, with populations persisting from 1984 to 2000 and spreading up to 4 km along a lake shore (Peschken and Derby, 1997). Releases were also made in Alberta in 1996 using galls from a population established in Oregon. At the 1996 Edmonton release site, 380 galls developed in the same season. In 1997, 34 galls were observed and in 1998, none. No gall formation was observed at the Lethbridge and Nanton release sites (A.S. McClay, C. Saunders, R. Butts, unpublished). *U. cardui* is well established in the Vancouver area, from which 25 redistribution releases have been made in British Columbia. To date one of these releases, near Nelson, is established (S. Turner, Kamloops, personal communication, 2001).

Evaluation of Biological Control

The bacterium *P. syringae* pv. *tagetis* was only able to infect *C. arvense* in the presence of an organosilicone surfactant (see also Johnson *et al.*, 1996). It caused chlorosis and stunting. Its effects were potentiated when applied together with a sublethal rate of glyphosate (Bailey *et al.*, 2000).

A. cirsinoxia infects primarily older, senescing leaves of *C. arvense* and requires at least 8 hours of leaf moisture for infection (Green and Bailey, 2000a, b), limiting its potential as a bioherbicide. Bailey *et al.* (2000) found that 18 of 71 fungal isolates caused significant reductions in shoot emergence and root weight, chlorosis and/or death of *C. arvense*. Most of the effective isolates were *Fusarium* spp. The efficacy of two of them was also tested using infested barley grains as a granular inoculant under greenhouse conditions, where they killed *C. arvense* in 4–6 weeks at an application rate of 250–500 g m^{-2}.

In Quebec, Forsyth and Watson (1986) determined that *T. ruficauda* attacked 70% of heads, reducing seed production by about 22%, and that defoliation by *C. rubiginosa* was rarely extensive enough to reduce plant vigour. Root mining by *C. pigra* sometimes killed plants, but regeneration of attacked plants was also observed. Main shoot galling by *U. cardui* reduced plant height, biomass and number of flowers, but side shoot galling had less impact.

Reports on the efficacy of *H. litura* are inconsistent. Peschken and Wilkinson (1981) concluded that larval mining produced no noticeable reduction in vigour of *C. arvense* plants. Attacked shoots were, in fact, taller on average than unattacked ones, possibly because the weevil attacks the earlier emerging rosettes, which later develop into taller shoots. They also found no evidence that *H. litura* aids in the spread of *P. punctiformis*. Rees (1990) reported that infestation by *C. litura* in Montana reduced *C. arvense* shoot production by 82% and that overwinter survival of infested plants was 12%, compared to 93% for uninfested plants. Interpretation of his results is difficult because the data are derived from unstructured field sampling rather than controlled experiments and it is not clear what is meant by a 'plant' in the study. Field experiments at Vegreville in 1990 and 1991 with *H. litura* on *C. arvense* plants growing in bare soil and with competition from a grass sward showed that infested plants growing in bare soil produced significantly fewer new shoots the following year, relative to the number produced in the year of establishment, than did adjacent untreated plants (Table 65.4).

At Indian Head, Saskatchewan, where *H. litura* was released in 1973, cultivation drastically reduced the *H. litura* population in 1979 (Peschken and Wilkinson, 1981) but on adjacent uncultivated land, *H. litura* continued to thrive without an apparent reduction of thistle density (D.P. Peschken, unpublished). Peschken and Derby (1992) found that mining by *H. litura*, together with galling by *U. cardui*, had no significant effect on dry weight, new shoot production or seed production of *C. arvense*.

L. cyanella appears capable of establishing on the Canadian prairies but seems unlikely to build up high densities. Because of its field preference for some native *Cirsium* spp. and its lack of impact on *C. arvense*, it is not recommended for further release.

Across Canada, rates of attack by *T. ruficauda* varied from 20 to 80%, and generally increased from east to west (Forsyth and Watson, 1985). In British Columbia, however, although up to 36% of heads were infested by *T. ruficauda*, attacked heads only contained on average one or two larvae. Levels of seed destruction were very low, up to 15 seeds m^{-2} from a total production of up to 1250 seeds m^{-2} (Lalonde and Roitberg, 1992).

In Quebec, *C. arvense* with *U. cardui* galls on the main shoot and on side shoots were significantly shorter than ungalled thistle shoots that had emerged before or during the laying period (Peschken *et al.*, 1982). In Ontario, *U. cardui* had spread up to 20 km from the original release site and was reducing *C. arvense* density (Alex, 1992). On the prairies, *U. cardui* persists only in Echo Valley Provincial Park, on a site near water and sheltered by trees, a habitat very well suited for *U. cardui*. Abiotic factors, namely temperature and moisture, regulate population levels (Peschken and Derby, 1997).

Biological control of *C. arvense* with introduced insects has had limited success, particularly in the prairies. This is due both to the vigorous nature of the weed, the poor adaptation of many agents to the prairie climate, and the lack of host specificity of most of the agents, which results in potential risks to native *Cirsium* spp. The approach proposed by Wan and Harris (1997) has potential for predicting the risks of non-target damage. However, as *A. carduorum* was not approved for release, it has not been possible to test these predictions under field conditions in Canada.

There appear to be few potential biological control agents from Europe left to be tested. A pesticide exclusion study suggested that, at least under agricultural conditions, *C. arvense* growth is not limited by invertebrate herbivory in western Europe (Edwards *et al.*, 2000). *C. arvense* has not been surveyed exhaustively for natural enemies in Asia, and other potential agents may be found there. Larvae of *Thamnurgus* sp. were reported feeding in *C. arvense* roots in China, but efforts to collect and rear this scolytid for host-specificity testing were unsuccessful (F.H. Wan and P. Harris, unpublished). A more precise identification of the ancestral range of *C. arvense* within Eurasia would be a useful guide to selection of areas for further surveys. Cladistic or phylogeographic methods (e.g. Bremer, 1992; Avise, 2000) may be useful for this purpose.

Because of the large number of native

Table 65.4. Shoot production by *Cirsium arvense* plants attacked or not attacked by *Hadroplontus litura* at Vegreville, 1990–1992 (A.S. McClay, unpublished).

		No. of shoots (mean ± SE)		
	n	In year of establishment	In following year	Rate of increase
Unattacked	8	23.3 ± 1.9	34.4 ± 7.1	1.41 ± 0.28[a]
Attacked	8	26.9 ± 3.2	22.1 ± 5.6	0.83 ± 0.23[a]

[a]Values significantly different, Wilcoxon signed-rank test, $P = 0.0273$.

Cirsium spp. in North America, reliable assessments of the potential for non-target damage are essential for future introductions of biological control agents against *C. arvense* (see Louda, 1999). The selection of test plants should be based on knowledge of their phylogenetic relationships with the target plant (McEvoy, 1996). Understanding the phylogenetic relationships among North American *Cirsium* spp., and between them, *C. arvense*, and other Eurasian *Cirsium* spp., is required.

Recommendations

Further work should include:

1. Further evaluation of the impact of *H. litura*;
2. Field validation of predicted host specificity of *A. carduorum*;
3. Increased focus on mycoherbicide development, possibly together with insect vectors;
4. Phylogenetic studies to determine relationships of *Cirsium* spp. in North America;
5. Biogeographic studies to locate the origin of *C. arvense* to guide further exploration for biological control agents.

Acknowledgements

We thank A.G. Wheeler and R. Lalonde for providing *Larinus planus*, T. Jessep for providing *Lema cyanella*, and E. Coombs for providing *Urophora cardui*. M. Sarazin, S. Turner, A. Watson, G. Sampson, C. Saunders and C. Maund provided unpublished release information. We are grateful for funding from the Canada–Alberta Environmentally Sustainable Agriculture Agreement and the Alberta Agricultural Research Institute.

References

Alex, J. (1992) Biological control of weeds in Ontario. In: Boiteau, G.J. (ed.) *Proceedings of the 38th Annual Meeting of the Canadian Pest Management Society, 27–31 July 1991, Fredericton, NB, Canada*, pp. 111–117.

Ali, S. (ed.) (1999) *Crop Protection 1999*. Alberta Agriculture, Food and Rural Development, Edmonton, Alberta.

Avise, J.C. (2000) *Phylogeography: the History and Formation of Species*. Harvard University Press, Cambridge, Massachussetts.

Bailey, K.L., Boyetchko, S.M., Derby, J., Hall, W., Sawchyn, K., Nelson, T. and Johnson, D.R. (2000) Evaluation of fungal and bacterial agents for biological control of Canada thistle. In: Spencer, N. (ed.) *Proceedings of the X International Symposium on Biological Control of Weeds*. USDA-ARS, Bozeman, Montana, pp. 203–208.

Berestetsky, A.O. (1997) Mycobiota of *Cirsium arvense* and allied species over the territory of the European part of Russia. *Mikologiya i Fitopatologiya* 31, 39–45.

Bousquet, Y. (ed.) (1991) *Checklist of Beetles of Canada and Alaska*. Research Branch Agriculture Canada, Ottawa, Ontario.

Bradley, J.D., Tremewan, W.G. and Smith, A.C. (1973) *British Tortricoid moths: Cochylidae and Tortricidae: Tortricinae*. Ray Society, London, UK.

Bremer, K. (1992) Ancestral areas: a cladistic reinterpretation of the center of origin concept. *Systematic Biology* 41, 436–445.

Brown, W.J. (1940) Notes on the American distribution of some species of Coleoptera common to the European and North American continents. *The Canadian Entomologist* 72, 5–88

Donald, W.W. (1990) Management and control of Canada thistle (*Cirsium arvense*). *Reviews of Weed Science* 5, 193–250.

Donald, W.W. (1994) The biology of Canada thistle (*Cirsium arvense*). *Reviews of Weed Science* 6, 77–101.

Donald, W.W. and Khan, M. (1996) Canada thistle (*Cirsium arvense*) effects on yield components of spring wheat (*Triticum aestivum*). *Weed Science* 44, 114–121.

Edwards, G.R., Bourdôt, G.W. and Crawley, M.J. (2000) Influence of herbivory, competition and soil fertility on the abundance of *Cirsium arvense* in acid grassland. *Journal of Applied Ecology* 37, 321–334.

Finnamore, D.B. (1984) Biological control of weeds update. *Adaptive Research Reports (New Brunswick Department of Agriculture and Rural Development)* 6, 190–192.

Forsyth, S.F. and Watson, A.K. (1985) Predispersal seed predation of Canada thistle. *The Canadian Entomologist* 117, 1075–1081.

Forsyth, S.F. and Watson, A.K. (1986) Stress inflicted by organisms on Canada thistle. In: Delfosse, E.S. (ed.) *Proceedings of the VI International Symposium on Biological Control of Weeds*. Agriculture Canada, Vancouver, British Columbia, pp. 425–431.

Freese, G. (1994) The insect complexes associated with the stems of seven thistle species. *Entomologia Generalis* 19, 191–207.

French, R.C., Nester, S.E. and Binder, R.G. (1994) Volatiles from germinating Canada thistle seed and root cuttings that stimulate germination of teliospores of the Canada thistle rust fungus, *Puccinia punctiformis*. *Journal of Agricultural and Food Chemistry* 42, 2937–2941.

Green, S. and Bailey, K.L. (2000a) Effects of leaf maturity, infection site, and application rate of *Alternaria cirsinoxia* conidia on infection of Canadian thistle (*Cirsium arvense*). *Biological Control* 19, 167–174.

Green, S. and Bailey, K.L. (2000b) Influence of moisture and temperature on infection of Canadian thistle by *Alternaria cirsinoxia*. *Plant Disease* 84, 1126–1132.

Heimann, B. and Cussans, G.W. (1996) The importance of seeds and sexual reproduction in the population biology of *Cirsium arvense* – a literature review. *Weed Research* 36, 493–503.

Hunter, J.H. (1996) Control of Canada thistle (*Cirsium arvense*) with glyphosate applied at the bud vs rosette stage. *Weed Science* 44, 934–938.

Johnson, D.R., Wyse, D.L. and Jones, K.J. (1996) Controlling weeds with phytopathogenic bacteria. *Weed Technology* 10, 621–624.

Lactin, D.J., Harris, P., Johnson, D.L., Wan, F.H. and Thomas, A.G. (1997) Modelling and mapping geographic ranges to evaluate weed biocontrol agents – a case study using *Altica carduorum* (Coleoptera, Chrysomelidae) and *Cirsium arvense* (Asteraceae). *Biocontrol Science and Technology* 7, 657–670.

La Ferla, A. (1939) Contributo alla conoscenza dei cleono del carciofo (*Cleonus piger* Scop.). *Bolletino del Laboratorio di Entomologia Agraria di Portici* 3, 25–33.

Lalonde, R.G. and Roitberg, B.D. (1992) Field studies of seed predation in an introduced weedy thistle. *Oikos* 65, 363–370.

Lalonde, R.G. and Shorthouse, J.D. (1985) Growth and development of larvae and galls of *Urophora cardui* (Diptera, Tephritidae) on *Cirsium arvense* (Compositae). *Oecologia* 65, 161–165.

Louda, S.M. (1999) Negative ecological effects of the musk thistle biological control agent, *Rhinocyllus conicus*. In: Follett, P.A. and Duan, J.J. (eds) *Nontarget Effects of Biological Control*. Kluwer Academic Publishers, Boston, Massachusetts, pp. 215–243.

Louda, S.M., Kendall, D., Connor, J. and Simberloff, D. (1997) Ecological effects of an insect introduced for the biological control of weeds. *Science* 277, 1088–1090.

Maund, C.M., McCully, K.V. and Sharpe, R. (1992) Biological control of selected weeds in pastures in New Brunswick during 1992. *Adaptive Research Reports (New Brunswick Department of Agriculture and Rural Development)* 14, 317–328.

Maund, C.M., McCully, K.V., Finnamore, D.B., Sharpe, R. and Parkinson, B. (1993) A summary of insect biological control agents released against weeds in NB pastures from 1990 to 1993. *Adaptive Research Reports (New Brunswick Department of Agriculture and Rural Development)* 15, 359–380.

Maund, C.M., Sharpe, R., Stairs, A. and McCully, K.V. (1994) Biological control of selected weeds with insects in New Brunswick pastures during 1994. *Adaptive Research Reports (New Brunswick Department of Agriculture)* 16, 385–397.

Maund, C.M., Sharpe, R. and McCully, K.V. (1995) Biological control of selected weeds with insects in New Brunswick pastures during 1995. *Adaptive Research Reports (New Brunswick Department of Agriculture)* 17, 227–240.

McClay, A.S. (1989) The potential of *Larinus planus* (F.) (Coleoptera: Curculionidae), an accidentally introduced insect in North America, for biological control of *Cirsium arvense* (L.) Scop. In: Delfosse, E.S. (ed.) *Proceedings of the VIII International Symposium on Biological Control of Weeds.* Istituto Sperimentale Patologia Vegetale (MAF), Rome, Italy, pp. 173–179

McClay, A.S. (2001) Canada thistle (*Cirsium arvense* [L.] Scop.). In: Van Driesche, R.G., Blossey, B., Hoddle, M. and Lyon, S. (eds) *Weed Biological Control in the Eastern United States.* United States Department of Agriculture Forest Service, Forest Health Technology Enterprise Team, Morgantown, West Virginia (in press).

McEvoy, P.B. (1996) Host specificity and biological pest control – how well is research on host specificity addressing the potential risks of biological control. *BioScience* 46, 401–405.

McLennan, B.R., Ashford, R. and Devine, M.D. (1991) *Cirsium arvense* (L.) Scop. competition with winter wheat (*Triticum aestivum* L.). *Weed Research* 31, 409–415.

Moore, R.J. (1975) The biology of Canadian weeds. 13. *Cirsium arvense* (L.) Scop. *Canadian Journal of Plant Science* 55, 1033–1048.

Moyer, J.R., Schaalje, G.B. and Bergen, P. (1991) Alfalfa (*Medicago sativa*) seed yield loss due to Canada thistle (*Cirsium arvense*). *Weed Technology* 5, 723–728.

O'Sullivan, P.A., Kossatz, V.C., Weiss, G.M. and Dew, D.A. (1982) An approach to estimating yield loss of barley due to Canada thistle. *Canadian Journal of Plant Science* 62, 725–731.

O'Sullivan, P.A., Weiss, G.M. and Kossatz, V.C. (1985) Indices of competition for estimating rapeseed yield loss due to Canada thistle. *Canadian Journal of Plant Science* 65, 145–149.

Peschken, D.P. (1984a) *Cirsium arvense* (L.) Scop., Canada thistle (Compositae). In: Kelleher, J.S. and Hulme, M.A. (eds) *Biological Control Programmes Against Insects and Weeds in Canada 1969–1980.* Commonwealth Agricultural Bureaux, Slough, UK, pp. 139–146.

Peschken, D.P. (1984b) Host range of *Lema cyanella* (Coleoptera: Chrysomelidae), a candidate for biocontrol of Canada thistle, and of four stenophagous, foreign thistle insects in North America. *The Canadian Entomologist* 116, 1377–1384.

Peschken, D.P. and Derby, J.L. (1992) Effect of *Urophora cardui* (L.) (Diptera: Tephritidae) and *Ceutorhynchus litura* (F.) (Coleoptera: Curculionidae) on the weed Canada thistle, *Cirsium arvense* (L.) Scop. *The Canadian Entomologist* 124, 145–150.

Peschken, D.P. and Derby, J.L. (1997) Establishment of *Urophora cardui* (Diptera: Tephritidae) on Canada thistle, *Cirsium arvense* (Asteraceae), and colony development in relation to habitat and parasitoids in Canada. In: Dettner, K., Bauer, G. and Völkl, W. (eds) *Vertical Food Web Interactions: Evolutionary Patterns and Driving Forces. Ecological Studies* 130, Springer, New York, New York, pp. 53–66.

Peschken, D.P. and Johnson, G.R. (1979) Host specificity and suitability of *Lema cyanella* (Coleoptera: Chrysomelidae) a candidate for the biological control of Canada thistle (*Cirsium arvense*). *The Canadian Entomologist* 111, 1059–1068.

Peschken, D.P. and Wilkinson, A.T. (1981) Biocontrol of Canada thistle (*Cirsium arvense*): releases and effectiveness of *Ceutorhynchus litura* (Coleoptera: Curculionidae) in Canada. *The Canadian Entomologist* 113, 777–785.

Peschken, D.P., Finnamore, D.B. and Watson, A.K. (1982) Biocontrol of the weed Canadian thistle (*Cirsium arvense*): releases and development of the gall fly *Urophora cardui* (Diptera: Tephritidae) in Canada. *The Canadian Entomologist* 114, 349–357.

Petanović, R., Boczek, J. and Stojnić, B. (1997) Taxonomy and bioecology of eriophyids (Acari, Eriophyoidea) associated with Canada thistle, *Cirsium arvense* (L.) Scop. *Acarologia* 38, 181–191.

Piper, G.L. and Andres, L.A. (1995) Canada thistle, *Cirsium arvense* (L.) Scop., Asteraceae. In: Nechols, J.R. (ed.) *Biological Control in the Western United States. Accomplishments and Benefits of Regional Research Project W-84, 1964–1989.* University of California, Division of Agriculture and Natural Resources, Oakland, California, pp. 233–236.

Promaine, A. (1999) Threatened species monitoring: Results of a 17-year survey of Pitcher's thistle, *Cirsium pitcheri*, in Pukaskwa National Park, Ontario. *Canadian Field-Naturalist* 113, 296–298.

Rees, N.E. (1977) Impact of *Rhinocyllus conicus* on thistles in southwestern Montana. *Environmental Entomology* 6, 839–842.

Rees, N.E. (1990) Establishment, dispersal and influence of *Ceutorhynchus litura* on Canada thistle (*Cirsium arvense*) in the Gallatin Valley of Montana. *Weed Science* 38, 198–200.

Root, R.B. (1973) Organization of a plant–arthropod association in simple and diverse habitats: the fauna of collards (*Brassica oleracea*). *Ecological Monographs* 43, 95–124.

Scherf, H. (1964) Die Entwicklungsstadien der mitteleuropaeischen Curculioniden (Morphologie, Bionomie, Ökologie). *Abhandlungen der Senckenbergischen Naturforschenden Gesellschaft* 506, 1–335.

Schroeder, D. (1980) The biological control of thistles. *Biocontrol News and Information* 1, 9–26.

Scoggan, H.G. (1979) *The Flora of Canada. Part 4. Dicotyledoneae (Loasaceae to Compositae).* National Museum of Canada, Ottawa, Ontario.

Skinner, K., Smith, L. and Rice, P. (2000) Using noxious weed lists to prioritize targets for developing weed management strategies. *Weed Science* 48, 640–644.

Thomas, R.F., Tworkoski, T.J., French, R.C. and Leather, G.R. (1994) *Puccinia punctiformis* affects growth and reproduction of Canada thistle (*Cirsium arvense*). *Weed Technology* 8, 488–493.

USDA Natural Resources Conservation Service (1999) The PLANTS database. http://plants.usda.gov/plants (15 November 2000)

Wan, F.H. and Harris, P. (1996) Host finding and recognition by *Altica carduorum*, a defoliator of *Cirsium arvense*. *Entomologia Experimentalis et Applicata* 80, 491–496.

Wan, F.H. and Harris, P. (1997) Use of risk analysis for screening weed biocontrol agents – *Altica carduorum* Guer. (Coleoptera, Chrysomelidae) from China as a biocontrol agent of *Cirsium arvense* (L.) Scop. in North America. *Biocontrol Science and Technology* 7, 299–308.

Wan, F.H., Harris, P., Cai, L.M. and Zhang, M.X. (1996a) Biology and ecology of *Altica carduorum* (Chrysomelidae, Coleoptera) from north-western China – a potential biocontrol agent for *Cirsium arvense* (Asteraceae) in Canada. *Biocontrol Science and Technology* 6, 509–519.

Wan, F.H., Harris, P., Cai, L.M. and Zhang, M.X. (1996b) Host specificity of *Altica carduorum* Guer. (Chrysomelidae, Coleoptera), a defoliator of *Cirsium arvense* (L.) Scop. (Asteraceae) from north-western China. *Biocontrol Science and Technology* 6, 521–530.

Ward, R.H. and Pienkowski, R.L. (1978) Biology of *Cassida rubiginosa*, a thistle-feeding shield beetle. *Annals of the Entomological Society of America* 71, 585–591.

Wheeler, A.G. Jr and Whitehead, D.R. (1985) *Larinus planus* (F.) in North America (Coleoptera: Curculionidae: Cleoninae) and comments on biological control of Canada thistle. *Proceedings of the Entomological Society of Washington* 87, 751–758.

White, D.J., Haber, E. and Keddy, C. (1993) *Invasive Plants of Natural Habitats in Canada: An Integrated Review of Wetland and Upland Species and Legislation Governing their Control.* Canadian Wildlife Service, Environment Canada, Ottawa, Ontario.

Winiarska, W. (1986) Roslinozerna entomofauna ostrozenia polnego, *Cirsium arvense* (L.) Scop. [Phytophagous insects found on *Cirsium arvense* (L.) Scop.]. *Polskie Pismo Entomologiczne* 56, 701–715.

Youssef, N.N. and Evans, E.W. (1994) Exploitation of Canada thistle by the weevil *Rhinocyllus conicus* (Coleoptera: Curculionidae) in northern Utah. *Environmental Entomology* 23, 1013–1019.

Zwölfer, H. (1965) Preliminary list of phytophagous insects attacking wild Cynareae (Compositae) in Europe. *Technical Bulletin, Commonwealth Institute of Biological Control* 6, 81–154.

Zwölfer, H. (1969) Experimental feeding ranges of species of Chrysomelidae (Col.) associated with Cynareae (Compositae) in Europe. *Technical Bulletin, Commonwealth Institute of Biological Control* 12, 115–130.

Zwölfer, H. (1988) Evolutionary and ecological relationships of the insect fauna of thistles. *Annual Review of Entomology* 33, 103–122.

Zwölfer, H. and Eichhorn, O. (1966) The host ranges of *Cassida* (Col.: Chrysomelidae) species attacking Cynareae (Compositae) in Europe. *Zeitschrift für Angewandte Entomologie* 58, 384–397.

Zwölfer, H. and Pattullo, W. (1970) Zur Lebensweise und Wirtsbindung des Distel-Blattkäfers *Lema cyanella* L. (*puncticollis* Curt.) (Col. Chrysomelidae). *Anzeiger für Schädlingskunde und Pflanzenschutz* 43, 53–59.

66 *Convolvulus arvensis* L., Field Bindweed (*Convolvulaceae*)

A.S. McClay and R.A. De Clerck-Floate

Pest Status

Field bindweed, *Convolvulus arvensis* L., a deep-rooted, climbing, herbaceous perennial native to Eurasia, is now widely distributed across North America. In Canada, it occurs in agricultural regions of all provinces except Newfoundland and Prince Edward Island (Weaver and Riley, 1982). In the prairie provinces, it occurs mainly in the south. *C. arvensis* has been viewed primarily as a weed of cropland. In the USA, crop losses were estimated at more than US$377 million per year (Boldt *et al.*, 1998). Reports of toxicity to horses and laboratory mice and the presence of tropane alkaloids in the plant suggest that it may also be of concern as a toxic plant to some livestock (Schultheiss *et al.*, 1995; Todd *et al.*, 1995). *C. arvensis* is a prohibited noxious weed under the Canada Seeds Act, and a noxious weed under the provincial Weed Control Acts of Alberta, Saskatchewan, Manitoba, Ontario and Quebec (Weaver and Riley, 1982).

Shoots of *C. arvensis* emerge from root buds when day temperatures reach about 14°C. Flowering occurs from late June. Seeds of *C. arvensis* can remain viable for up to 20 years in the soil and are the usual means of dispersal into new areas, while local spread occurs through lateral roots and rhizomes. Seedlings only 19 days old can regenerate from the root when the above-ground portion is removed (Weaver and Riley, 1982).

Background

C. arvensis cannot generally be controlled by chemicals alone. The only recommended herbicides in cereals are the Group 4 growth regulators such as 2,4-D (2,4-dichlorophenoxyacetic acid), dicamba and mecoprop, which provide only top growth suppression. Very few chemical control options exist for oilseeds (Ali, 1999). *C. arvensis* can be controlled in summer-fallow by repeated tillage every 3–4 weeks from June through September for two seasons, or by a combination of cultivation, crop rotation and herbicides (Dorrance, 1994). Biotypes of *C. arvensis* vary widely in their susceptibility to glyphosate (DeGennaro and Weller, 1984).

In Canada, biological control of *C. arvensis* has depended primarily on agents screened and introduced via the US programme, as recommended by Maw (1984). Extensive surveys for natural enemies were carried out in western Mediterranean Europe (Italy, France, Spain, Portugal, eastern Austria, Yugoslavia) from 1970 to 1977 (Rosenthal, 1981; Rosenthal and Buckingham, 1982). Two arthropods were approved for release, the defoliating moth *Tyta luctuosa* (Denis and Schiffermüller) and the gall mite *Aceria malherbae* Nuzzaci. The fungal pathogen *Phomopsis convolvulus* Ormeño-Núñez was also isolated in Canada and assessed as a possible biological control agent (Ormeño-Núñez *et al.*, 1988a; Morin *et al.*, 1989). Two other fungal pathogens, *Phoma proboscis* Heiny and *Stagonospora* sp., have been proposed as possible biological control agents in the USA and Europe, respectively (Heiny, 1994; Pfirter and Defago, 1998), but have not been studied in Canada.

No native *Convolvulus* spp. occur in Canada. In the closely related genus

Calystegia, *C. sepium* (L.) Robert Brown is widespread and common across Canada, *C. soldanella* (L.) Robert Brown ex Roemer and J.A. Schultes occurs on the coast of southern British Columbia, and *C. spithamaea* (L.) Pursh is found in open areas and thin woods from Ontario to Nova Scotia (Scoggan, 1979). In the USA, one native *Convolvulus* sp. and 16 *Calystegia* spp. occur (USDA Natural Resources Conservation Service, 1999). *Calystegia stebbinsii* Brummitt, from California, is listed as endangered under the US Endangered Species Act (US Fish and Wildlife Service, 1996).

Biological Control Agents

Pathogens

Fungi

P. convolvulus was isolated from diseased foliage of *C. arvensis* in Montreal, Quebec. Infected plants in the field showed rounded to irregular, light-brown leaf spots surrounded by a narrow, light-green zone. In pathogenicity tests, the first symptoms were pinpoint foliar lesions, followed by spots on leaves, petioles and stems, anthracnose-like symptoms and dieback of apices. Pycnidia were formed on lower parts of the plant, close to or directly in contact with the soil (Ormeño-Núñez *et al.*, 1988a, b). The fungus was maintained in culture on potato dextrose agar and was mass produced on barley grains for field and controlled-environment experiments (Vogelgsang *et al.*, 1998b).

Insects

T. luctuosa is one of the most frequently found insects feeding on *C. arvensis* in southern Europe (Rosenthal and Buckingham, 1982), where it also occurs on *C. sepium* and *Convolvulus althaeoides* L. (Rosenthal, 1978). This defoliator occurs throughout Europe from Scandinavia southwards, in Asia east to the Altai Mountains, Iraq, Afghanistan, Pakistan and northern India, and in North Africa. Eggs are laid on stems and foliage, larvae feed on leaves and flowers at night, and pupation occurs in the soil. There are five larval instars, and two or three generations per year in southern Europe (Rosenthal *et al.*, 1988). Short daylength induces pupal diapause, although some individuals enter diapause even at a 16 h photoperiod (Miller *et al.*, 2000). *T. luctuosa* was approved for release based on evaluation of an Italian population by Rosenthal (1978), although host specificity tests focused mainly on economic species; relatively few native North American *Convolvulaceae* were tested. Although *T. luctuosa* larvae fed on three out of five *Convolvulus* spp., *C. sepium*, three out of five *Ipomoea* spp., and *Dichondra repens* Förster, they completed development to the adult stage only on *C. arvensis*, *C. althaeoides* and *C. sepium*. Chessman *et al.* (1997) found that *T. luctuosa* larvae showed no feeding preference among four biotypes of *C. arvensis* and *C. sepium*, although development time was slightly slower on *C. sepium*.

Mites

A. malherbae, earlier referred to as *A. convolvuli* (Nalepa) (Rosenthal, 1983), was described as new by Nuzzaci *et al.* (1985). This gall mite feeds on *C. arvensis* leaves, inducing leaf distortion and galling. All life stages occur within the folded and distorted leaves. Heavily infested shoots become stunted and deformed (Rosenthal and Buckingham, 1982). The mite overwinters below ground on rhizome buds (Rosenthal, 1983). Its release in North America was approved following evaluation by Rosenthal and Platts (1990), although host-specificity tests showed that the mite would develop on several North American *Calystegia* spp. It was argued that native species would be less at risk than *C. arvensis* because of their low levels of abundance.

In greenhouse tests at Vegreville, *A. malherbae* induced some gall formation on potted *C. sepium* plants caged separately from *C. arvensis*. However, no breeding populations of *A. malherbae* were found in these galls, and they were probably induced by the feeding activity of the originally inoculated adults (A.S. McClay, unpublished). These results suggest that *A. malherbae* would be unable to establish field populations on *C. sepium*, in contrast to the conclusions of Rosenthal and Platts (1990); it is not clear whether their test plants were in contact with *C. arvensis* plants infested by *A. malherbae*. If this were the case, some of the galling observed on species other than *C. arvensis* may have been due to adult mites wandering on to the other test plants and feeding, without establishing breeding populations on those plants.

Releases and Recoveries

T. luctuosa was released in Canada four times (Table 66.1). At the 1991 release site near Irvine, Alberta, a few adults were seen in June 1992, confirming that the species had overwintered. However, permanent establishment has not been determined. One release, at Cluny, Alberta, was later discovered to have been made on *C. sepium* and not on *C. arvensis*. No sign of establishment was seen at this site the year after release. In Saskatchewan, no establishment was detected at the site at Weyburn and the site was later destroyed (P. Harris, Lethbridge, 2000, personal communication). In the USA, *T. luctuosa* was released in Texas, Oklahoma, Missouri, Kansas and Maryland, with no evidence of establishment to date (Miller *et al.*, 2000).

A. malherbae was released in British Columbia, Alberta, Saskatchewan, and Manitoba, on 24 occasions at 25 sites (Table 66.2). Most releases were from greenhouse colonies derived from mites originally collected near Thessaloniki, Greece. McClay *et al.* (1999) confirmed its establishment in Alberta. *A. malherbae* established successfully in Alberta and Montana, both from transplantation of infested *C. arvensis* plants into field sites and by attaching excised pieces of galled tissue to plants in the field. Additional releases in British Columbia, Saskatchewan, Manitoba and at Lethbridge, Alberta, are not known to have resulted in establishment (R.A. De Clerck-Floate, unpublished; P. Harris, Lethbridge, 2000, personal communication).

Evaluation of Biological Control

T. luctuosa is not known to be established anywhere in Canada or the USA, although it did survive one winter in Alberta. Overwinter survival, but no permanent establishment, was also reported in Maryland (Tipping and Campobasso, 1997). As a defoliator it is not expected to have a major impact on *C. arvensis*, a plant that can readily regenerate from stored reserves in the rhizomes, and no further releases of *T. luctuosa* are warranted. In Maryland, Tipping and Campobasso (1997) found that release of *T. luctuosa* on to *C. sepium* in maizefields did not increase defoliation above that caused by native herbivores, principally *Oidaematophorus monodactylus* (L.). Similar results should be expected on *C. arvensis* on the Canadian prairies,

Table 66.1. Releases and recoveries of *Tyta luctuosa* in Canada, 1989–1992.

Location	Release date	Number	Stage	Recoveries
Irvine, AB	4 July 1990	54	Larvae	None
Irvine, AB	16 August 1991	500	Larvae	Adults seen 1992
Cluny, AB[a]	15 August 1991	500	Larvae	None
Weyburn, SK	1989	300	Eggs	None

[a]On *Calystegia sepium*.
Alberta (AB), Saskatchewan (SK).

Table 66.2. Releases and recoveries of *Aceria malherbae* in Canada, 1989–2000.

Site	Release date	Stage, number	Land use	Recoveries
British Columbia				
Grand Forks	26 August 1994	77 galls	Edge of filbert orchard	1997 none
Kamloops	11 August 1992	1000 mites	Weighscale yard along highway	2000 none
Cawston	7 August 1998	443 galls	Within and on edge of orchard	2000 none
Alberta				
Dunmore (1)	26 August 1993	3 plants	Pasture	1994, 1995 very few
	30 June 1995	2 plants	None	
	13 May 1998	25 galls	None	
Dunmore (2)	30 June 1995	2 plants	Edge of pasture	1996 slight
Dunmore (3)	30 June 1995	1 plant	Roadside ditch	None
	13 May 1998	25 galls		1999 very few
Dunmore (4)	30 June 1995	100 stem pieces	Hayland	1995 very few
	17 June 1997	5 plants		1997 none
	13 May 1998	40 galls		1998 very few
Dunmore (5)	9 June 1999	30 galls	Dugout bank	1999 scattered galls, 2000 none
Dunmore (6)	9 June 1999	30 galls	Coulee slope	1999 moderate galling, 2000 none
Dunmore (7)	11 June 1999	30 galls	Waste land by railway tracks	1999 light galling, 2000 very few
Dunmore (8)	11 June 1999	30 galls	Dyke	1999 light galling, 2000 none
Lethbridge (1)	10 September 1994	30 galls	Landscaped area next to pond; under spruce trees	None
Lethbridge (2)	4 August 1998	100 galls	Edge of cultivated field	1999 and 2000 none
	10 August 1999	20 galls		
Medicine Hat (1)	30 June 1995	2 plants	Ditch bank	1996 very few
Medicine Hat (2)	13 May 1998	25 galls	Edge of irrigated field	2000 many galls over *c.* 10,000 m^2
Medicine Hat (3)	10 June 1999	Galls	Berm adjacent to highway in city	2000 good attack
Medicine Hat (4)	10 June 1999	Galls	City park	2000 good attack
Redcliff	30 June 1995	200 plant pieces	Waste ground	2000 heavy and widespread attack
Saskatchewan				
Cardross (1)	11 July 1995	Unknown	Farm shelterbelt	None
	25 June 1996	11 galls		
	11 June 1997	8 leaves		
Cardross (2)	11 July 1995	Unknown	Along abandoned road	None
	25 June 1996	6 galls		
Cardross (3)	25 June 1996	5 galls	Along dugout, in mowed field	None
Weyburn	14 July 1989	2 galls	Tree nursery	None
Assiniboia	26 July 1994	56 galls	Near shelterbelt in town	None
Spring Valley	24 June 1996	2–3 galls	Grain elevator yard	None
Manitoba				
Fannystelle	13 August 1992	1000 mites	Natural pasture	None?

where it is commonly heavily defoliated by native tortoise beetles, e.g. *Jonthonota nigripes* (Olivier) and *Deloyala guttata* (Olivier) (A.S. McClay, unpublished).

A. malherbae has shown good potential for effective *C. arvensis* control. Considerable variation exists among sites in its level of establishment and impact. At some sites there was no survival or only a few lightly galled leaves the year after release, while at others thriving mite populations and heavy damage were present up to 5 years after release. The most successful release was made in 1995 on wasteland around a disused greenhouse in the South Saskatchewan River valley near Redcliff, Alberta. By 1998, heavy damage to *C. arvensis* had occurred over an area of about 3000 m^2 (McClay *et al.*, 1999). In 1999, damage was even more extensive, with many plants completely galled and severely stunted. Variation in effectiveness among sites may be related to the amount of galled material released or the vigour of *C. arvensis* plants at the time of release. For instance, failed releases in Lethbridge, Alberta, were all made in late summer, when host vigour was low (R.A. DeClerck-Floate, unpublished; Table 66.2). Observations in 1999 also suggest that environmental conditions may play a role in variation among sites. Most sites at which strong mite populations developed were either close to the South Saskatchewan River, within the city of Medicine Hat (which lies in the river valley), or on irrigated farmland. On most non-irrigated upland sites away from the river valley only slight galling occurred (A.S. McClay, unpublished). This would be consistent with a requirement by the mites for high humidity, as suggested by Rosenthal (1983). All releases of *A. malherbae* to date have been made in uncultivated land (pastures, wasteland, roadsides, etc.). Its ability to survive in cropland and its effectiveness against *C. arvensis* there are unknown.

A granular barley formation of *P. convolvulus* applied to soil in field plots seeded with pre-germinated seeds or rootstocks of *C. arvensis* reduced its biomass by 98–100% (Vogelgsang *et al.*, 1998c). In field trials, surface application of the granular formulation was more effective than soil incorporation, although the opposite was observed under controlled environment conditions (Vogelgsang *et al.*, 1998a). In field plots, all rates of application down to 10 g of granular formulation per 0.25 m^2 plot gave close to 100% control (Vogelgsang *et al.*, 1998a). Accessions of *C. arvensis* from 11 localities in North America and Europe were all susceptible to *P. convolvulus*, although the degree of disease development differed among accessions (Vogelgsang *et al.*, 1999).

Recommendations

Further work should include:

1. Evaluating further the effectiveness of *A. malherbae*, with particular reference to the effects of environmental conditions, e.g. humidity, method and timing of release, and its ability to survive in annual cropping systems;
2. Active redistribution of *A. malherbae* to *C. arvensis* infested areas, using cost-effective release methods, i.e. attaching excised pieces of galled *C. arvensis* tissue to actively growing plants in the field in spring or early summer;
3. Further host range testing of *T. luctuosa* and *A. malherbae* against native *Convolvulaceae*, given the recent concerns for effects of weed biological control agents on native species, the limited number of native *Convolvulaceae* species used in pre-release testing with *T. luctuosa*, and uncertainties regarding the interpretation of test results with *A. malherbae*;
4. Optimizing production efficiency of *P. convolvulus* as a potential mycoherbicide, perhaps by including a powder to act as a diluent or extender, and increase the area that can be treated with a given amount of inoculum, particularly in high-value crops;
5. Commercializing *P. convolvulus* once production efficiency issues have been resolved.

Acknowledgements

We thank J. Littlefield for providing *A. malherbae*, P. Harris and M. Sarazin for information on earlier releases of the two arthropod agents, Alan Watson for information on *P. convolvulus*, and D. Henderson (Alberta), A. Sturko, S. Cesselli, E. Hogue, L. Edwards (British Columbia), G. Knight, G. Noble (Saskatchewan) and R. Kennedy (Manitoba) for assistance in locating and/or monitoring release sites.

References

Ali, S. (ed.) (1999) *Crop Protection 1999.* Alberta Agriculture, Food and Rural Development, Edmonton, Alberta.

Boldt, P.E., Rosenthal, S.S. and Srinivasan, R. (1998) Distribution of field bindweed and hedge bindweed in the USA. *Journal of Production Agriculture* 11, 377–381.

Chessman, D.J., Horak, M.J. and Nechols, J.R. (1997) Host plant preference, consumption, growth, development, and survival of *Tyta luctuosa* (Lepidoptera, Noctuidae) on biotypes of field bindweed and hedge bindweed. *Environmental Entomology* 26, 966–972.

DeGennaro, F.P. and Weller, S.C. (1984) Differential susceptibility of field bindweed (*Convolvulus arvensis*) biotypes to glyphosate. *Weed Science* 32, 472–476.

Dorrance, M.J. (ed.) (1994) *Practical Crop Protection.* Alberta Agriculture Food and Rural Development, Edmonton, Alberta.

Heiny, D.K. (1994) Field survival of *Phoma proboscis* and synergism with herbicides for control of field bindweed. *Plant Disease* 78, 1156–1164.

Maw, M.G. (1984) *Convolvulus arvensis* L., field bindweed (Convolvulaceae). In: Kelleher, J.S. and Hulme, M.A. (eds) *Biological Control Programmes against Insects and Weeds in Canada 1969–1980.* Commonwealth Agricultural Bureaux, Slough, UK, pp. 155–157.

McClay, A.S., Littlefield, J.L. and Kashefi, J. (1999) Establishment of *Aceria malherbae* (Acari: Eriophyidae) as a biological control agent for field bindweed (Convolvulaceae) in the northern Great Plains. *The Canadian Entomologist* 131, 541–547.

Miller, N.W., Nechols, J.R., Horak, M.J. and Loughin, T.M. (2000) Photoperiodic regulation of seasonal diapause induction in the field bindweed moth, *Tyta luctuosa* (Lepidoptera: Noctuidae). *Biological Control* 19, 139–148.

Morin, L., Watson, A.K. and Reeleder, R.D. (1989) Efficacy of *Phomopsis convolvulus* for control of field bindweed (*Convolvulus arvensis*). *Weed Science* 37, 830–835.

Nuzzaci, G., Mimmocchi, T. and Clement, S.L. (1985) A new species of *Aceria (Acari: Eriophyidae)* from *Convolvulus arvensis* L. *(Convolvulaceae)* with notes on other eriophyid associates of convolvulaceous plants. *Entomologica* 20, 81–89.

Ormeño-Núñez, J., Reeleder, R.D. and Watson, A.K. (1988a) A foliar disease of field bindweed (*Convolvulus arvensis*) caused by *Phomopsis convolvulus*. *Plant Disease* 72, 338–342.

Ormeño-Núñez, J., Reeleder, R.D. and Watson, A.K. (1988b) A new species of *Phomopsis* recovered from field bindweed (*Convolvulus arvensis*). *Canadian Journal of Botany* 66, 2228–2233.

Pfirter, H.A. and Defago, G. (1998) The potential of *Stagonospora* sp. as a mycoherbicide for field bindweed. *Biocontrol Science and Technology* 8, 93–101.

Rosenthal, S.S. (1978) Host specificity of *Tyta luctuosa* (Lep.: Noctuidae), an insect associated with *Convolvulus arvensis* (Convolvulaceae). *Entomophaga* 23, 367–370.

Rosenthal, S.S. (1981) European insects of interest in the biological control of *Convolvulus arvensis* in the United States. In: Del Fosse, E.S. (ed.) *Proceedings of the V International Symposium on Biological Control of Weeds.* Commonwealth Scientific and Industrial Research Organization, Brisbane, Australia, pp. 537–544.

Rosenthal, S.S. (1983) Current status and potential for biological control of field bindweed, *Convolvulus arvensis*, with *Aceria convolvuli.* In: Hoy, M.A., Knutson, L. and Cunningham, G.L. (eds) *Biological Control of Pests by Mites, Proceedings of a Conference, April 1982.* University of California, Berkeley, California, pp. 57–60.

Rosenthal, S.S. and Buckingham, G.R. (1982) Natural enemies of *Convolvulus arvensis* in western Mediterranean Europe. *Hilgardia* 50, 1–19.

Rosenthal, S.S. and Platts, B.E. (1990) Host specificity of *Aceria* (*Eriophyes*) *malherbae*, [*Acari: Eriophyidae*], a biological control agent for the weed, *Convolvulus arvensis* [*Convolvulaceae*]. *Entomophaga* 35, 459–463.

Rosenthal, S.S., Clement, S.L., Hostettler, N. and Mimmocchi, T. (1988) Biology of *Tyta luctuosa* [Lep.: Noctuidae] and its potential value as a biological control agent for the weed *Convolvulus arvensis*. *Entomophaga* 33, 185–192.

Schultheiss, P.C., Knight, A.P., Traubdargatz, J.L., Todd, F.G. and Stermitz, F.R. (1995) Toxicity of field bindweed (*Convolvulus arvensis*) to mice. *Veterinary and Human Toxicology* 37, 452–454.

Scoggan, H.G. (1979) *The Flora of Canada. Part 4. Dicotyledoneae (Loasaceae to Compositae).* National Museum of Canada, Ottawa, Ontario.

Tipping, P.W. and Campobasso, G. (1997) Impact of *Tyta luctuosa* (Lepidoptera, Noctuidae) on hedge bindweed (*Calystegia sepium*) in corn (*Zea mays*). *Weed Technology* 11, 731–733.

Todd, F.G., Stermitz, F.R., Schultheiss, P., Knight, A.P. and Traubdargatz, J. (1995) Tropane alkaloids and toxicity of *Convolvulus arvensis*. *Phytochemistry* 39, 301–303.

USDA Natural Resources Conservation Service (1999) The PLANTS database. http://plants.usda.gov/plants

US Fish and Wildlife Service (1996) Endangered and threatened wildlife and plants: determination of endangered status for four plants and threatened status for one plant from the central Sierran foothills of California. *Federal Register: 18 October 1996* 61(203), 54346–54358.

Vogelgsang, S., Watson, A.K. and DiTommaso, A. (1998a) Effect of soil incorporation and dose on control of field bindweed (*Convolvulus arvensis*) with the pre-emergence bioherbicide *Phomopsis convolvulus*. *Weed Science* 46, 690–697.

Vogelgsang, S., Watson, A.K. DiTommaso, A. and Hurle, K. (1998b) Effect of the pre-emergence bioherbicide *Phomopsis convolvulus* on seedling and established plant growth of *Convolvulus arvensis*. *Weed Research* 38, 175–182.

Vogelgsang, S., Watson, A.K., DiTommaso, A. and Hurle, K. (1998c) Field efficacy of *Phomopsis convolvulus* for control of *Convolvulus arvensis*. *Zeitschrift für Pflanzenkrankheiten und Pflanzenschutz* 16, 445–453.

Vogelgsang, S., Watson, A.K., DiTommaso, A. and Hurle, K. (1999) Susceptibility of various accessions of *Convolvulus arvensis* to *Phomopsis convolvulus*. *Biological Control* 15, 25–32.

Weaver, S.E. and Riley, W.R. (1982) The biology of Canadian weeds. 53. *Convolvulus arvensis* L. *Canadian Journal of Plant Science* 62, 461–472.

67 *Cynoglossum officinale* (L.), Houndstongue (*Boraginaceae*)

R.A. De Clerck-Floate and M. Schwarzländer

Pest Status

Houndstongue, *Cynoglossum officinale* (L.), is a noxious biennial or short-lived perennial weed of mountainous rangelands in north-western North America. Originally from Eurasia (Scoggan, 1978), the weed is thought to have been introduced to North America as a cereal seed contaminant in the 1800s (Knight *et al.*, 1984). Although it is reported from all Canadian provinces except Prince Edward Island and Newfoundland (presumably the northern territories too), it is particularly abundant in the interior of British Columbia (Upadhyaya *et al.*, 1988). The total area currently infested by *C. officinale* there is unknown, but a 1986 report estimated that over 2000 ha of forested rangeland, pasture and roadsides were infested and there were concerns over its increasing spread (Cranston and Pethybridge, 1986). The weed thrives particularly in forest openings created through logging activities, sometimes forming dense monocultures in these habitats (Upadhyaya and Cranston, 1991). It also is becoming a problem in the foothills of south-western Alberta, where it occurs in coulees, moist wooded draws, river/creek bottoms and along roadsides. Cattle are the main dispersers of seed to new sites, although deer and elk probably also contribute to its spread (De Clerck-Floate, 1997).

In British Columbia *C. officinale* is a major concern to cattlemen, second only to the knapweeds, *Centaurea* spp., as a priority for control (Upadhyaya and Cranston, 1991). The weed hinders establishment of forage on new pastures and its barbed seeds or 'burrs' attach to cattle, causing irritation, potential reductions in auction price of animals, and a negative impact on the rancher's reputation (Upadhyaya and Cranston, 1991). The market-related concerns are serious enough to prompt ranchers to spend time cleaning burrs off their cattle before they go to auction (Ranchers, Cranbrook, 1996, personal communication). It takes an estimated 5 man-days to clean burrs from 100 cows (Upadhyaya and Cranston, 1991). In England, Russia and the western USA, deaths of cattle (Greatorex, 1966; Baker *et al.*, 1991) and horses (Knight *et al.*, 1984; Stegelmeier *et al.*, 1996) have been attributed to consumption of *C. officinale*. The toxic substances involved are pyrrolizidine alkaloids, which occur at levels much higher than those found in another toxic range weed, tansy ragwort, *Senecio jacobaea* L. (Pfister *et al.*, 1992). Normally, livestock avoid feeding on green *C. officinale*, but problems arise when the plant senesces or is accidently dried in hay. Calves fed 1 kg of dried plants per kg body weight (60 mg of pyrrolizidine alkaloids kg^{-1} body weight) died within 48 hours due to severe liver damage, and even a chronic dose of one-quarter of this caused eventual death (Baker *et al.*, 1991).

Background

Current control options are limited. The herbicides picloram, dicamba, chlorsulfuron (Cranston and Pethybridge, 1986), and 2,4-D (2,4-dichlorophenoxyacetic acid) (Dickerson and Fay, 1982) will control *C. officinale*. However, use of picloram, the chemical of choice, is often not feasible because of cost and impact on non-target forages or tree species (Upadhyaya *et al.*, 1988). Cutting flowering plants at, or just above, the ground has also been suggested as a control method, but this usually reduces rather than eliminates seed production (Dickerson and Fay, 1982). If seeds have formed, but have not ripened at the time of cutting, they are still capable of germinating the following spring (R.A. De Clerck-Floate, unpublished). Both herbicide application and cutting are difficult and time consuming because of the large areas needing treatment and the uneven, obstacle-ridden terrain. Many ranchers and land managers believe that biological control is the only feasible control option.

European exploration for potential biological control agents began in 1988. Candidates subsequently studied included the root weevil *Mogulones* (*Ceutorhynchus*) *cruciger* Herbst, stem-boring weevil, *Mogulones trisignatus* Gyllendal, seed weevil *Mogulones borraginis* (Fabricius), and root flea beetle, *Longitarsus quadriguttatus* Pontoppidan (Freese, 1989). In 1992, preliminary host-specificity tests were conducted on two additional agents: the root weevil *Rhabdorhynchus varius* (Herbst) and the root fly, *Cheilosia pasquorum* Becker (Jordan and Schwarzländer, 1992). Initial screening showed that the host range of *R. varius* included *Echium vulgare* L. (Schwarzländer and Tosevski, 1993), a valued nectar-producing plant for honey bees, *Apis mellifera* L., in southern Ontario, so screening of this agent was stopped.

A recent shift in public attitude on the potential risks of biological control to native North American plants has affected this as well as other classical biological control programmes. Based on host-specificity tests using mostly European test plant species in the *Boraginaceae*, *M. cruciger*, the first candidate, was petitioned for release in 1993 (Jordan *et al.*, 1993). However, concerns were raised that certain native North American *Boraginaceae* had not been tested, particularly those in the same genus as *C. officinale* and in the North American genus *Amsinckia* (e.g. *A. carinata* A. Nelson and J.F. Macbridge is a species listed as threatened in Oregon). Canada also expressed concern over potential feeding on the European *Borago officinalis* L. (borage), grown to a limited degree as an alternative crop on the prairies. To address the concerns, additional host-specificity tests were conducted, which took another 3 years because of difficulties in obtaining and growing the native *Cynoglossum* spp. A supplemental petition was then submitted (De Clerck-Floate *et al.*, 1996) and *M. cruciger* was approved for release in Canada in 1997 and recommended for release in the USA. However, new concerns were raised by USDA-Fish and Wildlife Service over the safety of another species listed as threatened in the USA, *Cryptantha crassipes* I.M. Johnston. Currently, approval for release of *M. cruciger* there is pending further review. Meanwhile, Canada approved release of a second agent, *L. quadriguttatus*, (De Clerck-Floate *et al.*, 1997) in 1998. In Europe, screening of the remaining agents continues, using an expanded test plant list that includes several native North American *Boraginaceae*.

Several adventive or indigenous North American pathogens and insects have been found attacking *C. officinale* in British Columbia and Alberta. Some of these are being investigated for their distribution, ease of mass production, host specificity, efficacy and potential for integration into the current biological control programme. Of these agents, the pathogens show the most promise. The foliar fungus, *Phoma pomorum* Thumen, is not only host specific but is capable of reducing *C. officinale* biomass by 23% (Conner *et al.*, 2000). The powdery mildew fungus, *Erysiphe cynoglossi* (Wallroth) E. Braun is ubiquitous on *C. officinale* in British Columbia and Alberta and was found to significantly reduce seed production and quality (De Clerck-Floate, 1999). Other damaging pathogens include the root fungus, *Fusarium acuminatum* Ellis and Everhart, the bacterium, *Pseudomonas syringae* Van Hall, and several unidentified viruses (De Clerck-Floate *et al.*, 2000). Diapaused larvae of the indigenous moth, *Platyprepia virginalis* Boisduval, feed on *C. officinale* rosettes in early spring, but this defoliator did not have a significant impact on growth (Conner *et al.*, 2000) and has a broad host range (R.A. De Clerck-Floate, unpublished). Hence, it is not recommended for augmentative use.

Biological Control Agents

Insects

In Europe, and recently observed in Canada, diapaused *M. cruciger* adults emerge in spring (April–June) to feed on *C. officinale* shoots, mate and oviposit. Females emerging in summer also lay eggs, but at a lower rate. They tend to prefer bolting plants over rosettes, and large over small plants for oviposition (Prins *et al.*, 1992; Schwarzlaender, 1997). Oviposition in spring and autumn results in generation overlap in the field, such that larvae can be found within *C. officinale* roots throughout the year. There are three larval instars, after which the larvae exit host roots to pupate in the soil. Adults can live 1 year or longer, which also contributes to generational overlap. At several European sites more than 90% of plants were attacked in spring and the mean number of larvae per root reached 6.7 (Schwarzlaender, 1997). The weevil significantly reduced reproductive effort (Prins *et al.*, 1992) and biomass (Jordan *et al.*, 1993), and showed good potential as an effective agent.

M. cruciger is closely associated with, and highly specific on, *C. officinale* throughout the plant's range in central Europe (Jordan *et al.*, 1993; Schwarzlaender, 1997). Host-specificity tests indicated that *M. cruciger* prefers *C. officinale* over other species of *Boraginaceae*, but is still capable of developing to a lesser degree on other species and genera within the *Boraginaceae* (e.g. *Lappula deflexa* (Wahlenberg) Opiz, *Anchusa azurea* P. Mills, *Cynoglossum grande* Douglas ex Lehmann, *Borago officinalis, Hackelia floribunda* (Lehmann) I.M. Johnston, *Cryptantha* spp.) (Jordan *et al.*, 1993; De Clerck-Floate *et al.*, 1996).

Schwarzlander *et al.* (1997) and Jordan (1997) studied the life history of *L. quadriguttatus*. This univoltine flea beetle prefers attacking the rosette stage of its host. In Europe, adults emerge in May–June and, after 4–7 days of feeding, begin laying their eggs between the leaves or in the soil around the base of rosettes. Adults can be found feeding on the aerial parts of *C. officinale* throughout summer, whereas the larvae mine in rootlets and the outer cortex of tap roots during late summer and autumn. Larvae overwinter in the roots and emerge in spring to pupate in the soil.

European field records indicate a close association of *L. quadriguttatus* with *C. officinale*. Experiments confirmed that *L. quadriguttatus* has a host range mainly restricted to plant species within the genus *Cynoglossum*, but limited attack was found on species of other genera within *Boraginaceae* (e.g. *Anchusa, Echium, Lithospermum, Symphytum*) (Jordan, 1997; Schwarzlaender *et al.*, 1997; Schwarzländer, 2000).

Releases and Recoveries

Initial releases of *M. cruciger* from Hungary and Serbia were made in British Columbia in 1997 (Table 67.1). Some insects were kept at Lethbridge for laboratory rearing and the British Columbia Ministry of Forests also initiated propagation of the weevil within field cages. By 1998, a 50% mix of European-imported and Canadian laboratory/field-propagated adult weevils, in both post- and pre-diapause status, were being released. By 1999, 93% of the 8835 weevils released in British Columbia were laboratory- and field-propagated in Canada. Between 95 and 100% of the Alberta-released weevils were reared at Lethbridge in 1998 and 1999. Releases took place from early spring to autumn in both years. Recoveries have been made at most 1997 and 1998 release sites, regardless of location and month of release or the diapause status of adults at the time of release.

Table 67.1. Releases and recoveries of insects against *Cynoglossum officinale* in British Columbia (BC) and Alberta (AB).

Species	Province	Year	Total released	Number of releases	Recovery
Mogulones cruciger Herbst	BC	1997	1023	7	1998–2000
	BC	1998	3560[a]	17	1999–2000
	BC	1999	8835[b]	35	2000
	AB	1998	320	2	1999–2000
	AB	1999	2411	6	2000
Longitarsus quadriguttatus Pontoppidan	BC	1998	315	2	2000
	BC	1999	629	3	Not confirmed
	AB	1999	203	1	2000

[a]Of the total, 576 were reared in propagation plots at Kamloops, 1211 were laboratory reared at Lethbridge and 1773 came from Europe.
[b]Of the total, 3149 were reared at Kamloops, 5091 at Lethbridge and 595 were from Europe.

The ability to mass-propagate *M. cruciger* in the laboratory has allowed us to take a more experimental approach to initial releases. In April 1999, 5000 laboratory-reared adults were released at 20 sites in the East Kootenay area, British Columbia, as part of an experiment to determine the optimum number for release. The results will allow us to develop a prescription for effective use of the weevil.

Limited open and caged field releases of *L. quadriguttatus*, originally from Austria, were made in British Columbia in 1998, and in British Columbia and Lethbridge in 1999 (Table 67.1). In 2000, the beetle was recovered at both 1998 release locations (including caged propagation plots at Kamloops) and at the open propagation plot release made in 1999 at Lethbridge. Some of the beetles shipped are being laboratory reared at Lethbridge.

Evaluation of Biological Control

It is too early to fully evaluate the success of biological control attempts. However, initial indications are that *M. cruciger* is establishing well, increasing at release sites, dispersing to new sites and having an impact on *C. officinale*. In outdoor propagation plots at Lethbridge and Kamloops, the weevil has shown an excellent capacity for population increase and impact, to the point that it is now difficult to keep *C. officinale* available for *M. cruciger* in these plots. Some of the pathogens also show promise as biological control agents and, once investigated further, may be effectively integrated into the biological control programme for this weed.

Recommendations

Further work should include:

1. Continued monitoring of *M. cruciger* in British Columbia and Alberta, and developing release strategies that will ensure its predictable establishment, increase and impact;

2. Continued monitoring of *L. quadriguttatus* for establishment;

3. Monitoring non-target, native *Boraginaceae* (e.g. *Hackelia floribunda*, *Cryptantha celosioides* (Eastwood) Payson) for potential feeding by *M. cruciger* and *L. quadriguttatus*;

4. Continued screening of additional European candidate agents. Host-specificity tests should be completed on the root fly (*C. pasquorum*), stem weevil (*M. trisignatus*) and seed weevil (*M. borraginus*) using an expanded test-plant list, including *Boraginaceae* genera unique to North America (e.g. *Cryptantha*, *Plagiobothrys*, *Pectocarya*);

5. Continued studies on the biology, host specificity and efficacy of promising pathogens (e.g. *P. pomorum*, *P. syringae* and *F. acuminatum*).

Acknowledgements

Consortium funding for foreign screening of agents is acknowledged from the British Columbia Ministries of Forests, Agriculture and Food, the Wyoming Weed and Pest Districts, and Montana Noxious Weed Trust Fund. Support for research on *M. cruciger* in British Columbia is being provided by the British Columbia Beef Cattle Industry Development Fund (BCIDF) administered by the British Columbia Cattlemen's Association, British Columbia Hydro and Agriculture and Agri-Food Canada, Matching Investments Initiative (MII). BCIDF and MII provided support for research on indigenous/adventive pathogens and insects found on houndstongue in British Columbia. We also acknowledge the help of D. Brooke, S. Turner and V. Miller of British Columbia Ministry of Forests, B. Wikeem of Solterra Inc., L. Behne of the German Entomological Institute and I. Tosevski.

References

Baker, D.C., Pfister, J.A., Molyneux, R.J. and Kechele, P. (1991) *Cynoglossum officinale* toxicity in calves. *Journal of Comparative Pathology* 104, 403–410.

Conner, R.L., De Clerck-Floate, R.A., Leggett, F.L., Bissett, J.D. and Kozub, G.C. (2000) Impact of a disease and a defoliating insect on houndstongue (*Cynoglossum officinale*) growth; implications for weed biological control. *Annuals of Applied Biology* 136, 297–305.

Cranston, R.S. and Pethybridge, J.L. (1986) Report on houndstongue (*Cynoglossum officinale*) in British Columbia. Internal Report, British Columbia Ministry of Agriculture and Food, Victoria, British Columbia.

De Clerck-Floate, R. (1997) Cattle as dispersers of hound's-tongue on rangeland in southeastern British Columbia. *Journal of Range Management* 50, 239–243.

De Clerck-Floate, R. (1999) Impact of *Erysiphe cynoglossi* on the growth and reproduction of the rangeland weed *Cynoglossum officinale. Biological Control* 15, 107–112.

De Clerck-Floate, R., Schroeder, D. and Schwarzlaender, M. (1996) *Supplemental Information to the Petition (Can-93-4 and TAG 93-06) to release* Ceutorhynchus *(*Mogulones*)* cruciger *for the Biological Control of Hound's-tongue (*Cynoglossum officinale, *Boraginaceae) in Canada.* Agriculture and Agri-Food Canada Report.

De Clerck-Floate, R., Story, J. and Schwarzlaender, M. (1997) *Proposal to Introduce* Longitarsus quadriguttatus *Pont. (Col.: Chrysomelidae) for the Biological Control of Hound's-tongue (*Cynoglossum officinale *L.) in North America.* Agriculture and Agri-Food Canada Report.

De Clerck-Floate, R., Conner, R.L., Leggett, F.L., Hwang, S.F. and Yanke, L.J. (2000) Promising native/adventive pathogen and insect agents for the biological control of houndstongue in Canada. In: Spencer, N.R. (ed.) *Proceedings of the X International Symposium on Biological Control of Weeds, 4–14 July 1999, Bozeman, Montana, USA.* Montana State University, Bozeman, Montana, pp. 242–243.

Dickerson, J.R. and Fay, P.K. (1982) Biology and control of houndstongue (*Cynoglossum officinale*). *Proceedings of the Western Society of Weed Science* 35, 83–85.

Freese, A. (1989) *Weed projects for Canada; houndstongue (*Cynoglossum officinale *L.). Work in Europe in 1989.* European Station Report, International Institute for Biological Control.

Greatorex, J.C. (1966) Some unusual cases of plant poisoning in animals. *Veterinary Record* 78, 725–727.

Jordan, T. (1997) Host specificity of *Longitarsus quadriguttatus* (Pont., 1765) (Col., Chrysomelidae), an agent for the biological control of hound's-tongue (*Cynoglossum officinale* L., Boraginaceae) in North America. *Journal of Applied Entomology* 121, 457–464.

Jordan, T. and Schwarzländer, M. (1992) *Investigations on Potential Biocontrol Agents of Hound's-tongue* Cynoglossum officinale *L.* International Institute for Biological Control Annual Report.

Jordan, T., Schwarzländer, M., Tosevski, I. and Freese, A. (1993) Ceutorhynchus cruciger *Herbst (Coleoptera, Curculionidae): a Candidate for the Biological Control of Hound's-tongue (*Cynoglossum officinale *L., Boraginaceae) in Canada.* Final Report. International Institute of Biological Control.

Knight, A.P., Kimberling, C.V., Stermitz, F.R. and Roby, M.R. (1984) *Cynoglossum officinale* (Hound's-tongue) B A cause of pyrrolizidine alkaloid poisoning in horses. *Journal of the American Veterinary Medicine Association* 184, 647–650.

Pfister, J.A., Molyneux, R.J. and Baker, D.C. (1992) Pyrrolizidine alkaloid content of houndstongue (*Cynoglossum officinale* L.). *Journal of Range Management* 45, 254–256.

Prins, A.H., Nell, H.W. and Klinkhamer, P.G.L. (1992) Size-dependent root herbivory on *Cynoglossum officinale. Oikos* 65, 409–413.

Schwarzlaender, M. (1997) Bionomics of *Mogulones cruciger* (Coleoptera: Curculionidae), a below-ground herbivore for the biological control of hound's-tongue. *Environmental Entomology* 26, 357–365.

Schwarzländer, M. (2000) Host specificity of *Longitarsus quadriguttatus* Pont., a below-ground herbivore for the biological control of houndstongue. *Biological Control* 18, 18–26.

Schwarzländer, M. and Tosevski, I. (1993) *Investigations on Potential Biocontrol Agents of Hound's-tongue (*C. officinale *L.).* Annual Report, International Institute of Biological Control.

Schwarzlaender, M., Jordan, T. and Freese, A. (1997) *Investigations on* Longitarsus quadriguttatus *(Coleoptera, Chrysomelidae), a Below Ground Herbivore for the Biological Control of Hound's-tongue.* Revised Final Report, International Institute of Biological Control.

Scoggan, H.J. (1978) *The Flora of Canada. Part 4. Dicotyledonae (Loasaceae to Compositae).* National Museum of Natural Sciences, National Museums of Canada, Ottawa, Ontario, pp. 1282–1283.

Stegelmeier, B.L., Gardner, D.R., James, L.F. and Molyneux, R.J. (1996) Pyrrole detection and the pathologic progression of *Cynoglossum officinale* (houndstongue) poisoning in horses. *Journal of Veterinary Diagnostic Investigation* 8, 81–90.

Upadhyaya, M.K. and Cranston, R.S. (1991) Distribution, biology, and control of hound's-tongue in British Columbia. *Rangelands* 13, 103–106.

Upadhyaya, M.K., Tilsner, H.R. and Pitt, M.D. (1988) The biology of Canadian weeds. 87. *Cynoglossum officinale* L. *Canadian Journal of Plant Sciences* 68, 763–774.

68 *Cytisus scoparius* (L.) Link, Scotch Broom (*Fabaceae*)

R. Prasad

Pest Status

Scotch broom, *Cytisus scoparius* (L.) Link, native to Europe, was introduced from Hawaii into Sooke, British Columbia, in 1850 by William Grant. It greatly expanded its range along the Pacific (British Columbia) and Atlantic coasts (Nova Scotia) during the past century. In British Columbia, it has invaded forested, urban landscapes, rights-of-way and rangelands in the south-west (Vancouver, Victoria) and part of the interior east to Kootenay Lake and Castlegar (Peterson and Prasad, 1998). Human activities, e.g. planting along highways for beautification and prevention of soil erosion, have hastened its spread. *C. scoparius* rapidly invades disturbed areas, forming dense thickets that can suppress and inhibit mature vegetation, including conifer seedlings (Prasad, 2000). Its invasive features include stem photosynthesis, prolific seed production, longevity of seeds in the soil and nitrogen fixation (Prasad and Peterson, 1997). No solid data exist to evaluate its economic damage, which may be in the millions of dollars, particularly in urban land, where its infestations depreciate real estate values.

C. scoparius is a perennial, deciduous shrub that produces about 18,000 seeds per year per plant, although only half are viable. Seedlings begin flowering and setting seed at 2 years and continue to grow for 25–30 years, attaining a height of 3–6 m. A plant can propagate vegetatively after being cut or damaged.

Background

Chemical herbicides, e.g. 2,4,5-T (2,4,5-trichlorophenoxyacetic acid), 2,4-D (2,4-dichlorophenoxyacetic acid) (alone or combined with triclopyr or picloram), and tricholpyr, have provided effective control of *C. scoparius* (Miller, 1992a; Peterson and Prasad, 1998). Spraying with glyphosate in British Columbia gave somewhat inconsistent control (Zielke *et al.*, 1992). Fire can be used for vegetation control but seeds in the soil readily germinate after low to moderately severe burns (Peterson and Prasad, 1998).

A variety of mechanical methods has been used to control *C. scoparius*, with some having the undesired effect of actually increasing its spread and growth. Manual pulling is a popular and successful method of removal of young shrubs in urban and park areas, but is impractical in forests or inaccessible terrains. Pulling disturbs the soil, damaging desirable species and causing Scotch broom seeds to germinate. Manual cutting of older plants, especially to ground level during periods of moisture stress, is effective in preventing shrub regrowth (Miller, 1992b). Machinery is used to cut out high-density stands (Jones and Popenoe, 1996).

No single method effectively controls *C. scoparius*. A combination of strategies is required to reduce populations, e.g. depletion of the seed banks by disturbance, chemical treatment by herbicides, and manual cutting to reduce flowering and seed set. In Canada, *C. scoparius* has relatively few natural enemies. The occurrence of the native species, *Agonopterix ulicetella*, Stainton, on local *C. scoparius* and gorse, *Ulex europaeus* L., flowers was documented, but no attempt was made to use this as a biological control agent. In Europe, fungi and insects limit growth and distribution of *C. scoparius*.

Biological Control Agents

Vertebrates

Grazing by goats and sheep has been attempted to control *C. scoparius*, but field trials showed that sheep would not eat it (Zielke *et al.*, 1992). However, Lamancha goats effectively grazed *C. scoparius* on a small plot on southern Vancouver Island (Zielke *et al.*, 1992).

Insects

In southern Europe, several endemic seed feeders, e.g. *Apion fuscirostre* Fabricius and *Ceutorhynchus* spp., infest *C. scoparius*. Hosking (1992) reported several defoliators (e.g. *Gonioctena olivacea* Förster, *Sitona regensteinensis* Herbst, *Agonopterix* spp.), stem miners (e.g. *Apion immune* Kirby, *A. striatum* Kirby and *Leucoptera spartifoliella* Hübner) and small wood weevils found just below the dead branches. None of these agents has been released in Canada.

Pathogens

Punja and Ormrod (1979) reported foliage blight caused by *Alternaria alternata* Keissler and *Stemphylium* spp. under greenhouse conditions, but their bioherbicidal potentials were never tested. Prasad (1998, 2000) evaluated the potential of *Chondrostereum purpureum* Pouzar, *Fusarium tumidum* Sherbakoff and *Pleiochaeta setosa* L. under greenhouse conditions, and found that *F. tumidum* effectively reduced growth of *C. scoparius* by 50–70%, whereas the other two fungi had slight or variable effects. Subsequently, when 3-year-old *C. scoparius* stems were cut and treated with a new formulation of *C. purpureum*, a complete inhibition of resprouting was observed (Prasad and Naurais, 1999). The mycoherbicidal control by this fungus under field conditions is being tested. *Diaporthe inequalis* (Currey) Nitshke was found causing canker in stems and branches of *C. scoparius* in Nanaimo (R. Wall, Victoria, 2000, personal communication) but no attempt was made to use it for biological control.

Evaluation of Biological Control

The fungi *F. tumidum* and *C. purpureum* show promise against *C. scoparius* and could be developed as bioherbicides with improved formulation and virulence. The use of insects as potential biological control agents has not been exploited. These potential controls may be integrated with existing control techniques.

Recommendations

Further work should include:

1. Developing better formulations of fungi to improve inoculum viability and efficacy;
2. Developing affordable mass-production systems and application technology;
3. Host range testing of *F. tumidum* and *C. purpureum* to ensure their specificity;
4. Evaluating European insects for potential introduction.

References

Hosking, J.R. (1992) The impact of seed and pod eating insects on *Cytisus scoparius*. In: Delfosse, E. (ed.) *Proceedings of the 8th International Symposium on Biological Control of Weeds, 2–7 February, Canterbury*. Department of Scientific and Industrial Research Organizations, New Zealand, pp. 45–51.

Jones, C. and Popenoe, H. (1996) *Control Techniques of Scotch Broom*. National Park Service (USA), Redwood National Park, California.

Miller, G. (1992a) Chemical control of broom. *Oregon Department of Agriculture Weed Control Program, Broom/Gorse Quarterly* 1, 4.

Miller, G. (1992b) Manual control of broom. *Oregon Department of Agriculture, Weed Control Program, Broom/Gorse Quarterly* 1, 2–3.

Peterson, D. and Prasad, R. (1998) The biology of Canadian weeds. 109. *Cytisus scoparius* (l.) Link. *Canadian Journal of Plant Science* 78, 497–504.

Prasad, R. (1998) Evaluation of some fungi for bioherbicidal potential against Scotch broom (*Cytisus scoparius*) under greenhouse conditions. In: Wilcut, J. (ed.) *Abstracts and Proceedings of the Weed Science Society of America, 5–8 February, Chicago, Illinois*, pp. 38, 46.

Prasad, R. (2000) Some aspects of the impact of and management of the exotic weed, Scotch broom *Cytisus scoparius* in British Columbia. *Journal of Sustainable Forestry* 10, 339–345.

Prasad, R. and Naurais, S. (1999) Ecology, biology and control of alien plants (*Cytisus scoparius*) in British Columbia. In: Kelly, M., Howe, M. and Neill, B. (eds) *Proceedings of the California Exotic Plant Protection Council, 15–17 October, Sacramento, CA*. California Exotic Pest Plant Council, San Juan, Capistrano, vol. 5, pp. 23–25.

Prasad, R. and Peterson, D. (1997) Mechanisms of invasiveness of the exotic weed, Scotch broom (*Cytisus scoparius*) in British Columbia. In: *Proceedings of Expert Committee on Weeds, 9–12 December, Victoria, British Columbia*, pp. 197–198.

Punja, Z. and Ormrod, D.J. (1979) New or noteworthy plant diseases in coastal British Columbia 1975–77. *Canadian Plant Disease Survey* 59, 22–24.

Zielke, K., Boateng, J., Caldicott, N. and Williams, H. (1992) *Broom and Gorse: a Forestry Perspective Analysis*. British Columbia Ministry of Forests, Queens Printer, Victoria, British Columbia.

69 *Euphorbia esula* (L.), Leafy Spurge, and *Euphorbia cyparissias* (L.), Cypress Spurge (*Euphorbiaceae*)

R.S. Bourchier, S. Erb, A.S. McClay and A. Gassmann

Pest Status

Leafy spurge, *Euphorbia esula* L., was introduced into North America from Eurasia in the early 1800s (Gassmann *et al.*, 1996). It occurs in all Canadian provinces except Newfoundland and more than half of the US states (Alley and Messersmith, 1985). The most widely infested areas are in the prairie provinces and scattered areas in British Columbia, e.g. Thompson, Cariboo, Boundary, East Kootenay, Nechako and the North Okanagan and Bulkley valleys (Anonymous, 2000b). *E. esula* was first recorded from Huron County, Ontario, in 1889, followed by Manitoba in 1911, Saskatchewan in 1928, Alberta in 1933, and British Columbia in 1939 (Haber, 1997). It now infests more than 2 million ha in North America (Stelljes, 1997) including about 650,000 ha in North Dakota, South Dakota, Montana and Wyoming (Sell *et al.*, 1999). Infestations in Canada are estimated at about 8000 ha of pasture and native prairie in southern Saskatchewan (Anonymous, 2000c), about 141,000 ha in Manitoba (Manitoba leafy spurge stakeholders group, Brandon, 2000, personal communication) and more than 6000 ha in Alberta (McClay *et al.*, 1995). Combined economic losses have been estimated at US$130 million per year in North Dakota, South Dakota, Montana and Wyoming (Hansen *et al.*, 1997). *E. esula* has spread rapidly in rangeland, roadsides and non-crop riparian areas. An acrid, sticky white sap in stems causes direct toxicity to cattle, while displacement of rangeland due to competition from *E. esula* leads to reduced livestock production as well as secondary losses in other, associated industries (Leistritz *et al.*, 1992; Hansen *et al.*, 1997; Bangsund *et al.*, 1999). *Euphorbia cyparissias* L. is also native to Europe and contains sap that is toxic to livestock.

E. esula is a deep-rooted perennial that reproduces by seed and vegetative buds, and its stems can be more than 1 m tall (Best *et al.*, 1980). Seeds can persist in soil for up to 8 years (Selleck *et al.*, 1962). In the Canadian prairies, *E. esula* flowers from May to August. Seeds are explosively dispersed and carried by birds, insects and mammals, but the greatest spread of infestations is via vegetative root buds from individual plants (Haber, 1997).

The biology of *E. cyparissias* is similar to that of *E. esula* (Anonymous, 2000a). It is a perennial, reproducing both by seed and widely spreading, much-branched underground roots with numerous buds. *E. cyparissias* also forms dense stands, with stems attaining heights of 10–80 cm. Flowering begins in late spring or early summer and may continue until late autumn. Both fertile and non-fertile forms occur in Ontario, with the fertile form being the weed problem in abandoned cultivated land, woodland, roadsides and pastures.

Background

Biological control against *E. esula* was initiated in the 1960s in North America because of the difficulty and cost of controlling its populations on rangeland with herbicides, and because of the availability of its natural enemies in its native range (Harris *et al.*, 1985). Since 1970, 18 insects have been introduced (14 since 1980) into Canada (Julien and Griffiths, 1998; Harris, 2000a). These biological control agents are suited to particular habitats or combinations of dry and mesic as well as open and closed sites (Gassmann and Schroeder, 1995).

In North America, a taxonomic controversy remains as to whether *E. esula* is one species or an aggregate of two or more species (Crompton *et al.*, 1990; Gassmann *et al.*, 1996, and references therein; Rowe *et al.*, 1997; Geltman, 1998). Morphological and gas chromatographic studies suggest that North American *E. esula* is a single species (Crompton *et al.*, 1990; Evans *et al.*, 1991). These taxonomic problems have hindered selection of biological control agents; many of the European insects come from other *Euphorbia* spp. and thus may not be as well adapted to the North American spurge.

Biological Control Agents

Insects

Harris (1984) summarized the ecology and pre-1980 release data for *Hyles euphorbiae* (L.), *Chamaesphecia empiformis* (Esper), *Chamaesphecia tenthrediniformis* (Denis and Schiffermüller) and *Oberea erythrocephala* (Schrank).

Since 1978, five flea beetle species, *Aphthona cyparissiae* (Koch), *Aphthona flava* Guillebaume, *Aphthona nigriscutis* Foudras, *Aphthona czwalinae* Weise and *Aphthona lacertosa* Rosenhauer, have been released to control *E. esula* (Julien and Griffiths, 1998). These five are keyed in LeSage and Paquin (1996). They attack both *E. esula* and *E. cyparissias* (Gassmann and Schroeder, 1995). All are restricted to *Euphorbia* section *Esula*, with *A. czwalinae* having the narrowest and *A. nigriscutis* the widest host range (Gassmann *et al.*, 1996). They were introduced because of their different habitat preferences. Economically important spurges in North America, e.g. poinsettia, *Euphorbia pulcherrima* Willdenow, are not at risk because adults and larvae do not feed on them. None of the introduced *Aphthona* spp. occurs on annual spurges in their native range (Maw, 1981) and their larval biology excludes any sustained attack on annual spurges in nature. All are univoltine, overwinter as larvae in spurge roots, and have three larval instars. Pupation and adult emergence occur in late spring–early summer. Abiotic factors, e.g. temperature and/or humidity, are apparently the main mortality factors (Gassmann *et al.*, 1996). Adults are active throughout summer (June–September, depending on species), laying eggs on plant stems near the soil surface or in soil close to the plant. In the prairie provinces, *A. lacertosa* emerges and reaches peak abundance earlier than the other *Aphthona* spp., based on degree-day requirements (R. Hansen, Bozeman, 2000, personal communication). Newly hatched larvae aggregate and feed progressively on young to more mature roots. Adults feed on leaves of varying age from the lower part of the shoots up to the tips, including bracts, and produce feeding marks characteristic for each species group: the brown beetles (e.g. *A. cyparissiae*, *A. flava* and *A. nigriscutis*) start feeding from the leaf margin, whereas the black beetles (e.g. *A. czwalinae* and *A. lacertosa*) scrape the leaf surface, sometimes perforating it (Gassmann *et al.*, 1996). Adult leaf feeding reduces plant photosynthesis, and flower consumption reduces seed production. Larval feeding within the roots reduces a plant's ability to absorb water and nutrients, decreasing plant height, delaying flowering and weakening taproots (Rees *et al.*, 1996a).

A. czwalinae prefers mesic, loamy sites where the host plant grows with other vegetation, and is adapted to continental climates with cooler summer temperatures. *A. flava* prefers mesic to dry sites with sparse vegetation in areas with warm dry summers, as in subcontinental and submediterranean climates of south-eastern

Europe. It tolerates light shade, and is less likely to survive low temperatures than the other species. *A. lacertosa*, an eastern European species from the steppe biome, prefers loamy soils and can adapt locally to both dry and wet habitats. In Canada, *A. lacertosa* is expected to do well on sites that are too moist for *A. nigriscutis* and *A. cyparissiae*. *A. nigriscutis* is strongly associated with warm, open, very dry habitats with coarse soils, e.g. sandy knolls and hilltops, and is a semi-arid continental species with a very similar distribution in Europe to that of *A. lacertosa* but extending slightly further north and south. Generally, *A. nigriscutis* controls spurge in the open on coarse, dry prairie, but not in moister, shaded or mesic sites. *A. cyparissiae* is a subcontinental species adapted to slightly cooler summers and harsher winters; it prefers warm, open, sunny areas and slightly moister conditions than *A. nigriscutis* but less moist than *A. flava*. A sixth species of Mediterranean origin, *Aphthona abdominalis* Duftschmidt, was released in 1993 in the USA (Fornasari and Pecora, 1995).

Gassmann and Tosevski (1994) and Gassmann (1994) studied the ecology of the clearwing moths *Chamaesphecia hungarica* (Tomala), *Chamaesphecia astatiformis* (Herrich-Schaffer), and *Chamaesphecia crassicornis* Bartel. All are univoltine, overwinter in the roots of spurge plants, and pupate in early to late spring. *C. astatiformis* and *C. hungarica* overwinter as sixth- or seventh-instar larvae, whereas *C. crassicornis* overwinters as younger larvae and completes most larval development the following spring. Adult *C. hungarica* and *C. astatiformis* emerge from mid-May until the end of June in their native ranges, whereas *C. crassicornis* adults emerge in July. Females call by waving the ovipositor before mating. *C. hungarica* females lay, on average, 122 eggs singly on bracts, leaves and stems; *C. astatiformis* females oviposit mostly on vegetative shoots of young, small plants, with an average of 92 eggs being placed on the lower leaf surface or in the leaf axils on the upper part of the plant; and *C. crassicornis* females lay an average of 80 eggs singly in leaf axils and along stems. In all species, larvae hatch in 2–3 weeks. Larvae of *C. hungarica* penetrate the shoot just above the soil surface, and travel down the stem while mining the pith before entering the roots to feed, making a tunnel about 5 cm long (Lastuvka, 1982). In spring, larvae mine up to the base of the previous year's stem, exit, pupate and emerge as adults. Larvae of *C. crassicornis* and *C. astatiformis* drop to the ground and bore directly into the root. *C. crassicornis* larvae continue feeding the following spring and pupate in early June at the top of the exit tunnel. Both annual and biennial life cycles occur, although the latter is less common. Feeding by the larvae of all species destroys roots, depleting their reserves, causing loss of plant vigour and, eventually, plant death (Rees *et al.*, 1996b).

Since 1990, all three *Chamaesphecia* spp. have been introduced into Canada to control *E. esula* in different habitats (Tosevski *et al.*, 1996). In its native area, *C. hungarica* is found on plants growing in moist, loamy soils and in partly shaded habitats, e.g. riverbanks, swampy areas, and ditches. In contrast, *C. astatiformis* prefers mesic to dry loamy sites where the host plant is often mixed with other vegetation; it is adapted to a subcontinental climate with warm summers. *C. crassicornis* is best suited to mesic-dry to dry, open sites with coarse soils and a continental climate. All three species are restricted to *Euphorbia* section *Esula*, with *C. hungarica* primarily attacking *Euphorbia lucida* Waldstein and Kitaibel, *C. astatiformis* attacking *E. esula* (*s.s.*) and *C. crassicornis* attacking *Euphorbia virgata* Waldstein and Kitaibel in their native ranges (Gassmann, 1994; Gassmann and Tosevski, 1994). Larvae of the three species develop on North American *E. esula* but not on species in the sections *Chamaesyce*, *Agaloma* and *Poinsettia*, all of which contain economically important species. Of the three, *C. crassicornis* is considered the best biological control agent because leafy spurge acceptance is higher than for the other two species (Gassmann, 1994).

Harris and Soroka (1982) summarized

the biology of *Lobesia euphorbiana* (Freyer), which occurs from south and central Europe to the Ukraine. They studied a population originally collected from *E. lucida* and *Euphorbia seguieriana* Necker in northern Italy. It appears to be restricted to certain *Euphorbia* spp., in the sections *Galarhoeus*, *Esula* and *Chamaesyce*, the first two containing host plants in Europe. *L. euphorbiana* has two generations per year and possibly a third in Ontario (Harris, 2000b). Eggs are laid individually on lower leaf surfaces and larvae feed mainly on terminal buds by tying leaves or florets together into a tube and feeding from within the tube. The number of instars is thought to vary between four and five, depending on food quality. Laboratory tests suggested that larvae have a high temperature threshold, and may only survive in warm areas. Pupation occurs within the webbed tube about 26 days after oviposition, and adults emerge 10 days later and live for about a week. Overwintering occurs as pupae in leaf litter. The main damage to host plants is prevention of flowering rather than actual feeding damage. Harris and Soroka (1982) suggested that *L. euphorbiana* may reduce seed production of both spurge species but only in certain spurge stands, and will not likely, by itself, result in complete control of *E. esula*.

Harris (1985) summarized the biology of *Minoa murinata* (Scopoli) from central Europe, Spain, Corsica and Italy. In Europe, it is restricted to cooler areas of the spurge zone and larvae can tolerate prolonged cool periods. It has a lower temperature developmental threshold than *H. euphorbiae* (Harris, 1984, see below) and *L. euphorbiana*. *M. murinata* occurs in dry to moist sites in closed woods and is also the main species on *E. cyparissias* on sunny, dry chalk soil on heath-steppes, plains and highlands (Bergmann, 1955, as cited by Harris, 1985). It has 1–2 generations in its native range; adults emerge from May to June in continental areas where there are two generations, and later in June in areas with one generation. Two generations per year occurred in outdoor rearing cages at Vegreville, Alberta (McClay, 1996). Eggs are laid on the underside of *E. esula* and *E. cyparissias* leaves, with four instars feeding from the underside of the leaves. In the laboratory, larval duration was up to 20 days and pupal duration 16–57 days, with a minimum generation time of 33 days. Pupation occurs in the soil. *E. cyparissias* is the main host plant in the native range of *M. murinata*. In no-choice tests, larval feeding and pupation occurred on most *Euphorbia* spp., in the sections *Galarhoeus*, *Esulae*, *Chamaesyce* and *Petaloma*. In the laboratory, *E. esula* was found to be as good a host as *E. cyparissias* (Harris, 1985). The occurrence of *M. murinata* in a fairly broad range of habitats (especially cool, dry sites), as well as the fact that it is multivoltine, makes it an attractive potential biological control agent.

Spurgia esulae Gagné (formerly *Bayeria capitigena*) and *Spurgia capitigena* (Bremi) are bud-gall midges attacking *E. esula* in Europe. Gagné (1990), Pecora *et al.* (1991) and Nelson and Carlson (1999) reviewed their biology in native regions and the USA. Both were originally treated as *Bayeria capitigena* but Gagné (1990) separated them into two species and placed them in *Spurgia*. Both were introduced into North America because of their ability to infest spurge growing in shaded and moist areas (Fornasari, 1996), habitats that are not well colonized by existing biological control agents.

The larvae of both midges cause galls at the growing tips, which prevent host plant flowering and thus reduce seed production (Pecora *et al.*, 1991; Nelson and Carlson, 1999). The generation that overwinters does so as mature larvae in soil, pupating in spring, whereas larvae of spring and summer generations pupate in galls. Gall formation occurs from mid-April to late October in Europe. There are 3–5 generations, depending on weather (Harris, 2000b). First-generation galls produce the highest number of adults (Nelson and Carlson, 1999), with the number of galls present in the field declining as the season progresses (Mann *et al.*, 1996). Eggs are laid in groups on young leaves near growing tips, and larvae migrate to the tips to feed before spinning a silk cocoon and pupating (Pecora *et al.*, 1991). *S. esulae* appears to be restricted to *Euphorbia* spp. in the section *Esula*. Establishment success in North America varies among different *E. esula* genotypes (Lym *et al.*, 1996).

Pegomya curticornis (Stein) and *Pegomya euphorbiae* (Kieffer) were introduced to control *E. esula*. Initially they were thought to be one species, *P. argyrocephala* (Meigen), but Michelsen (1988) separated the group into five species. Gassmann (1987) and Gassmann and Tosevski (1993) studied their life histories in Europe, and Gassmann and Shorthouse (1990) described feeding strategies and gall induction. Both species are univoltine. Adults emerge in early spring from puparia that overwinter within galled shoots. Oviposition takes place 3–4 days after emergence and eggs are laid singly or in small groups on shoot tips. Larvae bore down the centre of the shoots and, upon reaching the base, induce gall formation on subterranean portions of the stem. There are three instars; the final instar is reached within 3 weeks and development is completed within 60–80 days. The plant is damaged early in the growing season as larvae mine the shoots, and galled shoots wilt and eventually die (Gassmann and Schroeder, 1995). The puparium is formed inside the gall in June. Both species belong to two feeding guilds: borers (first 4–5 weeks of larval development) and then gall inducers (6–8 weeks feeding within the lower part of the subterranean stem) (Gassmann and Shorthouse, 1990).

Identifying the host range of the two species has been compounded by the difficult taxonomy of European and North American *E. esula*. In Europe, *P. euphorbiae* is reared from *E. cyparissias*, *E. waldsteinii* [= *E. virgata* (Waldstein and Kitaibel)], *E. seguieriana* and rarely from *E. lucida* (Michelsen, 1988). *P. curticornis* is reared from several 'forms' of *E. esula*, in particular the hairy form of European *E. esula*, and larvae do not develop on the North American *E. esula*. In contrast, larvae of *P. euphorbiae* reared from *E. virgata* accept North American leafy spurge (Gassmann and Tosevski, 1993).

Releases and Recoveries

A. cyparissiae, *A. flava*, *A. nigriscutis* and *A. czwalinae* were released from 1982 to 1985 in mesic to very dry habitats and *A. lacertosa* was released from 1985 to 1990 in moist sites (Table 69.1). All are established in Canada (McClay *et al.*, 1995; Julien and Griffiths, 1998).

A. cyparissiae was first released in 1982 at two sites near Cardston, Alberta (McClay *et al.*, 1995; Julien and Griffiths, 1998) and from 1982 to 1986 in Saskatchewan and Alberta (Harris, 2000b). Up to 1994, 24 releases were made in Alberta, with establishment at a few sites, including Pincher Creek (McClay *et al.*, 1995). It is present in British Columbia, Alberta, Saskatchewan, Manitoba and Ontario. It controls *E. esula* in open, dry sites in Saskatchewan but not Alberta (Anonymous, 1997).

A. flava populations from Hungary and Italy were released from 1982 to 1983 (372 adults) near Cardston, Alberta, and yielded small numbers in 1986 (McClay *et al.*, 1995). Redistributions resulted in recoveries of beetles at 20 sites. It reduced spurge density at two sites in Alberta on coarse soil with high water tables (Harris, 2000c). The species is now considered to be established in British Columbia, Alberta and Ontario (Julien and Griffiths, 1998).

A. nigriscutis was first released near Cardston, Alberta, in 1983 from Hungarian populations. From 1988 to 1990, 24,860 adults were redistributed from the original site to 122 documented sites in Alberta (Table 69.1) (McClay *et al.*, 1995). It is considered established in British Columbia, Alberta, Saskatchewan, Manitoba, Ontario and Nova Scotia (Julien and Griffiths, 1998). Some releases, e.g. at Millet, Alberta, in 1988, did not result in establishment. However, more than 140,000 beetles were supplied for more than 260 releases by individual landowners, fieldmen and others from 1991 to 1994, and 50,000 more were supplied to other provinces and the USA for redistribution (McClay *et al.*, 1995). In Alberta, releases in 1997 resulted in establishment of beetles at several sites between 1998–2000 (R.S. Bourchier, unpublished).

Table 69.1. Number of releases (total number of insects) against *Euphorbia esula* and *Euphorbia cyparissias* by agent and province.

Agent	British Columbia	Alberta	Saskatchewan	Manitoba	Ontario	Quebec	Nova Scotia
A. cyparissiae (Koch)	8 (31,498)[a] 1989–95	38 (15,435)[a] 1982–95	107 (47,033)[b] 1982–92	94 (21,244)[j] 1982–94	9 (3437)[d] 1982–92	1 (133)[d] 1982	2 (2000) 1991–92
A. czwalinae Weise	See mixed releases	8 (2029) 1985–95	9 (1279)[a] 1985–95	8 (893) 1986–94	2 (63)[d] 1987	No releases	No releases
A. flava Guillebaume	18 (8165)[a] 1990–95	48 (16,472) 1982–92	10 (3650) 1983–91	No releases	4 (544)[d] 1982–87	No releases	1 (1000) 1992
A. lacertosa Rosenhauer	2 (1150) 1997	39 (29,092) 1990–2000	57 (2531)[l] 1987–96	37 (16,170)[k] 1991–2000	No releases	No releases	No releases
A. nigriscutis Foudras	274 (179,487) 1986–97	204 (359,632) 1986–97	135 (27,080)[g] 1983–96	133 (27,090)[h] 1983–97	9 (4982)[f] 1986–92	No releases	1 (1000) 1992
A. cyparissiae and *A. nigriscutis* (mixed)	No releases	2 (1114) 1983	No releases	No releases	No releases	No releases	No releases
A. lacertosa and *A. czwalinae* (mixed)	1 (~740) 1995	129 (252,100)[m] 1995–97	1 (600) 1995	2 (~900) 1995	No releases	No releases	No releases
A. lacertosa and *A. nigriscutis* (mixed)	No releases	No releases	No releases	1 (3000) 1997	No releases	No releases	No releases
L. euphorbiana (Freyer)	10 (630) 1987–98	2 (550) 1990–91	3 (417) 1991–94	19 (2042)[c] 1987–96	5 (397)[d] 1987–92	No releases	1 (300) 1991
M. murinata (Scopoli)	1 (500) 1994	13 (8626) 1991–95	No releases	3 (525)[a] 1988–91	No releases	No releases	No releases
O. erythrocephala (Schrank)	No releases	2 (143) 1980–86	1 (95) 1981	3 (102)[d] 1986–87	3 (69)[e] 1982–90	No releases	No releases
P. curticornis (Stein)	No releases	4 (143) 1988–93	No releases	2 (52)[a] 1988–90	No releases	No releases	No releases
P. euphorbiae (Kieffer)	No releases	4 (215) 1988–93	No releases	No releases	2 (145)[e] 1989–90	No releases	No releases
S. esulae Gagné	4 (1375) 1990–93	4 (1675) 1989–93	No releases	3 (500) 1989–92	2 (400) 1992	No releases	2 (985)[d] 1990–91
H. euphorbiae (L.)	2 (2200) year unknown	No releases	2 (338) 1984	1 (746) 1985	No releases	No releases	No releases
S. capitigena (Bremi)	No releases	No releases	No releases	1 (50) 1987	No releases	2 (800)[e] 1990	3 (1575)[i] 1990–91

Continued

Table 69.1. *Continued.*

Agent	British Columbia	Alberta	Saskatchewan	Manitoba	Ontario	Quebec	Nova Scotia
C. empiformis (Esper)	No releases	No releases	No releases	No releases	3 (596)[f] 1989	No releases	No releases
Totals	320 (~225,745)	497 (~687,226)	325 (~ 83,023)	305 (~73,314)	39 (~10,633)	3 (~933)	10 (~6860)

[a]Information on number of insects released missing for 1–2 releases.
[b]Information on number of insects released missing for 3–5 releases.
[c]Information on number of insects released missing for 5–10 releases.
[d]1 release for *E. cyparissias*.
[e]2 releases for *E. cyparissias*.
[f]3 releases for *E. cyparissias*.
[g]Information on number of insects released missing for 69 releases.
[h]27,090 insects released in 84 releases (unknown number of insects released for 49 releases).
[i]All 3 releases for *E. esula* and *E. cyparissias*.
[j]Information on number of insects released missing for 34 releases.
[k]Information on number of insects released missing for 14 releases.
[l]Information on number of insects released missing for 48 releases.
[m]Mixed releases were primarily *A. lacertosa* with a small proportion of *A. czwalinae*.

A. czwalinae was first released in small numbers at Spring Coulee, Alberta, in 1985 and again in 1993 (McClay *et al.*, 1995; Julien and Griffiths, 1998). It is established in Manitoba where it reduced weed flowering on a moist clay riverbank site subject to flooding. Establishment has not been confirmed in Saskatchewan (Julien and Griffiths, 1998) although recently a small number of beetles have been recovered at some release sites in Alberta in 1999–2000 (A.R. Kalischuk and R.S. Bourchier, unpublished).

A. lacertosa from populations collected in Hungary and Yugoslavia was first released in 1990 near Spruce Grove, Alberta (Julien and Griffiths, 1998). Release sites are located near sites where both *A. nigriscutis* and *A. flava* failed to establish. Consistent with observations in Europe, *A. lacertosa* prefers more mesic and loamy sites than the other species (McClay *et al.*, 1995). The beetle is considered established in Alberta, Saskatchewan and Manitoba (Julien and Griffiths, 1998).

In 1997, releases of a mixed *A. lacertosa* and *A. czwalinae* population, collected from North Dakota, were made in Alberta, Saskatchewan and Manitoba. Populations from these mixed releases established in all provinces and, by 1999, the dominant species in Alberta was *A. lacertosa* (A.R. Kalischuk and R.S. Bourchier, unpublished). Releases on the Blood Reserve, southern Alberta, resulted in outbreak densities of beetles in 1999–2000. *A. lacertosa* had a significant, visible impact on spurge densities at several release sites within 1 year of the releases (Table 69.2) (R.S. Bourchier, unpublished).

C. hungarica and *C. astatiformis* were released in 1991 and 1993, respectively, from populations collected in Yugoslavia (Julien and Griffiths, 1998) and *C. crassicornis* collected from Hungary was released in field cages in 1994. None of these species have established in open releases on the prairies; however, larvae of all three species have overwintered successfully in cages at Lethbridge (P. Harris, Lethbridge, 2000, personal communication).

L. euphorbiana from Italian populations was first released in 1983 (Julien and Griffiths, 1998; Harris, 2000b). Most of the releases since then have taken place in Manitoba (Table 69.1). The moth is considered established in British Columbia, Manitoba, Saskatchewan and Ontario but not in Alberta or Nova Scotia (Harris, 2000b). Densities high enough to enable redistribution occur in British Columbia and Manitoba (S. Turner, Kamloops, and P. Harris, Lethbridge, 2000, personal communication).

M. murinata was first released in Manitoba in 1988 from German populations (Table 69.1) (Julien and Griffiths, 1998). It has survived in field cages in Alberta and Saskatchewan, but is not considered established in any western province.

S. capitigena and *S. esulae* from Italy (via USA) were released together in 1987 (Julien and Griffiths, 1998). *S. capitigena* is considered established in Alberta and Saskatchewan whereas *S. esulae* is established in Alberta, Saskatchewan, Manitoba, Nova Scotia and Ontario. No major impact on spurge populations has yet been recorded (Julien and Griffiths, 1998; Harris, 2000b).

P. euphorbiae and *P. curticornis* from Hungarian populations were released in

Table 69.2. *Aphthona* spp. release sites at Blood Indian Reserve, Alberta, 1997–1998.

	A. nigriscutis	*A. lacertosa*	Total
Release sites, 1997	92	33	125
Confirmed establishments, 1998	81 (88%)	33 (100%)	114
Sites with visible halos	18/20 (90%)	21/26 (81%)	39/46 (87%)
Mean halo size around release point (m^2)	2.29	0.86	
Beetles released, 1997	338,000	71,500	409,500

1988 (Julien and Griffiths, 1998). *P. euphorbiae* survived 4 years in cages at Millet but redistribution failed and currently there are no established field populations (McClay *et al.*, 1995). A *Pegomya* sp. from this initial population, identified as *P. curticornis* (P. Harris, Lethbridge, 2000, personal communication), established and overwintered at a field site at Regina. Given the results of the host screening trials in Europe, these individuals were likely *P. euphorbiae*. This population needs to be re-examined because, if confirmed as *P. curticornis*, this establishment suggests that it can sometimes attack North American *E. esula*. Regardless of the species, no control at the release sites up to 1992 occurred (P. Harris, Lethbridge, 2000, personal communication).

The status of insects released before 1980 (Harris, 1984) is updated here. *H. euphorbiae* by itself is not an extremely effective agent, which may be a function of its susceptibility to predation and disease (Harris and Soroka, 1982; Hansen, 1996). There have been a few releases in Saskatchewan and Manitoba since 1980 (Table 69.1) and, currently, the moth is considered established in Ontario, where larvae have been collected and overwintered in the laboratory and re-released in spring, and in southern Alberta, where temperatures are high enough for larval development (Harris, 2000b).

O. erythrocephala, first released in 1979, was released again in 1986 (20 adults) in Alberta but yielded no adults up to 1992 (McClay *et al.*, 1995). Releases were also made in Saskatchewan during 1990 from a cage colony (Julien and Griffiths, 1998). The beetle is established at a few North American sites, but persists only at low numbers (Rees *et al.*, 1986, in Gassmann and Schroeder, 1995). It is established in Alberta, but its population has not increased sufficiently to have an impact (Rees *et al.*, 1996c).

C. tenthrediniformis, originally released in 1971 from populations of *E. esula* (*s.l.*) collected in Austria and Greece (Julien and Griffiths, 1998), is not considered established in Canada (Harris, 2000b). Similarly, *C. empiformis*, first released in 1970, is not considered established, although one additional release was made in 1989 in Ontario (Table 69.1). *C. tenthrediniformis* is now believed to have too narrow a host range to attack the North American *E. esula* complex (Harris, 1984).

Evaluation of Biological Control

Biological control of *E. esula* has been successful in terms of agent establishment and because outbreaks of *Aphthona* spp. are providing control in some habitats (McClay *et al.*, 1995; Julien and Griffiths, 1998; Lym, 1998; Kirby *et al.*, 2000; R.S. Bourchier, unpublished). In Edmonton, where *A. nigriscutis* was released in dense stands in 1988 and 1989, *E. esula* cover was reduced to less than 1% and above-ground biomass was reduced to less than 1 g m^{-2} 5 years after release (McClay *et al.*, 1995).

The principal requirement is now to quantify the impact of the existing biological control agents and assess their behaviour. Most *A. lacertosa* releases in Alberta were made in 1997 and beetle outbreaks were already observed by 2000 (I.D. Jonsen and R.S. Bourchier, unpublished). Many of the predictions about habitat preferences and behaviour of the insects are based on observations at low densities in the country of origin. Of particular interest is what happens to outbreak beetle populations when local spurge populations collapse. Impact data have only recently been published for the USA (Kirby *et al.*, 2000) and are being collected for *Aphthona* spp. in Alberta (R.S. Bourchier, unpublished), Saskatchewan (G. Bowes, Saskatoon, 2000, personal communication) and Manitoba (P. McCaughey, Brandon, 2000, personal communication). Recent observations of an *A. lacertosa* outbreak suggest that it may be able to suppress spurge in a broader range of habitats than expected (I.D. Jonsen and R.S. Bourchier, unpublished). Given the US results, there will likely still be problems in controlling spurge in shrubby riparian areas and under full forest canopy, e.g. in Manitoba and some spurge infestations in British Columbia (D. Brooke, Kamloops, 2000, personal communication).

Additional European *Aphthona* spp. that may be more effective in shaded environments are available (Gassmann, 1996); however, a petition submitted in 1996 for release of *A. venustula* was returned for additional non-target host screening. This testing is critical to address general concerns that have been raised about non-target effects of biological control agents (Louda *et al.*, 1997; Pemberton, 2000; Strong and Pemberton, 2000). The host-range testing is complicated because it is difficult to obtain and cultivate the species of concern or suitable surrogates.

There is still considerable potential to evaluate agents that have already been released in North America and have remained at low density. There is a need to determine the reasons for the failure of their populations to increase; some biological control agents may simply require a long period at low density to adapt to new conditions. In addition, impact assessment should be linked to habitat and climate attributes to determine their role in the success or failure of control.

Recommendations

Further work should include:

1. Determining the status of *E. esula* control and the impact of established agents at previous release sites, especially in Manitoba and Saskatchewan, to identify sites where biological control is not working and why;
2. Temporal and spatial studies of the population dynamics of outbreaking *Aphthona* populations;
3. Assessment of the interactions between *Aphthona* spp. and other available control methods, e.g. herbicides, grazing;
4. Establishing nursery sites for *Aphthona* spp., particularly *A. lacertosa*, as sources for re-distribution;
5. Studies of DNA of original *A. lacertosa* populations and those released in 1997 to determine if outbreaking populations are genetically different from original populations;
6. Conducting non-target host screening for additional *Aphthona* spp. (*A. venustula*, *A. ovata*) in Europe;
7. Determining the environmental impact of *E. esula* outbreaks on native flora and fauna to enable risk assessments of further biological control releases for control of this invasive species;
8. Assessing reasons for failure of some biological control agents to establish, or for populations to increase, e.g. why *L. euphorbiana* and *O. erythrocephala* persist only at low densities.

Acknowledgements

Funds for the ongoing insect research programme on leafy spurge have been provided by the Southern Applied Research Association (Alberta), Blood Tribe Lands Department, and the Matching Investments Initiative of Agriculture and Agri-Food Canada.

References

Alley, H.P. and Messersmith, C.G. (1985) Chemical control of leafy spurge. In: Watson, A.K. (ed.) *Leafy Spurge. Monograph Series of the Weed Science Society of America* 3, 65–78.

Anonymous (1997) Biological Weed Control Agents – Leafy spurge. Alberta Agriculture, Food and Rural Development. http://www.agric.gov.ab.ca/sustain/biolog2.html#7 (January 2001)

Anonymous (2000a) Spurge, cypress. Publication 505. Ontario Weeds. Ontario Vegetation Management Association, Ontario Ministry of Agriculture and Food. http://www.ovma.on.ca/Weeds/spurge.htm (January 2001)

Anonymous (2000b) Summary of Biological Control Releases – Leafy Spurge. British Columbia Ministry of Forests. Forests Practices Branch. http://www.for.gov.bc.ca/hfp/pubs/interest/noxious/nox06.htm (January 2001)

Anonymous (2000c) Biological Control of Leafy Spurge. Saskatchewan Agriculture and Food. http://www.agr.gov.sk.ca/DOCS/crops/integrated_pest_management/weed_control/Biocon.asp?firstPick=Crops&secondpick=Integrated%20Pest%20Management&thirdpick=Weed%20Control (January 2001)

Bangsund, D.A., Leistritz, F.L. and Leitch, J.A. (1999) Assessing economic impacts of biological control of weeds: the case of leafy spurge in the northern Great Plains of the United States. *Journal of Environmental Management* 56, 35–43.

Bergmann, A. (1955) Die Grossschmetterlinge Mitteldeutschlands. *Urania-Verlag.* 5(1), 560pp.

Best, K.F., Bowes, G.G., Thomas, A.G. and Maw, M.G. (1980) The biology of Canadian weeds. 39. *Euphorbia esula* L. *Canadian Journal of Plant Science* 60, 651–663.

Crompton, C.W., Stahevitch, A.E. and Wojtas, W.A. (1990) Morphometric studies of the *Euphorbia esula* group (Euphorbiaceae) in North America. *Canadian Journal of Botany* 68, 1978–1988.

Evans, J.O., Torell, J.M., Valcarce, R.V. and Smith, G.G. (1991) Analytical pyrolysis-pattern recognition for the characterisation of leafy spurge (*Euphorbia esula* L.) biotypes. *Annals of Applied Biology* 119, 47–58.

Fornasari, L. (1996) Biology and ethology of *Aphthona* spp. (Coleoptera: Chrysomelidae, Alticinae) associated with *Euphorbia* spp. (Euphorbiaceae). *Chrysomelidae Biology* 3, 293–313.

Fornasari, L. and Pecora, P. (1995) Host specificity of *Aphthona abdominalis* Duftschmid (Coleoptera: Chrysomelidae), a biological control agent for *Euphorbia esula* L. (leafy spurge, Euphorbiaceae) in North America. *Biological Control* 5, 353–360.

Gagné, R.J. (1990) Gall midge complex (Diptera: Cecidomyiidae) in bud galls of Palearctic *Euphorbia* (Euphorbiaceae). *Annals of the Entomological Society of America* 83, 335–345.

Gassmann, A. (1987) *Investigations on the* Pegomya argyrocephala *Complex of Species (Diptera: Anthomyiidae) to Select Candidate Biological Control Agents for Leafy and Cypress Spurge.* Final Report, CABI-European Station, Delémont, Switzerland.

Gassmann, A. (1994) Chamaesphecia crassicornis *Bartel 1912 (Lepidoptera: Sesiidae), a Suitable Agent for the Biological Control of Leafy Spurge (*Euphorbia esula *L.) (Euphorbiaceae) in North America.* Final Report, CABI-European Station, Delémont, Switzerland.

Gassmann, A. (1996) Life history and host specificity of *Aphthona venustula* Kutsch. (Col., Chrysomelidae), a candidate for the biological control of leafy spurge (*Euphorbia esula* L.) in North America. *Journal of Applied Entomology* 120, 405–411.

Gassmann, A. and Schroeder, D. (1995) The search for effective biological control agents in Europe: history and lessons from leafy spurge (*Euphorbia esula* L.) and cypress spurge (*Euphorbia cyparissias* L.). *Biological Control* 5, 466–477.

Gassmann, A. and Shorthouse, J.D. (1990) Structural damage and gall induction by *Pegomya curticornis* and *Pegomya euphorbiae* (Diptera: Anthomyiidae) within the stems of leafy spurge (*Euphorbia* × *pseudovirgata*) (Euphorbiaceae). *The Canadian Entomologist* 122, 429–439.

Gassmann, A. and Tosevski, I. (1993) *Investigations on Additional Biocontrol Agents of Leafy Spurge (*Euphorbia esula *s.l.).* Annual Report, CABI-European Station, Delémont, Switzerland.

Gassmann, A. and Tosevski, I. (1994) Biology and host specificity of *Chamaesphecia hungarica* and *Ch. astatiformis* (Lep.: Sesiidae), two candidates for the biological control of leafy spurge, *Euphorbia esula* (Euphorbiaceae) in North America. *Entomophaga* 39, 237–245.

Gassmann, A., Schroeder, D., Maw, E. and Sommer, G. (1996) Biology, ecology, and host specificity of European *Aphthona* spp. (Coleoptera, Chrysomelidae) used as biocontrol agents for leafy spurge, *Euphorbia esula* (Euphorbiaceae), in North America. *Biological Control* 6, 105–113.

Geltman, D.V. (1998) Taxonomic notes on *Euphorbia esula* (Euphorbiaceae) with special reference to its occurrence in the east part of the Baltic region. *Annales Botanici Fennici* 35, 113–117.

Haber, E. (1997) *Invasive Exotic Plants of Canada.* Fact Sheet No. 9, *Leafy Spurge.* National Botanical Services, Ottawa, Ontario. April 1997. http://infoweb.magi.com/~ehaber/factsprg.html (January 2001)

Hansen, R. (1996) *Hyles euphorbiae* (Lepidoptera: Sphingidae). Leafy spurge hawk moth. http://www.nysaes.cornell.edu/ent/biocontrol/weedfeeders/hyles.html (January 2001)

Hansen, R.W., Richard, R.D., Parker, P.E. and Wendel, L.E. (1997) Distribution of biological control agents of leafy spurge (*Euphorbia esula* L.) in the United States: 1988–1996. *Biological Control* 10, 129–142.

Harris, P. (1984) *Euphorbia esula–virgata* complex, leafy spurge and *E. cyparissias* L., cypress spurge (Euphorbiaceae). In: Kelleher, J.S. and Hulme, M.A. (eds) *Biological Control Programmes*

Against Insects and Weeds in Canada 1969–1980. Commonwealth Agriculture Bureaux, Slough, UK, pp. 159–169.

Harris, P. (1985) Minoa murinata *(Scop.), (Lepidoptera: Geometridae) a Candidate for the Biocontrol of Leafy Spurge (*Euphorbia esula-virgata *complex) and Cypress Spurge in Canada.* Information report, Agriculture and Agri-Food Canada Research Station, Regina, Saskatchewan.

Harris, P. (2000a) Leafy and cypress spurge, *Euphorbia esula* L. and *E. cyparissias* L. Lethbridge Research Centre. Biology of Target Weeds. http://res2.agr.ca/lethbridge/weedbio/hosts/blfysprg.htm (January 2001)

Harris, P. (2000b) Biocontrol agents. Lethbridge Research Centre. Agents Tried in Biocontrol. http://res2.agr.ca/lethbridge/weedbio/agents/.htm (January 2001)

Harris, P. (2000c) Lethbridge Research Centre. Classical Biocontrol of Weeds. *Aphthona flava.* http://res2.agr.ca/lethbridge/weedbio/hosts/slfysprg.htm (January 2001)

Harris, P. and Soroka, J. (1982) Lobesia (Lobesoides) euphorbiana *(Frr.) (Lepidoptera: Oleuthreutinae): a Candidate for the Biological Control of Leafy Spurge in North America.* Information Report, Agriculture and Agri-food Canada. Research Station, Regina, Saskatchewan.

Harris, P., Dunn, P.H., Schroeder, D. and Vonmoos, R. (1985) Biological control of leafy spurge in North America. In: Watson, A.K. (ed.) *Leafy Spurge.* Monograph Series of the Weed Science Society of America, No. 3, pp. 79–92.

Julien, M.H. and Griffiths, M.W. (eds) (1998) *Biological Control of Weeds. A World Catalogue of Agents and their Target Weeds*, 4th edn. CAB International, Wallingford, UK.

Kirby, D.R., Carlson, R.B., Krabbenhoft, K.D., Mundal, D. and Kirby, M.M. (2000) Biological control of leafy spurge with introduced flea beetles (*Aphthona* spp.). *Journal of Range Management* 53, 305–308.

Lastuvka, Z. (1982) A contribution to morphology and biology of the clear-wing moths *Chamaesphecia tenthrediniformis* (Den. et Schiff.) s.l. and *Chamaesphecia hungarica* (Tom.) (Lepidoptera, Sesiidae). *Acta Universitatis Agriculturae* 4, 69–83.

LeSage, L. and Paquin, P. (1996) Identification keys for *Aphthona* flea beetles (Coleoptera: Chrysomelidae) introduced in Canada for the control of spurge (*Euphorbia* spp., Euphorbiaceae). *The Canadian Entomologist* 128, 593–603.

Leistritz, F.L., Thompson, F. and Leitch, J.A. (1992) Economic impact of leafy spurge (*Euphorbia esula*) in North Dakota. *Weed Science* 40, 275–280.

Louda, S.M., Kendall, D., Connor, J. and Simberloff, D. (1997) Ecological effects of an insect introduced for the biological control of weeds. *Science* 277, 1088–1090.

Lym, R.G. (1998) The biology and integrated management of leafy spurge (*Euphorbia esula*) on North Dakota rangeland. *Weed Technology* 12, 367–373.

Lym, R.G., Nissen, S.J., Rowe, M.L., Lee, D.J. and Masters, R.A. (1996) Leafy spurge (*Euphorbia esula*) genotype affects gall midge (*Spurgia esulae*) establishment. *Weed Science* 44, 629–633.

Mann, K., Sobhian, R., Littlefield, J. and Cristofaro, M. (1996) *Petition for the Introduction and Release of the Gall Midge* Spurgia capitigena *(Bremi) (Diptera: Cecidomyiidae) into the United States for the Biological Control of Leafy Spurge.* Information Report, United States Department of Agriculture, Agriculture Research Service.

Maw, E. (1981) Biology of some *Aphthona* spp. (Col.: Chrysomelidae) feeding on *Euphorbia* spp. (Euphorbiaceae), with special reference to leafy spurge (*Euphorbia* sp. near *esula*). MSc thesis, University of Alberta, Edmonton, Alberta.

McClay, A.S. (1996) Biological control in a cold climate: temperature responses and climatic adaptation of weed biocontrol agents. In: Moran, V.C. and Hoffmann, J.H. (eds) *Proceedings of the IX International Symposium on Biological Control of Weeds.* University of Cape Town, Stellenbosch, South Africa, pp. 377–383.

McClay, A.S., Cole, D.E., Harris, P. and Richardson C.J. (1995) *Biological Control of Leafy Spurge in Alberta: Progress and Prospects.* Alberta Environmental Centre, Vegreville, Alberta.

Michelsen, V. (1988) Taxonomy of the species of *Pegomya* (Diptera: Anthomyiidae) developing in the shoots of spurges (*Euphorbia* spp). *Entomologica Scandinavica* 18, 425–435.

Nelson, J.A. and Carlson, R.B. (1999) Observations on the biology of *Spurgia capitigena* Bremi on leafy spurge in North Dakota. *Biological Control* 16, 128–132.

Pecora, P., Pemberton, R.W., Stazi, M. and Johnson, G.R. (1991) Host specificity of *Spurgia esulae* Gagné (Diptera: Cecidomyiidae), a gall midge introduced into the United States for control of leafy spurge (*Euphorbia esula* L. 'complex'). *Environmental Entomology* 20, 282–287.

Pemberton, R.W. (2000) Predictable risk to native plants in weed biological control. *Oecologia* 125, 489–494.
Rees, N.E., Pemberton, R.W., Rizza, A. and Pecora, P. (1986) First recovery of *Oberea erythrocephala* on the leafy spurge complex in the United States. *Weed Science* 34, 395–397.
Rees, N.E., Spencer, N.R., Knutson, L.V., Fornasari, L., Quimby, P.C. Jr, Pemberton, R.W. and Nowierski, R.M. (1996a) *Aphthona cyparissias*. In: Rees, N.E., Quimby P.C. Jr, Piper, G.L, Coombs, E.M., Turner, C.E., Spencer, N.R. and Knutson, L.V. (eds) *Biological Control of Weeds in the West*. Western Society of Weed Science Publishers, Bozeman, Montana.
Rees, N.E., Spencer, N.R., Knutson, L.V., Fornasari, L., Quimby, P.C. Jr, Pemberton, R.W. and Nowierski, R.M. (1996b) *Chamaesphecia hungarica*. In: Rees, N.E., Quimby P.C. Jr, Piper, G.L, Coombs, E.M., Turner, C.E., Spencer, N.R. and Knutson, L.V. (eds) *Biological Control of Weeds in the West*. Western Society of Weed Science Publishers, Bozeman, Montana.
Rees, N.E., Spencer, N.R., Knutson, L.V., Fornasari, L., Quimby, P.C. Jr, Pemberton, R.W. and Nowierski R.M. (1996c) *Oberea erythrocephala*. In: Rees, N.E., Quimby P.C. Jr, Piper, G.L, Coombs, E.M., Turner, C.E., Spencer, N.R. and Knutson, L.V. (eds) *Biological Control of Weeds in the West*. Western Society of Weed Science Publishers, Bozeman, Montana.
Rowe, M.L., Lee, D.J., Nissen, S.J., Bowditch, B.M. and Masters, R.A. (1997) Genetic variation in North American leafy spurge (*Euphorbia esula*) determined by DNA markers. *Weed Science* 45, 446–454.
Sell, R.S., Bangsund, D.A. and Leistritz, F.L. (1999) *Euphorbia esula*: perceptions by ranchers and land managers. *Weed Science* 47, 740–749.
Selleck, G.W., Coupland, R.T. and Frankton, C. (1962) Leafy spurge in Saskatchewan. *Ecological Monographs* 32, 1–29.
Stelljes, K.B. (1997) Project to Target Leafy Spurge. United States Department of Agriculture, Agricultural Research Service. http://alembic.nal.usda.gov/is/pr/1997/970903.spurge.htm (January 2001)
Strong, D.R. and Pemberton, R.W. (2000) Biological control of invading species: risk and reform. *Science* 288, 1969–1971.
Tosevski, I., Gassmann, A. and Schroeder, D. (1996) Description of European *Chamaesphecia* spp. (Lepidoptera: Sesiidae) feeding on *Euphorbia* (Euphorbiaceae), and their potential for biological control of leafy spurge (*Euphorbia esula*) in North America. *Bulletin of Entomological Research* 86, 703–714.

70 *Galium spurium* L., False Cleavers (*Rubiaceae*)

A.S. McClay, R. Sobhian and W. Zhang

Pest Status

False cleavers, *Galium spurium* L., an annual plant native to Europe, is a widespread, introduced species in Canada. It occurs primarily in the prairie provinces and locally in British Columbia, Ontario and Quebec. In much of the literature, false cleavers is not distinguished from cleavers, *Galium aparine* L. However, the most abundant and troublesome annual *Galium* sp. in arable land on the prairies is *G.*

spurium (Malik and Vanden Born, 1987a, 1988).

G. spurium is a major weed of canola, *Brassica napus* L. and *B. rapa* L., and other crops. During the 1990s, in each of the prairie provinces, it increased its abundance more rapidly than any other cropland weed. In Alberta, for example, it occurred in less than 1% of cereal and oilseed fields surveyed in 1973–1977, and 18% of fields surveyed in 1997 (Thomas *et al.*, 1998a, b, c). Heavy infestations cause yield losses by competing with crops; a population of 100 plants m^{-2} reduced canola yield by 18% (Malik and Vanden Born, 1987b). Its seed cannot be separated easily from canola seed, leading to crop contamination. In 1994 the average level of cleavers contamination in export canola cargoes was 14.64 seeds per 25 g (D.R. DeClercq, Winnipeg, 1995, personal communication), equivalent to 0.16% *G. spurium* contamination by weight across the prairies. Contamination of 1% or more leads to downgrading and consequent price reductions. Contamination of crop seed also results in new infestations. Under the Canada Seeds Act, *G. spurium* is a primary noxious weed seed and there is zero tolerance for its seed in all grades of pedigreed seed of cereals, forage crops, and oilseeds (Malik and Vanden Born, 1987a). The clinging stems can tangle up equipment, causing delays and difficulty in harvesting (Stromme, 1995).

G. spurium is a slender, branched plant with whorled leaves and straggling or climbing stems up to 200 cm long. All parts of the plant, including the fruits, are 'sticky' due to a covering of short, hooked spines or bristles (Malik and Vanden Born, 1988). In Alberta, seed sown in May produced plants that flowered from early July to late August and developed fruits from mid-July to early September. Seedlings that emerged in August and September did not flower in the first season, but were able to overwinter and resume growth the following spring. Potted plants produced up to 3500 seeds per plant (Malik and Vanden Born, 1987a).

Background

Prior to the introduction of herbicide-tolerant canola, no effective herbicides were available to control *G. spurium* in canola. Multiple herbicide resistance has now been detected in a population in Alberta. This biotype is cross-resistant to quinclorac and ALS (acetolactate synthase)-inhibiting herbicides, including imazethapyr, one of the products for which herbicide-tolerant canola has been developed. With increasing use of these varieties, it can be predicted that ALS-resistant *G. spurium* will continue to be selected for, and that herbicide resistance will become more common in this species (Hall *et al.*, 1998). Classical biological control was therefore pursued.

Batra (1984) surveyed the phytophagous insects feeding on *Galium* spp. in Europe and identified two possible biological control agents for use against *G. aparine* or *G. spurium*: the stem-galling midge, *Geocrypta galii* (H. Loew), and the leaf-rolling mite, *Cecidophyes galii* (Karpelles).

Biological Control Agents

Mites

In 1994, a gall mite was discovered causing heavy damage to a population of *G. aparine* at Carnon, southern France. It was originally identified as *C. galii*, a species associated with several European *Galium* spp. (Karpelles, 1884; Nalepa, 1889, 1893), but was later described as a new species, *Cecidophyes rouhollahi* Craemer, on the basis of host preference and slight but consistent morphological differences (Craemer *et al.*, 1999). Infested leaves roll up around the midvein; heavily attacked plants become brown and stunted and their seed production is severely reduced. Seedlings with as few as four leaves were infested in the field and showed typical leaf rolling, but the cotyledons were not affected. The mite was found in the field near Montpellier as early as February, causing deformation of *G. aparine* plants. It multiplies actively on this host throughout spring and early summer.

Host specificity testing of *C. rouhollahi* in Europe showed that it would accept *G. spurium* readily as a host, and would attack only a few closely related annual European *Galium* spp. in the subgenus *Kolgyda* (R. Sobhian, unpublished). All of these occur in North America as introduced weeds. None of the perennial native North American *Galium* spp. and no plants outside the genus *Galium* were attacked. A petition for release of *C. rouhollahi* in Canada is in preparation.

In greenhouse experiments in France, *C. rouhollahi* caused severe damage to *G. spurium*. Inoculated plants suffered 40% mortality after 78 days, surviving plants produced no seed, and their biomass was reduced by 60% compared to uninoculated controls (R. Sobhian, unpublished). Field-collected mites survived 3 days in a freezer at −19.5°C, suggesting that the mite has good cold tolerance. *C. rouhollahi* has good potential as a biological control agent; its effectiveness in the field will depend on its ability to survive under the climatic conditions and cropping practices on the prairies.

Pathogens

Fungi

In Canada, indigenous fungi are being evaluated to control *G. spurium* (W. Zhang, unpublished). In 1998 and 1999, diseased leaves, stems, flowers and seeds were collected from crop fields in Alberta (near Peace River, Edmonton, Lamont, Vegreville and Vermilion) and Saskatchewan (Saskatoon). A total of 163 fungal isolates were obtained, 74 of which were shown to be pathogenic to *G. spurium* by Koch's postulates. Pathogenic isolates were further assessed for weed control efficacy (virulence) using a 0–3 scale (0, no symptoms; 1, light infection; 2, moderate infection; and 3, severe infection to death). Forty-seven isolates showed a virulence rating of 2 or 3 to *G. spurium*. Virulent isolates proceeded to crop safety tests (preliminary host range) on nine major crops (wheat, *Triticum aestivum* L., barley, *Hordeum vulgare* L., oats, *Avena sativa* L., canola, flax, *Linum usitatissimum* L., safflower, *Carthamus tinctorius* L., field pea, *Pisum sativum* L., lentil, *Lens culinaris* Medikus, and lucerne, *Medicago sativa* L.). To date, one very promising isolate (CL98–103) has been identified.

Preliminary laboratory and greenhouse studies demonstrated that CL98–103 can kill *G. spurium* with a 12–16 h dew period and is non-pathogenic to canola and eight other major crops. Further host specificity tests on 41 plant species or cultivars demonstrated that CL98–103 is sufficiently safe to use in western Canada. Large quantities of spores were easily produced in a liquid medium within 48–72 h, suggesting that CL98–103 has potential as a bioherbicide to control *G. spurium*. Its field effectiveness will depend on development of formulations to overcome its dew requirement and other environmental limitations.

Recommendations

Further work should include:

1. Release of *C. rouhollahi* in Canada;
2. Post-release monitoring of *C. rouhollahi* to estimate its development rate, population increase, dispersal and overwinter survival under various cultural conditions;
3. Evaluation of the impact of the mite on growth and reproduction of *G. spurium* in the field when applied at different growth stages of the weed;
4. Development of formulations and application methods for isolate CL98–103.

Acknowledgements

We are grateful to the Canola Council of Canada, the Alberta Agricultural Research Institute, the Canadian Seed Growers' Association, and the Saskatchewan Agriculture Development Fund for financial support.

References

Batra, S.W.T. (1984) Phytophages and pollinators of *Galium* (Rubiaceae) in Eurasia and North America. *Environmental Entomology* 13, 1113–1124.
Craemer, C., Sobhian, R., McClay, A.S. and Amrine, J.W. (1999) A new species of *Cecidophyes* (Acari: Eriophyidae) from *Galium aparine* (Rubiaceae) with notes on its biology and potential as a biological control agent for *Galium spurium*. *International Journal of Acarology* 25, 255–263.
Hall, L.M., Stromme, K.M., Horsman, G.P. and Devine, M.D. (1998) Resistance to acetolactate synthase inhibitors and quinclorac in a biotype of false cleavers (*Galium spurium*). *Weed Science* 46, 390–396.
Karpelles, L. (1884) Über Gallmilben (*Phytoptus* Duj.). *Sitzungsberichte der kaiserlichen Akademie der Wissenschaften. Mathematisch-naturwissenschaftliche Classe. Abtheilung 1 (Vienna)* 90, 46–55, f. 41–11.
Malik, N. and Vanden Born, W.H. (1987a) Growth and development of false cleavers (*Galium spurium* L.). *Weed Science* 35, 490–495.
Malik, N. and Vanden Born, W.H. (1987b) False cleavers (*Galium spurium* L.) competition and control in rapeseed. *Canadian Journal of Plant Science* 67, 839–844.
Malik, N. and Vanden Born, W.H. (1988) The biology of Canadian weeds. 86. *Galium aparine* L. and *Galium spurium* L. *Canadian Journal of Plant Science* 68, 481–499.
Nalepa, A. (1889) Beiträge zur Systematik der Phytopten. *Sitzungsberichte der kaiserlichen Akademie der Wissenschaften. Mathematisch-naturwissenschaftliche Classe. Abtheilung 1 (Vienna)* 98, 112–156.
Nalepa, A. (1893) Katalog der bisher beschriebenen Gallmilben, ihrer Gallen und Nährpflanzen, nebst Angabe der einschlägigen Literatur und kritischen Zusätzen. *Zoologische Jahrbücher. Abtheilung für Systematik, Geographie und Biologie der Thiere (Jena)* 7, 274–328.
Stromme, K. (1995) *Biology and Control of False Cleavers.* Agronomy Unit, Alberta Agriculture, Food and Rural Development, Edmonton, Alberta.
Thomas, A.G., Frick, B. and Hall, L. (1998a) *Weed Population Shifts in Alberta.* Agriculture and Agri-Food Canada, Saskatoon, Saskatchewan.
Thomas, A.G., Frick, B., van Acker, R. and Joosse, D. (1998b) *Weed Population Shifts in Manitoba.* Agriculture and Agri-Food Canada, Saskatoon, Saskatchewan.
Thomas, A.G., Frick, B., Wise, R.F. and Juras, L.T. (1998c) *Weed Population Shifts in Saskatchewan.* Agriculture and Agri-Food Canada, Saskatoon, Saskatchewan.

71 *Hypericum perforatum* L., St John's Wort (*Clusiaceae*)

K.I.N. Jensen, P. Harris and M.G. Sampson

Pest Status

St John's wort, *Hypericum perforatum* L., is a cosmopolitan weed native to Eurasia that is common in all provinces, except those in the prairies (Crompton *et al.*, 1988). However, in Manitoba it has recently invaded the tall grass prairie region where it is displacing native species. *H. perforatum* can exceed 1 m in height, is deep-rooted and is particularly adapted to regions with hot, dry summers, where it occurs in both open and semi-open habitats. It is commonly found along roadsides, waste areas, disturbed or burned sites, and it is a weed of rangelands, pastures and perennial fruit crops, e.g. strawberries, *Fragaria* × *ananassa* Duchesne, and lowbush blueberries, *Vaccinium angustifolium* Aiton, in eastern Canada. In Quebec and Manitoba, *H. perforatum* is listed as a noxious weed, but it is not listed in the Weed Seeds Order nor is there restriction on its importation and sale in Canada. In British Columbia and Ontario, biological control programmes have successfully reduced its importance (Harris and Maw, 1984). In the Atlantic provinces, infestations are generally small and scattered, due perhaps to cooler, moister conditions and competition from native species, but it may occur as an important weed locally.

Campbell and Delfosse (1984) and Crompton *et al.* (1988) reviewed the biology of *H. perforatum*. It is a highly variable, short-lived perennial that propagates by seed and short rhizomes and overwinters as a procumbent, basal rosette. Black glands on flowers, leaves and stems contain the naphthodianthrone hypericin, a photodynamic, reddish pigment that can induce Type I photosensitization in non-pigmented skin of livestock on exposure to bright sunlight. Symptoms range from blistering and loss of performance to tissue necrosis and, in severe cases, death (Giese, 1980). Photosensitization has been associated with a narrow-leaved subspecies, *H. perforatum* var. *angustifolium* De Candolle, from southern Europe that contains high levels of hypericin, and not with the northern, round-leaved forms (Southwell and Campbell, 1991). In Canada, *H. perforatum* has not been classified into subspecies, but populations differ widely in their foliar characteristics and hypericin content. Hypericin levels of Nova Scotia selections of the weed were about one-half and one-third of those in selections from western North America and Australia, respectively (Jensen *et al.*, 1995). Therefore, the status of *H. perforatum* as a phototoxic weed in Atlantic Canada is questionable.

Mitich (1994) reviewed the role of *H. perforatum* in folklore and folk medicine. Its pre-1800 introduction and widespread distribution in North America are partly due to its use as a garden and medicinal plant. Several biomedically active naphthodianthrones, flavinoids, and phloroglucanols have been extracted from *H. perforatum* (Nahrstedt and Butterweck, 1997). Interest in its pharmacological properties has accelerated since the late 1980s, particularly as an antidepressant, and sales of *H. perforatum* products in Canada exceeded Can$2 million in 1998 (Englemeyer and Brandle, 1999). Some harvesting of *H. perforatum* from 'wild' stands occurs, and recommendations for its commercial production are being developed. This must now be weighed in future biological control programmes against this weed.

Background

H. perforatum was first recognized as a serious weed in the 1940s in the southern interior of British Columbia. Chemical control in rangelands proved expensive and ineffective due to the weed's tolerance to many herbicides and its ability to rapidly re-infest treated sites (Crompton *et al.*, 1988). Hence, Canada's first biological weed control programme was initiated against *H. perforatum* in British Columbia in 1951, modelled on successful programmes undertaken in Australia in the 1920s and 1930s and in California in the 1940s (see references in Delfosse and Cullen, 1984). The evolution of the Canadian programme is well documented (Harris *et al.*, 1969; Harris and Maw, 1984). In its native range, 37 insects are known to feed on *H. perforatum*. Some of these have specialized feeding behaviour or physiological or physical mechanisms for avoiding the phototoxic effects of hypericin (Fields *et al.*, 1989, 1991). Seven of ten species released worldwide as biological control agents against the weed (Julien, 1992) have also been released in Canada: *Agrilus hyperici* (Creutzer), *Aplocera plagiata* L., *Aphis chloris* (Koch), *Chrysolina hyperici* (Förster), *Chrysolina quadrigemena* (Suffrian), *Chrysolina varians* (Shaller), and *Zeuxidiplosis giardi* (Kieffer) (Harris and Peschken, 1971; Harris and Maw, 1984). Of these, *C. varians* and the gall-forming midge, *Z. giardi*, did not survive in British Columbia (Harris and Peschken, 1971). There is no evidence that established insects have attacked any native *Hypericum* sp.

Successful biological control of *H. perforatum* in Canada and elsewhere has largely been dependent on the performance of *C. quadrigemina* and *C. hyperici*, each having distinct climatic limitations that affect their relative effectiveness (Harris, 1962; Williams, 1985). *C. quadrigemina*, originally from southern France, is best adapted to, and dominates on, warmer, drier, open sites having late fall frosts (Harris, 1962). In British Columbia, it has been successful on open and semi-open sites below 1000 m elevation dominated by Ponderosa pine, *Pinus ponderosa* D. Douglas ex Lawson and Lawson, and having a humidity index of 24–30 (Harris *et al.*, 1969), and there the weed has been reduced to less than 2% of its pre-release levels (Harris and Maw, 1984). *C. quadrigemina* is also well established in southern Ontario (Alex, 1981; Fields *et al.*, 1988) and its success there has resulted in *H. perforatum* being removed from the Noxious Weed List. The beetle has not survived in the Maritimes (Harris and Maw, 1984) and it performs poorly in moister regions of British Columbia, e.g. the East Kootenays (Williams, 1985). In contrast, *C. hyperici*, which initially originated from England, performs best in moister, cooler montane and maritime regions and it is well established in the Atlantic provinces (Sampson, 1987; Sampson and MacSween, 1992; Maund *et al.*, 1993; Morrison *et al.*, 1998) and areas of British Columbia with a humidity index of 30–40 and on sites dominated by Douglas fir, *Pseudotsuga menziesii* (Mirabel) Franco (Harris *et al.*, 1969). Five to 13 years were required for these insects to adapt their behaviour and life cycle to overwinter in sufficient numbers to provide control. Control of *H. perforatum* by *Chrysolina* spp. is augmented by additional stresses placed on the plant, including drought stress (Williams, 1985), competition from other plant species (Cullen *et al.*, 1997) and disease (Morrison *et al.*, 1998). There is still a need for supplementary biological control agents, particularly those that are effective in moister habitats (Williams, 1985). Three other insects have been established in Canada that are of minor or secondary importance.

Biological Control Agents

Insects

The current status of insects initially studied, released and reported by Harris and colleagues prior to 1980 is summarized here. No new species have been introduced since.

Early attempts to establish the root-boring beetle *A. hyperici* in British Columbia from California were unsuccessful (Harris and Peschken, 1971; Harris and Maw, 1984). More recently in the USA, it has adapted and expanded its range northward. In northern Idaho, the numbers per plant remain low but at two of four study sites more than 50% of dead plants showed signs of feeding or had exit holes (Campbell and McCaffey, 1991). *A. hyperici* imported from Idaho in the late-1980s has survived in British Columbia, but so far populations remain low and cause negligible damage (Harris, 1999).

A. chloris from Germany, released and established in 1979 near Cranbrook (Harris and Maw, 1984), was redistributed and established widely in British Columbia in the 1990s (Table 71.1). The aphid did not establish in New Brunswick, possibly due to destruction of the site, nor in Manitoba. The agent is well established in Nova Scotia. It is best adapted to humid, cooler montane and maritime regions; it appears that predation restricts its effectiveness in warmer regions (Briese and Judd, 1995). Nymphs and adults feed on stems and leaves, and high densities can desiccate or

Table 71.1. Summary of the number of insect releases against *Hypericum perforatum* from 1980 to 1997, recorded in *Insect Liberations in Canada Bulletins*.

	Number of releases (province[a])	
Species	1981–1989	1990–1997
Agrilus hyperici (Creutzer)	3 (BC)	None
Aphis chloris (Koch)	None	14(BC), 1(MB), 1(NB), 3(NS)
Aplocera plagiata L.	5(NS), 1(SK)	None
Chrysolina hyperici (Förster)	1(ON)	1(MB), 2(NB), 1(NS)
Chrysolina quadrigemina (Suffrian)	3(BC), 3(ON), 1(NB)	None

[a](BC) British Columbia, (SK) Saskatchewan, (MB) Manitoba, (ON) Ontario, (NB) New Brunswick, (NS) Nova Scotia.

kill plants. In British Columbia, *H. perforatum* was controlled within 200 m of one release site and had spread 10 km (Harris, 1999). In Nova Scotia, the aphid has spread 60 km in 8 years from releases on the mainland and Cape Breton Island. At two sites, *H. perforatum* density was reduced by more than 90% and mortality was observed when aphids fed on roots (Sampson and MacSween, 1992).

Harris (1967) discussed the biology of *A. plagiata*, and Harris and Maw (1984) and Harris and Peschken (1971) summarized results of early releases. This defoliator has established over a 300 km^2 area of south-central British Columbia from releases made in the late 1970s. It disperses readily but populations remain low and do minimal damage to *H. perforatum* (Harris, 1999). Overwintering larvae are susceptible to fungal diseases, which may account for poor establishment on moister sites (Harris, 1967). Later releases in New Brunswick (Maund *et al.*, 1993) and Nova Scotia (Sampson, 1987) have not established.

Harris and Peschken (1971) discussed the biology of *C. hyperici*. This defoliator is widely established in New Brunswick, Nova Scotia and Ontario, but in Ontario *Chrysolina* populations are dominated by *C. quadrigemina* (Alex, 1981; Fields *et al.*, 1988). It is also the most common species in the cooler, moister regions of British Columbia (Williams, 1985). After release in eastern Ontario in 1969, *Chrysolina* spp. have been dispersing about 5 km year^{-1} (Fields *et al.*, 1988) and were found in Quebec along the Ottawa River in 1993 (LeSage, 1996). *C. hyperici* was first observed in Cape Breton in 1985, suggesting that dispersal from release sites in 1969 in Nova Scotia may approach 10 km year^{-1}. *C. hyperici* will likely disperse throughout the range of *H. perforatum* in eastern Canada. It has not yet been released or reported in Newfoundland. In addition to natural dispersal, considerable attempts have been made to redistribute beetles in Ontario (Alex, 1981), New Brunswick (Maund *et al.*, 1993) and Nova Scotia (Sampson, 1987), and *C. hyperici* was introduced into Prince Edward Island near Montague in the early 1990s (Sampson and MacSween, 1992). The long-term effect of *C. hyperici* herbivory on *H. perforatum* in Atlantic Canada is minimal. Although stands or individual plants do occur with extensive defoliation and high numbers of adults, adult densities typically range from less than 1–5 per plant (Sampson, 1987; Sampson and MacSween, 1992; Morrison *et al.*, 1998). Larvae and adults feed for 2–2½ months of the weed's 6–7-month growing season, and healthy plants fully recover after adults aestivate in early August. Mortality during aestivation must be high in Atlantic Canada because few adults are observed in autumn. Control of *H. perforatum* by *C. hyperici* has also been unsatisfactory in moister regions of British Columbia (Williams, 1985) presumably because plants recover in the absence of drought stress.

Harris and Peschken (1971) discussed the biology of *C. quadrigemina* in Canada. By the 1970s it had reduced *H. perforatum* to less than 2% of the pre-release levels in most arid regions in south-central British Columbia and it was effective in reducing the weed to negligible levels at release sites in southern Ontario (Harris and Maw, 1984). Populations of *C. quadrigemina* have also dispersed to the lower Fraser Valley. This region, although moist, tends to have dry summers that would allow the beetle to complete its obligatory summer aestivation. Similarly, the recent presence of *C. quadrigemina* in southern coastal regions of British Columbia does not indicate an adaptation to moister conditions as this region also has dry summers, similar to Italy, which is within its native range. Alex (1981) reported successful efforts to redistribute *C. quadrigemina* within south-western Ontario and Fields *et al.* (1988) reported on its natural dispersal in eastern Ontario. In the 1980s, beetles from the Fraser Valley were redistributed to New Brunswick but these failed to establish, likely due to excessively wet summers.

Pathogens

Fungi

Colletotrichum gloeosporioides (Penzig) Penzig & Saccardo f. sp. *hypericum* is an endemic fungus causing anthracnose on *H. perforatum*, first observed controlling the weed in lowbush blueberry fields in Nova Scotia. It occurs widely in Nova Scotia (Crompton *et al.*, 1988; Hildebrand and Jensen, 1991) and also in New Brunswick and Prince Edward Island. The fungus effectively controls all growth stages of *H. perforatum* when applied as a foliar spray consisting of an aqueous suspension of conidia (Hildebrand and Jensen, 1991; Jensen and Doohan, 1994). Regrowth is controlled by secondary disease cycles. Templeton (1992) correctly categorized *C. gloeosporioides* f. sp. *hypericum* as an 'orphan' mycoherbicide, that is, an unlikely candidate for commercial development despite its effectiveness.

In its native range, *C. gloeosporioides* f. sp. *hypericum* provides significant control of *H. perforatum*, often making other control measures unnecessary. It has potential as a 'classical' agent. In non-arable habitats in Nova Scotia, e.g. pastures and riverbanks, mortality of mature plants ranged from 36 to 96% during the growing season, and 50% of surviving infected plants did not survive the winter. Seedling mortality approached 100% and no infected seedlings survived the winter. The fungus overwinters within infected plants, seed and old plant material. In Nova Scotia, stem lesions are first observed in early May and cycles of secondary infection occur thereafter. Infected plants become reddish-yellow and are easy to identify (Morrison *et al.*, 1998). Stem lesions become sunken, with dark-brown centres and red–purple margins that expand or coalesce, girdling the stem and withering the distal portions. Crown infection kills the basal rosette and the mature plant. Under moist conditions, setose acervuli produce masses of conidia in a gelatinous matrix that are disseminated by rain-splash or other physical means. The sexual stage has not been observed on field-collected material or on artificial media (Hildebrand and Jensen, 1991).

Both larvae and adults of *C. hyperici* have been observed to feed in lesions on infected plants, and further infection may be enhanced by feeding injury. Field-collected adults were shown to be contaminated with fungal conidia, and healthy plants became infected when fed on (Morrison *et al.*, 1998). In several field studies (Jensen and Doohan, 1994), the rapid, random spread of disease appeared to be associated with immigration of *C. hyperici* adults to the plots. In controlled studies, adults that had fed on diseased plants, or contacted sporulating cultures of the pathogen, effectively disseminated the disease and controlled the weed (Morrison *et al.*, 1998).

Mycoherbicide applications of the fungus have been virulent on all *H. perforatum* biotypes tested (Jensen and Doohan, 1994;

Shepherd, 1995), including those from eastern and western Canada, Oregon and Australia. Native Australian *Hypericum* spp. were not susceptible, but several *Hypericum* spp. native to eastern Canada were, suggesting perhaps that the pathogen may have originated from native species. A wide range of crop and related species did not develop symptoms when sprayed with conidia suspensions. The host range appears narrow and restricted to *H. perforatum* and related North American species, but further testing is warranted. The fungus has the potential to augment *H. perforatum* control elsewhere, particularly in moist, shady habitats or cooler, wetter regions where *Chrysolina* spp. do not provide adequate control.

Evaluation of Biological Control

This success in classical biological control continues. In many areas *C. hyperici* and *C. quadrigemina* are the dominant biological control agents and their range continues to expand. Although the following agents are established, their effect has been negligible: *A. hyperici* is not common in any part of its northern range and any further release is not warranted; *A. chloris* appears adapted only to Nova Scotia and the interior of British Columbia and attempts to introduce it elsewhere would likely not be successful; and *A. plagiata* has only established in British Columbia, where it provides negligible control, so further redistribution is not warranted.

C. gloeosporioides f. sp. *hypericum* could potentially improve the overall control of *H. perforatum*, particularly in habitats where *Chrysolina* spp. have not been effective.

Recommendations

Future work should include:

1. Releasing *C. hyperici* in the tall-grass prairie regions of Manitoba recently invaded by *H. perforatum* and monitoring the possible expansion of the weed into the prairies;
2. Determining the geographic range of *C. gloeosporioides* f. sp. *hypericum*, to facilitate regulatory approval for its distribution within Canada;
3. Determining the possible effects of the disease caused by *C. gloeosporioides* f. sp. *hypericum* on the host–herbivore dynamics prior to any release;
4. Integrating *C. gloeosporioides* f. sp. *hypericum* with *C. hyperici* to improve biological control where the insect alone has not proven effective.

References

Alex, J.F. (1981) St John's wort. *Canadian Agricultural Insect Pest Review*, p. 68.

Briese, D.T. and Judd, P.W. (1995) Establishment, spread and initial impact of *Aphis chloris* Koch (Homoptera: Aphididae) introduced into Australia for the biological control of St John's wort. *Biocontrol Science and Technology* 5, 271–285.

Campbell, C.L. and McCaffrey, J.P. (1991) Population trends, seasonal phenology, and impact of *Chrysolina quadriegimina*, *C. hyperici* (Coleoptera: Chrysomelidae), and *Agrilus hyperici* (Coleoptera: Buprestdae) associated with *Hypericum perforatum* in northern Idaho. *Environmental Entomology* 20, 303–315.

Campbell, M.H. and Delfosse, E.S. (1984) The biology of Australian weeds. 13. *Hypericum perforatum* L. *Journal of the Australian Institute of Agricultural Science* 50, 63–73.

Crompton, C.W., Hall, I.V., Jensen, K.I.N. and Hildebrand, P.D. (1988) The biology of Canadian weeds. 83. *Hypericum perforatum* L. *Canadian Journal of Plant Science* 68, 149–162.

Cullen, J.M., Briese, D.T. and Groves, R.H. (1997) Towards the integration of control methods for St John's wort: workshop summary and recommendations. *Plant Protection Quarterly* 12, 103–106.

Delfosse, E.S. and Cullen, J.M. (1984) New activities in biological control of weeds in Australia. III. St John's wort: *Hypericum perforatum*. In: Delfosse, E.S. (ed.) *Proceedings of the Fifth International Symposium on Biological Control of Weeds*. Commonwealth Scientific and Industrial Research Organization, Melbourne, Australia, pp. 575–581.

Englemeyer, C.E. and Brandle, J.E. (1999) St John's wort – *Hypericum perforatum*. http://res.agr.ca/lond/pmrc/study/newcrops/stjohnswort.html (4 January 2000)

Fields, P.G., Arnason, J.T. and Philogène, B.J.R. (1988) Distribution of *Chrysolina* spp. (Coleoptera: Chrysomelidae) in eastern Ontario, 18 years after their initial release. *The Canadian Entomologist* 120, 937–938.

Fields, P.G., Arnason, J.T. and Philogène, B.J.R. (1989) Behavioral and physical adaptions of three insects that feed on the phototoxic plant *Hypericum perforatum*. *Canadian Journal of Zoology* 68, 339–346.

Fields, P.G., Arnason, J.T., Philogène, B.J.R., Aucoin, R.R., Morand, P. and Sousy-Breau, C. (1991) Phototoxins as insecticides and natural plant defences. *Memoirs of the Entomological Society of Canada* 159, 29–38.

Giese, A.C. (1980) Hypericism. *Photochemistry and Photobiology Reviews* 5, 229–255.

Harris, P. (1962) Effect of temperature on fecundity and survival of *Chrysolina quadrigemina (Suffr.)* and *C. hyperici* (Först.) (Coleoptera: Chrysomelidae). *The Canadian Entomologist* 94, 774–780.

Harris, P. (1967) Suitability of *Anaitis plagiata* (Geometridae) for biocontrol of *Hypericum perforatum* in dry grassland of British Columbia. *The Canadian Entomologist* 99, 1304–1310.

Harris, P. (1999) Status of introduced and main indigenous organisms on weeds targeted for biocontrol in Canada. http://res.agr.ca/leth/weedbio/table.htm (6 January 2000)

Harris, P. and Maw, M. (1984) *Hypericum perforatum* L., St John's wort (Hypericaceae). In: Kelleher, J.S. and Hulme, M.A. (eds) *Biological Control Programmes Against Insects and Weeds in Canada 1969–1980*. Commonwealth Agriculture Bureaux, Slough, UK, pp. 171–177.

Harris, P. and Peschken, D.P. (1971) *Hypericum perforatum* L., St John's wort (Hypericaceae). In: *Biological Control Programmes against Insects and Weeds in Canada 1959–1968*. Technical Communication No. 4, Commonwealth Institute of Biological Control, Trinidad, Commonwealth Agricultural Bureaux, Farnham Royal, UK, pp. 89–94.

Harris, P., Peschken, D.P. and Milroy, J. (1969) The status of biological control of the weed *Hypericum perforatum* in British Columbia. *The Canadian Entomologist* 101, 1–15.

Hildebrand, P.D. and Jensen, K.I.N. (1991) Potential for the biological control of St John's-wort (*Hypericum perforatum*) with an endemic strain of *Colletotrichum gloeosporioides*. *Canadian Journal of Plant Pathology* 13, 60–70.

Jensen, K.I.N. and Doohan, D.J. (1994) *Potential for Control of St John's Wort in Nova Scotia Pastures Using a Native, Host-specific* Colletotrichum gloeosporioides. Final Project Report, Canada/Nova Scotia Livestock Feed Initiative Agreement, ALFI-TT9429.

Jensen, K.I.N., Gaul, S.O., Specht, E.G. and Doohan, D.J. (1995) Hypericin content of Nova Scotia biotypes of *Hypericum perforatum* L. *Canadian Journal of Plant Science* 75, 923–926.

Julien, M.H. (1992) *Biological Control of Weeds – a World Catalogue of Agents and their Target Weeds*, 3rd edn. CAB International, Wallingford, UK.

LeSage, L. (1996) Expansion de l'aire de répartition de *Chrysolina hyperici* (Forster) dupuis son introduction en Ontario (Coleoptera: Chrysomelidae). *Proceedings of the Entomological Society of Ontario* 127, 127–130.

Maund, C.M., McCully, K.V. and Sharpe, R. (1993) A summary of insect biological agents released against weeds in pastures in New Brunswick from 1990 to 1993. *New Brunswick Department of Agriculture, Adaptive Research Report* 15, 359–380.

Mitich, L.W. (1994) Intriguing world of weeds – common St John's wort. *Weed Technology* 8, 658–661.

Morrison, K.D., Reekie, E.G. and Jensen, K.I.N. (1998) Biocontrol of common St Johnswort (*Hypericum perforatum*) with *Chrysolina hyperici* and a host-specific *Colletotrichum gloeosporioides*. *Weed Technology* 12, 426–435.

Nahrstedt, A. and Butterweck, V. (1997) Biologically active and other chemical constituents of *Hypericum perforatum* L. *Pharmacopsychiatry* 30, 129–134.

Sampson, M.G. (1987) *Biological Control of Weeds in Nova Scotia*. Final Project Report, Canada/Nova Scotia Agri-Food Development Agreement, TDP 1987–19.

Sampson, M.G. and MacSween, T. (1992) *Biological Control of Weeds in Nova Scotia*. Final Project Report, Canada/Nova Scotia Livestock Feed Initiative Agreement, TDP-63.

Shepherd, R.C.H. (1995) A Canadian isolate of *Colletotrichum gloeosporioides* (Penzig) Penzig and Saccardo as a potential biological control agent for St John's wort in Australia. *Plant Protection Quarterly* 10,148–151.

Southwell, I.A. and Campbell, M.H. (1991) Hypericin content variation in *Hypericum perforatum* in Australia. *Phytochemistry* 30, 475–478.
Templeton, G.E. (1992) Use of *Colletotichum* as mycoherbicides. In: Bailey, J.A. and Jeger, J.E. (eds) Colletotrichum: *Biology, Pathology and Control.* CAB International, Wallingford, UK, pp. 358–380.
Williams, K.S. (1985) Climatic influences on weeds and their herbivores: biological control of St John's wort in British Columbia. In: Delfosse, E.S. (ed.) *Proceedings of the Sixth International Symposium on Biological Control of Weeds.* Agriculture Canada, Ottawa, Ontario, pp. 127–132.

72 *Linaria dalmatica* (L.) Miller, Dalmatian Toadflax (*Scrophulariaceae*)

R.A. De Clerck-Floate and P. Harris

Pest Status

Dalmatian toadflax, *Linaria dalmatica* (L.) Miller, is an invasive, perennial weed of grasslands, open forests and rights-of-way in western North America that was introduced as an ornamental from eastern Europe in the early 1900s (Alex, 1962; Vujnovic and Wein, 1997). Two forms of the species occur in North America, broad-leaved and narrow-leaved; the former is more important. Since 1980, *L. dalmatica* has become a serious problem in the southern interior of British Columbia and contiguous areas of south-west Alberta, where it currently infests thousands of hectares of range and forest land and is still spreading (R.A. De Clerck-Floate and V. Miller, unpublished). Although the weed also occurs in Saskatchewan, Manitoba, Ontario, Quebec and Nova Scotia (Vujnovic and Wein, 1997), it currently is not considered a major problem in those provinces.

Strong, early season vegetative growth from an extensive root system and lateral stems allow *L. dalmatica* to compete successfully with surrounding rangeland vegetation, particularly winter annuals, biennials and shallow-rooted perennials (Robocker, 1974; Lajeunesse *et al.*, 1993). On coarse-textured soils where it typically grows (Alex, 1962; Robocker, 1974; Vujnovic and Wein, 1997), *L. dalmatica* can form dense stands that displace valued forage and native plant species. The weed is also a prolific seed producer; a large, multistemmed plant may shed up to 500,000 seeds (Robocker, 1970) that can remain viable in soil for up to 10 years (Robocker, 1974). Although *L. dalmatica* contains toxic chemicals (Vujnovic and Wein, 1997), cattle and wildlife generally avoid grazing on it. However, because of significant losses in grazing potential on infested lands, cattlemen in British Columbia have listed *L. dalmatica* as their third control priority after knapweeds, *Centaurea* spp., and houndstongue, *Cynoglossum officinale* L.

Background

Control of *L. dalmatica* is difficult. Chemical treatment is uneconomical and potentially environmentally damaging when applied to large weed stands on grasslands. Although picloram alone or with fluroxypyr or 2,4-D (2,4-dichlorophenoxyacetic acid) can effectively control *L. dalmatica* (Lajeunesse *et al.*, 1993; Vujnovic and Wein, 1997), it leaches readily through the coarse soils, is not as effective under dry conditions, and at high application rates will kill many broad-leaved, non-target species (Lajeunesse *et al.*, 1993). Even if successful, the chemicals require reapplication every 3–4 years for up to 12 years for long-term control. Where *L. dalmatica* grows close to water, chemical control is not an option. Mechanical control, e.g. pulling or mowing, is also not feasible in most cases (Lajeunesse *et al.*, 1993).

Biological control against *L. dalmatica* was initiated together with that for *L. vulgaris* Miller, in the 1960s, with release of the defoliating moth, *Calophasia lunula* (Hufnagel) (Harris and Carder, 1971; Harris, 1984). European agents released in Canada since 1991 to control *L. dalmatica* include the stem-boring weevil, *Mecinus janthinus* Germar, the root moth, *Eteobalea intermediella* (Treitschke), the root-galling weevil, *Gymnetron linariae* Panzer, and an *L. dalmatica* strain of the seed weevil, *Gymnetron antirrhini* (Paykull). In addition, the European flower-feeding beetle, *Brachypterolus pulicarius* (L.) occurs adventively on broad-leaved *L. dalmatica* in Saskatchewan and British Columbia. In British Columbia, the seed-feeding weevil *Gymnetron netum* (Germar) is adventive on both forms of *L. dalmatica* near Creston and was recently introduced accidently on the broad-leaved form in Kamloops (R.A. De Clerck-Floate, unpublished). Macedonian and German populations of an *L. dalmatica* strain of *G. netum* are being screened for host specificity. Recent emphasis is on acquiring and testing representative species of some important native North American genera of *Scrophulariaceae*, e.g. *Antirrhinum*, *Castilleja* and *Pedicularis*.

Biological Control Agents

Insects

The biology of agents shared with the programme for *L. vulgaris* are covered under that species (see McClay and De Clerck-Floate, Chapter 73 this volume) and are not discussed here unless host-related differences exist.

Adult *M. janthinus* emerge as early as late March to early April on *L. dalmatica*, which grows in sunny, south-facing microhabitats, (R.A. De Clerck-Floate, unpublished), in contrast to May emergence from *L. vulgaris* (Jeanneret and Schroeder, 1992).

E. intermediella attacks both *L. dalmatica* and *L. vulgaris*. Unlike *E. serratella*, *E. intermediella* is bivoltine, with the possibility of overlapping generations in Europe (Saner *et al.*, 1994). Eggs are deposited in clusters on the lower stems of toadflax. Larvae tunnel down into the central root where they complete most of their development. Penultimate-instar larvae return to the upper root or the base of stems to pupate. Typically, 3–7 larvae develop per plant and, depending on plant size, can cause considerable damage (Saner *et al.*, 1994). Because *E. intermediella* has a Mediterranean distribution in Europe, a restricted establishment in southern areas of Canada is probable. On *L. dalmatica*, *E. intermediella* prefers vegetative to reproductive plants (Saner *et al.*, 1994). Host records (Riedl, 1969) and host-specificity tests (Saner *et al.*, 1994) indicate that *E. intermediella* is host specific, only attacking species within the tribe *Antirrhineae*. Approval for release of *E. intermediella* in Canada was obtained in 1991.

G. antirrhini adults emerge in late spring to mate and oviposit into developing seed capsules of *L. dalmatica* (Groppe, 1992). The three larval instars feed on seeds. Pupation occurs within the capsules and adults typically emerge in late summer to overwinter in soil litter. Late-developing weevils may diapause within capsules. *G. antirrhini* is univoltine. It is thought to have been introduced to North America from its native Eurasia in the early 1900s (Smith, 1959). The adventive populations occur on *L. vulgaris* in the north-eastern and the north-western USA (Smith, 1959), wherever this weed grows in Canada (R.A. De Clerck-Floate, unpublished), and also on the narrow-leaved form of *L. dalmatica* in British Columbia and Washington (Smith, 1959). Host specificity tests on an *L. dalmatica* strain of *G. antirrhini* from Yugoslavia showed a narrow host range, and complete development only occurred on *L. dalmatica* and occasionally on *L. vulgaris* (Groppe, 1992).

Releases and Recoveries

Several releases of *C. lunula* were made on *L. dalmatica* in Canada since 1980. Most occurred from 1985 to 1989 in the southern interior of British Columbia (total of 4860 *C. lunula* mostly in the larval stage; 11 releases). The northernmost release site was Kamloops (50°40′N) and the southernmost was Grand Forks (49°02′N). Three releases (total of 566 *C. lunula*) were also made in southern Alberta in 1991, 1995 and 1997.

Establishment of *C. lunula* on *L. dalmatica* in Canada was thought to have been unsuccessful at the time of the first report of the moth's establishment on this weed in Missoula, Montana (McDermott *et al.*, 1990). McClay and Hughes (1995) indicated that all but the southernmost areas of Canada are unsuitable for *C. lunula* on the basis of insufficient degree-days for larval development. In British Columbia, *C. lunula* larvae were found near Trail (49°06′N) on *L. dalmatica* in 1995 where the degree-days are sufficient. Larvae were also reported during monitoring of *L. dalmatica* biological control sites in summer, 2000, near Castlegar (49°12′N), Trail, Christina Lake, Grand Forks (49°02′N) and Creston (49°06′N) (R.A. De Clerck-Floate and V. Miller, unpublished). Of 32 sites monitored between the East and West Kootenay Mountains, *C. lunula* larvae were found at 13 sites (40%). These either came from the original releases made in 1985 and 1989 at Grand Forks and Castlegar, respectively, or are founder populations originating from the USA. In southern Alberta, establishment has been confirmed as unsuccessful at two of the three release sites (Lethbridge, 49°42′N, and Scandia, 50°13′N), but the third site (Del Bonita, 49°02′N) has yet to be checked. According to McClay and Hughes (1995), Scandia should have enough degree-days to allow completion of a full generation of the moth.

M. janthinus was initially released at five sites in British Columbia and Alberta against *L. dalmatica* in 1991 and 1992 (Table 72.1). Initial releases were small (29–65 individuals), yet only the Pincher Creek release was unsuccessful. One of the initial releases in Kamloops became the source population for 19 releases in 1994 that ranged from 49°02′N (Grand Forks) to 52°08′N (William's Lake). Most of these releases successfully established (Table 72.1). In southern Alberta, *M. janthinus* has only established at Scandia, one of six sites where releases were made on *L. dalmatica* from 1992 to 1998 (R.A. De Clerck-Floate, unpublished).

At some of the 1994 release sites, 100% attack of *L. dalmatica* stems by *M. janthinus* was achieved within 3 years of release, and some large, reproductive stems of *L. dalmatica* produced over 100 adults, based on spring stem dissections (R.A. De Clerck-Floate and V. Miller, unpublished). Despite more than 95% adult overwinter mortality at some sites and in some years, outbreak numbers of the weevil were noted 3–5 years after release at most 1994 sites; even at the northernmost site, William's Lake. Weevil redistribution from selected 1994 sites to new *L. dalmatica* infestations began in 1996. Only the initial releases are listed in Table 72.1 because of the large number of releases in recent years, e.g. in British Columbia 27,294 adults were collected and redistributed to 129 new sites in 1999 (S. Turner, Kamloops, 2000, personal communication).

E. intermediella was only recently established on *L. dalmatica* in propagation plots at Kamloops (Table 72.2). Initial releases (1991–1996) were made using eggs shipped

Table 72.1. Initial releases and subsequent recoveries of *Mecinus janthinus* against broad-leaved *Linaria dalmatica* in Canada. All releases were of adults in spring and were uncaged except where indicated. Recoveries indicate years the sites were monitored and *M. janthinus* was found.

Location	Year of release	Number released	Recoveries
Alberta			
Pincher Creek	1992	30	None
Scandia	1992	29	1993–1999
British Columbia			
Cranbrook (1)	1994	300	1995–2000
Cranbrook (2)	1994	100	None (site destroyed 1994)
Grand Forks (1)	1994	300	1995–2000
Grand Forks (2)	1994	300	1995–2000
Grand Forks (3)	1994	300	1995–2000
Heffly Creek	1994	90	1999
Kamloops (1)	1991	40 (caged)	1992–2000
Kamloops (2)	1991	38	1996, 1999
Kamloops (3)	1994	150	1993–2000
Lillooet (1)	1992	65	1994
	1994	450	1995, 1996, 1998, 1999
Lillooet (2)	1994	450	1996, 1998
Monte Lake	1992	92	1993, 1996, 1998
Needles	1994	183	1995–1999
Princeton (1)	1994	300	None
Princeton (2)	1994	300	None
Princeton (3)	1994	300	None
Salmon Arm	1994	200	1995 and 1997 (numbers low in 1997)
Trail	1994	500	1995–2000 (control achieved by 1999)
Vernon	1994	100	1995–1998 (site destroyed in 1998)
William's Lake (1)	1994	530	1995, 1998–2000
William's Lake (2)	1994	530	1995, 1998–2000
William's Lake (3)	1994	530	1995 (very small patch of toadflax)

from Europe. In addition to the problems with mould during transit, it is suspected that high mortality was suffered during and after transfer of eggs and neonate larvae to the base of plants using fine paintbrushes. However, in 1998, late-instar larvae and pupae within field-collected *L. dalmatica* roots were shipped and quarantined at Lethbridge until adult emergence. Adults were then released into propagation plots at Lethbridge and Kamloops. The presence of an established colony in Kamloops has been confirmed through the recovery of new-generation adults from caged plots in 1998–2000 (S. Turner, Kamloops, 2000, personal communication; Table 72.2). No open-field releases have yet been made.

Releases of *G. linariae* on *L. dalmatica* in propagation plots at Kamloops and Lethbridge failed to establish. Although galls with pupae and adults were retrieved within the same year of releases in 1996 and 1997 at Lethbridge and in 1996 in Kamloops, recovery of new adults did not persist beyond 1 year. Many of the root galls formed by *G. linariae* on *L. dalmatica* were occluded with no evidence of insect survival. A hypersensitive plant response may be involved in causing mortality of early stages of *G. linariae* (see Fernandes, 1990). It was not until the 1997 releases of *G. linariae* on *L. vulgaris* in Kamloops that successful establishment was achieved. No open-field releases of *G. linariae* have been made in Canada.

The first releases of the *L. dalmatica* strain of *G. antirrhini* were made in Canada in 1993 within caged propagation plots at Kamloops (Table 72.3).

Table 72.2. Releases and recovery of *Eteobalea intermediella* on *Linaria dalmatica* in propagation plots at Lethbridge, Alberta and Kamloops, British Columbia. All releases were within cages except for the 1998 release at Lethbridge.

Location	Year of release	Number and stage	Recoveries
Alberta			
Lethbridge	1992	33 larvae in potted plants	None
	1998	7 adults	None
British Columbia			
Kamloops	1991	389 neonate larvae	None
	1992	360 neonate larvae	None
	1993	480 eggs/neonate larvae	None
	1994	133 eggs/neonate larvae	None
	1996	559 eggs/neonate larvae	None
	1998	94 adults	1998–2000

Subsequent releases within the same plots produced a surviving colony (S. Turner, Kamloops, 1997–2000, personal communication). Beginning in 1994, some plots were uncaged at Kamloops and adult *G. antirrhini* were found both outside and inside cages beginning in 1998. Using weevils collected from the plots, five open-field releases of *G. antirrhini* have been made on *L. dalmatica* in British Columbia, from 1998 to 2000 (Table 72.3). At the Kamloops open-field release site, no evidence of weevil attack was found in 1999 (D. Brooke, Kamloops, 2000, personal communication).

Evaluation of Biological Control

Currently, *M. janthinus* is showing the most promise in controlling *L. dalmatica*. At several 1994 release sites, e.g. Grand Forks, Kamloops, William's Lake, a complete suppression of *L. dalmatica* flowering and severe stunting of shoot growth is evident (R.A. De Clerck-Floate, unpublished). Most of this impact is attributed to feeding on stem apices by mass-emerging adults in spring, something not predicted by European studies (Jeanneret and Schroeder, 1992; Saner *et al.*, 1994). Because *L. dalmatica* produces its flowering stems in one

Table 72.3. Releases and recoveries of the *Linaria dalmatica* strain of *Gymnetron antirrhini* in Lethbridge, Alberta and Kamloops, British Columbia. All releases were of post-diapaused adults.

Location	Year of release	Number and method	Recoveries
Alberta			
Lethbridge (plots)	1994	210, caged	None
	1997	13, open	None
British Columbia			
Kamloops (plots)	1993	300, caged	Unknown
	1994	200, caged	Unknown (site destroyed)
	1995	4, caged	Unknown
	1996	240, caged	1998–2000
Kamloops (field)	1996	80, open	None 1999
Penticton (field)	1999	728, open	Yet to be monitored
Princeton (field)	1999	200, open	Yet to be monitored
Merritt (field)	2000	200, open	Yet to be monitored
Penticton (field)	2000	331, open	Yet to be monitored

spring flush (Saner *et al.*, 1994) it does not have the within-season flexibility to compensate later for feeding by *M. janthinus* adults. Complete control has been achieved at the 1994 Trail site where winter temperatures were consistently mild, thus allowing high overwinter survival and a rapid build-up of the weevil. Although *M. janthinus* on *L. dalmatica* in British Columbia is parasitized by Ichneumonidae, Pteromalidae and Torymidae (G. Gibson, A. Bennett, and R.A. De Clerck-Floate, unpublished), parasitism levels are typically less than 5% at most sites.

Although *C. lunula* has established in the southernmost regions of British Columbia on *L. dalmatica*, its range is expected to remain restricted, based on degree-day requirements (McClay and Hughes, 1995). Its occurrence is sporadic within the climatic area suitable for its development and, although it can completely defoliate plants (V. Miller, Nelson, 2000, personal communication), its densities are generally too low for it to be effective in controlling *L. dalmatica* on its own.

The flower- and seed-feeding agents *B. pulciarius* and *G. netum*, found sporadically on *L. dalmatica*, appear to be too rare to have a major impact on seed production.

The remaining available agents, *E. intermediella*, *G. linariae* and the *L. dalmatica* strain of *G. antirrhini*, are too recently established for an accurate evaluation of their impact. Until we get *G. linariae* to establish on *L. dalmatica*, it is premature to suggest that it has potential as a biological control agent.

Recommendations

Further work should include:

1. Continuing *M. janthinus* redistribution to new *L. dalmatica* infestations and developing release protocols, e.g. optimum number for release in different biogeoclimatic areas;
2. Continued monitoring of previous *M. janthinus* releases for establishment, population change and impact on target and non-target plant species;
3. Increasing *E. intermediella* colony size at Kamloops and attempting field releases;
4. Attempting further field releases of *G. antirrhini* and determining factors that may affect its establishment;
5. Attempting further releases of *G. linariae* under varying field conditions to determine factors affecting plant suitability for gall development and insect survival;
6. Completing the screening of the Macedonian and German Rhine Valley populations of *G. netum*, comparing their population attributes and host specificity to populations already occurring adventively on broad-leaved *L. dalmatica* in southern British Columbia, and obtaining release approval;
7. Investigating host specificity of other potential European agents, e.g. the thrips, *Taeniothrips linariae* Priesner, and gall midge, *Diodaulus linariae* (Winnertz) Rübsaamen.

Acknowledgements

We gratefully acknowledge D. Brooke, V. Miller and S. Turner of the British Columbia Ministry of Forests for their efforts in propagating, releasing and monitoring agents. G. Gibson and A. Bennett identified the parasitoids. The British Columbia Ministry of Agriculture, Food and Fisheries, the British Columbia Ministry of Forests, Montana Noxious Weed Trust Fund, USDA-APHIS and the Wyoming Weed and Pest Districts funded overseas screening of agents. The British Columbia Cattlemen's Association, the British Columbia Beef Cattle Industry Development Council, the British Columbia Grazing Enhancement Fund, Canadian Pacific Railway, the Pest Management Alternatives Office and the Agriculture and Agri-Food Canada Matching Investments Initiative funded research in British Columbia and Alberta.

References

Alex, J.F. (1962) The taxonomy, history, and distribution of *Linaria dalmatica*. *Canadian Journal of Botany* 40, 295–307.

Fernandes, G.W. (1990) Hypersensitivity: a neglected plant resistance mechanism against insect herbivores. *Environmental Entomology* 19, 1173–1182.

Groppe, K. (1992) *Final Report.* Gymnetron anthirrhini *Paykull (Col.: Curculionidae). A Candidate for Biological Control of Dalmatian Toadflax in North America.* International Institute of Biological Control, European Station, Delémont, Switzerland.

Harris, P. (1984) *Linaria vulgaris* Mill., yellow toadflax, and *L. dalmatica* (L.) Mill., broad-leaved toadflax (Scrophulariaceae). In: Kelleher, J.S. and Hulme, M.A. (eds) *Biological Control Programmes Against Insects and Weeds in Canada 1969–1980.* Commonwealth Agriculture Bureaux, Slough, UK, pp. 179–182.

Harris, P. and Carder, A.C. (1971) *Linaria vulgaris* Mill., yellow toadflax, and *L. dalmatica* (L.) Mill., broad-leaved toadflax (Scrophulariaceae). In: *Biological Control Programmes Against Insects and Weeds in Canada 1959–1968.* Technical Communication No. 4, Commonwealth Institute of Biological Control, Trinidad, Commonwealth Agricultural Bureaux, Farnham Royal, UK, pp. 94–97.

Jeanneret, P. and Schroeder, D. (1992) Biology and host specificity of *Mecinus janthinus* Germar (Col.: Curculionidae), a candidate for the biological control of yellow and Dalmatian toadflax, *Linaria vulgaris* (L.) Mill. and *Linaria dalmatica* (L.) Mill. (Scrophulariaceae) in North America. *Biocontrol Science and Technology* 2, 25–34.

Lajeunesse, S.E., Fay, P.K., Cooksey, D., Lacey, J.R., Nowierski, R.M. and Zamora, D. (1993) *Dalmatian and Yellow Toadflax: Weeds of Pasture and Rangeland.* Extension Service, Montana State University, Bozeman, Montana.

McClay, A.S. and Hughes, R.B. (1995) Effect of temperature on developmental rate, distribution, and establishment of *Calophasia lunula* (Lepidoptera: Noctuidae), a biological agent for toadflax (*Linaria* spp.). *Biological Control* 5, 368–377.

McDermott, G.J., Nowierski, R.M. and Story, J.M. (1990) First report of establishment of *Calophasia lunula* Hufn. (Lepidoptera: Noctuidae) on Dalmatian toadflax, *Linaria genistifolia* subsp. *dalmatica* Maire and Petitmengin, in North America. *The Canadian Entomologist* 122, 767–768.

Riedl, T. (1969) Matériaux pour la connaissance des Momphidae paléarctiques (Lepidoptera). Partie IX. Revue des Momphidae européennes, y compris quelques espèces d'Afrique du Nord et du Proche-Orient. *Poskie Pismo Entomologiczne* 39, 635–919.

Robocker, W.C. (1970) Seed characteristics and seedling emergence of Dalmatian toadflax. *Weed Science* 18, 720–725.

Robocker, W.C. (1974) *Life History, Ecology, and Control of Dalmatian Toadflax.* Technical Bulletin 79, Washington Agricultural Experiment Station, Washington State University, Pullman, Washington.

Saner, M.A., Jeanneret, P. and Müller-Schärer, H. (1994) Interaction among two biological control agents and the developmental stage of their target weed, Dalmatian toadflax, *Linaria dalmatica* (L.) Mill. (Scrophulariaceae). *Biocontrol Science and Technology* 4, 215–222.

Smith, J.M. (1959) Notes on insects, especially *Gymnaetron* spp. (Coleoptera: Curculionidae), associated with toadflax, *Linaria vulgaris* Mill. (Scrophulariaceae), in North America. *The Canadian Entomologist* 91, 116–121.

Vujnovic, K. and Wein, R.W. (1997) The biology of Canadian weeds. 106. *Linaria dalmatica* (L.) Mill. *Canadian Journal of Plant Science* 77, 483–491.

73 *Linaria vulgaris* Miller, Yellow Toadflax (*Scrophulariaceae*)

A.S. McClay and R.A. De Clerck-Floate

Pest Status

Yellow toadflax, *Linaria vulgaris* Miller, is a herbaceous perennial European weed that spreads vigorously both by seed and by creeping roots. It is widespread in uncultivated and cultivated land, particularly under reduced tillage, throughout Canada up to 60°N. Its abundance and impact on the prairies declined in the late 1950s, possibly due to the effects of two European insects that became established at that time (Harris, 1984). However, it is still considered a significant problem in parts of central Alberta, the Peace River district, and north-western Saskatchewan. In New Brunswick, *L. vulgaris* is a serious problem in fields of strawberries, *Fragaria* × *ananassa* Duchesne, and raspberries, *Rubus idaeus* L., in orchards, and in some fields of alfalfa, *Medicago sativa* L., hay and grain (Maund *et al.*, 1992).

L. vulgaris is distasteful to cattle and avoided by them when grazing (Mitich, 1993). It competes with crops, reducing yield. O'Donovan and Newman (1989) found that a natural infestation of *L. vulgaris* in a wheat field reduced wheat yield by 11% for each 50 shoots m^{-2}. Actual densities in the centre of the patch were up to about 200 shoots m^{-2}. At Lacombe, Alberta, barley yield was reduced by about 90 g m^{-2} for each 100 *L. vulgaris* shoots m^{-2} in both reduced-tillage and zero-tillage plots (Fig. 73.1) (A.S. McClay, R.A. De Clerck-Floate and K.N. Harker, unpublished).

Root spread from small transplants of *L. vulgaris* can be up to 1 m year^{-1} in fallow land or 0.5 m year^{-1} in a barley crop (Nadeau *et al.*, 1991). Root pieces taken from seedlings as young as 3 weeks old can produce new shoots when transplanted (Nadeau *et al.*, 1992). Nadeau and King (1991) found that the amount of seeds shed, from mid-August to mid-October, could be up to 210,000 seeds m^{-2}, but most fell within 0.5 m of the parent plants. Seed viability and dormancy were major factors affecting establishment.

Background

Few effective herbicides for *L. vulgaris* exist, although preharvest applications of glyphosate at 0.9 kg ha^{-1} reduced densities by over 80% the following year, resulting in a significant increase in crop yields of barley, *Hordeum vulgare* L., canola, *Brassica napus* L. and *B. rapa* L., and flax, *Linum usitatissimum* L. (Baig *et al.*, 1999). Chemical control possibilities are limited due to resistance of *L. vulgaris* to common herbicides (Saner *et al.*, 1995).

Previous work on biological control of *L. vulgaris*, summarized by Harris and Carder (1971) and Harris (1984), began in the 1960s with the release of the defoliating moth, *Calophasia lunula* (Hufnagel). In the 1980s, renewed interest in the control of *L. vulgaris* and Dalmatian toadflax, *L. dalmatica* (L.) Miller, revived the biological control programmes against both of these weeds, and several more European insect agents were screened and approved for release against *L. vulgaris*.

No native *Linaria* spp. occur in North America; the three North American species have been transferred to *Nuttallanthus*

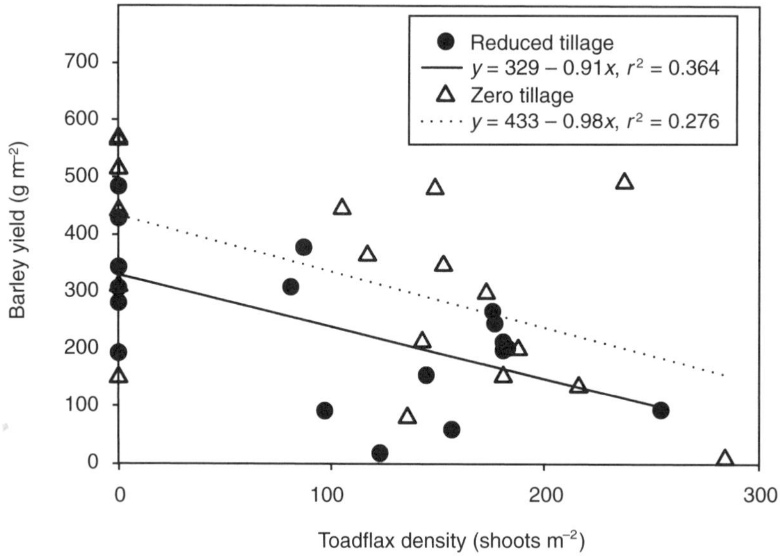

Fig. 73.1. Effect of *Linaria vulgaris* shoot density on yield of barley under zero tillage and reduced tillage, Lacombe, Alberta, 1994.

(Sutton, 1988; USDA Natural Resources Conservation Service, 1999). Thus, the risks of non-target damage appear relatively low.

Biological Control Agents

Insects

Brachypterolus pulicarius (L.), a European flower-feeding beetle, accidentally introduced into Canada before 1961, feeds extensively on shoot tips, flower buds and anthers of *L. vulgaris*, and is now widespread throughout its range. *Gymnetron antirrhini* (Paykull), a seed-feeding weevil adventive to North America, is also widespread, but is parasitized in Wisconsin by an introduced European pteromalid, *Pteromalus microps* Graham (Volenberg and Krauth, 1996), which may reduce its effectiveness. Harris (1984) suggested that *C. lunula*, by then established in Ontario, could be established elsewhere in Canada. It is established on *L. dalmatica* in Montana (McDermott *et al.*, 1990).

Mecinus janthinus Germar is a univoltine stem-mining weevil native to central and southern Europe and southern Russia. Females oviposit into the stems of *L. vulgaris*, where the larvae feed in tunnels and pupate. Adults eclose from the pupae in late summer but remain within the stems over winter, emerging the following spring to feed on the foliage, mate, and oviposit. Larval tunnelling in the stems causes premature wilting and suppresses flowering. Host-specificity tests showed that it would develop only on some *Linaria* spp. (Jeanneret and Schroeder, 1992). The weevil was approved for release in Canada in 1991.

Eteobalea serratella Treitschke is a univoltine moth widely distributed from southern and central Europe to Mongolia (Riedl, 1975a, b, 1978). Eggs are deposited close to the stem base and newly hatched larvae bore into the plant through leaf axils or other suitable entry points. Larvae feed in silk-lined tunnels in all parts of the root system, but mainly in the cortex and root crown. They pupate in the tunnel and adults emerge through an exit hole near ground level, about 2 cm below the upper end of the mine. Larvae overwinter in the roots but there is no obligate diapause (Saner *et al.*, 1990). Host testing of an *E.*

serratella population from Rome, Italy, showed that development was restricted to some perennial *Linaria* spp. and that *L. vulgaris* was the preferred host. In the field, plants in dry habitats were killed by *E. serratella* (Saner *et al.*, 1990). The Rome population of *E. serratella* was approved for field release in Canada in 1991.

Gymnetron linariae Panzer, a univoltine root-galling weevil, was collected during 1987–1993 from central and southern Europe and southern Russia (Jordan, 1994). Adults emerge in April and May to feed and oviposit. Eggs are laid singly into shallow pockets chewed into the root surface by females, and the galls develop within 2 weeks. There are three larval instars. New-generation adults emerge from July to late summer, but a portion of the population may diapause within the galls. Host-specificity tests showed that only a few *Linaria* spp., including *L. vulgaris* and *L. dalmatica*, were acceptable for gall induction and weevil development (Jordan, 1994). *G. linariae* was approved for release in Canada in 1995.

Releases and Recoveries

In Alberta and Saskatchewan, numerous releases of *C. lunula* failed to result in establishment, probably due to insufficient degree-day accumulation for complete development (McClay and Hughes, 1995). In New Brunswick, a release of 1025 larvae at Nashwaaksis in 1990 resulted in establishment (Maund *et al.*, 1993). This release (referred to as 'Fredericton'), incorrectly reported as not established by McClay and Hughes (1995), was in the area predicted to be suitable on the basis of degree-days. In Nova Scotia, *C. lunula* became established from releases made in 1984–1991 and can now be found throughout the province (G. Sampson, Truro, 2000, personal communication).

M. janthinus releases were not made against *L. vulgaris* until 1994. Overwinter survival of adult *M. janthinus* in *L. vulgaris* plants was tested at Vegreville and Lethbridge. At each site, groups of five females and 3–4 males were placed on each of 30 potted plants in a greenhouse cage and allowed to oviposit for 4 days in May, 1994. All plants were set out in field plots in mid-July, 1994. In autumn, 1994, half of the plants were brought into the laboratory and stems were dissected for *M. janthinus*. The remaining plants were left in the plots over winter and dissected in early April, 1995, to determine the numbers of adults surviving. Percentage survival at Vegreville and Lethbridge was 68% and 18%, respectively (Table 73.1). This difference may have been related to greater snow cover at the Vegreville site, providing better thermal insulation for overwintering adults in the stems.

M. janthinus has been released at 42 locations, mostly in Alberta, up to 2000 (Table 73.2). Most monitoring was conducted by taking stem samples from release sites towards the end of the growing season in September and dissecting to check for *M. janthinus*. At most sites, breeding was confirmed within the release year. The dissections showed mixtures of larval stages, pupae and adults. A similar result was found during autumn sampling at the Wilbert, Saskatchewan, site in 1996–1998 (R.A. De Clerck-Floate and A.G. Thomas, unpublished). As only the adult stage overwinters, the mixture of stages suggests that

Table 73.1. Overwinter survival of two insects in *Linaria vulgaris* at Vegreville and Lethbridge, Alberta, 1994–1995.

Location	Species	Autumn	Spring	% Survival
Vegreville	*Mecinus janthinus* Germar (adults per plant)	11.9	8.1	68.0
	Eteobalea serratella Treitschke (larvae + pupae per plant)	3.7	1.76	47.6
Lethbridge	*Mecinus janthinus* (adults per plant)	3.5	1.1	17.7
	Eteobalea serratella (larvae + pupae per plant)	0.1	0.0	0.0

Table 73.2. Releases and recoveries of *Mecinus janthinus* against *Linaria vulgaris* in Canada, 1994–2000. All releases were of adults in spring or early summer, and were uncaged unless otherwise indicated.

Location	Year	Number	Site description	Recoveries
Alberta				
Lafond	1994	84	Seeded pasture	None
Wetaskiwin	1994–1996	370	Field margin – caged	None
Nisku	1995	50	Hayland	None
Mannville	1995	50	Pasture	None
Kinsella	1995–1997	770	Fallow	1996–1997
Edmonton	1995	62	Park	1995
Derwent	1996	200	Conservation area	1996
Rivercourse	1996	200	Old road bed	1996
Rosalind	1996	194	Rough pasture	1997–2000
Kinsella	1997	200	Hayland	1998
Tofield	1997	200	Roadside, creek bank	None
Edmonton	1997	533	Park	1997–2000
Edmonton	1998	200	Freeway embankment	1998–2000
Edmonton	1998	200	Freeway embankment	1998–2000
Edmonton	1998	200	Freeway embankment	1998
Kinsella	1998	200	Pasture	1998
Kinsella	1998	200	Pasture	1998
Fairview	1998	200	Hay pasture	1998–1999
Fairview	1998	200	Grass seed	1998–1999
Fairview	1998	200	Pasture	1998–1999
Fairview	1998	200	Hay pasture	1999
Brownvale	1999	60	Pasture	1999
Edmonton	1999	100	Park	1999–2000
Edmonton	1999	100	Park	1999–2000
Bashaw	1999	60	Nature reserve	None
Lacombe	1999	60	Pasture	Unknown
Langdon	1999	60	Recreation area	2000
Breton	1999	60	Roadside	2000
Kinsella	2000	200	Old railway line	2000
Kinsella	2000	200	Field margin	None
Derwent	2000	200	Nature reserve	2000
Pine Lake	2000	200	Nature reserve	2000
Pine Lake	2000	200	Nature reserve	2000
Fairview	2000	200	Canola field	Unknown
Whitelaw	2000	200	Wheat, underseeded to lucerne	Unknown
Grande Prairie	2000	200	Industrial park	2000
Grande Prairie	2000	200	Industrial park	2000
Saskatchewan				
Wilbert	1996–1998	2696	Seeded pasture	1997–1998
Last Mt. Lake	1997	77	Native mixed grass prairie	None
Manitou Sand Hills	1998	100	Beach of saline lake	Unknown
Marsden	1998	200	Grazing	1998
Nova Scotia				
St Croix	1995 and 1997	253	Roadside/streamside	1999

M. janthinus may be approaching its climatic limits in Alberta, and that only eggs laid early in the season will result in complete development through to the adult stage, implying that releases should be made as early as possible in the season to maximize the chances of establishment. Survival for at least one winter has been

Table 73.3. Releases and recoveries of *Eteobalea serratella* against *Linaria vulgaris* in Canada, 1992–1996. All releases were made between early June and mid-July, and were open unless otherwise noted.

Location	Year	Stage	Number	Site description and notes	Recoveries
British Columbia					
Kamloops	1992 and 1995	Eggs, larvae	1629	Propagation plots: eggs and neonate larvae transferred to plant base	None
Alberta					
Duvernay	1992	Eggs, larvae	575	Improved pasture: eggs and neonate larvae transferred to plant base	1993 – larvae found late June, none since then
Kinsella	1995	Adults	92	Pasture: caged release	1996 – larvae and pupae found in late September, none since then
Edmonton	1995	Adults	40	Park	1996 – larvae found in September, none since then
Lethbridge	1995	Eggs, larvae	4323	Propagation plots: eggs and neonate larvae transferred to plant base	None
Mannville	1995	Adults	140	Pasture	None
Derwent	1996	Larvae, pupae	Unknown	Meadow, conservation area: transplanted plants containing larvae and pupae	None
Saskatchewan					
Senlac	1993	Larvae	101	Native pasture: larvae within roots of potted plants. Pots sunk into toadflax patch	None
Nova Scotia					
St Croix (1)	1992	Larvae	114	Open release, abandoned field	None
St Croix (2)	1992	Eggs	53	Park, open release	None
St Croix (3)	1995	Eggs	1494	Roadside: eggs transferred to plant base	1999 – moths seen in low numbers

confirmed at 14 sites in Alberta, one site in Saskatchewan and the one release site in Nova Scotia (Table 73.2). Although population densities have generally remained low, they have increased annually at one 1996 release site near Rosalind in central Alberta, with 68% of stems attacked and a mean of 2.03 adults and pupae per stem by 1999. *Pteromalus microps* was reared from *M. janthinus* at several field release sites in Alberta.

In Alberta, six field releases of *E. serratella* were made from 1992 to 1996 (Table 73.3), by transferring eggs or neonate larvae on to stem bases of plants in the field, by transplanting infested plants containing larvae and/or pupae, and by releasing adults in the open or in field cages. The number of releases was limited by the difficulty of maintaining a viable laboratory colony as a source of material for field release – rearing is very labour-intensive, adult emergence is often spread over a long period, and survival rates are fairly low. *E. serratella* bred and survived through one winter at three of the six Alberta release sites (Table 73.3). Subsequent monitoring by dissection of root samples showed no definite evidence of established populations, although occasional old tunnels suggesting possible larval feeding were found up to 4 years after release at the Kinsella site (A.S. McClay, unpublished). Overwinter survival studies in transplanted plants were carried out at Vegreville and Lethbridge in 1994–1995 using methods similar to those described for *M. janthinus*. Very little establishment was obtained on the plants at Lethbridge, but at Vegreville overwinter survival of larvae and pupae was 47.6% (Table 73.1). Success of *E. serratella* in other parts of Canada has also been poor. In Nova Scotia, three releases were made (Table 73.3) and establishment has been confirmed at one site, but in low numbers (G. Sampson, Truro, 2000, personal communication). In Saskatchewan, however, a release did not result in establishment. In Alberta and British Columbia, multiple attempts to establish a colony within propagation plots at Lethbridge and Kamloops failed (Table 73.3).

Releases of *G. linariae* have only been made on *L. vulgaris* in propagation plots at Lethbridge, Alberta and Kamloops, British Columbia. Initial attempts to establish *G. linariae* on *L. dalmatica* at Kamloops failed, but when introduced in 1997 and 1998 to caged plots of *L. vulgaris*, a surviving colony was obtained (S. Turner, Kamloops, 2000, personal communication). Similar releases of *G. linariae* in Lethbridge in 1996 and 1997 on a caged, mixed stand of *L. vulgaris* and *L. dalmatica* did not result in a sustained colony. Galls with pupae and adults were found in August of both years, but no overwinter survival occurred (R.A. De Clerck-Floate, unpublished).

Evaluation of Biological Control

Feeding by *B. pulicarius* delays *L. vulgaris* flowering and reduces seed production by 74% (Nadeau and King, 1991; McClay, 1992) but the weevil's presence has not sufficiently curtailed the weed.

No detailed studies on the impact of *C. lunula* exist. In Nova Scotia, late-instar larvae are found on *L. vulgaris* only in September (G. Sampson, Truro, 2000, personal communication), consistent with the observation that this area has barely sufficient degree-days for *C. lunula* to complete development (McClay and Hughes, 1995). Under these circumstances it is unlikely to have much impact. In New Brunswick, mature larvae appear by mid-July and in some years may have a partial second generation. Damage levels varied widely but at some sites up to 30% defoliation was observed (Maund *et al.*, 1994, 1995), leading these authors to rate *C. lunula* as being of good potential effectiveness (Maund *et al.*, 1993).

In 1999, studies at the Rosalind, Alberta, site suggested that attack by *M. janthinus* reduces flowering and seed production and increases mortality of attacked stems (A.S. McClay, unpublished). The relatively low establishment rate and slow population build-up of *M. janthinus* on *L. vulgaris* in Alberta contrasts with the successful establishment and promising impact observed on *L. dalmatica* in British Columbia (see De Clerck-Floate and Harris, Chapter 72 this volume). This difference may be due to differences in the microclimate. Typically, *L. dalmatica* thrives on sunny, dry, south-facing slopes, which heat up sooner than the typical *L. vulgaris* habitats. Emergence of *M. janthinus* at *L. dalmatica* sites in British Columbia can be as early as late March, compared to late April into May at *L. vulgaris* sites, thus giving the insects plenty of time to complete development by fall (R.A. De Clerck-Floate, unpublished).

In the absence of definite establishment for most releases, it is not yet possible to evaluate the impact of *E. serratella* in the field. In greenhouse experiments, Volenberg *et al.* (1999) found that feeding by three larvae of *E. serratella* per *L. vulgaris* plant reduced root biomass by 20%. Saner and Müller-Schärer (1994) found that attacked plants had a shorter flowering season and produced seeds of lower weight. Hence, if the establishment problems can be overcome, *E. serratella* may be an effective agent.

It also is not yet possible to evaluate the impact of *G. linariae* on *L. vulgaris* because of limited establishment and field releases. However, a thriving colony of the weevil established on *L. vulgaris* in propagation plots in Kamloops suggests that *G. linariae* can survive in the field. No published studies on the impact of this agent on *L. vulgaris* exist but in Europe some plants yielded over 20 galls (Jordan, 1994), probably causing a severe drain on plant growth and reproduction.

Recommendations

Further work should include:

1. Introducing *C. lunula* only in areas where sufficient degree-days are available for its development;
2. Improving rearing and monitoring methods for *E. serratella* and continuing efforts to establish it;
3. Evaluating the impact of *M. janthinus* and factors affecting its establishment, including microclimate;
4. Field releasing *G. linariae* on *L. vulgaris* and closely monitoring it for establishment;
5. Introducing populations of previously approved agents from areas with colder climates, e.g. eastern Europe or southern Russia, with additional host-specificity screening as needed;
6. Evaluating *Eteobalea intermediella* Riedl, already released against *L. dalmatica* in British Columbia, against *L. vulgaris*;
7. Screening other potential agents from Europe, including the thrips *Taeniothrips linariae* Priesner and the moth *Eupithecia linariata* (Denis and Schiffermüller).

Acknowledgements

We thank the Alberta Agricultural Research Institute and the Canada Alberta Environmentally Sustainable Agriculture Agreement for funding. G. Gibson identified *Pteromalus microps*. The release and/or monitoring efforts of G. Sampson, C. Saunders, J. Loland, M. Baert, E. Johnson, T. Jorgenson, and S. Turner are gratefully acknowledged.

References

Baig, M.N., Darwent, A.L., Harker, K.N. and O'Donovan, J.T. (1999) Preharvest applications of glyphosate for yellow toadflax (*Linaria vulgaris*) control. *Weed Technology* 13, 777–782.

Harris, P. (1984) *Linaria vulgaris* Mill., yellow toadflax, and *L. dalmatica* (L.) Mill., broad-leaved toadflax (Scrophulariaceae). In: Kelleher, J.S. and Hulme, M.A. (eds) *Biological Control Programmes Against Insects and Weeds in Canada 1969–1980.* Commonwealth Agricultural Bureaux, Slough, UK, pp. 179–182.

Harris, P. and Carder, A.C. (1971) *Linaria vulgaris* Mill., yellow toadflax, and *L. dalmatica* (L.) Mill., broad-leaved toadflax (Scrophulariaceae). In: *Biological Control Programmes Against Insects and Weeds in Canada 1959–1968.* Commonwealth Agricultural Bureaux, Slough, UK, pp. 94–97.

Jeanneret, P. and Schroeder, D. (1992) Biology and host specificity of *Mecinus janthinus* Germar (Col.: Curculionidae), a candidate for the biological control of yellow and Dalmatian toadflax, *Linaria vulgaris* (L.) Mill. and *Linaria dalmatica* (L.) Mill. in North America. *Biocontrol Science and Technology* 2, 25–34.

Jordan, K. (1994) Gymnetron linariae *Panzer (Col.: Curculionidae): a Candidate for Biological Control of Dalmatian and Yellow Toadflax in North America.* International Institute of Biological Control, European Station, Delémont, Switzerland, p. 36.

Maund, C.M., McCully, K.V. and Sharpe, R. (1992) Biological control of selected weeds in pastures in New Brunswick during 1992. *Adaptive Research Reports (New Brunswick Department of Agriculture)* 14, 317–328.

Maund, C.M., McCully, K.V., Finnamore, D.B., Sharpe, R. and Parkinson, B. (1993) A summary of insect biological control agents released against weeds in NB pastures from 1990 to 1993. *Adaptive Research Reports (New Brunswick Department of Agriculture)* 15, 359–380.

Maund, C.M., Sharpe, R., Stairs, A. and McCully, K.V. (1994) Biological control of selected weeds with insects in New Brunswick pastures during 1994. *Adaptive Research Reports (New Brunswick Department of Agriculture)* 16, 385–397.

Maund, C.M., Sharpe, R. and McCully, K.V. (1995) Biological control of selected weeds with insects in New Brunswick pastures during 1995. *Adaptive Research Reports (New Brunswick Department of Agriculture)* 17, 227–240.
McClay, A.S. (1992) Effects of *Brachypterolus pulicarius* (L.) (Coleoptera: Nitidulidae) on flowering and seed production of common toadflax. *The Canadian Entomologist* 124, 631–636.
McClay, A.S. and Hughes, R.B. (1995) Effects of temperature on developmental rate, distribution, and establishment of *Calophasia lunula* (Lepidoptera, Noctuidae), a biocontrol agent for toadflax (*Linaria* spp.). *Biological Control* 5, 368–377.
McDermott, G.J., Nowierski, R.M. and Storey, J.M. (1990) First report of establishment of *Calophasia lunula* Hufn. (Lepidoptera: Noctuidae) on Dalmatian toadflax, *Linaria genistifolia* ssp. *dalmatica* (L.) Maire and Petitmengin, in North America. *The Canadian Entomologist* 122, 767–768.
Mitich, L.W. (1993) Yellow toadflax. *Weed Technology* 7, 791–793.
Nadeau, L.B. and King, J.R. (1991) Seed dispersal and seedling establishment of *Linaria vulgaris* Mill. *Canadian Journal of Plant Science* 71, 771–782.
Nadeau, L.B., Dale, M.R.T. and King, J.R. (1991) The development of spatial pattern in shoots of *Linaria vulgaris* (Scrophulariaceae) growing on fallow land or in a barley crop. *Canadian Journal of Botany* 69, 2539–2544.
Nadeau, L.B., King, J.R. and Harker, K.N. (1992) Comparison of growth of seedlings and plants grown from root pieces of yellow toadflax (*Linaria vulgaris*). *Weed Science* 40, 43–47.
O'Donovan, J.T. and Newman, J.C. (1989) Influence of toadflax on yield of wheat. *Expert Committee on Weeds, Research Report (Western Canada)* 3, 201.
Riedl, T. (1975a) Brève révision des espèces du groupe d'*Eteobalea beata* (Walsingham) (Insecta, Lepidoptera, Cosmopterygidae). *Bulletin du Muséum National d'Histoire Naturelle* 335, 1293–1302.
Riedl, T. (1975b) Sur la répartition de quelques espèces françaises de Momphidae (s.l.). *Alexanor* 9, 185–191.
Riedl, T. (1978) Sur la répartition de certains Momphidae s.l. dans la region Méditerranéenne (Lepidoptera). *Mitteilungen der Entomologische Gesellschaft, Basel* 28, 72–75.
Saner, M.A. and Müller-Schärer, H. (1994) Impact of root mining by *Eteobalea* spp. on clonal growth and sexual reproduction of common toadflax, *Linaria vulgaris* Mill. *Weed Research* 34, 199–204.
Saner, M., Groppe, K. and Harris, P. (1990) Eteobalea intermediella *Riedl and* E. serratella *Treitschke (Lep., Cosmopterigidae), Two Suitable Agents for the Biological Control of Yellow and Dalmatian Toadflax in North America.* Final report. International Institute of Biological Control, Delémont, Switzerland.
Saner, M.A., Clements, D.R., Hall, M.R., Doohan, D.J. and Crompton, C.W. (1995) The biology of Canadian weeds. 105. *Linaria vulgaris* Mill. *Canadian Journal of Plant Science* 75, 525–537.
Sutton, D.A. (1988) *A revision of the tribe Antirrhineae.* Oxford University Press, London, UK.
USDA Natural Resources Conservation Service (1999) The PLANTS database. http://plants.usda.gov/plants (4 May 2000)
Volenberg, D.S. and Krauth, S.J. (1996) First record of *Pteromalus microps* (Hymenoptera, Pteromalidae) in the New World. *Entomological News* 107, 272–274.
Volenberg, D.S., Hopen, H.J. and Campobasso, G. (1999) Biological control of yellow toadflax (*Linaria vulgaris*) by *Eteobalea serratella* in peppermint (*Mentha piperita*). *Weed Science* 47, 226–232.

74 *Lythrum salicaria* L., Purple Loosestrife (*Lythraceae*)

C.J. Lindgren, J. Corrigan and R.A. De Clerck-Floate

Pest Status

Purple loosestrife, *Lythrum salicaria* L., is a Eurasian wetland perennial, likely introduced to North America in the early 1800s (Thompson *et al.*, 1987). Cultivated varieties of *L. salicaria*, developed as early as 1937 (Harp and Collicut, 1983), have been widely used across North America by gardeners and landscapers and have further contributed to its spread (Ottenbreit, 1991; Lindgren and Clay, 1993). *L. salicaria* is capable of forming continuous stands that can displace native vegetation, which provides food, cover and breeding areas for wildlife. Thompson *et al.* (1987) estimated that controlling this plant across the invaded wetlands of 19 American states would cost US$45.9 million per year.

L. salicaria has invaded every Canadian province (White *et al.*, 1993). In British Columbia, it can be found along the Fraser River, Iona Island, Westham Island, Vancouver Island, Jericho Park (Vancouver), the Ladner Marsh, the Okanagan Valley, Chilliwack and Nelson (Myers and Denoth, 1999). In Alberta, the first infestation was reported in 1990 near Medicine Hat. Ali and Verbeek (1999) reported more than 315,000 plants in 1994 and infestations in as many as 185 individual wetlands in 1999. In Saskatchewan, *L. salicaria* is found mostly in urban settings, e.g. Saskatoon, Moose Jaw, Regina, Swift Current and Yorkton (A. Salzl, Saskatoon, 1999, personal communication). In Manitoba, *L. salicaria* was first reported in 1896, and has since spread to every major river system in southern Manitoba, with large infestations in the south basins of lakes Winnipeg and Manitoba. In Ontario, *L. salicaria* has a long history of residency (100+ years), and many extensive populations are established south of the 49th parallel (White *et al.*, 1993). In Quebec, large populations exist in the Eastern Townships, and along the lower Ottawa and St Lawrence River valleys (White *et al.*, 1993). Although *L. salicaria* has been present in Quebec since the 1800s, farmers became concerned in 1949 when loosestrife began replacing forage crops in riparian pastures (Templeton and Stewart, 1999). In New Brunswick, *L. salicaria* is a concern in most of the lower marsh in the Saint John flood plain. Prior to the 1960s, botanical surveys revealed none in this region (J. Wile, Amherst, 1999, personal communication). In Nova Scotia, *L. salicaria* is widespread, with large infestations reported on Cape Breton and on the mainland (G. Sampson, Truro, 1999, personal communication). In Prince Edward Island, *L. salicaria* can be found throughout the province, with larger infestations found around larger towns and villages. It is also present in salt marshes on the upper Hillsborough River (T. Duffy, Charlottetown, 2000, personal communication). In Newfoundland, *L. salicaria* is present in western, central and eastern regions of the island. However, its distribution is patchy and it is not common anywhere. *L. salicaria* has not been recorded from Labrador (P. Dixon, St John's, 2000, personal communication).

L. salicaria, including all cultivated varieties, has been designated a noxious weed in Prince Edward Island (1991), Alberta

(1992) and Manitoba (1996). Provincial working groups formed to combat this weed include the Alberta Purple Loosestrife Eradication Program, Saskatchewan Purple Loosestrife Eradication Project, the Manitoba Purple Loosestrife Project and Project Purple in Ontario.

Background

Malecki *et al.* (1993) stated: 'No effective method is available to control *L. salicaria*, except where it occurs in small localized stands and can be intensively managed.' Control methods attempted include water-level manipulation, physical removal, mowing, burning and herbicide application, but these are costly, localized and short-term. Biological control represents the only option, given the geographical and temporal scales of the problem (Malecki *et al.*, 1993).

Biological Control Agents

Insects

Diehl *et al.* (1997) collected 51 species of resident herbivorous insects on *L. salicaria* in Manitoba, but concluded that they are not effective in reducing its density there. Based on the history of the spread of this plant across Canada (White *et al.*, 1993), we believe this conclusion applies nationally.

In Europe, over 100 species of phytophagous insects have been associated with *L. salicaria* (Batra *et al.*, 1986). De Clerck-Floate (1992) recommended that the European root-mining weevil, *Hylobius transversovittatus* (Goeze), and the leaf beetles, *Galerucella calmariensis* L. and *Galerucella pusilla* Duftschmid, be released against *L. salicaria*. These agents have narrow host ranges, climatic origins compatible with those of Canada, and potential for causing extensive damage to *L. salicaria*. These three species were approved for release in 1992.[1] Two other European weevils, *Nanophyes marmoratus* Goeze and *Nanophyes brevis* Boheman, were approved for release in 1994. Releases of four of the agents were made in Canada from 1992 to 1999. European screening prior to agent importation revealed populations of *N. brevis* to be infected with an unidentified nematode, so this agent was not released in Canada.

H. transversovittatus adults are mainly nocturnal, feed on foliage and stem tissue, and can live for several years (Blossey, 1993). Eggs are laid into the lower part of the main shoot or on to the root, with larval development taking 1–2 years. In the field, long wet periods will delay larval development.

G. calmariensis and *G. pusilla* adults emerge from winter diapause in late May to early June and begin feeding on young foliage. Oviposition begins in early June and peaks about mid-June. Larvae feed on shoot tips, foliage and flowers. Peak numbers of larvae occur from late June to early July. Mature larvae pupate in soil around the host plants. First-generation adults occur in August, and in some years well into October. A second generation has been observed in British Columbia, Manitoba and Ontario.

N. marmoratus is univoltine. In Europe, overwintered adults start feeding on young foliage in late May, moving to the upper parts of flower spikes to feed on unopened flowers as flower buds develop (Blossey and Schroeder, 1995). Eggs are laid from June to September, with the female usually depositing one egg into the tip of a young flower bud. Larvae consume the stamens and ovary; attacked buds do not flower and are aborted. New-generation adults appear in August, feeding on foliage prior to overwintering.

Releases and Recoveries

Biological control programmes have been initiated in every province except Newfoundland. A summary of releases is given in Table 74.1.

[1] Starter populations of *H. transversovittatus, G. calmariensis* and *G. pusilla* were obtained from Europe via the USA in 1992 and reared at the University of Guelph and the Agriculture and Agri-Food Canada Lethbridge Research Centre for initial Canadian distribution.

Table 74.1. Known liberations of biological control agents against *Lythrum salicaria* in Canada,1992–1999. Total number of each species released is followed by life stage (^{A}, adult; ^{L}, larva; ^{P}, pupa; ^{E}, eggs) and (number of releases).

Province	Year	*Galerucella calmariensis* L.	*Galerucella pusilla* Duftschmid	*Galerucella* spp.	*Hylobius transversovittatus* Goeze	*Nanophyes marmoratus* Goeze	Total
British Columbia	1993	1308^{A,L} (7)					1308
	1994	1430^{A} (4)	400^{A} (2)		180^{E,L} (1)		2010
	1995	1218^{A} (4)		475^{A} (1)			1693
	1996	453^{A} (2)					456
	1997	3550^{A} (12)	150^{A} (1)				3700
	1998			100^{A} (1)			100
	1999			133^{A} (2)			133
Alberta	1993	388^{A} (2)					388
	1994	100^{A} (1)					100
	1996			75^{A} (1)			75
	1997			175^{A} (1)			175
	1998			200^{A} (2)			200
Saskatchewan	1999	5150^{A} (4)					5150
Manitoba	1992				40^{E} (1)		40
	1993	1981^{A,L} (6)	366^{A} (2)				2347
	1994	1037^{A} (12)	448^{A} (6)		140^{L} (1)		1625
	1995	5883^{A} (12)			1500^{E} (3)		7383
	1996	7650^{A} (15)			550^{E} (1)		8200
	1997	32,500^{A} (15)			1600^{E} (5)	720^{A} (3)	34,820
	1998	50,750^{A} (15)					50,750
	1999	57,190^{A,L} (28)			110^{A} (1)		57,300
Ontario	1992			2800^{L} (6)			2800
	1993			15,700^{A} (50)	300^{L} (2)		16,000
	1994			22,100^{A} (38)	553^{L} (1)		22,653
	1995			30,600^{A} (45)			30,600
	1996			27,950^{A,L} (27)			27,950
	1997			218,965^{A,L,P} (55)			218,965
	1998			80,000^{L} (16)			80,000
	1999			90,000^{L} (12)			90,000
Quebec	1996	1200^{A} (2)	1200^{A} (2)				2400
	1997			8000^{A,L} (8)			8000
	1998			2000^{L} (3)			2000
Nova Scotia	1994			100^{A} (1)	189^{L} (2)		289
	1995			300^{A} (1)			300
	1996			975^{A} (1)			975
	1997			4600^{A} (4)			4600
	1998			31,000^{A,L} (4)			31,000
	1999			100,000^{L} (3)			100,000
New Brunswick	1993		148^{A} (1)				148
	1994	990^{A} (2)	250^{A} (2)				1240
	1995	1000^{A} (2)	800^{A} (2)				1800
	1996			500^{A} (2)			500
	1997			3600^{A} (2)			3600
	1998			20,000^{A} (5)			20,000
	1999			77,000^{L} (5)			77,000
Prince Edward Island	1993	390^{A} (4)	950^{A} (5)				1340
	1994	150^{A}					150
	1996			1400^{A} (4)			1400
	1997			2300^{A} (2)			2300
	1998			20,000^{L,P} (9)			20,000
	1999			50,000^{A} (6)			50,000
Grand total 1992–1999							995,963

H. transversovittatus has been released in British Columbia, Alberta, Manitoba, Ontario and Nova Scotia. At Iona, British Columbia, it is believed that the weevil did not establish due to high tides. In Alberta, larvae were released (within roots of transplanted plants) in 1994 in an open garden plot at Lethbridge, and adults were recovered in 1998 and 1999. In Manitoba, *H. transversovittatus* was released in October, 1992, in the Spruce Woods/Cypress area. Larvae overwintered but no adult weevils have been found to date. In 1996, eggs implanted into cut stems developed and adults were found in 1999. Adults obtained from Cornell University also were released in Manitoba in 1999, near the Libau Marsh. In Ontario, *H. transversovittatus* was released in 1993 and 1994. Releases were discontinued after 1994 because the species was difficult and expensive to rear. It did not establish at any of the Ontario release sites. In Nova Scotia, the status of *H. transversovitattus*, released as larvae in 1994, is uncertain.

N. marmoratus adults were released[2] in 1997 in the Libau Marsh, Manitoba. The population successfully overwintered and reproduced in 1998.

Portions of the initial European importations of *G. calmariensis* and *G. pusilla* were distributed to programmes in Alberta, Manitoba and Ontario in 1992 (Hight *et al.*, 1995). All subsequent Canadian releases of these two species are descended from these populations.

In British Columbia, releases were done annually from 1993 to 1999, with both *Galerucella* spp. being released at 37 sites. It is estimated that 50 to 83% of these have established (R. Cranston, Abbotsford, 1999, personal communication). In Alberta, at one of the three original (1993–1994) release sites near Lethbridge, the beetles established along one side of Gaeol Lake. Releases of *Galerucella* spp. were made at Fort Macleod from 1996 to 1998. Establishment has been confirmed there but beetle numbers are low. The Saskatchewan Purple Loosestrife Eradication Project obtained *G. calmariensis* brood stock (from Manitoba) in 1999 and began mass rearing and releases near Saskatoon and Moose Jaw.

In Manitoba, initial releases of *Galerucella* occurred in 1993. The Manitoba Purple Loosestrife Program has mass-reared *G. calmariensis* from 1994 to 1999, and released this species at over 100 sites from 1993 to 1999. *G. pusilla* was released at eight Manitoba sites in 1993–1994. In an effort to increase agent production, a satellite mass-rearing project was initiated in 1999, involving local stakeholder groups, e.g. the Manitoba Weed Supervisors Association, to rear and release *G. calmariensis* in their local areas.

In Ontario, initial releases of *Galerucella* adults were made at the Speed River, Guelph, in 1992. From 1993 to mid-1996, laboratory-reared *Galerucella* spp. were released at 151 sites into the following general areas: the Grand River watershed around Kitchener–Waterloo and Cambridge, several wetlands in the Mississauga–Burlington area, the Lake St Clair–Detroit River area, the Niagara region, around the lower Bruce Peninsula, the lower Trent watershed, and the Rideau valley watershed. After mid-1996, all Ontario releases were done by redistributing adults and larvae collected from well-established field populations containing both species. In 1996–1997, releases were concentrated in the Grand River watershed as part of a watershed-wide management plan. After termination of the Ontario Program in 1997 (due to lack of funding), a private company continued to make releases with field-collected larvae of both species in 1998 and 1999.

In Quebec, initial releases of adult *Galerucella* spp. in 1996 were along the St Lawrence River and rivers in the Outaouais region, but no establishment occurred (Templeton and Stewart, 1999). In 1997,

[2]The Manitoba Purple Loosestrife Project partnered with Cornell University and the Minnesota Purple Loosestrife Program in autumn, 1996, to collect and import *N. marmoratus* and *N. brevis* from Europe.

adults and larvae, and, in 1998, larvae were released at Lac St François National Wildlife Reserve, near Nicolet, in Hull near the Champlain bridge, and at Cap Tourmente National Wildlife Reserve. In spring, 1998, overwintered adults were found at these release sites (Templeton and Stewart, 1999).

In the Maritimes, *Galerucella* spp. were released from 1993 to 1999 at 23 sites in New Brunswick, by the provincial Department of Agriculture and Rural Development and Ducks Unlimited Canada. The Nova Scotia Agricultural College reared and released beetles from 1994 to 1999. They have established at over 50 sites (G. Sampson, Truro, 1999, personal communication). In Prince Edward Island, beetles have been released at 31 sites since 1993, including Bothwell, Souris, Stratford and Southport (J. Stewart, Charlottetown, 1999, personal communication), and have established at most release sites. From 1997 to 1999, release programmes were intensified in the three Maritime provinces, with over 300,000 *Galerucella* spp. being released at 39 sites.

Evaluation of Biological Control

The biological control programme against *L. salicaria* appears to be developing into a major success. Based upon initial data and observations from across Canada (and the USA), it is apparent that the *Galerucella* spp. alone may be able to effectively control *L. salicaria* in a variety of habitats. In the following discussion, 'control' is considered to mean: (i) over 95% suppression of *L. salicaria* biomass; (ii) over 99% suppression of flowering and seed production; and (iii) substantial replacement of *L. salicaria* with other plant species.

In British Columbia, herbivory damage by *G. calmariensis* released near Chilliwack and at Jericho Park in 1999 was estimated at 90–100% (Myers and Denoth, 1999). In Alberta, populations of *L. salicaria* were suppressed along one side of Gaeol Lake as a result of *G. calmariensis* releases in 1993 and, by 1998, the beetles had dispersed across the lake and established in a new *L. salicaria* stand. In Nova Scotia, *G. calmariensis* had reduced flowering by 80–90% in at least one release site in 1999 (G. Sampson, Truro, 1999, personal communication).

Results from Canada's two largest provincial programmes merit further discussion. In Manitoba, close to 100% control of *L. salicaria* has been achieved at many release sites, including Delta Marsh, areas within the Libau Marsh, Winnipeg River at Great Falls, Red Rock Lake in the Whiteshell, along Highway #317, and sites in the City of Winnipeg. Fixed monitoring stations were established at two release sites in the Libau Marsh and one site in the Delta Marsh, with data collected from 30 randomly tagged stems per site at 10-day intervals from late spring to early autumn. Populations of *G. calmariensis* increased significantly in the third (Delta), fourth or fifth years (Libau sites) after release. In the Libau Marsh, herbivory resulted in all stems being destroyed between 5 and 6 years after release of *G. calmariensis*. The Delta Marsh received the fewest beetles (250), with all *L. salicaria* stems being destroyed by mid-July of each year since 3 years after release. Within a year of explosion of beetle populations, high levels of herbivory resulted in death of all stems at these sites by July to early August. To obtain significant control of *L. salicaria* in Manitoba, *Galerucella* egg densities approaching 600 eggs m^{-2} need to be attained (Diehl, 1999). At the Delta Marsh site, Diehl (1999) reported a 2537% increase in the number of eggs m^{-2} between the second and third years after release. This resulted in a reduction in numbers of stems from 32 to 0 m^{-2}. Diehl (1999) also reported that there was no difference in overwintering survival between the two *Galerucella* spp., that both can tolerate prolonged periods of spring flooding, and that initial dispersal was largely limited to within 5 m of the point of release. An integrated vegetation management strategy is being developed in Manitoba, integrating *G. calmariensis* with herbicide applications (Lindgren *et al.*, 1998, 1999).

Integration of herbicide use with beetles resulted in the most effective suppression of *L. salicaria* stem densities. In herbicide-alone trials, stem densities at the end of the study were greater than before treatment (Henne, 2000).

In Ontario, large populations of the two *Galerucella* spp. (>50 egg masses m^{-2}) were beginning to control *L. salicaria* by 1995 at three of the initial (1992–1993) release sites. By 1999, *L. salicaria* was under control in seven areas of southern Ontario. Densities of 300–600 egg masses m^{-2} have been found in all these areas, and these sites were virtually unrecognizable as *L. salicaria* infestations by 1999 (Bowen, 1998). Effective beetle populations are established in most of the heavily infested areas of southern Ontario, including the Detroit River below Windsor, the western end of Lake Ontario (Bowen, 1998), through much of the Grand River watershed, the Sydenham River in Owen Sound, the Otonabee River in Peterborough, and the Rideau River watershed. Beetles have spread from several release sites (Grand River, Speed River, Etobicoke Creek and Lake Ontario) to occupy at least 100 km of shoreline. The rate of spread is estimated to be 5–10 km $year^{-1}$ from the best release sites. A comprehensive watershed-wide control strategy, initiated in 1996 by the Grand River Watershed Management Plan for Purple Loosestrife, was highly successful. It is anticipated that control of *L. salicaria* will be achieved through most of this watershed in the next 5–10 years. Beetles continue to spread in Ontario, and we believe that they will eventually be found in all of the *L. salicaria* populations in the province.

Of the biological control agents available for *L. salicaria*, *G. calmariensis* has proved highly reproductive, easy to mass-rear, effective and has been the most widely released agent across Canada. Monitoring indicates an *L. salicaria–G. calmariensis* interaction model as follows: significant increases in the *G. calmarienis* population occur as early as the third or fourth year after release, followed by suppression or elimination of *L. salicaria* sexual reproduction, a decline in overall stem height, a reduction in stem number and, finally, a change in the *G. calmariensis* population growth curve from positive to negative as *L. salicaria* is suppressed (Lindgren, 2000). Observations from Ontario further suggest that *G. calmariensis* and *G. pusilla* can coexist and provide effective weed control. At the Ontario sites, the *Galerucella* spp. were released less than 1 km from each other. Populations of the two species subsequently overlapped within 2 years. The coalescence of the two *Galerucella* species at these sites promoted both control and rapid, long-distance dispersal from the original release sites. Finally, in Ontario, effective redistribution of *Galerucella* spp. from successful field sites has been done, with a high rate of establishment and weed control.

Limited feeding by *G. calmariensis* was observed on the native, non-target species *Lythrum alatum* Pursh and *Decodon verticillatus* (L.) Elliott at the Royal Botanical Gardens in Burlington, Ontario (Corrigan *et al.*, 1998). Both of these had been attacked in 'no-choice' host-specificity testing prior to beetle importation into North America (Kok *et al.*, 1992). We believe that the feeding observed at the Botanical Gardens is a short-term, spillover effect, and that these species are not at long-term risk from the biological control agents (Corrigan *et al.*, 1998). The impact of *L. salicaria* on two endangered plant species, *Sidalcea hendersonii* Watson and *Caltha palustris* L., is also under investigation in British Columbia (Myers and Denoth, 1999).

Historically, biological control programmes targeted agricultural weeds. Because *L. salicaria* is a weed of aquatic habitats, it has resulted in new audiences being introduced to biological control of weeds (Blossey *et al.*, 1996). To build support, it is essential that programme objectives and results be communicated to them. The importance of fostering community awareness and involving community partners cannot be overlooked, especially for weeds invading natural areas.

The effort to control *L. salicaria* has been immense, with the involvement of numerous stakeholder groups and contributions from a equally large number of funding agencies across Canada. While *L. salicaria* is an exotic species recognized as a primary invader of natural habitats (White *et al.*, 1993), it is unfortunate that programme funding has restricted and, in some cases, eliminated provincial biological weed control initiatives. Despite the encouraging control results so far, it may be premature to restrict our biological control toolbox to only the *Galerucella* spp. Long-term funding (15–20 years) is needed to further the biological control efforts against *L. salicaria*.

Recommendations

Further work should include:

1. Assessing the establishment and performance of *H. transversovittatus* and *N. marmoratus*;
2. Long-term monitoring of the biological control agents and associated changes in *L. salicaria* populations;
3. Documenting the response of native plant communities;
4. Further developing integrated vegetation management strategies.

Acknowledgements

J. Meyers, M. Denoth, R. Cranston, S. Ali, C. Verbeek, A. Salzl, J. Diehl, G. Sampson, J. Wile, T. Duffy, J. Stewart, K. Templeton, J. Laing, D. Mackenzie, K. McCully, R. Langevin and B. Blossey provided important information. G. Lee initiated the Canadian programme development. Canadian efforts would not have been possible without the screening and host-specificity testing conducted by American and European cooperators.

References

Ali, S. and Verbeek, C. (1999) *The Alberta Purple Loosestrife Eradication Program 1999 Status Report.* Alberta Agriculture, Food and Rural Development, Edmonton, Alberta.

Batra, S.W.T., Schroeder, D., Boldt, P.E. and Mendl, W. (1986) Insects associated with purple loosestrife (*Lythrum salicaria*) in Europe. *Proceedings of the Entomological Society of Washington* 88, 748–759.

Blossey, B. (1993) Herbivory below ground and biological weed control: life history of a root-boring weevil on purple loosestrife. *Oecologia* 94, 380–387.

Blossey, B. and Schroeder, D. (1995) Host specificity of three potential biological weed control agents attacking flowers and seeds of *Lythrum salicaria* (Purple Loosestrife). *Biological Control* 5, 47–53.

Blossey, B., Malecki, R.A., Schroeder, D. and Skinner, L. (1996) A biological weed control programme using insects against purple loosestrife, *Lythrum salicaria*, in North America. In: Moran, V.C. and Hoffmann, J.H. (eds) *Proceedings of the IX International Symposium on Biological Control of Weeds, 19–26 January 1996, Stellenbosch, South Africa.* University of Cape Town, Cape Town, South Africa, pp. 351–355.

Bowen, K. (1998) Beetles offer hope for purple loosestrife control. *Pappus* 17, 21–27.

Corrigan, J.E., MacKenzie, D.L. and Simser, L. (1998) Field observations of non-target feeding by *Galerucella calmariensis* [Coleoptera: Chrysomelidae], an introduced biological control agent of purple loosestrife, *Lythrum salicaria* [Lythraceae]. *Proceedings of the Entomological Society of Ontario* 129, 99–106.

De Clerck-Floate, R. (1992) *The Desirability of Using Biocontrol Against Purple Loosestrife in Canada.* Agriculture Canada, Lethbridge, Alberta.

Diehl, J.K. (1999) Biological control of purple loosestrife, *Lythrum salicaria* L. (Lythraceae) with *Galerucella* spp. (Coleoptera: Chrysomelidae): dispersal, population change, overwintering ability, and predation of the beetles, and impact on the plant in southern Manitoba wetland release sites. MSc thesis. University of Manitoba, Winnipeg, Manitoba.

Diehl, J.K., Holliday, N.J., Lindgren, C.J. and Roughley, R.E. (1997) Insects associated with purple loosestrife, *Lythrum salicaria* L., in southern Manitoba. *The Canadian Entomologist* 129, 937–948.

Harp, H.F. and Collicutt, L.M. (1983) *Lythrums for Home Gardens.* Publication 1285E, Communications Branch, Agriculture Canada, Ottawa, Ontario.

Henne, D.C. (2000) Evaluation of an integrated management approach for the control of purple loosestrife, *Lythrum salicaria* L., in southern Manitoba: biological control and herbicides. MSc thesis, University of Manitoba, Winnipeg, Manitoba.

Hight, S.D., Blossey, B., Laing, J. and De Clerck-Floate, R. (1995) Establishment of insect biological control agents from Europe against *Lythrum salicaria* in North America. *Environmental Entomology* 24, 967–977.

Kok, L.T., McAvoy, T.J. , Malecki, R.A., Hight, S.D., Drea, J.J. and Coulson, J.R. (1992) Host specificity tests of *Galerucella calmariensis* (L.) and *G. pusilla* (Duft.) (Coleoptera: Chrysomelidae), potential biological control agents of purple loosestrife, *Lythrum salicaria* L. (Lythraceae). *Biological Control* 2, 282–290.

Lindgren, C.J. (2000) Performance of a biological control agent, *Galerucella calmariensis* L. (Coleoptera: Chrysomelidae) on Purple Loosestrife *Lythrum salicaria* L. in southern Manitoba (1993–1998). In: Spencer, N.R. (ed.) *Proceedings of the X International Symposium on Biological Control of Weeds, 4–14 July 1999, Bozeman, Montana USA.* Montana State University, Bozeman, Montana, pp. 367–382.

Lindgren, C.J. and Clay, R.T. (1993) Fertility of 'Morden Pink' *Lythrum virgatum* in Manitoba. *HortScience* 28, 954.

Lindgren, C.J., Gabor, T.S. and Murkin, H.R. (1998) Impact of triclopyr amine on *Galerucella calmariensis* L. (Coleoptera: Chrysomelidae) and a step toward integrated management of purple loosestrife *Lythrum salicaria* L. *Biological Control* 12, 14–19.

Lindgren, C.J., Gabor, T.S. and Murkin, H.R. (1999) Compatibility of glyphosate with *Galerucella calmariensis*; a biological control agent for purple loosestrife (*Lythrum salicaria*). *Journal of Aquatic Plant Management* 37, 44–48.

Malecki, R.A., Blossey, B., Hight, S.D., Schroder, D., Kok, L.T. and Coulson, J.R. (1993) Biological control of purple loosestrife. *BioScience* 43, 680–686.

Myers, J. and Denoth, M. (1999) *Endangered Species Recovery Fund Report, 31 November, 1999.* University of British Columbia, Vancouver, British Columbia.

Ottenbreit, K. (1991) The distribution, reproductive biology, and morphology of *Lythrum* species, hybrids and cultivars in Manitoba. MSc thesis, University of Manitoba, Winnipeg, Manitoba.

Templeton, K. and Stewart, R.K. (1999) *Pilot Project on the Biological Control of Purple Loosestrife in Quebec.* MacDonald College, McGill University, Montreal, Quebec, Canadian Wildlife Service and Ontario Royal Botanical Gardens.

Thompson, D.Q., Stuckey, R.L. and Thompson, E.B. (1987) Spread impact and control of purple loosestrife (*Lythrum salicaria*) in North American wetlands. *United States Fish and Wildlife Service, Fish Wildlife Research* 2, 1–55.

White, D.J., Haber, E. and Keddy, C. (1993) *Invasive Plants of Natural Habitats in Canada.* Canadian Wildlife Service, Environment Canada, Ottawa, Ontario.

75 *Malva pusilla* Smith, Round-leaved Mallow (*Malvaceae*)

K. Mortensen and K.L. Bailey

Pest Status

Round-leaved mallow, *Malva pusilla* Smith, also called *M. rotundifolia* L., was introduced from Eurasia and occurs in every province except Newfoundland, but is most common in the prairie provinces. It has long been considered a weed of farmyards, gardens and waste areas (Frankton and Mulligan, 1987). In Saskatchewan and Manitoba, surveys indicated that *M. pusilla* has become more common in cultivated land (Thomas, 1978a, b; Thomas and Wise, 1988; Thomas *et al.*, 1995). In Alberta, *M. pusilla* doubled in abundance on cultivated land from 1980 to 1985, according to the Alberta weed alert reporting system. High infestations of *M. pusilla* were found mainly in eastern Saskatchewan and Manitoba, and are more prevalent on dark soils (Makowski, 1995). *M. pusilla* can be a serious weed in less competitive crops, e.g. flax, *Linum usitatissimum* L., and lentils, *Lens culinaris* Medicus. Yield losses up to 90% have been reported in flax (Makowski and Morrison, 1989), and up to 80% in lentils (Makowski, 1995). *M. pusilla* causes fewer problems in competitive cereal crops but, if it gets a head start, yield losses up to 30% may occur in wheat, *Triticum aestivum* L. (Makowski, 1995). *M. pusilla* can cause serious problems in harvest equipment and large amounts of seeds are left in stubble after harvest.

M. pusilla is an annual that emerges throughout summer and grows well into autumn. It has a long tap root, a prostrate growth habit, and a stem with many branches that can extend over 1 m long and produce large amounts of seeds. Due to the hard seed coat, seeds exhibit low germination if not scarified, and thus can persist for a long time in soil. Seed capsules are about the size of cereal kernels, the individual seeds are slightly smaller than canola, *Brassica napus* L. and *B. rapa* L., seeds, making it difficult to screen out *M. pusilla* seeds using standard methods. Thus, it can be a serious contaminant in seed of many crops.

Background

Some herbicides, e.g. bromoxynil (3,5-dibromo-4-hydroxybenzonitrite) plus MCPA (4-chloro-2-methylphenoxyacetic acid), appear to give good control, with larger leaves turning completely necrotic after 7–10 days. However, new growth initiated within 1 week at the centre of surviving plants appeared normal. None of the tested herbicides gave consistent control (Makowski and Morrison, 1989). Cultivation can kill *M. pusilla* if the tap root is severed below the crown, otherwise regrowth will occur. Mowing and grazing will delay growth for a short time, but rapid recovery with increased branching below the injured area usually takes place.

Makowski and Morrison (1989) reported several insects on *Malva* spp. from various areas of the world. *M. rotundifolia* was described as a host for many of the insects reported, but confusion in the taxonomy of *M. pusilla* and common mallow, *Malva neglecta* Wallroth, have resulted in both weed species being referred to as *M. rotundifolia*. In Saskatchewan, *Vanessa cardui* (L.) larvae fed on leaves of *M. pusilla*. The potato aphid, *Macrosiphum euphorbiae* (Thomas), was found on *M. rotundifolia* in eastern Washington (Landis *et al.*, 1972). *Calycomyza malvae* (Burgess) larvae form leaf mines on *M. rotundifolia* in the USA (Spencer and Steyskal 1986). *Systena blanda* (Melsheimer) adults fed on *M. rotundifolia* in onion, *Allium cepa* L., fields in Ohio (Drake and Harris, 1931). The nematode, *Ditylenchus dipsaci* (Kuhn) Filip, was found on *M. pusilla* in Italy (Greco, 1976). Several fungi have been reported on *M. pusilla* in Canada and the USA: a rust, *Puccinia malvacearum* Montagne; leaf spots caused by *Cercospora* spp., *Septoria malvicola* Ellis and Martin, *Colletotrichum malvarum* (Braun and Caspary) Southworth, and *Colletotrichum gloeosporioides* (Penzig) Saccardo f. sp. *malvae* (Mortensen, 1988; Farr *et al.*, 1989). *C. gloeosporioides* f. sp. *malvae* is the only agent that showed sufficient impact on *M. pusilla* under prairie conditions.

Biological Control Agents

Pathogens

Fungi

C. gloeosporioides f. sp. *malvae* (sexual stage unknown) causes anthracnose of *M. pusilla* and was first observed in 1982 on its seedlings in a greenhouse. Later, the disease was found at various locations in Saskatchewan and Manitoba. Sticky masses of conidia are produced in acervuli on infected leaves and stems. Conidia suspend readily in water and spread by rain-splash to neighbouring healthy plants, where germination and new infection take place. The fungus overwinters in infected *M. pusilla* debris but, under natural conditions, not in sufficient amount to give adequate control (Mortensen, 1988). *C. gloeosporioides* f. sp. *malvae* was shown to be sufficiently host specific, could be produced on artificial media, and was effective in controlling *M. pusilla* when applied as a spore suspension. Thus, *C. g. malvae* was deemed to have the characteristics required for a successful biological herbicide (TeBeest and Templeton, 1985; Charudattan, 1991).

In 1985, an agreement to commercialize *C. g. malvae* was signed between Philom Bios Inc., Saskatoon, Saskatchewan, and Agriculture and Agri-Food Canada. Commercialization included registration and successful marketing.

Registration

In Canada, biological control products are regulated under the Canadian Pest Control Products Act. When the first registration package for *C. g. malvae* was submitted in 1987, there were no well-defined guidelines or regulations for safety testing of microbial pest control products. The requirements at that time were loosely based on those used to determine hazards associated with chemical pesticides, and on the microbial guidelines developed by the Environmental Protection Agency (EPA) under the US Federal Insecticide, Fungicide, and Rodenticide Act (FIFRA). Canadian regulatory agencies at that time did not accept the US registration requirements and insisted on additional information. As a result very few of the initial safety tests conducted in the mid-1980s with *C. g. malvae* were accepted. Therefore, a series of meetings was held with Agriculture Canada, Health and Welfare Canada, and Environment Canada to review and agree on types of safety tests and the protocols to be used to generate the data. Human and environmental toxicity, infectivity, irritation and residue protocols were determined, based on expansions of the EPA-approved protocols for microbial pest control agents. Consultation with the EPA confirmed that the results generated with the Canadian-approved tests would be acceptable for EPA's review of product registrability under the US FIFRA. The costs for conducting the Canadian-approved protocols (four acute mammalian infectivity/ toxicity tests, two mammalian irritation studies, and three environmental toxicity studies) were triple those quoted for the EPA-approved protocols (Cross and Polonenko, 1996). Environmental toxicity included crop tolerance, infectivity and efficacy tests on eight field crops (Mortensen and Makowski, 1997; Makowski and Mortensen, 1998, 1999). Subsequently, a complete *C. g. malvae* registration application was prepared and resubmitted for regulatory review by the end of 1990. The regulatory review process was completed within 13 months and a full registration was granted in February 1992. *C. g. malvae* (tradename: BioMal) was the first bioherbicide product to receive registration under the Canadian Pest Control Products Act.

Marketing

Bioherbicides need to be fast acting, predictable, easy to use and provide a level of weed control comparable to chemical herbicides before they are generally accepted by industry and producers (Bowers, 1982; Charudattan, 1990). Many plant pathogens are quite host specific, which allows a bioherbicide to be used to control a weed in a closely related crop. The disadvantage is that they will only control a single weed species. The problem weed must therefore be of significant economic importance for a private company to invest in commercialization of a bioherbicide (Charudattan, 1990; Cross and Polonenko, 1996). *C. gloeosporioides* f. sp. *malvae* provides satisfactory control of *M. pusilla* but forms only sublethal lesions on closely related species, e.g. *M. neglecta*, *M. parviflora* and *Abutilon theophrasti* (Mortensen, 1988). Early market assessment, based on producer responses in the mid-1980s, indicated that the incidence of *M. pusilla* in Saskatchewan alone was about 160,000 ha. Later more detailed market research by Philom Bios in the early 1990s showed that the number of 'treatable' hectares, i.e. areas for which producers would buy the product, was no greater than 40,000 ha in western Canada. Further, *M. pusilla* occurs in patches, so producers would likely 'spot-spray' and therefore preferred to receive the product in packages sufficient to treat 0.8 ha (2 acres). Philom Bios determined that the production and packaging costs would have driven the BioMal retail price beyond what producers would be willing to pay, and would not have provided any return on their investment in the product development process or any margins to their marketing partners and distribution system. Therefore, a decision was made in 1994 not to pursue commercial sales of BioMal (Cross and Polonenko, 1996). The licence to commercialize and market BioMal was then terminated.

Evaluation of Biological Control

The registration of BioMal in 1992 only included control of *M. pusilla* in eight field crops (Makowski and Mortensen, 1992). However, as discussed, this was not sufficient from a marketing perspective. *C. g. malvae* can be safely used in many vegetable crops (Mortensen, 1988) and has effectively controlled *M. pusilla* and increased yield in strawberries, *Fragaria* × *ananassa* Duschene (Mortensen and Makowski, 1995). Extending the licence to vegetable crops, small fruits and gardens, where *M. pusilla* is often a serious problem (Makowski and Morrison, 1989), would increase the market potential considerably. Although *C. g. malvae* does not adequately control related weeds, recent experiments showed that it may be possible to increase the effectiveness of *C. g. malvae* on the marginal host *A. theophrasti* through improvement in application methods, and application together with reduced amounts of chemical herbicides (Kutcher and Mortensen, 1999). *A. theophrasti* is a serious weed in maize, *Zea mays* L., and soybean, *Glycine max* (L.) Merrill, in eastern Canada and the USA, and is difficult to control due to its biology and tolerance for many herbicides used in these crops (Spencer, 1984; Warwick and Black, 1988). This larger potential market for *C. g. malvae* should be of interest to industry.

In 1998, negotiations were initiated between Agriculture and Agri-Food Canada, Philom Bios Inc., and a US company, Encore Technologies (Carlson Business Centre, Minnetonka, Minnesota 55305, USA), for re-registration and commercialization of *C. g. malvae*. An agreement has been reached, and Encore Technologies is in the process of re-registering *C. g. malvae* for control of *M. pusilla*. Release of the product on to the market was planned for 2001.

Recommendations

Further work should include:

1. Increasing the effectiveness of *C. g. malvae* on *A. theophrasti* through improvement in application methodology, and application together with reduced rates of chemical herbicides.

References

Bowers, R.C. (1982) Commercialization of microbial biological control agents. In: Charudattan, R. and Walker, H.L. (eds) *Biological Control of Weeds with Plant Pathogens*. John Wiley & Sons, New York, New York, pp. 157–173.

Charudattan, R. (1990) Assessment of efficacy of mycoherbicide candidates. In: Delfosse, E.S. (ed.) *Proceedings of the VII International Symposium on Biological Control of Weeds (1988), Rome, Italy*. Istituto Sperimentale per la Patologia Vegitale, Ministero dell'Agricoltura e dell Foreste, Rome, Italy, pp. 455–464.

Charudattan, R. (1991) The mycoherbicide approach with plant pathogens. In: TeBeest, D.O. (ed.) *Microbial Controls of Weeds*. Chapman and Hall, New York, New York, pp. 24–57.

Cross, J.V. and Polonenko, D.R. (1996) An industry perspective on registration and commercialization of biocontrol agents in Canada. *Canadian Journal of Plant Pathology* 18, 446–454.

Drake, C.J. and Harris, H.M. (1931) The palestriped flea beetle, a pest of young seedling onions. *Journal of Economical Entomology* 24, 1132–1137.

Farr, D.F., Bills, G.F., Chamuris, G.P. and Rossman, A.Y. (1989) *Fungi on Plants and Plant Products in the United States*. APS Press, St Paul, Minnesota.

Frankton, C. and Mulligan, G.A. (1987) *Weeds of Canada*. Publication 948, Agriculture Canada, Ottawa, Ontario.

Greco, N. (1976) Weed host of *Ditylenchus dipsaci* in Puglia. *Nematology of Mediterranean* 4, 99–102.

Kutcher, H.R. and Mortensen, K. (1999) Genotypic and pathogenic variation of *Colletotrichum gloeosporioides* f. sp. *malvae*. *Canadian Journal of Plant Pathology* 21, 37–41.

Landis, B.J., Powell, D.M. and Fox, L. (1972) Overwintering and winter dispersal of the potato aphid (*Macrosiphum euphorbiae*: Hem., Hom., Aphididae) in Eastern Washington. *Enviromental Entomology* 1, 68–71.

Makowski, R.M.D. (1995) Round-leaved mallow (*Malva pusilla*) interference in spring wheat (*Triticum aestivum*) and lentil (*Lens culinaris*) in Saskatchewan. *Weed Science* 43, 381–388.

Makowski, R.M.D. and Morrison, I.N. (1989) The biology of Canadian weeds. 91. *Malva pusilla* Sm. (= *M. rotundifolia* L.). *Canadian Journal of Plant Science* 69, 861–879.

Makowski, R.M.D. and Mortensen, K. (1992) The first mycoherbicide in Canada: *Colletotrichum gloeosporioides* f. sp. *malvae* for round-leaved mallow control. In: Richardson, R.G. (ed.) *Proceedings of the First International Weed Congress 2*. Monash University, Melbourne, Australia, pp. 298–300.

Makowski, R.M.D. and Mortensen, K. (1998) Latent infections and penetration of the bioherbicide agent *Colletotrichum gloeosporioides* f. sp. *malvae* on non-target field crops under controlled environmental conditions. *Mycological Research* 102, 1545–1552.

Makowski, R.M.D. and Mortensen, K. (1999) Latent infections and residues of the bioherbicide agent *Colletotrichum gloeosporioides* f. sp. *malvae*. *Weed Science* 47, 589–595.

Mortensen, K. (1988) The potential of an endemic fungus, *Colletotrichum gloeosporioides*, for control of round-leaved mallow (*Malva pusilla*) and velvetleaf (*Abutilon theophrasti*). *Weed Science* 36, 473–478.

Mortensen, K. and Makowski, R.M.D. (1995) Tolerance of strawberries to *Colletotrichum gloeosporioides* f. sp. *malvae*, a mycoherbicide for control of round-leaved mallow (*Malva pusilla*). *Weed Science* 43, 429–433.
Mortensen, K. and Makowski, R.M.D. (1997) Effects of *Colletotrichum gloeosporioides* f. sp. *malvae* on plant development and biomass of non-target field crops under controlled and field conditions. *Weed Research* 37, 351–360.
Spencer, N.R. (1984) Velvetleaf, *Abutilon theophrasti* (Malvaceae). History and economic impact in United States. *Economic Botany* 38, 406–416.
Spencer, K.A. and Steyskal, G.C. (1986) *Manual of the Agromyzidae (Diptera) of the United States.* Agricultural Handbook 638, United States Department of Agriculture, Agriculture Research Service, Washington, DC, pp. 140–149, 235.
TeBeest, D.O. and Templeton, G.E. (1985) Mycoherbicides. Progress in the biological control of weeds. *Plant Disease* 69, 6–10.
Thomas, A.G. (1978a) *The 1978 Weed Survey of Cultivated Land in Saskatchewan.* Weed Survey Series. Publication 78–2, Agriculture Canada, Regina, Saskatchewan.
Thomas, A.G. (1978b) *The 1978 Weed Survey of Cultivated Land in Manitoba.* Weed Survey Series. Publication 78–3, Agriculture Canada, Regina, Saskatchewan.
Thomas, A.G. and Wise, R.F. (1988) *Weed Survey of Manitoba Cereal and Oilseed Crops 1986.* Publication 88–1, Weed Survey Series. Agriculture Canada, Regina, Saskatchewan.
Thomas, A.G., Wise, R.F., Frick, B.L. and Juras, L.T. (1995) *Saskatchewan Weed Survey, Cereal, Oilseed and Pulse Crops 1995.* Publication 96–1, Weed Survey Series, Agriculture and Agri-Food Canada, Saskatoon Research Centre, Saskatoon, Saskatchewan.
Warwick, S.I. and Black, L.D. (1988) The biology of Canadian weeds. 90. *Abutilon theophrasti.* *Canadian Journal of Plant Science* 68, 1069–1085.

76 *Matricaria perforata* Mérat, Scentless Chamomile (*Asteraceae*)

A.S. McClay, H.L. Hinz, R.A. De Clerck-Floate and D.P. Peschken

Pest Status

Scentless chamomile, *Matricaria perforata* Mérat,[1] an introduced summer annual, winter annual or short-lived perennial native to Europe and Asia, has become a widely distributed weed of disturbed and cultivated land in Canada, particularly in the prairie provinces. It is common in roadsides, drainage ditches, cropland, hayland, wasteland (Woo *et al.*, 1991) and industrial areas. In agricultural land it is associated with slough margins and transition areas, such as field edges and rights-of-way (Bowes *et al.*, 1994). It occurs particularly in low-lying areas that are poorly drained and difficult to cultivate in spring (Douglas, 1989; Woo *et al.*, 1991).

[1]In the North American literature, scentless chamomile has mostly been referred to as *Matricaria perforata* Mérat. In Europe it is usually referred to as *Tripleurospermum inodorum* (L.) Schultz-Bipontinus or *T. perforatum* (Mérat) Laínz.

M. perforata spreads rapidly because of its profuse seed production, up to 256,000 seeds per plant. Plants emerging by early July usually flower and produce seed, whereas those emerging from mid-July onwards overwinter as rosettes that bolt and flower the following year (Blackshaw and Harker, 1997). In farmers' fields in Saskatchewan, *M. perforata* at a density of 25 plants m^{-2} in spring wheat, *Triticum aestivum* L., caused yield losses ranging from 30 to 80%. Actual densities of *M. perforata* in these fields reached up to 70 plants m^{-2}. The winter annual form is particularly competitive and yield losses due to *M. perforata* were greater in moist years (Douglas *et al.*, 1991, 1992). *M. perforata* can act as a host for several insect pests of crops (Woo *et al.*, 1991) and for one pathogen, aster yellows phytoplasma, which attacks a wide range of crop species (Khadhair *et al.*, 1999).

Background

Several herbicides are available to control *M. perforata* in cereals; however, most are only effective against seedlings. In canola, clopyralid (which has recropping restrictions for other crops such as pulses), glufosinate ammonium (only for tolerant varieties of canola) and diquat (used as a crop desiccant) are currently used. In most forage legumes, pulses and special crops, no chemical control is available (Ali, 1999).

Because few native plants in Canada are closely related to *M. perforata*, and because over 70 insects and fungi were recorded from it in the literature, of which two insects and two fungi were considered to have a narrow enough host range to be worth further study, Peschken (1989) and Peschken *et al.* (1990) proposed it as a target for biological control.

Biological Control Agents

Insects

Freese and Günther (1991) conducted field surveys for insects associated with *M. perforata* in Europe. The first agent selected, *Omphalapion hookeri* (Kirby), was tested for host specificity at Regina, beginning in 1988. Beginning in 1991, further agents were studied in Switzerland. The root-mining weevils, *Diplapion confluens* Kirby and *Coryssomerus capucinus* (Beck), and the stem-mining weevil, *Microplontus rugulosus* (Herbst), had too broad a host range within the tribe Anthemideae and posed a potential risk to possible future cultivation of German chamomile, *Chamomilla recutita* L., in Canada (Hinz and Leiss, 1996; Hinz and Müller-Schärer 2000a), so were rejected for introduction.

O. hookeri (Kirby) (previously placed in *Apion*) is a small, univoltine weevil, distributed widely across Europe. Females oviposit in young flower buds of *M. perforata* and larvae feed on developing seeds. Pupation occurs in the capitulum and adult weevils emerge in late summer and overwinter (Freese, 1991). Apart from *M. perforata*, *O. hookeri* develops only on *Matricaria maritima* L. subspp. *maritima* and *phaeocephala* (Ruprecht) Rauschert (Peschken and Sawchyn, 1993). The population used for screening, and for the initial releases, originated from southern Germany. While screening tests were in progress, an adventive population of *O. hookeri* was discovered in Nova Scotia; field observations on this population confirmed its host specificity (Peschken *et al.*, 1993). *O. hookeri* was approved for release in Canada in 1992.

Napomyza sp. near *lateralis* Fallén[2] and *Botanophila* sp. near *spinosa* (Rondani) are two stem-mining flies currently under study in Switzerland. Extensive testing on the former showed that it

[2]The name *N. lateralis* has been applied to morphologically identical insects from a wide range of host plants in the *Asteraceae* (Spencer, 1976), but host-specificity tests on the population from *M. perforata* suggest that there may be several sibling species with more restricted host ranges included under this name.

strongly preferred *M. perforata* over all 66 test plant species and varieties offered. Although occasional oviposition and development occurred on 14 non-target species under no-choice conditions, its field host range seems to be very restricted (Hinz, 1999). Larvae of *Botanophila* sp. near *spinosa* were found mining in developing shoots of *M. perforata* in Switzerland. Survival and oviposition have so far been very limited under confined conditions, and it has not yet been possible to start host-specificity tests.

Microplontus edentulus (Schultze), a univoltine stem-mining weevil, occurs in eastern Europe and southern Ukraine. Females lay eggs in the upper parts of stems of bolted *M. perforata*. Larvae tunnel in stems and also mine up branches to feed in flower-head bases (A.S. McClay, unpublished). In late summer the mature larva cuts an exit hole in the stem, drops to the ground and quickly burrows into the soil, where it forms a pupation cell, develops to an adult and overwinters (Hinz *et al.*, 1996). In screening tests, *M. edentulus* showed a high level of specificity for *M. perforata*. Occasional oviposition and development to the adult stage occurred under laboratory conditions on a few other species of *Matricaria*, *Chamomilla* and *Anthemis*, but these did not appear to be normal hosts in the field (Hinz *et al.*, 1996). *M. edentulus* was approved for release in 1997.

Rhopalomyia tripleurospermi Skuhravá was discovered in eastern Austria during surveys for potential biological control agents for *M. perforata*. Host range tests showed that it was restricted to *M. perforata* (Skuhravá and Hinz, 2001). It produces four generations per year in the field in Europe, and induces galls in various meristematic tissues, including apical meristems of rosettes and bolting plants, leaf axils, buds and flowers. Galls contain up to 80 chambers, each containing one larva, and females in culture produced an average of 61 offspring (Hinz, 1998). The galls appear externally as proliferations of very short shoots that stunt the plant and reduce flowering along the axis on which they occur (Hinz and Müller-Schärer, 2000b), and at high attack levels may kill overwintering rosettes (H.L. Hinz, unpublished). *R. tripleurospermi* was approved for release in 1999.

Releases and Recoveries

The first releases of *O. hookeri* were made in 1992 in Saskatchewan and Alberta, and establishment occurred immediately. Adults released from both the German and Nova Scotia populations have since become established at numerous sites in British Columbia, Alberta, Saskatchewan and Manitoba. Releases of as few as 38 adults resulted in establishment (McClay and De Clerck-Floate, 1999). Two redistribution releases have also been made in Nova Scotia (G. Sampson, Truro, 2000, personal communication). In 1992, 450 *O. hookeri* were released at Hillsborough, New Brunswick, but no weevils were observed in 1993 (Maund *et al.*, 1993). At Vegreville, Alberta, *O. hookeri* has been found up to 7 km from the release site 7 years after release (A.S. McClay, unpublished). At some sites monitored in Saskatchewan in 1998–1999 the weevil had reached considerable numbers, with attack as high as 85% in Wapella in the south-east and 95% at Tisdale in the north-east (Table 76.1). Although showing good dispersal capabilities on its own, *O. hookeri* is being redistributed in Alberta and Saskatchewan. In Alberta, it has been mass reared on potted *M. perforata* plants in outdoor field cages, and stored over winter at 0°C and 100% relative humidity, with good survival (McClay, 1999).

M. edentulus has been released nine times in four provinces (Table 76.2), using progeny of weevils collected in eastern Austria. These were open releases of 25–75 adults, except for the release at Vegreville, Alberta, in 1997, in which 16 infested plants were transplanted into a field plot before emergence of larvae from the stems. Based on larval emergence from other plants in the same rearing cage, about 2000 larvae are thought to have emerged from the transplanted plants. No

Table 76.1. Recoveries and population levels of *Omphalapion hookeri* in Saskatchewan and British Columbia, 1998–1999.

Location	Release year	Monitoring year	% Seed heads infested (proportion)	Mean number of weevils/ infested head ± SE	Maximum number of weevils/head
Saskatchewan					
Balcarres	1995	1999	71 (144/204)	3.4 ± 0.2	11
Bankend	1995	1999	33 (69/210)	2.9 ± 0.3	10
Bethune (#1)[a]	1995	1998	0 (0/202)	0	0
		1999	5 (10/199)	2.1 ± 0.3	4
Bethune (#2)[a]	1995	1998	1 (3/221)	1.7 ± 0.6	3
		1999	0.5 (1/203)	1.0	1
Canwood (#1)[a]	1995	1999	51 (106/210)	3.7 ± 0.3	20
Dubuc	1995	1999	46 (99/217)	5.1 ± 0.4	14
Edenwold (#1)	1996	1999	27 (54/204)	2.6 ± 0.3	9
Edenwold (#2)	1996	1999	0	–	–
Hafford (#1)[a]	1996	1998	3 (5/198)	2.2 ± 0.6	4
Hafford (#2)[a]	1996	1998	10 (22/210)	1.9 ± 0.2	4
Holdfast[a]	1995	1998	1 (2/202)	1	1
		1999	9 (17/199)	2.3 ± 0.3	5
McLean	1995	1999	39 (78/200)	3.0 ± 0.3	14
Melville	1995	1999	50 (103/205)	4.4 ± 0.3	13
Qu'Appelle	1995	1999	53 (107/202)	3.5 ± 0.3	13
Rocanville	1996	1999	14 (28/207)	1.9 ± 0.2	4
Tantallon	1995	1998	2 (3/148)	1	1
		1999	5 (10/206)	2.7 ± 0.6	7
Tisdale (#1)[a]	1995	1999	95 (191/202)	5.1 ± 0.2	15
Tisdale (#4)[a]	1996	1999	41 (84/205)	3.1 ± 0.2	7
Wapella (#2)[a]	1996	1998	68 (125/184)	3.9 ± 0.2	8
		1999	85 (172/203)	4.6 ± 0.2	14
Whitewood (#1)[a]	1995	1998	71 (22/31)	3.3 ± 0.4	7
		1999	70 (143/203)	5.0 ± 0.3	14
Whitewood (#2)[a]	1995	1998	10 (5/50)	1.8 ± 0.4	3
Whitewood (#3)[a]	1996	1998	9 (5/54)	1.8 ± 0.6	4
British Columbia					
Ft St John (#2)[a]	1992/93	1999	1 (1/92)	1	1
Ft St John (#3)	1998	1999	0	–	–
Ft St John (#4)	1998	1999	2 (2/98)	2.5 ± 0.5	3

[a]Sites that are the same as listed in McClay and De Clerck-Floate (1999), where monitoring information goes back to 1996. Sampled seed heads from all sites were collected randomly within 40 m of each release point.

signs of establishment have been found so far at any of the adult release sites. However, at the site of the larval release, adults and attacked stems were found from 1998 to 2000. In 1999, 62% of *M. perforata* stems sampled from a large, naturally occurring patch about 100 m from the release area showed mining by *M. edentulus* larvae, with a mean of 1.97 mines per attacked stem. Attacked stems were also common in 2000, indicating that a well-established population has persisted for 3 years at this site.

In Alberta, British Columbia, Saskatchewan and Manitoba, 55 releases of *R. tripleurospermi* were made in 1999. Because adult midges live for only a few hours at room temperature (Hinz, 1998), most releases were made by transplanting infested plants containing mature larvae or pupae into field sites. Some releases near Vegreville were made by releasing

Table 76.2. Releases and recoveries of *Microplontus edentulus* against *Matricaria perforata*, 1997–2000.

Location	Release date	Stage	Number	Recoveries
Alberta				
Vegreville	22 July 1997	Larvae in plants	*c.* 2000	1998–2000
Spirit River	3 June 1998	Adults	50	None
Edmonton	4 June 1998	Adults	50	None
Bruce	5 June 1998	Adults	75	None
Beaverlodge	6 June 1998	Adults	25	None
British Columbia				
Hudson's Hope	13 June 1997	Adults	50	None
Fort St John	4 June 1998	Adults	50	None
Manitoba				
Winnipeg Beach	4 June 1998	Adults	75	None
Saskatchewan				
Hafford	5 June 1998	Adults	50	None

adult midges from the greenhouse colony into 1 m^3 field cages placed over field stands of *M. perforata*. Initial establishment occurred readily in the field, both from adult releases and from transplantation of galled plants. Releases made from April to early August 1999, in Alberta and British Columbia, resulted in 74% gall formation (excluding sites that were subsequently destroyed or not monitored). At the Vegreville site, where releases were made from 23 April to 29 June 1999, galls were found up to 500 m from the release plot by late September. A vigorous *R. tripleurospermi* population was present at this site in 2000 and overwinter survival was also confirmed at many other sites in Alberta and Saskatchewan (A.S. McClay and G. Bowes, unpublished).

Evaluation of Biological Control

The only agent that has been established long enough for any evaluation of control is *O. hookeri*. The reduction in seed production it caused was detectable in field samples collected in Vegreville in 1996. Each individual of *O. hookeri* completing development reduced seed production in a head by 11.2 seeds, and it was estimated that a density of 15 weevils per head would be needed to approach complete seed destruction (McClay and De Clerck-Floate, 1999). On this basis, there are some sites in Saskatchewan where *O. hookeri* is having an impact on seed production (Table 76.1). A site in Tisdale, for instance, had on average five weevils per attacked seed head and reached a maximum of 15 per head. Several sites reached maximum numbers of 14+ per head (Table 76.1). In Nova Scotia, the best establishment sites now have around two weevils per head, but this is still insufficient to have an impact on *M. perforata* populations (G. Sampson, Truro, 2000, personal communication). The rapid dispersal of *O. hookeri* may lead to low initial rates of population build-up, until it has become generally distributed throughout areas infested with *M. perforata*. Three adults of *Pteromalus anthonomi* (Ashmead) emerged from several thousand field-collected *M. perforata* seed heads at Vegreville in 1999; it is not yet known if they were parasitic on *O. hookeri*.

The apparent lack of establishment of *M. edentulus* at most sites may be a reflection of dispersal rather than true failure to establish. The establishment at Vegreville shows that it is well able to persist and increase under the climatic and soil conditions of at least some parts of the Canadian prairies. Hinz *et al.* (1996) reported that *M. edentulus* significantly reduced the biomass and number of seeds produced by potted *M. perforata* plants. Its impact under field conditions in Canada is unknown.

R. tripleurospermi survived well over

winter in Alberta and caused severe galling on some plants. Heavy galling stunts growth of flowering branches and appears to reduce or delay flowering. The impact of this species on *M. perforata* in the field will depend on its phenology, the degree to which it is affected by native parasitoids, and the plant's ability to regrow after gall damage. *Aprostocetus* n. sp., found parasitizing up to 70% of larvae and pupae in Europe (Hinz, 1998), was eliminated from the culture sent to Canada. Another parasitoid, *Mesopolobus* sp., was reared from *M. perforata* plants infested with *R. tripleurospermi* that had been kept in an outdoor rearing cage at Vegreville. *Mesopolobus* sp. was not found in culture cages that had been kept in a greenhouse, and is presumably native.

M. perforata is likely to be a difficult target for biological control. Infestations can increase rapidly when uncontrolled, due to its profuse seed production, and decline over 2–3 years in the presence of competition from perennial plants. It may thus be difficult for agents to track the spatial and temporal variability of the weed population, although all three agents released to date appear to have good dispersal capabilities. Although parasitism of the introduced agents by native chalcids is so far very low, this may become a problem in future, particularly for *R. tripleurospermi*.

Recommendations

Further work should include:

1. Continued rearing and redistribution of *O. hookeri*, *M. edentulus* and *R. tripleurospermi* within areas where *M. perforata* is a problem;
2. Evaluating their impact, separately and in combination in controlled small plot studies;
3. Completing the screening of the two stem-mining flies, *Napomyza* sp. near *lateralis* and *Botanophila* sp. near *spinosa*;
4. Elucidating the *N. lateralis* sibling species complex;
5. Developing a release strategy that includes targeting relatively persistent infestations, e.g. those along rights-of-way and in abandoned gravel pits, redistributing agents so they are uniformly established over large infested areas, and using multivoltine agents, e.g. *R. tripleurospermi*.

Acknowledgements

Financial support for research on biological control of *M. perforata* was provided by the Canada–Alberta Environmentally Sustainable Agriculture Agreement, Alberta Agricultural Research Institute, Saskatchewan Agriculture Development Fund, Manitoba Sustainable Development Innovation Fund, Peace River Agriculture Development Fund, and Nova Gas Transmission Ltd. Parasitoid identifications were provided by G. Gibson. G. Bowes and G. Sampson provided information on biological control programmes against scentless chamomile in Saskatchewan and Nova Scotia, respectively.

References

Ali, S. (ed.) (1999) *Crop Protection 1999.* Alberta Agriculture, Food and Rural Development, Edmonton, Alberta.

Blackshaw, R.E. and Harker, K.N. (1997) Scentless chamomile (*Matricaria perforata*) growth, development, and seed production. *Weed Science* 45, 701–705.

Bowes, G.G., Spurr, D.T., Thomas, A.G., Peschken, D.P. and Douglas, D.W. (1994) Habitats occupied by scentless chamomile (*Matricaria perforata* Mérat) in Saskatchewan. *Canadian Journal of Plant Science* 74, 383–386.

Douglas, D.W. (1989) *The Weed Scentless Chamomile* (Matricaria perforata *Mérat) in Saskatchewan: Farmers' Perspectives, History and Distribution, Habitats, Biology, Effects on Crop Yield and Control.* Agriculture Canada, Saskatoon, Saskatchewan.

Douglas, D.W., Thomas, A.G., Peschken, D.P., Bowes, G.G. and Derksen, D.A. (1991) Effects of summer and winter annual scentless chamomile (*Matricaria perforata* Mérat) interference on spring wheat yield. *Canadian Journal of Plant Science* 71, 841–850.

Douglas, D.W., Thomas, A.G., Peschken, D.P., Bowes, G.G. and Derksen, D.A. (1992) Scentless chamomile (*Matricaria perforata* Mérat) interference in winter wheat. *Canadian Journal of Plant Science* 72, 1383–1387.

Freese, A. (1991) *Apion hookeri* Kirby (Col., Curculionidae), a potential agent for the biological control of *Tripleurospermum perforatum* (Mérat) Wagenitz [= *T. inodorum* (L.) C.H. Schultz, *Matricaria perforata* Mérat, *M. inodora* L.] (Asteraceae, Anthemideae) in Canada. *Journal of Applied Entomology* 112, 76–88.

Freese, A. and Günther, W. (1991) The insect complex associated with *Tripleurospermum perforatum* (Asteraceae: Anthemideae). *Entomologia Generalis* 16, 53–68.

Hinz, H.L. (1998) Life history and host specificity of *Rhopalomyia* n. sp. (Diptera : Cecidomyiidae), a potential biological control agent of scentless chamomile. *Environmental Entomology* 27, 1537–1547.

Hinz, H.L. (1999) *Investigations on Potential Biocontrol Agents of Scentless Chamomile,* Tripleurospermum perforatum *(Mérat) Laínz. Annual Report 1999.* CABI Bioscience Centre Switzerland, Delémont, Switzerland.

Hinz, H.L. and Leiss, K. (1996) *Investigations on Potential Biocontrol Agents of Scentless Chamomile* (Tripleurospermum perforatum *(Mérat) Wagenitz). Annual Report.* International Institute of Biological Control, Delémont, Switzerland.

Hinz, H.L. and Müller-Schärer, H. (2000a) Suitability of two root-mining weevils for the biological control of scentless chamomile, *Tripleurospermum perforatum*, with special regard to potential non-target effects. *Bulletin of Entomological Research* 90, 497–508.

Hinz, H.L. and Müller-Schärer, H. (2000b) Influence of host condition on the performance of *Rhopalomyia* n. sp. (Diptera: Cecidomyiidae), a biological control agent for scentless chamomile, *Tripleurospermum perforatum. Biological Control* 18, 147–156.

Hinz, H., Bacher, S., McClay, A.S. and De Clerck-Floate, R. (1996) Microplontus *(*Ceutorhynchus*)* edentulus *(Schltz.) (Col.: Curculionidae), a Candidate for the Biological Control of Scentless Chamomile in North America.* International Institute of Biological Control, Delémont, Switzerland.

Khadhair, A.H., McClay, A., Hwang, S.F. and Shah, S. (1999) Aster yellows phytoplasma identified in scentless chamomile by microscopical examinations and molecular characterization. *Journal of Phytopathology* 147, 149–154.

Maund, C.M., McCully, K.V., Finnamore, D.B., Sharpe, R. and Parkinson, B. (1993) A summary of insect biological control agents released against weeds in NB pastures from 1990 to 1993. *Adaptive Research Reports (New Brunswick Department of Agriculture)* 15, 359–380.

McClay, A.S. (1999) *Biological Control of Scentless Chamomile: Final Report. AARI project number 97M165.* Alberta Research Council, Vegreville, Alberta.

McClay, A.S. and De Clerck-Floate, R.A. (1999) Establishment and early effects of *Omphalapion hookeri* (Kirby) (Coleoptera: Apionidae) as a biological control agent for scentless chamomile, *Matricaria perforata* Mérat (Asteraceae). *Biological Control* 14, 85–95.

Peschken, D.P. (1989) *Petition for the Approval of the Weed Scentless Chamomile as a Target for Classical Biological Control in Canada.* Agriculture Canada Research Station, Regina, Saskatchewan.

Peschken, D.P. and Sawchyn, K.C. (1993) Host specificity and suitability of *Apion hookeri* Kirby (Coleoptera: Curculionidae), a candidate for the biological control of scentless chamomile, *Matricaria perforata* Mérat (Asteraceae) in Canada. *The Canadian Entomologist* 125, 619–628.

Peschken, D.P., Thomas, A.G., Bowes, G.G. and Douglas, D.W. (1990) Scentless chamomile (*Matricaria perforata*) – a new target weed for biological control. In: DelFosse, E.S. (ed.) *Proceedings of the VII International Symposium on Biological Control of Weeds.* Istituto Sperimentale per la Patologia Vegetale, Rome, Italy, pp. 411–416.

Peschken, D.P., Sawchyn, K.C. and Bright, D.E. (1993) First record of *Apion hookeri* Kirby (Coleoptera: Curculionidae) in North America. *The Canadian Entomologist* 125, 629–631.

Skuhravá, M. and Hinz, H.L. (2000) *Rhopalomyia tripleurospermi* sp. n. (Diptera: Cecidomyiidae), a new gall midge species on *Tripleurospermum perforatum* (Asteraceae : Anthemideae) in Europe, and a biological control agent in Canada. *Acta Societatis Zoologicae Bohemicae* 64, 425–435.

Spencer, K.A. (1976) *The Agromyzidae (Diptera) of Fennoscandia and Denmark*. Vol. 5 part 2, *Fauna Entomologica Scandinavica*. Scandinavica Science Press, Klampenborg, Denmark.
Woo, S.L., Thomas, A.G., Peschken, D.P., Bowes, G.G., Douglas, D.W., Harms, V.W. and McClay, A.S. (1991) The biology of Canadian weeds. 99. *Matricaria perforata* Mérat (Asteraceae). *Canadian Journal of Plant Science* 71, 1101–1119.

77 *Myriophyllum spicatum* L., Eurasian Water Milfoil (*Haloragaceae*)

R.A. Ring, N.N. Winchester and I.V. MacRae

Pest Status

Eurasian water milfoil, *Myriophyllum spicatum* L., native to Eurasia, is an important weed in aquatic ecosystems in southern British Columbia (Aiken *et al.*, 1979). Among unwanted or nuisance plants that cause various problems through excessive growth, e.g. native water lilies, *Nuphar* spp., pondweeds, *Potamogeton* spp., and coontail, *Ceratophyllum* sp., *M. spicatum* is usually the most severe. Nine *Myriophyllum* spp. are known in British Columbia, but the rapid, dense growth that often results in mats and clumps at the surface characterizes *M. spicatum* (Ceska, 1977; Ceska and Ceska, 1986). This perennial plant reproduces vegetatively mainly by fragmentation or propagation from root crowns. Although seeds are produced, seedlings are not considered important in its reproduction and spread (Newroth, 1990). *M. spicatum* displaces native vegetation by re-growing from root crowns early in spring and, in summer, grows up to 5 cm per day, reaching the surface in water up to 4–5 m deep (Anonymous, 1986). It can also grow in almost all substrates from rocks to gravel, sand, silt or clay (Warrington, 1983). Because it seems to prefer habitats frequented by humans, or areas modified for public use, it is often perceived as a major threat to water use. Since 1971, *M. spicatum* has adversely affected recreational use of infested waters and beaches by fragment accumulation along the water's edge, spoiling the aesthetic quality of offshore water, and increasing the risk for swimmers. The dense growth of untreated *M. spicatum* may also have contributed to drowning tragedies, and has been associated with 'swimmers itch' problems.

In the Okanagan Valley region, beach use by residents and tourists is an important recreational activity (Phipps and James, 1981). From 1970 to 1980 aquatic weeds became one of the main problems for residents and visitors, despite ongoing control programmes (Anonymous, 1986). Historically, most areas in the Okanagan Valley lakes, Shuswap Lake and Cultus Lake did not have nuisance aquatic plants before *M. spicatum* became established. In some areas motorboats, sailboats with keels, and water skiing were curtailed until *M. spicatum* was removed or controlled. Shore-based angling was also adversely affected and trollers in mid-lake encountered mats of floating milfoil that entangled their fishing lines. In addition, *M. spicatum* can reach densities that interfere with some shore and river salmonids. The plants interfere with spawning by covering spawning gravels and, possibly, accumulation of organic matter and gravel compaction could cause further deterioration (Anonymous, 1986). Other adverse effects include clogged agricultural, industrial and power generation water intakes, lower dissolved oxygen concentrations and increased populations of permanent pool mosquitoes (Smith and Barko, 1990).

A 10% reduction in values of lakefront property due to heavy weed infestation amounts to a loss in value of at least Can$3.7 million for the entire Okanagan basin. Furthermore, if research continues together with a surveillance and plant removal programme in the interior of British Columbia, the total cost of the programme (from 1976 to 1980) was estimated to be Can$3.22 million (Buchanan, 1976). Various aspects of this option have continued until 2000. No dollar figures are available for the costs of research and surveillance, but the estimated operating costs and treatment rates for selected mechanical control methods were, in 1986: harvester = Can$1200 ha^{-1}; rototiller = Can$400–1300 ha^{-1}; shallow water tillage = Can$125–400 ha^{-1}; diver-operated dredge = Can$2500–19,000 ha^{-1}; bottom barriers = Can$8000–26,000 ha^{-1}.

From 1976 to 1985 in the Okanagan lakes system, about 150 ha infested with *M. spicatum* were controlled annually by mechanical methods. In Shuswap Lake in 1985, 38.82 ha were treated at a cost of Can$4500 ha^{-1}, and in Cultus Lake in 1985, 4.65 ha were treated at Can$4000 ha^{-1}, amounting to Can$800,000 per annum, excluding equipment rental or depreciation of capital costs of machinery, expenses incurred from transport/launching of machines, or administrative costs (Anonymous, 1986). Nor does this consider the costs of experimental treatments, such as using the herbicide 2,4-D in the Okanagan Lakes during the late 1970s and early 1980s.

Presently, in southern British Columbia, *M. spicatum* occupies about 1500 ha, of which about 300 ha are managed, mainly by mechanical controls (Kangasniemi *et al.*, 1993). Apparently, these are becoming increasingly effective as technical improvements to machines are made. Consequently, mechanical harvesting, derooting, and rototilling are currently the methods of choice in high-use areas. However, biological control remains an option in areas where intensive mechanical methods are environmentally inappropriate or too expensive (Kangasniemi *et al.*, 1993) or, perhaps, where biological control could be integrated more effectively with mechanical methods.

Background

On-going attempts to control *M. spicatum* using 2,4-D, rototilling and harvesting have not effectively solved the problem. Biological control of aquatic weeds has been attempted for water hyacinth, *Eichhornia crassipes* (Martius) Solms-Laubach (Center *et al.*, 1984), has been used successfully for alligatorweed, *Alternantha philoxeroides* (Martius) Grisebach (Cofrancesco, 1984), and was suggested for *M. spicatum* (Buckingham *et al.*, 1981).

In the Okanagan valley several infestations of *M. spicatum* were found to be affected by insect damage in surveys undertaken in the late 1970s and early 1980s. Retarded shoot elongation and failure to flower resulted from the larval feeding activities of a non-biting midge, *Cricotopus myriophylli* Oliver, a caddis-fly, *Triaenodes tarda* Milne, and a weevil, probably *Eurhychiopsis lecontei* (Dietz) (Kangasniemi *et al.*, 1993).

In 1979 the British Columbia Ministry of Environment, Water Investigations Branch, reviewed the potential of biological control agents against *M. spicatum* (Anonymous, 1979). Among the more 'promising' organisms identified were herbivorous fish, snails, *Physa* sp., crayfish, *Cambrus* sp., insects (over 25 species were identified that feed on *M. spicatum* in Eurasia), fungi and bacteria (Balciunas, 1982).

Biological Control Agents

Insects

Triaenodes larvae occur in plant beds in both lotic and lentic waters, where they swim readily with their characteristic cases, formed from green plants (Wiggins, 1977). *Triaenodes tarda* Milne, native to North America, can use *M. spicatum* to build its case. Larvae were the primary agent that suppressed growth in several hectares of *M. spicatum* in Magic Lake, a shallow, eutrophic lake on Pender Island. All five instars feed heavily on growing tips and foliage of *M. spicatum*, incorporating them into their cases, which results in a significant impact on *M. spicatum* growth and development. Densities of 3–10 larvae per plant cause a significant decrease in plant growth. Higher densities produce greater cropping, but the optimum number was six larvae per plant. Feeding damage is more severe in later instars. Larvae swim actively to other plants and floating fragments when one food source is exhausted. Pupae anchor on the plants themselves, with both ends of the case sealed and cemented to the plant. Emerging adults were successfully mated in a rearing tent. Males are short lived but females live for about 2 weeks. Egg masses were recovered and the F_1 generation was subsequently reared throughout the winter. *T. tarda* feeds throughout the growing season (May–7October) and all life stages are present. Some synchrony exists in the population at Magic Lake, a large pulse of adults appearing in late July–early August.

The duration of each life-cycle stage can be manipulated, so mass production is feasible. Larvae survive in a wide range of temperatures and early instars overwintering in the lakeshore sediments become active as the water temperature increases to 4.0°C. They are also tolerant of anoxic conditions. These attributes should enable introduction of *T. tarda* into infested areas where it will cause the most feeding damage, and will allow co-ordinated introduction with the next agent.

C. myriophylli (Oliver, 1984) was found damaging several well-established weed beds of *M. spicatum* in the Okanagan Valley lakes system in the late 1970s and early 1980s (Kangasniemi, 1983; Kangasniemi and Oliver, 1983). Larvae of *C. myriophylli* establish on the apical portions of stems, construct cases, and feed on the meristematic tissue (Anonymous, 1981; Oliver, 1984). When *C. myriophylli* densities are sufficient they impact *M. spicatum* populations by reducing overall height and preventing surfacing and flowering. Plants remain a metre or more below the surface throughout the year (Kangasniemi *et al.*, 1993). Denuding the plant of growing tissue in this manner suppresses *M. spicatum* to an economically acceptable level. Laboratory trials determined the number of *C. myriophylli* larvae per meristem necessary to suppress *M. spicatum* growth, how quickly growth could be suppressed, and the midge's host preference (MacRae, 1988; MacRae *et al.*, 1990). One larva can eat all the meristematic tissue from an apical tip of stem, inhibiting growth. Consumption is so rapid that no significant difference in the new growth of apical tips occurs when one, two or three larvae feed on them. The rapidity with which one larva can completely strip a meristematic region, well within the time period to complete the second or third larval instar, implies that each larva requires more than one meristem to complete development. Feeding damage by *C. myriophylli* was assessed using varying larval densities (1–4 larvae per plant). All larval densities had a significant impact on plant growth, with no significant differences among them.

Host-preference studies (Ring, 1988) showed that *C. myriophylli* preferred *M. spicatum* and had a marked inability to feed on any of the 12 native species tested, except for the closely related *M. sibiricum* Kamarov (= *M. exalbescens* Fernald). *C. myriophylli* showed a significant preference for *M. spicatum* over *M. sibiricum*. Larvae on culled meristems placed into an aquarium planted with *M. spicatum* had no difficulty becoming established on the fresh plants. They also readily relocated after this food source was depleted, thus indicating that *C. myriophylli* larvae can relocate to lateral growing tips once apical tips have been browsed. *C. myriophylli* was not introduced with *M. spicatum*, as was initially assumed, but is native to British Columbia, the original host plant being *M. sibiricum* (Kangasniemi *et al.*, 1993).

Because its life cycle is about 30 days and *C. myriophylli* is multivoltine, it is attractive for mass-rearing. However, when adult males emerge, they swarm over visible markers that can be quite high (>3 m) and the vertical mating swarms may be difficult to see. This behaviour makes laboratory simulations very difficult, so mass-rearing of this chironomid has not yet been successful.

Evaluation of Biological Control

The life-cycle features of *T. tarda*, combined with a wide environmental tolerance, should ensure a successful mass-rearing programme. However, its success as a biological control agent for *M. spicatum* in other lakes may be limited by the presence of predatory fish and additional factors relating to habitat suitability for *Triaenodes,* e.g. eutrophication. Both *T. tarda* and *C. myriophylli* have the potential to be integrated into a control programme because infestations of *M. spicatum* are spreading and mechanical control techniques in British Columbia are limited to high-priority areas, e.g. public beaches and marinas. Biological control techniques may prove to be valuable, inexpensive alternatives and provide for expansion of currently treated areas.

Potential disruption of ecosystems and further complication or exacerbation of existing problems must be resolved if these agents are to be used elsewhere. For instance, *C. myriophylli* belongs in the *Cricotopus sylvestris* group (Oliver, 1984). In many rice-producing areas of the world, *C. sylvestris* Wulp, often referred to as the 'rice midge' or 'rice seed midge', is a pest (Berczic, 1979; Gigarick, 1984). Its propensity to attack rice, *Oryza sativa* L., may also extend to its close relative, *C. myriophylli,* so this must be tested if *C. myriophylli* is to be exported.

Recommendations

Further work should include:

1. Determining optimal culturing requirements and mass rearing methodologies for *T. tarda* and *C. myriophylli*;
2. Investigating how large populations of these two insects can be accumulated and stored at low temperature;
3. Testing for the ideal transporting methods and conditions;
4. Integrating these biological control agents into existing management controls for *M. spicatum*;
5. Evaluating Eurasian species associated with *M. spicatum* for their suitability as biological control agents.

Acknowledgements

We thank the staff of the Water Management Branch of the British Columbia Ministry of Environment for valuable assistance and logistical support. This work was supported by a grant from the Science Council of British Columbia.

References

Aiken, S.G., Newroth, P.R. and Wile, I. (1979) The biology of Canadian weeds. 34. *Myriophyllum spicatum* L. *Canadian Journal of Plant Science* 59, 201–215.

Anonymous (1979) *The Feasibility of Using Biological Control Agents for Control of Eurasian Water Milfoil in British Columbia.* Aquatic Plant Management Program Vol. V. Canada Information Bulletin, Province of British Columbia, Water Investigations Branch, Victoria, British Columbia.

Anonymous (1981) *A Summary of Biological Research on Eurasian Watermilfoil in British Columbia.*

Aquatic Plant Management Program Vol. XI. Canada Information Bulletin, Province of British Columbia, Water Investigations Branch, Victoria, British Columbia.

Anonymous (1986) *A Review of Aquatic Plant Management Methods and Programs in British Columbia.* Aquatic Plant Management Program Volume XII. Canada Information Bulletin, Ministry of Environment, Victoria, British Columbia.

Balciunas, J.K. (1982) *Insects and Other Macroinvertebrates Associated with Eurasian Watermilfoil in the United States.* Technical Report A-82-5, United States Army Engineer Waterways Experiment Station, Vicksburg, Mississippi.

Berczic, A. (1979) Animal pests of rice in Hungary and the problem of their control. *Opuscula Zoologica (Budapest)* 15, 61–74.

Buchanan, R.J. (1976) *Briefing paper on Eurasian Watermilfoil (*Myriophyllum spicatum *L.).* Report no. 2463, Canada Water Investigations Branch, British Columbia Ministry of Environment, Victoria, British Columbia.

Buckingham, G.R., Bennett, C.A. and Ross, B.M. (1981) *Investigation of Two Insect Species for Control of Eurasian Watermilfoil.* Technical Report A-81-4, United States Army Engineer Waterways Experimental Station, Vicksburg, Mississippi.

Center, T.D., Durden, W.C. and Corman, D.A. (1984) *Efficacy of* Sameodes albiguttalis *as a Biocontrol of Waterhyacinth.* Aquatic Plant Management Laboratory, United States Department of Agriculture, Fort Lauderdale, Florida, Technical Report A-84-2, for United States Army Engineer Waterways Experimental Station, Vicksburg, Mississippi.

Ceska, O. (1977) *Studies on Aquatic Macrophytes. Part XVII. Phytochemical Differentiation of* Myriophyllum *Taxa Collected in British Columbia.* Prepared by University of Victoria, Victoria, British Columbia, for Water Investigations Branch, British Columbia Ministry of Environment, Victoria, British Columbia.

Ceska, A. and Ceska, O. (1986) *Myriophyllum* Haloragaceae species in British Columbia: Problems with identification. In: *Proceedings of the First International Symposium on Watermilfoil* (Myriophyllum spicatum) *and Related Haloragaceae Species, 23–24 July 1985, Vancouver, British Columbia, Canada.* The Aquatic Plant Management Society Incorporated.

Cofrancesco, A.F. (1984) *Alligatorweed and its Biocontrol Agents.* Information Exchange Bulletin A-84-3, Environmental Resources Division, Engineering Laboratory, United States Army Engineer Waterways Experimental Station, Vicksburg, Mississippi.

Gigarick, A.A. (1984) General problems with rice invertebrate pests and their control in the USA. Fifteenth Pacific Science Congress on Rice Pest Management, Dunedin, New Zealand, 1983. *Protection Ecology* 7, 105–128.

Kangasniemi, B.J. (1983) Observations on herbivorous insects that feed on *Myriophyllum spicatum* in British Columbia. In: Taggart, J. (ed.) *Lake Restoration, Protection and Management. Proceedings of the Second Annual Conference, North American Lake Management Society, October, 1982, Vancouver, British Columbia, Canada.* United States Environmental Protection Agency, Washington, DC, pp. 214–218.

Kangasniemi, B.J. and Oliver, D.R. (1983) Chironomidae (Diptera) associated with *Myriophyllum spicatum* in Okanagan Valley lakes, British Columbia. *The Canadian Entomologist* 115, 1545–1546.

Kangasniemi, B., Speier, H. and Newroth, P. (1993) Review of Eurasian watermilfoil biocontrol by the milfoil midge. In: *Proceedings of the Twenty-seventh Annual Meeting of the Aquatic Plant Control Research Program, 16–19 November 1992, Bellevue, Washington. Miscellaneous Paper A-93-2.* United States Army Corps of Engineers, Waterways Experimental Station, Vicksburg, Mississippi, pp. 19–22.

MacRae, I.V. (1988) Evaluation of *Cricotopus myriophylli* Oliver (Diptera: Chironomidae) as a potential biocontrol agent for Eurasian water milfoil, *Myriophyllum spicatum.* MSc thesis, University of Victoria, Victoria, British Columbia.

MacRae, I.V., Winchester, N.N. and Ring, R.A. (1990) Feeding activity and host preference of the milfoil midge, *Cricotopus myriophylli* Oliver (Diptera: Chironomidae). *Journal of Aquatic Plant Management* 28, 89–92.

Newroth, P.R. (1990) Prevention of the spread of Eurasian water milfoil. In: *Proceedings, National Conference on Enhancing the States' Lake and Wetland Management Programs.* United States Environmental Protection Agency, North American Lake Management Society, Chicago, Illinois, pp. 93–100.

Oliver, D.R. (1984) Description of a new species of *Cricotopus* Van Der Wulp (Diptera: Chironomidae) associated with *Myriophyllum spicatum*. *The Canadian Entomologist* 116, 1287–1292.
Phipps, S.A. and James, S.A. (1981) *Water-based Recreation in the Okanagan Basin, 1980 Review*. Canada–British Columbia Okanagan Basin Implementation Agreement, Victoria, British Columbia.
Ring, R.A. (1988) *Biocontrol of Eurasian Watermilfoil*. Final Report, Science Council of British Columbia, Vancouver, British Columbia.
Smith, L. and Barko, J.W. (1990) Ecology of Eurasian watermilfoil. *Journal of Aquatic Plant Management* 28, 55–64.
Warrington, P.D. (1983) *An Introduction to Life Histories of* Myriophyllum *spp. in South Western British Columbia*. Water Management Branch, British Columbia Ministry of Environment, Victoria, British Columbia.
Wiggins, G.L. (1977) *Larvae of the North American Caddisfly Genera (*Trichoptera*)*. University of Toronto Press, Toronto, Ontario, pp. 161–177.

78 *Setaria viridis* (L.) Beauvois, Green Foxtail (*Poaceae*)

S.M. Boyetchko

Pest Status

Green foxtail, *Setaria viridis* (L.) Beauvois, a weed of European origin and one of the world's most common weeds (Fernald, 1950; Douglas *et al.*, 1985), is found in temperate zones but has also been reported in higher elevations in the cooler subtropics of South and North America, Australia and Asia (Holm *et al.*, 1977). It is economically important in several countries, including Canada, because of its prolific seed production, dense stands and strong ability to compete well with spring-sown crops (Holm *et al.*, 1977, 1979; Douglas *et al.*, 1985). The weed is found in cultivated fields cropped to barley, *Hordeum vulgare* L., maize, *Zea mays* L., flax, *Linum usitatissimum* L., rapeseed, *Brassica napus* L. and *B. rapa* L., soybean, *Glycine max* (L.) Merrill, sunflower, *Helianthus annuus* L., tomato, *Lycopersicon esculentum* L., and wheat, *Triticum aestivum* L., in addition to gardens, waste places and roadsides (Frankton and Mulligan, 1970). *S. viridis* was reported in 46% of fields on the prairies (Thomas *et al.*, 1996, 1998a, b). In Saskatchewan, it was estimated that competition from *S. viridis* in wheat amounts to 7.8% in yield loss (Hume, 1989). The value of annual losses due to grass weeds, including *S. viridis*, from reductions in crop yield, dockage, cleaning costs, lower crop grade and quality, and costs associated with chemical and cultural control have been estimated at Can$120–$500 million.

S. viridis often emerges late in spring, because it requires higher soil temperatures (20–30°C) for germination and emergence than most cereal crops, but is more competitive in early spring (Blackshaw *et al.*, 1981). Soil moisture appears to have a greater effect on seed germination than soil temperature. Shallow seeding depths of

1.5–2.5 cm are preferred by *S. viridis*, and emergence decreases with increasing seeding depth (Dawson and Bruns, 1962).

Background

A variety of chemical herbicides are used to control *S. viridis* (Douglas *et al.*, 1985; Beckie *et al.*, 1999) but recent surveys revealed that at least one in every 20 fields in Saskatchewan (about 1 million ha) have Group 1 (acetyl-CoA carboxylase [ACCase] inhibitor) resistant *S. viridis* (Beckie *et al.*, 1999). Resistance to Group 3 (dinitroanilines) and cross-resistance to Group 1 and 3 herbicides have also been reported, but with much lower incidence (Beckie and Morrison, 1993; Morrison and Devine, 1994; Morrison *et al.*, 1995; Retzinger and Mallory-Smith, 1997; Beckie *et al.*, 1999).

Several insects and pathogenic fungi, bacteria and viruses have been associated with *S. viridis* (Douglas *et al.*, 1985). In Saskatchewan, insects associated with *S. viridis* include *Lygus borealis* (Kelton), *Stenodema vicinum* (Provancher), *Hebecephalus occidentalis* Beamer and Tuthill, *H. rostratus* Beamer and Tuthill, *Helochara communis* Fitch, *Latalus personatus* Beirne, along with various beetles (Chrysomelidae, Melyridae), flies (Agromyzidae, Anthomyiidae, Chloropidae) and parasitic wasps (Chalcidoidea). Fungi reported on *S. viridis* include *Fusarium equiseti* (Corda) Saccardo, *Pyricularia grisea* (Cooke) Saccardo, *Pythium debaryanum* Hesse, *P. graminicola* Subramaniam, and *Sclerospora graminicola* (Saccardo) Schroeter (Conners, 1967). Many of these are also pathogens of cereals and other crops. The potential of these organisms for biological control has not been pursued.

During the past 20–30 years, research on plant pathogens for biological control of weeds has been greatly intensified (Charudattan, 1991; Boyetchko, 1999). Most of the organisms used have been foliar-applied fungi but, more recently, use of deleterious rhizobacteria has shown promise to control several weed species, particularly weedy grasses. Foliar pathogens have historically shown less than adequate control of weedy grasses because the meristem of grasses is covered by a leaf sheath, thereby protecting the growing point from infection. In addition, many fungal pathogens of weeds are often found on crops. However, soil-borne bacteria, e.g. *Pseudomonas*, *Flavobacterium* and *Xanthomonas* spp., show tremendous potential as pre-emergent biological control agents, by inhibiting or suppressing weed seed germination and/or root growth and development (Kremer and Kennedy, 1996).

Biological Control Agents

Pathogens

Bacteria

Several hundred weed-suppressive soil bacteria have been evaluated as biological control agents against *S. viridis*, many of which show at least 80% suppression to root growth and/or seed germination in laboratory bioassays (Boyetchko, 1997, 1998). Two bacterial strains with significant deleterious effects on *S. viridis* were field tested for 3 years in Saskatoon. Formulation plays a key role in their survival during the growing season and some formulations, such as peat-based granules, may provide slow release of bacteria (similar to slow-release fertilizers) for biological control, particularly for weeds such as *S. viridis* that emerge later in the growing season (Boyetchko, 1996). Use of granular formulations, e.g. peat-based granules, reduced weed emergence and above-ground biomass by 45–60%, depending on rate of application. Bacterial survival in the field over the growing season depended on the type of formulation used and bacterial strain. Peat prills provided slow release of the bacteria, resulting in season-long weed control. In 1999 and 2000, field results using a pesta formulation indicated that this may also have potential for stabilizing the bacteria and being highly effective in the field (S.M. Boyetchko, D. Daigle and W. Connick Jr, unpublished). Up to 90% weed control was achieved using the pesta formulation. Clay-based formulations were ineffective in the field. Nutritional factors also are significant in enhancing biological control activity of the bacterial strains tested. Fermentation media and incorporation of precursors for bacterial secondary metabolites can enhance biological control.

Fungi

Three fungal pathogens, *Drechslera gigantea* (Heald and Wolf) Ito, *Exserohilum rostratum* (Drechsler) Leonard and Suggs and *Exserohilum longirostratum* (Subramaniam) Sivan, alone or in mixtures, showed bioherbicidal activity against a variety of weedy grasses in Florida (Chandramohan and Charudattan, 1997). Preliminary results demonstrated that they can control 1-week-old *S. viridis* seedlings 3 days after inoculation, indicating their strong potential for biological control. Extensive survey and screening activities for additional foliar and soil-borne fungal biological control agents showed that a variety of fungi, including *Alternaria*, *Cephalosporium*, *Colletotrichum*, *Fusarium*, *Phoma* and *P. grisea*, are pathogenic to *S. viridis* (Boyetchko *et al.*, 1998). These fungi continue to be assessed for their potential. However, more effective delivery systems and inoculum levels that reflect practical application rates will dictate their suitability for biological control.

Evaluation of Biological Control

Despite the variety of native insects that feed on *S. viridis*, biological control with bacteria and fungi appears to be more promising. Bacteria applied as pre-emergent biological control agents provide a viable method for reducing the competitive nature of the weed while not being constrained by the amount of leaf wetness or dew often required by foliar applied pathogens. These bacteria are easy to mass-produce through liquid fermentation, and discovery of new granular formulations will ensure their ease of application by farmers. Use of highly virulent and fast-acting foliar fungal pathogens, e.g. *D. gigantea*, *E. rostratum* and *E. longirostratum*, can offset the requirement for long periods of leaf wetness, particularly for the grass weeds growing in the prairies, where long dew periods are infrequent.

Recommendations

Further work should include:

1. Evaluating soil bacteria for biological control, stabilizing them through fermentation and formulation, and understanding the underlying mechanisms of action to enhance efficacy;
2. Evaluating the three fungal pathogens, originally from Florida;
3. More extensive surveys for foliar fungal pathogens in ecoregions where *S. viridis* is a problem, to discover ecotypes or isolates that can infect *S. viridis* and significantly suppress it;
4. Developing formulations for application of foliar and soil-borne biological control agents, particularly formulations that reduce the dew period requirements, important where moisture is often a limiting factor, e.g. the prairies, and formulations, e.g. granules, for pre-emergent agents.

References

Beckie, H.J. and Morrison, I.N. (1993) Effective kill of trifluralin-susceptible and -resistant green foxtail (*Setaria viridis*). *Weed Technology* 7, 15–22.

Beckie, H.J., Thomas, A.G. and Legere, A. (1999) Nature, occurrence, and cost of herbicide-resistant green foxtail (*Setaria viridis*) across Saskatchewan ecoregions. *Weed Technology* 13, 626–631.

Blackshaw, R.E., Stobbe, E.H., Shaykewich, C.F. and Woodbury, W. (1981) Influence of soil temperature and soil moisture on green foxtail (*Setaria viridis*) establishment in wheat (*Triticum aestivum*). *Weed Science* 29, 179–184.

Boyetchko, S.M. (1996) Formulating bacteria for use as biological control agents. In: *Proceedings of the 1996 National Meeting of Expert Committee on Weeds, Victoria, British Columbia, 9–12 December 1996*, pp. 85–88.

Boyetchko, S.M. (1997) Efficacy of rhizobacteria as biological control agents of grassy weeds. In: *Proceedings, Soils and Crops Workshop '97, Saskatoon, Saskatchewan.* Extension Division, College of Agriculture, University of Saskatchewan, Saskatoon, Saskatchewan, pp. 460–465.

Boyetchko, S.M. (1998) Evaluation of deleterious rhizobacteria for biological control of grassy weeds. In: *Proceedings of the IV International Bioherbicide Workshop, University of Strathclyde, Glasgow, Scotland, 6–7 August 1998,* p. 16.

Boyetchko, S.M. (1999) Innovative applications of microbial agents for biological weed control. In: Mukerji, K.G., Chamola, B.P. and Upadhyay, K. (eds) *Biotechnological Approaches in Biocontrol of Plant Pathogens.* Kluwer Academic/Plenum Publishers, London, UK, pp. 73–97.

Boyetchko, S.M., Wolf, T.M., Bailey, K.L., Mortensen, K. and Zhang, W.M. (1998) Survey and evaluation of fungal pathogens for biological control of grass weeds. In: *Proceedings, Soils and Crops Workshop '98, Saskatoon, Saskatchewan.* Extension Division, College of Agriculture, University of Saskatchewan, Saskatoon, Saskatchewan, pp. 424–429.

Chandramohan, S. and Charudattan, R. (1997) Bioherbicidal control of grassy weeds with a pathogen mixture. *Weed Science Society of America Abstracts* 37, 56.

Charudattan, R. (1991) The mycoherbicide approach with plant pathogens. In: TeBeest, D.O. (ed.) *Microbial Control of Weeds.* Chapman and Hall, New York, New York, pp. 24–57.

Conners, I.L. (1967) *An Annotated Index of Plant Diseases in Canada.* Publication 1251, Canada Department of Agriculture, Ottawa, Ontario.

Dawson, J.H. and Bruns, V.F. (1962) Emergence of barnyardgrass, green foxtail and yellow foxtail seedlings from various soil depths. *Weeds* 10, 136–139.

Douglas, B.J., Thomas, A.G., Morrison, I.N. and Maw, MG. (1985) The biology of Canadian weeds. 70. *Setaria viridis* (L.) Beauv. *Canadian Journal of Plant Science* 65, 669–690.

Fernald, M.L. (1950) *Gray's Manual of Botany*, 8th edn. American Book Company, New York, New York.

Frankton, C. and Mulligan, G.A. (1970) *Weeds of Canada.* Publication 948, Canada Department of Agriculture, Ottawa, Ontario.

Holm, L., Pancho, J.V., Herberger, J.P. and Plucknett, D.L. (1979) *A Geographical Atlas of World Weeds.* John Wiley and Sons, New York.

Holm, L.G., Plucknett, D.L., Pancho, J.V. and Herberger, J.P. (1977) *The World's Worst Weeds.* The University Press of Hawaii, Honolulu, Hawaii.

Hume, L. (1989) Yield losses in wheat due to weed communities dominated by green foxtail (*Setaria viridis* [L.] Beauv.): A multispecies approach. *Canadian Journal of Plant Science* 69, 521–529.

Kremer, R.J. and Kennedy, A.C. (1996) Rhizobacteria as biocontrol agents of weeds. *Weed Technology* 10, 601–609.

Morrison, I.N. and Devine, M.D. (1994) Herbicide resistance in the Canadian prairie provinces: five years after the fact. *Phytoprotection* 75 (Suppl.), pp. 5–16.

Morrison, I.N., Bourbeois, L., Friesen, L. and Kelner, D. (1995) Betting against the odds: The problem of herbicide resistance. In: Roberts, T.L. (ed.) *Proceedings of the 1995 Western Canada Agronomy Workshop.* Potash and Phosphate Institute of Canada, Red Deer, Alberta, pp. 159–164.

Retzinger, E.J. and Mallory-Smith, C. (1997) Classification of herbicides by site of action for weed resistance management strategies. *Weed Technology* 11, 384–393.

Thomas, A.G., Frick, B.L. and Hall, L.M. (1998a) *Alberta Weed Survey: Cereal and Oilseed Crops 1997.* Weed Survey Series Publication 98-2, Agriculture and Agri-Food Canada, Saskatoon, Saskatchewan.

Thomas, A.G., Frick, B.L., Van Acker, R.C., Knezevic, S.Z. and Joosse, D. (1998b) *Manitoba Weed Survey: Cereal and Oilseed Crops 1997.* Weed Survey Series Publication, Agriculture and Agri-Food Canada, Saskatoon, Saskatchewan.

Thomas, A.G., Wise, R.F., Frick, B.L. and Juras, L.T. (1996) *Saskatchewan Weed Survey: Cereal, Oilseed and Pulse Crops 1995.* Weed Survey Series Publication 96-1, Agriculture and Agri-Food Canada, Saskatoon, Saskatchewan.

79 *Silene vulgaris* (Moench) Garcke, Bladder Campion (*Caryophyllaceae*)

D.P. Peschken, A.S. McClay and R.A. De Clerck-Floate

Pest Status

Bladder campion, *Silene vulgaris* (Moench) Garcke, an introduced, persistent, deep-rooted perennial weed that reproduces mainly by seed (Wall and Morrison, 1990), is a primary noxious weed under the Canada Seeds Act (Anonymous, 1987) and occurs in the north-eastern and central USA and in every Canadian province to latitude 54°N (Scoggan, 1979). *S. vulgaris* is primarily a weed of roadsides, gravel pits and waste places. It thrives on sandy, coarse sandy and light soils. In Manitoba, field-wide infestations were reported on 1245 ha in 1984, primarily in hayfields, pastures and lucerne, *Medicago sativa* L., seed fields (M. Goodwin, 1999, Saskatoon, personal communication). According to weed surveys in the three prairie provinces, *S. vulgaris* infested 21 of 14,026 annually cultivated fields surveyed from 1976 to 1997 (Thomas and Wise, 1983a, 1984, 1985, 1987, 1988; Thomas *et al.*, 1997, 1998a, b). Most of the infested fields (10) were found in the Aspen Parkland (Black Soils) ecoregion, and in the Interlake Plain (3) and Lake Manitoba (4) ecoregions of Manitoba (Ecological Stratification Working Group, 1995; A.G. Thomas, 1999, Saskatoon, personal communication). In the Peace River region, British Columbia, only 2 of 372 fields in forage crops were infested with *S. vulgaris* (Thomas and Wise, 1983b). In Manitoba and Saskatchewan, none of 241 lucerne seed fields surveyed was infested (Goodwin *et al.*, 1985; Loeppky and Thomas, 1998; Malik *et al.*, 1991).

Cattle eat *S. vulgaris*, but its fodder value is low (Caputa, 1983). Clean-out losses can be as high as 30% in contaminated lucerne seed (Goodwin, 1985). Contaminated hay or seed cannot be sold legally. On Red River clay, lucerne and barley, *Hordeum vulgare* L., compete successfully with *S. vulgaris* (Wall and Morrison, 1990), but whether that is the case on poorer soils is not known. In Ontario and Quebec, *S. vulgaris* and *Vicia cracca* L. are the most important reservoir hosts of the lucerne mosaic virus (Paliwal, 1982).

Background

Once established, *S. vulgaris* is difficult and expensive to control (Manitoba Agriculture, 1985). Intensive summer fallow for 2 years is required to starve out *S. vulgaris*, but this may lead to soil erosion, especially on the light soils where it thrives. Infested fields should not be seeded to perennial forage crops (Dorrance, 1994). No herbicides are registered for within-crop control in any of the three prairie provinces (Ali, 1999; Manitoba Agriculture, 1999; Saskatchewan Agriculture and Food, 1999). Imazapyr at the rate of 3 l ha^{-1} controls *S. vulgaris*, but this use is registered only in non-cropped/non-grazed areas such as industrial sites or railroad ballast (Ali, 1999). Biological control was attempted to aid in control of *S. vulgaris* and to prevent the spread of severe infestations.

Biological Control Agents

Insects

Peschken and Derby (1990) investigated the host specificity of the seed feeder, *Hadena perplexa* (Denis and Schiffermüller). Although this moth has been reported only from *S. vulgaris* in Europe, the laboratory host range included four different genera. Therefore, *H. perplexa* was not recommended for release.

Cassida azurea Fabricius (mistakenly identified as *Cassida hemisphaerica* Herbst by Maw and Steinhausen, 1980a, b) occurs in much of Europe, in Algeria, and in Siberia (Bibolini, 1975), but it is absent from Great Britain, Holland and Scandinavia. Bibolini (1975) and Maw (1976) described its biology. In northern Italy, adults of this univoltine beetle appear in April and feed on young shoots of *S. vulgaris*, followed by mating and oviposition until early August. There are five larval instars. Young larvae tend to feed within the clusters of young apical leaves. Later stages also feed on succulent leaves and within buds and flowers. Older larvae may empty one flower every 24 h, leaving only the calyx. The larvae pupate inside or, rarely, on the outside of flowers, and on leaves. Adults overwinter in the upper layer of soil, where 88% of buried adults survived the winter of 1989–1990, and 91% that of 1990–1991 (Peschken *et al.*, 1997). The sites where the beetles had been buried were covered by snow (D.P. Peschken, unpublished).

Maw (1976) screened *C. azurea* using a breeding colony from stock collected in southern France and supplemented in 1986 with beetles collected near Brig, Switzerland. Peschken *et al.* (1997) conducted further host-specificity tests. *C. azurea* is restricted to *Silene* spp. It was able to complete development on native and introduced *Silene* spp., although development was slower and survival less on the native species. In contrast to laboratory results, *C. azurea* has been recorded only from *S. vulgaris* in the field in Europe, where seven *Silene* spp. co-occur in the same geographic area (Bibolini, 1975). Permission for field releases was granted in 1989 (Peschken *et al.*, 1997).

Pathogens

The rust *Uromyces behenis* (de Candolle) Unger occurs on *S. vulgaris* in Germany (Ale-Agha, 1994), but has not been studied as a candidate for biological control in Canada.

Releases and Recoveries

Breeding adults of *C. azurea* were released in spring, and sexually inactive beetles in autumn, in Manitoba, Saskatchewan and Alberta, beginning in 1989 (Table 79.1). In Alberta, an additional 1998 sexually active beetles were released at three unmonitored sites in 1995 and 1996. Colonies at 25 release sites were monitored: at five sites the colonies survived for at least 1 year; at ten sites, for at least 2–8 years; at one site the colony initially died out but subsequent releases survived for 3 years; at five sites the colonies did not survive one full year; at one site the colony survived for 6 years, but then was not recovered; and three sites were destroyed.

Evaluation of Biological Control

Monitoring of *C. azurea* establishment and spread at several sites and formal monitoring of *S. vulgaris* population changes at the Manitoba release sites (Peschken *et al.*, 1997; Table 79.1) showed that populations of *C. azurea* on most sites were too small to have an impact on *S. vulgaris* density. The most successful release appears to be the 1991 release at Fort Assiniboine, Alberta (Table 79.1). By 1996, feeding damage on all *S. vulgaris* plants in this pasture occurred and little of the weed was left around the release point. Canada thistle, *Cirsium arvense* (L.) Scoparius, was becoming abundant in 1996 and the site was mowed in 1997. It is not clear whether

Table 79.1. Releases and recoveries of *Cassida azurea* against *Silene vulgaris*.

Locality	Year released	No. of *C. azurea* released and stage[a]	Years recovered[b]	Remarks
Manitoba				
Vassar site 1	1989	1086 BA, 264 L	1990–1996	Cage and open-field releases; pasture with scattered lucerne; coarse, sandy soil
	1990	1580 DA		
Vassar site 2	1989	50 BA	1992, 1994, 1995	Cage releases, lucerne field; coarse, sandy soil; site destroyed in 1996
	1991	300 DA		
Fishing River	1990	732 DA	1991–1998	Hayfield, mixed forages; sandy soil
Valley River	1991	513 DA	1992–1998	Alfalfa field; sandy soil
Arborg site 1	1993	250 A	Not recovered	Uncultivated land , sandy soil. Beetles did not overwinter, perhaps because there was very little snow cover all winter
	1994	125 BA	Not recovered	
Arborg site 2	1993	250 A	Not recovered	Uncultivated land, sandy soil. Beetles did not overwinter, perhaps because there was very little snow cover all winter.
	1994	125 BA	Not recovered	
	1996	321 BA	1997–1999	Release site was protected with flax straw for the winter of 1996–1997. Excellent winter survival. In 1997 defoliation of *S. vulgaris* over about 50 m^2. In 1998 and 1999 beetles thinly spread over about 0.8 ha. Only individual plants defoliated
	1997	100 BA		
Grandview site 1	1994	300 BA	1995–1996	Hayfield on sandy ridge
Grandview site 2	1994	100 BA	1995–1996	Highway ditch, cut for hay
Whitemouth site 1	1994	100 BA	1995	Edge of lucerne field. No *C. azurea* found in 1996 or 1997
Whitemouth site 2	1994	100 BA	1995	Release site in fence line between pasture and lucerne field. No *C. azurea* found in 1996 or 1997
Saskatchewan				
Regina	1989	61 BA, 20 L	1990–1994	Cage release on Research Station; dense *S. vulgaris*; gravelly soil, site destroyed 1994
Maple Creek	1991	975 DA	1992	On railway bed
Alberta				
Redwater site 1	1990	320 BA	1991–1999	In 1996 had spread 110 m from release point but *C. azurea* population sparse
Redwater site 2	1992	200 BA	1992–1997	*C. azurea* population sparse
Olds	1990	217 DA	1990–1996	Pasture; sandy soil. Site flooded in 1996

Continued

Table 79.1. *Continued.*

Locality	Year released	No. of *C. azurea* released and stage[a]	Years recovered[b]	Remarks
Lethbridge	1993	About 50 L, E	1994–1999	Open garden plot at Lethbridge Research Centre
Fort Assiniboine	1991	100 DA	1992–1997	Pasture. *C. azurea* abundant in 1997, spread to 106 m from release point. In 1996 extensive defoliation, reduction in *S. vulgaris*. Mowing in 1997. No *C. azurea* seen in 1999
Millet	1991	150 DA	1991–1996	Pasture; sandy soil
Morley	1991	100 DA	Not recovered	Dry gravelly roadside; plants dusty
Nisku	1991	100 DA	1992–1993	Coarse gravel on railway bank; site sprayed in 1994
Bassano	1992	200 DA	Not recovered	Gravel pile
	1993	200 DA	Not recovered	Gravel pile
Pincher Creek	1993	200 DA	1994	Dry rocky slope
Claresholm	1993	200 DA	1994	Disused railway bank
Drayton Valley	1993	200 DA	1994	Roadside; grey wooded soil somewhat sandy
Rimbey	1993	200 DA	1994	Farmyard and garden; black loam soil

[a] A, adult beetles, sexual stage not recorded; BA, breeding adults; DA, adults in sexual diapause; L, larvae; E, eggs.
[b] The most recent year indicates when the site was last monitored.

the decrease in *S. vulgaris* was due to biological control, competition with *C. arvense* or mowing. At the two sites near Arborg, Manitoba (Table 79.1), colony overwintering failed, perhaps due to lack of snow cover.

Recommendations

Further work should include:

1. Investigating *U. behenis* and the seed feeders *Delia flavifrons* (Hufnagel) and *Hadena* spp. other than *H. perplexa*;
2. Continued monitoring of populations of *C. azurea* to determine the reasons for its success or failure to control *S. vulgaris*.

Acknowledgements

The following people provided assistance in locating release sites, making releases and monitoring them: J. Booth, D. Cole, A. Dearborn, C. Dearborn, P. Drebnisky, D. Henderson, R. Kennedy, B. Kuypers, R. McTavish, M. Moore, K. Patzer, F. Paulson, C. Pouteau, B. Ralston-Chalmers, E. Richardson, T. Seitz, R. Tarrant, M. Weiss and S. Wylie.

References

Ale-Agha, N. (1994) Ein kurzer Bericht zur Darstellung einiger Rostarten auf *Silene* im Duisburger Raum. *Mededelingen Faculties Landbouwkundige en Toegepaste Biologische Wetenschappen, Universiteit Gent* 59, 3a, 847–852.

Ali, S. (ed.) (1999) *Crop Protection 1999.* AGDEX 606–1, Alberta Agriculture and Food and Rural Development.

Anonymous (1987) *Seeds Act* 1959, *c.* 35, s. 1. Minister of Supply and Services Canada, Ottawa, Ontario.

Bibolini, C. (1975) Contributo alla conoscenza dei crisomelidae italiani (Coleoptera-Chrysomelidae). III. Osservazioni sulla etologia di *Cassida denticollis* Suffr., *Cassida prasina* Illig. e *Cassida ornata* Creutz e loro distribuzione geografica. *Frustula Entomologica* 13, 1–91.

Caputa, J. (1983) Weeds of meadows (*Silene vulgaris, Silene Flos-cuculi,* description, control). Les mauvaises herbes des prairies. *Revue Suisse d'Agriculture* 15, 214–215.

Dorrance, M.J. (ed.) (1994) *Practical Crop Protection.* Alberta Agriculture, Food and Rural Development, Edmonton, Alberta.

Ecological Stratification Working Group (1995) *A National Ecological Framework for Canada.* Agriculture and Agri-Food Canada, Research Branch, Centre for Land and Biological Resources Research and Environment Canada, State of the Environment Directorate, Ecozone Analysis Branch, Ottawa/Hull, Canada, Report and national map at 1:75000,000 scale.

Goodwin, M. (1985) Weed alert – bladder campion. In: *1985 Manitoba Weed Fair, Brandon, 17–18 January 1985*, Brandon, Manitoba, pp. 38–39.

Goodwin, M.S., Thomas, A.G., Morrison, I.N. and Wise, R.F. (1985) *Weed Survey of Alfalfa Seed Fields in Manitoba.* Weed Survey Series Publication No. 85-1, Agriculture Canada, Regina, Saskatchewan.

Loeppky, H.A. and Thomas, A.G. (1998) Weed survey of Saskatchewan alfalfa seed fields. In: Goerzen, D.W. (ed.) *Proceedings of 16th Annual Canadian Alfalfa Seed Conference, Saskatoon, Saskatchewan.* Saskatchewan Alfalfa Seed Producers Association, Saskatoon, Saskatchewan, pp. 53–57.

Malik, N.G., Bowes, G. and Waddington, J. (1991) *Weed Management Strategies in Lucerne Grown for Seed.* Final Report for Saskatchewan Agriculture Development Fund Project # V860050017. Agriculture Canada, Melfort, Saskatchewan.

Manitoba Agriculture (1985) *How to Control Bladder Campion.* Weed Facts Agdex No. 641.

Manitoba Agriculture (1999) *Guide to Crop Protection.* Manitoba Agriculture, Winnipeg, Manitoba.

Maw, M.G. (1976) Biology of the tortoise beetle, *Cassida hemisphaerica* (Coleoptera: Chrysomelidae), a possible biological control agent for the bladder campion, *Silene cucubalus* (Caryophyllaceae), in Canada. *The Canadian Entomologist* 108, 945–954.

Maw, M.G. and Steinhausen, W.R. (1980a) Corrigendum for 'Biology of the tortoise beetle, *Cassida hemisphaerica,* (Coleoptera: Chrysomelidae), a possible biological control agent for bladder campion, *Silene cucubalus* (Caryophyllaceae), in Canada' [*The Canadian Entomologist* 108, 945–954, 1976]. *The Canadian Entomologist* 112, 639.

Maw, M.G. and Steinhausen, W.R. (1980b) *Cassida azurea* (Coleoptera: Chrysomelidae) – not *C. hemisphaerica* – as a possible biological control agent of bladder campion, *Silene cucubalus* (Caryophyllaceae) in Canada. *Zeitschrift für Angewandte Entomologie* 90, 420–422.

Paliwal, Y.C. (1982) Virus diseases of alfalfa and biology of alfalfa mosaic virus in Ontario and western Quebec. *Canadian Journal of Plant Pathology* 4, 175–178.

Peschken, D.P. and Derby, J.L. (1990) Evaluation of *Hadena perplexa* [Lepidoptera: Phalaenidae] as a biological control agent of bladder campion *Silene vulgaris* [Caryophyllaceae] in Canada: rearing and host specificity. *Entomophaga* 35, 653–657.

Peschken, D.P., De Clerck-Floate, R. and McClay, A.S. (1997) *Cassida azurea* Fab. (Coleoptera: Chrysomelidae): Host specificity and establishment in Canada as a biological control agent against the weed *Silene vulgaris* (Moench) Garcke. *The Canadian Entomologist* 129, 949–958.

Saskatchewan Agriculture and Food (1999) *Guide to Crop Protection.* Saskatchewan Agriculture and Food, Regina, Saskatchewan.

Scoggan, H.J. (1979) The Flora of Canada. Part 4. Dicotyledoneae (Losaceae to Compositae). *National Museum of Natural Sciences (Ottawa) Publications in Botany* 7, 1117–1711.

Thomas, A.G. and Wise, R.F. (1983a) *Weed Surveys of Saskatchewan Cereal and Oilseed Crops from 1976 to 1979.* Weed Survey Series Publication No. 83-6, Agriculture Canada, Regina, Saskatchewan.

Thomas, A.G. and Wise, R.F. (1983b) *Peace River Region of British Columbia Weed Survey of Forage Crops – 1978, 1979 and 1980.* Weed Survey Series, Publication No. 83-5, Agriculture Canada, Regina, Saskatchewan.

Thomas, A.G. and Wise, R.F. (1984) *Weed Surveys of Manitoba Cereal and Oilseed Crops from 1978, 1979 and 1981.* Weed Survey Series Publication No. 84-1, Agriculture and Agri-Food Canada, Regina, Saskatchewan.

Thomas, A.G. and Wise, R.F. (1985) *Dew's Alberta Weed Survey (1973–1977)*. Weed Survey Series Publication No. 85–3, Agriculture Canada, Regina, Saskatchewan.
Thomas, A.G. and Wise, R.F. (1987) *Weed Survey of Saskatchewan Cereal and Oilseed Crops (1986)*. Weed Survey Series Publication No. 87-1, Agriculture and Agri-Food Canada, Regina, Saskatchewan.
Thomas, A.G. and Wise, R.F. (1988) *Weed Survey of Manitoba Cereal and Oilseed Crops (1987)*. Weed Survey Series Publication No. 88-1, Agriculture and Agri-Food Canada, Regina, Saskatchewan.
Thomas, A.G., Frick, B.L. and Hall, L.M. (1998a) *Alberta Weed Survey of Cereal and Oilseed Crops in 1997*. Weed Survey Series Publication No. 98-2, Agriculture and Agri-Food Canada, Saskatoon, Saskatchewan.
Thomas, A.G., Frick, B.L., Van Acker, R.C., Knezevic, S.Z. and Joosse, D. (1998b) *Manitoba Weed Survey of Cereal and Oilseed Crops in 1997*. Weed Survey Series Publication No. 98-1, Agriculture and Agri-Food Canada, Saskatoon, Saskatchewan.
Thomas, A.G., Kelner, D.J., Wise, R.F. and Frick, B.L. (1997) *Manitoba Weed Survey Comparing Zero and Conventional Tillage Crop Production Systems (1994)*. Weed Survey Series Publication No. 97-1, Agriculture and Agri-Food Canada, Saskatoon, Saskatchewan.
Wall, D.A. and Morrison, I.N. (1990) Competition between *Silene vulgaris* (Moench) Garcke and alfalfa (*Medicago sativa* L.). *Weed Research* 30, 145–151.

80 *Sonchus arvensis* L., Perennial Sow-thistle (*Asteraceae*)

A.S. McClay and D.P. Peschken

Pest Status

Perennial sow-thistle, *Sonchus arvensis* L.,[1] native to Europe and western Asia, occurs throughout Canada, and is a significant weed of agricultural crops across the prairies. It grows best in saturated soils and at relatively cool temperatures (Zollinger and Kells, 1991). In Michigan, Zollinger and Kells (1993) found that natural infestations of *S. arvensis* at densities from 61 to 96 shoots m^{-2} reduced yields of soybean, *Glycine max* (L.) Merrill, by up to 87% and dry edible bean, *Phaseolus vulgaris* L., by up to 84%. In Saskatchewan and Manitoba, Peschken *et al.* (1983) estimated crop losses in canola at Can$4.1 million per year. Current total losses in all crops and provinces would be many times this amount.

S. arvensis is a vigorous, deep-rooted, perennial herb up to 150 cm tall. All parts of the plant contain latex. It reproduces by windblown seed and spreads by means of horizontal spreading roots. Vertical roots can penetrate 2 m into the soil and can produce vegetative buds up to 50 cm below the soil surface. New shoots develop in late April from overwintering buds on roots or stem bases. Flowering begins in July and fruit maturation takes about 10 days (Lemna and Messersmith, 1990).

[1]Two forms occur in Canada, *S. arvensis* L. subsp. *arvensis* and *S. arvensis* L. subsp. *uliginosus* (von Bieberstein) Nyman, the latter distinguished mainly by the presence of glandular hairs on the peduncles and involucres.

Background

Several herbicides control *S. arvensis* in cereal crops; most, however, give top-growth control only (Ali, 1999). *S. arvensis* densities can be reduced in a canola–barley rotation by in-crop applications of clopyralid in canola, *Brassica napus* L. and *B. rapa* L. (with an additional pre-seeding application of glyphosate in the first year), followed by clopyralid + MCPA (4-chloro-2-methylphenoxyacetic acid) in barley, *Hordeum vulgare* L. (Darwent *et al.*, 1998).

In reduced tillage systems, *S. arvensis* has sometimes been reported to increase; however, Blackshaw *et al.* (1994) and Derksen *et al.* (1994) found that it responded inconsistently to tillage treatments. Stevenson and Johnston (1999) showed that *S. arvensis* densities tend to increase in crop rotations with a high frequency of broadleaf crops, e.g. canola, pea, *Pisum sativum* L., or flax, *Linum usitatissimum* L., possibly due to a shortage of herbicide options for its control in these crops.

In Europe, Schroeder (1974) reported 53 insects feeding on *S. arvensis* and recommended 11 as potential biological control agents, most of them seed- or flower-feeding species. Three of these have now been screened and released in Canada. Shurobenkov (1983) listed some insects associated with *S. arvensis* in Russia but did not add any new candidate species. Peschken (1984) suggested the root-boring moth, *Celypha roseana* (Schläger), as an additional possible candidate.

Two other European *Sonchus* spp., spiny annual sow-thistle, *S. asper* (L.) Hill, and annual sow-thistle, *S. oleraceus* L., occur in Canada and are significant weed problems. Some of the biological control agents released against *S. arvensis* will also attack one or both of these. According to the PLANTS database (USDA Natural Resources Conservation Service, 1999) no native *Sonchus* spp. occur in North America and there is only one species of the subtribe Sonchinae, as defined by Bremer (1994). Thus non-target risks appear to be of minor concern.

Biological Control Agents

Insects

Cystiphora sonchi (Bremi), a gall midge, attacks *S. arvensis*, and to a lesser extent other *Sonchus* spp., throughout Europe (Peschken, 1982). Females lay their eggs through the stomatal openings on the lower surface of leaves towards the end of the leaf expansion period (De Clerck and Steeves, 1988; De Clerck-Floate and Steeves, 1995). As larvae hatch, they form a single-chambered pustule gall protruding from the upper surface of the leaf. Pupation occurs either in a cocoon in the gall or in the soil after emergence of the mature larva. In Europe three generations per year occur (Peschken, 1982). Female *C. sonchi* produce single-sexed broods (McClay, 1996).

Tephritis dilacerata (Loew), a gall-forming fly, is most frequently found attacking *S. arvensis* in Europe and can only be reared reliably on that species, although there are some records from *S. oleraceus* and *S. asper* (Bérubé, 1978a). It oviposits into young flower buds where the larvae induce a button-shaped gall that prevents flower opening. Larvae feed on developing florets and receptacle tissue and pupate in the flower head, emerging in late summer as adults that overwinter (Bérubé, 1978b; Shorthouse, 1980). The insect thus spends about 10 months of the year as an adult, including 2–3 months after the likely time of emergence from overwintering sites until mid-July, when *S. arvensis* buds become available for oviposition. Attacked heads usually contain 1–8 puparia, although up to 20 can sometimes be found (A.S. McClay, unpublished). Peschken (1979) confirmed the host specificity of *T. dilacerata*. In Europe *T. dilacerata* is parasitized by *Pteromalus sonchi* Janzon (Janzon, 1983). A few individuals of a *Pteromalus* sp. were reared from *S. arvensis* heads galled by *T. dilacerata* at Vegreville, Alberta, in 1992 (A.S. McClay, unpublished).

Liriomyza sonchi Hendel, a leaf-mining fly, is widespread in Europe and extends to central Asia (Hendel, 1931–1936; Spencer, 1976). Females lay up to 140 eggs through the upper epidermis of the leaf and larvae form blotch mines, sometimes with several larvae to a mine. Pupation occurs in the soil or occasionally on the leaf surface (Peschken and Derby, 1988). Two generations per year occur in the field (Hendel, 1931–1936). Host-specificity testing of a population from lower Austria showed that in no-choice tests *L. sonchi* would breed readily on *S. arvensis* and at a low rate on *S. asper*, *S. oleraceus*, *Aetheorrhiza bulbosa* (L.) Cassini, and *Taraxacum officinale* Weber. Ten cultivars of lettuce, *Lactuca sativa* L., were tested using 837 female *L. sonchi*; a single adult emerged from one plant (Peschken and Derby, 1988).

Releases and Recoveries

C. sonchi was released at 19 sites across Canada from 1981 to 1991. It established in Alberta, Saskatchewan, Manitoba, Nova Scotia and probably New Brunswick, but not in British Columbia or Quebec (Table 80.1). *C. sonchi* is now widely distributed in Saskatchewan. At Vegreville, Alberta, it initially increased rapidly after a release in 1984, completing three generations per year (Peschken *et al.*, 1989), but in 1987 the density declined to less than half of its peak value and parasitic Hymenoptera emerged from a high percentage of galls collected in July. In 1988, the *C. sonchi* population at Vegreville collapsed: no galls were observed until early August, when five were found in a search of the entire 1000 m^2 plot. On an adjacent creek bank galls were still fairly numerous. Similar declines occurred at some Saskatchewan release sites. One reason for the population collapses may be parasitism. The larval endoparasitoid *Aprostocetus* sp. near *atticus* Graham[2] was the most abundant parasitoid at both the Alberta and Saskatchewan sites. This species also attacks the introduced dandelion leaf-gall midge, *Cystiphora taraxaci* (Kieffer), in Saskatchewan (Peschken *et al.*, 1993). Three other parasitoid species, *Neochrysocharis formosa* (Westwood), *Chrysonotomyia* sp. and *Zatropis* sp. near *justica* (Girault), occurred on *C. sonchi* in very small numbers. Samples were collected at Vegreville in 1988 and 1989, and at Outlook and Pike Lake, Saskatchewan, in 1990 and 1991, to evaluate levels of mortality from parasitism and other causes (Table 80.2).

T. dilacerata did not establish. Peschken (1984) described its early release (Table 80.3). In Alberta, from 1991 to 1995, further attempts to establish *T. dilacerata* from eastern Austria were made. Nine open and field-cage releases of a total of 3870 adults were made at Lethbridge, Sherwood Park and Vegreville (Table 80.4). Flies were released either in July when *S. arvensis* flower buds began to appear, in September to allow dispersal of flies to find overwintering sites, or in November by placing open cages of flies directly into possible overwintering sites. Flies released included adults emerged from galls collected in Austria, field-collected adults directly imported from Austria, and flies reared on potted plants or in field cages at Vegreville and overwintered as described below. All July releases resulted in good breeding success, with adult progeny emerging from galls by September. However, no overwinter survival was observed from any release, except for a single male seen in May 1995 at the 1994 release site.

The effects of shelter and snow cover on overwinter survival of *T. dilacerata* were investigated at Vegreville from 1991 to 1994 in 30 × 30 × 30 cm screened cages under various conditions: outdoors under snow cover; in a growth chamber at 6°C; with and without a layer of leaf litter in the cage; with and without monthly feeding at room temperature with a honey/yeast extract/mineral salts solution; and along a

[2]This is either a colour variant of *A. atticus* or a closely related, undescribed species (J. LaSalle, Riverside, 1989, personal communication). *Aprostocetus atticus* was originally described from Greece, where its possible host is *Cystiphora* sp. (Graham, 1987).

Table 80.1. Releases and recoveries of *Cystiphora sonchi* against *Sonchus arvensis*, 1981–1999. All releases were of galls containing larvae and/or pupae. Because of dispersal, it is not always clear which releases were responsible for currently established populations.

Location	Year	Number	Recoveries
British Columbia			
Abbotsford	1984	4,070	None
Telkwa	1992	150	None
Alberta			
Ribstone	1981	5,000	None
Ribstone	1982	4,500	None
Ribstone	1983	3,000	None
Vegreville	1983	6,111	None
Vegreville	1984	2,700	1985–1999
Saskatchewan			
Regina	1981	2,900	Established around Regina, at Last Mountain Lake and Echo Valley Provincial Park, probably from these releases
Regina	1983	10,203	
Regina	1984	500	
Regina	1987	61,510	
Melfort	1981	7,500	
Outlook	1981	8,000	1991. Also at Douglas Provincial Park, possibly from this release
Wishart	1981	3,500	
Saskatoon	1984	800	Galls seen near Saskatoon 1998
Saskatoon	1985	600	
Pike Lake	1985	5,750	1991
Estlin	1986	6,500	
Manitoba			
Deloraine	1982	2,000	Established at Rossburn, MB, possibly from these releases
Deloraine	1983	2,234	
Deloraine	1984	4,079	
Quebec			
Sainte-Anne-de Bellevue	1981	5,000	Not established
Sainte-Anne-de Bellevue	1982	2,100	
New Brunswick			
St Quentin	1991	3,789	Galls formed but no overwinter survival
St Quentin	1992	500	
St Quentin	1993	450	Galls formed but no overwinter survival
Lincoln	1993	61	No gall formation
St Quentin	1994	500	Galls seen in August 1995
Nova Scotia			
Great Village	1984	4,500	
Great Village	1985	500	
Great Village	1986	13,246	
Bible Hill	1985	500	Established and now distributed through Truro area
Colchester County	1985	1,100	
Truro	1985	100	
Truro	1986	14,754	
Windsor	1985	1,636	Not established

Table 80.2. Estimated mortality of *Cystiphora sonchi* due to parasitism by *Aprostocetus* sp. nr. *atticus* and other causes in Alberta and Saskatchewan, 1988–1991.

Location	Sampling date	Method	Total larvae	Mortality (%) due to	
				Parasitism	Unknown
Alberta					
Vegreville	9 Aug. 1988	Emergence	219	50	–
Vegreville	Jul–Aug. 1989	Dissection	1150	20[a]	–
Vegreville	23 Aug. 1990	Dissection	672	72[a]	–
Saskatchewan					
Pike Lake	June–Aug. 1990	Emergence	1382	18	65[b]
Pike Lake	July–Aug. 1990	Dissection	463	14	22[c]
Pike Lake	June–Aug. 1991	Dissection	648	28	42[c]
Outlook	June–Aug. 1990	Emergence	2487	13	64[b]
Outlook	June–Aug. 1990	Dissection	635	9	23[c]
Outlook	June–Aug. 1991	Dissection	2362	8	25[c]

[a]All paralysed larvae were assumed to be parasitized.
[b]Larvae that exited the galls but failed to develop to adults.
[c]Larvae paralysed but no parasitoid eggs or larvae found on dissection.

Table 80.3. Releases and recoveries of *Tephritis dilacerata* adults against *Sonchus arvensis*, 1979–1984; 'fall' refers to releases in autumn of recently emerged adults, while 'spring' refers to releases of overwintered adults ready to breed.

Location	Year	Season	Number	Recoveries
Alberta				
Ribstone	1981	Fall	2000	None
Saskatchewan				
Regina and Estevan	1979	Spring	810 (total)	Bred well in 1979 in Regina, poorly in Estevan
Wishart	1981	–	38	None
Melfot	1981	–	860	None
Outlook	1981	–	195	None
Regina	1982	–	278	None
Quebec				
Ste Anne de Bellevue	1981	–	1947	None
Nova Scotia				
North River	1984	–	1899	None
Prince Edward Island				
Lauretta	1981	–	2000	None

Table 80.4. Field releases and recoveries of adult *Tephritis dilacerata* against *Sonchus arvensis* in Alberta, 1991–1994.

Location	Release date	Number	Source	Cage/open	Recoveries and notes
Sherwood Park	September 1991	1600	Galls from Austria	Open	No recoveries 1992
Lethbridge	September 1991	1370	Galls from Austria	Open	No recoveries 1992
Vegreville	July 1992	15	Overwintered progeny from flies received 1991	Caged	Galls collected Sept., 73 flies emerged
Vegreville	July 1992	65	Field-collected adults from Austria	Caged	Galls collected Sept., 773 flies emerged
Vegreville	July 1993	174	Flies reared 1992 and overwintered in cages	Caged	Galls collected Sept., 750 flies emerged
Vegreville	July 1993	23	Flies reared 1992 and overwintered in cages	Caged	Galls collected Sept., 298 flies emerged
Vegreville	November 1993	300	Reared 1993 in cages	Caged	Flies placed in open cage in perennial sow-thistle stand No recoveries 1994
Vegreville	November 1993	300	Reared 1993 in cages	Both	Flies placed in open cage among bushes near perennial sow-thistle stand. No recoveries 1994
Vegreville	July 1994	23	Reared 1993 and overwintered in cages	Caged	1 male seen May 1995

42 m transect running from inside a stand of trembling aspen into an open mowed area. Also, overwinter survival of the progeny of flies that had overwintered once in Alberta was compared to that of progeny from imported flies. Survival varied very widely among years and, to a much lesser extent, among treatments. In 1991/1992, there was no survival of 2400 flies in the growth chamber, probably due to desiccation; survival of 3600 flies outdoors was increased from 1.3 to 4.2% by providing a layer of litter in the cages, but was not enhanced by monthly feeding. In 1992/1993, survival of 901 flies overwintered outdoors under snow cover with a 15 cm litter layer in the cages was much higher than in 1991/1992; survival of progeny from flies that had successfully overwintered was 75.2%, while survival of progeny of flies imported that summer from Austria was 72.0%. In 1993/1994, total survival of 800 flies along the transect was only 0.9%, with no significant location effect along the transect. These results, overall, suggest that microhabitat variability and year-to-year weather variations affect the rate of overwinter survival. There was no evidence that survival for one winter in Alberta had selected for increased cold-hardiness.

These studies showed that *T. dilacerata*

Table 80.5. Releases of *Liriomyza sonchi* against *Sonchus arvensis*, 1987–1991.

Location	Year	Number	Stage	Recoveries
Saskatchewan				
Outlook	1987	103	Adults	None
Regina	1987	107	Adults	None
Indian Head	1988	349	Adults	None
Outlook	1988	354	Adults	None
Pike Lake	1988	816	Adults	None
Regina	1988	24	Adults	None
Regina	1988	132	Pupae	None
Regina	1988	45	Larvae	None
Pike Lake	1989	171	Adults	None
Regina	1989	150	Adults	None
New Brunswick				
St Quentin	1990	2118	Pupae	Mines observed later in summer: no overwinter survival
St Quentin	1991	1268	Pupae	None
Nova Scotia				
Colchester County	1989	748	Pupae	None
Colchester County	1989	1137	Adults	None
Garland	1989	468	Adults	None
Garland	1989	679	Pupae	None
North West River	1991	546	Pupae	None

will readily accept Canadian *S. arvensis* plants as hosts; that it is able to complete development and emerge by September, when conditions should still be favourable to allow the flies to seek overwintering habitats; that, under certain conditions of shelter and snow cover, the flies can successfully overwinter in the field in Alberta; and that these overwintered flies can successfully breed the following summer. The fly's wide distribution in Europe (Bérubé, 1978b) also suggests that it should survive on the Canadian prairies. It is not clear, therefore, why releases of *T. dilacerata* have so far failed to establish. Possibly, during the time before the appearance of *S. arvensis* flower buds, overwintered adults cannot find suitable food in the field, suffer excessive losses from predation, or become too widely dispersed to find mates.

Field releases of *L. sonchi* began in 1987 (Table 80.5) but it has not established. Although Julien and Griffiths (1998) reported its establishment in New Brunswick, this report appears to be in error. Leaf mines were observed later the same summer after the 1990 release at St Quentin, New Brunswick, but there was no overwinter survival (Maund *et al.*, 1993; C. Maund, Fredericton, 2000, personal communication).

Evaluation of Biological Control

In Alberta or Saskatchewan, the only agent established to date, *C. sonchi*, has not had any noticeable effect on the vigour or population density of *S. arvensis*. This is in contrast to the significant impact that *Cystiphora schmidti* (Rübsaamen) has had on skeletonweed, *Chondrilla juncea* L., in the USA and Australia (Julien and Griffiths, 1998). The fact that *C. sonchi* oviposits only into leaves towards the end of their expansion period (De Clerck-Floate and Steeves, 1995), and does not damage meristematic tissue, may reduce its effectiveness. The high rates of parasitism observed on this species from 1989 onwards may also have decreased its effectiveness. In Nova Scotia, the population of *S. arvensis* at the original release site of *C. sonchi* has levelled off at about 60% of its former density (G. Sampson, Truro, 2000, personal communication). This situation requires further study to determine whether *C. sonchi* is responsible for the apparent reductions. Further releases of *C. sonchi* do not appear to be justified at present, until it can be determined whether it is responsible for any impact on *S. arvensis*.

It should be possible to establish *T. dilacerata*. However, the impact of this species is likely to be limited, as its effect is only on seed production. Harris and Shorthouse (1996) suggested that the galls of *T. dilacerata* are not nutrient sinks and that this is likely to further limit its effectiveness.

Recommendations

Further work should include:

1. Assessing whether *C. sonchi* is reducing *S. arvensis* populations in Nova Scotia;
2. Attempting to establish *L. sonchi*;
3. Screening of *C. roseana* for its suitability.

Acknowledgements

We thank M. Sarazin, G. Sampson, C. Maund, K. Brown, R. Cranston, J. Lischka, G. Davis, A. Watson, and the late A.T.S. Wilkinson for information on agent releases and G. Scheibelreiter for collecting *T. dilacerata* in Austria.

References

Ali, S. (ed.) (1999) *Crop Protection 1999*. Alberta Agriculture, Food and Rural Development, Edmonton, Alberta.

Bérubé, D.E. (1978a) The basis for host plant specificity in *Tephritis dilacerata* and *T. formosa* [Dip.: Tephritidae]. *Entomophaga* 23, 331–337.

Bérubé, D.E. (1978b) Larval descriptions and biology of *Tephritis dilacerata* [Dip.: Tephritidae], a candidate for the biocontrol of *Sonchus arvensis* in Canada. *Entomophaga* 23, 69–82.

Blackshaw, R.E., Larney, F.O., Lindwall, C.W. and Kozub, G.C. (1994) Crop rotation and tillage effects on weed populations on the semi-arid Canadian prairies. *Weed Technology* 8, 231–237.

Bremer, K. (1994) *Asteraceae: Cladistics and Classification*. Timber Press, Portland, Oregon.

Darwent, A.L., Harker, K.N. and Clayton, G.W. (1998) Perennial sowthistle control with sequential herbicide treatments applied under minimum and zero tillage systems. *Canadian Journal of Plant Science* 78, 505–511.

De Clerck, R.A. and Steeves, T.A. (1988) Oviposition of the gall midge *Cystiphora sonchi* (Bremi) (Diptera: Cecidomyiidae) via the stomata of perennial sow-thistle (*Sonchus arvensis* L.). *The Canadian Entomologist* 120, 189–193.

De Clerck-Floate, R.A. and Steeves, T.A. (1995) Patterns of leaf and stomatal development explain ovipositional patterns by the gall midge *Cystiphora sonchi* (Diptera, Cecidomyiidae) on perennial sow thistle (*Sonchus arvensis*). *Canadian Journal of Zoology* 73, 198–202.

Derksen, D.A., Thomas, A.G., Lafond, G.P., Loeppky, H.A. and Swanton, C.J. (1994) Impact of agronomic practices on weed communities: fallow within tillage systems. *Weed Science* 42, 184–194.

Graham, M.W.R. de V. (1987) A reclassification of the European Tetrastichinae (Hymenoptera: Eulophidae) with a revision of certain genera. *Bulletin of the British Museum (Natural History), Entomology Series* 55, 1–392.

Harris, P. and Shorthouse, J.D. (1996) Effectiveness of gall inducers in weed biological control. *The Canadian Entomologist* 128, 1021–1055.

Hendel, F. (1931–1936) 59. Agromyzidae. In: Lindner, E. (ed.) *Die Fliegen der palaearktischen Region*. Schweizerbart'sche Verlag, Stuttgart, Germany, pp. 1–570.

Janzon, L.A. (1983) *Pteromalus sonchi* n. sp. (Hymenoptera: Chalcidoidea), a parasitoid of *Tephritis dilacerata* (Loew) (Diptera: Tephritidae), living in flower-heads of *Sonchus arvensis* L. (Asteraceae) in Sweden. *Entomologica Scandinavica* 14, 309–315.

Julien, M.H. and Griffiths, M.W. (1998) *Biological Control of Weeds: a World Catalogue of Agents and Their Target Weeds*, 4th edn. CAB International, Wallingford, UK.

Lemna, W.K. and Messersmith, C.G. (1990) The biology of Canadian weeds. 94. *Sonchus arvensis* L. *Canadian Journal of Plant Science* 70, 509–532.

Maund, C.M., McCully, K.V., Finnamore, D.B., Sharpe, R. and Parkinson, B. (1993) A summary of insect biological control agents released against weeds in NB pastures from 1990 to 1993. *Adaptive Research Reports (New Brunswick Department of Agriculture)* 15, 359–380.

McClay, A.S. (1996) Unisexual broods in the gall midge *Cystiphora sonchi* (Bremi) (Diptera: Cecidomyiidae). *The Canadian Entomologist* 128, 775–776.

Peschken, D.P. (1979) Host specificity and suitability of *Tephritis dilacerata* [*Dip.: Tephritidae*]: a candidate for the biological control of perennial sow-thistle (*Sonchus arvensis*) [*Compositae*] in Canada. *Entomophaga* 24, 455–461.

Peschken, D.P. (1982) Host specificity and biology of *Cystiphora sonchi* (Dip.: Cecidomyiidae), a candidate for the biological control of *Sonchus* species. *Entomophaga* 27, 405–416.

Peschken, D.P. (1984) *Sonchus arvensis* L., perennial sow-thistle, *S. oleraceus* L., annual sow-thistle and *S. asper* (L.) Hill, spiny annual sow-thistle (Compositae). In: Kelleher, J.S. and Hulme, M.A. (eds) *Biological Control Programmes Against Insects and Weeds in Canada 1969–1980*. Commonwealth Agricultural Bureaux, Slough, UK, pp. 205–209.

Peschken, D.P. and Derby, J.A.L. (1988) Host specificity of *Liriomyza sonchi* Hendel. (Diptera: Agromyzidae), a potential biological agent for the control of weedy sow-thistles, *Sonchus* spp., in Canada. *The Canadian Entomologist* 120, 593–600.

Peschken, D.P., Thomas, A.G. and Wise, R.F. (1983) Loss in yield of rapeseed (*Brassica napus*, *Brassica campestris*) caused by perennial sowthistle (*Sonchus arvensis*) in Saskatchewan and Manitoba. *Weed Science* 31, 740–744.

Peschken, D.P., McClay, A.S., Derby, J.L. and De Clerck, R.A. (1989) *Cystiphora sonchi* (Diptera: Cecidomyiidae), a new biological control agent established on the weed perennial sow-thistle (*Sonchus arvensis*) (Compositae) in Canada. *The Canadian Entomologist* 121, 781–791.

Peschken, D.P., Gagné, R.J. and Sawchyn, K.C. (1993) First record of the dandelion leaf-gall midge, *Cystiphora taraxaci* (Kieffer, 1888) (Diptera: Cecidomyiidae), in North America. *The Canadian Entomologist* 125, 913–918.

Schroeder, D. (1974) The phytophagous insects attacking *Sonchus* spp. (Compositae) in Europe. In: Wapshere, A.J. (ed.) *Proceedings of the Third International Symposium on Biological Control of Weeds*. Commonwealth Agricultural Bureaux, Slough, UK, pp. 89–96.

Shorthouse, J.D. (1980) Modification of the flower heads of *Sonchus arvensis* (family Compositae) by the gall former *Tephritis dilacerata* (order Diptera, family Tephritidae). *Canadian Journal of Botany* 58, 1534–1540.

Shurobenkov, B.G. (1983) Phytophages of the field sow thistle. *Zashchita Rastenii* 11, 22–23.

Spencer, K.A. (1976) *The Agromyzidae (Diptera) of Fennoscandia and Denmark*. Vol. 5 part 2, *Fauna Entomologica Scandinavica*. Scandinavica Science Press, Klampenborg, Denmark.

Stevenson, F.C. and Johnston, A.M. (1999) Annual broadleaf crop frequency and residual weed populations in Saskatchewan Parkland. *Weed Science* 47, 208–214.

USDA Natural Resources Conservation Service (1999) The PLANTS database. http://plants.usda.gov/plants (5 April 2000)

Zollinger, R.K. and Kells, J.J. (1991) Effect of soil pH, soil water, light intensity, and temperature on perennial sowthistle (*Sonchus arvensis* L.). *Weed Science* 39, 376–384.

Zollinger, R.K. and Kells, J.J. (1993) Perennial sowthistle (*Sonchus arvensis*) interference in soybean (*Glycine max*) and dry edible bean (*Phaseolus vulgaris*). *Weed Technology* 7, 52–57.

81 *Tanacetum vulgare* L., Common Tansy (*Asteraceae*)

D.J. White

Pest Status

Common tansy, *Tanacetum vulgare* L., was introduced from eastern Europe and the British Isles as early as 1638. Steady increases in populations and in habitats colonized have resulted in its designation as a noxious weed in Quebec, Manitoba, Alberta and British Columbia. Roadsides, railways, fence lines, field margins, permanent seeded pasture, lake shores and river and creek banks had the highest densities and area. The importance of *T. vulgare* in European folk medicine has prompted extensive research on its phytochemistry, and pharmacology (Nemeth *et al.*, 1994). In contrast, before 1993, limited research was done to understand factors that regulate its populations. In Alberta, the problem for agricultural producers is the persistent and increasing colonization of pastures and hay fields by *T. vulgare*, and possible toxicity in cattle. The proportionally greater, high-density areas of *T. vulgare* in riparian habitats serve as continued sources of re-infestation and result in serious native habitat displacement. The north central region is the centre of *T. vulgare* infestations and plant density. A 1993 survey estimated that 26,384 ha, in 58 municipal districts, were infested and the total estimated annual cost to municipalities and private landowners for controlling *T. vulgare* was Can\$256,612 (Can\$9.70 ha^{-1}) (White, 1997).

T. vulgare is a fast-growing perennial that flowers early in its life cycle, produces many easily dispersed seeds, reproduces vegetatively and is a good competitor (Baker, 1965). Stems often remain erect for 2 years in undisturbed habitats and retain seed with high germination rates after dispersal in the second year (White, 1997). A small percentage of plants flowering in July produce viable seed, with a 10–20% germination rate by mid-August. Seed collected from erect stems, following overwintering, germinate at a rate of 70%, with further increases to 90% following additional cold treatment. Seed weight varied markedly among habitats, e.g. average weights of 50 seeds along stream banks was 6.66 mg, and along roadsides, 8.47 mg. In contrast, plant height over time varied more within sites than between sites (White, 1997).

Background

The limited effectiveness of conventional herbicide and cultural control methods, and the environmental risks associated with these methods (toxicity and erosion in areas of high infestation), prompted recommendations to develop alternative control measures.

White (1997) showed that establishment of *T. vulgare* was greatest in pastures seeded with species such as meadow foxtail, *Alopecurcus pratensis* L., and streambank wheatgrass, *Agropyron riparium* Scribner and Smith, that did not quickly produce high levels of ground cover. Grazing decreased *T. vulgare* populations but also promoted continued seedling establishment on bare ground. Although heavy grazing resulted in decreased populations, the presence of plants in surrounding ungrazed vegetation resulted in the persistent establishment of new seedlings on readily available bare ground. Mid-season levels of non-structural carbohydrates in the roots and rhizomes of *T. vulgare* are higher in ungrazed than in grazed habitats. Maintenance of a vigorous perennial rootstock appeared essential for producing large amounts of seed. However, vegetative spread did not appear to be as important to seed dispersal as seedling establishment in undisturbed habitats. Simulated herbivory experiments within natural habitats demonstrated the highly conditional responses of *T. vulgare* to the type of defoliation, natural habitat and moisture and light availability. These response characteristics provide information for selection of potentially successful biological control agents (White, 1997).

Biological Control Agents

Insects

Few insects feed on *T. vulgare* in north-central Alberta. None are abundant enough or inflict damage at a level capable of adversely affecting *T. vulgare* populations. In contrast, the diversity of insect species and amount of plant damage reported in Europe suggests a high potential for successful introduction of biological control agents into North America. Although most of the species on *T. vulgare* in Europe are oligophagous and polyphagous, several appear to be monophagous and could be suitable for introduction. Of particular interest is the root-feeding guild, e.g. *Dicrorampha* spp., *Celyphya rufana* Scopoli and *Phytoecia nigricornis* (Fabricius) (Friese and Schroeder, 1997; Schmitz, 1998).

Pathogens

Fungi

In Alberta, a stem rust, *Puccinia tanaceti* de Candolle var. *tanaceti*, and a powdery mildew, *Erysiphe cichoracearum* de Candolle, were found in isolated situations on mature and senescent stems and leaves of *T. vulgare* (White, 1997).

Recommendations

Further work should include:

1. Screening of European root-feeding insects for specificity;
2. Evaluating the effectiveness of potential agents in light of the complex infraspecific chemotype variation of *T. vulgare* and its persistence under heavy vertebrate and simulated insect herbivory.

References

Baker, H.G. (1965) Characteristics and modes of origin of weeds. In: Baker, H.G. and Stebbins, C.L. (eds) *The Genetics of Colonizing Species.* Academic Press, New York, pp. 147–169.

Friese, J. and Schroeder, D. (1997) *Field Surveys for Phytophagous Insects Associated with* Tanacetum vulgare *in Northern Europe.* Annual Report, European Station, International Institute of Biological Control, Delémont, Switzerland.

Nemeth, E.Z., Hethelyi, E. and Bernath, J. (1994) Comparison studies on *Tanacetum vulgare* L. Chemotypes. *Journal of Herbs, Spices and Medicinal Plants* 2, 85–92.

Schmitz, G. (1998) The phytophagous insect fauna of *Tanacetum vulgare* L. (Asteraceae) in central Europe. *Beiträge zur Entomologie* 48, 219–236.

White, D.J. (1997) *Tanacetum vulgare* L.: weed potential, biology, response to herbivory, and prospects for classical biological control in Alberta. MSc thesis, Department of Entomology, University of Alberta, Edmonton, Alberta.

82 *Taraxacum officinale* (Weber), Dandelion (*Asteraceae*)

S.M. Stewart-Wade, S. Green, G.J. Boland, M.P. Teshler, I.B. Teshler, A.K. Watson, M.G. Sampson, K. Patterson, A. DiTommaso and S. Dupont

Pest Status

Dandelion, *Taraxacum officinale* Weber, is a herbaceous perennial native to Europe that now occurs in over 60 countries worldwide (Holm *et al.*, 1997). It is a weed in pastures, forages, orchards, vineyards, vegetable gardens, turf in golf courses, municipal parks and home gardens (Burpee, 1992; Holm *et al.*, 1997). Although its presence may not cause economic losses, it is an aesthetic problem, especially during flowering and seed production periods (Holm *et al.*, 1997). It is also an increasing problem in annual crops in western Canada (Derksen and Thomas, 1996).

T. officinale is an autumn–spring germinating perennial that reproduces apomictically by seed or vegetatively via root segments (Holm *et al.*, 1997; Moerkerk and Barnett, 1998).

Background

Several herbicides are registered to control *T. officinale* (Daniel and Freeborg, 1987) but there is concern about their potential negative effects on humans, animals and the environment (Meyer and Allen, 1994). There has been increasing legislation to restrict the use of certain herbicides in numerous municipalities (Riddle *et al.*, 1991). Alternative methods, e.g. biological control, have therefore been investigated.

Biological Control Agents

Insects

The weevil *Ceutorhynchus punctiger* Gyllenhall attacks flower buds, seeds and leaves of *T. officinale*, but host specificity and key mortality factors must first be studied (McAvoy *et al.*, 1983). Another weevil, *Barypeithes pellucidus* (Boheman), feeds lightly on *T. officinale* leaves and moderately on the epidermis of the scapes (Galford, 1987). The black vine weevil, *Otiorhynchus sulcatus* (Fabricius), feeds on *T. officinale* (Masaki *et al.*, 1984). The potato leafhopper, *Empoasca fabae* (Harris), survives and reproduces on *T. officinale* (Lamp *et al.*, 1984). Root-feeding larvae of the Japanese beetle, *Popillia japonica* Newman, and the southern masked chafer, *Cyclocephala lurida* Bland, feed upon and reduce root biomass of *T. officinale* (Crutchfield and Potter, 1995). The cynipid wasp, *Phanacis taraxaci* (Ashmead), forms galls on the abaxial surface of maturing *T. officinale* leaves, which influences the partitioning of photoassimilates by actively redirecting carbon resources from unattacked leaves (Paquette *et al.*, 1993; Bagatto *et al.*, 1996). The first record of European dandelion leaf-gall midge, *Cystiphora taraxaci* Kieffer, in north-central Saskatchewan was by Peschken *et al.* (1993). This midge induces purple–red pustule galls on the upper surface of leaves (Neuer–Markmann and Beiderbeck, 1990).

Pathogens

Viruses

In the Okanagan Valley, British Columbia, a *Carlavirus* with the proposed name of dandelion latent virus (DaLV), was isolated from naturally infected *T. officinale* exhibiting no visible symptoms (Johns, 1982).

Fungi

Using fungi as mycoherbicides is a control option for many weeds (Charudattan, 1991; TeBeest, 1996; Mortensen, 1998), including *T. officinale*. At least 15 fungi have been recorded on *T. officinale* in Canada but only a few have been considered for biological control (Anonymous, 1957; Conners, 1967; Ginns, 1986; Riddle *et al.*, 1991). Riddle *et al.* (1991) and Brière *et al.* (1992) evaluated isolates of *Sclerotinia sclerotiorum* (Libert) De Bary and *Sclerotinia minor* Jagger (see Huang *et al.*, Chapter 99 this volume) for their virulence on *T. officinale* under growth room and field conditions. Riddle *et al.* (1991) found significant negative correlations between isolate virulence and dry weights of inoculated plants in a controlled environment, and positive correlations between isolate virulence and reduction in the number of *T. officinale* plants in inoculated turfgrass swards. However, concern exists about using this virulent polyphagous plant pathogens as mycoherbicides.

A collaborative project involving three academic institutions and three industrial partners (University of Guelph (UG), McGill University (MU), Nova Scotia Agricultural College (NSAC), Dow AgroSciences Inc., BioProducts Centre Inc., and Saskatchewan Wheat Pool) was established with the aim of developing a bioherbicide to control *T. officinale* in turfgrass, targeting home garden use as the primary potential market. Numerous fungi, pathogenic on *T. officinale*, were collected and screened, and those with the highest potential were selected for further study. Eight isolates were evaluated in June, July and September 1996 for their efficacy to control dandelion under growth room and field conditions in Ontario, Quebec and Nova Scotia. These were spore and/or mycelial liquid formulations of *Phoma herbarum* Westendorp (G5/2), *Phoma exigua* Desmazières (GIII) and *Phoma* sp. (G961.16) produced by UG; *Myrothecium roridum* Tode Fries (AC133) and *Plectosphaerella cucumerina* (Lindfors) W. Gams (AC9530) produced by NSAC; and *Curvularia inaequalis* Boedjin (Mac2) and *Colletotrichum* sp. Corda (Mac4/H) produced by MU. Two solid formulations of *S. minor* (Mac1), produced by MU, comprising mycelium in sodium alginate granules (Brière *et al.*, 1992) and mycelial-colonized barley grits (a modification of the barley-based formulation used by Ciotola *et al.*, 1991) were also evaluated. Mac1 was the most consistently effective isolate at controlling *T. officinale* under growth room and field conditions, despite varying location, season and formulation. Based on the results and considering that sodium alginate is more expensive than barley, isolate Mac1 as a barley-based formulation was selected for further study.

Field trials were conducted in June, July and September 1997, and in May and September 1998, at all three locations, using both transplanted and natural stands of *T. officinale*, with isolate Mac1 formulated as both barley grits and kaolin clay granules (Teshler *et al.*, 1998). Field efficacy trials were designed to assess the effect of: (i) dose using spot application (0.2, 0.4 or 0.8 g per plant) and broadcast application (10, 20, 40, 60, 120 g m^{-2}); (ii) timing of application (morning, noon, afternoon and evening applications); (iii) single versus split application; (iv) irrigation regime; (v) mowing regime; (vi) storage of inoculum (stored for 5, 14, 18 or 21 weeks prior to application); and (vii) *T. officinale* growth stage (seedling, bud, flowering). Safety issues were addressed by testing the pathogenicity of Mac1 on turfgrasses, survival in soil, turf and compost, and potential for dissemination. Laboratory trials were also conducted at all three locations to determine the optimal medium and conditions for growth and storage of Mac1, to develop quality assurance assays and to improve the solid substrate production system.

The success of the field efficacy trials depended on dew or rainfall for the establishment of infection. Efficacy was low if prolonged hot, dry conditions prevailed during a trial. The barley formulation of Mac1 had greater efficacy than the kaolin clay formulation at all locations, with optimum application rates of 0.4–0.8 g per plant when spot applied and 60 g m^{-2} when broadcast. Under favourable weather conditions (cool to moderate temperatures and sufficient moisture), Mac1 formulated as barley grits usually produced visible disease symptoms within 1–3 days after inoculation, and significant disease development and plant mortality by 7–14 days after inoculation. In general, efficacy was not affected by timing of application, single versus split application, mowing regime, length of storage, or *T. officinale* growth stage. Irrigation only increased the efficacy of Mac1 at the Ontario site during dry conditions in 1997. Mac1 did not infect any of the turfgrass species tested. Sclerotia formed on the inoculum in some field plots, but sclerotial degradation in the field was rapid, with no viable sclerotia found after 4 months. Sclerotia were also killed within 5 h when exposed to compost temperatures of 50°C. Mycelial transfer from infected *T. officinale* to lettuce, *Lactuca sativa* L. (a highly susceptible species), only occurred when plants were in direct contact with each other. The potential for infection of common garden plants such as petunia, *Petunia* sp., via the use of inoculated lawn clippings as a mulch, was minimal. When stored at room temperature, Mac1 on barley grits rapidly lost viability on potato dextrose agar (generally within 3 weeks). However, at 4°C viability of inoculum was maintained up to 25 weeks, although it declined progressively.

Evaluation of Biological Control

Isolate Mac1 as a barley grit formulation showed good efficacy on *T. officinale*, provided dew or rainfall occurred shortly after inoculation. Strict user guidelines concerning timing of application (to coincide with forecast precipitation), survival and transfer of this fungus should optimize its efficacy and minimize the potential risks of carry-over to susceptible, non-weed hosts.

Such intensive collaboration among public and private research organizations in developing a potential bioherbicide is unique. Within 4 years, the project progressed from collection and screening of numerous fungal isolates, to field evaluation and formulation of a single candidate isolate, to initiation of the government registration process. However, many factors contributed to the subsequent discontinuation of the project, including changes in research direction and priorities among the industrial sponsors; insufficient international market size; poor performance of Mac1 under prolonged, dry weather conditions; sclerotia formation in the field; costs of large-scale production; and the need for refrigeration during storage and distribution of the barley grit formulation.

Recommendations

Further work should include:

1. Production of Mac1 on a small, local scale, e.g. a made-to-order basis, to avoid the costs and degradation of quality associated with large-scale production and long-term storage;

2. Investigation of integrated pest-management strategies to complement Mac1, including other biological control agents, e.g. insects, cultural methods and chemicals.

References

Anonymous (1957) *Report of the Minister of Agriculture for Canada for the year ended 3 March 1955*. Agriculture and Agri Food Canada, Queens Printer, Ottawa, Ontario.

Bagatto, G., Paquette, L.C. and Shorthouse, J.D. (1996) Influence of galls of *Phanacis taraxaci* on carbon partitioning within common dandelion, *Taraxacum officinale*. *Entomologia Experimentalis et Applicata* 79, 111–117.

Brière, S.C., Watson, A.K. and Paulitz, T.C. (1992) Evaluation of granular sodium alginate formulations of *Sclerotinia minor* as a potential biological control agent for turfgrass weed species. *Phytopathology* 82, 1081.

Burpee, L.L. (1992) A method for assessing the efficacy of a biocontrol agent on dandelion (*Taraxacum officinale*). *Weed Technology* 6, 401–403.

Charudattan, R. (1991) The mycoherbicide approach with plant pathogens. In: TeBeest, D.O. (ed.) *Microbial Control of Weeds.* Chapman and Hall, New York, pp. 24–57.

Ciotola, M., Wymore, L. and Watson, A. (1991) *Sclerotinia*, a potential mycoherbicide for lawns. *Weed Science Society of America Abstracts* 31, 81.

Conners, I.L. (1967) *An Annotated Index of Plant Diseases in Canada.* Publication 1251, Research Branch, Canada Department of Agriculture.

Crutchfield, B.A. and Potter, D.A. (1995) Feeding by Japanese beetle and southern masked chafer grubs on lawn weeds. *Crop Science* 35, 1681–1684.

Daniel, W.H. and Freeborg, R.P. (1987) *Turf Managers Handbook.* Harcourt Brace Jovanovich, Duluth, Minnesota.

Derksen, D.A. and Thomas, A.G. (1996) Dandelion control in cereal and oilseed crops. *Expert Committee on Weeds (ECW) Proceedings.* Expert Committee on Weeds, Victoria, British Columbia, pp. 63–69.

Galford, J.R. (1987) Feeding habits of the weevil *Barypeithes pellucidus* (Coleoptera: Curculionidae). *Entomological News* 98, 163–164.

Ginns, J.H. (1986) *Compendium of Plant Disease and Decay Fungi in Canada 1960–1980.* Publication 1813, Research Branch, Canada Department of Agriculture.

Holm, L., Doll, J., Holm, E., Pancho, J. and Herberger, J.P. (1997) *World Weeds: Natural Histories and Distribution.* John Wiley and Sons, New York, New York.

Johns, L.J. (1982) Purification and partial characterization of a carlavirus from *Taraxacum officinale. Phytopathology* 72, 1239–1242.

Lamp, W.O., Morris, M.J. and Armbrust, E.J. (1984) Suitability of common weed species as host plants for the potato leafhopper, *Empoasca fabae. Entomologica Experimentalis et Applicata* 36, 125–131.

Masaki, M., Ohmura, K. and Ichinohe, F. (1984) Host range studies of the black vine weevil, *Otiorhynchus sulcatus* (Fabricius) (Coleoptera: Curculionidae). *Applied Entomology and Zoology* 19, 95–106.

McAvoy, T.J., Kok, L.T. and Trumble, J.T. (1983) Biological studies of *Ceutorhynchus punctiger* (Coleoptera: Curculionidae) on dandelion in Virginia. *Annals of the Entomological Society of America* 76, 671–674.

Meyer, M.H. and Allen, P. (1994) Dandelion dilemma: a decision case in turfgrass management. *Horticulture Technology* 4, 190–193.

Moerkerk, M.R. and Barnett, A.G. (1998) *More Crop Weeds.* R.G. and F.J. Richardson, Melbourne, Australia.

Mortensen, K. (1998) Biological control of weeds using microorganisms. In: Boland, G.J. and Kuykendall, L.D. (eds) *Plant–Microbe Interactions and Biological Control.* Marcel Dekker, New York, New York, pp. 223–248.

Neuer-Markmann, B. and Beiderbeck, R. (1990) Biology and host range of the gall midge species *Cystiphora taraxaci* under growth chamber conditions (Diptera: Cecidomyiidae). *Entomologia Generalis* 15, 209–216.

Paquette, L.C., Bagatto, G. and Shorthouse, J.D. (1993) Distribution of mineral nutrients within the leaves of common dandelion (*Taraxacum officinale*) galled by *Phanacis taraxaci* (Hymenoptera: Cynipidae). *Canadian Journal of Botany* 71, 1026–1031.

Peschken, D.P., Gagne, R.J. and Sawchyn, K.C. (1993) First record of the dandelion leaf-gall midge, *Cystiphora taraxaci* (Kieffer, 1888) (Diptera: Cecidomyiidae), in North America. *The Canadian Entomologist* 125, 913–918.

Riddle, G.E., Burpee, L.L. and Boland, G.J. (1991) Virulence of *Sclerotinia sclerotiorum* and *S. minor* on dandelion (*Taxacum officinale*). *Weed Science* 39, 109–118.

TeBeest, D.O. (1996) Biological control of weeds with plant pathogens and microbial pesticides. In: Sparks, D.L. (ed.) *Advances in Agriculture,* Vol. 56. Academic Press, Toronto, Ontario, pp. 115–137.

Teshler, I., Teshler, M., DiTommaso, A. and Watson, A. (1998). Application of multifactorial experimental design to optimize a fungal formulation for biocontrol of dandelion (*Taraxacum officinale*). *Expert Committee on Weeds (ECW) Proceedings.* Expert Committee on Weeds, Winnipeg, Manitoba, p. 76.

83 *Ulex europaeus* L., Gorse (*Fabaceae*)

R. Prasad

Pest Status

Gorse, *Ulex europaeus* L., is a shrub native to Mediterranean Europe (Misset and Gourret, 1995) that arrived in Canada in the last century via Oregon (Isaacson, 1992a). It is found mainly in British Columbia (Vancouver, Vancouver Island, Gulf Islands and Queen Charlotte Islands) at low elevations in the coastal western hemlock, *Tsuga heterophylla* (Rafinesque-Schmaltz) Sargent, and coastal Douglas fir, *Pseudotsuga menziesii* (Mirbel), biogeoclimatic zones (Meidinger and Pojar, 1991) and is classed as a noxious weed. It also invaded the east coast of North America as far north as Massachusetts but its low frost tolerance may limit its spread further north. *U. europaeus* is a serious weed in many coastal areas worldwide (Richardson and Hill, 1996), suppressing tree growth in forested landscapes. It invades dry and disturbed sites, forming thickets that suppress and retard native vegetation, probably including conifer seedlings (Prasad, 2000). Although gorse can occupy the same site as Scotch broom, *Cytisus scoparius* (L.) Link (see Prasad, Chapter 68 this volume), it prefers drier sites and can persist longer, thus posing a greater threat. *U. europaeus* is invasive due to specialized stem photosynthesis, prolific seed production, longevity of seeds in soil and nitrogen fixation (Zielke *et al.*, 1992). Once established, the *U. europaeus* canopy architecture prohibits growth of other plants (Richardson and Hill, 1996). *U. europaeus* threatens native plant diversity because it establishes large, dense thickets, creating conditions that inhibit their growth (Lee *et al.*, 1986). Of particular concern is the Garry oak, *Quercus garryana* Douglas, ecosystem (Nuszdorfer *et al.*, 1991). *U. europaeus* is also a fire hazard because of the high concentration of oil within its branches (Zielke *et al.*, 1992). In some areas, its spread has been linked to agriculture, where it has been occasionally planted as hedgerows and subsequently invaded pastures and road verges. Although no data exist on the value of economic losses, it is believed to be considerable as real estate values decline due to severe infestations in urban landscapes.

Ulex europaeus germinates from seeds produced by young and old plants. Seedlings begin to flower after 2 years and continue to flower in winter. A mature plant produces large numbers of seeds that survive in the soil for several years. Vegetative propagation after cutting or wounding is profuse. Some plants attain a height of 4–5 m and survive 25–30 years.

Background

Chemical herbicides have been effectively used to control *U. europaeus* and *C. scoparius* (Peterson and Prasad, 1998). Historically, the most widely used compound was 2,4–5 trichlorophenoxy acetic acid (2,4,5-T) applied as a foliar spray or to the stump (Balneaves and Perry, 1982) but it is now banned in British Columbia (Zielke *et al.*, 1992). Glyphosate combined with an organosilicone surfactant is equally effective (Balneaves and Perry, 1982). Triclopyr (Garlon-3) applied as foliar spray gives almost complete control of gorse seedlings and resprouts (Hartley and Popay, 1982).

Fire has been used to control *U. europaeus*. Rolston and Talbot (1980) reported 62% reduction in seed numbers in the top 10 cm of soil following a fire. After burning, grazing by goats for 2–3 years reduced gorse populations to negligible levels (Radcliffe, 1985).

Although manual cutting is another control option, it is difficult in well-established populations because of the spiny nature of the plant.

U. europaeus is attacked by a range of insects and mites (Syrett *et al.*, 1999); however, none has been introduced into Canada for biological control.

Biological Control Agents

Vertebrates

Goats and sheep have been employed to control *U. europaeus* populations, particularly in New Zealand. An intensive level of goat stocking (25–30 goats ha^{-1}) was very effective in reducing its populations (Radcliffe, 1985). Krause *et al.* (1988) noted that goats were more economical than conventional herbicides.

Insects

Exapion ulicis Förster, a seed-feeding weevil, was introduced into the USA and has spread throughout major *U. europaeus*-infested areas on the west coast (Isaacson, 1992b). In Washington state, *E. ulicis* has reduced seed production by as much as 96% on some sites. Adults lay eggs on the pods in early spring and larvae feed on developing seeds within (Isaacson, 1992b). Adults also feed on foliage, possibly making *U. europaeus* more susceptible to the pathogenic fungus, *Colletotrichum* sp. (Markin *et al.*, 1996).

Apion scutellare Kirby, a gall-forming weevil, has also been considered for biological control but attempts to introduce it into Hawaii, where *U. europaeus* is a problem, have been unsuccessful.

Agonopterix ulicetella Stainton, a North American moth, often colonizes *U. europaeus* in Oregon, Hawaii and British Columbia (Markin *et al.*, 1996) but is unlikely to be used in inundative releases because of its potential spread to native plants.

Mites

Tetranychus lintearis Dufour colonizes *U. europaeus* and feeds on the cell contents of spines and stems (Hill and O'Donnell, 1991; Isaacson, 1992b). Since 1989, populations from New Zealand have been released and became established in Hawaii and Oregon (Markin *et al.*, 1996) where they gave good control of *U. europaeus*. Even though aggressive and successful, *T. lintearis* has not yet been released in Canada.

Pathogens

Fungi

Many fungi have been isolated from *U. europaeus* but few promising biological control candidates have been found (Johnston, 1990). In New Zealand, research is in progress to develop *Gibberella tumida* Broad (Brende) as a mycoherbicide (Johnston and Park, 1994) but its performance is erratic under field conditions.

In Canada, *Chondrostereum purpureum* (Persoon ex Fries Pouzar) has been developed to control resprouting in hardwood weeds, and work is being done to adopt it for *U. europaeus* and *C. scoparius* (Prasad and Naurais, 1999). However, because *U. europaeus* rapidly resprouts from cut stems, *C. purpureum* efficacy is not consistent under field conditions.

Evaluation of Biological Control

Growth and reproduction of *U. europaeus* is generally too vigorous to be adequately controlled by insects or mites. Even if seed production is reduced by 96%, each plant could still add about 300 seeds to the persistent seed bank. The tap root allows the plant to recover from serious herbivory and even a severely reduced seed production may favour establishment of new stands.

No single strategy can completely control/eradicate *U. europaeus* once it is established. All types of management, including biological control, should be attempted early, right after seedling emergence, to prevent extensive proliferation and colony establishment. An integrated approach using manual cutting and herbicide or bioherbicide treatments coupled with burning is likely to be more effective than any one control measure alone. Control measures should aim at depleting/flushing out seedbanks by spraying herbicides on seedlings before flowering, preferably using systemic herbicides that destroy the root/underground parts as well. Biological control using insects, mites, vertebrates or pathogens should be complementary, especially in environmentally sensitive areas, because these agents are best suited for reducing the infestation by cutting down seed production.

Recommendations

Further work should include:

1. Refining *C. purpureum* formulations and testing at different times of the year to improve control;
2. Evaluating and introducing suitably adapted populations of *T. lintearis* and *E. ulicis*;
3. Developing an integrated management programme.

References

Balneaves, J.M. and Perry, C. (1982) Long term control of gorse–bracken mixtures for forest establishment in Nelson, N.Z. *New Zealand Journal of Forestry* 27, 219–225.

Hartley, M.J. and Popay, A.I. (1982) Control of gorse seedlings by low rates of herbicides. In: Hartley, M.J. (ed.) *Proceedings of the 35th New Zealand Weed and Pest Control Conference, Palmerston North, New Zealand*, pp. 138–140.

Hill, R.L. and O'Donnell, D.J. (1991) The host range of *Tetranychus lintearis* (Acarina: Tetranychidae). *Experimental and Applied Acarology* 11, 253–269.

Isaacson, D. (1992a) Distribution and status of gorse. *Oregon Department of Agriculture Weed Control Program, Broom/Gorse Quarterly* 1(1), 1–2.

Isaacson, D. (1992b) Status of biocontrol agents for control of gorse. *Oregon Department of Agriculture Weed Control Program, Broom/Gorse Quarterly* 1(1), 3–4.

Johnston, P.R. (1990) Potential fungi for the biological control of some New Zealand weeds. *New Zealand Journal of Agricultural Research* 33, 1–14.

Johnston, P.R. and Park, S.L. (1994) Evaluation of the mycoherbicidal potential of fungi found on broom and gorse in New Zealand. In: Popay, A. (ed.) *Proceedings of the 47th New Zealand Plant Protection Conference*. New Zealand Plant Protection Society, Hamilton, New Zealand, pp. 121–124.

Krause, M.A., Beck, A.C. and Dent, J.B. (1988) Control of gorse in hill country: an assessment of chemical and biological methods. *Agricultural Systems* 26, 35–49.

Lee, W.G., Allen, R.B. and Johnson, D.N. (1986). Succession and dynamics of gorse (*Ulex europaeus* L.) communities in the Dunedin Ecological District, South Island, N.Z. *New Zealand Journal of Botany* 24, 279–292.

Markin, G.P., Yashioka, E.R. and Conant, P. (1996) Biological control of gorse in Hawaii. In: Moran, V. and Hoffman, J. (eds) *Proceedings of the X International Symposium on Biological Control of Weeds*. University of Capetown, Capetown, South Africa, pp. 371–375.

Meidinger, D. and Pojar, J. (1991) *Ecosystems of British Columbia*. Special Report Series #6, British Columbia Ministry of Forests, Victoria, British Columbia, pp. 81–111.

Misset, M.T. and Gourret, J.P. (1995) Flow cytometric analysis of different ploidy levels observed in the genus *Ulex* L. in Brittany, France. *Botanica Acta* 109, 72–79.

Nuszdorfer, F.C., Klinka, K. and Demarchi, D.A. (1991) Coastal Douglas-fir zone. In: Meidinger, D.

and Pojar, J. (eds) *Ecosystems of British Columbia.* Special Report Series #6, British Columbia Ministry of Forests, Victoria, British Columbia, pp. 81–93.

Peterson, D. and Prasad, R. (1998) The biology of Canadian weeds. 109. *Cytisus scoparius* L. (Link). *Canadian Journal of Plant Science* 78, 497–504.

Prasad, R. (2000) Some aspects of the impact and management of the exotic weed, Scotch broom (*Cytisus scoparius*) in British Columbia. *Journal of Sustainable Forestry* 15, 339–345.

Prasad, R. and Naurais, S. (1999) Invasiveness of alien plants: impact of Scotch broom on Douglas-fir seedlings and its control. In: Kelly, M., Howe, M. and Neill, B. (eds) *Proceedings of the California Exotic Plant Protection Council, Sacramento, CA, USA, 15–17 Oct.* California Exotic Pest Plant Council, San Juan Capistrano, California, Vol. 5, pp. 23–25.

Radcliffe, J.E. (1985) Grazing management of goat and sheep for gorse control. *New Zealand Journal of Experimental Agriculture* 13, 181–190.

Richardson, R.G. and Hill, R.L. (1996) The biology of Australian weeds. 34. *Ulex europaeus* L. *Plant Protection Quarterly* 13, 46–58.

Rolston, M. and Talbot, J. (1980) Soil temperatures and regrowth of gorse burnt after treatment with herbicides. *New Zealand Journal of Experimental Agriculture* 8, 55–61.

Syrett, P., Fowler, S.V., Coombs, E.M., Hosking, J.R., Marking, G.P., Paynter, Q. and Shepherd, A.W. (1999) The potential for biological control of Scotch broom (*Cytisus scoparius*) and related weedy species. *Biocontrol News and Information* 20(1), 33 N.

Zielke, K., Boateng, J., Caldicott, N. and Williams, H. (1992) *Broom and Gorse: a Forestry Perspective Analysis.* British Columbia Ministry of Forests, Queens Printer, Victoria, British Columbia.

84 *Alternaria panax* Whetzel, Alternaria Blight (Pleosporaceae)

J.A. Traquair

Pest Status

Alternaria panax Whetzel, causal agent of Alternaria blight, is a ubiquitous pathogen of American ginseng, *Panax quinquefolius* L., in all areas of its commercial production and/or natural occurrence in North America and Asia. The major sites of commercial production in Canada in both mulched, artificial shade gardens and woodland sites are southern Ontario and southern British Columbia. Alternaria blight was first noticed in New York state, USA (Whetzel and Rosenbaum, 1912; Whetzel *et al.*, 1930) and has since been reported in Ontario and British Columbia in artificial gardens and woodland sites (Howard *et al.*, 1994; Reeleder and Fisher, 1995; Punja, 1997). The disease is most severe in artificial shade gardens where plant density is high and where cool, moist foliar canopies provide ideal conditions for production and spread of conidia.

Symptoms include stem spot and damping-off of seedlings, stem and foliar spot or blight of mature plants, and mould of stratified seed. Spots are characterized by central water-soaked tissue that quickly dries and turns brown in a target-board pattern with yellow–brown margins. *A. panax* is thought to overwinter as conidia and mycelium in mulch and infested crop residue (David, 1988; Parke and Shotwell, 1989; Howard *et al.*, 1994). Based on experiences in the Orient, which probably apply to Canada and the USA, crop loss assessments range from minor leaf and stem spot or foliar blight in 10–20% of stands to major epidemics involving extensive defoliation and blight, with 100% loss of crop in shade gardens (Proctor and Bailey, 1987; Reeleder and Fisher, 1995; Proctor 1996). The current export value of Canadian ginseng is Can$60 million. Necrosis of leaf tissue certainly reduces photosynthetic surface and causes reduced root growth and marketable yield.

Background

Regular and frequent applications of foliar fungicides are recommended in Canada and the USA (Parke and Shotwell, 1989; Howard *et al.*, 1994; Oliver, 1996; Proctor, 1996). Current non-chemical approaches to control of Alternaria blight include removal of infected plants, careful attention to sanitation and avoidance of excessive nitrogen fertilization in order to limit overdevelopment of the ginseng canopy, which impedes air circulation around plants (Howard *et al.*, 1994).

Removal of crop residue and straw mulch from ginseng gardens is not practical or economical. Growers and buyers of American ginseng as a medicinal crop are very interested in non-chemical disease control and the guaranteed absence of fungicide residue. Therefore, biological control is well-worth pursuing.

Biological Control Agents

Bacteria, Fungi

In the USA, Joy and Parke (1995) reported that foliar applications of the Gram-negative bacterium, *Burkholderia* (= *Pseudomonas*) *cepacia* (Palleroni and Holmes) Yabuuchi, Kasako, Oyaizu, Yano, Hotta, Hashimoto, Ezahi and Arahawa, suppressed Alternaria leaf blight of *P. quinquefolius*. However, use of *B. cepacia* has been halted because of reports of certain strains being opportunistic human pathogens of cystic fibrosis patients (Holmes *et al.*, 1998). In Canada, experimental drench applications to straw mulch and soil, and seed coating with actinomycetous bacteria such as *Streptomyces* spp., are effective for the biological control of overwintering conidial and mycelial inoculum of *A. panax* in straw mulch, soil, and stratified ginseng seed *in vitro* and in pot cultures. Antagonism is based mainly on the production of antifungal compounds and antibiosis. Similarly, the fungi *Trichoderma harzianum* Rifai, *Gliocladium virens* Miller, Giddens and Foster, *Trametes versicolor* (L.: Fries) Pilat, *Irpex lacteus* (Fries: Fries) Fries, and *Chondrostereum purpureum* (Persoon: Fries) Pouzar are effective biological control agents for suppression of *Alternaria* diseases of ginseng (J.A. Traquair and G.J. White, unpublished). However, extensive hyperparasitism has been observed *in vitro* as the mechanism of inhibition on nutrient agar and straw substrates and on various mulch materials in ginseng pots under controlled environmental conditions.

Evaluation of Biological Control

Biological control is a promising approach to the eradicative and preventive control of Alternaria blight and spot diseases of perennial *P. quinquefolius* crops because of the potential to destroy soil-, crop debris- and mulch-borne inoculum over a 4–5-year production cycle.

Recommendations

Further work should include:

1. Determining field efficacy of bacterial and fungal biological control agents;
2. Developing formulation and delivery of bacterial and fungal agents.

References

David, J.C. (1988) *Alternaria panax*. CMI Descriptions of Pathogenic Fungi and Bacteria. Set 96, Nos 951–960. *Mycopathologia* 103, 105–124.
Holmes, A., Govan, J. and Goldstein, R. (1998) Agricultural use of *Burkholderia* (*Pseudomonas*) *cepacia*: A threat to human health? *Emerging Infectious Diseases* 4, 221–227.
Howard, R.J., Garland, J.A. and Seaman, W.L. (eds) (1994) *Diseases and Pests of Vegetable Crops in Canada*. Canadian Phytopathology Society and Entomological Society of Canada, Ottawa, Ontario.
Joy, A.E. and Parke, J.L. (1995) Biocontrol of Alternaria leaf blight on American ginseng by *Burkholderia cepacia* AMMD. In: Bailey, W.G., Whitehead, C., Proctor, J.T.A. and Kyle, J.T. (eds) *Proceedings of the International Ginseng Conference, Vancouver 1994*. Simon Fraser University, Burnaby, British Columbia, pp. 93–100.
Oliver, A. (ed.) (1996) *Ginseng Production Guide for Commercial Growers*. Province of British Columbia, Ministry of Agriculture, Fisheries and Food, Kamloops, British Columbia.
Parke, J.L. and Shotwell, K.M. (1989) *Diseases of Cultivated Ginseng*. Bulletin A3465, University of Wisconsin-Extension and United States Department of Agriculture, pp. 10–12.
Proctor, J.T.A. (1996) Ginseng: old crop, new directions. In: Janick, J. (ed.) *Progress in New Crops*. American Society of Horticultural Sciences, Alexandria, Virginia, pp. 565–577.
Proctor, J.T.A. and Bailey, W.G. (1987) Ginseng: industry, botany, and culture. *Horticulture Reviews* 9, 187–236.
Punja, Z.K. (1997) Fungal pathogens of American ginseng (*Panax quinquefolium*) in British Columbia. *Canadian Journal of Plant Pathology* 19, 301–306.
Reeleder, R.D. and Fisher, P. (1995) *Diseases of Ginseng*. Factsheet No. 95-003, Ontario Ministry of Agriculture, Food and Rural Affairs, pp. 1–4.
Whetzel, H.H. and Rosenbaum, J. (1912) *Diseases of Ginseng and Their Control*. Bulletin 250, United States Bureau of Plant Industry, pp. 1–40.
Whetzel, H.H., Rosenbaum, J., Braun, J.W. and McClintoch, J.A. (1930) *Ginseng Diseases and Their Control*. Farmers' Bulletin 736, United States Department of Agriculture, pp. 1–7.

85 *Botryotinia fuckeliana* (de Bary) Whetzel, Grey Mould and Botrytis Blight (Sclerotiniaceae)

J.T. Calpas, J.P. Tewari and J.A. Traquair

Pest Status

The worldwide fungus, *Botryotinia fuckeliana* (de Bary) Whetzel [anamorph, *Botrytis cinerea* (Persoon) Fries], causal agent of grey mould or Botrytis blight, causes serious losses to a wide range of greenhouse crops (Howard *et al.*, 1994; Hausbeck and Moorman, 1996) including vegetables, bedding plants, bulbs, cut flowers, potted plants and perennials. These and other field crops, e.g. American ginseng, *Panax quinquefolius* L., herbal crops, market vegetables and small fruits are high in value and constitute a fast-growing component of the Canadian agriculture/horticulture sector.

Although the biology of *B. cinerea* on many host plants is well understood, the disease it causes continues to cause significant losses in greenhouse crops and field-grown horticultural crops. Prolonged leaf wetness, high humidity and cool temperatures favour the rapid development and spread of Botrytis blight and grey mould in densely planted greenhouse and field-grown horticultural crops (Parke and Shotwell, 1989; Howard *et al.*, 1994; Reeleder and Fisher, 1995; Hausbeck and Moorman, 1996; Punja, 1997b). Continued significant losses occur, even though this disease can be one of the easiest to control through proper environmental management (Jarvis, 1992; Howard *et al.*, 1994). Strict control of the environment, to prevent conditions that favour development of grey mould, can be very difficult in the field and during early months of the greenhouse cropping season (January through March).

Background

Fungicides are commonly employed to control grey mould and Botrytis blight; however, strains of the fungus are now resistant to several of them (Howard *et al.*, 1994; Elad *et al.*, 1995). Further, consumer demand has placed additional pressure on producers of market vegetables, small fruits, ornamentals and medicinal crops to reduce pesticide use and employ integrated disease-management strategies. Therefore, increased demand and funding has occurred for the development of biological control agents for greenhouse crop pests and diseases. Greenhouse growers have responded in Alberta, for example, by typically spending about Can\$15,000–20,000 ha^{-1} $year^{-1}$ to produce their vegetable crops without insecticide use.

Environmental control is the basis for optimism in the development of biological control for diseases of greenhouse crops (Andrews, 1990; Punja, 1997a). For *B. cinerea* in greenhouses, biological control is attractive because the environment can be manipulated to increase agent efficacy. However, for high-value, field-grown horticultural crops, environmental manipulation is more difficult (Yu and Sutton, 1998). In these circumstances, control of Botrytis blight and grey mould can be limited by contamination with wind-blown conidial inoculum from other crops and weeds. In the case of perennial horticultural crops such as ginseng and berries, persistence of sclerotial inoculum in the soil and mulch is an added constraint (Parke and Shotwell, 1989; Howard *et al.*, 1994). In Ontario, infections from overwintering sclerotial inoculum and polycyclical infections from wind-blown conidial inoculum from diseased leaves and fruit in the current crop are also serious problems in grey mould control in dense plantings of strawberry, *Fragaria* × *ananassa* (L.) Duchesne, and raspberry, *Rubus idaeus* L.

Several commercial biological control products based on *Trichoderma* spp. are available in the USA and other countries (D. Fravel, Beltsville, 1999, personal communication[1]) but none are registered in Canada. Examples of these products included Trichodex®, RootShield® or Bio-Trek, T-22G or T-22 Planter Box, Promote® and Trichoseal®. Because *Trichoderma* spp. are endemic to all Canadian soils, they are excellent candidates for biological control of *B. cinerea*, subject to local testing and development for registration in Canada.

Biological Control Agents

Fungi, Bacteria

Research into developing biological controls for *B. cinerea* has been undertaken for several crops, including apple, *Malus pumila* Miller (= *M. domestica* Borkhausen)

[1]http://www.barc.usda.gov/ars/Beltsville/barc/psi/bpdl/bioprod.htm

(Tronsmo and Ystaas, 1980), rose, *Rosa* spp. (Redmond *et al.*, 1987), snap bean *Phaseolus vulgaris* L. (Nelson and Powelson, 1988), black spruce seedlings, *Picea mariana* (Miller) Britton, Sterns and Poggenburg (Zhang *et al.*, 1994), strawberry (Peng and Sutton, 1991; Sutton and Peng, 1993) and raspberry (Yu and Sutton, 1998). *Trichoderma* spp. (Dik and Elad, 1999) and *Gliocladium* spp. (Sutton *et al.*, 1997) are among the most promising fungal biological control agents against *B. cinerea*, and different strains have the ability to control a range of pathogens under a variety of environmental conditions (Lorito *et al.*, 1993; Punja, 1997b). Research has also been directed at development of biological control for *B. cinerea* in greenhouse crops, including the use of *Trichoderma harzianum* Rifai against *B. cinerea* in greenhouse tomato, *Lycopersicon esculentum* Miller (O'Neill *et al.*, 1996).

Trichoderma spp. have a high degree of adaptability, are common throughout the world under a variety of environmental conditions and substrates (Hjeljord and Tronsmo, 1998), and can be used as antagonists in combination with fungicides (Elad *et al.*, 1993). *Trichoderma* isolates that are fast-growing saprophytes and can establish high populations on the crop plant compete with *B. cinerea* in the phyllosphere, and colonize potential infection sites to the exclusion of *B. cinerea* (Hjeljord and Tronsmo, 1998). *Trichoderma* spp. are also known to be aggressive mycoparasites that directly attack fungal pathogens such as *B. cinerea* (Bélanger *et al.*, 1995; Hjeljord and Tronsmo, 1998).

The adaptability of *Trichoderma* spp. also raises concern that certain strains could be plant pathogens. Although reports of *Trichoderma* spp. causing plant disease exist (Menzies, 1993; Hjeljord and Tronsmo, 1998), considering the amount of work done on *Trichoderma* spp. as potential biological control agents, the risk appears slight. The possibility that *Trichoderma* spp. could themselves become introduced pathogens is an integral component of the ecological research involving these fungi. Environmental fate and the risk of these biological control agents as pests are important concerns in developing a biological control strategy.

Another concern is that *Trichoderma* spp., particularly *T. harzianum,* can cause serious disease problems in commercial mushroom, *Agaricus bisporus* (Lange) Imbach, culture (Hjeljord and Tronsmo, 1998). However, Muthumeenakshi *et al.* (1998) indicated that strains of *T. harzianum* useful for biological control are not likely to be aggressive pathogens of mushrooms, and this can be confirmed by genotyping and commercial trials.

Boyle (1999) also demonstrated that presence of mushroom-aggressive strains of *T. harzianum* in mushroom compost is, in itself, not enough to cause a disease epidemic. The disease process is complex and depends on a number of additional factors, including the condition of the spawn and the full microbiota of compost (Boyle, 1999). Use of *T. harzianum* as a biological control organism does not inherently pose any greater threat to mushroom culture than it does to the actual crop to which it is applied. Certainly, it does not pose any greater threat to mushroom culture than the widespread agricultural use of chemical fungicides.

In Canada, research is focused on testing biological control agents developed and registered in other countries to manage Botrytis blight and grey mould. These evaluations are being undertaken together with development of new Canadian products. Practical studies on formulation and delivery under local environmental conditions, with ecological research aimed at optimized activity and environmental impact, are being undertaken.

In Alberta, the *Botrytis* biological control program (mitigated in 1998 at the Crop Diversification Centre-South in Brooks and at the Department of Agriculture, Food and Nutrition Science, University of Alberta in Edmonton) is responding to the need for alternatives to chemical controls for *B. cinerea*, and the desire to reduce pesticide use in greenhouse vegetable crop production. The use of *Trichoderma* spp. as a biological control for *B. cinerea* in greenhouse crops is being undertaken to select and assess potential isolates that are effective under commercially relevant conditions. The tomato model system was chosen because of the particular problems greenhouse tomato growers were experiencing with the disease.

A molecular biology component allows for characterization of the biological control agents as well as for identification and tracking of candidates. One hundred and sixty isolates of *B. cinerea* from 32 locations throughout Alberta were characterized based on random polymorphic DNA (RAPD) analysis and their virulence on tomato. Genetic characterization of 100 isolates of *Trichoderma* spp. was completed in 1999. Screening of these isolates against *B. cinerea* using a tomato-stem-piece assay was started in late 1999 and the most promising isolates are being evaluated in greenhouse trials.

In Ontario, sclerotial, mycelial and conidial inocula of *B. cinerea* on field-grown, horticultural crops were targeted for biological control. Control of foliar and seedling blight and seed mould of American ginseng were studied using selected wood-decay basidiomycetes, e.g. *T. harzianum*, *Trichoderma virens* Miller, Giddens, and Foster, and *Streptomyces* spp., including *S. griseoviridis* (Anderson, Ehrlich, Sun, and Burkholder), in the commercial product Mycostop® (Kemira Agro Oy, Finland). Of 26 assorted agaricoid and polyporaceous basidiomycetes screened *in vitro* and in pots under controlled environment conditions, using sand and straw as delivery systems, *Irpex tulipiferae* Schwein and *Coriolus versicolor* (Fries) Quélet [syn. *Trametes versicolor* (L.) Fries] were the most effective antagonists. They were capable of degrading melanin in fungal walls of *B. cinerea*, hyperparasitizing sclerotia, mycelium, conidiophores and conidia, and suppressing disease (White, 1999). *Trichoderma* spp. and *Streptomyces* spp. killed sclerotia and were very suppressive to *B. cinerea* on seeds during the 18-month stratification period (Waite and Traquair, 1998; J.A. Traquair, unpublished). Having generated marked strains of *B. cinerea* as nitrogen non-utilizing (nit) mutants and hygromycin-resistant transformants (White *et al.*, 1998), we are now investigating the epidemiological impact of overwintering inoculum on straw mulch and dead plant material relative to wind-blown inoculum from neighbouring crops and weeds. Infections from overwintering sclerotial inoculum and polycyclical infections from wind-blown conidial inoculum from diseased leaves and fruit in the current crop are also serious problems in *B. cinerea* control in dense plantings of strawberry and raspberry crops. Peng and Sutton (1991), Sutton and Peng (1993) and Sutton *et al.* (1997) reported biological protection of foliage and fruits with *Gliocladium* spp. sprayed on the phylloplane and further distributed by bees. Yu and Sutton (1998) determined the environmental manipulations (temperature and moisture) necessary to optimize biological control by these antagonists.

Recommendations

Future work should include:

1. Canadian registration of a commercial biological control product for *B. cinerea* based on *Trichoderma* spp.;
2. Development of new biological control agents and Canadian commercial biological control products to increase the range and diversity of biological controls available to the Canadian horticulture industry, establishing sustainable biological control of *B. cinerea* in greenhouse and field environments.

Acknowledgements

The financial support of the Alberta Agriculture Research Institute is acknowledged for the Alberta studies on greenhouse crops.

References

Andrews, J.H. (1990) Biological control in the phyllosphere: Realistic goal or false hope? *Canadian Journal of Plant Pathology* 12, 300–307.

Bélanger, R.R., Dufour, N., Caron, J. and Benhamou, N. (1995) Chronological events associated with the antagonistic properties of *Trichoderma harzianum* against *Botrytis cinerea*: indirect evidence for sequential role of antibiosis and parasitism. *Biocontrol Science and Technology* 51, 41–53.

Boyle, D. (1999) Why mushrooms are not wiped out by green mould. *Mushroom World* 10, 5–10.

Dik, A.J. and Elad, Y. (1999). Comparison of antagonists of *Botrytis cinerea* in greenhouse-grown cucumber and tomato under different climatic conditions. *European Journal of Plant Pathology* 105, 123–137.

Elad, Y., Zimand, G., Zaqs, Y., Zuriel, S. and Chet, I. (1993) Use of *Trichoderma harzianum* in combination or alternation with fungicides to control cucumber grey mold (*Botrytis cinerea*) under commercial greenhouse conditions. *Plant Pathology* 42, 324–332.

Elad, Y., Gullino, M.L., Shteinberg, D. and Aloi, C. (1995) Managing *Botrytis cinerea* on tomato in greenhouses in the Mediterranean. *Crop Protection* 14, 105–109.

Hausbeck, M.K. and Moorman, G.W. (1996) Managing *Botrytis* in greenhouse-grown flower crops. *Plant Disease* 80, 1212–1219.

Hjeljord, L. and Tronsmo, A. (1998) *Trichoderma* and *Gliocladium* in biological control: an overview. In: Harman, G.E. and Kubicek, C.P. (eds) *Trichoderma* and *Gliocladium,* Vol. 2. *Enzymes, Biological Control and Commercial Applications.* Taylor & Francis, London, pp. 131–145.

Howard, R.J., Garland, J.A. and Seaman, W.L. (1994) *Diseases and Pests of Vegetable Crops in Canada.* The Canadian Phytopathological Society and Entomological Society of Canada, Ottawa, Ontario.

Jarvis, W.R. (1992) *Managing Diseases in Greenhouse Crops*, 1st edn. American Phytopathological Society Press, St Paul, Minnesota.

Lorito, M., Harman, G.E., Hayes, C.K., Broadway, R.M., Tronsmo, A., Woo, S.L. and DiPietro, A. (1993) Chitinolytic enzymes produced by *Trichoderma harzianium*: Antifungal activity of purified endochitinase and chitobiosidase. *Phytopathology* 83, 302–307.

Menzies, J.G. (1993) A strain of *Trichoderma viride* pathogenic to germinating seedlings of cucumber, pepper and tomato. *Plant Pathology* 42, 784–791.

Muthumeenakshi, S., Brown, A.E. and Mills, P.R. (1998) Genetic comparison of the aggressive weed mould strains of *Trichoderma harzianum* from mushroom compost in North America and the British Isles. *Mycological Research* 102, 385–390.

Nelson, M.E. and Powelson, M.L. (1988) Biological control of grey mold of snap beans by *Trichoderma hamatum. Plant Disease* 72, 727–729.

O'Neill, T.M., Niv, A., Elad, Y. and Shteinberg, D. (1996) Biological control of *Botrytis cinerea* on tomato stem wounds with *Trichoderma harzianum. European Journal of Plant Pathology* 102, 635–643.

Parke, J.L. and Shotwell, K.M. (1989) *Diseases of Cultivated Ginseng.* Bulletin A3465, University of Wisconsin-Extension and United States Department of Agriculture, pp. 10–12.

Peng, G. and Sutton, J.C. (1991) Evaluation of microorganisms for biocontrol of *Botrytis cinerea* in strawberry. *Canadian Journal of Plant Pathology* 13, 247–257.

Punja, Z.K. (1997a) Comparative efficacy of bacteria, fungi, and yeasts as biological control agents for diseases of vegetable crops. *Canadian Journal of Plant Pathology* 19, 315–323.

Punja, Z.K. (1997b) Fungal pathogens of American ginseng (*Panax quinquefolius*) in British Columbia. *Canadian Journal of Plant Pathology* 19, 301–306.

Redmond, J.C., Marois, J.J. and MacDonald, J.D. (1987) Biocontrol of *Botrytis cinerea* on roses with epiphytic microorganisms. *Plant Disease* 71, 799–802.

Reeleder, R.D. and Fisher, P. (1995) *Diseases of Ginseng.* Factsheet No. 95-003, Ontario Ministry of Agriculture, Food and Rural Affairs. pp. 1–4.

Sutton, J.C. and Peng, G. (1993) Biocontrol of *Botrytis cinerea* in strawberry leaves. *Phytopathology* 83, 615–621.

Sutton, J.C., Li, D., Peng, G., Yu, H. and Zhang, P. (1997) *Gliocladium roseum*: a versatile adversary of *Botrytis cinerea* in crops. *Plant Disease* 81, 316–328.

Tronsmo, A. and Ystaas, J. (1980) Biological control of *Botrytis cinerea* on apple. *Plant Disease* 64, 1009–1011.

Waite, D. and Traquair, J.A. (1998) *In vitro* antagonism of ginseng seed mold (*Botrytis cinerea*). *Canadian Journal of Plant Pathology* 20, 342.
White, G.J. (1999) Biological control of Botrytis blight of American ginseng using wood-decay basidiomycetes. MSc Thesis, University of Western Ontario, London, Ontario.
White, G.J., Dobinson, K. and Traquair, J.A. (1998) Selection of nitrate-nonutilizing mutants in *Verticillium, Alternaria* and *Botrytis*. *Canadian Journal of Plant Pathology* 20, 340.
Yu, H. and Sutton, J.C. (1998) Effects of inoculum density, wetness duration and temperature on control of *Botrytis cinerea* by *Gliocladium roseum* in raspberry. *Canadian Journal of Plant Pathology* 20, 243–252.
Zhang, P.G., Sutton, J.C. and Hopkin, A.A. (1994) Evaluation of microorganisms for biocontrol of *Botrytis cinerea* in container-grown black spruce seedlings. *Canadian Journal of Forest Research* 24, 1312–1316.

86 *Cochliobolus sativus* (Ito and Kuribayashi) Drechsler ex Dastur, Common Root Rot (*Pleosporaceae*)

S.M. Boyetchko and J.P. Tewari

Pest Status

Cochliobolus sativus (Ito and Kuribayashi) Drechsler ex Dastur [anamorph *Bipolaris sorokiniana* (Saccardo) Shoemaker (= *Helminthosporium sativum* Pammel, C.M. King and Bakke)] causes common root rot, one of the most widespread diseases of cereals. The disease occurs primarily on spring and winter wheat, *Triticum aestivum* L., and barley, *Hordeum vulgare* L., and occasionally on tall or meadow fescue grass, *Festuca elatior* L. (Trevathan, 1992). *C. sativus* affects any below-ground and above-ground part of the plant. The most common disease symptoms are root rot, spot blotch or leaf blight, and blackpoint of seeds (Conner, 1990). Although plants are not necessarily killed, economic losses are generally attributed to reductions in tiller number and kernels per tiller, resulting in lower seed quality, including increased seed discoloration and reduced grain yield (Ledingham *et al.*, 1973; Piening *et al.*, 1976; Duczek, 1989; Trevathan, 1992; Duczek and Jones-Flory, 1993).

In the prairies, annual yield losses in spring barley from 1970 to 1972 averaged 10.3% (Piening *et al.*, 1976), while yield losses of 5.7% have been reported for hard red spring wheat (Ledingham *et al.*, 1973). In Ontario, 26% reduction in barley grain yield has been attributed to spot blotch (Clark, 1979). In south-western Quebec, seedling blight and common root rot intensities of 25% and 70%, respectively, occurred (Pua *et al.*, 1985). Seedling blight and root rot severity were directly correlated with yield losses, while spot blotch intensity was not. Resistance of cultivars to common root rot disease has been related to the level of discoloration on the subcrown internodes (Tinline and Ledingham, 1979; Duczek *et al.*, 1985). In the slight, moderate, and severe disease rating categories, the mean losses in grain yield were 29%, 38% and 59%, respectively, compared to the clean control (Verma and Morrall, 1976). Bailey *et al.* (1997) showed that grain yield losses in wheat and barley were 16–29%, thus suggesting that yield losses may have been underestimated in previous studies or that *C. sativus* is only one of the factors affecting root growth.

Inoculum of *C. sativus* can be seed-borne and is often soil-borne, with very high inoculum potential in the field, often from 8 to 253 conidia g^{-1} of soil (Chinn *et al.*, 1962; Duczek, 1981). Duczek *et al.* (1985) reported that disease severity reached 75% in wheat and barley when inoculum in soil was 10–60 and 50–120 conidia cm^{-3} of soil, respectively. Conidia and mycelia can also survive in crop residues retained under minimum and zero tillage (Ledingham, 1961; Chinn, 1976a, b; Reis and Wunsche, 1984). Butler (1959) reported that *C. sativus* conidia remained viable in straw for up to 2 years and that inoculum survival was reduced in moist soils compared to dry soils. In addition, sporulation continued longer under minimum and zero tillage than under conventional tillage, which promotes decomposition of residues containing *C. sativus* inoculum (Duczek and Wildermuth, 1992). Under moist and warm conditions, most sporulation occurs within 20 days; burial of residues by incorporation through tillage decreases this type of sporulation.

Background

A variety of control measures to reduce common root rot severity exists. Resistant or tolerant wheat cultivars occur (Wildermuth and McNamara, 1987; Stack, 1994; Bailey *et al.*, 1997) but still exhibit disease symptoms. Duczek and Wildermuth (1992) found that tolerance to common root rot was more prevalent in barley than in wheat.

Tillage and crop rotation affect disease incidence and severity. Severity of common root rot with cereals grown under reduced tillage decreases, while leaf spot disease increases (Conner *et al.*, 1987; Bailey *et al.*, 1992). Tinline and Spurr (1991) reported that intensity of common root rot, frequency of isolation of *C. sativus* from plants, and level of inoculum in the upper 8 cm of soil were lower under zero-tillage than under conventional tillage. Inoculum of *C. sativus* has also been associated with non-cereal crops, e.g. soybean, *Glycine max* (L.) Merill, lupin, *Lupinus* spp., canola, *Brassica napus* L. and *B. rapa* L., lucerne, *Medicago sativa* L., vetch, *Vicia* spp., and clover, *Trifolium* spp., grown in rotation with wheat (Spurr and Kiesling, 1961; Gourley, 1968; Wildermuth and McNamara, 1987; Heimann *et al.*, 1989). Survival of *C. sativus* in crop residues, including non-host plants, could therefore result in carryover of inoculum from year to year (Fernandez, 1991). Verma *et al.* (1975) reported that common root rot developed more rapidly in wheat grown in low-phosphorus soils compared to high ones. The application of phosphorus fertilizer to stubble field resulted in a significant reduction in incidence of barley common root rot (Piening *et al.*, 1983).

Fungicide treatment of seeds has been used to control seed-borne disease, but has limited application for cereal root rot control (Verma *et al.*, 1986). Although Verma (1983) reported effective control with some chemicals, e.g. triadimenol, some phytotoxicity was noted.

Perforation and lysis of *C. sativus* conidia and hyphae by soil bacteria and mycophagous amoebae was reported (Old and Patrick, 1976; Old, 1977; Anderson and Patrick, 1980; Duczek, 1983, 1986; Fradkin and Patrick, 1985). Annular depressions and perforations 1–7 μm in diameter in the conidial wall were produced by giant amoebae resembling *Leptomyxa reticulata* Goodey (Old, 1977), while two other soil amoebae, *Theratromyxa weberi* Zwillenberg and *Vampyrella vorax* Cienkowski, caused perforations less than 1 μm in diameter in the fungal spore wall (Anderson and Patrick, 1980). Duczek (1983, 1986) discovered populations of *Thecamoeba granifera minor*, as the dominant hyphal-feeding and spore-perforating amoeba in Saskatchewan. In some cases, perforation and lysis of

C. sativus conidia were the result of bacterial activity (Old and Patrick, 1976; Fradkin and Patrick, 1985). *C. sativus* conidia showed various degrees of inhibition (particularly on germination) when exposed to cell-free culture filtrates and washed bacterial cells of different bacterial strains. The authors concluded that soil microflora may play an important role in the survival of soil-borne pathogens, e.g. *C. sativus*, and indicated the biological control potential of these bacteria.

Biological Control Agents

Bacteria

Hanson (2000) evaluated *Burkholderia* (= *Pseudomonas*) *cepacia* (Palleroni and Holmes) Yabuuchi, Kosako, Oyaizu, Yano, Hotta, Hashimoto, Ezahi, and Arakawa, strain Ral-3, and *Pseudomonas fluorescens* Trevisan (Migula) strain 63–49 as potential biological control agents of *C. sativus. In vitro* studies evaluating the impact of abiotic factors on pathogen suppression by the bacteria showed strong inhibitory effects on fungal growth. Fungal inhibition was significantly affected by pH (pH 6.0 provided optimal control) while nutritional amendments, particularly a carbon source, had a major impact on suppressing fungal growth through antibiosis. However, field results were inconsistent. Seed treatment of spring wheat with the bacteria did not result in significant disease suppression or enhanced crop yield.

Fungi

Idriella bolleyi (R. Sprague) von Arx [= *Microdochium bolleyi* (R. Sprague) de Hoog and Herm Nijh] reduced common root rot disease by 16% (Duczek, 1997). Seed treatment of barley with this fungus also resulted in an increase in grain yield but similar results were not found with *C. sativus* in wheat. Its further development as a biological control agent has not been pursued.

A reduction in common root rot disease in cereals colonized by the symbiotic arbuscular–mycorrhizal fungi has been reported (Boyetchko and Tewari, 1988; Thompson and Wildermuth, 1989; Rempel and Bernier, 1990; Boyetchko, 1991). Thompson and Wildermuth (1989) showed an inverse relationship between arbuscular–mycorrhizal fungus root colonization and infection of roots by *C. sativus* in winter and summer field crops. However, Wani *et al.* (1991) reported no relationship between incidence of common root rot and root colonization by arbuscular–mycorrhizal fungi under controlled environment and field conditions. Levels of *C. sativus* inoculum were not quantified, and the variation in inoculum density in the field was unknown. Rempel and Bernier (1990) reported that *Glomus intraradices* Schenck and Smith reduced severity of common root rot in wheat and protected it against any yield reduction that may have been attributed to *C. sativus.* Three arbuscular–mycorrhizal fungal species effectively controlled common root rot severity at different *C. sativus* inoculum densities in barley in greenhouse experiments, with *Glomus intraradices* and *Glomus mosseae* (Nicolson and Gerdemann) Gerdemann and Trappe being more effective at suppressing the disease than *Glomus dimorphicum* Boyetchko and Tewari (Boyetchko and Tewari, 1988; Boyetchko, 1991). A concomitant application of the arbuscular–mycorrhizal fungi and phosphorus fertilizer reduced disease severity greater than an application of phosphorus alone, indicating the mediation of improved nutrient uptake as one mode of action. However, further studies indicated that enhanced phosphorus nutrition was not solely responsible for disease suppression and that the mechanisms for biological control may be multicomponent.

Evaluation of Biological Control

Arbuscular–mycorrhizal fungi are the most promising of the agents studied. Unfortunately, the inability to mass-produce these beneficial symbiotic fungi, mainly due to their biotrophic nature, does not allow for their production and application in large-scale agricultural production systems, but may work under glasshouse agricultural systems. Exploitation of mycorrhizal diversity and functioning under natural field conditions may prove to be a viable alternative for using these fungi for biological control of soil-borne diseases.

Recommendations

Further work should include:

1. Determining the feasibililty of developing microbial-based biological control of *C. sativus* in cereals;

2. Determining the diversity, ecology and functioning of indigenous arbuscular–mycorrhizal fungi and whether their natural populations could be enhanced through different crop-production systems, e.g. conventional versus low inputs, or soil amendments to suppress *C. sativus.*

References

Anderson, T.R. and Patrick, Z.A. (1980) Soil vampyrellid amoebae that cause small perforations in conidia of *Cochliobolus sativus. Soil Biology and Biochemistry* 12, 159–167.

Bailey, K.L., Mortensen, K. and Lafond, G.P. (1992) Effects of tillage systems and crop rotations on root and foliar diseases of wheat, flax, and peas in Saskatchewan. *Canadian Journal of Plant Science* 72, 583–591.

Bailey, K.L., Duczek, L.J. and Potts, D.A. (1997) Inoculation of seeds with *Bipolaris sorokiniana* and soil fumigation methods to determine wheat and barley tolerance and yield losses caused by common root rot. *Canadian Journal of Plant Science* 77, 691–698.

Boyetchko, S.M. (1991) Biological control of the common root rot of barley through the use of vesicular–arbuscular mycorrhizal fungi. PhD thesis, University of Alberta, Edmonton, Alberta.

Boyetchko, S.M. and Tewari, J.P. (1988) The effect of VA mycorrhizal fungi on infection by *Bipolaris sorokiniana* in barley. *Canadian Journal of Plant Pathology* 10, 361.

Butler, F.C. (1959) Saprophytic behaviour of some cereal root-rot fungi. IV. Saprophytic survival in soils of high and low fertility. *Annals of Applied Biology* 47, 28–36.

Chinn, S.H.F. (1976a) Influence of rape in crop rotation on prevalence of *Cochliobolus sativus* conidia and common root rot of wheat. *Canadian Journal of Plant Science* 56, 199–201.

Chinn, S.H.F. (1976b) *Cochliobolus sativus* conidia populations in soil following various cereal crops. *Phytopathology* 66, 1082–1084.

Chinn, S.H.F., Sallans, B.J. and Ledingham, R.J. (1962) Spore populations of *Helminthosporium sativum* in soils in relation to the occurrence of common root rot of wheat. *Canadian Journal of Plant Science* 42, 720–727.

Clark, R.V. (1979) Yield losses of barley cultivars caused by spot blotch. *Canadian Journal of Plant Pathology* 1, 113–117.

Conner, R.L. (1990) Interrelationship of cultivar reactions to common root rot, black point, and spot blotch in spring wheat. *Plant Disease* 74, 224–227.

Conner, R.L., Lindwall, C.W. and Atkinson, T.G. (1987) Influence of minimum tillage on severity of common root rot in wheat. *Canadian Journal of Plant Pathology* 9, 56–58.

Duczek, L.J. (1981) Number and viability of conidia of *Cochliobolus sativus* in soil profiles in summerfallow in Saskatchewan. *Canadian Journal of Plant Pathology* 3, 12–14.

Duczek, L.J. (1983) Populations of mycophagous amoebae in Saskatchewan soils. *Plant Disease* 67, 606–608.

Duczek, L.J. (1986) Populations in Saskatchewan soils of spore-perforating amoebae and an amoeba (*Thecamoeba granifera* s.sp. *minor*) which feeds on hyphae of *Cochliobolus sativus. Plant and Soil* 92, 295–298.

Duczek, L.J. (1989) Relationship between common root rot (*Cochliobolus sativus*) and tillering in spring wheat. *Canadian Journal of Plant Pathology* 11, 39–44.

Duczek, L.J. (1997) Biological control of common root rot in barley by *Idriella bolleyi. Canadian Journal of Plant Pathology* 19, 402–405.

Duczek, L.J. and Jones-Flory, L.L. (1993) Relationship between common root rot, tillering, and yield loss in spring wheat and barley. *Canadian Journal of Plant Pathology* 15, 153–158.

Duczek, L.J. and Wildermuth, G.B. (1992) Effect of temperature, freezing period, and drying on the sporulation of *Cochliobolus sativus* on mature stem bases of wheat. *Canadian Journal of Plant Pathology* 14, 130–136.

Duczek, L.J., Verma, P.R. and Spurr, D.T. (1985) Effect of inoculum density of *Cochliobolus sativus* on common root rot of wheat and barley. *Canadian Journal of Plant Pathology* 7, 382–386.

Fernandez, M.R. (1991) Recovery of *Cochliobolus sativus* and *Fusarium graminearum* from living and dead wheat and nongramineous winter crops in southern Brazil. *Canadian Journal of Botany* 19, 1900–1906.

Fradkin, A. and Patrick, Z.A. (1985) Interactions between conidia of *Cochliobolus sativus* and soil bacteria as affected by physical contact and exogenous nutrients. *Canadian Journal of Plant Pathology* 7, 7–18.

Gourley, C.O. (1968) *Bipolaris sorokiniana* on snap beans in Nova Scotia. *Canadian Plant Disease Survey* 48, 34–36.

Hanson, K.G. (2000) Characterization of potential biological control agents antagonistic to soilborne fungal pathogens. MSc thesis, University of Saskatchewan, Saskatoon, Saskatchewan.

Heimann, M.G., Stevenson, W.R. and Raud, R.E. (1989) *Bipolaris sorokiniana* found causing lesions on snapbean in Wisconsin. *Plant Disease* 73, 701.

Ledingham, R.J. (1961) Crop rotations and common root rot in wheat. *Canadian Journal of Plant Science* 41, 479–486.

Ledingham, R.J., Atkinson, T.G., Horricks, J.S., Mills, J.T., Piening, L.J. and Tinline, R.D. (1973) Wheat losses due to common root rot in the prairie provinces of Canada, 1969–1971. *Canadian Plant Disease Survey* 53, 113–122.

Old, K.M. (1977) Giant soil amoebae cause perforation of conidia of *Cochliobolus sativus*. *Transactions of the British Mycological Society* 68, 277–320.

Old, K.M. and Patrick, Z.A. (1976) Perforation and lysis of spores of *Cochliobolus sativus* and *Thielaviopsis basicola* in natural soils. *Canadian Journal of Botany* 54, 2798–2809.

Piening, L.J., Atkinson, T.G., Horricks, J.S., Ledingham, R.J., Mills, J.T. and Tinline, R.D. (1976) Barley losses due to common root rot in the prairie provinces of Canada, 1970–72. *Canadian Plant Disease Survey* 56, 41–45.

Piening, L.J., Walker, D.R. and Dagenais, M. (1983) Effect of fertilizer on root rot of barley on stubble and fallowland. *Canadian Journal of Plant Pathology* 5, 136–139.

Pua, E.C., Pelletier, R.L. and Klinck, H.R. (1985) Seedling blight, spot blotch, and common root rot in Quebec and their effect on grain yield in barley. *Canadian Journal of Plant Pathology* 7, 395–401.

Reis, E.M. and Wunsche, W.A. (1984) Sporulation of *Cochliobolus sativus* on residues of winter crops and its relationship to the increase of inoculum density in soil. *Plant Disease* 68, 411–412.

Rempel, C.B. and Bernier, C.C. (1990) *Glomus intraradices* and *Cochliobolus sativus* interactions in wheat grown under two moisture regimes. *Canadian Journal of Plant Pathology* 12, 338.

Spurr, H.W. Jr and Kiesling, R.L. (1961) Field and host studies of parasitism by *Helminthosporium sorokinianum*. *Plant Disease Reporter* 45, 941–943.

Stack, R.W. (1994) Susceptibility of hard red spring wheats to common root rot. *Crop Science* 34, 276–278.

Thompson, J.P. and Wildermuth, G.B. (1989) Colonization of crop and pasture species with vesicular–arbuscular mycorrhizal fungi and negative correlation with root infection by *Bipolaris sorokiniana*. *Canadian Journal of Botany* 69, 687–693.

Tinline, R.D. and Ledingham, R.J. (1979) Yield losses in wheat and barley cultivars from common root rot in field tests. *Canadian Journal of Plant Science* 59, 313–320.

Tinline, R.D. and Spurr, D.T. (1991) Agronomic practices and common root rot in spring wheat: Effect of tillage on disease and inoculum density of *Cochliobolus sativus* in soil. *Canadian Journal of Plant Pathology* 13, 258–266.

Trevathan, L.E. (1992) Seedling emergence, plant height, and root mass of tall fescue grown in soil infested with *Cochliobolus sativus*. *Plant Disease* 76, 270–273.

Verma, P.R. (1983) Effect of triadimenol, imazalil, and nuarimol seed treatment on common root rot and grain yields in spring wheat. *Canadian Journal of Plant Pathology* 5, 174–176.

Verma, P.R. and Morrall, R.A.A. (1976) The epidemiology of common root rot in Manitou wheat. 4. Appraisal of biomass and grain yield in naturally infected crops. *Canadian Journal of Botany* 54, 1656–1665.

Verma, P.R., Morrall, R.A.A., Randell, R.L. and Tinline, R.D. (1975) The epidemiology of common

root rot in Manitou wheat. III. Development of lesions on subcrown internodes and the effect of added phosphate. *Canadian Journal of Botany* 53, 2568–2580.
Verma, P.R., Spurr, D.T. and Sedun, F.S. (1986) Effect of triadimenol, imazalil, and nuarimol seed treatment on subcrown internode length, coleoptile-node-tillering and common root rot in spring wheat. *Plant and Soil* 91, 133–138.
Wani, S.P., McGill, W.B. and Tewari, J.P. (1991) Mycorrhizal and common root-rot infection, and nutrient accumulation in barley grown on Breton loam using N from biological fixation or fertilizer. *Biology and Fertility of Soils* 12, 46–54.
Wildermuth, G.B. and McNamara, R.B. (1987) Susceptibility of winter and summer crops to root and crown infection by *Bipolaris sorokiniana*. *Plant Pathology* 36, 481–491.

87 *Cronartium ribicola* J.C. Fischer, White Pine Blister Rust (*Cronartiaceae*)

J.A. Bérubé

Pest Status

Cronartium ribicola J.C. Fischer, white pine blister rust, native to Asia, was introduced from Europe into Canada in the early 1900s and rapidly spread throughout the country, affecting five-needle pines such as eastern white pine, *Pinus strobus* L., western white pine, *P. monticola* Douglas Don, whitebark pine, *P. albicaulis* Engelmann, limber pine, *P. flexilis* James, and sugar pine, *P. lambertiana* Douglas. It is one of the most important forest diseases in North America, where it causes mortality and an annual loss of more than 20 million m^3 and, if not controlled, makes growing white pine impossible or unprofitable (Benedict, 1967). *C. ribicola* attacks pines of all ages and sizes, killing smaller pines quickly whereas larger pines may develop cankers that girdle, retard growth, weaken stems and finally kill the tree. Infection rates on planted seedlings between 15 and 50% are common in zones where white pine is still common, and can easily reach up to 75% in areas at the distribution limit of white pine.

C. ribicola has a complex life cycle, with two hosts and five kinds of spores. The aeciospores and spermagonia are found on the pine host in late spring and early summer. The urediospores, teliospores and basidiospores are found on wild and cultivated currant and gooseberry bushes, *Ribes* spp. Aeciospores can travel long distances, spreading the disease to far-away *Ribes* bushes. Infection with basidiospores is localized (several hundred metres) and occurs on pine needles in late summer. Cankers may take years to develop and kill the tree.

Climate, altitude, slope, aspect, topographic position, site richness and drainage are documented to have an impact on infection severity (Van Arsdel *et al.*, 1961). In general, cool and wet weather favours the disease, as *C. ribicola* spores require water to germinate. Trees grown above the morning dew zone escape the disease. There is also historical (Piché, 1917;

Lachmund, 1926) and genetic evidence of restricted gene flow between the *C. ribicola* populations of eastern and western Canada (Hamelin *et al.*, 2000), which may induce differences in virulence.

Background

Natural forests are nearly impossible to protect with reasonable means. Plantations or intensely managed sites can be treated in various ways to minimize impact. Site selection and preparation, *Ribes* eradication, branch pruning and use of a sterol-synthesis-inhibiting fungicide (Bérubé, 1996) are control options available.

Various fungal biological control agents have been proposed against pine rusts, e.g. *Scytalidium uredinicola* Kuhlman (Hiratsuka *et al.*, 1979), *Darluca filum* (Bivona-Bernardi: Fries) M.J. Berkeley (Kendrick, 1985), *Tuberculina maxima* Rostkovius (Bergdahl and French, 1978; Fairbairn *et al.*, 1983), *Cladosporium gallicola* Sutton (Tsuneda and Hiratsuka, 1979), and *Monocillium nordii* (Bourchier) Gams (Tsuneda and Hiratsuka, 1980), but none of these has shown efficacy under controlled laboratory experiments or in field trials.

Biological Control Agents

Fungi

The fungus *Microsphaeropsis arundinis* (Ahmad) Sutton, effective in controlling apple scab, *Venturia inaequalis* (Cooke) Winter (Bernier *et al.*, 1996), demonstrated effectiveness against *C. ribicola* at the uredial stage (Bérubé *et al.*, 1998). Nearly complete destruction of the uredial stage was observed, thus inhibiting the following telial stage and infectious basidiospores. Bérubé *et al.* (1998) demonstrated in laboratory experiments that *M. arundinis* strains P-176 and P-130 caused from 83.5 to 93.8% and 94.2 to 98.7% uredial mortality, respectively, when inoculated after infection with the disease. In contrast, no mortality occurred in controls up to 14 days after inoculation with *C. ribicola*.

We have collected and screened more specific fungal biological control agents targeting *C. ribicola* on its white pine host under nursery conditions (Bérubé *et al.*, 1998). Sixty-three white pine needle fungal endophytes were tested and seven species showed various levels of inhibition of *C. ribicola*.

Evaluation of Biological Control

An ascomycete temporarily labelled as Species A by Bérubé *et al.* (1998) has been field tested in white pine plantations in Newfoundland and in Quebec since 1998. Due to the length of disease development and symptom expression (up to 5 years), it is too early to evaluate field efficacy.

Recommendations

Further work should include:

1. Evaluating the potential of *M. arundinis* to control *C. ribicola* on *Ribes* sp., because cultivation of currants has a high economic potential that is presently limited by pesticide regulations;
2. Clarifying the host range, distribution and mode of action of promising biological control agents;
3. Describing formally the promising agents.

References

Benedict, W.V. (1967) White pine blister rust. In: *Important Forest Insects and Diseases of Mutual Concern to Canada, the United States and Mexico.* Canadian Department of Forestry and Rural Development, pp. 185–198.

Bergdahl, D.R. and French, D.W. (1978) Occurrence of *Tuberculina maxima* on *Cronartium* and *Endocronartium* rusts in Minnesota. *Plant Disease Reporter* 62, 811–812.

Bernier, J., Carisse, O. and Paulitz, T.C. (1996) Fungal communities isolated from dead apple leaves from orchards in Quebec. *Phytoprotection* 77, 129–134.
Bérubé, J.A. (1996) Use of triadimefon to control white pine blister rust. *The Forestry Chronicle* 72, 637–638.
Bérubé, J.A., Trudelle, J.G., Carisse, O. and Dessureault, M. (1998) Endophytic fungal flora from eastern white pine needles and apple tree leaves as a means of biological control for white pine blister rust. In: Proceedings of the First IUFRO Rusts of Forest Trees WP Conference, 2–7 August 1998, Saariselka, Finland. *Finnish Forest Research Institute, Research Papers* 712, 305–309.
Fairbairn, N., Pickard, M.A. and Hiratsuka, Y. (1983) Inhibition of *Endocronartium harknessii* spore germination by metabolites of *Scytalidium uredinicola* and *S. album* and the influence of growth medium on inhibitor production. *Canadian Journal of Botany* 61, 2147–2152.
Hamelin, R.C., Hunt, R.S., Geils, B.W., Jensen, G.D., Jacobi, V. and Lecours, N. (2000) Barrier to gene flow between eastern and western populations of *Cronartium ribicola* in North America. *Phytopathology* 90, 1073–1078.
Hiratsuka, Y., Tsuneda, A. and Sigler, L. (1979) Occurrence of *Scytalidium uredinicola* on *Endocronartium harknessii* in Alberta, Canada. *Plant Disease Reporter* 63, 512–513.
Kendrick, B. (1985) *The Fifth Kingdom.* Mycologue Publications, Waterloo, Ontario.
Lachmund, H.G. (1926) Studies of white pine blister rust in the west. *Journal of Forestry* 24, 874–884.
Piché, G.C. (1917) Notes sur la rouille vésiculeuse du pin blanc. *Ministère des Terres et Forêts, Province de Québec, Circulaire* 1, 1–10.
Tsuneda, A. and Hiratsuka, Y. (1979) Mode of parasitism of a mycoparasite *Cladosporium gallicola* on western gall rust *Endocronartium harknessii. Canadian Journal of Plant Pathology* 1, 31–36.
Tsuneda, A. and Hiratsuka, Y. (1980) Parasitization of pine stem rust fungi by *Monocillium nordii. Phytopathology* 70, 1101–1103.
Van Arsdel, E.P., Riker, A.J., Kouba, T.F., Suomi, V.E. and Bryson, R.A. (1961) *The Climatic Distribution of Blister Rust on White Pine in Wisconsin.* Station Paper 39, United States Department of Agriculture, Forest Service, Lake States Forest Experimental Station, St Paul, Minnesota.

88 *Erwinia amylovora* (Burrill) Winslow, Broadhurst, Buchanan, Krumwiede, Rogers and Smith, Fire Blight (Enterobacteriaceae)

A.M. Svircev, J.J. Gill and P. Sholberg

Pest Status

The bacterium, *Erwinia amylovora* (Burrill) Winslow, Broadhurst, Buchanan, Krumwiede, Rogers and Smith, is the causal agent of fire blight, a major disease of pear, *Pyrus communis* L., and apple, *Malus pumila* Miller (= *Malus domestica* Borkhausen). Commercial pear cultivars currently grown in Canada are highly susceptible to infection by *E. amylovora*. Although resistant pear cultivars are available (Hunter, 1999), consumer demand favours the planting of susceptible pear cultivars such as Bartlett, Flemish Beauty and Bosc. In commercially grown apples, varying levels of fire blight resistance occur. Cultivation of scions such as Fuji and Gala on fire blight susceptible M9 and M26 dwarfing rootstocks are popular.

E. amylovora begins its annual infection cycle in early spring with activation of the bacterial population residing in the overwintering cankers. Cankers are necrotic regions established in woody tissues of susceptible pear or apple trees. The actively growing bacterial cells are located in the canker margins and are extruded on to the bark surface. The bacterial droplets on the canker surface, commonly known as bacterial ooze, may be disseminated by insects, wind and rain to the newly opened blossoms, which act as primary infection sites. Invasion of blossoms by *E. amylovora* may lead to further necrosis of the blossoms and adjacent shoots. In susceptible cultivars, bacteria will migrate down the shoots and colonize the main body of the tree.

Background

In Canada, streptomycin, applied as an aerial spray, is the only product registered to control blossom blight. Control options for advanced fire blight infections are limited to removal of diseased wood. Streptomycin resistance had been documented in several locations worldwide (McManus and Jones, 1994; Chou and Jones, 1995) and was first identified in Canada in 1993 (Sholberg and Bedford, 1993). In British Columbia, after the planting of many new, high density orchards on susceptible rootstocks and scions, and the occurrence of weather conducive to fire blight in 1997 and 1998, streptomycin resistance became widespread (Sholberg *et al.*, 2000).

Several biological control agents are commercially available to control blossom blight infections caused by *E. amylovora* (Johnson and Stockwell, 1998), e.g. Blight Ban® A506 (*Pseudomonas fluorescens* (Travisan) Migula A506, Plant Health Technologies, Boise, Idaho, USA) and Blight Ban® C9-1 (*Erwinia herbicola* (Lönis) Day C9-1, Plant Health Technologies, US experimental permit). Blight Ban® has not yet been registered for use in Canada. In Washington state, screening trials on caged apple trees treated with strain E325 of *Pantoea agglomerans* Gavini, Mergaert, Beji, Mielcarek, Izard, Kersters and DeLey provided 42% or better control than A506 and 24% better control than C9–1 (Pusey, 1999). The biofungicide, Serenade® (*Bacillus subtilis* (Ehrenberg) Cohn (Q ST713 strain), AgraQuest, Davis, California, USA) is effective against *E. amylovora*, according to company information, and research trials are in progress. Both Serenade® and E325 are being considered for joint registration in Canada and the USA.

Biological control agents prevent infection of the flower surface in various ways. They may colonize the flower surface and subsequently prevent epiphytic growth of *E. amylovora* on the stigma, hypanthium or nectarthodes (Wilson and Lindow, 1993). Antibiosis and competition for resources have also been demonstrated as a mechanism of action for strains of *E. herbicola* (Erskine and Lopatecki, 1975; Ishimaru *et al.*, 1988; Vanneste *et al.*, 1992; Wilson *et al.*, 1992; Wodzinski *et al.*, 1994).

Control of plant pathogens by bacteriophages was investigated sporadically, with mixed results (Vidaver, 1976; Munsch *et al.*, 1995; Jones *et al.*, 1998). Erskine (1973) and Ritchie and Klos (1977) studied bacteriophages of *E. amylovora* and postulated their possible role in the epidemiology of fire blight, but their potential for biological control was not examined further.

The current trends in the apple orchards towards high-density plantings of susceptible cultivars and rootstocks point to the necessity of developing new and innovative control strategies.

Biological Control Agents

Viruses

Bacteriophages of *E. amylovora* were isolated from soil surrounding blighted trees

(Gill *et al.*, 1999). Their presence in soil surrounding apple or pear trees appears to be associated with the presence of fire blight disease symptoms. Forty-five bacteriophage isolates were recovered from the field, purified and enriched in culture, and DNA was extracted. Thirty-seven of the isolates were placed into one of six groups (named restriction fragment length polymorphism (RFLP) groups) based on the patterns obtained by digestion of the bacteriophage DNA with four restriction endonucleases. Some of the isolates were identified as PEa1-type bacteriophages using polymerase chain reaction (PCR). Of the 45 bacteriophages evaluated, only 13 (29%) were able to produce visible plaques on all 13 *E. amylovora* strains tested. Bacteriophages in Group 3, similar to PEa1, and its relatives, showed little or no lytic activity against some isolates of *E. amylovora* from British Columbia orchards, and against two strains isolated from Harrow, Ontario. The exception to this pattern was phage isolate PEa 31–3, which formed plaques on all *E. amylovora* strains. Certain isolates exhibited consistent ability to inhibit the development of disease symptoms in the form of bacterial exudate or ooze, when evaluated in the immature pear plug system. When arranged by RFLP group, the bacteriophages in Groups 3 and 6 exhibited the highest levels of overall biological control activity on the pear assay. Most bacteriophages in Groups 1, 2, 4 and 5 tested using this system exhibited minimal biological control activity. In the absence of a control agent, the bacterial population on the plug surface increased by 100-fold or more, from 1×10^6 colony-forming units (cfu) at the time of application to between 1×10^8 and 1×10^{10} cfu at the time of evaluation. Bacteriophage treatment was able to reduce this population increase, by as much as 97% in the case of phage PEa 51–2. Significant control ($P \leq 0.05$) of *E. amylovora* population on the plug surface was obtained in some instances.

In the immature pear fruit bioassay, bacteriophages were able to inhibit the ability of *E. amylovora* to produce bacterial ooze. Bacteriophages of Groups 3 and 6 exhibited the greatest overall ability to suppress ooze formation. Although reductions in bacterial populations were significant, the population surviving bacterial phage treatment was large, numbering from 6×10^7 to 2×10^9 cfu.

Bacteria

In British Columbia, a trial was conducted on Jonagold, Golden Delicious and Elstar apple trees. The treatments were *Pseudomonas fluorescens* (Travisan) Migula strain A506, *P. agglomerans* strain E325 and streptomycin. The biological control agents and streptomycin were applied at early and full bloom. Blossoms were inoculated with *E. amylovora* 48 h later, followed by wetting for 4 h or longer. As expected, streptomycin was the most effective material on all three cultivars. E325 and A506 both reduced the number of infected blossoms on Elstar but were ineffective on Golden Delicious. E325 also reduced infected blossoms on Jonagold although A506 was ineffective on this cultivar.

Evaluation of Biological Control

Research on biological control agents such as E325 and A506 indicated that they would be useful for disease control in Canada, especially where streptomycin resistance is known to occur. The multifaceted approach to fire blight control, which incorporates the use of disease forecasting models, streptomycin and biological control agents, can lead to successful control of fire blight in orchards. Research on the use of bacteriophages and other biological control agents, while in its very early stages, holds promise.

Recommendations

Further work should include:

1. Optimizing the biological control activity of bacteriophages by field-testing systems that will increase their stability on the flower surface;

2. Further testing of biological control agents that have shown promise in preliminary trials.

Acknowledgements

A.L. Jones, Michigan State University, donated the PCR primers used for identifying the PEa1-type bacteriophages.

References

Chou, C.S. and Jones, A.L. (1995) Molecular analysis of high-level streptomycin resistance in *Erwinia amylovora*. *Phytopathology* 85, 324–328.

Erskine, J.M. (1973) Characteristics of *Erwinia amylovora* bacteriophage and its possible role in the epidemiology of fire blight. *Canadian Journal of Microbiology* 19, 837–845.

Erskine, J.M. and Lopatecki, L.E. (1975) *In vitro* and *in vivo* interactions between *Erwinia amylovora* and related saprophytic bacteria. *Canadian Journal of Microbiology* 21, 35–41.

Gill, J.J., Svircev, A.M., Myers, A.L. and Castle, A.J. (1999) Biocontrol of *Erwinia amylovora* using bacteriophages. *Phytopathology* 89, S27.

Hunter, D.M. (1999) Update on Harrow fire blight-resistant pear cultivars and selections. *Compact Fruit Tree* 32, 59–62.

Ishimaru, C.A., Klos, E.J. and Brubaker, R.R. (1988) Multiple antibiotic production by *Erwinia herbicola*. *Phytopathology* 78, 746–750.

Johnson, K.B. and Stockwell, V.O. (1998) Management of fire blight: a case study in microbial ecology. *Annual Review of Phytopathology* 36, 227–248.

Jones, J.B., Somodi, G.C., Jackson, L.E. and Harbaugh, B.K. (1998) Control of bacterial spot on tomato in the greenhouse and field with bacteriophages. *Seventh International Conference on Plant Pathology*, Paper Number 5.2.14.

McManus, P.S. and Jones, A.L. (1994) Epidemiology and genetic analysis of streptomycin-resistant *Erwinia amylovora* from Michigan and evaluation of oxytetracycline for control. *Phytopathology* 84, 627–633.

Munsch, P., Olivier, J.M. and Elliott, T.J. (1995) Biocontrol of bacterial blotch of the cultivated mushroom with lytic phages: some practical considerations. In: *Science and Cultivation of Edible Fungi, Volume 2: Proceedings of the 14th International Congress, Oxford, 17–22 September 1995*, pp. 595–602.

Pusey, P.L. (1999) Selection and field testing of *Pantoea agglomerans* strain E325 for biocontrol of fire blight of apple and pear. *Phytopathology* 89, S62.

Ritchie, D.F. and Klos, E.J. (1977) Isolation of *Erwinia amylovora* bacteriophage from aerial parts of apple trees. *Phytopathology* 67, 101–104.

Sholberg, P. and Bedford, K. (1993) Streptomycin resistant *Erwinia amylovora* (fire blight) in British Columbia. In: Smirle, M.J. (ed.) *Research Highlights 1993*. Agriculture Canada, Summerland, British Columbia, pp. 48–49.

Sholberg, P., Bedford, K. and Haag, P. (2000) Occurrence and control of streptomycin-resistant *Erwinia amylovora* in British Columbia. *Canadian Journal of Plant Pathology* 22, 179.

Vanneste, J.L., Yu, J. and Beer, S.V. (1992) Role of antibiotic production by *Erwinia herbicola* Eh252 in biological control of *Erwinia amylovora*. *Journal of Bacteriology* 174, 2785–2796.

Vidaver, A.K. (1976) Prospects for control of phytopathogenic bacteria by bacteriophages and bacteriocins. *Annual Review of Phytopathology* 14, 451–465.

Wilson, M. and Lindow, S.E. (1993) Interactions between the biological control agent *Pseudomonas fluorescens* A506 and *Erwinia amylovora* in pear blossoms. *Phytopathology* 83, 117–123.

Wilson, M., Epton, H.A.S. and Sigee, D.C. (1992) Interactions between *Erwinia herbicola* and *E. amylovora* on the stigma of hawthorn blossoms. *Phytopathology* 82, 914–918.

Wodzinski, R.S., Umholtz, T.E., Rundle, J.R. and Beer, S.V. (1994) Mechanisms of inhibition of *Erwinia amylovora* by *Erw. herbicola in vitro* and *in vivo*. *Journal of Applied Bacteriology* 76, 22–29.

89 *Fusarium oxysporum* Schlechtendahl f. sp. *cyclaminis* Gerlach, Fusarium Wilt of Cyclamen (*Hyphomycetes*)

J.A. Gracia-Garza

Pest Status

Fusarium oxysporum Schlechtendahl f. sp. *cyclaminis* Gerlach causes the serious disease cyclamen wilt of *Cyclamen persicum* Miller. It was first observed in Europe around the 1930s (Barthelet and Gaudineau, 1936). Since then, the disease has been reported from all parts of the world where cyclamen is produced, e.g. Germany, France, Belgium, Netherlands, Italy, Brazil, USA and Canada (Tompkins and Snyder, 1972; Pitta and Teranishi, 1979; Grouet, 1985; Rattink, 1986; Copeman, 1993; Minuto and Garibaldi, 1998). It was first reported in Canada in 1988 (Matteoni, 1988); however, the disease was present long before that (W. Brown, Vineland, July 2000, personal communication). Plants infected with *F. o. cyclaminis* can appear healthy for months before showing symptoms. It is often when flowering begins that most infected plants will show yellowing leaves, wilting and eventually total collapse. Examination of corms of infected plants show a typical brown–red discoloration of the vascular vessels.

With the implementation of recirculating nutrient solutions for irrigation, concerns about disease dispersal in large greenhouse operations are growing. *F. o. cyclaminis* can survive for long periods of time in water without losing its viability, and as a saprophyte growing under benches or other areas in greenhouses. Estimates of about 100 colony-forming units (cfu) ml^{-1} of nutrient solution are found in reservoirs used for recirculating. *F. o. cyclaminis* can be carried through the recirculating nutrient solutions and infect healthy plants. Initial introduction of *F. o. cyclaminis* to commercial operations is most likely by either infected seedlings or infested seed, although the proportion of seed carrying the pathogen has been estimated to be very low (<1%). Also, the pathogen has been isolated from Mycetophilidae and Ephydridae collected in commercial operations, increasing the potential for disease dissemination.

Losses of the entire crop (40,000–50,000 plants) due to cyclamen wilt have been reported (Tompkins and Snyder, 1972). Growers in the Niagara region, southern Ontario, have reported losses as high as 40–50% of their crop, although losses of 10% are more common.

Background

Control of *F. o. cyclaminis* has not been successful with current chemicals or cultural practices, although these practices are still an important component of an integrated disease management strategy to maintain crop losses at low levels. Several studies have been conducted to evaluate the efficacy of chemical fungicides to control this pathogen (Grouet, 1985; Minuto, 1995; O'Neill, 1995; Chase, 1998; Minuto and Garibaldi, 1998); however, in most instances, the results have not been satisfactory. In the USA, drenching the potting medium with Benlate was recommended in the 1980s (Powell, 1982; Tayama, 1987). Efficacy of Benlate against *F. o. cyclaminis* on cyclamen is questionable and it is not registered for that purpose in Canada. Grouet (1985) reported a significant reduction of the disease with applications of maneb in combination with chlorothalonil. More recently, the application of fungicides such as Carbendazim (Benizimidazole) and fludioxonil (Phenylpyrroles) (Medallion®, Novartis Crop Protection, Inc., Greenboro, North Carolina, USA) has resulted in moderate to good control (Minuto, 1995; O'Neill, 1995; Chase, 1998). At present, neither of these products is registered in Canada for use against *F. o. cyclaminis*.

Biological control agents against *Fusarium* spp. have been studied extensively in recent years. Among those organisms that have been reported as having potential for control of *F. o. cyclaminis* are several non-pathogenic *Fusarium* strains (Garibaldi, 1988; Eparvier *et al.*, 1991; Rattink, 1993; Minuto, 1995; Minuto and Garibaldi, 1998). Other organisms that have shown antagonistic effects against *Fusarium* spp. are *Pseudomonas* spp. (Xu *et al.*, 1987; van Peer *et al.*, 1990; Eparvier *et al.*, 1991), *Bacillus subtilis* (Cohen) Prazmowski as a seed colonizer (Zhang *et al.*, 1996), the mycoparasites *Trichoderma* spp. and *Gliocladium* spp. (Sivan *et al.*, 1985; Rattink, 1993; Datnoff *et al.*, 1995; Zhang *et al.*, 1996), the vesicular–arbuscular mycorrhizal fungus *Glomus intraradices* Schenk and Smith (Datnoff *et al.*, 1995), and the bacterium *Streptomyces griseoviridis* Anderson, Ehrlich, Sun and Burkholder (Rattink, 1993). Several of these organisms are already available in markets of the USA and Europe. In Canada, no products are currently available, although some products sold as plant growth promoters contain one or several of these organisms.

An integrated disease-management programme using commercial products containing organisms with known biological control activity and organisms with potential to control *F. o. cyclaminis* is being developed.

Biological Control Agents

Bacteria, Fungi

The following products and/or organisms were evaluated either alone or in combination using cyclamen cultivar Laser White: BTM humic acid and several genera of beneficial bacteria (*Bacillus*, *Clostridium*, *Enterobacter*, *Pseudomonas* and *Rhizobium*) (Earth Corp Environmental Ltd, Calgary, Alberta); Hungavit earthworm castings (BioLife 2000 Ltd, Budapest, Hungary); Modicell mixture of enzymes extracted from several genera of fungi (DeruNed bv, Bergschenhoek, The Netherlands); Mycorise endomycorrhizal fungus (Premier Tech, Rivière-du-Loup, Quebec); and RootShield® drench (Bioworks Inc., Geneva, New York, USA) containing *Trichoderma harzianum* Rifai, strain KRL-AG2. The following organisms were also included: non-pathogenic *Fusarium* (isolates CS1 and CS20) (D.R. Fravel, collection), *Pseudomonas corrugata* Roberts and Scarlett, strain 13, *P. fluorescens* (Trevisan) Migula, strain 15 (T.C. Paulitz, collection) and *T. hamatum* Rifai, strain TMCS 3 (J.A. Garcia-Garza, collection). Combinations of some of the products/organisms were also included. For methods and frequency of application see Table 89.1.

Each treatment was applied to the growing medium before planting, as well as prior to seedling transplant. Plants were inoculated through sub-irrigation with *F. o. cyclaminis*, isolate Fo9, originally obtained from a diseased cyclamen plant. Populations of *F. o. cyclaminis* in the recirculating tanks were monitored throughout the experiments.

At the end of the experiment, the number of surviving plants in each treatment was recorded. Surviving plants were then removed from the soil and their corms were cut in half to check for discoloration due to *F. o. cyclaminis*. Infected plants may look healthy but if colonization of the vascular system has taken place the plant will eventually die.

Preliminary results showed that combinations of Mycorise and BTM, Mycorise and Modicell, or RootShield® and CS1

Table 89.1. Products and/or organisms tested[a] against *Fusarium oxysporum* f.sp. *cyclaminis*.

Treatment	Application technique	Application frequency	Per cent survival	Per cent of discolored corms
Control (non-inoculated, Fo9 −)			95	5
Control (inoculated, Fo9 +)			18	97
BTM	Drench	3	18	77
Hungavit	Drench	3	14	84
Modicell	Drench	3	11	90
Mycorise	Soil mix	1	18	80
Non-pathogenic Fusarium CS1	Drench	1	15	89
Non-pathogenic Fusarium CS20	Drench	1	21	74
Pseudomonas corrugata	Drench	1	11	95
Pseudomonas fluorescens	Drench	1	15	85
RootShield®	Drench	1	3	93
TMCS 3 (*Trichoderma hamatum*)	Drench	1	0	98
Mycorise + BTM	Soil mix/drench	1/3	36	67
Mycorise + Modicell	Soil mix/drench	1/3	41	70
RootShield® + BTM	Drench	1/3	15	88
RootShield® + CS1	Drench	1/1	42	59
Daconil®	Drench	3	7[b]	87[b]
Medallion®	Drench	3	0[b]	93[b]

[a]Test products/organisms were compared for efficacy against the fungicides Daconil 2787 (aromatics) (ISK, Biosciences Corp. Mentor, Ohio, USA) and Medallion® (fludioxonil).
[b]Data from only one test.

reduced severity of the disease and colonization of the corms (Table 89.1). Although percentage of surviving plants was low from the growers' point of view, reduction of losses was significant. In initial experiments, application of Mycorise appeared to slow the colonization process; plant mortality was retarded by 3–4 weeks. There are several possible mechanisms involved in the reduction of plant mortality and/or colonization of the vascular system. The composition of products such as BTM can have positive effects against deleterious organisms in the rhizosphere. Control of Fusarium wilt on carnations by *Pseudomonas* spp., which are present in BTM, has been attributed to the production of siderophores and antibiotics (Xu *et al.*, 1987; van Peer *et al.*, 1990). BTM also contains *Bacillus* spp. Some strains of *B. subtilis* have been found to reduce root colonization by *Fusarium* spp. in cotton (Zhang *et al.*, 1996), possibly as a result of antibiosis. Colonization of the root system with an endomycorrhizal fungus prior to exposure to *F. o. cyclaminis*, which can be easily achieved in the floriculture greenhouse industry, may help the host by protecting potential entry points for *F. o. cyclaminis*. Suppression of Fusarium wilt of cyclamen, seen in the combination of RootShield® (containing *T. harzianum* strain T22) and the non-pathogenic *Fusarium* isolate CS1, may involve competition from both biological control agents and mycoparasitic activity of *T. harzianum*, and induced resistance response triggered by the non-pathogenic *Fusarium* isolate.

Populations of *F. o. cyclaminis* in tanks used to irrigate control plants (Fo9-inoculated) were higher (1400 cfu ml^{-1}) throughout the experiment than any other treatment ($\bar{x}$ = 200 cfu ml^{-1}). The effect of the organisms contained in the products tested has not been evaluated in the laboratory against *F. o. cyclaminis*. However, population densities in most treatments remained between 100 and 200 cfu ml^{-1}.

Recommendations

Further work should include:

1. Determining the nature of the interaction between the organisms contained in Mycorise and BTM and the possible mechanisms of action of these organisms when suppressing *F. o. cyclaminis*;

2. Evaluating other cultural practices to enhance the disease suppression due to biological control agents.

Acknowledgements

The author wishes to thank W. Brown and T.J. Blom for their collaboration in this project. Thanks to D.R. Fravel and T.C. Paulitz for providing cultures of the non-pathogenic *Fusarium* and the pseudomonads bacteria, respectively. Appreciation is also extended to all the companies and their representatives who provided samples of products used in this research and to Flowers Canada (Ontario) Inc. for partly supporting this research.

References

Barthelet, J. and Gaudineau, M. (1936) Les maladies des cyclamens. *Revue de Pathologie Végétale et Entomologie Agricole de France* 23, 101–122.

Chase, A.R. (1998) Fusarium diseases of some ornamentals. Hal, J. and Robb, K. (eds) *Proceedings of the 14th Conference on Insect and Disease Management of Ornamentals*, Del Mar, California, 21–23 February 1998, Society of American Florists, pp. 53–57.

Copeman, R.J. (1993) Program toward an integrated approach to managing *Fusarium* wilt of cyclamen. *Cecil Delworth Bulletin*, 37–38.

Datnoff, L.E., Nemec, S. and Pernezny, K. (1995) Biological control of *Fusarium* crown and root rot of tomato in Florida using *Trichoderma harzianum* and *Glomus intraradices*. *Biological Control* 5, 427–431.

Eparvier, A., Lemanceau, P. and Alabouvette, C. (1991) Population dynamics of non-pathogenic *Fusarium* and fluorescent *Pseudomonas* strains in rockwool, a substratum for soilless cultures. *Microbiology Ecology* 86, 177–184.

Garibaldi, A. (1988) Research on substrates suppressive to *Fusarium oxysporum* and *Rhizoctonia solani*. *Acta Horticulturae* 221, 271–277.

Grouet, D. (1985) Vascular Fusarium disease of cyclamen. *Phytoma* 372, 49–51.

Matteoni, J. A. (1988) Diseases of cyclamen in Ontario from 1983 to 1987. *Canadian Plant Disease Survey* 68, 84.

Minuto, A. (1995) Evaluation of antagonistic strains of *Fusarium* spp. in the biological and integrated control of *Fusarium* wilt of cyclamen. *Crop Protection* 14, 221–226.

Minuto, A. and Garibaldi, A. (1998) Evaluation of the spread of *Fusarium oxysporum* f. sp. *cyclaminis* in cyclamen crop grown using ebb and flow irrigation. *Colture-Protette* 27, 21–26.

O'Neill, T. (1995) Evaluation of fungicides against *Fusarium* wilt (*Fusarium oxysporum* f. sp. *cyclaminis*) of cyclamen. *Annals of Applied Biology* 126, 20–21.

Peer, R. van, Kuik, A.J. van, Rattink, H. and Schippers, B. (1990) Control of *Fusarium* wilt in carnation grown on rockwool by *Pseudomonas* sp. strain WCS417r and by Fe-EDDHA. *Netherlands Journal of Plant Pathology* 96, 119–132.

Pitta, G.P.B. and Teranishi, J. (1979) Occurrence of wilt (*Fusarium oxysporum* Schl. f. *cyclaminis* n.f.) on *Cyclamen persicum* Mill. *Biológico* 45, 213–215.

Powell, C. (1982) *Fusarium* wilt on pot mums and cyclamen. *Ohio State Flower Grower's Hotline* 2, 1–2.

Rattink, H. (1986) Some aspects of the etiology and epidemiology of *Fusarium* wilt on cyclamen. *International Symposium on Crop Protection* 51, 617–624.
Rattink, H. (1993) Biological control of *Fusarium* crown and root rot of tomato on a recirculation substrate system. *Mededelingen van de Faculteit Landbouwwetenschappen Rijksuniversiteit Gent* 58, 1329–1334.
Sivan, A., Chet, I., Zeidan, O. and Oko, O. (1985) Application of *Trichoderma harzianum* for biological control of *Fusarium* wilt on melons and of *Fusarium* crown rot on tomatoes. *Phytoparasitica* 13, 1.
Tayama, H.K. (1987) Control of cyclamen Fusarium wilt – A preliminary report. *Ohio Florists' Association Bulletin*, 693, 1–3.
Tompkins, C.M. and Snyder, W.C. (1972) Cyclamen wilt in California and its control. *Plant Disease Reporter* 56, 493–497.
Xu, T., Peer, R. van, Rattink, H. and Schippers, B. (1987) The potential use of fluorescent *Pseudomonas* in the protection of carnations against *Fusarium* wilt in hydroponics. *Acta Horticulturae* 216, 93–100.
Zhang, J.X., Howell, C.R. and Starr, J.L. (1996) Suppression of *Fusarium* colonization of cotton roots and *Fusarium* wilt by seed treatments with *Gliocladium virens* and *Bacillus subtilis*. *Biocontrol Science and Technology* 6, 175–187.

90 *Fusarium oxysporum* Schlechtendahl f. sp. *lycopersici*, Tomato Wilt (*Hyphomycetes*)

J. Bao and G. Lazarovits

Pest Status

Fusarium oxysporum Schlechtendahl f. sp. *lycopersici* Snyder and Hansen strain (race 1, designated as *Fol*), the causal agent of tomato wilt, is an important pathogen of tomato, *Lycopersicon esculentum* L. Fusarium wilts are destructive vascular diseases of many economically important crops worldwide. The diseases are caused by a wide diversity of pathogenic forms (forma speciales and races) within the species *Fusarium oxysporum* (Armstrong and Armstrong, 1981; Gordon and Martyn, 1997). Fusarium wilt pathogens are typically soilborne, very persistent in soil, and difficult to control using conventional methods such as chemical fungicides or crop rotation.

Background

The two most effective methods to control Fusarium wilts are soil disinfestation with broad-spectrum biocidal chemicals and the use of resistant cultivars. In instances when fumigants cannot be used and resistant cultivars are not available, growers have few alternatives for managing these diseases. Biological control, using disease-suppressive microorganisms to improve plant health, has been advocated as a promising vehicle for plant disease control (Alabouvette, 1990; Cook, 1993; Lumsden *et al.*, 1995; Handelsman and Stabb, 1996). Control of Fusarium wilts using non-pathogenic *Fusarium* (NPF) strains has, in fact, been shown to be very successful in many crops under controlled environmental conditions (Ogawa and Komada, 1984; Paulitz *et al.*, 1987; Alabouvette, 1990; Larkin *et al.*, 1996; Postma and Luttikholt, 1996; Larkin and Fravel, 1999). Under field conditions, however, disease control with biological agents has generally not met expectations. This can be attributed mostly to our poor understanding of what conditions need to be established in the field for the desired interactions among the host, the pathogen, and the NPF strain. This chapter summarizes tests on several NPF strains, together with other biological control agents, in the laboratory as well as in the field, to investigate their disease control efficacy and interactions.

Biological Control

Fungi, Bacteria

Tomato Fusarium wilt was used as a model pathogen–host system. The pathogen used was *F. oxysporum* f. sp. *lycopersici* (race 1, designated as *Fol*). The host plant was the disease-susceptible tomato cultivar Bonny Best. Fungi were grown on potato dextrose agar (PDA, or its broth, PDB, Difco) and bacteria on nutrient broth (or its agar-amended solid medium, Difco), respectively. Komada's (K) medium (Komada, 1975) was used to determine *Fusarium* populations in tomato roots. Tomato transplants were generated by planting a seed into a plug tray (2.5 $cm^2 \times 4.5$ cm) containing about 1 g dry Promix per cell (Premier Horticulture Inc., Quebec). Seedlings were allowed to grow for 2 weeks in a growth chamber at 25/20°C (light/dark) with 14 h of fluorescent light/10 h dark and watered daily. Each seedling was inoculated by adding either 5×10^6 fungal spores or 10^9 bacterial cells into the plug medium at either seeding or transplantation time. Inoculated tomato seedlings were transplanted into 7 cm^3 pots filled with pathogen (*Fol*)-inoculated soil. *Fol*-inoculated soil was prepared by mixing a sandy loam soil with ground wheat bran culture of *Fol* to give 10^5 colony-forming units (cfu) g^{-1} dried soil. This inoculum level was found to provide the most consistent disease incidence on tomato plants. Disease development was rated using disease incidence as a percentage of total plants infected and as disease severity index (DSI), where disease severity of infected plants was assessed on a 0–4 scale with 0 = healthy plant, 1 = leaves curved or lower leaves yellow with no apparent plant stunting, 2 = all leaves curved with apparent plant stunting, 3 = all lower leaves dead, and 4 = entire plant killed. Reduced disease incidence or DSI compared to control plants was used as an indicator of biological control efficacy.

A total of 308 microbial isolates from various taxonomic groups were collected from soils sampled from south-western Ontario. From these isolates, 152 (39 of *Fusarium* spp., eight of *Gliocladium* spp., 14 of *Trichoderma* spp., five of *Talaromyces* spp., one *Stilbella* sp., 48 of unidentified fungi, 21 of *Streptomyces* spp., and 16 of unidentified bacteria) were bioassayed on tomato plants to control tomato wilt. Of the 152 isolates, 47 displayed some antibiosis to the pathogen on dual culture plates but did not provide significant disease suppression in the actual bioassay with tomato plants. Repeated screening experiments on tomato plants indicated that *Fusarium* isolates provided the best disease control of all organisms tested. *Fusarium* isolate SA70 reduced disease severity by 80% or more, and incidence by 90%. The control obtained with SA70 was consistent from experiment to experiment and therefore it was selected as a model biological control agent for further study.

Disease control efficacy by SA70 was affected by the inoculum density of SA70 applied in the seedling plug in comparison to inoculum level of the pathogen *Fol* already present in the soil. Plants inoculated with 2×10^5 and 2×10^7 SA70 spores per plug had a disease incidence of 60% and 12%, respectively, after 20 days of post-transplantation, compared to the plants treated with water that had 100% disease incidence. When plants were inoculated with SA70 at 5×10^6 spores per seedling and transplanted into soil containing increasing levels of pathogen densities, disease protection became progressively less effective. At a *Fol* inoculum level of 2.5×10^4 cfu g^{-1} soil, disease reduction was about 90%, but with *Fol* at 5×10^4 cfu g^{-1} it was 60%, and at 1×10^5 cfu g^{-1} only 25%. Nevertheless, all SA70-treated plants had a significantly lower disease severity index ($P < 0.05$). Disease control efficacy was also affected by timing of inoculation with SA70. Seedlings inoculated with SA70 at seeding time or 7 days after seeding had a significantly lower DSI than those inoculated at transplantation time, which was at 14 days after seeding.

To boost growth of the SA70 strain in Promix plugs as a means of improving disease control efficacy, we added soybean meal, wheat bran, a combination of broth and melted PDA into the Promix before inoculation with the fungus and seeding. Neither soybean meal nor wheat bran enhanced disease control, but both reduced seed germination and seedling growth. PDA amendment, however, improved disease control by SA70. Improvement of disease control was found to be related to an increase in the initial rhizosphere colonization by SA70, suggesting that the interaction between the pathogen and NPF in/on the root was important.

A benomyl-resistant mutant (70B10), generated from strain SA70, was used to examine root colonization (initial, rhizosphere or internal) and how this process relates to disease control. Pre-inoculation of 70B10 tomato roots reduced the colonization by *Fol*, suggesting possible exclusion of the pathogen from root tissues. The presence of *Fol* prior to inoculation by 70B10, in contrast, increased root colonization by 70B10. These results are similar to those reported by Larkin *et al.* (1996) but differ from those of Steinberg *et al.* (1999), who found that colonization by one *Fusarium* strain always reduced root colonization by another.

A genetically tagged biological control strain was developed and studies of the interactions among the non-pathogen, pathogen and the plant were undertaken. Strain SA70 used as a biological control agent was morphologically identical to the pathogenic isolate *Fol*. To differentiate and track the two *Fusarium* strains when they were introduced into the same root system, strain SA70 was genetically marked with both β-glucuronidase (GUS) and hygromycin B (HmB) resistance genes (*gusA* and *hph*) using DNA-mediated transformation (Bao *et al.*, 2000). One transformant, 70T01, was selected from more than 100 transformants as it had a stable single copy of each *gusA* and *hph* gene integrated into the SA70 fungal genomic DNA and the gene expressions were shown to be consistent. A simple and reliable detection technique was developed to measure GUS activity from the fungal mycelium using FastPrep® equipment to extract GUS, which disrupts cell integrity in a reproducible manner (Bao *et al.*, 2000). This process gave us a consistent method for measuring GUS activity from infected plant tissues and from fungal mycelium. GUS activity was determined from mycelium samples and also from supernatants of protoplasts derived from 70T01 mycelium. By plating non-lysed protoplasts on to agar medium we determined the population that could form colonies (colony-forming protoplast, CFP). GUS activity was highly correlated with mycelia dry weight ($r^2 > 0.9$) or CFP numbers ($r^2 > 0.8$). The CFP–GUS activity relationship provided the first attempt to measure absolute biomass for a filamentous fungus based on the single cell concept (true cfu). As little as 300–500 CFP ml^{-1} extract of 70T01-inoculated roots was detectable, and it showed the presence of 6–50 times more fungal biomass than found using the cfu plating method. The enzyme marker not only provided a powerful tool to monitor a specific fungus in an ecological niche, but also a means to quantify the fungus in plant root tissues. The level of fungal biomass found using the CFP-GUS relationship method, however, did not always agree with that found using the cfu method and we remain unsure as to which procedure was more correct. Quantification using the cfu method was, in general, considered to be affected by many factors, including the presence of high numbers of fungal spores that could give high cfu counts without any great level of tissue colonization; uneven maceration of tissues; and reduced growth of the fungus due to fungal ageing or production of toxic components in plant roots during maceration. We, and others, have accepted the assumption that mycelia are the dominant fungal forms in plant roots and that they play a more important role in disease suppression than spores. Obtaining a more accurate picture of the extent of mycelium colonization is seen as a requirement for understanding the plant–biological control agent relationship. The CFP-GUS technique overcomes several limitations seen with the cfu method and may provide a new tool for quantification of filamentous fungi in a root ecosystem.

70T01 mycelia in tomato roots were localized using the X-Gluc histochemical staining method, based on the fungal GUS expression. The mycelia were found to colonize primarily the epidermis or the outer cortex cell layers along tomato roots. The colonization was discontinuous and uneven. This was the first time that the pathogen and the NPF strain were visually differentiated simultaneously in the same root system. *Fol* mycelia were rarely observed at sites colonized by 70T01, suggesting that pre-colonization by 70T01 could reduce infection by *Fol* and lead to disease reduction. In contrast, where abundant *Fol* mycelia were found, 70T01 mycelia were not observed, suggesting that the two organisms were likely competing for root space. *Fol* was localized in the vascular tissues of the root, an area which the NPF strain rarely penetrated, even though the NPF mycelia were found colonizing the outer epidermis root cell layers in the same root segment. Thus, direct interaction between the two organisms does not likely occur in the inner root tissues but is restricted to the surface. Colonization by 70T01 prior to invasion by *Fol* is thus considered as a prerequisite for disease control. Host defence reactions, typically noted as increased cell wall thickness or formation of papillae on the cell wall, were also observed on sections from 70T01-inoculated roots. These plant cell defence reactions to 70T01 colonization may also be involved in preventing invasion by *Fol* or triggering induced-resistance responses.

The 70T01 population densities (GUS activity or cfu number) in two or three different root zones (in-plug, intermediate, and distal root segments) were determined at various times after transplantation of the plug seedlings into *Fol*-inoculated soil. Root segments from the seedling plug (the inoculation zone) had much higher 70T01 densities than the non-70T01-inoculated sites (intermediate or distal root zones that grew into the soil from the seedling plug). In contrast, the *Fol* cfu population densities were low at the 70T01-inoculated zones, but very high (often >10 times higher) at the non-inoculated zones. Root colonization by 70T01 in the intermediate or distal root zones was usually very low, indicating that the NPF strain did not actively move with the growing roots. This colonization pattern result obtained using GUS detection further confirmed the results obtained using histochemical localization, indicating that colonization by 70T01 decreased with distance away from the 70T01-inoculated zone. Thus, newly elongated root areas where tissues were not colonized by 70T01 are available for infection by *Fol*. The pathogen then can enter into the vascular bundle, and spread upward into the plant unimpeded.

In 1997 and 1998, several biological control agents were tested to control Fusarium wilt of muskmelon, *Cucumis melo* var. *reticulatus* Naudin, in the Delhi area of Ontario. In addition to SA70, we tested Fo7, CS20 (both *F. oxysporum*); and Fs-7, CS-1 (both *F. solani* (Martin) Saccardo); G-4, G-37 G-6, G-10, and Gv (*Trichoderma virens* Miller, Giddens, and Foster); and bacterial strains Bc-F (*Burkholderia vietnamiensis* Gillis, Van, Bardin, Goor, Hebbar, Willems, Segers, Kersters, Heulin, and Fernandez), M3 (unidentified) and B-B1-4-1 (*Streptomyces* sp.) (Bao *et al.*, 1999). Cordelle, a muskmelon cultivar highly susceptible to most races of the *Fusarium* pathogen, was used in both years of the trials. Transplants were generated in Promix plugs in a greenhouse and transplanted to the field in June. We observed significant differences among the treatments for disease incidence and disease severity within the first 7 weeks post-transplantation but not later. Several of the treatments increased disease severity and some delayed disease. None, however, provided sufficient protection to be recommended as a practical disease-control strategy.

Evaluation of Biological Control

With the appropriate tools, the opportunity exists to examine the workings of biological control in the ecological setting used by the pathogen and the control agents. However, this study pointed out that we also need to develop much more information for selecting, testing, formulating and delivering biological control agents. We selected the control organism by screening a large number of microorganisms from soil. In retrospect, a more competitive Fusarium strain than SA70 may have been found by using the root or the rhizosphere as the source for potential candidates. Even then, our selection process may still have been less than optimal. We do not know in what part of the root the organism of choice resides or how many isolates should be tested. If we test 100 isolates, do we know that those 100 isolates are not clonal and that we are really looking at one single isolate? Is the time and location of the plants we use for the source of the control agents important? The arrival of effective biological control will be accelerated by a better understanding and a more systematic approach for studying the activities of biological control agents as they exist in nature.

Recommendations

Further work should include:

1. Obtaining a better understanding of, and using a more systematic approach for, selecting and testing potential biological control agents;
2. Developing effective formulations and delivery systems.

Acknowledgements

We thank Nightingale Farms, Environment Canada, and Agriculture and Agri-Food Canada Matching Investment Initiatives programme for funding this project.

References

Alabouvette, C. (1990) Biological control of *Fusarium* wilt pathogens in suppressive soils. In: Hornby, D. (ed.) *Biological Control of Soil-borne Plant Pathogens.* CAB International, Wallingford, UK, pp. 27–43

Armstrong, G.M. and Armstrong, J.K. (1981) Formae speciales and races of *Fusarium oxysporum* causing wilt diseases. In: Nelson, P.E., Toussoun, T.A. and Cook, J.R. (eds) Fusarium: *Diseases, Biology, and Taxonomy.* Pennsylvania State University Press, University Park, Pennsylvania, pp. 391–399.

Bao, J.R., Hill, J., Lazarovits, G., Fravel, D. and Howell, C.R. (1999) Biological control of *Fusarium* wilt of muskmelon using nonpathogenic *Fusarium* spp. and other biological agents, 1997–1998. *Biological and Cultural Tests for Control of Plant Diseases* 14, 160.

Bao, J.R., Velema, J., Dobinson, K.F. and Lazarovits, G. (2000) Using GUS expression in a nonpathogenic *Fusarium oxysporum* strain to measure fungal biomass. *Canadian Journal of Plant Pathology* 22, 70–78.

Cook, R.J. (1993) Making greater use of introduced microorganisms for biological control of plant pathogens. *Annual Review of Phytopathology* 31, 53–80.

Gordon, T.R. and Martyn, R.D. (1997) The evolutionary biology of *Fusarium oxysporum. Annual Review of Phytopathology* 35, 111–128.

Handelsman, J. and Stabb, E.V. (1996) Biocontrol of soilborne plant pathogens. *The Plant Cell* 8, 1855–1869.
Komada, H. (1975) Development of a selective medium for quantitative isolation of *Fusarium oxysporum* from natural soil. *Review of Plant Protection Research* 8, 115–125.
Larkin, R.P. and Fravel, D.R. (1999) Mechanisms of action and dose–response relationships governing biological control of *Fusarium* wilt of tomato by nonpathogenic *Fusarium* spp. *Phytopathology* 89, 1152–1161.
Larkin, R.P., Hopkins, D.L. and Martin, F.N. (1996) Suppression of *Fusarium* wilt of watermelon by nonpathogenic *Fusarium oxysporum* and other microorganisms recovered from a disease-suppressive soil. *Phytopathology* 86, 812–819.
Lumsden, R.D., Lewis, J.A. and Fravel, D.R. (1995) Formulation and delivery of biocontrol agents for use against soilborne plant pathogens. In: Hall, F.R. and Barry, J.W. (eds) *ACS Symposium Series 595: Biorational Pest Control Agents.* American Chemical Society, Washington, DC, pp. 165–182.
Ogawa, K. and Komada, H. (1984) Biological control of *Fusarium* wilt of sweet potato by nonpathogenic *Fusarium oxysporum. Annals of the Phytopathology Society of Japan* 50, 1–9.
Paulitz, T.C., Park, C.S. and Baker, R. (1987) Biological control of *Fusarium* wilt of cucumber with nonpathogenic isolates of *Fusarium oxysporum. Canadian Journal of Microbiology* 33, 349–353.
Postma, J. and Luttikholt, A.J.G. (1996) Colonization of carnation stems by a nonpathogenic isolate of *Fusarium oxysporum* and its effect on *Fusarium oxysporum* f. sp. *dianthi. Canadian Journal of Botany* 74, 1841–1851.
Steinberg, C., Whipps, J.M., Wood, D., Fenlon, J. and Alabouvette, C. (1999) Mycelial development of *Fusarium oxysporum* in the vicinity of tomato roots. *Mycological Research* 103, 769–778.

91 *Heterobasidion annosum* (Fries) Brefeld, Annosus Root Rot (*Polyporaceae*)[1]

G. Laflamme

Pest Status

Heterobasidion annosum (Fries) Brefeld (= *Fomes annosus* (Fries) Karsten), causal agent of annosus root rot, is found on all continents. Because *H. annosum* causes extensive damage worldwide, it is considered to be one of the most destructive pathogens in evergreen forests. Forest pathologists have classified the species in different groups based on their host and, more recently, the continent where it is found. At least five intersterility groups exist: the European P, S and F groups, and the North American P and S groups (Mitchelson and Korhonen, 1998). The letters stand for **P**ine, **S**pruce and **Fi**r. After 20 years of research, the three European groups are now divided into three different species, *H. annosum* being restricted to the P group. Up to now in eastern Canada, only the P group has been identified. The dis-

[1]Hawksworth *et al.* (1995) classified this pathogen in *Polyporaceae*, but Niemelä and Korhonen (1998) reported that it was considered to be more closely related to species in the *Bondarzewiaceae*.

ease occurs in over 170 tree species worldwide. Although the extent of damage varies with species, the greatest damage occurs in conifers. In eastern Canada, red pine, *Pinus resinosa* Aiton, is the most-affected species. Other species growing near red pines infected with *H. annosum* have also been found to be infected with the pathogen, but these do not seem to be hosts for primary infection. In eastern Canada, the disease was first detected in southern Ontario in 1955, and 15 years later in Larose Forest near Ottawa (Laflamme, 1994). In Quebec, the first case of this disease was discovered in 1989 about 40 km from the Larose Forest (Laflamme and Blais, 1993). Since then, it has spread to other red pine plantations.

The disease was first identified by Hartig (1874) who demonstrated (Hartig, 1900) that it is transmitted from tree to tree by root contact, creating characteristic 'circles of mortality'. Rishbeth (1951) discovered that the fungus becomes established in a stand by spores that colonize freshly cut stumps. The discovery of this key element in the propagation of *H. annosum* finally made it possible to develop methods aimed at controlling its introduction into forests by treating stumps. Various chemical products were then tested and Rishbeth (1963) was the first to use biological control, with promising results.

H. annosum basidiospores can travel long distances. Rishbeth (1959) found viable spores over the ocean more than 300 km from the closest possible source of infection. Thus, after being transported by wind, basidiospores settle on freshly cut stump surfaces and germinate. Such surfaces are selective for a number of microorganisms, including *H. annosum*. Therefore, spores must colonize the stump surface soon after the tree is felled and before other microorganisms move in. The window of opportunity varies, depending on host and climate, and can range from a few days to 3 or 4 weeks. However, infection rarely occurs more than 2 weeks after felling (Hodges, 1969).

Background

Mechanical eradication of infected trees has been the sole method to control *H. annosum*. As of 1999, no commercial formulations, chemical or biological, were registered for use in Canada. Rishbeth (1963) observed that untreated stumps were often colonized by the saprophytic fungus *Phlebiopsis gigantea* (Fries) Jülich (= *Peniophora gigantea* (Fries) Massee). Once established, this fungus prevented *H. annosum* from infecting the stump. *P. gigantea* has the additional advantage of producing large quantities of spores when cultivated in the laboratory. Like many other wood-rotting fungi, *P. gigantea* spreads by spores produced on fruiting bodies made of a thin and porous layer on the surface of the substrate colonized by the fungus. This resupinate form of fruiting body produces millions of spores. Although other potential microorganisms have been tested (Holdenrieder and Greig, 1998), *P. gigantea* is the only one that has been commercialized (Korhonen *et al.*, 1994).

The isolates of *P. gigantea* used for the Kemira formulation Rotstop®, registered in a few European countries (Korhonen *et al.*, 1994), are considered quite different from our North American isolates (Vainio and Hantula, 2000). Thus, it could be very difficult to obtain a registration for this commercial formulation for use in eastern Canada unless the original isolate is replaced by a North American one.

Biological Control Agents

Fungi

In western Quebec, Bussières *et al.* (1996) evaluated the potential of *P. gigantea* for use in red pine plantations to control *H. annosum*. Their results showed that *P. gigantea* colonized most red pine stumps 12 months following application of the inoculum. Natural colonization of stumps by *P. gigantea* did not provide adequate protection against infection by *H. annosum*. However, the application of *P. gigantea* on fresh stumps ensures its presence there and results in a more extensive colonization of this saprophyte.

Recommendations

Future work should include:

1. Formulation and commercialization of a Canadian isolate of *P. gigantea* for use on red and Scots pine;
2. Testing the susceptibility of other *Pinus* spp. to infection by *H. annosum*;
3. Developing techniques to apply the biological control product; manual treatment should be evaluated for small-scale operations and devices that can be fitted on a tree harvester should be tested, making the treatment completely mechanized for large-scale operations (Thor, 1997);
4. Further studying other antagonistic fungal species, e.g. *Phaeotheca dimorphospora* DesRochers et Ouellette, for their potential as additional biological agents for other tree species (Roy, 1999).

References

Bussières, G., Dansereau, A., Dessureault, M., Roy, G., Laflamme, G. and Blais, R. (1996) *Lutte Contre la Maladie du Rond dans l'Ouest du Québec. Projet No. 4023. Essais, Expérimentations et Transfert Technologique en Foresterie.* Ressources naturelles Canada, Service canadien des forêts, Ottawa, Ontario.

Hartig, R. (1874) *Wichtige Krankheiten der Waldbäume. Beiträge zur Mycologie und Phytopathologie für Botaniker und Forstmänner.* J. Springer, Berlin, Germany.

Hartig, R. (1900) *Lehrbuch der Pflanzenkrankheiten. 3rd Auftreten des Lehrbuches des Baumkrankheiten, 1882, 1889.* Springer, Berlin, Germany.

Hawksworth, D.L., Kirk, P.M., Sutton, B.C. and Pegler, D.N. (1995) *Ainsworth and Bisby's Dictionary of Fungi*, 8th edn. CAB International, Wallingford, UK.

Hodges, C.S. (1969) Modes of infection and spread of *Fomes annosus. Annual Review of Phytopathology* 7, 247–266.

Holdenrieder, O. and Greig, B.J.W. (1998) Biological method of control. In: Woodward, S., Stenlid, J., Karjalainen, R. and Hüttermann, A. (eds) Heterobasidion annosum*: Biology, Ecology, Impact and Control.* CAB International, Wallingford, UK, pp. 235–258.

Korhonen, K., Lipponen, K., Bendz, M., Johansson, M., Ryen, L., Venn, K., Seiskari, P. and Niemi, M. (1994) Control of *Heterobasidion annosum* by stump treatment with 'Rotstop', a new commercial formulation of *Phlebiopsis gigantea.* In: Johansson, M. and Stenlid, J. (eds) *Proceedings of the Eighth International Conference on Root and Butt Rot, Sweden and Finland, 9–16 August 1993.* CAB International, Wallingford, UK, pp. 675–685.

Laflamme, G. (1994) *Annosus Root Rot Caused by* Heterobasidion annosum. Information Leaflet LFC 27, Natural Resources Canada, Canadian Forest Service, Quebec Region.

Laflamme, G. and Blais, R. (1993) Première mention de *Heterobasidion annosum* au Québec. *Phytoprotection* 74, 171.

Mitchelson, K. and Korhonen, K. (1998) Diagnosis and differentiation of intersterility groups. In: Woodward, S., Stenlid, J., Karjalainen, R. and Hüttermann, A. (eds) Heterobasidion annosum*: Biology, Ecology, Impact and Control.* CAB International, Wallingford, UK, pp. 71–92.

Niemelä, T. and Korhonen, K. (1998) Taxonomy of the genus *Heterobasidion.* In: Woodward, S., Stenlid, J., Karjalainen, R. and Hüttermann, A. (eds) Heterobasidion annosum: *Biology, Ecology, Impact and Control.* CAB International, Wallingford, UK, pp. 27–33.

Rishbeth, J. (1951) Observations on the biology of *Fomes annosus* with particular reference to East Anglia pine plantations. II. Spore production, stump infection, and saprophytic activity in stumps. *Annals of Botany* 15, 1–21.

Rishbeth, J. (1959) Dispersal of *Fomes annosus* and *Peniophora gigantea. Transactions of the British Mycological Society* 42, 243–260.

Rishbeth, J. (1963) Stump protection against *Fomes annosus* III. Inoculations with *Peniophora gigantea*. *Annals of Applied Biology* 52, 63–77.
Roy, G. (1999) Développement d'un agent de lutte biologique contre *Heterobasidion annosum*. Thèse de Doctorat, Université Laval, Québec, Canada.
Thor, M. (1997) Stump treatment against *Heterobasidion annosum*: Techniques and biological effect in practical forestry. Licentiate's dissertation, Swedish University of Agricultural Sciences, Department of Forest Mycology and Pathology, Uppsala, Sweden.
Vainio, E.J. and Hantula, J. (2000) Genetic differentiation between European and North American populations of *Phlebiopsis gigantea*. *Mycologia* 92, 436–446.

92 *Leptosphaeria maculans* (Desmazières) Cesati and De Notaris, Blackleg of Canola (*Leptosphaeriaceae*)

P.D. Kharbanda, J. Yang, P.H. Beatty, J.P. Tewari and S.E. Jensen

Pest Status

A virulent strain of *Leptosphaeria maculans* (Desmazières) Cesati and De Notaris [conidial state: *Phoma lingam* (Tode: Fries) Desmazières], causal agent of the blackleg disease, has become one of the most important diseases of canola, *Brassica napus* L. and *B. rapa* L., in several temperate countries during the past 20 years. It is a serious yield-limiting factor in canola/rapeseed production. In Australia it caused a serious epidemic in 1971 and 1972 and nearly destroyed the rapeseed industry (Bokor *et al.*, 1975). Blackleg was the major disease of rapeseed in parts of France, England and Germany (Gladders and Musa, 1980). In Canada, the virulent strain was found in Saskatchewan in 1975 and has since spread rapidly in the west (Kharbanda, 1992; Petrie, 1994; Chigogora and Hall, 1995; Juska *et al.*, 1997). Annual canola crop losses caused by blackleg are estimated to be nearly Can$50 million dollars.

L. maculans is seed-borne and also survives on infected canola stubble. It produces sexual fruiting bodies, pseudothecia, containing asci and ascospores. Rain-splashed pycnidiospores and air-borne ascospores serve as primary inocula that are dispersed to new crops and initiate disease. In nature, *L. maculans* persists in a saprophytic mode, colonizing dead tissues. Pseudothecia are formed continuously on host stubble and discharge ascospores. The production of ascospores is greatly affected by temperature, moisture, light and nutrients (Petrie, 1994). Ascospores are formed in the same year on winter canola stems in Ontario whereas, in western Canada, they are discharged the next spring and early summer. Ascospores continue to discharge from the stubble for 3–5 years. Secondary inoculum mainly consists of pycnidiospores that are produced on infected canola plants and ascospores from infected stubble of previous years. Pycnidiospores are primarily distributed by rain-splash within short distances and cause secondary infections under suitable conditions.

Lesions develop on leaves, stems and pods, and produce more pycnidia. Ascospores are effectively dispersed over a few kilometres by wind. Ascospores appear to be more infective than pycnidiospores. Infection by ascospores is affected by temperature and wetness duration (Biddulph *et al.*, 1999). Primary infection of seedlings from the ascospore inoculum results in latent infection on 'Westar', a susceptible cultivar, and the period of latent infection is much shorter than on 'Cresor', a resistant cultivar. Latent infection was also found in other commercial canola varieties and stinkweed, *Thlaspi arvense* L., infected by different *L. maculans* strains.

Background

Fungicidal seed treatments and foliar applications of fungicides such as propiconazole do not effectively control blackleg disease (Kharbanda, 1992). Tolerant cultivars combined with cultural management and seed testing have been used to manage the disease. Completely resistant cultivars may soon become available. Nevertheless, alternative disease control methods are required. Biological agents to enhance control of blackleg disease are needed.

Biological Control Agents

Fungi

Petrie (1982) reported partial suppression of the virulent *L. maculans* with the weakly virulent strain of the pathogen *in vitro* and *in vivo*. Tewari and Briggs (1995), Tewari *et al.* (1997) and Shinners and Tewari (1997, 1998) investigated the fungi *Cyathus olla* Batsch: Peres and *Cyathus striatus* (Hudson: Peres) Peres for their role in increasing decomposition of canola residues, and consequently reducing inoculum of *L. maculans* present on the stubble. In the laboratory, Starzycki *et al.* (1998) tested strains of *Trichoderma viride* Peres: Fries and *Trichoderma harzianum* Rifai for their protective ability against *L. maculans* and *Sclerotinia sclerotiorum* (Libert) de Bary and found that various strains of *Trichoderma* spp. had different inhibitory effects against the two pathogens.

Bacteria

Kharbanda and Dahiya (1990) found a strain of *Penicillium verrucosum* Dierckx that produced a metabolite toxic to *L. maculans*. Chakraborty *et al.* (1994) tested *in vitro* antagonism of *Erwinia herbicola* (Löhnis) Dye, a phyllosphere microorganism on canola, against *L. maculans* and found a partially thermolabile antifungal substance in the bacterial culture that significantly reduced the severity of blackleg disease.

A strain of *Paenibacillus polymyxa* (Prazmowski) Ash *et al.* PKB1 (previously *Bacillus polymyxa* Prazmowski), isolated from canola roots, was found to be highly inhibitory to the growth of *L. maculans* and some other pathogenic fungi *in vitro*. Since 1994, we have explored the use of this strain, alone or in combination with fungicides, to control blackleg and some other diseases of canola. Molecular probes and specific primers developed by Yang *et al.* (1997, 1998) were used to detect *P. polymyxa*. Other strains are also being investigated (de Freitas *et al.*, 1999). The antifungal agent produced by PKB1 appears to be a combination of cyclic depsipeptide compounds of 883 Da and 897 Da that are very similar or identical to fusaricidins A and B, respectively (Beatty *et al.*, 1998; Beatty, 2000).

Yang *et al.* (1996) and Kharbanda *et al.* (1997) tested the effectiveness of *P. polymyxa* PKB1 against *L. maculans* and other disease-causing fungi, e.g. *Sclerotinia sclerotiorum*, *Rhizoctonia solani* Kühn, *Alternaria* spp., *Pythium* spp., *Botrytis* spp., *Ascochyta* spp., *Pyrenophora teres* Drechsler, *P. tritici-repentis* (Diedicke) Drechsler, *Didymella* sp. and *Fusarium* spp., by measuring fungal inhibition zones on potato-dextrose agar and nutrient agar plates and by determining mycelium dry weight from potato-dextrose broth shake-cultures containing the bacterial filtrate. In both Petri plate and liquid-culture tests, *P. polymyxa* PKB1 was found to have significant inhibitory effects on all fungi tested. It reduced blackleg incidence and severity on the susceptible cultivar 'Westar' but there was no significant difference between treatments on the resistant cultivar 'Quantum' (Yang *et al.*, 1996).

Kharbanda *et al.* (1997) compared the performance of PKB1 and the fungicide propiconazole on survival of *L. maculans* on canola stubble. Infected canola stubble in pots sprayed with either propiconazole (125 g active ingredient (a.i.) ha^{-1}) or *P. polymyxa* PKB1 suspension (7.4×10^7 cells ml^{-1}) and incubated at temperatures ranging from −5°C to 20°C showed that, 10 weeks after inoculation, propiconazole significantly reduced the number of pycnidia under most temperature regimes (except at 20°C, and at various temperatures on buried samples). Although *P. polymyxa* PKB1 was not effective in reducing the number of pycnidia on the stem surface, it significantly reduced *L. maculans* survival under most conditions compared with untreated or propiconazole-treated stubble. Kharbanda *et al.* (1997) and Yang *et al.* (1996) determined that *P. polymyxa* PKB1 could be used in combination with these chemicals in an integrated pest-management system.

Kharbanda *et al.* (1998) and Yang *et al.* (1999) evaluated compost as a carrier of *P. polymyxa* PKB1 for large-scale application. The viability of *L. maculans* was significantly reduced in stubble treated with propiconazole, propiconazole + PKB1, and PKB1 + compost. There were significant differences in pseudothecia production in response to treatments and burial methods in samples retrieved after 18 months. Compost + PKB1 and propiconazole + PKB1 had a significant inhibitory effect on ascospore formation on canola stubble.

Canola seeds coated with *P. polymyxa* PKB1 spores were tested in the laboratory for its effect on disease reduction. In Petri plate tests, canola seeds coated with *P. polymyxa* PKB1 had higher germination on *L. maculans* culture plates than uncoated seeds. In a growth-chamber test, *P. polymyxa* PKB1-coated 'Westar' canola seeds had significantly lower cotyledon infection than uncoated seeds when the seeds were planted in *L. maculans*-infested soil (J. Yang, unpublished).

Evaluation of Biological Control

P. polymyxa PKB1 is capable of inhibiting growth of several fungi that cause important diseases on canola, and other field and greenhouse crops. Most chemicals used on canola do not have deleterious effects on the growth of *P. polymyxa* PKB1. Propiconazole significantly reduced the number of pycnidia on stubble and *P. polymyxa* PKB1 significantly reduced survival of *L. maculans* under growth chamber and field conditions. Compost could be a useful carrier for delivery of *P. polymyxa* PKB1.

Recommendations

Further work should include:

1. Investigating the application of *P. polymyxa* PKB1 for disease control of other field and greenhouse crops;
2. Further experimentation on separation of the antifungal compounds fusaricidins A and B and on application of the compounds in disease control;
3. Screening additional bacterial isolates for biological control of *L. maculans*.

References

Ash, C., Prist, F.G. and Collins, M.D. (1994) Validation List No. 51. *International Journal of Systematic Bacteriology* 44, 852.

Beatty, P.H. (2000) Investigation of an antifungal antibiotic production by an environmental isolate of *Paenibacillus polymyxa*. PhD thesis, University of Alberta, Edmonton, Alberta.

Beatty, P.H., Kharbanda, P.D. and Jensen, S.E. (1998) Purification and partial characterization of an antifungal antibiotic produced by *Bacillus polymyxa* PKB1. In: *The Annual Meeting of the American Society for Microbiology, 17–21 May, Atlanta, Georgia, USA*. American Society of Microbiology, Washington, DC, Abstract.

Biddulph, J.E., Fitt, B.D.L., Leech, P.K. and Gladders, P. (1999) Effects of temperature and wetness duration on infection of oilseed rape leaves by ascospores of *Leptosphaeria maculans* (stem canker). *European Journal of Plant Pathology* 105, 769–781.

Bokor, A., Barbetti, M.J., Brown, A.G.P., MacNish, G.C. and Wood, P.McR. (1975) Blackleg of rapeseed. *Journal of Agriculture in Western Australia* 16, 7–10.

Chakraborty, B.N., Chakraborty, U. and Basu, K. (1994) Antagonism of *Erwinia herbicola* towards *Leptosphaeria maculans* causing blackleg disease of *Brassica napus*. *Letters in Applied Microbiology* 18, 74–76.

Chigogora, J.L. and Hall, R. (1995) Relationship among measures of blackleg in winter oilseed rape and infection of harvested seed by *Leptosphaeria maculans*. *Canadian Journal of Plant Pathology* 17, 25–30.

de Freitas, J.R., Boyetchko, S.M., Germida, J.J. and Khachatourians, G.G. (1999) Development of natural microbial metabolites as biocontrol products for canola pathogens. *Canadian Journal of Plant Pathology* 21, 193–194.

Gladders, P. and Musa, T.M. (1980) Observations on the epidemiology of *Leptosphaeria maculans* stem canker in winter oilseed rape. *Plant Pathology* 29, 28–37.

Juska, A., Busch, L. and Tanaka, K. (1997) The blackleg epidemic in Canadian rapeseed as a 'normal agricultural accident'. *Ecological Society of America* 7, 1350–1356.

Kharbanda, P.D. (1992) Performance of fungicides to control blackleg of canola. *Canadian Journal of Plant Pathology* 14, 169–176.

Kharbanda, P.D. and Dahiya, J.S. (1990) A metabolite of *Penicillium verrucosum* inhibitory to growth of *Leptosphaeria maculans* and *Rhizoctonia solani*. *Canadian Journal of Plant Pathology* 12, 335.

Kharbanda, P.D., Yang, J., Beatty, P.H., Jensen, S.E. and Tewari, J.P. (1997) Potential of a *Bacillus* sp. to control blackleg and other diseases of canola. *Phytopathology* 87, S51.

Kharbanda, P.D., Clark, T., Yang, J. and Tewari, J.P. (1998) Suppression of *Leptosphaeria maculans* with *Bacillus polymyxa* amended compost and agronomic benefits of using compost. *Canadian Journal of Plant Pathology* 21, 195.

Petrie, G.A. (1982) Blackleg of rapeseed (canola) caused by *Leptosphaeria maculans*: interaction of virulent and weakly virulent strains and implications for biological control. *Canadian Journal of Plant Pathology* 4, 309.

Petrie, G.A. (1994) 1994 survey for blackleg and other diseases of canola. *Canadian Journal of Plant Pathology* 75, 142–144.

Shinners, T.C. and Tewari, J.P. (1997) Diversity in crystal production by some birds nest fungi (Nidulariaceae) in culture. *Canadian Journal of Chemistry* 75, 850–856.

Shinners, T.C. and Tewari, J.P. (1998) Morphological and RAPD analysis of *Cyathus olla* from crop residue. *Mycologia* 90, 980–989.

Starzycki, M., Starzycka, E. and Matuszczak, M. (1998) Fungi of the genus *Trichoderma* spp. and their protective ability against the pathogens *Phoma lingam* (Tode ex Fr.) Desm. and *Sclerotinia sclerotiorum* (Lib.) de Bary. *Review of Plant Pathology* 77, 1411.

Tewari, J.P. and Briggs, K.G. (1995) Field infestation of canola stubble by a bird nest fungus. *Canadian Journal of Plant Pathology* 17, 291.

Tewari, J.P., Shinners, T.C. and Briggs, K.G. (1997) Production of calcium oxalate crystals by two species of *Cyathus* in culture and infested plant debris. *Zeitschrift für Naturforschung* 52c, 421–425.

Yang, J., Kharbanda, P.D. and Tewari, J.P. (1996) Inhibitory effect of a biocontrol agent (*Bacillus* sp.) against *Leptosphaeria maculans* and DNA fingerprinting of the biocontrol agent using PCR-RAPD. *Proceedings of the International Workshop on Biological Control of Plant Diseases*, China Agricultural University Press, Beijing, China 21, 206, p. 99.

Yang, J., Kharbanda, P.D. and Tewari, J.P. (1997) Detection of a biocontrol agent (*Bacillus* sp.) against *Leptosphaeria maculans* using Dig-labeled probes. *Canadian Journal of Plant Pathology* 20, 218.

Yang, J., Kharbanda, P.D. and Tewari, J.P. (1998) Development of specific primers to a biocontrol agent against *Leptosphaeria maculans*. *Phytopathology* 88, S101.

Yang, J., Mooney, H.D., Clark, T. and Kharbanda, P.D. (1999) Development of compost as a delivery medium for a bacterial biocontrol agent. *Canadian Journal of Plant Pathology*.

93 *Monilinia fructicola* (Winter) Honey, Brown Rot (*Hyphomycetes*)

T. Zhou and P. Sholberg

Pest Status

The fungus *Monilinia fructicola* (Winter) Honey is the causal agent of brown rot, the most severe disease of stone fruits, including apricot, *Prunus armeniaca* Marsh, cherry, *P. avium* L., peach, *P. persica* (L.) Batsch, and plum, *P. domestica* Link, in Canada. Although the disease may affect blossoms and twigs, it is highly destructive to fruits, and can ruin half or more of the crop before harvest, with the remaining fruit subject to post-harvest decay. In a 2-year survey conducted in southern Ontario, 20–80% of commercially ripe peaches collected from local orchards developed brown rot decay after only 4–5 days' incubation at room temperature (Zhou *et al.*, 1997).

M. fructicola overwinters in two ways: (i) in mummified fruit; and (ii) in twig cankers resulting primarily from the previous season's rotted fruit. In spring, mycelium of *M. fructicola* in mummified fruit on the tree and on the ground and in the twig cankers produces chains of elliptical conidia, while the mycelium in mummied fruit buried in the ground produces several small, brownish, cup-shaped apothecia, which form asci and ascospores. Both conidia and ascospores can cause blossom infection. Although ascospores are relatively rare in Ontario, in years when apothecia are found severe blossom blight has been noted. Conidia from infected blossoms may contribute to infections of small green fruit, and ripening fruit later that year. Fruit infection also takes place after harvest, in storage and in transit. On fruit, brown rot starts with small, circular brown spots. The spots spread rapidly, and are sooner or later covered with ash-coloured tufts of conidia. One large or several small rotten areas may be present on the fruit, which finally becomes completely rotted.

Background

Currently, control of *M. fructicola* still relies on preharvest application(s) of fungicide(s) such as captan and iprodione. Public health concerns about the presence of chemical residues in the food supply have led to the restriction or withdrawal of most postharvest fungicide treatments in Canada and the USA (Wilson *et al.*, 1994). Although iprodione was registered in the USA for the postharvest treatment of peaches against brown rot before 1996, no such registration was obtained in Canada. In fact, no fungicide is currently available in Canada for postharvest treatment against *M. fructicola*.

During the past two decades substantial efforts have been made to find alternatives to synthetic fungicides to control postharvest diseases of fruits. Pusey and Wilson (1984) and Smilanick *et al.* (1993) reported that numerous microorganisms inhibited *Monilinia* spp. on peach fruits. Pusey *et al.* (1986) investigated *Bacillus subtilis* (Ehrenberg) Cohn, and McKeen *et al.* (1986) showed that it produced an antibiotic substance toxic to *M. fructicola*. Strains of *Pseudomonas corrugata* Roberts and Scarlett and *Pseudomonas cepacia*

Roberts and Scarlet ex Burkholder greatly reduced peach decay when applied up to 12 h after inoculation with *M. fructicola*, but they controlled brown rot poorly when applied to peaches with natural preharvest infections of *M. fructicola* (Smilanick *et al.*, 1993). Although some microorganisms have shown great potential for controlling postharvest diseases, no biological product is currently available commercially to control postharvest diseases of peach.

Biological Control Agents

Bacteria

In British Columbia, Utkehede and Sholberg (1986) tested 21 isolates of *Bacillus subtilis* and one isolate of *Enterobacter aerogenes* Hormaeche and Edwards on agar for antagonism to several pathogenic fungi, including *M. fructicola*. All inhibited *M. fructicola*. However, when the bacteria were tested on mature cherry fruit, 15 isolates of *B. subtilis* were effective but the isolate of *E. aerogenes* did not control *M. fructicola*. One isolate of *B. subtilis* reduced brown rot to 9% compared to 84% in the untreated control and was as effective as iprodione, the fungicide most commonly used by orchardists to control *M. fructicola*. Bechard *et al.* (1998) purified an antimicrobial compound from an isolate of *B. subtilis* and partially characterized it as a lipopeptide (Bechard *et al.*, 1998). It does not appear to be the same compound as that found by McKeen *et al.* (1986) but it is antibacterial and antifungal.

In southern Ontario, Zhou and DeYoung (1996) isolated several microorganisms from apple leaves, including saprophytic isolates of *Pseudomonas syringae* van Hall, *Pseudomonas* spp. and yeasts, and showed that these isolates suppressed apple scab during the growing season. Some of these microorganisms also inhibited isolates of *Penicillium expansum* Link and *Botrytis cinerea* Persoon: Fries, and effectively controlled blue mould and grey mould of apple under cold storage and controlled atmosphere storage (Zhou *et al.*, 1998, 2001). These promising microorganisms were evaluated as pre- and postharvest applications for their suppression of brown rot of peach.

In preharvest treatments, *P. syringae* isolate MA-4 was evaluated for 2 years for controlling *M. fructicola* in peach orchards in Vineland, Ontario. In 1997, the experiment was conducted in a 6-year-old 'Loring' peach orchard. Peach trees were sprayed with water, as a control, or cell suspensions of the isolate MA-4, once at 3 weeks prior to harvest or twice, at 3 weeks and 1 week prior to harvest, respectively. A foliar calcium fertilizer (Cab'y: 10% Ca^{2+} and 0.5% boron) at a final concentration of 1% was added to the bacterial suspensions, with a final concentration of 10^7 colony-forming units (cfu) ml^{-1}. Brown rot development was monitored by counting the number of peaches with brown rot, both on and under each tree, every 2–3 days after the first application.

Development of peach brown rot during the 3 weeks prior to harvest was significantly different among the treatments (P = 0.05). Brown rot in the treatment with two applications of *P. syringae* isolate MA-4 developed more slowly than that in the water check, and at harvest the incidence of peaches with brown rot was 5.4%, about 70% lower than that in the water check (17.2%). In the treatments with one application of isolate MA-4, brown rot was only slightly reduced. Application of the foliar fertilizer Cab'y alone did not give significant brown rot control as compared with the water check (Zhou and Schneider, 1998). Similar results were obtained in the two experiments conducted in 'Redhaven' and 'Loring' peach orchards in 1998. At harvest, two applications of *P. syringae* isolate MA-4 (10^7 cfu ml^{-1}) with 1% Cab'y reduced brown rot by 48–70% as compared to the water controls. These were as effective as two applications of captan fungicide.

In postharvest treatments, commercially ripe 'Redhaven' peaches were wounded and coinoculated with isolates of *P. syringae* (MA-4 and NSA-6), *P. fluorescens* (Trevisan) Migula (BAP-3) at a concentration of 1×10^7 cfu ml^{-1} or an isolate of *Candida* sp. (NSD-4) at a concentration of 1×10^6 cfu ml^{-1} in combination with a spore suspension of *M. fructicola* at 1×10^4 conidia ml^{-1}. After 5 days' incubation at 22°C, *P. syringae* isolates NSA-6 and MA-4 reduced brown rot to 28% and 73%, respectively, from 98% in the inoculated check. The isolates of BAP-3 and NSD-4 were not effective in controlling brown rot.

In another experiment, 'Loring' peaches harvested from an orchard with high incidence of preharvest fruit rot were soaked for 2 min in cell suspensions of *P. syringae* isolates. After 3 days' incubation at room temperature, the incidence of brown rot in the water check reached 65%, but was only 29% and 30% for peaches treated with *P. syringae* isolates MA-4 and NSA-6, respectively. In a similar experiment, addition of 0.5% $CaCl_2$ in the cell suspensions significantly improved the activity of *P. syringae* (Zhou *et al.*, 1999).

Evaluation of Biological Control

Biological control of *M. fructicola* was effective both before and after harvest, as shown by the above experiments. There is a great need for postharvest control of brown rot in stone fruits and the biological controls as identified above would serve the purpose well.

Recommendations

Further work should include:

1. Facilitating the registration of *P. syringae* isolates as environmentally sound control agents.

Acknowledgements

The Canada Agricultural Adaptation Council, Ontario Tender Fruit Producers' Marketing Board, Nabisco, Ltd, and the Matching Investment Initiative grant from Agriculture and Agri-Food Canada provided funding for the research conducted in Ontario.

References

Bechard, J., Eastwell, K.C., Sholberg, P.L., Mazza, G. and Skura, B. (1998) Isolation and partial chemical characterization of an antimicrobial peptide produced by a strain of *Bacillus subtilis*. *Journal of Agricultural and Food Chemistry* 46, 5355–5361.

McKeen, C.D., Reilly, C.C. and Pusey, P.L. (1986) Production and partial characterization of antifungal substances antagonistic to *Monilinia fructicola* from *Bacillus subtilis*. *Phytopathology* 76, 136–139.

Pusey, P.L. and Wilson, C.L. (1984) Postharvest biological control of stone fruit brown rot by *Bacillus subtilis*. *Plant Disease* 68, 753–756.

Pusey, P.L., Wilson, C.L., Hotchkiss, M.W. and Franklin, J.D. (1986) Compatibility of *Bacillus subtilis* for postharvest control of peach brown rot with commercial fruit waxes, dicloran, and cold-storage conditions. *Plant Disease* 70, 587–590.

Smilanick, J.L., Denisarrue, R., Bosch, J.R., Gonzalez, A.R., Henson, D. and Janisiewicz, W.J. (1993) Control of postharvest brown rot of nectarines and peaches by *Pseudomonas* species. *Crop Protection*. 12, 513–520.

Utkhede, R.S. and Sholberg, P.L. (1986) *In vitro* inhibition of plant pathogens by *Bacillus subtilis* and *Enterobacter aerogenes* and *in vivo* control of postharvest cherry diseases. *Canadian Journal of Microbiology* 32, 963–967.

Wilson, C.L., El-Ghaouth, A., Chalutz, E., Droby, S., Stevens, C., Lu, J.Y., Khan, V. and Arul, J. (1994) Potential of induced resistance to control postharvest diseases of fruits and vegetables. *Plant Disease* 78, 837–844.

Zhou, T. and DeYoung, R. (1996) Control of apple scab with applications of phyllosphere microorganisms. In: Tang, W., Cook, R.J. and Rovira, A. (eds) *Advances in Biocontrol of Plant Diseases*. Beijing China Agricultural University Press, Beijing, China, pp. 369–399.
Zhou, T. and Schneider, K. (1998) Control of peach brown rot by preharvest applications of an isolate of *Pseudomonas syringae*. *Abstracts of the 7th International Congress of Plant Pathology*, Abstract 3, 5.2.20, British Society for Plant Pathology, Birmingham, UK.
Zhou, T., Schneider, K. and Walker, G. (1997) Peaches: to wax or not to wax. *The Tender Fruit Grape Vine* 2(2), 10–12.
Zhou, T., Northover, J. and Schneider, K. (1998) Control of postharvest diseases of apple with saprophytic isolates of *Pseudomonas syringae*. *Canadian Journal of Plant Pathology* 20, 343.
Zhou, T., Northover, J. and Schneider, K.E. (1999) Biological control of postharvest diseases of peach with phyllosphere isolates of *Pseudomonas syringae*. *Canadian Journal of Plant Pathology* 21, 375–381.
Zhou, T., Schneider, K.E., Chu, C. and Liu, W.T. (2001) Postharvest control of blue mold and grey mold on apples using isolates of *Pseudomonas syringae*. *Canadian Journal of Plant Pathology* 23(3) (in press).

94 *Penicillium expansum* Link, Blue Mould of Apple (*Hyphomycetes*)

T. Zhou and P. Sholberg

Pest Status

Penicillium expansum Link causes blue mould, a destructive fruit rot of apple, *Malus pumila* Miller (= *M. domestica* Borkhausen), and occurs in most apple-growing areas of the world. Other names for this disease are soft rot and penicillium rot. In North America, blue mould is the most important postharvest disease of apples. *P. expansum* not only causes fruit decay, but also produces the carcinogenic mycotoxin patulin. This toxin may rise to unacceptable levels in fruit destined for processing. Generally, losses are 2–5%, depending on cultivar and length of storage for fruit kept in controlled atmosphere storage.

P. expansum is a common saprophyte that sporulates profusely. It is present almost everywhere and can survive long periods of unfavourable conditions. Bulk bins, packing lines and storage rooms are usually contaminated. The pathogen invades fruit mainly through wounds or bruises, but under favourable conditions it can also infect fruit through lenticels. Symptoms of blue mould appear as soft, light-brown, watery spots. When the relative humidity is high, conidia are produced on the spots in coremia that are initially white and then become blue–green, giving rise to the description 'blue mould'. Under favourable conditions, the entire fruit can rot in 2 weeks. During storage, *P. expansum* spreads by contact between infected and sound fruit.

Background

In commercial practice, thiabendazole (TBZ), a benzimidazole, and captan are the only fungicides registered for postharvest use on apple. Control of *P. expansum* relies on the use of TBZ as a drench treatment before cold storage and/or spray treatment on the packing line (Koffmann and Penrose, 1987). However, benzimidazole-resistant isolates are now present in most packing houses (Jones and Aldwinkle, 1990; Sholberg and Haag, 1996). Research in the mid-1980s showed that most of the benzimidazole-resistant strains of *P. expansum* were sensitive to the antioxidant chemical diphenylamine (DPA) used to control storage scald on apple. The combination of TBZ with DPA effectively controls both TBZ-sensitive and TBZ-resistant *P. expansum* in most apple storage situations (Rosenberger and Meyer, 1985). However, future use of DPA is under evaluation because of its possible undesirable biological degradation products that may be carcinogenic. Searches for alternative control strategies, particularly biological control, have increased greatly, due to the development of fungicide-resistant pathogens and public demand for fungicide-free produce. Currently, two biofungicides, BioSave110™ and Aspire™, have been registered in the USA for postharvest use on apple, but no biological product is available in Canada to control *P. expansum* of apple.

In the past decade, substantial progress has been made in finding alternatives to synthetic fungicides to control postharvest diseases of fruits. Several antagonistic microorganisms have been discovered to reduce postharvest fungal decay of apple and other pome fruits. Strains of *Pseudomonas syringae* van Hall are effective in controlling blue mould of citrus and pome fruit (Janisiewicz and Jeffers, 1997), and have been commercialized as Bio-Save® biofungicides. Other bacteria, e.g. *Burkholderia* (= *Pseudomonas*) *cepacia* (Palleroni and Holmes) Kabuuchi, Kosako, Oyaiza, Yano, Hotta, Hashimoto, Ezaki and Arakawa (Janisiewicz and Roitman, 1988), *Pseudomonas gladioli* Severini (Mao and Cappellini, 1989), *B. pumilus* Meyer and Gottheil, and *Bacillus amyloliquefaciens* (ex Fukumoto) Priest (Mari *et al.*, 1996) were reported to reduce blue mould and/or grey mould on apple or pear. The yeast *Candida oleophila* Montrocher (Aspire®) effectively controls blue mould on citrus and pome fruit (Wilson *et al.*, 1994; Lurie *et al.*, 1995). Other yeasts, e.g. *Cryptococcus laurentii* (Kufferath) C.E. Skinner, *Rhodotorula glutinis* (Fresen) Harrison (Chand-Goyal and Spotts, 1997), *Pichia anomala* (Hansen) Kurtzman and *Candida sake* (Saito and Ota) van Uden and Buckley (Jijakli *et al.*, 1993), effectively control *P. expansum* on apple.

Biological Control Agents

Bacteria

In British Columbia, *in vitro* tests showed that both *Enterobacter aerogenes* Hormaeche and Edwards and *Bacillus subtilis* (Ehrenberg) Cohn were effective biological control agents against *P. expansum* (Utkehede and Sholberg, 1986). Experiments conducted in 1990 on stored apples showed that *E. aerogenes, B. subtilis* and *P. syringae* prevented decay (Sholberg *et al.*, 1990). However, difficulties associated with the registration process discouraged efforts to register biological control agents for postharvest use. Interest in biological control was again revived when potential biological control organisms were discovered in the tissue of harvested apples (Sholberg *et al.*, 1995). The isolates, predominantly *B. subtilis*, were found to be effective. Several of the isolates reduced, by about half, the diameter of blue mould lesions in apples stored at 5°, 10° and 20°C when compared to the control. One *B. subtilis* isolate was effective against a wide range of fungi and bacteria, probably because it produced an antibiotic, recently characterized by Bechard *et al.* (1998).

In Ontario, microorganisms isolated from apple fruits and leaves collected from eastern Ontario were screened for apple scab control during the growing season and some of them, including isolates of *P. syringae* and *Candida* sp., suppressed apple scab by up to 70% (Zhou and

DeYoung, 1996). Two of the isolates, NSA-6 and MA-4, identified as non-pathogenic *P. syringae,* showed some effect in suppressing major postharvest diseases of peach, e.g. brown rot and rhizopus rot, caused by *Monilinia fructicola* (Winter) Honey (see Zhou and Sholberg, Chapter 93 this volume) and *Rhizopus stolonifer* (Ehrenberg: Fries) Vuillemin, respectively (Zhou *et al.*, 1999). These isolates also inhibited spore germination of *P. expansum* and *B. cinerea in vitro* (T. Zhou, unpublished) and were further developed as biological agents to control blue mould of apple.

Zhou *et al.* (1998) evaluated four isolates of *P. syringae* – MA-4, MB-4, MD-3b, and NSA-6 – as biological control agents. 'McIntosh' apples treated with individual isolates and incubated at 4°C showed significant reductions in the incidence of blue mould. A subsequent experiment to test various concentrations (10^5–10^8 colony-forming units (cfu) ml^{-1}) of the agents showed that while the incidence of blue mould in controls was 100%, it was 83, 69, 22 and 6% in treatments with isolate MA-4 at concentrations of 10^5, 10^6, 10^7 and 10^8 cfu ml^{-1}, respectively.

Zhou *et al.* (2001) evaluated spray treatments consisting of water suspensions of *P. syringae* MA-4, *P. expansum* or a mixture of the two suspensions. Application of *P. syringae* MA-4 greatly reduced blue mould of both 'Empire' and 'Red Delicious' apples inoculated with *P. expansum.* After incubation at 4°C for 42 days, incidence of blue mould in both apple varieties in the treatment with *P. syringae* MA-4 were 4.5% and 7.5%, respectively, significantly lower than 12% and 25%, respectively, in the corresponding water controls. For apples not inoculated with *P. expansum, P. syringae* MA-4 reduced blue mould of 'Red Delicious' apples to 5%, compared to 10.5% in the water control. However, statistically, *P. syringae* MA-4 did not reduce blue mould incidence on 'Empire' apples. When apples were incubated under 18°C, all treatments with *P. syringae* MA-4 had significantly lower incidence of blue mould compared to the water controls. On *P. expansum*-inoculated 'Empire' apples, *P. syringae* MA-4 reduced blue mould incidence to 10% after 12 days' incubation, significantly lower than 34% in the water control. Incidence of blue mould of non-inoculated 'Empire' apples was also significantly reduced by *P. syringae* MA-4 to 2%, compared to 16.5% in the water control. Because of the slow development of blue mould, 'Red Delicious' apples were incubated at 18°C for 20 days. By the end of the incubation period, blue mould incidence of inoculated 'Red Delicious' apples reached 20% in the water control, but only 10% in the treatment with *P. syringae* MA-4. For non-inoculated apples, blue mould incidence in the treatment of *P. syringae* MA-4 was reduced to 0.5%, compared to 7.5% in the water control (Zhou *et al.*, 2001).

In storage trials, 'Empire' and 'Red Delicious' apples artificially wounded and soaked in a suspension of *P. expansum* at a final concentration of 10^3 conidia ml^{-1} were treated as follows: (i) water (control); (ii) 450 µl active ingredient ml^{-1} of thiabendazole plus 1000 µl ml^{-1} diphenylamine; (iii) biofungicide BioSave1000 (freeze dried formula) at a concentration equivalent to 5×10^8 cfu ml^{-1}; and (iv) *P. syringae* isolate MA-4 at 5×10^8 cfu ml^{-1}. The treated apples were separated into two groups: one was incubated in a cold room at 1°C and the other in a controlled atmosphere room (1°C, 2.5% O_2 and 2.5% CO_2). Incidence of blue mould was determined after 4 months. On 'Red Delicious' apple, the treatments of BioSave and *P. syringae* MA-4 greatly reduced blue mould incidence, to 51% and 4%, respectively, compared to 88% in the control under cold storage, and to 25% and 4%, respectively, compared to 95% in the control under controlled atmosphere storage. There was no disease in the fungicide treatments. A very similar trend was found on 'Empire' apple. Treatment with BioSave and MA-4 reduced blue mould incidence, under cold storage, to 10% and 2%, respectively, compared to 38% in the control and, under controlled atmosphere storage, to 45% and 9%, respectively, compared to 69% in the control. The reduction by the isolate MA-4 was similar to the fungicide treatments, which reduced blue mould incidence to 0% and 9% in cold storage and controlled atmosphere storage, respectively.

Evaluation of Biological Control

Research on biological control of postharvest pathogens of apple continues to show promise. Isolates of several species of microorganisms are effective in controlling blue mould and other postharvest diseases of apple under controlled conditions.

Recommendations

Further work should include:

1. Comparing and evaluating the most promising isolates on a commercial scale;
2. Facilitating registration of these biological control agents.

Acknowledgements

The authors would like to thank O.L. (Sam) Lau, Okanagan Federated Shippers Association, and C.L. Chu, University of Guelph, for providing fruit, storage facilities and other resources for use in conducting postharvest apple trials.

References

Bechard, J., Eastwell, K.C., Sholberg, P.L., Mazza, G. and Skura, B. (1998). Isolation and partial chemical characterization of an antimicrobial peptide produced by a strain of *Bacillus subtilis*. *Journal of Agricultural Food Chemistry* 46, 5355–5361.

Chand-Goyal, T. and Spotts, R.A. (1997) Biological control of postharvest diseases of apple and pear under semi-commercial conditions using three saprophytic yeasts. *Biological Control* 10,199–206.

Janisiewicz, W.J. and Jeffers, S.N. (1997) Efficacy of commercial formulation of two biofungicides for control of blue mold and gray mold of apples in cold storage. *Crop Protection* 16, 629–633.

Janisiewicz, W.J. and Roitman, J. (1988) Biological control of blue mold and gray mold on apple and pear with *Pseudomonas cepacia*. *Phytopathology* 78, 1697–1700.

Jijakli, M., Lepoivre, H., Tossut, P. and Thonard, P. (1993) Biological control of *Botrytis cinerea* and *Penicillium* sp. on postharvest apples by two antagonistic yeasts. *Mededelingen van de Faculteit Landbouwwetenschappen Universiteit Gent* 58, 1349–1358.

Jones, A. and Aldwinckle, H. (1990) *Compendium of Apple and Pear Diseases*. APS Press, St Paul, Minnesota.

Koffmann, W. and Penrose, L.J. (1987) Fungicides for the control of blue mold (*Penicillium* spp.) in pome fruits. *Scientia Horticulturae* 31, 225–232.

Lurie, S., Droby, S., Chalupowicz, L. and Chalutz, E. (1995) Efficacy of *Candida oleophila* strain 182 in preventing *Penicillium expansum* infection of nectarine fruits. *Phytoparasitica* 23, 231–234.

Mao, G.H. and Cappellina, R.A. (1989) Postharvest biocontrol of gray mold of pear by *Pseudomonas gladioli*. *Plant Pathology* 79, 1153.

Mari, M., Lori, R., Leoni, O. and Marchi, A. (1996) Bioassays of glucoinolate-derived isothiocyanates against postharvest pear pathogens. *Plant Pathology* 45, 753–760.

Rosenberger, D.A. and Meyer, F.W. (1985) Negatively correlated cross-resistance to diphenylamine in benomyl-resistant *Penicillium expansum*. *Phytopathology* 75, 74–79.

Sholberg, P.L. and Haag, P.D. (1996) Incidence of postharvest pathogens of stored apples in British Columbia, BC, Canada. *Canadian Journal of Plant Pathology* 18, 81–85.

Sholberg, P.L., Haag, P. and Utkhede, R.S. (1990) Use of bacteria to control postharvest diseases of stored apples. In: Utkhede, R.S. (ed.) *Research Highlights, 1990*. Agriculture Canada, Summerland, British Columbia.

Sholberg, P.L., Marchi, A. and Bechard, J. (1995) Biocontrol of postharvest diseases of apple using *Bacillus* spp. isolated from stored apples. *Canadian Journal of Microbiology* 41, 247–252.

Utkhede, R.S. and Sholberg, P.L. (1986) *In vitro* inhibition of plant pathogens by *Bacillus subtilis* and

Enterobacter aerogenes and *in vivo* control of postharvest cherry diseases. *Canadian Journal of Microbiology* 32, 963–967.
Wilson, C.L., El Ghaouth, E., Droby, S., Stevens, C., Lu, J.Y., Khan, V. and Arul, J. (1994) Potential on induced resistance to control postharvest diseases of fruits and vegetables. *Plant Disease* 78, 837–844.
Zhou, T. and DeYoung, R. (1996) Control of apple scab with applications of phyllosphere micro-organisms. In: Tang, W., Cook, R.J. and Rovira, A. (eds) *Advances in Biocontrol of Plant Diseases.* Beijing China Agricultural University Press, Beijing, China, pp. 369–399.
Zhou, T., Northover, J. and Schneider, K. (1998) Control of postharvest diseases of apple with saprophytic isolates of *Pseudomonas syringae. Canadian Journal of Plant Pathology* 20, 343.
Zhou, T., Northover, J. and Schneider, K.E. (1999) Biological control of postharvest diseases of peach with phyllosphere isolates of *Pseudomonas syringae. Canadian Journal of Plant Pathology* 21, 375–381.
Zhou, T., Schneider, K.E., Chu, C. and Liu, W.T. (2001). Postharvest control of blue mold and grey mold on apples using isolates of *Pseudomonas syringae. Canadian Journal of Plant Pathology* 23(3) (in press).

95 *Phytophthora cactorum* (Lebert and Cohn) Schröter, Crown and Root Rot (*Pythiaceae*)

R.S. Utkhede

Pest Status

Phytophthora cactorum (Lebert and Cohn) Schröter is the causal agent of crown and root rot, a serious disease of apple trees, *Malus pumila* Miller (= *M. domestica* Borkhausen), worldwide. It may also affect cherry, *Prunus avium* L., peach, *Prunus persica* (L.) Batsch, plum, *Prunus spinosa* L., and apricot, *Prunus armeniaca* L., trees. The Commonwealth Mycological Institute prepared a world distribution map of *P. cactorum* (Anonymous, 1965). In North America, the disease was first reported as early as 1858 when dying apple trees were discovered in Michigan (Baines, 1939). In Canada, the disease was first reported in 1928 in the Okanagan Valley, British Columbia, and about Can$2 million per year is lost due to it. Losses have been reported on all ages of fruit trees of the major species.

Phytophthora crown and root rot often results in the death of affected trees. About 3% of trees are affected by *P. cactorum* in any orchard in the Okanagan valley. Blackwell (1943) reviewed the life history of *P. cactorum*. The first visible sign of an infected apple tree is usually foliar chlorosis followed by purplish-red colour of leaves in late summer and autumn. Infection of apple trees by *P. cactorum* occurs at the root crown, with invasion extending distally along the main roots. It takes about 2–3 years before the tree dies.

Background

Chemical pesticides such as metalaxyl and fosetyl Al are registered to control crown and root rot of apple trees. Because pesticide safety, ground water contamination and sustainable agriculture are currently important public concerns, biological control products need to be developed. Moreover, the prospects for biological control have never been better, as recent research on plant–microbe interactions and biotechnology is showing real potential for new and effective approaches.

Biological Control Agents

Bacteria

Enterobacter agglomerans (Beijernck) Ewing and Fife, strain B8,[1] was isolated from soil in an Okanagan Valley orchard. It was shown to be antagonistic to *P. cactorum* on cornmeal agar, producing an antibiotic inhibitory to mycelial growth (Utkhede, 1983). Neither the growth of *E. agglomerans* nor its antagonistic effect on *P. cactorum* were affected by any of the six herbicides tested (Utkhede, 1982), which suggested that herbicides may not be a limiting factor on the use of a bacterial antagonist for biological control of *P. cactorum.* Under greenhouse conditions, *E. agglomerans* significantly reduced infections of apple seedlings caused by three isolates of *P. cactorum* in sterile field soil (Utkhede, 1984a). *E. agglomerans* also significantly reduced the population of viable *P. cactorum* oospores in the top 30 mm of soil where oospores generally survive. Complete inhibition of *P. cactorum* growth was observed with 40% concentration of autoclaved *E. agglomerans* extract (Utkhede and Gaunce, 1983). The growth was significantly reduced by low pH alone (4.5 or less) but even when the pH of *E. agglomerans* extract was raised to 6, *P. cactorum* growth was completely inhibited.

In a short-term orchard trial *E. agglomerans,* applied as a soil drench, significantly reduced the percentage of crown rot infection (Utkhede, 1987). In a long-term orchard trial, biological control of *P. cactorum* was achieved by application of *E. agglomerans* (strain B8) as soil and trunk drenches (Utkhede and Smith, 1991). Growth and antagonistic ability of *E. agglomerans* were not significantly affected over a 4-week period on cornmeal agar containing 50 or 100 mg l^{-1} of metalaxyl, fosetyl-AL or mancozeb. This suggested that it may be possible to use this bacterial isolate together with chemical fungicides to control crown and root rot of apple trees. Metalaxyl, alternated with *E. agglomerans,* significantly reduced disease incidence and increased fruit yield under orchard conditions (Utkhede and Smith, 1993). Strain B8 of *E. agglomerans* and its method of application were patented (Patent No. 1,316,856) in Canada.

A powder formulation of *E. agglomerans* (developed by Lallemond Inc., 15130, Saint-Simon, France) was applied in spring and autumn over a 3-year period as soil and trunk drenches, at the rate of 1×10^{10} colony-forming units per tree, to control *P. cactorum* at two locations in the Okanagan Valley (Utkhede and Smith, 1997). This significantly reduced disease severity and increased trunk cross-sectional area and fruit yield of Macspur trees on MM106 rootstock when compared with the untreated control. Genetic transformation of *E. agglomerans* with salicylate-utilizing gene was achieved to improve its biological control activity under orchard conditions (Utkhede *et al.*, 2000). This biological control agent is not yet registered for use by growers in Canada.

Attempts were also made to identify additional biological control agents. Twenty-one isolates of *Bacillus subtilis*

[1]This biological agent was identified by Dr J.F. Bradbury, Commonwealth Mycological Institute, Kew, Surrey, England, in 1985 as *Enterobacter aerogenes* (Kruse) Hormaeche and Edwards. In 1993, the strain was re-identified as *Enterobacter agglomerans* by Microbial ID, Inc., Burksdale Professional Centre, Newark, Delaware, USA, based on fatty acid analysis.

(Ehrenberg) Cohn were antagonistic to growth of six *P. cactorum* isolates on cornmeal agar (Utkhede, 1984b). Six bacterial antagonists – AB6, AB9, AB3, EBW3, EBW1 and BACT-X – provided significant reductions of infection with *P. cactorum* on 'McIntosh' apple seedlings under greenhouse conditions.

Evaluation of Biological Control

Strain B8 of *E. agglomerans* increased tree growth and fruit production, and reduced root and crown rot of apple trees caused by *P. cactorum*. Strain B8 has potential as a biological control agent of *P. cactorum*, particularly among organic apple growers in the Okanagan Valley.

The success of this biological control agent, like others, will also depend on lower consumer quality criteria, e.g. visual perfection of agricultural and horticultural products, the grower accepting lower disease control, the registration requirements for biological agents clearly defined and their cost not prohibitively expensive, realistic registration requirements for biological agents (different from pesticides), and appropriate legislation to implement integrated disease management practices.

Recommendations

Future work should include:

1. Registration of *E. agglomerans* as a biological control agent against *P. cactorum*;
2. Finding a commercial partner to manufacture the biological product.

References

Anonymous (1965) *Distribution Map of Plant Diseases*, Map No. 280. Edition 2, Phytophthora cactorum. Commonwealth Mycological Institute.

Baines, R.C. (1939) Phytophthora trunk canker or collar rot of apple trees. *Journal of Agricultural Research* 59, 159–184.

Blackwell, E. (1943) The life history of *Phytophthora cactorum* (Leb. & Cohn) Schroet. *Transactions of the British Mycological Society* 26, 71–89.

Utkhede, R.S. (1982) Effects of six herbicides on the growth of *Phytophthora cactorum* and a bacterial antagonist. *Pesticide Science* 13, 693–695.

Utkhede, R.S. (1983) Inhibition of *Phytophthora cactorum* by bacterial isolates and effects of chemical fungicides on their growth and antagonism. *Zeitschrift für Pflanzenkrankheiten und Pflazenschutz* 90, 140–145.

Utkhede, R.S. (1984a) Effect of bacterial antagonist on *Phytophthora cactorum* and apple crown rot. *Journal of Phytopathology* 109, 169–175.

Utkhede, R.S. (1984b) Antagonism of isolates of *Bacillus subtilis* to *Phytophthora cactorum*. *Canadian Journal of Botany* 62, 1032–1035.

Utkhede, R.S. (1987) Chemical and biological control of crown and root rot of apple caused by *Phytophthora cactorum*. *Canadian Journal of Plant Pathology* 4, 295–300.

Utkhede, R.S. and Gaunce, A.P. (1983) Inhibition of *Phytophthora cactorum* by a bacterial antagonist. *Canadian Journal of Botany* 61, 3343–3348.

Utkhede, R.S. and Smith, E.M. (1991) Biological and chemical treatments for control of *Phytophthora cactorum* in a high density apple orchard. *Canadian Journal of Plant Pathology* 13, 267–270.

Utkhede, R.S. and Smith, E.M. (1993) Long-term effects of chemical and biological treatments on crown and root rot of apple trees caused by *Phytophthora cactorum*. *Soil Biology and Biochemistry* 25, 383–386.

Utkhede, R.S. and Smith, E.M. (1997) Effectiveness of dry formulations of *Enterobacter agglomerans* for control of crown and root rot of apple trees. *Canadian Journal of Plant Pathology* 19, 397–401.

Utkhede, R., Nie, J., Xu, H., Eastwell, K. and Wiersma, P. (2000) Transformation of biocontrol agent *Enterobacter agglomerans* with salicylate utilizing gene and its monitoring in orchard soil. *Journal of Horticultural Science and Biotechnology* 75, 50–54.

96 *Pythium* spp., Damping-off, Root and Crown Rot (*Pythiaceae*)

T.C. Paulitz, H.C. Huang and J.A. Gracia-Garza

Pest Status

Pythium spp. are the causal agents of damping-off in seedlings and root and crown rot, important worldwide diseases of field and greenhouse crops, vegetables and turfgrass. *Pythium* spp. have a wide host range, attacking almost all greenhouse crops. The disease is especially devastating in highly susceptible young plants in greenhouses, because growing conditions, e.g. high densities and peat-based planting media lacking the normal biological buffering of soil, make it easy for *Pythium* spp. to spread and colonize. Damping-off is also one of the major factors limiting production of field crops in western Canada. *Pythium* spp. isolated from crops in the prairies include *P. debaryanum* Heese, *P. hypogynum* Middleton, *P. irregulare* Buisman, *P. paroecandrum* Drechsler, *P. salpingophorum* Drechsler, *P. sylvaticum* Campbell and Hendrik, *P. torulosum* Trow, *P. ultimum* Trow, and *Pythium* sp. 'group G'.[1] Non-fruiting strains of *Pythium* occur on various hosts in southern and central Alberta (Cormack, 1951; Stelfox and Williams, 1980; Huang *et al.*, 1992; Hou *et al.*, 1997). The major hosts of *Pythium* sp. 'group G' are safflower, *Carthamus tinctorius* L., canola, *Brassica napus* L. and *B. rapa* L., dry field pea, *Pisum sativum* var. *arvense* (L.), sugar beet, *Beta vulgaris* L., lettuce, *Lactuca sativa* L., cucumber, *Cucumis sativus* L., muskmelon, *Cucumis melo* L. var. *reticulatus* Naudin, spinach, *Spinacia oleracea* L., marigold, *Tagetes* spp., tomato, *Lycopersicon esculentum* Miller, carrot, *Daucus carota sativus* (Hoffman) Arcangeli, sunflower, *Helianthus annuus* L. (Huang *et al.* 1992), cicer milkvetch, *Astragulus cicer* L. (Hou *et al.*, 1997) and lucerne, *Medicago sativa* L. (Stelfox and Williams, 1980; Hou *et al.*, 1997). *Pythium irregulare*, a pathogenic species on cicer milkvetch and lucerne in southern Alberta (Hou *et al.*, 1997), was also found on lucerne in eastern Ontario (Basu, 1983).

The disease can be severe on canola in the Peace River Region, Alberta (Harrison, 1989), and on sugarbeet and safflower in southern Alberta, resulting in thin stands of these crops. In Alberta, field incidence of damping-off reaches 99% in canola (Harrison, 1989; Turkington and Harrison,

[1] *Pythium* sp. 'group G' is an asexual form incapable of producing oogonia and antheridia in culture and was proven to be a divergent form of *Pythium ultimum* (Huang *et al.*, 1992).

1994), 83% in cicer milkvetch (Hou *et al.*, 1997), 30% in cucumber (Chang *et al.*, 1994) and 82% in safflower (Howard *et al.*, 1990; Huang *et al.*, 1992; Muendel *et al.*, 1995). Pythium damping-off and root rot is also prevalent on lucerne (Stelfox and Williams, 1980), processing peas (Sumar *et al.*, 1982) and dry field peas (Howard *et al.*, 1995). In central Saskatchewan, Pythium root rot was widespread on dry field peas (Hwang and Chakravarty, 1993).

In greenhouses, *Pythium* spp. are among the most important root and seedling pathogens, on both vegetables and horticulture crops. In British Columbia, *P. aphanidermatum* (Edson) Fitzpatrick, *P. irregulare* and *Pythium* sp. 'group G' were responsible for root disease and crown rot of greenhouse cucumbers (Favrin *et al.*, 1988). In Quebec, *P. aphanidermatum* and *P. ultimum* were the most commonly isolated species from greenhouse cucumber (Paulitz *et al.*, 1992). In addition, recirculating hydroponic systems such as rockwool, ebb and flow and nutrient film are especially susceptible to the introduction and spread of *Pythium* spp. via zoospores in the water (Paulitz, 1997). Under greenhouse conditions, Pythium damping-off is a potential problem because disease incidence may reach 95–100%. Given that, in 1998, the total value of greenhouse sales was Can$1.19 billion and vegetables were valued at Can$285 million (Statistics Canada, 1998), losses due to Pythium damping-off may be considerable.

In field-grown vegetable crops, *Pythium* spp. cause root rot and damping-off in carrot, beet, crucifers, cucurbits, lettuce, sweet corn, *Zea mays* L., pea, bean, *Phaseolus vulgaris* L., tomato, aubergine, *Solanum melongena* L. var. *esculentum* Nees, and pepper, *Capsicum annuum* L. (Howard *et al.*, 1994) and postharvest rots (leaks) in cucurbits and potato, *Solanum tuberosum* L. On turfgrass, *Pythium* spp. cause a summer blight or patch disease (Couch, 1995) and cool season dieback (Hsiang *et al.*, 1995). In 1998, 113,720 ha were planted with vegetables, for a total value of Can$513 million dollars (Statistics Canada, 1999).

Martin and Loper (1999) reviewed the biology, ecology and epidemiology of *Pythium* spp., which function in a similar way to other plant pathogenic soil-borne fungi. Temperature and soil moisture are important factors affecting the outbreak of Pythium damping-off. For example, in southern Alberta high soil moisture (near field capacity) and high temperature (>10°C) are conducive factors for Pythium damping-off in safflower (Muendel *et al.*, 1995). *Pythium* spp. can survive in the soil as thick-walled sexual resting spores called oospores. Oospores can remain dormant in soil and germinate to form hyphae or sporangia, thin-walled structures that asexually give rise to motile flagellated zoospores. These 'swimming' spores are chemotactically attracted to plant exudates from roots or seeds, attach to the plant, encyst by forming a cell wall around the spore, and infect the plant via a germ tube. Young plant tissues such as radicles and hypocotyls of seedlings and root tips are especially vulnerable. Susceptibility to damping-off (seedling rot) generally decreases with age. A film of water around soil particles is required for the production and dispersal of zoospores. Therefore, disease is more severe in wet, poorly drained soils. *Pythium* is more tolerant of higher CO_2 and low O_2 than other soil microbes. Damping-off caused by *P. ultimum* is more severe at cool soil temperatures (15–20°C), whereas that caused by *P. aphanidermatum* is more severe at high temperatures (above 25°C).

Background

In field crops only one fungicide seed treatment, Thiram 75 WP, is registered to control damping-off in sugarbeet, mustard, *Brassica* spp., grasses, bean, pea, soybean, *Glycine max* (L.) Merrill, corn and safflower (Anonymous, 1999). Other seed treatment fungicides, e.g. Apron (metalaxyl), are registered to control seed rots and seedling blights of alfalfa, clover, *Trifolium* spp., birdsfoot trefoil, *Lotus corniculatus* L., canola, pea, bean and sugarbeet, caused by *Pythium* spp. (Anonymous, 1999). However, the use of chemical fungicides has become an important environmental issue.

In greenhouses, ornamental crops are treated with etridiazol (Truban 25ED or Truban 30WP). Metalaxyl (Subdue 2G) is a systemic granular fungicide that can also be used. Until recently, no fungicides were registered to control *Pythium* on greenhouse vegetable crops, but propamocarb hydrochloride (Previcur N) has received a minor use registration in Canada (PCP#26288).

There are no disease-resistant crop cultivars or varieties. Instead, cultural strategies are used (Menzies and Bélanger, 1996; Paulitz, 1997). Sanitation prevents introduction and spread of the pathogen. If soil is used, it must be sterilized. Most soil-less substrates, e.g. peat and rockwool, usually do not contain the pathogen. Accidental introduction of *Pythium* spp. into recirculating hydroponic systems can be devastating. Treatment of recirculating water with UV, heat or ozone to kill inoculum is used extensively in the UK and Europe, and growers in Ontario are testing some of these systems. Filtration of hydroponic solutions with membranes or slow sand filtration is another way of reducing the inoculum load in hydroponic systems.

In vegetable crops, seeds are routinely treated with captan in addition to the fungicides used for field crops. Cultural management includes tillage methods that reduce soil compaction and planting seeds in well-drained soil when soil temperature is optimum for germination.

Biological Control Agents

Bacteria

Liang *et al.* (1996) tested 665 strains of rhizosphere bacteria isolated from plant roots collected in Alberta and found 23 that were antagonistic to *Pythium* sp. 'group G'. Fifteen of these were identified to species and were tested for efficacy as seed treatments to control damping-off of safflower in soil naturally infested with the pathogen. Two strains of *Erwinia carotovora* (Jones) Bergey, Harrison, Breed, Hammer and Huntoon, one strain of *Pantoea agglomerans* (Beijerinck) Gavini (= *E. herbicola* (Lohnis) Dye), four strains of *E. rhapontici* (Millard) Burkholder, one strain of *Pseudomonas putida* (Trevisan) Migula, and three strains of *Pseudomonas fluorescens* Migula significantly ($P < 0.05$) reduced pre-emergence damping-off and increased seedling emergence of safflower. In addition, treatment of safflower seeds with *P. agglomerans* and *P. fluorescens* also resulted in a significant increase in seedling height.

Some of the selected strains have been tested to control damping-off of safflower, canola, dry field pea and sugarbeet in fields naturally infested with *Pythium* spp., predominantly *Pythium* sp. 'group G'. These preliminary trials indicated that seed treatment with indigenous strains, e.g. *P. agglomerans, E. rhapontici* and *P. fluorescens,* effectively reduced incidence of Pythium damping-off and thereby increased seedling emergence (H.C. Huang *et al.*, unpublished).

In greenhouse cucumber, Paulitz *et al.* (1992) screened bacteria against zoospores of *P. aphanidermatum*. From over 600 bacteria isolated from the rhizosphere of cucumbers grown in different soils from Quebec, two isolates of *Pseudomonas corrugata* Roberts and Scarlett and three isolates of *P. fluorescens* were selected and tested under simulated commercial conditions in rockwool inoculated or not with *P. aphanidermatum* (Rankin and Paulitz, 1994). Two of these isolates increased fruit production under inoculated and non-inoculated conditions. *P. fluorescens* isolates 63–49 and 63–28 (developed by Agrium Inc., Saskatoon, Saskatchewan) were tested in Quebec and British Columbia under simulated commercial conditions, and increases in yields up to 18% under inoculated conditions were obtained (McCullagh *et al.*, 1996). Isolate 63–28 has also been tested on tomato and it increased fruit yield and fruit weight by 13% and 18%, respectively (Gagné *et al.*,

1993). Investigation into the mechanisms of these bacteria suggested that they interfered with the germination, attraction and distribution of encysted zoospores on the rhizoplane of cucumber roots (Zhou and Paulitz, 1993). Experiments with split roots suggested that the bacteria could induce a systemic resistance throughout the root system of cucumber (Zhou and Paulitz, 1994). Chen *et al.* (1998) confirmed the mechanism of induced resistance with *P. corrugata* 13 and *P. fluorescens* 63–28 (later identified as *P. aureofaciens* Kluyver). Inoculation of cucumber roots with either isolate resulted in elevated levels of salicylic acid, which is involved in the systemic signalling process (Chen *et al.*, 1999). Chen *et al.* (2000) detected elevated levels of phenylalanine ammonia lyase (PAL), peroxidase (PO) and polyphenoloxidase (PPO), enzymes involved in defence reactions, in roots treated with these bacteria. Gamard *et al.* (1997) and Paulitz *et al.* (2000) found that *P. aureofaciens* isolate 63–28 also produced three unique furanone or butyrolactone antibiotics with activity against *Pythium*, *Phytophthora* and *Rhizoctonia* spp. Benhamou *et al.* (1996) demonstrated the antifungal activity of the bacteria against *P. ultimum* in pea roots.

Fungi

In field crops, several indigenous species of fungi antagonistic to soil-borne pathogens were tested for control of *Pythium* spp. Preliminary results showed that seed treatment with *Trichoderma viride* Persoon: Fries, *Trichoderma harzianum* Rifai, *Talaromyces flavus* (Klöcker) Stolk and Samson, and *Penicillium aurantiogriseum* Dierckx were effective in reducing Pythium damping-off of sugarbeet under controlled environments (H.C. Huang *et al.*, unpublished).

In greenhouse crops, several fungal biological control agents are commercially available worldwide for use against *Pythium* spp., including *T. harzianum* (RootShield®), *Gliocladium virens* Miller, Giddens and Foster (SoilGard®) and *Streptomyces griseoviridis* (Krainsky) Waksman and Henrici (Mycostop®). These are currently being tested on poinsettia and other floricultural crops in recirculating systems (J. Gracia-Garza, unpublished).

Evaluation of Biological Control

The use of microbial seed treatment to control damping-off appears feasible for field crops such as sugarbeet, pulses, oilseeds, forages and perhaps vegetables in the prairies. However, the effectiveness of disease control varies with species and strains of microorganism.

Biological control treatments are well suited for greenhouse crops, where there is a lack of biological buffering in the near-sterile substrates, where the environment can be controlled to favour the biological control agent, where the economic value of the crop is high, and where there is a lack of registered fungicides because of the small potential market. While worldwide, six *Trichoderma* and two *Gliocladium* products have become available in the past 5 years, some *Pseudomonas* strains, developed by Canadian companies and universities, have not been commercialized. Current research on bacterial biological control agents focuses on improving seed-treatment techniques, shelf-life, and ecological studies on interactions between agents and other natural populations of microorganisms in soil.

Recommendations

Further work should include:

1. Selecting indigenous strains that are not only effective but also adapted to prairie conditions for use as seed-treatment agents;
2. Improving seed-treatment techniques for maintaining efficacy and shelf-life of biological control agents;
3. Understanding mechanisms of competition between biological control agents and other natural microbial populations under field conditions;
4. Developing organic soil amendments

that may further improve efficacy and consistency of biological control agents for protection of field crops;
5. Testing available commercial products under Canadian conditions;
6. Testing other isolates that may not have the potential to be commercialized but could be further developed for small markets.

References

Anonymous (1999) Fungicides. In: *Crop Protection 1999. AGDEX 606–1.* Alberta Agriculture, Food and Rural Development, Edmonton, Alberta, pp. 334–376.

Basu, P.K. (1983) Survey of eastern Ontario alfalfa fields to determine common fungal diseases and predominant soil-borne species of *Pythium. Canadian Plant Disease Survey* 63, 51.

Benhamou, N., Bélanger, R. and Paulitz, T. (1996) Pre-inoculation of Ri T-DNA transformed pea roots with *Pseudomonas fluorescens* inhibits colonization by *Pythium ultimum* Trow: an ultrastructural and cytochemical study. *Planta* 199, 105–117.

Chang, K.F., Chen, W., Choban, B. and Mirza, M. (1994) Pythium root rot of field grown cucumbers in central Alberta in 1994. *Canadian Plant Disease Survey* 74, 111.

Chen, C., Bélanger, R.R., Benhamou, N. and Paulitz, T.C. (1998) Induced systemic resistance (ISR) by *Pseudomonas* spp. impairs pre- and post-infection development of *Pythium aphanidermatum* on cucumber roots. *European Journal of Plant Pathology* 104, 877–886.

Chen, C., Bélanger, R., Benhamou, N. and Paulitz, T. (1999) Role of salicylic acid in systemic resistance induced by *Pseudomonas* spp. against *Pythium aphanidermatum* in cucumber roots. *European Journal of Plant Pathology* 105, 477–486.

Chen, C., Bélanger, R., Benhamou, N. and Paulitz, T. (2000) Defense enzymes induced in cucumber roots by treatment with plant growth-promoting rhizobacteria (PGPR) and *Pythium aphanidermatum. Physiological and Molecular Plant Pathology* 56, 13–23.

Cormack, M.W. (1951) Root rot or wilt of safflower. In: Conners, I.L. and Savile, D.B.O. (compilers) *30th Annual Report of Canadian Plant Disease Survey 1950.* Canada Department of Agriculture, Science Service, Division of Botany and Plant Pathology, Ottawa, Ontario.

Couch, H.B. (1995) *Diseases of Turfgrass*, 3rd edn. Krieger Publishing, Malabar, Florida.

Favrin, R.J., Rahe, J.E. and Mauza, B. (1988) *Pythium* spp. associated with crown rot of cucumbers in British Columbia greenhouses. *Plant Disease* 72, 683–687.

Gagné, S., Dehbi, L., Le Quéré, D., Cayer, F., Morin, J.-L., Lemay, R. and Fournier, N. (1993) Increase of greenhouse tomato fruit yields by plant growth-promoting rhizobacteria (PGPR) inoculated into the peat-based growing media. *Soil Biology and Biochemistry* 25, 269–272.

Gamard, P., Sauriol, F., Benhamou, N., Bélanger, R. and Paulitz, T. (1997) Novel butyrolactones with antifungal activity produced by *Pseudomonas aureofaciens* strain 63–28. *Journal of Antibiotics* 50, 742–749.

Harrison, L.M. (1989) Canola disease survey in the Peace River region in 1988. *Canadian Plant Disease Survey* 69, 59.

Hou, T.J., Huang, H.C. and Acharya, S.N. (1997) A preliminary study on damping-off of cicer milkvetch in southern Alberta. *Acta Prataculturae Sinica* 6, 47–50.

Howard, R.J., Moskaluk, E.R. and Sims, S.M. (1990) Survey for seedling blight of safflower. *Canadian Plant Disease Survey* 70, 82.

Howard, R.J., Garland, J.A. and Seaman, W.L. (eds) (1994) *Diseases and Pests of Vegetable Crops in Canada.* Canadian Phytopathology Society and Entomological Society of Canada, Ottawa, Ontario.

Howard, R.J., Briant, M.A. and Sims, S.M. (1995) Pea root rot survey in southern Alberta in 1994. *Canadian Plant Disease Survey* 75, 153–154.

Hsiang, T., Wu, C., Yang, L. and Liu, L. (1995) Pythium root rot associated with cool-season dieback of turfgrass in Ontario and Quebec. *Canadian Plant Disease Survey* 75, 191–195.

Huang, H.C., Morrison, R.J., Muendel, H.-H., Barr, D.J.S., Klassen, G.R. and Buchko, J. (1992) *Pythium* sp. 'group G', a form of *Pythium ultimum* causing damping-off of safflower. *Canadian Journal of Plant Pathology* 14, 229–232.

Hwang, S.F. and Chakravarty, P. (1993) Root rot disease complex of field pea in central Saskatchewan in 1990. *Canadian Plant Disease Survey* 73, 98–99.

Liang, X.Y., Huang, H.C., Yanke, L.J. and Kozub, G.C. (1996) Control of damping-off of safflower by bacterial seed treatment. *Canadian Journal of Plant Pathology* 18, 43–49.

Martin, F.N. and Loper, J.E. (1999) Soilborne plant diseases caused by *Pythium* spp.: ecology, epidemiology, and prospects for biological control. *Critical Reviews in Plant Sciences* 18, 111–181.

McCullagh, M., Utkhede, R., Menzies, J., Punja, Z. and Paulitz, T.C. (1996) Evaluation of plant growth-promoting rhizobacteria for biological control of Pythium root rot of cucumber grown in rockwool and effects on yield. *European Journal of Plant Pathology* 102, 747–755.

Menzies, J.G. and Bélanger, R.R. (1996) Recent advances in cultural management of diseases of greenhouse crops. *Canadian Journal of Plant Pathology* 18, 186–193.

Muendel, H.-H., Huang, H.C., Kozub, G.C. and Barr, D.J.S. (1995) Effect of soil moisture and temperature on seedling emergence and incidence of *Pythium* damping-off in safflower. *Canadian Journal of Plant Science* 75, 505–509.

Paulitz, T.C. (1997) Biological control of root pathogens in soilless and hydroponic systems. *HortScience* 32, 193–196.

Paulitz, T.C., Zhou, T. and Rankin, L. (1992) Selection of rhizosphere bacteria for biological control of *Pythium aphanidermatum* on hydroponically grown cucumber. *Biological Control* 2, 226–237.

Paulitz, T.C., Nowak-Thompson, B., Gamard, P., Tsang, E. and Loper, J. (2000) A novel antifungal furanone from *Pseudomonas aureofaciens*, a biocontrol agent of fungal plant pathogens. *Journal of Chemical Ecology* 26, 1515–1524.

Rankin, L. and Paulitz, T.C. (1994) Evaluation of rhizosphere bacteria for biological control of Pythium root rot of greenhouse cucumbers in hydroponic culture. *Plant Disease* 78, 447–451.

Statistics Canada (1998) *Greenhouse, Sod and Nursery Industries*. Catalogue no. 22–202-XIB, pp. 14–15.

Statistics Canada (1999) *Fruit and Vegetable Production*. Catalogue no. 22-003SXIB.

Stelfox, D. and Williams, J.R. (1980) *Pythium* species in alfalfa fields in central Alberta. *Canadian Plant Disease Survey* 60, 35.

Sumar, S.P., Mohyuddin, M. and Howard, R.J. (1982) Diseases of pulse crops in Alberta, 1978–79. *Canadian Plant Disease Survey* 62, 33–38.

Turkington, T.K. and Harrison, L.M. (1994) Survey of canola diseases in the Peace River region of Alberta, 1993. *Canadian Plant Disease Survey* 74, 94–95.

Zhou, T. and Paulitz, T.C. (1993) *In vitro* and *in vivo* effects of *Pseudomonas* spp. on *Pythium aphanidermatum*: Zoospore behavior in exudates and on the rhizoplane of bacteria-treated cucumber roots. *Phytopathology* 83, 872–876.

Zhou, T. and Paulitz, T.C. (1994) Induced resistance in the biological control of *Pythium aphanidermatum* by *Pseudomonas* spp. on European cucumber. *Journal of Phytopathology* 142, 51–63.

97 *Rhizoctonia solani* Kühn, Damping-off and Seedling Blight (*Hyphomycetes*)

J.A. Traquair, H.C. Huang, S.M. Boyetchko and S. Jabaji-Hare

Pest Status

Rhizoctonia solani Kühn causes seedling damping-off and blight, root rot, leaf spot, stem rot and black scurf or stem canker in a wide range of field crops, vegetables and ornamentals throughout Canada and worldwide (Martens *et al.*, 1984; Ginns, 1986; Howard *et al.*, 1994; Turkington and Harrison, 1994). Root rot occurs on pre-emergent seedlings, whereas damping-off occurs on post-emergent seedlings and often as a leaf spot and girdling stem (crown) rot in older canola, *Brassica napus* L. and *B. rapa* L., seedlings, and green-house-grown tomato, *Lycopersicon esculentum* Miller, and cabbage, *Brassica oleracea* L., transplants (Martens *et al.*, 1984; Tewari, 1985; Howard *et al.*, 1994).

In Ontario, Rhizoctonia damping-off and root rot are major diseases (10–50% incidence in plug trays) affecting the production and marketability of tomato plug transplants grown in greenhouses (Howard *et al.*, 1994), leading to reduced stand establishment in both greenhouse and field and problems for mechanical planting systems. In field seeding, *Rhizoctonia* can significantly reduce emergence and, in severe cases, can cause complete seedling mortality. Sippell *et al.* (1985) reported yield losses in canola of 23–36%. In Quebec, annual losses caused by black scurf and canker of potato amount to Can$4 million (Banville, 1989).

Rhizoctonia solani exists mainly as the sterile anamorph of a corticoid basidiomycete, producing hyaline to brownish-coloured vegetative mycelium with characteristic hyphal branching at right angles, and producing brown to blackish-coloured, rudimentary sclerotia that consist of compact aggregations of moniliform cells on the plant surface and microsclerotial aggregations of thick-walled hyphae between and within infected root and hypocotyl cells (Carling and Sumner, 1992; Howard *et al.*, 1994). Strains are not readily distinguishable based on morphological characters, but different subspecies groups can be recognized on the basis of anastomosis group and nucleic acid fingerprints (Carling and Sumner, 1992). AG-2-1 is the common designation for seedling canola and sugarbeet isolates from the field, even though AG-4 isolates have been obtained from mature plants, e.g. tomato, cabbage and other transplanted vegetable crops grown in greenhouses (Hwang *et al.*, 1986; Gugel *et al.*, 1987; Sabaratnam, 1999), whereas AG-3 isolates are prevalent on potato, *Solanum tuberosum* L. (Hooker, 1981; Xue *et al.*, 1998). AG-4 and AG-2 isolates are characteristic of root rot in beans (Howard *et al.*, 1994). AG-2-1 isolates are the predominant ones from American ginseng, *Panax quinquefolius* L., which is also susceptible to AG-3 isolates from potato (Reeleder and Brammall, 1994). *Rhizoctonia solani* on canola in the field prefers cooler temperatures and high soil moisture for disease development (Teo *et al.*, 1988).

Background

Curative chemical control after infection is difficult. Fungicidal seed treatments and chemical drenches of the potting medium for greenhouse-grown transplants are recommended (Howard *et al.*, 1994; Anonymous, 1996). Fungicides such as a mixture of iprodione + thiram + lindane (Foundation®) or iprodione + thiram (Foundation Lite®) are registered to control Rhizoctonia damping-off of canola and mustard by seed treatment (Anonymous, 1999). Thiabendazole (Mertect) is registered to control Rhizoctonia storage rot of potato and sugarbeet, *Beta vulgaris* L., caused by *R. solani* (Anonymous, 1999).

Cultural methods in greenhouses that are effective include strict sanitation, sterilization of potting medium and plug trays or, preferably, the use of new plug trays (Howard *et al.*, 1994). To control *R. solani* in canola a firm, moist seedbed and shallow seeding are recommended (Teo *et al.*, 1988). A seeding depth of 1.5–2.5 cm will result in higher seedling emergence compared to seeding deeper (3.0–4.0 cm) (Gugel *et al.*, 1987; Kharbanda and Tewari, 1996). Crop rotation and controlling crucifer volunteers and weeds are additional control measures for the disease (Kharbanda and Tewari, 1986). On potato and bean, *Phaseolus vulgaris* L., integrated disease management includes sanitation (disease-free seed tubers), shallow planting, crop rotation with cereals, grasses or buckwheat, and fungicidal treatment of seed and tubers (Howard *et al.*, 1994).

Cultivar resistance to Rhizoctonia diseases is lacking for tomato and very limited for cruciferous transplants and other vegetables (Howard *et al.*, 1994). No resistant field crop cultivars exist, although they can differ in their susceptibility to *R. solani* (Kharbanda and Tewari, 1996).

Biological Control Agents

Bacteria

Various rhizobacteria have been investigated[1] as potential biological control agents for *R. solani* on field crops and greenhouse transplants (De Freitas *et al.*, 1999; J.R. De Freitas, S.M. Boyetchko, J.J. Germida and G.G. Kachatourians, unpublished). Rhizosphere and endophytic bacteria were isolated from *B. napus*, cultivars 'Legend', 'Excel' and 'Quest', and *B. rapa*, cultivar 'Parkland', and antifungal activity *in vitro* was assessed after 7 days using dual plate cultures on one-half strength potato dextrose agar (PDA). Out of 1223 bacterial strains evaluated, 9.7% inhibited *R. solani* AG-4 and 11.4% inhibited AG-2-1 strains. Fatty acid methyl ester profiles (FAME) and analysis by gas chromatography using the MIDI system (Microbial Identification System, Inc., Newark, USDA) indicated that most of the bacteria with antifungal activity were *Pseudomonas*, *Xanthomonas*, *Burkholderia* and *Bacillus* spp. Other bacterial genera identified included *Arthrobacter, Curtobacterium, Cytophaga, Flavobacterium, Hydrenophaga, Sphingobacteria, Micrococcus* and *Variovorax*. A significant portion of potential bacterial biological control agents were unknown species that could not be found in the current MIDI library. Further detailed characterization of secondary metabolites produced by the bacterial strains is under way.

Streptomycetous rhizobacteria from tomato are effective antagonists of *R. solani* when applied as seed treatments or amendment to artificially infested, peat-based potting media, suppressing seedling damping-off by 94% and 93%, respectively, compared to 71% suppression by plug drenching (Sabaratnam, 1999). Seed coating of lyophilized, living bacterial filaments in wettable powders is the most effective delivery method in plug-tray systems of transplant production (J.A. Traquair and S. Sabaratnam, unpublished). Suppression of Rhizoctonia damping-off on tomato seedlings with seed-coatings or plug drenches with the recommended wettable powder formulation and rates of the dried spores and cells of *Streptomyces*

[1] A collaborative study between Agriculture and Agri-Food Canada (Saskatoon Research Centre) and the University of Saskatchewan.

griseoviridis (Mycostop®), originally isolated from sphagnum moss in Finland and registered in European countries (Kemira) and in USA (AgBio), have not been as effective as tomato Streptomycetes (J.A. Traquair and S. Sabaratnam, unpublished).

Fungi

Trichoderma and *Gliocladium* spp. are the most studied fungal biological control agents for Rhizoctonia damping-off in numerous crops (Lumsden *et al.*, 1993). When applied as a seed treatment or soil amendment, *Trichoderma harzianum* Rifai reduced severity of symptoms on canola seedlings by 46.4% in *R. solani* infested soils (Calman, 1990). The hyphae of *R. solani* showed extensive hyperparasitic coiling by *T. harzianum*. Although *T. harzianum* was considered a potentially good candidate for biological control of *R. solani*, further work on its registration in Canada has not been pursued. In Quebec, Benyagoub *et al.* (1994, 1996) studied *Stachybotrys elegans* (Pidopl) W. Gams, as a destructive mycoparasite of hyphae and sclerotia of *R. solani* (AG-3) infecting potato. Xue *et al.* (1998) showed that several binucleate, non-pathogenic *Rhizoctonia* species (AG-G) also induce peroxidases, glucanases and chitinases that lead to systemic host resistance to *R. solani* (AG-4) in beans.

Competitive Interactions

Composted agricultural and industrial wastes have shown considerable promise as soil amendments to control soil-borne plant pathogens (Huang and Huang, 1993). They can contain allelochemicals that inhibit pathogens directly or they can stimulate the activity of natural soil-borne microbial antagonists (Patrick, 1986; H.C. Huang *et al.*, unpublished). For example, amendment of soil with 160 ppm of CF-5, a liquid compound containing extracts from fermented agricultural wastes and 10% (v/v) allyl alcohol (Huang and Huang, 1993), was not only effective in reducing incidence of damping-off of kale, *Brassica oleracea* var. *acephala* De Candolle, and pea, *Pisum sativum* L., caused by *R. solani*, but also effective in increasing populations of antagonistic microorganisms such as *Trichoderma* spp. and *Bacillus* spp. (Huang *et al.*, 1993). Another study showed that at 150–400 ppm, the CF-5 compound effectively controlled apothecial production of *Sclerotinia sclerotiorum* (Libert) de Bary and stimulated growth and sporulation of *Trichoderma* spp. (Huang *et al.*, 1997). In American ginseng, various organic mulches, composts and *Trichoderma* spp. (R.D. Reeleder and R.A. Brammall, unpublished) are being investigated as biological control agents in artificial shade gardens.

Evaluation of Biological Control

Management of Rhizoctonia diseases in soil or soilless culture is based on thorough understanding of population dynamics of *R. solani* and its biological control agents in a given crop environment (Huang, 1992). Development of effective organic amendment technologies and successful control of Rhizoctonia damping-off of field and containerized crops by organic amendment and microbial activity must be based on sound ecological principles. Much-needed information on the environmental fate of biological control agents can be approached more easily with the advent of recent biotechnologies and molecular biology. Genetical insertion of bioluminescent markers is a useful approach to monitoring stability and distribution of streptomycetous biological control agents on tomato roots (Sabaratnam *et al.*, 1999; S. Sabaratnam and J.A. Traquair, unpublished). PCR markers for *Stachybotrys* spp. and *Rhizoctonia* spp. will also facilitate ecological and environmental fate studies on bean (Bounou *et al.*, 1999; Wang *et al.*, 1999).

Recommendations

Further work should include:

1. Determining survival, ecology and mechanisms of activity of microbial biological control agents and amendments that enhance them;
2. Improving formulations and delivery methods for preventive biological control in diverse agricultural and horticultural systems;
3. Lengthening shelf-life and improving activity of formulations, e.g. by adding nutrients (amendments) that support adequate growth and rapid dispersal of biological control agents in the rhizosphere.

References

Anonymous (1996) *Growing Vegetable Transplants in Plug Trays.* Publication 250/22, Ontario Ministry of Agriculture, Food and Rural Affairs.

Anonymous (1999) Fungicides. In: *Crop Protection 1999.* AGDEX 606–1. Alberta Agriculture, Food and Rural Development, Edmonton, Alberta, pp. 334–376.

Banville, G. (1989) Yield losses and damage to potato plants caused by *Rhizoctonia solani* Kühn. *American Potato Journal* 66, 821–834.

Benyagoub, M., Jabaji-Hare, S.H., Banville, G. and Charest, P.M. (1994) *Stachybotrys elegans*: a destructive mycoparasite of *Rhizoctonia solani. Mycological Research* 98, 493–505.

Benyagoub, M., Jabaji-Hare, S.H., Chamberland, H. and Charest, P.M. (1996) Gold cytochemistry of the mycoparasitic interaction between *Stachybotrys elegans* and its host *Rhizoctonia solani* (AG-3). *Mycological Research* 100, 79–86.

Bounou, S., Jabaji-Hare, S.H., Hogue, R. and Charest, P.M. (1999) Polymerase chain reaction-based assay for specific detection of *Rhizoctonia solani. Mycological Research* 103, 1–8.

Calman, A.I. (1990) Canola seedling blight in Alberta: pathogens, involvement of *Pythium* spp. and biological control of *Rhizoctonia solani.* MSc thesis, University of Alberta, Edmonton, Alberta.

Carling, D.E. and Sumner, D.R. (1992) *Rhizoctonia.* In: Singleton, L.L., Mihail, J.D. and Rush, C.M. (eds) *Methods for Research on Soilborne Phytopathogenic Fungi.* American Phytopathological Society Press, St Paul, Minnesota, pp. 157–165.

De Freitas, J.R., Boyetchko, S.M., Germida, J.J. and Khachatourian, G.G. (1999) Development of natural microbial metabolites as biocontrol products for canola pathogens. *Canadian Journal of Plant Pathology* 21, 193–194.

Ginns, J.H. (1986) *Compendium of Plant Disease and Decay Fungi in Canada 1960–80.* Research Branch Publication No. 1816. Canadian Government Publishing Centre, Ottawa, Ontario.

Gugel, R.K., Yitbarek, S.M., Verma, P.R., Morrall, R.A.A. and Sadasivaiah, R.S. (1987) Etiology of the *Rhizoctonia* root rot complex in the Peace River region of Alberta. *Canadian Journal of Plant Pathology* 9, 119–128.

Hooker, W.J. (ed.) (1981) *Compendium of Potato Diseases.* American Phytopathological Society Press, St Paul, Minnesota.

Howard, R.J., Garland, J.A. and Seaman, W.L. (eds) (1994) *Diseases and Pests of Vegetable Crops in Canada.* Canadian Phytopathology Society and Entomological Society of Canada, Ottawa, Ontario.

Huang, H.C. (1992) Ecological basis of biological control of soil-borne plant pathogens. *Canadian Journal of Plant Pathology* 14, 86–91.

Huang, H.C. and Huang, J.W. (1993) Prospects for control of soil-borne plant pathogens by soil amendment. *Current Topics in Botanical Research* 1, 223–235.

Huang, J.W., Yang, S.H. and Huang, H.C. (1993) Effect of allyl alcohol and soil microorganisms on *Rhizoctonia solani. Plant Pathological Bulletin* (*Taiwan*) 2, 259.

Huang, H.C., Huang, J.W., Saindon, G. and Erickson, R.S. (1997) Effect of allyl alcohol and agricultural wastes on carpogenic germination of sclerotia of *Sclerotinia sclerotiorum* and colonization by *Trichoderma* spp. *Canadian Journal of Plant Pathology* 19, 43–46.

Hwang, S.F., Swanson, T.A. and Evans, I.R. (1986) Characterization of *Rhizoctonia solani* isolates from canola in west central Alberta. *Plant Disease* 70, 681–687.

Kharbanda, P.D. and Tewari, J.P. (1996) Integrated management of canola diseases using cultural methods. *Canadian Journal of Plant Pathology* 18, 168–175.
Lumsden, R.D., Lewis, J.A. and Locke, J.C. (1993) Managing soil-borne plant pathogens with fungal antagonists. In: Lumsden, R.D. and Vaughn, J.L. (eds) *Pest Management: Biologically-based Technologies*. American Chemical Society, Washington, DC, pp. 196–203.
Martens, J.W., Seaman, W.L. and Atkinson, T.G. (eds) (1984) *Diseases of Field Crops in Canada*. Canadian Phytopathological Society, Ottawa, Ontario.
Patrick, Z.A. (1986) Allelopathic mechanisms and their exploitation for biological control. *Canadian Journal of Plant Pathology* 8, 225–228.
Reeleder, R.D. and Brammall, R.A. (1994) Pathogenicity of *Pythium* species, *Cylindrocarpon destructans*, and *Rhizoctonia solani* to ginseng seedlings in Ontario. *Canadian Journal of Plant Pathology* 16, 311–316.
Sabaratnam, S. (1999) Biological control of Rhizoctonia damping-off of tomato with a rhizosphere actinomycete. PhD thesis, University of Western Ontario, London, ON, Canada.
Sabaratnam, S., Cuppels, D.A. and Traquair, J.A. (1999) Insertion of a luciferase gene cassette into a streptomycetous biocontrol agent. *Phytopathology*, 89, S67.
Sippell, D.W., Sadasivaiah, R.S. and Cox, M. (1985) Factors affecting severity of root rot of canola in the Peace River region. *Canadian Journal of Plant Pathology* 8, 354.
Teo, B.K., Yitbarek, S.M., Verma, P.R. and Morrall, R.A.A. (1988) Influence of soil moisture, seeding date, and *Rhizoctonia solani* isolates (AG 2-1 and AG 4) on disease incidence and yield in canola. *Canadian Journal of Plant Pathology* 10, 151–158.
Tewari, J.P. (1985) *Diseases of Canola Caused by Fungi in the Canadian Prairies*. Agriculture and Forestry Bulletin 8, University of Alberta, Edmonton, AB, Canada, pp. 13–20.
Turkington, T.K. and Harrison, L.M. (1994) Survey of canola diseases in the Peace River region of Alberta, 1993. *Canadian Plant Disease Survey* 74, 94–95.
Wang, X., Leclerc-Potvin, C., Charest, P.M. and Jabaji-Hare, S.H. (1999) Generation of species-specific marker for the identification of *Stachybotrys elegans*. *Phytopathology* 89, S83.
Xue, L., Charest, P.M. and Jabaji-Hare, S.H. (1998) Systemic induction of peroxidases, 1,3-beta-glucanases, chitinases, and resistance in bean plants by binucleate *Rhizoctonia* species. *Phytopathology* 88, 359–365.

98 *Sclerotinia homoeocarpa* F. T. Bennett, Dollar Spot of Turfgrass (*Sclerotiniaceae*)

G.J. Boland, T. Zhou and J.I. Boulter

Pest Status

Sclerotinia homeocarpa F.T. Bennett[1] is the causal agent of dollar spot, one of the most important plant diseases that affects turfgrasses. It can cause disease in at least 40 plant hosts throughout North and Central America, Europe, Australia, New Zealand

[1]Although the pathogen is currently classified in *Sclerotinia*, most authorities believe it will eventually be reclassified in *Lanzia*, *Moellerodiscus* or *Rutstroemia* (Vargas and Powell, 1997; Kohn, 1979a, b; Walsh *et al.*, 1999).

and Japan (Fenstermacher, 1980; Vargas, 1994; Couch, 1995; Walsh *et al.*, 1999). Most hosts are grasses (*Poaceae*) but some are *Cyperaceae*, *Caryophyllaceae*, *Convolvulaceae* and *Fabaceae* (Walsh *et al.*, 1999). Dollar spot can cause considerable damage to highly maintained golf-course putting greens, closely mown fairways and bowling greens (Goodman and Burpee, 1991); and less intensively managed turfgrass such as home lawns, recreational and athletic facilities, and educational or industrial properties. Dollar spot reduces the aesthetic and playing quality of infected turf, and disease can also contribute to weed encroachment and plant death (Smith *et al.*, 1989). Except for western Canada and the US Pacific north-west, dollar spot is the most common turf disease in North America (Couch, 1995). More money is spent on managing dollar spot than any other turfgrass disease on golf courses (Goodman and Burpee, 1991).

Symptoms of dollar spot on turfgrass swards vary according to the turfgrass species and management practices, although disease symptoms are particularly severe on creeping bentgrass, *Agrostis palustris* Hudson. On closely mown turf, such as on golf-course putting greens, the disease develops into sunken, circular, straw-colored patches that range in size from a few blades of grass to the size of a silver dollar (5–7.5 cm diameter), hence the disease name (Vargas, 1994; Couch, 1995). Necrotic patches are noticeable because they contrast sharply with adjacent healthy turfgrass.

S. homoeocarpa is reported to overwinter as darkly pigmented stromata and as dormant mycelium in the crowns and roots of infected plants. It primarily infects leaves through mycelial growth into cut leaf tips and stomata, but direct penetration also occurs. Sporulation by *S. homoeocarpa* is rare in field conditions and, therefore, these structures are considered to have a minor role in the epidemiology of the disease. Local infection results when mycelium grows from diseased to healthy leaves that are close together. Over larger areas, the pathogen is distributed primarily through physical displacement of infested and diseased tissues, e.g. grass clippings on machinery and shoes. Environment has a strong influence on development of dollar spot and this disease primarily occurs during warmer weather (Couch, 1995; Walsh *et al.*, 1999).

Background

Dollar spot is primarily managed through the use of regular applications of fungicides and cultural practices. Fungicides have been the primary method of disease control for at least 40 years. Often, multiple applications of fungicides are required to maintain disease-free turf throughout a growing season and, as a result, resistance to fungicides in *S. homoeocarpa* has posed an ongoing challenge to the turfgrass industry (Walsh *et al.*, 1999).

Cultural controls include any practices that reduce the amount and duration of leaf wetness on turf, e.g. irrigation during the day to promote rapid drying of leaves, removal of infested and/or moist grass clippings that will not dry during the day, and pruning or removal of trees and shrubs to increase aeration and minimize shade so that dew evaporates more quickly. Applications of nitrogen are known to be effective for reducing disease severity, although the manner in which this occurs has not been clarified.

The use of composts and other organic amendments for disease suppression has potential to be beneficial both ecologically and economically. Although compost use may not control turfgrass diseases such as dollar spot to a level that may replace fungicide use, its integration with current disease management practices may reduce fungicide use and associated problems. Naturally suppressive composts can be incorporated into normal golf-course maintenance by replacing sphagnum peat or other organic materials used in topdressing mixtures or soil amendments. Composts suppress plant diseases through a combination of physico-chemical and biological characteristics. Physico-chemical characteristics include any physical or chemical aspects of composts that reduce disease severity by directly or indirectly affecting the pathogen or host capacity for growth, such as nutrient levels, organic matter, moisture, pH and other factors. In North America, work on biologically based control of *S. homoeocarpa* on creeping bentgrass has been emphasized because the disease on this host is severe and this interferes with the playabilty of golf-course putting greens.

Further attempts to develop biological control agents for dollar spot of turf are warranted because of the severity and economic importance of this disease, the prevalence of fungicide resistance in populations of *S. homoeocarpa*, the suitability of turfgrass as an environment for establishment and maintenance of microbial biological control agents, and the potential for commercial development and use of biological control products.

Biological Control Agents

Compost-inhabiting microbial populations are important biological control agents because they compete with pathogens for nutrients, produce antibiotics, lytic and other extracellular enzymes, are parasites or predators, induce host-mediated resistance in plants and interact in other ways that decrease disease development (Nelson, 1991; Nelson and Craft, 1991, 1992; Hoitink *et al.*, 1997a, b; Whipps, 1997; Boulter *et al.*, 2000).

In field trials, Boulter *et al.* (1999) assessed the efficacy of composts in suppressing dollar spot. Overall, there were relatively few consistent differences among treatments, but there were significant (P = 0.05) differences between treatments and the pathogen-treated control. Compost-rate treatments applied once per season did not suppress disease compared to a pathogen-treated control. However, compost-rate treatments applied every 3 weeks did suppress disease severity compared to a pathogen-treated control, and were as effective as bi-weekly applications of the fungicide chlorothalonil (applied at the manufacturer's recommended preventive rates) in suppressing disease. These results indicate that reductions in dollar spot severity by applications of compost every 3 weeks were comparable to applications of a fungicide every 2 weeks.

Significant differences were not detected among most compost treatments in field experiments. This may have been because feedstock compositions were not sufficiently different to elicit distinctive results. Individual composts in these experiments were based on selected ratios of known but similar feedstocks and, therefore, nutrient and microbial activity may be more similar than anticipated. Variability among compost batches in disease suppression may not be as important as previously thought. The efficacy of all composts in suppression of dollar spot may reflect an underlying principle that activity is associated more with resident microbial activity and nutrient availability than the presence of a specific microbial microflora or feedstock combination. Differences among composts may also have remained undetected because all rates of application may have exceeded a critical threshold for efficacy. Lower application rates may have revealed differences among the composts.

Fungi

Goodman and Burpee (1991) examined inundative applications of selected biological control agents. In controlled environments, colonized sand–cornmeal top-dressings were compared for disease suppression, and four of 24 potential antagonists suppressed disease by 25–90%. In field trials, maximum disease intensities following treatment by isolates of *Fusarium heterosporum* Nees ex Fries, an *Acremonium* sp. and an unidentified bacterium were 5%, 14% and 44%, compared to 84% in plots that were not top-dressed and 64% in plots that were top-dressed with non-infested, autoclaved sand–cornmeal. Subsequent field trials with *F. heterosporum* compared living with heat-killed sand–cornmeal treatments, and indicated that heating did not reduce efficacy of the treatment. The results established that treatment of turf with sand–cornmeal top-dressings colonized by *F. heterosporum* could significantly suppress dollar spot, and that the mechanism of action may be production of heat-stable substances toxic to *S. homoeocarpa*.

Boland and Smith (2000) subsequently compared *F. heterosporum* with several other fungal and bacterial antagonists in 2 years of field trials in naturally and artificially infested swards of creeping bentgrass. Under high inoculum concentrations of *S. homoeocarpa*, none of the biological control agents were particularly effective compared to a fungicide control. Of the microorganisms tested, *F. heterosporum* was the only species that provided significant disease suppression in more than one trial.

Hypovirulent isolates of **S. homoeocarpa**

Hypovirulence is a phenotypic response of selected isolates within a population of a plant pathogen characterized by reduced virulence, but it may also be associated with characters such as reduced growth rate, sporulation and/or survival. Although hypovirulence has been associated with several modes of action, most often it has been associated with the presence of double-stranded ribonucleic acid (dsRNA) (Nuss and Koltin, 1990). The potential of using hypovirulent isolates of a fungal pathogen in a biological control strategy resides in the ability to transfer hypovirulence from hypovirulent isolates to virulent isolates, and thereby reduce the mean disease severity of the population through overall reductions in virulence, growth, sporulation and/or survival (Zhou and Boland, 1998b).

To obtain hypovirulent isolates, 132 isolates of *S. homoeocarpa* were evaluated for virulence on detached leaves and swards of creeping bentgrass and for the presence of dsRNA. Thirteen of 132 isolates (9.8%) did not initiate dollar spot lesions in inoculated swards 4 weeks after inoculation, and were considered to be hypovirulent. dsRNA was detected in six of the 13 hypovirulent isolates (46.2%) (Zhou and Boland, 1997). It was also found that, compared to typical isolates of *S. homoeocarpa,* these six isolates often grew slowly on potato dextrose agar (PDA), formed thin colonies with atypical colony margins, and failed to produce a typical black stroma. In *in vitro* experiments, hypovirulence and dsRNA were transmitted from hypovirulent isolate Sh12B to a virulent DMI (sterol demethylation inhibitor)-fungicide-resistant isolate, Ky-7, and the converted isolate was hypovirulent, contained dsRNA, and grew on medium amended with 2 μg active ingredient ml^{-1} tebuconizole (BayHWG 1608). Hypovirulence and dsRNA were also transferred to at least four other isolates of *S. homoeocarpa*. The characterization of transmissible hypovirulence and dsRNA in *S. homoeocarpa* provided potential for using hypovirulent isolates in management of dollar spot of turfgrass.

Zhou and Boland (1998a) evaluated selected hypovirulent isolates of *S. homoeocarpa* for efficacy in suppressing dollar spot of turfgrass under growth-room and field conditions. Under growth-room conditions, hypovirulent isolates Sh12B, Sh09B or Sh08D of *S. homoeocarpa* caused 3.4–30.4% diseased turf, in comparison to virulent isolates Sh48B and Sh14D, which caused 80.2–90.2% disease. In treatments that received both virulent and hypovirulent isolates, only hypovirulent isolate Sh12B significantly reduced dollar spot severity compared to the pathogen-treated control.

In a field experiment conducted in 1993 on swards of creeping bentgrass, experimental plots were artificially inoculated with a virulent isolate of *S. homoeocarpa*, and then treated with a hypovirulent isolate in various formulations. Ten days after inoculation, the percentage diseased turf for each formulation of hypovirulent isolate Sh12B was 6.3%, 12.5% and 20.8%, for treatments applied as a mycelial suspension (80 ml m^{-2}), granular mix (8 g m^{-2}) and alginate pellets (8 g m^{-2}), respectively, and were significantly lower than their respective formulation controls (31.2%, 23.8% and 30.0%, respectively). Suppression of dollar spot by the mycelial suspension of hypovirulent isolate Sh12B was still evident 45 days after treatment, and residual disease suppression persisted until the next growing season (Zhou and Boland, 1998a). Similarly, significant suppression of dollar spot by isolate Sh12B was observed when this experiment was repeated the following year.

Zhou and Boland (1998a) determined the effects of a hypovirulent isolate on suppressing naturally occurring dollar spot. Treatments with a mycelial suspension and alginate pellets of hypovirulent isolate Sh12B significantly reduced dollar spot up to 58%, compared to their respective formulation controls. With few exceptions, there were no statistical differences between treatments with hypovirulent isolate Sh12B and the fungicide Daconil 2787. Multiple applications of the hypovirulent isolate did not result in greater suppression of dollar spot as compared to a single application.

Evaluation of Biological Control

All of the strategies examined to date have provided effective results under defined experimental conditions.

Recommendations

Further work should include:

1. Addressing the comparative efficacy and commercial potential of these biological control strategies, and providing increased emphasis on identification of mechanisms of action responsible for the observed efficacy;
2. Comparing the biological control agents and strategies with those being developed in other regions to identify those most effective for continued development.

References

Boland, G.J. and Smith, E.A. (2000) Influence of biological control agents on dollar spot of creeping bentgrass, 1999. *Biological and Cultural Tests for Control of Plant Disease* 15, 50.

Boulter, J.I., Boland, G.J. and Trevors, J.T. (1999) Evaluation of compost for biological control of dollar spot (*Sclerotinia homoeocarpa*) on creeping bentgrass (*Agrostris palustris*). *Phytopathology* 89, S8.

Boulter, J.I., Boland, G.J. and Trevors, J.T. (2000) Compost: A study of the development process and end-product potential for suppression of turfgrass disease. *World Journal of Microbiology and Biotechnology* 16, 115–134.

Couch, H.B. (1995) *Diseases of Turfgrasses*, 3rd edn. Krieger Publishing, Malabar, Florida.

Fenstermacher, J.M. (1980) Certain features of dollar spot disease and its causal organism, *Sclerotinia homoeocarpa*. In: Joyner, B.G. and Larsen, P.O. (eds) *Advances in Turfgrass Pathology: Proceedings of the Symposium on Turfgrass Diseases, 15–17 May 1979, Columbus, Ohio*. B.G. Harcourt Brace Jovanovich, Duluth, Minnesota.

Goodman, D.M. and Burpee, L.L. (1991) Biological control of dollar spot disease of creeping bentgrass. *Phytopathology* 81, 1438–1446.

Hoitink, H.A.J., Han, D.Y., Krause, M.S., Zhang, W., Stone, A.G. and Dick, W.A. (1997a) *How to Optimize Disease Control Induced by Composts*. Ohio Agricultural Research and Development Center, Ohio State University, Wooster, Ohio.

Hoitink, H.A.J., Stone, A.G. and Han, D.Y. (1997b) Suppression of plant disease by composts. *HortScience* 32, 184–187.

Kohn, L.M. (1979a) A monographic revision of the genus *Sclerotinia*. *Mycotaxon* 9, 365–444.

Kohn, L.M. (1979b) Delimitation of the economically important plant pathogenic *Sclerotinia* species. *Phytopathology* 69, 881–886.

Nelson, E.B. (1991) Introduction and establishment of strains of *Enterobacter cloacae* in golf course turf for the biological control of dollar spot. *Plant Disease* 75, 510–514.

Nelson, E.B. and Craft, C.M. (1991) Suppression of dollar spot with topdressings amended with composts and organic fertilizers. 1989. *Biological and Cultural Tests for Control of Plant Disease* 6, 93.

Nelson, E.B. and Craft, C.M. (1992) Suppression of dollar spot on creeping bentgrass and annual bluegrass turf with compost-amended topdressings. *Plant Disease* 76, 954–958.

Nuss, D.L. and Koltin, Y. (1990) Significance of dsRNA genetic elements in plant pathogenic fungi. *Annual Review of Phytopathology* 28, 37–58.

Smith, J.D., Jackson, N. and Woolhouse, A.R. (1989) Dollar spot disease. In: *Fungal Diseases of Amenity Turf Grasses*, 3rd edn. E. & F.N. Spon, New York, New York.

Vargas, J.M. Jr (1994) *Management of Turfgrass Diseases*, 2nd edn. Lewis Publishers, Boca Raton, Florida.

Vargas, J.M. Jr and Powell, J.F. (1997) Mycelial compatibility and systematics of *Sclerotinia homoeocarpa*. *Phytopathology* 87, S79.

Walsh, B., Ikeda, S.S. and Boland, G.J. (1999) Biology and management of dollar spot (*Sclerotinia homoeocarpa*); an important disease of turfgrass. *HortScience* 34, 13–21.

Whipps, J.M. (1997) Ecological considerations involved in commercial development of biological control agents for soil-borne diseases. In: Dirk van Elsas, J., Trevors, J.T. and Wellington, E.M.H. (eds) *Modern Soil Microbiology*. Marcel Dekker, New York, New York.

Zhou, T. and Boland, G.J. (1997) Hypovirulence and double-stranded RNA in *Sclerotinia homoeocarpa*. *Phytopathology* 87, 147–153.

Zhou, T. and Boland, G.J. (1998a) Suppression of dollar spot by hypovirulent isolates of *Sclerotinia homoeocarpa*. *Phytopathology* 88, 788–794.

Zhou, T. and Boland, G.J. (1998b) Biological control strategies for *Sclerotinia* species. In: Boland, G.J. and Kuykendall, L.D. (eds) *Plant–Microbe Interactions and Biological Control*. Marcel Dekker, New York, New York, pp. 127–156.

99 *Sclerotinia sclerotiorum* (Libert) de Bary and *Sclerotinia minor* Jagger, Sclerotinia Diseases (*Sclerotiniaceae*)

H.C. Huang, S.D. Bardin, G.J. Boland, R.D. Reeleder and S.M. Boyetchko

Pest Status

Sclerotinia spp. comprise a group of fungi pathogenic to higher plants. Most hosts of the main pest species, *S. sclerotiorum* (Libert) de Bary, are herbaceous plants in the *Asteraceae*, *Fabaceae*, *Brassicaceae*, *Solanaceae*, *Apiaceae* and *Ranunculaceae* (Boland and Hall, 1994; Huang, 1997). The host range of *S. sclerotiorum* consists of 408 species (Boland and Hall, 1994). The host range for *S. minor* is considerably smaller and includes 94 species (Melzer *et al.*, 1997). Kohn (1979) revised the *Sclerotiniaceae* and limited the genus to three species: *S. sclerotiorum*, *S. minor* Jagger and *S. trifoliorum* Ericksson. Two additional species have been added since: *S. asari* Wu and Wang (Wu and Wang, 1983) and *S. nivalis* Saito (Saito, 1997; Li *et al.*, 2000). *S. sclerotiorum* and *S. minor*, the two species found in Canada, are reviewed here.

Sclerotinia diseases can cause serious losses in yield and quality of important field and vegetable crops. In western Canada, Sclerotinia diseases of canola/rapeseed, *Brassica napus* L. and *B. rapa* L., caused an estimated loss of more than Can$15 million in 1982 (Martens *et al.*, 1984). In Saskatchewan, canola/rapeseed stem rot caused by *S. sclerotiorum* occurred in 62% of fields (Morrall *et al.*, 1976). Wilt of sunflower, *Helianthus annuus* L., due to *S. sclerotiorum* reduced seed yield by more than 70% when wilting occurred within 4 weeks of flowering (Dorrell and Huang, 1978). In southern Alberta, white mould of dry bean, *Phaseolus vulgaris* L., was found in 80–100% of the fields, with 0–90% of plants infected by *S. sclerotiorum* in each field (Huang *et al.*, 1988). In Ontario, white mould significantly reduced seed yields of dry bean in field trials when disease incidence was higher than 40% (Haas and Bolwyn, 1973). In Alberta (Xue and Burnett, 1994) and Manitoba (Xue *et al.*, 1995), stem rot of dry pea, *Pisum sativum* L., caused by *S. sclerotiorum*, was ranked as the third most common disease. Blossom blight of alfalfa, *Medicago sativa* L., caused by *S. sclerotiorum* and/or *Botrytis cinerea* Persoon ex Fries, is prevalent in Alberta, Saskatchewan and Manitoba (Gossen *et al.*, 1997). In Quebec, Devaux (1991) recorded only one field of soybean, *Glycine max* (L.) Merrill, severely infected by *S. sclerotiorum*.

In Ontario, lettuce, *Lactuca sativa* L., drop caused by *S. minor* and *S. sclerotiorum* was present in 71% and 57% of the fields, respectively (Melzer *et al.*, 1993), with *S. minor* causing yield losses of more than 35% (Melzer and Boland, 1994). In Quebec, lettuce drop due to *S. sclerotiorum* caused 1.7% of crop loss, and losses in transplanted crops were consistently higher than in seeded crops (Reeleder and Charbonneau, 1987). Sclerotinia rot of carrot, *Daucus carota sativus* Arcangeli, was occasionally observed in Quebec but did not cause significant yield losses (Arcelin and Kushalappa, 1991). However, it is an important disease of stored carrots (Pritchard *et al.*, 1992). In Prince Edward Island, *S. sclerotiorum* incidence in tobacco, *Nicotiana tabacum* L., fields increased from 40% in 1985 to 76% in 1986, and yield losses in some fields were estimated as high as 10% (Martin and Arsenault, 1987).

Research on biology and epidemiology of *S. sclerotiorum* in Canada was reviewed by Bardin and Huang (2001). In soil, *S. sclerotiorum* survives mainly as black sclerotia, which are the primary source of inoculum for the disease. However, dormant mycelium in stored seeds can play an important role in pathogen dissemination and disease epidemiology in bean (Tu, 1988).

Depending on environmental and physiological conditions, e.g. temperature, moisture and exogenous source of nutrients, sclerotia can germinate carpogenically to produce apothecia and ascospores or myceliogenically to produce mycelia (Bardin and Huang, 2001). The pathogen produces white, fluffy mycelia on the surface of invaded tissues or causes plant wilt, depending whether the above-ground or underground tissues are infected. Ascospores are the primary source of inoculum for infection of above-ground tissues, causing diseases, e.g. white mould of bean, stem blight of canola, pod rot of pea, head rot of sunflower and blossom blight of alfalfa. Mycelium from myceliogenic germination of sclerotia of *S. sclerotiorum* in soil is the primary source of inoculum for infection of root tissues in sunflower wilt and carrot root rot (Bardin and Huang, 2001).

Secondary spread of Sclerotinia diseases can occur by direct contact between diseased and healthy tissues (Huang and Hoes, 1980). New sclerotia are produced in and on infected tissues. They may survive in or on the soil, remain with crop residues or persist in harvested tissues, e.g. pods, seeds and roots. The longevity of *S. sclerotiorum* sclerotia is affected by environmental conditions and the presence or absence of natural enemies. Melanins of normal sclerotia may be important for increasing resistance of *S. sclerotiorum* to adverse environmental conditions and attack by microorganisms (Huang, 1983). Of epidemiological significance, the pathogen can spread when *Sclerotinia*-contaminated pollen grains are transported by pollinating insects (Stelfox *et al.*, 1978).

Background

Chemical methods have been the preferred method to control Sclerotinia diseases (see Bardin and Huang, 2001). Fungicides commonly used are benomyl, vinclozolin, iprodione, chlorothalonil and DCT (diazinon 6%, captan 18%, thiophanate-methyl 14%). However, benomyl and iprodione delayed plant maturation by about 1 week when used to control white mould of bean. Other compounds, including urea, calcium cyanamide, formulated compounds, e.g. S-H mixture (Huang and Sun 1991) and CF-5 (Huang *et al.*, 1997), and the herbicides chlorsulfuron, cyanazine, metribuzin, triallate and trifluralin (Teo *et al.*, 1992) inhibited carpogenic germination of *S. sclerotiorum* sclerotia, whereas the triazine herbicides, simazine and atrazine, did not influence carpogenic germination of sclerotia but inhibited the normal differentiation and development of apothecia (Huang and Blackshaw, 1995).

Ozone, ultraviolet-C and modified atmospheres also provide some control of Sclerotinia rot of carrot and celery in storage (Reyes, 1988; Reeleder *et al.*, 1989; Ouellette *et al.*, 1990; Mercier *et al.*, 1993; Liew and Prange, 1994).

Breeding crops for *Sclerotinia* resistance, and cultural methods, e.g. use of pathogen-free seeds, seeding rate and row spacing, tillage, flooding, irrigation and crop rotation, have been tried (see Bardin and Huang, 2001). Cultural practices are only effective when used as part of an integrated pest-management strategy. Additionally, the increasing concern over use of chemical pesticides has increased the need to examine alternative control strategies.

Biological Control Agents

Pathogens

Bacteria

Bacillus cereus Frankland and Frankland, strain alf-87A, sprayed on to pea plants at blossom stage, significantly reduced incidence of basal pod rot caused by *S. sclerotiorum* ascospores (Huang *et al.*, 1993). Antibiosis was involved in pathogen suppression because ascospore germination and vegetative growth of *S. sclerotiorum* were inhibited by secreted metabolites of *B. cereus*. In field experiments, *Bacillus subtilis* (Ehrenberg) Cohn significantly decreased white mould incidence and severity on bean, and its effectiveness appeared to be cumulative over the years (Tu, 1997). However, reduction of white mould by *B. subtilis* was not consistent from one field trial to another (Boland, 1997). De Freitas *et al.* (1999) screened rhizosphere and endophytic bacteria from canola and selected strains that produce novel metabolites with antibiosis activity against *S. sclerotiorum*. Some of the antagonistic strains belonged to species of *Pseudomonas*, *Xanthomonas*, *Burkholderia* and *Bacillus* but a significant portion of the strains were not found in the current MIDI library, indicating that they may be new species.

Fungi

Most of the fungal biological control agents that have been evaluated to date were isolated from sclerotia of *S. sclerotiorum* and from the phylloplane (leaf surface) of susceptible hosts, e.g. rapeseed and bean petals, and lettuce leaves. Some agents, e.g. *Coniothyrium minitans* Campbell (Huang, 1977; Tu, 1984; Huang and Kokko, 1987, 1988), *Gliocladium catenulatum* Gilman and Abbott (Huang, 1978, 1980), *G. virens* Miller and Foster (Tu, 1980), *Talaromyces flavus* (Klöcker) Stolk and Samson (McLaren *et al.*, 1986, 1989), *Trichoderma viride* Persoon ex Fries (Huang, 1980) and *Trichothecium roseum* (Persoon: Fries)

Link (Huang and Kokko, 1993), are mycoparasites of *S. sclerotiorum* sclerotia.

Soil treatments with mycoparasites, e.g. *C. minitans*, effectively reduced the number of sclerotia (Huang, 1979, 1980) as well as apothecia produced from sclerotia (McLaren *et al.*, 1996; Huang and Erickson, 2000). Although *C. minitans* is a destructive parasite of *S. sclerotiorum*, killing sclerotia and hyphae (Huang and Hoes, 1976), it appeared ineffective in controlling the pathogen in an actively growing state, and thus failed to reduced the pathogen's spread (Huang, 1980).

Foliar application of spore suspensions of *C. minitans*, *T. flavus*, *T. roseum* and *Trichoderma virens* (Miller, Giddens and Foster) von Arx effectively reduced white mould incidence of dry bean under field conditions (Huang *et al.*, 2000b). In southern Alberta, *C. minitans* was the most effective agent and reduced the number of infected plants by an average of 56% but was not as efficient as benomyl. In Ontario, Boland (1997) found that another strain of *C. minitans* was effective in 1 of 4 trials but was no more effective than other antagonists tested. The difference in efficacy of *C. minitans* in these reports may be due to differences in strain, dosage or formulation of the agents, time and method of application, and the particular agro-ecological environment that affects the population dynamics of the pathogen and its biological control agents.

Fungi isolated from the anthoplane (flower surface) of bean and rapeseed and the phylloplane of lettuce were saprophytes highly competitive at colonizing senescent plant tissues. *Alternaria alternata* (Fries) Keissler and *Cladosporium cladosporioides* (Fries) de Vries were the most prevalent fungi recovered from bean and rapeseed petals (Boland and Hunter, 1988; Boland and Inglis, 1989; Inglis and Boland, 1990). These organisms, sprayed on bean plants, rapidly colonized flower petals and prevented white mould development in the greenhouse (Boland and Inglis, 1989) but did not provide consistent control in field trials (Inglis and Boland, 1990, 1992). Competition for nutrients by these fungi appeared to be the main suppressive mechanism of *S. sclerotiorum*. Other *Sclerotinia*-suppressive fungi include *Drechslera* sp., *Epicoccum purpurascens* Ehrenberg and Schlechtendahl (*E. nigrum* Link), *Fusarium graminearum* Schwabe (*Gibberella zeae* (Schwabe) Petch), *Fusarium heterosporum* Nees, *Myrothecium verrucaria* (Albertini and Schweinitz) Ditmar, and *T. viride* (Mercier and Reeleder, 1987a, b; Boland and Inglis, 1989; Inglis and Boland, 1990, 1992). In contrast with other fungi, control of white mould by *E. purpurascens* was independent of environmental changes for control of white mould and acted against *S. sclerotiorum* via antibiosis (Zhou *et al.*, 1991; Hannusch and Boland, 1996). New biotypes of *E. purpurascens*, tolerant to iprodione and with improved sporulation, were created from wildtype isolates exposed to shortwave UV light (Zhou and Reeleder, 1989, 1990). Biological control activity of these new biotypes *in vitro* and in the field was also improved compared to the wild type (Zhou and Reeleder, 1989). When tested to control white mould of bean in the field, all fungal treatments became less effective as environmental conditions became more conducive for the disease (Boland, 1997). In the USA, Adams and Fravel (1990) reported successful control of lettuce drop caused by *S. minor* using the mycoparasite *Sporidesmium sclerotivorum* Ueker, Ayers and Adams.

Insects

Bradysia coprophila Lintner larvae were associated with sclerotia and suppressed *S. sclerotiorum* populations in soil (Anas and Reeleder, 1987). *In vitro* tests showed that sclerotia damaged by larval feeding had greatly reduced levels of mycelial germination (0–30%), whereas undamaged sclerotia germinated at a rate of 95%. Larvae were shown to produce salivary gland secretions that contain chitinase, which further reduced the ability of sclerotia to germinate (Anas *et al.*, 1989). Sclerotia that had been grazed by the larvae were more susceptible to colonization by *Trichoderma* spp. (Gracia-Garza *et al.*, 1997a, b). Fungus gnats are often regarded as greenhouse pests (see Gillespie *et al.*, Chapter 10 this volume), so objections are sometimes raised when encouragement of gnat populations in field soils is proposed. However, when effects of gnats on greenhouse-grown plants were evaluated, larvae failed to survive on healthy plants (Anas and Reeleder, 1988). In contrast, when selected plant species were inoculated with various plant pathogens it was found that all diseased plants supported larval development.

There has been interest in using honeybees, *Apis mellifera* L., as biological couriers to control blossom-mediated diseases, by placing a biological control agent in a dispenser in such a way that bees departing from the hive must walk through the inoculum (Israel and Boland, 1992). Additional information on the influence of biological control agents and their formulations on honeybee health is required. Similarly, leafcutter bees, *Megachile rotundata* (Fabricius), used as pollinators for commercial production of alfalfa seed (Goplen *et al.*, 1980), should be investigated as a potential delivery system for biological control of blossom blight of alfalfa caused by *S. sclerotiorum* and *Botrytis cinerea* (Gossen *et al.*, 1997; Huang *et al.*, 2000a).

Soil amendments

Organic soil amendments affect microbial population dynamics by intensifying microbial activity and enhancing competition among soil microorganisms, which can lead to control of soil-borne pathogens and promotion of plant growth (Huang and Huang, 1993). Soil amended with formulated products, e.g. S-H mixture (Sun and Huang, 1985) and CF-5 (Huang and Huang, 1993), both made from organic and inorganic waste materials, controlled apothecial production from *S. sclerotiorum* (Huang and Sun, 1991; Huang *et al.*, 1997). Disease suppression by these formulated compounds was due to a combination of toxic effects on the pathogens and stimulating effects on antagonistic microorganisms in the rhizosphere. In addition to increasing competition among soil microorganisms to manage soil-borne pathogens, formulated amendments can also improve soil fertility and plant growth.

Evaluation of Biological Control

The use of mycoparasitic and antagonistic microorganisms to control *S. sclerotiorum* appears feasible. For example, *C. minitans* is promising as a spray and as a soil amendment and *E. purpurascens* is promising as a spray. However, progress in developing biological control products is slow due to difficulties in inoculum production and inconsistent field efficacy. Biological control of Sclerotinia diseases has potential as part of integrated pest management.

Recommendations

Further work should include:

1. Improving formulation of biological control agents to increase shelf-life and efficacy, and promote growth and colonization of the agents in soil or on plants;
2. Improving application methods (soil amendments, spray formulation), timing of application and delivery method, e.g. use of bees, of the biological control agents;
3. Determining how soil factors, e.g. structure and chemical composition, and environmental factors, e.g. temperature and moisture, can promote survival and proliferation of regionally adapted, biological control agents instead of pathogens;
4. Selecting bacteria or fungi adapted to low temperatures that could have potential to control Sclerotinia diseases in stored crops, e.g. carrots and celery;
5. Developing biological control programmes for *S. minor*;
6. Studying control mechanisms by various agents, e.g. mycoparasites, and antagonistic fungi and bacteria;
7. Integrating non-chemical control methods to enhance survival or build-up of popula-

tions of beneficial organisms in soil and reduce populations of *Sclerotinia* spp. while reducing pesticide use;
8. Carefully selecting crops for rotation so that those cultivated prior to a susceptible crop can enhance natural or artificially inoculated populations of microorganisms useful for biological control of *S. sclerotiorum*;
9. Determining the effect of decomposition of soil amendments on potential biological control agents.

References

Adams, P.B. and Fravel, D.R. (1990) Economical biological control of Sclerotinia lettuce drop by *Sporidesmium sclerotivorum*. *Phytopathology* 80, 1120–1124.

Anas, O. and Reeleder, R.D. (1987) Recovery of fungi and arthropods from sclerotia of *Sclerotinia sclerotiorum* in Quebec muck soils. *Phytopathology* 77, 327–331.

Anas, O. and Reeleder, R.D. (1988) Feeding habits of larvae of *Bradysia coprophila* on fungi and plant tissue. *Phytoprotection* 69, 73–78.

Anas, O., Alli, I. and Reeleder, R.D. (1989) Inhibition of germination of sclerotia of *Sclerotinia sclerotiorum* by salivary gland secretions of *Bradysia coprophila*. *Soil Biology and Biochemistry* 21, 47–52.

Arcelin, R. and Kushalappa, A.C. (1991) A survey of carrot diseases on muck soils in the southern part of Quebec. *Canadian Plant Disease Survey* 71, 147–153.

Bardin, S.D. and Huang, H.C. (2001) Research on biology and control of Sclerotinia diseases in Canada. *Canadian Journal of Plant Pathology* 23, 88–98.

Boland, G.J. (1997) Stability analysis for evaluating the influence of environment on chemical and biological control of white mold (*Sclerotinia sclerotiorum*) of bean. *Biological Control* 9, 7–14.

Boland, G.J. and Hall, R. (1994) Index of plant hosts of *Sclerotinia sclerotiorum*. *Canadian Journal of Plant Pathology* 16, 93–108.

Boland, G.J. and Hunter, J.E. (1988) Influence of *Alternaria alternata* and *Cladosporium cladosporioides* on white mold of bean caused by *Sclerotinia sclerotiorum*. *Canadian Journal of Plant Pathology* 10, 172–177.

Boland, G.J. and Inglis, G.D. (1989) Antagonism of white mold (*Sclerotinia sclerotiorum*) of bean by fungi from bean and rapeseed flowers. *Canadian Journal of Botany* 67, 1775–1781.

De Freitas, J.R., Boyetchko, S.M., Germida, J.J. and Khachatourians, G.G. (1999) Development of natural microbial metabolites as biocontrol products for canola pathogens. *Canadian Journal of Plant Pathology* 21, 193–194.

Devaux, A. (1991) Incidence of soybean diseases in the St-Hyacinth region in 1990. *Canadian Plant Disease Survey* 71, 109.

Dorrell, D.G. and Huang, H.C. (1978) Influence of Sclerotinia wilt on seed yield and quality of sunflower wilted at different stages of development. *Crop Science* 18, 974–976.

Goplen, B.P., Baenzier, H., Bailey, L.D., Gross, A.T.H., Hanna, M.R., Michaud, R., Richards, K.W. and Waddington, J. (1980) *Growing and Managing Alfalfa in Canada*. Agriculture Canada Publication #1705, Agriculture Canada, Ottawa, Ontario.

Gossen, B.D., Lan, Z., Harrison, L.M., Holley, J. and Smith, S.R. (1997) Survey of blossom blight of alfalfa on the Canadian prairies in 1996. *Canadian Plant Disease Survey* 77, 91–92.

Gracia-Garza, J.A., Bailey, B.A., Paulitz, T.C., Lumsden, R.D., Reeleder, R.D. and Roberts, D.P. (1997a) Effect of sclerotial damage of *Sclerotinia sclerotiorum* on the mycoparasitic activity of *Trichoderma hamatum*. *Biocontrol Science and Technology* 7, 401–413.

Gracia-Garza, J.A., Reeleder, R.D. and Paulitz, T.C. (1997b) Degradation of sclerotia of *Sclerotinia sclerotiorum* by fungus gnats (*Bradysia coprophila*) and the biocontrol fungi *Trichoderma* spp. *Soil Biology and Biochemistry* 29, 123–129.

Haas, J.H. and Bolwyn, B. (1973) Predicting and controlling white mold epidemics in white beans. *Canada Agriculture* 18, 28–29.

Hannusch, D.J. and Boland, G.J. (1996) Influence of air temperature and relative humidity on biological control of white mold of bean (*Sclerotinia sclerotiorum*). *Phytopathology* 86, 156–162.

Huang, H.C. (1977) Importance of *Coniothyrium minitans* in survival of sclerotia of *Sclerotinia sclerotiorum* in wilted sunflower. *Canadian Journal of Botany* 55, 289–295.

Huang, H.C. (1978) *Gliocladium catenulatum*: hyperparasite of *Sclerotinia sclerotiorum* and *Fusarium* species. *Canadian Journal of Botany* 56, 2243–2246.

Huang, H.C. (1979) Biological control of Sclerotinia wilt in sunflower. *Canada Agriculture* 24, 12–14.

Huang, H.C. (1980) Control of Sclerotinia wilt of sunflower by hyperparasites. *Canadian Journal of Plant Pathology* 2, 26–32.

Huang, H.C. (1983) Pathogenicity and survival of the tan-sclerotial strain of *Sclerotinia sclerotiorum*. *Canadian Journal of Plant Pathology* 5, 245–247.

Huang, H.C. (1997) *Sclerotinia sclerotiorum* (Lib.) de Bary. In: *Crop Protection Compendium*. CAB International, Wallingford, UK, CD-ROM, module 1.

Huang, H.C. and Blackshaw, R.E. (1995) Influence of herbicides on the carpogenic germination of *Sclerotinia sclerotiorum* sclerotia. *Botanical Bulletin of Academia Sinica* 36, 59–64.

Huang, H.C. and Erickson, R.S. (2000) Soil treatment with fungal agents for control of apothecia of *Sclerotinia sclerotiorum* in bean and pea crops. *Plant Pathology Bulletin* 9, 53–58.

Huang, H.C. and Hoes, J.A. (1976) Penetration and infection of *Sclerotinia sclerotiorum* by *Coniothyrium minitans*. *Canadian Journal of Botany* 54, 406–410.

Huang, H.C. and Hoes, J.A. (1980) Importance of plant spacing and sclerotial position to development of Sclerotinia wilt of sunflower. *Plant Disease* 64, 81–84.

Huang, H.C. and Huang, J.W. (1993) Prospects for control of soilborne plant pathogens by soil amendment. *Current Topics in Botanical Research* 1, 223–235.

Huang, H.C. and Kokko, E.G. (1987) Ultrastructure of hyperparasitism of *Coniothyrium minitans* on sclerotia of *Sclerotinia sclerotiorum*. *Canadian Journal of Botany* 65, 2483–2489.

Huang, H.C. and Kokko, E.G. (1988) Penetration of hyphae of *Sclerotinia sclerotiorum* by *Coniothyrium minitans* without the formation of appressoria. *Journal of Phytopathology* 123, 133–139.

Huang, H.C. and Kokko, E.G. (1993) *Trichothecium roseum*, a mycoparasite of *Sclerotinia sclerotiorum*. *Canadian Journal of Botany* 71, 1631–1638.

Huang, H.C. and Sun, S.K. (1991) Effects of S-H mixture or Perlka^{TM} on carpogenic germination and survival of sclerotia of *Sclerotinia sclerotiorum*. *Soil Biology and Biochemistry* 23, 809–813.

Huang, H.C., Kokko, M.J. and Phillippe, L.M. (1988) White mold of dry bean (*Phaseolus vulgaris* L.) in southern Alberta, 1983–87. *Canadian Plant Disease Survey* 68, 11–13.

Huang, H.C., Kokko, E.G., Yanke, L.J. and Phillippe, R.C. (1993) Bacterial suppression of basal pod rot and end rot of dry peas caused by *Sclerotinia sclerotiorum*. *Canadian Journal of Microbiology* 39, 227–233.

Huang, H.C., Huang, J.W., Saindon, G. and Erickson, R.S. (1997) Effect of allyl alcohol and fermented agricultural wastes on carpogenic germination of sclerotia of *Sclerotinia sclerotiorum* and colonization by *Trichoderma* spp. *Canadian Journal of Plant Pathology* 19, 43–46.

Huang, H.C., Acharya, S.N. and Erickson, R.S. (2000a) Etiology of alfalfa blossom blight caused by *Sclerotinia sclerotiorum* and *Botrytis cinerea*. *Plant Pathology Bulletin* 9, 11–16.

Huang, H.C., Bremer, E., Hynes, R.K. and Erickson, R.S. (2000b) Foliar application of fungal biocontrol agents for the control of white mold of dry bean caused by *Sclerotinia sclerotiorum*. *Biological Control* 18, 270–276.

Inglis, G.D. and Boland, G.J. (1990) The micro-flora of bean and rapeseed petals and the influence of the microflora of bean petals on white mold. *Canadian Journal of Plant Pathology* 12, 129–134.

Inglis, G.D. and Boland, G.J. (1992) Evaluation of filamentous fungi isolated from petals of bean and rapeseed for suppression of white mold. *Canadian Journal of Microbiology* 38, 124–129.

Israel, M.S. and Boland, G.J. (1992) Influence of formulation on efficacy of honey bees to transmit biological controls for management of Sclerotinia stem rot of canola. *Canadian Journal of Plant Pathology* 14, 244.

Kohn, L.M. (1979) A monographic revision of the genus *Sclerotinia*. *Mycotaxon* 4, 365–444.

Li, G.Q., Wang, D.B., Jiang, D.H., Huang, H.C. and Laroche, A. (2000) First report of *Sclerotinia nivalis* on lettuce in central China. *Mycological Research* 104, 232–237.

Liew, C.L. and Prange, R.K. (1994) Effect of ozone and storage temperature on postharvest diseases and physiology of carrots (*Daucus carota* L.). *Journal of the American Society for Horticultural Science* 119, 563–567.

Martens, J.W., Seaman, W.L. and Atkinson, T.G. (eds) (1984) *Diseases of Field Crops in Canada*. The Canadian Phytopathological Society.

Martin, R.A. and Arsenault, W.J. (1987) Prevalence and severity of Sclerotinia stalk rot of tobacco on Prince Edward Island, 1985 and 1986. *Canadian Plant Disease Survey* 67, 41–43.

McLaren, D.L., Huang, H.C. and Rimmer, S.R. (1986) Hyperparasitism of *Sclerotinia sclerotiorum* by *Talaromyces flavus*. *Canadian Journal of Plant Pathology* 8, 43–48.

McLaren, D.L., Huang, H.C., Rimmer, S.R. and Kokko, E.G. (1989) Ultrastructural studies on infection of sclerotia of *Sclerotinia sclerotiorum* by *Talaromyces flavus*. *Canadian Journal of Botany* 67, 2199–2205.

McLaren, D.L., Huang, H.C. and Rimmer, S.R. (1996) Control of apothecial production of *Sclerotinia sclerotiorum* by *Coniothyrium minitans* and *Talaromyces flavus*. *Plant Disease* 80, 1373–1378.

Melzer, M.S. and Boland, G.J. (1994) Epidemiology of lettuce drop caused by *Sclerotinia minor*. *Canadian Journal of Plant Pathology* 16, 170–176.

Melzer, M.S., Smith, E.A. and Boland, G.J. (1993) Survey of lettuce drop at Holland Marsh, Ontario. *Canadian Plant Disease Survey* 73, 105.

Melzer, M.S., Smith, E.A. and Boland, G.J. (1997) Index of plant hosts of *Sclerotinia minor*. *Canadian Journal of Plant Pathology* 19, 272–280.

Mercier, J. and Reeleder, R.D. (1987a) Effect of pesticides maneb and carbaryl on the phylloplane microflora of lettuce. *Canadian Journal of Microbiology* 33, 212–216.

Mercier, J. and Reeleder, R.D. (1987b) Interactions between *Sclerotinia sclerotiorum* and other fungi on the phylloplane of lettuce. *Canadian Journal of Botany* 65, 1633–1637.

Mercier, J., Arul, J., Ponnampalam, R. and Boulet, M. (1993) Induction of 6-methoxymellein and resistance to storage pathogens in carrot slices by UV-C. *Journal of Phytopathology* 137, 44–54.

Morrall, R.A.A., Dueck, J., McKenzie, D.L. and McGee, D.C. (1976) Some aspects of *Sclerotinia sclerotiorum* in Saskatchewan, 1970–75. *Canadian Plant Disease Survey* 56, 56–62.

Ouellette, E., Raghavan, G.S.V. and Reeleder, R.D. (1990) Volatile profiles for disease detection in stored carrots. *Canadian Agricultural Engineering* 32, 255–261.

Pritchard, M.K., Boese, D.E. and Rimmer, S.R. (1992) Rapid cooling and field-applied fungicides for reducing losses in stored carrots caused by cottony soft rot. *Canadian Journal of Plant Pathology* 14, 177–181.

Reeleder, R.D. and Charbonneau, F. (1987) Incidence and severity of diseases caused by *Botrytis cinerea*, *Pythium tracheiphilum* and *Sclerotinia* spp. on lettuce in Quebec, 1985–1986. *Canadian Plant Disease Survey* 67, 45–46.

Reeleder, R.D., Raghavan, G.S.V., Monette, S. and Gariepy, Y. (1989) Use of modified atmospheres to control storage rot of carrot caused by *Sclerotinia sclerotiorum*. *International Journal of Refrigeration* 12, 159–163.

Reyes, A.A. (1988) Suppression of *Sclerotinia sclerotiorum* and watery soft rot of celery by controlled atmosphere storage. *Plant Disease* 72, 790–792.

Saito, I. (1997) *Sclerotinia nivalis*, sp. nov., the pathogen of snow mold of herbaceous dicots in northern Japan. *Mycoscience* 38, 227–236.

Stelfox, D., Williams, J.R., Soehngen, U. and Topping, R.C. (1978) Transport of *Sclerotinia sclerotiorum* ascospores by rapeseed pollen in Alberta. *Plant Disease Reporter* 62, 576–579.

Sun, S.K. and Huang, J.W. (1985) Formulated soil amendment for controlling Fusarium wilt and other soilborne diseases. *Plant Disease* 69, 917–920.

Teo, B.K., Verma, P.R. and Morrall, R.A.A. (1992) The effects of herbicides and mycoparasites at different moisture levels on carpogenic germination in *Sclerotinia sclerotiorum*. *Plant and Soil* 139, 99–107.

Tu, J.C. (1980) *Gliocladium virens*, a destructive mycoparasite of *Sclerotinia sclerotiorum*. *Phytopathology* 70, 670–674.

Tu, J.C. (1984) Mycoparasitism by *Coniothyrium minitans* on *Sclerotinia sclerotiorum* and its effect on sclerotial germination. *Phytopathologische Zeitschrift* 109, 261–268.

Tu, J.C. (1988) The role of white mold-infected white bean (*Phaseolus vulgaris* L.) seeds in the dissemination of *Sclerotinia sclerotiorum* (Lib.) de Bary. *Journal of Phytopathology* 121, 40–50.

Tu, J.C. (1997) Biological control of white mould in white bean using *Trichoderma viride*, *Gliocladium roseum* and *Bacillus subtilis* as protective foliar spray. Proceedings of the 49th International Symposium on Crop Protection, Gent, Belgium, 6 May, 1997, Part IV. *Mededelingen Faculteit Landbouwkundige en Toegepaste Biologische Wetenschappen, Universiteit Gent* 62, 979–986.

Wu, Y.S. and Wang, C.G. (1983) *Sclerotinia asari* Wu and Wang: a new species of *Sclerotiniaceae*. *Acta Phytopathologica Sinica* 13, 9–14.

Xue, A.G. and Burnett, P.A. (1994) Diseases of field pea in central Alberta in 1993. *Canadian Plant Disease Survey* 74, 102–103.
Xue, A.G., Warkentin, T.D., Rashid, K.Y., Kennaschuk, E.O. and Platford, R.G. (1995) Diseases of field pea in Manitoba in 1994. *Canadian Plant Disease Survey* 75, 156–157.
Zhou, T. and Reeleder, R.D. (1989) Application of *Epicoccum purpurascens* spores to control white mold of snap bean. *Plant Disease* 73, 639–642.
Zhou, T. and Reeleder, R.D. (1990) Selection of strains of *Epicoccum purpurascens* for tolerance to fungicides and improved biocontrol of *Sclerotinia sclerotiorum*. *Canadian Journal of Microbiology* 36, 754–759.
Zhou, T., Reeleder R.D. and Sparace S.A. (1991) Interactions between *Sclerotinia sclerotiorum* and *Epicoccum purpurascens*. *Canadian Journal of Botany* 69, 2503–2510.

100 *Sphaerotheca* and *Erysiphe* spp., Powdery Mildews (*Erysiphaceae*)

R.R. Bélanger, W.R. Jarvis and J.A. Traquair

Pest Status

Powdery mildew fungi, *Sphaerotheca* spp. and *Erysiphe* spp., are ubiquitous phyllosphere pathogens of numerous field and greenhouse crops. Their epidemiology and pathogenesis have been studied extensively but the diseases they cause remain among the most important plant diseases worldwide.

In greenhouses, powdery mildew diseases are particularly aggressive because of the constant, favourable environmental conditions that accelerate their development (Elad *et al.*, 1996). They attack most plant species and are prominent on the three most important greenhouse crops in Canada: roses, *Rosa* spp., cucumber, *Cucumis sativus* L., and tomato, *Lycopersicon esculentum* L. In roses, the disease is caused by *Sphaerotheca pannosa* (Wallroth: Fries) Léveille var. *rosae* Woronichin, now classified as *Podosphaera pannosa* (Wallroth: Fries) de Bary. On long English cucumber, the pathogen has been recently redefined from *Sphaerotheca fuliginea* (Schlechtendahl: Fries) Pollacci to *Podosphaera xanthii* (Castagne) U. Braun and N. Shishkoff. On both roses and long English cucumber, powdery mildew is the single most limiting factor in greenhouse production.

Tomato crops were long thought to be exempt from powery mildew attacks. However, greenhouse tomato has recently become a prominent host of *Erysiphe* spp. and this disease has reached epidemic proportions in certain parts of Canada (Bélanger and Jarvis, 1994) and the USA within a few years of its discovery. At the same time, greenhouse tomato has become increasingly more susceptible to the disease all over Europe.

Taken together, these three crops account for more than 50% of the total value of greenhouse sales, estimated at roughly Can$1.2 billion (Statistics Canada, 1998). The cost for their control can reach Can$10,000 ha^{-1} year^{-1}.

Background

Powdery mildews are largely controlled by regular applications of fungicides. In roses, dodemorph-acetate (Meltatox®) is probably the most efficient and the most commonly used product. In cucumber, myclobutanil (Nova 40W) has been recently registered for powdery mildew control under greenhouse conditions. In tomato, only Microfine Wettable Sulphur (sulphur 92%) is registered against powdery mildew. The latter product is often used on roses as well.

While no cultivars of roses, long English cucumber or tomato are known to be completely resistant to powdery mildews, some are more tolerant than others. However, as it is often the case, the most productive cultivars are also the most susceptible and growers will usually favour productivity even if it implies more fungicide treatments.

Biological Control Agents

Fungi

Considering the ubiquity of powdery mildews and their devastating impact, it does not appear that they have received a proportionate research effort over the years in the field of biological control. This is rather surprising as one would expect these fungi to be easy targets for hyperparasites because of their ectotrophic growth. If this assumption is undeniable, achieving complete control of powdery mildew with natural enemies remains elusive. Over the years, several natural antagonists have been described and all agents are fungi (Bélanger *et al.*, 1998).

Ampelomyces quisqualis Cesati was the first fungus to be reported as a parasite of powdery mildews (Yarwood, 1932). Since then it has been shown to parasitize several species of powdery mildew (Sundheim, 1982; Sundheim and Tronsmo, 1988). Under greenhouse or field conditions, most workers have reported that this antagonist was effective only under very high humidity (Jarvis and Slingsby, 1977).

Verticillium lecanii (A. Zimmermann) Viégas is polyphagous and parasitizes arthropods, rusts and powdery mildews (Sundheim and Tronsmo, 1988). Askary *et al.* (1998) tested various strains for their ability to parasitize potato aphid, *Macrosiphum euphorbiae* Thomas, and *S. fuliginea*. One of them, *V. lecanii* strain 198499, was found to be virulent on both organisms, although its activity against *S. fuliginea* was not as good as that of *Pseudozyma flocculosa* (see below). Investigations into the mode of action of this strain by electron microscopy suggested that antibiosis was an important component of its virulence (Askary *et al.*, 1997).

Tilletiopsis spp. have often been associated with biological control against powdery mildew (Hijwegen and Buchenauer, 1984). Urquhart *et al.* (1994) isolated several *Tilletiopsis* spp. from powdery mildew-infected leaves sampled in the lower Fraser Valley, British Columbia. They showed that two species, *T. washingtonensis* Nyland and *T. pallescens* Gokhale, when applied at a rate of 1×10^8 conidia ml^{-1}, could reduce the incidence of cucumber powdery mildew under greenhouse conditions. They originally suggested that glucanases were involved in activity of the antagonists (Urquhart *et al.*, 1994) but recent evidence indicates that antibiosis is the main mode of action (Z.K. Punja, Burnaby, 1998, personal communication).

Pseudozyma flocculosa (Traquair, L.A. Shaw and Jarvis) Boekhout and Traquair is the most recent and probably the most efficient natural antagonist of powdery mildew to be identified. It was discovered along with another closely related species, *P. rugulosa* (Traquair, L.A. Shaw and Jarvis) Boekhout and Traquair (Traquair *et al.*, 1988).[1] Jarvis *et al.* (1989) were the first to

[1]At the time, Traquair *et al.* (1988) described both species as yeast-like fungi in the *Endomycetaceae*, *Stephanoascus flocculosus* Traquair, Shaw and Jarvis (anamorph: *Sporothrix flocculosa* Traquair, Shaw and Jarvis) and *S. rugulosus* Traquair, Shaw and Jarvis (anamorph: *S. rugulosa* Traquair, Shaw and Jarvis). However, they were later redefined as basidiomycetous yeasts related to anamorphs of Ustilaginales belonging to the genus *Pseudozyma* Bandoni emend. Boekhout (Boekhout, 1995).

report that both fungi were powerful antagonists of cucumber powdery mildew, *S. fuliginea*, with *P. flocculosa* apparently more active under different environmental conditions. Subsequently, Hajlaoui and Bélanger (1991, 1993) demonstrated that the same two antagonists were equally effective against *S. pannosa* var. *rosae* and *Erysiphe graminis* de Candolle (= *Blumeria graminis* (de Candolle) E.O. Speer f. sp. *tritici* Émile Marchal), responsible for rose and wheat powdery mildew, respectively. In controlled experiments, *P. flocculosa* was found to be less demanding than *P. rugulosa* or *T. washingtonensis* Nyland in terms of temperature and humidity requirements.

Cytological and microscopical studies indicated that *P. flocculosa* did not penetrate its host but rather induced a rapid plasmolysis of powdery mildew cells (Hajlaoui *et al.*, 1992). These results suggested that the antagonist acted by antibiosis rather than by parasitism. Furthermore, when culture filtrates of the fungus were extracted and bioassayed against target fungi, it was possible to reproduce the same cell reactions as observed when powdery mildew fungi were confronted with *P. flocculosa* (Hajlaoui *et al.*, 1994).

Chemical analysis of the culture filtrates revealed the presence of at least four molecules with antifungal activity, three of them being closely related fatty acids (Choudhury *et al.*, 1994; Benyagoub *et al.*, 1996a). Avis *et al.* (2000) were able to synthesize two of the three fatty acids and demonstrated that they account for the antagonistic activity of *P. flocculosa*. These molecules act by interfering with membrane fluidity and, as a result, membrane composition would determine the level of specificity. Indeed, the resistance of *P. flocculosa* to its own antibiotics versus the relative sensitivity of other fungi appears to be linked to the sterol composition in fungal membranes (Benyagoub *et al.*, 1996b). This hypothesis has been further confirmed and was proposed as a model of activity of the antibiotics in the membranes (T.J. Avis and R.R. Bélanger, in press). Based on this model, it becomes easy to determine the sensitivity of fungi to *P. flocculosa* and to evaluate the possibility of development of resistant strains. So far, in spite of repeated exposures to the synthesized antibiotics, it has not been possible to obtain a resistant strain of *S. fuliginea*, which would indicate that resistance development in the field is unlikely, considering that the antibiotics degrade very rapidly in nature.

On the other hand, mutants of *P. flocculosa* that have lost their ability to produce the antibiotics have recently been obtained (Y. Cheng and R.R. Bélanger, unpublished). Bioassays with these mutants have confirmed that they have lost their antagonistic properties. These mutants will be extremely valuable in pursuing studies into the mode of action of *P. flocculosa*.

When tested under commercial conditions under a restrictive research permit, fresh preparations of *P. flocculosa* offered as good a control of rose powdery mildew as the commonly used fungicides dodemorph-acetate (Meltatox®) and microfine sulphur (Bélanger *et al.*, 1994). In addition, for some cultivars, the biological treatment improved flower quality by eliminating the stress (phytotoxicity) caused by fungicides.

These results prompted the commercial development of a formulation based on *P. flocculosa* conidia (Sporodex®) for use against powdery mildew on greenhouse crops. In two large-scale trials Sporodex® achieved the best level of powdery mildew control on long English cucumber when compared to AQ-10® (a commercial product based on *A. quisqualis*) and fresh preparations of *V. lecanii* (Dik *et al.*, 1998).

An improved formulation leaving no residues was further developed and tested under commercial conditions in The Netherlands, Canada and Colombia. In The Netherlands, treatment of a semitolerant long English cucumber cultivar with Sporodex® allowed the crop to be grown pesticide-free for a complete season (16 weeks). In Canada, R. Cerkauskas (Harrow, 2000, personal communication) compared Sporodex® to myclobutanil in a commercial greenhouse. While absolute control of powdery mildew with Sporodex® was not as good as with the fungicide, cucumber yield had been improved by as much as 15% under the biological treatment. Finally, Bureau (1999) evaluated the efficacy of Sporodex® against rose powdery mildew in standard commercial greenhouses for rose production in Colombia. In two separate trials, Bureau reported that the product was as efficient as fungicides used for powery mildew control, and flower quality was improved.

Evaluation of Biological Control

Sporodex® is effective for control of powdery mildew in greenhouse crops and offers a safe, efficient and chemical-free means of control. A submission for registration has been filed in Canada and the USA.

Biological control of powdery mildews remains a challenge in spite of the different agents that have been identified as their natural enemies. For optimal success, the ecology of both the pathogen and the biological control agent(s) should be respected when carrying out a biological control programme.

Recommendations

Future work should include:

1. Improving delivery and formulations of biological control agents to alleviate the high humidity requirements that most require for maximum efficacy.

Acknowledgements

Plant Products Co. Ltd (Brampton, Ontario) supported research and development of Sporodex®.

References

Askary, H., Benhamou, N. and Brodeur, J. (1997) Ultrastructural and cytochemical investigations of the antagonists effect of *Verticillium lecanii* on cucumber powdery mildew. *Phytopathology* 87, 359–368.

Askary, H., Carrière, Y., Bélanger, R.R. and Brodeur, J. (1998) Pathogenicity of the fungus *Verticillium lecanii* to aphids and powdery mildew. *Biocontrol Science and Technology* 8, 23–32.

Avis, T.J., Boulanger, R.R. and Bélanger, R.R. (2000) Synthesis and biological characterization of (*Z*)-9-heptadecenoic and (*Z*)-6-methyl-9-heptadecenoic acids, fatty acids with antibiotic activity produced by *Pseudozyma flocculosa*. *Journal of Chemical Ecology* 26, 987–1000.

Bélanger, R.R. and Jarvis, W.R. (1994) Occurrence of powdery mildew on greenhouse tomatoes in Canada. *Plant Disease* 78, 640.

Bélanger, R.R., Labbé, C. and Jarvis, W.R. (1994) Commercial-scale control of rose powdery mildew with a fungal antagonist. *Plant Disease* 78, 420–424.

Bélanger, R.R., Dik, A.J. and Menzies, J.G. (1998) Powdery mildews – Recent advances toward integrated control. In: Boland, G.J. and Kuykendall, L.D. (eds) *Plant–Microbe Interactions and Biological Control*. Marcel Dekker, New York, pp. 89–109.

Benyagoub, M., Willemot, C. and Bélanger, R.R. (1996a) Influence of a subinhibitory dose of antifungal fatty acids from *Sporothrix flocculosa* on cellular lipid composition in fungi. *Lipids* 31, 1077–1082.

Benyagoub, M., Bel Rhlid, R. and Bélanger, R.R. (1996b) Purification and characterization of new fatty acids with antibiotic activity produced by *Sporotrhix flocculosa*. *Journal of Chemical Ecology* 22, 405–413.

Boekhout, T. (1995) *Pseudozyma bandoni* emend. Boekhout, a genus for yeast-like anamorphs of Ustilaginales. *Journal of General and Applied Microbiology* 41, 355–366.

Bureau, A. (1999) Évaluation du biofongicide Sporodex contre le blanc poudreux de la rose cultivée sous serres colombiennes. Thèse de maîtrise no. 18017, Université Laval, Quebéc.

Choudhury, S.R., Traquair, J.A. and Jarvis, W.R. (1994) 4-Methyl-7,11-heptadecadenal and 4-methyl-7,11-heptadecadienoic acid: New antibiotics from *Sporothrix flocculosa* and *Sporothrix rugulosa*. *Journal of Natural Products* 57, 700–704.

Dik, A.J., Verhaar, M.A. and Bélanger, R.R. (1998) Comparison of three biological control agents against cucumber powdery mildew (*Sphaerotheca fuliginea*) in semi-commercial-scale glasshouse trials. *European Journal of Plant Pathology* 104, 413–423.

Elad, Y., Malathrakis, N.E. and Dik, A.J. (1996) Biological control of Botrytis-incited diseases and powdery mildews in greenhouse crops. *Crop Protection* 15, 229–240.

Hajlaoui, M. and Bélanger, R.R. (1991) Comparative effects of temperature and humidity on the activity of three potential antagonists of rose powdery mildew. *Netherlands Journal of Plant Pathology* 97, 203–208.

Hajlaoui, M. and Bélanger, R.R. (1993) Antagonism of the yeast-like phylloplane fungus *Sporothrix flocculosa* against *Erysiphe graminis* var. *tritici. Biocontrol Science and Technology* 3, 427–434.

Hajlaoui, M.R., Benhamou, N. and Bélanger, R.R. (1992) Cytochemical study of the antagonistic activity of *Sporothrix flocculosa* on rose powdery mildew, *Sphaerotheca pannosa* var. *rosae. Phytopathology* 82, 583–589.

Hajlaoui, M.R., Traquair, J.A., Jarvis, W.R. and Bélanger, R.R. (1994) Antifungal activity of extracellular metabolites produced by *Sporothrix flocculosa. Biocontrol Science and Technology* 4, 229–237.

Hijwegen, T. and Buchenauer, H. (1984) Isolation and identification of hyperparasitic fungi associated with Erysiphaceae. *Netherlands Journal of Plant Pathology* 90, 70–82.

Jarvis, W.R. and Slingsby, K. (1977) The control of powdery mildew of greenhouse cucumber by water sprays and *Ampelomyces quisqualis. Plant Disease Reporter* 61, 728–730.

Jarvis, W.R., Shaw, L.A. and Traquair, J.A. (1989) Factors affecting antagonism of cucumber powdery mildew by *Stephanoascus flocculosus* and *S. rugulosus. Mycological Research* 92, 162–165.

Statistics Canada (1998) *Greenhouse, Sod and Nursery Industries.* Catalogue no. 22-202-XIB, pp. 14–15.

Sundheim, L. (1982) Control of cucumber powdery mildew by the hyperparasite *Ampelomyces quisqualis* and fungicides. *Plant Pathology* 31, 209–214.

Sundheim, L. and Tronsmo, A. (1988) Hyperparasites in biological control. In: Mekerji, K.G. and Garg, K.L. (eds) *Biocontrol of Plant Diseases,* Vol. I. CRC Press, Boca Raton, Florida, pp. 53–69.

Traquair, J.A., Shaw, L.A. and Jarvis, W.R. (1988) New species of *Stephanoascus* with *Sporothrix* anamorphs. *Canadian Journal of Botany* 66, 926–933.

Urquhart, E.J., Menzies, J.G. and Punja, Z.K. (1994) Growth and biological control activity of *Tilletiopsis* species against powdery mildew (*Sphaerotheca fuliginea*) on greenhouse cucumber. *Phytopathology* 84, 341–351.

Yarwood, C.E. (1932) *Ampelomyces quisqualis* on clover mildew. *Phytopathology* 22, 31.

101 *Venturia inaequalis* (Cooke) Winter, Apple Scab (*Venturiaceae*)

O. Carisse, J. Bernier and V. Philion

Pest Status

Venturia inaequalis (Cooke) Winter (anamorph *Spilocea pomi* Fries), causal agent of apple scab, is distributed worldwide but it is more prevalent in regions with cold and wet spring conditions, e.g. Ontario, Quebec and the Maritimes. It is considered to be the single most important disease of apple *Malus pumila* Miller (= *M.*

domestica Borkhausen) in eastern Canada and in several other apple production areas such as the USA and Europe. Most apple cultivars are susceptible to *V. inaequalis*. Fungicides are presently the only control method, with 8–16 fungicide sprays applied yearly. In Quebec, these fungicides represent 9.8% of all pesticides used in agriculture. This is an important input cost to growers, e.g. in Quebec, where scab control may cost up to Can$3 million or about 10% of all production costs. These costs vary depending on weather pattern, control programmes and products used.

In autumn, when apple leaves have fallen, *V. inaequalis* becomes saprophytic. On infected leaves, two compatible mating types come together to form a pseudothecium initial through fertilization and formation of the ascogonium. The fungus overwinters as pseudothecial initials. In early spring, the pseudothecia mature and, when leaves are wetted by rain, ascospores are ejected. Ascospores will germinate on susceptible leaves if there is enough free water. Once the appressoria and infection pegs are formed, the hyphae move subcutically, and lesions produce conidia, which will be splash dispersed to new leaves and fruit throughout the season.

Background

Control if *V. inaequalis* is mostly achieved by applying fungicides, despite the cost, risk of resistance development and environmental and health concerns. Development of fungicide resistance had an impact on apple production in Canada. In Ontario, about 50% of the *V. inaequalis* isolated from samples were resistant to some eradicant fungicides currently used, e.g. *V. inaequalis* developed resistance to Benlate (benomyl) within only 3 years (Ontario Ministry of Agriculture and Food, 1993). The fungus is becoming increasingly resistant to dodine and there is concern about resistance to even the most recently developed families of fungicides, such as DMI® (sterol demethylation inhibitors) (Braun and McRae, 1992; Carisse and Pelletier, 1994), and the strobilurins and anilino pyrimidines. Almost all fungicide applications are applied to control primary infection and, depending on the region and level of control of primary infections, secondary infections. In some cases, fungicides are applied in late summer and autumn to control storage scab or to reduce primary inoculum the next year.

So far, very little success has occurred in reducing the number of fungicide applications needed to control *V. inaequalis*, mainly because growers know that inadequate control can cause rapid disease development, resulting in serious losses. Further, concerns that a reduced spray programme against *V. inaequalis* could increase the risk of secondary diseases slowed the adoption of reduced spray strategies. As a result, growers tend to spray large quantities of fungicides on all cultivars, including those with a known low susceptibility and in orchards with a very low inoculum level.

Biological Control Agents

Fungi

Biological control targeting both primary and secondary leaf infection has been tested with little success (Andrews *et al.*, 1983; Cullen *et al.*, 1984; Boudreau and Andrews, 1987; Burr *et al.*, 1996; Carisse, 1999). Because ascospores are the main source of primary inoculum in several production areas, targeting ascospore reduction using biological control was more successful (Miedtke and Kennel, 1990; Young and Andrews, 1990; Carisse *et al.*, 2000). Heye (1982) screened 57 organisms for their ability to inhibit pseudothecia formation. The results showed that *Athelia bombacina* Persoon completely inhibited pseudothecial formation in controlled laboratory experiments and in the field. Moreover, subsequent reports (Young and Andrews, 1990) showed that this approach was encouraging, even more so if the potential synergism of other compatible methods was also considered. These methods include urea applications (Burchill and Cook, 1970) and chemical applications of etephon or Ethrel® (2-chloroethylphosphonic acid) in autumn to promote defoliation (Heye, 1982). The results were incomplete because *A. bombacina* was not evaluated over the course of an entire ascospore ejection season and in large field trials (Miedtke and Kennel, 1990). Although promising, this biological control agent was not developed to a commercial level. Research was undertaken to develop a biological control agent that would interfere with overwintering of *V. inaequalis*. Because ascospores are produced in pseudothecia that overwinter in dead apple leaves, organisms sharing this very specific ecological niche were collected and tested for their potential to inhibit pseudothecia development and consequently ascospore production. To do so, dead apple leaves were collected in early spring and late autumn, 1993, in six abandoned orchards located in the different apple-growing regions of Quebec. A total of 189 fungal isolates were recovered from leaves collected in spring and 156 from those collected in autumn. Most of the isolates (75%) were deuteromycetes and 15 had never been recorded previously as apple-leaf colonizers in North America (Bernier *et al.*, 1996). The orchard saprophytes and a known antagonist, *A. bombacina*, were evaluated *in vitro* to determine their ability to degrade apple leaves and to inhibit pseudothecia and ascospore production (Philion *et al.*, 1997a). From this evaluation, five fungal isolates, *Microsphaeropsis* sp., *M. arundinis* (Ahmad), *Ophiostoma* sp., *Diplodia* sp. and *Trichoderma* sp., were selected, based on their capacity to inhibit ascospore production. These potential biological control agents were further tested under orchard conditions. The most consistent reduction in ascospore production was obtained with *Microsphaeropsis* sp., strain P130A (Carisse *et al.*, 2000). Study of the mode of action of this isolate revealed that it is a mycoparasite (Benyagoub *et al.*, 1998).

The strategy developed consists of applying the biological control agent to apple leaves in autumn to inhibit sexual-stage development and thus reduce ascospore potential. The following spring, ascospore density is monitored with spore traps and the decision whether to apply a fungicide is made on the basis of inoculum potential reduction (Carisse *et al.*, 1999), actual number of ascospores present in air (Philion *et al.*, 1997b) and infection risk based on weather conditions, tree phenology and fungicide residue level from previous sprays. This scab management strategy was evaluated in Quebec, in a mature orchard of 0.41 ha planted with 'McIntosh' and 'Lobo' cultivars. The biological control agent was applied, in mid-October, at a rate of 10^{11} spores ha^{-1}, as a postharvest, pre-leaf-fall treatment. The effect of strain P130A on ascospore production was evaluated the next spring by measuring the concentration of *V. inaequalis* ascospores in the air during each rain event during the primary infection period from the end of April until late June. In 1997 and 1998, the application of strain P130A resulted in a 70.7% and 79.8% reduction, respectively, in the total amount of air-borne ascospores trapped compared to the control plots. In other similar trials, in 1998–1999, the biological control agent reduced ascospore production by 70–85% depending on the inoculum potential in the orchards. This reduction of inoculum allowed about a 40% reduction of fungicide sprays. A better reduction of inoculum was obtained when the biological control agent was mixed with 5% urea (46% N). Trials were conducted in orchards with different levels of inoculum. In low-inoculum orchards, application of the biofungicide alone or mixed with urea resulted in a substantial reduction in the number of fungicide sprays required (five as compared to nine in the untreated plot). However, in orchards with high inoculum potential, autumn application of the biofungicide alone or mixed with urea resulted in a small reduction in the number of fungicides required (five as compared to six in the untreated plot). The incidence of scab was substantially reduced from 12% in untreated plot to 2.21% and 1.18% in the treated plots.

Evaluation of Biological Control

The results of field trials clearly demonstrated that biological control of ascospores to reduce inoculum is successful and resulted in reduced fungicide use. Consequently biological control should be incorporated into apple scab management programmes. Continued research on mass production and formulation of microbial fungicides will facilitate commercialization and reduce the investment required by companies to develop biological control agents. However, the requirements for evaluating the environmental impacts of the release of microbial fungicides should be clarified to accelerate registrations in Canada.

Recommendations

Further work should include:

1. Development of a precise method to quantify air-borne inoculum in order to increase the benefit provided by application of *Microsphaeropsis* sp., strain P130A;
2. Evaluation of the impact of reduced spray programmes on secondary diseases development;
3. Development of a stable formulation that would allow application earlier in autumn;
4. Evaluation of the best application techniques and timing;
5. Integrating the biological control agent with other products that enhance defoliation and leaf decomposition;
6. Selection of strains of *Microsphaeropsis* sp., based on their efficacy and fitness;
7. Searching for other biological control agents.

References

Andrews, J.H., Berbee, F.M. and Nordheim, E.V. (1983) Microbial antagonism to the imperfect stage of the apple scab pathogen, *Venturia inaequalis*. *Phytopathology* 73, 228–234.

Benyagoub, M., Benhamou, N. and Carisse, O. (1998) Cytochemical investigation of the antagonistic interaction between *Microsphaeropsis* sp. (isolate P130A) and *Venturia inaequalis*. *Phytopathology* 88, 605–613.

Bernier, J., Carisse, O. and Paulitz, T.C. (1996) Fungal communities isolated from dead apple leaves from orchards in Quebec. *Phytoprotection* 77, 129–134.

Boudreau, M.A. and Andrews, J.H. (1987) Factors influencing antagonism of *Chaetomium globosum* to *Venturia inaequalis*: A case study in failed biocontrol. *Phytopathology* 77, 1470–1475.

Braun, P.G. and McRae, K.B. (1992) Composition of a population of *Venturia inaequalis* resistant to myclobutalanil. *Canadian Journal of Plant Pathology* 14, 215–220.

Burchill, R.T. and Cook, R.T.A. (1970) The interaction of urea and micro-organism in suppressing the development of perithecia of *Venturia inaequalis* (Cke) Wint. In: Preece, T.F. and Dickinson, C.H. (eds) *Ecology of Leaf Surface Micro-organisms*. Academic Press, New York, New York, pp. 471–483.

Burr, T.J., Matteson, M.C., Smith, C.A., Corral-Garcia, M.R. and Huang, T. (1996) Effectiveness of bacteria and yeast from apple orchards as biological control agents of apple scab. *Biological Control* 6, 151–157.

Carisse, O. (1999) 50 years of research on biological control of apple scab. *International Organization for Biological Control/Western Palaearctic Regional Section, Bulletin* 23, 5–10.

Carisse, O. and Pelletier, J.R. (1994) Sensivity distribution of *Venturia inaequalis* to fenarimol in Quebec apple orchards. *Phytoprotection* 75, 35–43.

Carisse, O., Svircev, A. and Smith, R. (1999) Integrated biological control of apple scab. *International Organization for Biological Control/Western Palaearctic Regional Section, Bulletin* 23, 23–28.

Carisse, O., Philion, V., Rolland, D. and Bernier, J. (2000) Effect of fall application of fungal antagonists on spring ascospore production of the apple scab pathogen, *Venturia inaequalis*. *Phytopathology* 90, 31–37.

Cullen, D., Barbee, F.M. and Andrews, J.H. (1984) *Chaetomium globosum* antagonizes the apple scab pathogen, *Venturia inaequalis*, under field conditions. *Canadian Journal of Botany* 62, 1814–1818.

Heye, C.C. (1982) Biological control of the perfect stage of the apple scab pathogen, *Venturia inaequalis* (Cke) Wint. PhD thesis, University of Wisconsin, Madison, Wisconsin.
Miedtke, U. and Kennel, W. (1990) *Athelia bombacina* and *Chaetomium globosum* as antagonists of the perfect stage of the apple scab pathogen (*Venturia inaequalis*) under under field conditions. *Journal of Plant Diseases* 97, 24–32.
Ontario Ministry of Agriculture and Food (1993) *1994–1995 Fruit Production Recommendations.* Ontario Ministry of Agriculture and Food Publication 360, pp. 18, 24–31.
Philion, V., Carisse, O. and Paulitz, T. (1997a) *In vitro* evaluation of fungal isolates for their ability to influence leaf rheology, production of pseudothecia, and ascospores of *Venturia inaequalis. European Journal of Plant Pathology* 103, 441–452.
Philion, V., Carisse, O., Garcin, A. and Vanesson, S. (1997b) Monitoring airborne ascospore of *Venturia inaequalis* scab. *International Organization for Biological Control/Western Palaearctic Regional Section, Bulletin* 20(9), 180–184.
Young, C.S. and Andrews, J.H. (1990) Inhibition of pseudothecial development of *Venturia inaequalis* by the basidiomycete *Athelia bombacina* in apple leaf litter. *Phytopathology* 80, 536–542.

102 *Verticillium dahliae* Klebahn, Verticillium Wilt (*Moniliaceae*), and *Streptomyces scabies* (Thaxter) Lambert and Loria, Potato Scab (*Streptomycetaceae*)

G. Lazarovits, M. Tenuta, K.L. Conn and N. Soltani

Pest Status

Verticillium dahliae Klebahn, causal agent of Verticillium wilt, causes severe yield reductions in various important crops worldwide (Powelson and Rowe, 1993). In Canada, *V. dahliae* is an important pathogen on potato, *Solanum tuberosum* L. and tomato, *Lycopersicon esculentum* L. In Ontario, the vast majority of fields sampled (>50) near Alliston were found to have more than 80% disease incidence (G. Lazarovits, unpublished). A survey of five tomato fields near Leamington revealed that more than 50% of the plants were infected by race 2 of *V. dahliae*, for which there are no resistant cultivars (Dobinson and Lazarovits, 1994). The economic value of loss due to this disease in Canada has never been clarified, but in the USA Powelson and Rowe (1993) rated early dying as the most important disease of both seed and commercial potato crops and as the second most important yield constraint to potato production.

Infection is initiated from microsclerotia that overwinter in soil or in infected plant debris. Microsclerotia are highly adapted for survival in soil, where they can remain viable for more than a decade (Wilhelm, 1955). The incidence and severity of Verticillium wilt is directly related to microsclerotia density (Pullman and DeVay, 1982; Xiao and Subbarao, 1998) and, because these are the primary source of inoculum, they need to be targeted for disease control. Because various plant pathogenic nematode species enhance Verticillium wilt (Powelson and Rowe, 1993), reduction in their populations can also be targeted to reduce wilt severity.

Streptomyces scabies (Thaxter) Lambert and Loria, a Gram-positive bacterium, is the predominant causal agent of potato scab, an economically important disease in North America and Europe (Lambert and Loria, 1989; Goyer *et al.*, 1996; Loria *et al.*, 1997). In Canada, this disease is often a limiting production factor in all provinces where potatoes are grown. Growers can lose up to 50% of the value of the tubers delivered to processors due to scab. Unsightly tubers are not marketable for table stock. Depending on the *S. scabies* strain and soil conditions, bacterial invasion can lead to shallow, raised or deep-pitted lesions (Goyer *et al.*, 1996; Loria *et al.*, 1997). Pathogenic *S. scabies* strains produce phytotoxins (thaxtomins) that are an excellent indicator of virulence (King *et al.*, 1991; Loria *et al.*, 1995; Conn *et al.*, 1998). *S. scabies* poses a long-term threat to potato production because spores and mycelium can survive in soil or on plant residues for over a decade (Kritzman and Grinstein, 1991).

Background

No effective disease control strategy is available to growers to control Verticillium wilt or potato scab. In more intense agricultural settings, fumigation with chemical sterilants such as methyl bromide, Vapam and Chloropicrin can kill either nematodes, *V. dahliae*, or both, and thus reduce disease incidence (Easton *et al.*, 1974; Ben Yephet *et al.*, 1983). However, these pesticides are unavailable or too costly to be widely used by Canadian growers.

The development of biological control concepts have been ongoing for over a century, and driven mainly by entomologists observing control of insect pests by predators and parasitoids. As a result, the classical definition of biological control is 'a population-level process in which one species population lowers the numbers of another species by mechanisms such as predation, parasitism, pathogenicity, or competition' (Van Driesche and Bellows, 1996). Concepts for biological control of plant pathogens have a much shorter history and begin in the first decades of the 20th century. Sanford (1926) observed that addition of grass clippings to soil reduced the incidence of potato scab of potato due to displacement of the pathogen by saprophytic organisms. Because the activity of plant pathogens, particularly those resident in soil, is altered by a variety of mechanisms in addition to predation, parasitism and competition by antagonists, the definition of biological control applied to plant pathogens is broader than that applied to other pests (Van Driesche and Bellows, 1996). In their overview of the concepts of biological control of plant pathogens, Cook and Baker (1983) stated that 'biological control is the reduction of the amount of inoculum or disease-producing activity of a pathogen accomplished by or through one or more organisms (antagonists) other than man'. They elaborated that 'antagonists are biological agents with the potential to interfere in the life processes of plant pathogens'. Mechanisms by which antagonists interfere with or suppress plant diseases are many and include parasitism, competition, toxin or antibiotic production and acquired resistance of the plant host. The compounds generated as a result of microorganism activity do not have to be directly toxic to pathogens but can include compounds that stimulate their premature germination or increase the activity of microbial antagonists.

The addition of organic amendments to soil supplies a rich source of energy and nutrients to microorganisms and the amendments themselves alter the physical and chemical environment of soil. As a result, addition of amendments can change the populations and activities of soil organisms. This suggests that one approach to achieving biological control of soil-borne plant pathogens is to 'feed' soil the right substrate to promote antagonists to pathogens. The use of organic amendments that are converted into biologically active products in soil is rapidly expanding in many countries, either as a stand-alone process or together with other treatments such as solarization (Gamliel, 2000).

Biological Control

Soil amendments

The use of organic amendments to control plant pathogens was initiated in the hope of using amendments, e.g. bloodmeal and soymeal, as carriers of biological control agents that would also help to establish them in soil. However, we found that the amendments alone, without addition of biological control agents, suppressed the incidence of Verticillium wilt of aubergine, *Solanum melongena* var. *esculentum* Nees, as Wilhelm (1955) found. Because of the demonstrated efficacy of amendments, research was therefore directed towards determining if they suppress plant disease by increasing the population and activity of microbial antagonists of pathogenic organisms.

To undertake studies on biological control of *V. dahliae* and *S. scabies*, techniques to quantitatively add and recover the pathogens from soil were first developed. Hawke and Lazarovits (1994) developed and M. Tenuta and G. Lazarovits (unpublished) modified a rapid bioassay that permitted quantitative determination of survival of microsclerotia added to amended soil. Conn *et al.* (1998) developed a semiselective agar medium to isolate *S. scabies* from amended soil and, when used in conjunction with determination of which recovered isolates produced thaxtomin, was capable of quantifying pathogenic *S. scabies* in soil at populations greater than 10^3 colony-forming units (cfu) g^{-1} of soil. We have also developed a more rapid means to detect and quantify *S. scabies* in soil using the polymerase chain reaction (PCR) (Lazarovits *et al.*, 1998). This method detects the *nec1* gene sequence in *S. scabies*, which is highly linked to pathogenicity (Bukhalid and Loria, 1997).

Lazarovits and Conn (1997) developed a soil microcosm assay to facilitate testing the survival of *V. dahliae* microsclerotia and, recently, *S. scabies*, added to soil-amendment mixtures placed in test tubes. The impact of amendments on soil pH, ion concentrations and numbers of selected soil microbial groups is routinely done using soil microcosms. Conn and Lazarovits (1999) and Lazarovits *et al.* (1999) compared the incidence of Verticillium wilt in potato plants to the survival of *V. dahliae* microsclerotia buried in amended soil in the field and in the same amended soil tested in soil microcosms done in the laboratory. Various amendments, including soymeal meat and bonemeal, solid cattle manure, liquid swine and poultry manures and organic fertilizers were tested. The impact of amendments in reducing microsclerotia survival in microcosms accurately predicted the efficacy of these amendments in reducing Verticillium wilt in the field. Development of the microcosm assay has led to the ability to manipulate and measure many parameters in soil that potentially influence survival of *V. dahliae* microsclerotia and *S. scabies*, advancing our understanding of the mode of action of amendments in controlling plant diseases.

Various types of amendments added to soil reduced disease incidence and levels of many pathogens and pests (Lazarovits *et al.*, 2000). In Ontario, studies on commercial potato fields near Alliston showed reduced incidence of Verticillium wilt and potato scab and the numbers of plant pathogenic nematodes after addition of nitrogenous organic amendments such as meat, bonemeal and soymeal (37 tonnes ha^{-1}), and poultry manure (66 tonnes ha^{-1}) to soil (Conn and Lazarovits, 1999; Lazarovits *et al.*, 1999). Laboratory studies showed that within weeks of adding bloodmeal and soymeal to soil, *V. dahliae* microsclerotia were killed (Hawke, 1994). Addition of the same materials to autoclaved soils did not result in microsclerotia death (Hawke,

1994). It was concluded that soil microorganisms acting on nitrogenous organic amendments were responsible for killing the microsclerotia.

Lazarovits and Conn (1997) showed that addition of swine manure to soils from commercial potato fields near Alliston killed *V. dahliae* microsclerotia within days under both laboratory and greenhouse conditions, but only in some soils tested. In field trials, 55 hl of swine manure ha^{-1} reduced severity of Verticillium wilt, potato scab and the numbers of plant parasitic nematodes for 3 years, but at only one of two sites examined (Conn and Lazarovits, 1999). Disease levels at this site were reduced by 60–80% compared to the control treatment.

In field trials in Ontario, addition of 10 and 20 hl ha^{-1} of ammonium lignosulphonate, a by-product of the pulp and paper industry with a high nitrogen and carbon content, consistently resulted in a 30–70% decrease in incidence of Verticillium wilt compared to controls (Soltani *et al.*, 2000; N. Soltani, K.L. Conn and G. Lazarovits, unpublished). The incidence of potato scab was also reduced six- to 11-fold, with marketable yield (less than 5% surface covered with scab lesions) increasing three- to 20-fold compared to controls (Soltani *et al.*, 2000; N. Soltani, K.L. Conn and G. Lazarovits, unpublished). In Prince Edward Island in 1999, field studies using 10 hl of ammonium lignosulphonate ha^{-1} applied to two commercial potato farms showed little effect on incidence of Verticillium wilt at either site, though the incidence of disease was generally low (G. Lazarovits, K.L. Conn and W. Kelly, unpublished). However, incidence of potato scab was decreased by 60% and marketable yield increased sixfold at both sites (G. Lazarovits, K.L. Conn and W. Kelly, unpublished). In laboratory experiments, 10 and 20 hl of ammonium lignosulphonate ha^{-1} reduced nematode numbers in soil by 60% and 95%, respectively (Soltani and Lazarovits, 1998).

Antagonism of pathogens due to transformation of soil amendments has been investigated as a biological control mechanism. Using soil microcosms, M. Tenuta and G. Lazarovits (unpublished) identified ammonia (NH_3) and nitrous acid (HNO_2) as the primary toxic products released during decomposition of high nitrogen amendments. The concentrations of the toxicants achieved were controlled by soil pH, with NH_3 requiring pH to be in excess of 8.5 and HNO_2 requiring pH to be below 5.5. An excess of 10 mmol of NH_3 and 0.05 mmol of HNO_2 maintained over a 4-day period were sufficient to kill 95% of the microsclerotia buried in soil. Levels of NH_3 and HNO_2 found in amended soil were also sufficient to kill propagules of other plant pathogens, including *S. scabies*, in toxicity assays done in solution (M. Tenuta and G. Lazarovits, unpublished). Generation of NH_3 and HNO_2 following amendment is soil specific (M. Tenuta and G. Lazarovits, unpublished). NH_3 does not form in soils with organic carbon levels above 1.7%. HNO_2 does not form when buffering capacity for organic soils and initial soil pH for mineral soils are high. While NH_3 and HNO_2 are toxic to some organisms, the numbers of other soil microorganisms increase by 100–1000-fold in amended soils. Thus, these products stimulate general microbial activity and populations. In the case of toxicity due to NH_3 and HNO_2, the antagonists responsible for their production are ammonifying and nitrifying bacteria and fungi, respectively.

Using soil microcosms, Conn and Lazarovits (2000) showed that swine manure killed microsclerotia within days of addition and this occurred only when soil pH was below 5.5. In solution studies, the mechanism of their rapid death under acid conditions was identified to be the presence of volatile fatty acids in the manure (M. Tenuta, K.L. Conn and G. Lazarovits, unpublished). Such materials are produced during anaerobic fermentation by *Eubacterium* spp. and *Clostridium* spp. (Zhu and Jacobson, 1999). Adding chemically pure volatile fatty acids to soil also killed microsclerotia (K.L. Conn, M. Tenuta and G. Lazarovits, unpublished). Swine manure also killed microsclerotia 1–3 weeks after addition to soil (K.L. Conn,

M. Tenuta and G. Lazarovits, unpublished). In alkaline soils this effect was due to NH_3 generation, and in poorly buffered acid soils to HNO_2 generation (K.L. Conn, M. Tenuta and G. Lazarovits, unpublished).

Adding any of the above-mentioned amendments increased the overall soil microbial population by 1–3 cfu g^{-1} soil, including individual groups such as Gram-positive and Gram-negative bacteria, fluorescent bacteria, proteolytic and ammonifying bacteria, ammonifying fungi and total fungi (Conn and Lazarovits, 1999; Lazarovits *et al.*, 1999; Soltani *et al.*, 2000; M. Tenuta and G. Lazarovits, unpublished). A shift in the predominant species present also occurred. A single application of ammonium lignosulphonate increased the fungal population for three seasons at one site (N. Soltani, K.L. Conn and G. Lazarovits, unpublished). In addition, the metabolic activity of organisms increased in response to addition of amendments as determined by soil respiration (M. Tenuta and G. Lazarovits, unpublished) and fluorescein diacetate (M. Tenuta, unpublished). Some amendments, e.g. swine manure and ammonium lignosulphonate, were found to increase the levels of biological control agents, including *Trichoderma* spp. and *Talaromyces flavus* (Klöcker) Stolk and Samson, and this may have been responsible for the observed efficacy up to 3 years after application to field soil (Conn and Lazarovits, 1999; Soltani *et al.*, unpublished). Microcosm studies indicate that ammonium lignosulphonate is not directly toxic to *V. dahliae* microsclerotia and that no toxic transformation products are produced in soil (N. Soltani, K.L. Conn and G. Lazarovits, unpublished). In field studies using ammonium lignosulphonate, control of Verticillium wilt was achieved without an apparent reduction in pathogen populations in soil. Similar results were obtained with addition of sudangrass, *Sorghum bicolor* (L.) Moench, as a green manure to a potato field, resulting in reduced incidence of Verticillium wilt but not in levels of *V. dahliae* propagules or those of pathogenic nematodes (Davis *et al.*, 1996). Thus, we suspect a reduction in disease severity resulted from some mechanism other than the toxic components described above.

Recommendations

Further work should include:

1. Developing a plant bioassay to identify effects of amendments or biological control agents that suppress plant disease but do not reduce levels of pathogens in soil;
2. Optimizing the use of amendments to lower the amounts needed and increasing efficacy in a broader range of soils through formulations customized for soil type and disease pressure;
3. Examining how, or if, biological control agents are contributing to long-term disease suppression capabilities of ammonium lignosulphonate and swine manure.

Acknowledgements

Financial support was provided by The Fats and Proteins Research Foundation Inc., Ontario Potato Growers' Marketing Board, South Simcoe Potato Growers Association, the Canada–Ontario Agriculture Green Plan, Ontario Pork, Agricultural Adaptation Council, Canadapt Program, Prince Edward Island Producers Yield Club and the Matching Investment Initiative of Agriculture and Agri-Food Canada.

References

Ben Yephet, Y., Siti, E. and Frank, Z. (1983) Control of *Verticillium dahliae* by metam-sodium in loess soil and effect on potato tuber yields. *Plant Disease* 67, 1223–1225.

Bukhalid, R.A. and Loria, R. (1997) Cloning and expression of a gene from *Streptomyces scabies* encoding a putative pathogenicity factor. *Journal of Bacteriology* 179, 7776–7783.

Conn, K.L. and Lazarovits, G. (1999) Impact of animal manures on verticillium wilt, potato scab, and soil microbial populations. *Canadian Journal of Plant Pathology* 21, 81–92.

Conn, K.L. and Lazarovits, G. (2000) Soil factors influencing the efficacy of liquid swine manure added to soil to kill *Verticillium dahliae*. *Canadian Journal of Plant Pathology* 22 400–406.

Conn, K.L., Leci, E., Kritzman, G. and Lazarovits, G. (1998) A quantitative method for determining soil populations of *Streptomyces* and differentiating potential potato scab-inducing strains. *Plant Disease* 82, 631–638.

Cook, R.J. and Baker, K.F. (1983) *The Nature and Practice of Biological Control of Plant Pathogens*. APS Press, St Paul, Minnesota, 539 pp.

Davis, J.R., Huisman, O.C., Westermann, D.T., Hafez, S.L., Everson, D.O., Sorensen, L.H. and Schneider, A.T. (1996) Effects of green manures on verticillium wilt of potato. *Phytopathology* 86, 444–453.

Dobinson, K. and Lazarovits, G. (1994) Incidence of *Verticillium dahliae* infection in processing tomatoes in southern Ontario. *Canadian Plant Disease Survey* 74, 113–114.

Easton, G.D., Nagle, M.E. and Bailey, D.L. (1974) Fumigants, rates, and application methods affecting Verticillium wilt incidence and potato yields. *American Potato Journal* 51, 71–77.

Gamliel, A. (2000) Soil amendments: a non chemical approach to the management of soilborne pest. *Acta Horticulturae* 532, 39–47.

Goyer, C., Otrysko, B. and Beaulieu, C. (1996) Taxonomic studies on *Streptomycetes* causing potato common scab: a review. *Canadian Journal of Plant Pathology* 18, 107–113.

Hawke, M.A. (1994) The survival of microsclerotia of *Verticillium dahliae*. PhD thesis, University of Western Ontario, London, Ontario.

Hawke, M.A. and Lazarovits, G. (1994) Production and manipulation of individual microsclerotia of *Verticillium dahliae* for use in studies of survival. *Phytopathology* 84, 883–890.

King, R.R., Lawrence, C.H. and Clark, M.C. (1991) Correlation of phytotoxin production with pathogenicity of *Streptomyces scabies* isolates from scab infected potato tubers. *American Potato Journal* 68, 675–680.

Kritzman, G. and Grinstein, A. (1991) Formalin application against soil-borne *Streptomyces*. *Phytoparasitica* 19, 248.

Lambert, D.H. and Loria, R. (1989) *Streptomyces scabies* sp. nov., nom. rev. *International Journal of Systematic Bacteriology* 39, 387–392.

Lazarovits, G. and Conn, K.L. (1997) *Assessment of the Influence of Manures for the Control of Soil-borne Pests Including Fungi, Bacteria, and Nematodes*. COESA Report No.: RES/MAN-010/97, Canada–Ontario Agriculture Green Plan. http://res.agr.ca/lond/gpres/reporlst.html

Lazarovits, G., Yang, Z., Conn, K.L., Bukhalid, R.A. and Loria, R. (1998) Detection of pathogenic *Streptomyces scabies* from soil using PCR and primers from Nec1 virulence locus. *Canadian Journal of Plant Pathology* 20, 335.

Lazarovits, G., Conn, K.L. and Potter, J. (1999) Reduction of potato scab, verticillium wilt, and nematodes by soymeal and meat and bone meal in two Ontario potato fields. *Canadian Journal of Plant Pathology* 21, 345–353.

Lazarovits, G., Conn, K.L. and Tenuta, M. (2000) Control of *Verticillium dahliae* with soil amendments: efficacy and mode of action. In: Tjamos, E.C., Rowe, R.C., Heale, J.B. and Fravel, D.R. (eds) Advances in Verticillium Research and Disease Management. *Proceedings of the Seventh International Verticillium Symposium, Athens, Greece, 1997*. APS Press, St Paul, Minnesota, pp. 274–291.

Loria, R., Bukhalid, R.A., Creath, R.A., Leiner, R.H., Olivier, M. and Steffens, J.C. (1995) Differential production of thaxtomins by pathogenic *Streptomyces* species *in vitro*. *Phytopathology* 85, 537–541.

Loria, R., Bukhalid, R.A., Fry, B.A. and King, R.R. (1997) Plant pathogenicity in the genus *Streptomyces*. *Plant Disease* 81, 836–846.

Powelson, M.E. and Rowe, R.C. (1993) Biology and management of early dying of potatoes. *Annual Review of Phytopathology* 31, 111–126.

Pullman, G.S. and DeVay, J.E. (1982) Epidemiology of Verticillium wilt of cotton: a relationship between inoculum density and disease progression. *Phytopathology* 72, 549–554.

Sanford, G.B. (1926) Some factors affecting the pathogenicity of *Actinomyces scabies*. *Phytopathology* 16, 525–547.

Soltani, N. and Lazarovits, G. (1998) *Effects of Ammonium Lignosulfonate on Soil Microbial*

Population, Verticillium Wilt, and Potato Scab. Annual International Research Conference on Methyl Bromide Alternatives and Emission Reductions, Orlando, Florida, pp. 20-1–20-4.

Soltani, N., Brown, A., Conn, K. and Lazarovits, G. (2000) Control of verticillium wilt and potato scab with ammonium lignosulfonate. *Phytopathology* 90, S73.

Van Driesche, R.G. and Bellow, T.S. (1996) *Biological Control.* Chapman and Hall, New York, New York.

Wilhelm, S. (1955) Longevity of Verticillium wilt fungus in the laboratory and the field. *Phytopathology* 45, 180–181.

Xiao, C.L. and Subbarao, K.V. (1998) Relationships between *Verticillium dahliae* inoculum density and wilt incidence, severity, and growth of cauliflower. *Phytopathology* 88, 1108–1115.

Zhu, J. and Jacobson, L.D. (1999) Correlating microbes to major odorous compounds in swine manure. *Journal of Environmental Quality* 28, 737–744.

Appendix I: Noteworthy Publications[1] on Biological Control 1981–2000

M.J. Sarazin

A CD-ROM search covering the literature since 1980 revealed that a substantial number of books were published on biological control in general, or some aspect of it. Listed below are 130 publications considered to be of use to the biological control community at large. The references are listed alphabetically by author or editor. In addition to general treatments, many can be divided into the following categories: biological control agent (i.e. insects, pathogens, nematodes and mites), target host (i.e. insects, weeds, pathogens, mites and nematodes) and by system other than agricultural (i.e. medical/veterinary, forestry and glasshouses). By listing the major references in this way, gaps become apparent, such as an overview of a certain topic, although references published before 1980 may have covered these areas. Peripheral areas such as integrated pest management, rearing, host–plant interactions and taxonomy of important biological control agents are available but not listed here, except where they provide an important overview of the subject (e.g. Quicke 1997, parasitic wasps).

The following references (listed by first author/editor and year) treat *insects as biological control agents* (*the publication involves predators either solely or in addition to parasitoids): Anderson (1982), Beckage (1993), *Bellows (1999), *Boethel (1986), *Coll (1998), *Coulson (2000), *Crawley (1992), *Croft (1990), Flint (1998), Fry (1989), Godfray (1994), Grenier (1988), Hawkins (1994), Hunter (1997), Kauffman (1992), LaSalle (1993), Noyes (1994), Pickett (1998), Poinar (1984), Quicke (1997), Ridgway (1998), *Sarazin (1981–1991, 1988–1995, 1992–2000); Shaw (1997), *Taylor (1984), Toft (1991), *Van Driesche (1996), *Vincent (1992), Waage (1986), Wajnberg (1991, 1994).

The following references (listed by first author/editor and year) treat *pathogens as biological control agents* (*fungal pathogen and **viral pathogen): Beckage (1993), Bellows (1999), Boland (1998), *Burge (1988), Burges (1981), Charudattan (1982), Cheng (1984), Coulson (2000), Crawley (1992), Croft (1990), Fuxa (1987), **Granados (1986), *Hall (1982), *Ignoffo (1988), Jackson (1992), **Kurstak (1982), Laird (1990), **Maramorosch (1985), McClay (1990), Navon (2000), Poinar (1984, 1988), Sarazin (1988–1995), Tanada (1993), TeBeest (1991), Van Driesche (1996), Vincent (1992).

The following references (listed by first author/editor and year) treat *nematodes as biological control agents*: Akhurst (1993), Bellows (1999), Coulson (2000), Eidt (1994), Evans (1993), Gaugler (1990), Navon (2000), Nickle (1991), Sarazin (1988–1995), Van Driesche (1996), Vincent (1992).

The following references (listed by first author/editor and year) treat *mites as biological control agents*: Bellows (1999), Coulson (2000), Gerson (1990), Habersaat (1989), Hoy (1987), Kostiainen (1996), Lindquist (1996), Sarazin (1988–1995), Van Driesche (1996), Vincent (1992).

[1]The concentration being on books useful as references.

The following references (listed by first author/editor and year) treat *insects as target hosts*: Akhurst (1993), American Mosquito Control Association (1985), Baker (1990), Beckage (1993), Bellows (1999), Ben-Dov (1997), Boethel (1986), Burges (1981), Cameron (1989), Cheng (1984), Coulson (2000), DeBach (1991), Eidt (1994), Fry (1989), Fuxa (1987), Gaugler (1990), Granados (1986), Gunasekaran (1996), Hall (1982), Hawkins (1994, 1999), Hoffmann (1993), Ignoffo (1988), Jackson (1992), Jervis (1996), Kelleher (1984), Kostiainen (1996), Laird, (1990), Loomans (1995), Mahr (1993), Minks (1989), Navon (2000), Nechols (1995), Noyes (1994), Patterson (1986), Pickett (1998), Poinar (1984), Raupp (1993), Rice Mahr (1993), Robinson (1989), Rosen (1990), Rutz (1990), Samways (1981), Sarazin (1981–1991, 1988–1995, 1992–2000), Schaefer (1983), Tanada (1993), Van den Bosch (1982), Van der Geest (1991), Van Driesche (1992), Van Driesche (1996), Van Lenteren (1992), Vincent (1992), Wajnberg (1991, 1994), Waterhouse (1987), Wood (1988).

The following references (listed by first author/editor and year) treat *weeds as target hosts*: Bellows (1999), Boland (1998), Cameron (1989), Charudattan (1982), Coulson (2000), DeBach (1991), Harley (1992), Harris (1991), Hokkanen (1985), Julien (1997, 1998), Kelleher (1984), McClay (1989, 1990), Nechols (1995), Powell (1994), Rosenthal (1984), Samways (1981), Sarazin (1981–1991, 1988–1995, 1992–2000), TeBeest (1991), Van Driesche (1996), Vincent (1992), Waterhouse (1987, 1994), Watson (1993), Wood (1988).

The following references (listed by first author/editor and year) treat *pathogens as target hosts*: Baker (1982, 1990), Bailey (1992), Bellows (1999), Burges (1981), Campbell (1989), Cook (1983), Coulson (2000), Gunasekaran (1996), Hawkins (1999), Hornby (1990), Mukerji (1988, 1999), Sarazin (1988–1995), Tjamos (1992), Van Driesche (1996), Vincent (1992), Wilson (1994), Windels (1985), Wood (1988).

The following references (listed by first author/editor and year) treat *mites as target hosts*: Bellows (1999), Coulson (2000), Helle (1986), Kostiainen (1996), Lindquist (1996), Mahr (1993), Raupp (1993), Sarazin (1988–1995).

The following references (listed by first author/editor and year) treat *nematodes as target host*: Bellows (1999), Coulson (2000), Poinar (1988), Stirling (1991).

The following references (listed by first author/editor and year) treat *systems other than agricultural* (*indicates glasshouse system, ** indicates forestry system and *** indicates medical/veterinary system): **Eidt (1994), ***Hall (1982), **Hulme (1982), *Hussey (1985), ***Laird (1981), *Malais (1992), *Steiner (1987).

General List

Akhurst, R., Bedding, R. and Kaya, H. (eds) (1993) *Nematodes and the Biological Control of Insect Pests.* CSIRO, Melbourne, Australia, 178pp.

Allen, G. and Rada, A. (1984) *The Role of Biological Control in Pest Management.* University of Ottawa Press, Ottawa, Ontario, 173pp.

American Mosquito Control Association (1985) *Biological Control of Mosquitoes.* American Mosquito Control Association, Fresno, California, 218pp.

Anderson, R.M. and Canning, E.U. (eds) (1982) *Parasites as Biological Control Agents.* Cambridge University Press, Cambridge, New York, New York, 298pp.

Andow, D.A., Nyvall, R.F. and Ragsdale, A. (eds) (1997) *Ecological Interactions and Biological Control.* Westview, Boulder, Colorado, 350pp.

Baker, K.F. and Cook, R.J. (1982) *Biological Control of Plant Pathogens.* American Phytopathological Society, St Paul, Minnesota, 433pp.

Baker, R.R. and Dunn, P.E. (eds) (1990) *New Directions in Biological Control: Alternatives for Suppressing Agricultural Pests and Diseases.* Alan R. Liss, New York, New York, 837pp.

Bailey, J.A. and Jeger, M.J. (eds) (1992) Colletotrichum: *Biology, Pathology and Control.* CAB International, Wallingford, UK, 388pp.

Barbosa, P. (1998) *Conservation Biological Control.* Academic Press, San Diego, California, 396pp.
Beckage, N.E., Thompson, S.N. and Federici, B.A. (eds) (1993) *Parasites and Pathogens of Insects.* Academic Press, New York, New York, 740pp.
Bellows, T.S. and Fisher, T.W. (eds) (1999) *Handbook of Biological Control.* Academic Press, San Diego, California, 1046pp.
Ben-Dov, Y. and Hodgson, C.J. (eds) (1997) *Soft Scale Insects: Their Biology, Natural Enemies and Control.* Elsevier, Amsterdam, The Netherlands, 476pp.
Boethel, D.J. and Eikenbary, R.D. (1986) *Interactions of Plant Resistance and Parasitoids and Predators of Insects.* Ellis Horwood Limited, Chichester, UK, 224pp.
Boland, G.J. and Kuykendall, L.D. (eds) (1998) *Plant-Microbe Interactions and Biological Control.* Marcel Dekker, New York, New York, 442pp.
Burge, M.N. (ed.) (1988) *Fungi in Biological Control Systems.* Manchester University Press, Manchester, UK, 269pp.
Burges, H.D. (ed.) (1981) *Microbial Control of Pests and Plant Diseases, 1970–1980.* Academic Press, London, UK, 949pp.
Cameron, P.J., Hill, R.L., Bain, J. and Thomas, W.P. (eds) (1989) *A Review of Biological Control of Invertebrate Pests and Weeds in New Zealand 1874 to 1987.* CAB International, Wallingford, UK, 424pp.
Campbell, R.E. (1989) *Biological Control of Microbial Plant Pathogens.* Cambridge University Press, Cambridge, UK, 218pp.
Cavalloro, R. (ed.) (1987) *Integrated and Biological Control in Protected Crops.* A.A. Balkema, Rotterdam, The Netherlands, 251pp.
Charudattan, R. and Walker, H.L. (eds) (1982) *Biological Control of Weeds With Plant Pathogens.* John Wiley and Sons, New York, New York, 293pp.
Cheng, T.C. (1984) *Pathogens of Invertebrates: Application in Biological Control and Transmission Mechanisms.* Plenum Press, New York, New York, 278pp.
Coll, M. and Ruberson, J.R. (eds) (1998) *Predatory Heteroptera.* Entomological Society of America, Lanham, Maryland, 233pp.
Cook, R.J. and Baker, K.F. (1983) *The Nature and Practice of Biological Control of Plant Pathogens.* The American Pathological Society, St Paul, Minnesota, 539pp.
Coombs, J. and Hall, K.E. (1998) *Dictionary of Biological Control and Integrated Pest Management.* CPL Scientific, Newbury, UK, 196pp.
Coulson, J.R., Vail, P.V., Dix, M.E., Nordlund, D.A. and Kauffman, W.C. (eds) (2000) *110 Years of Biological Control Research and Development in the United States Department of Agriculture – 1883–1993.* Administrative Report No. 2000–1, United States Department of Agriculture, Agricultural Research Service, 645pp.
Crawley, M.J. (ed.) (1992) *Natural Enemies: The Population Biology of Predators, Parasites, and Diseases.* Blackwell Scientific Publications, Oxford, UK, 576pp.
Croft, B.A. (1990) *Arthropod Biological Control Agents and Pesticides.* John Wiley and Sons, New York, New York, 723pp.
DeBach, P. and Rosen, D. (1991) *Biological Control by Natural Enemies.* 2nd edn. Cambridge University Press, Cambridge, UK, 440pp.
Dent, D. (ed.) (1995) *Integrated Pest Management: Principles and Systems Development.* Chapman and Hall, London, UK, 356pp.
Eidt, D.C. and Thurston, G.S. (1994) *Entomopathogenic Nematodes for Insect Pest Management in a Cold Climate.* Canadian Forest Service, Fredericton, New Brunswick, Canada, 66pp.
Evans, K., Trudgill, D.L. and Webster, J.M. (eds) (1993) *Plant Parasitic Nematodes in Temperate Agriculture.* CAB International, Wallingford, UK, 648pp.
Flint, M.L. and Dreistadt, S.H. (1998) *Natural Enemies Handbook.* UC Division of Agriculture and Natural Sciences, Oakland, California, 154pp.
Follett, P.A. and Duan, J.J. (1999) *Non Target Effects of Biological Control.* Kluwer Academic Publishers, Hingham, Massachusetts, 336pp.
Fry, J.M. (1989) *Natural Enemy Databank.* CAB International, Wallingford, UK, 192pp.
Fuxa, J.R. and Tanada, Y. (eds) (1987) *Epizootiology of Insect Diseases.* John Wiley and Sons, New York, New York, 555pp.
Gaugler, R. and Kaya, H.R. (eds) (1990) *Entomopathogenic Nematodes in Biological Control.* CRC Press, Boca Raton, Florida, 365pp.

Gerson, U. and Smiley, R.L. (1990) *Acarine Biocontrol Agents.* Chapman and Hall, London, UK, 174pp.

Godfray, H.C.J. (1994) *Parasitoids: Behavioural and Evolutionary Ecology.* Princeton University Press, Princeton, New Jersey, 473pp.

Granados, R.R. and Federici, B.A. (eds) (1986). *The Biology of Baculoviruses.* CRC Press, Boca Raton, Florida, 2 volumes, 275 + 276pp.

Grenier, S. (1988) Applied biological control with tachinid flies (Diptera, Tachinidae): a review. *Anzeiger für Schädlingskunde, Pflanzenschutz, Umweltschutz* 61, 49–56.

Gunasekaran, M. and Weber, D.J. (eds) (1996) *Molecular Biology of the Biological Control of Pests and Diseases of Plants.* CRC Press, Boca Raton, Florida, 219pp.

Habersaat, U. (1989) *The Importance of Predatory Soil Mites as Predators of Agricultural Pests, with Special Reference to* Hypoaspis angusta *Karg, 1965 (Acari: Gamasina).* Eidgenossische Technische Hochschule, Zurich, Switzerland, 205pp.

Hall, F.R. and Menn, J.J. (1998) *Biopesticides: Use and Delivery.* Humana Press, Totowa, New Jersey, 626pp.

Hall, R.A. and Papierok, B. (1982) Fungi as biological control agents of arthropods of agricultural and medical importance. *Parasitology* 84, 205–240.

Harley, K.L.S. and Forno, I.W. (1992) *Biological Control of Weeds: a Handbook for Practitioners and Students.* Inkata Press, Melbourne, Australia, 74pp.

Harris, P. (1991) Classical biocontrol of weeds: its definition, selection of effective agents, and administrative-political problems. *The Canadian Entomologist* 123, 827–849.

Hawkins, B.A. (1994) *Pattern and Process in Host–Parasitoid Interactions.* Cambridge University Press, Cambridge, UK, 190pp.

Hawkins, B.A. and Cornell, H.V. (eds) (1999) *Theoretical Approaches to Biological Control.* Cambridge University Press, New York, New York, 424pp.

Helle, W. and Sabelis, M.W. (eds) (1986) *Spider Mites: Their Biology, Natural Enemies and Control, Vol. 1B.* Elsevier, Amsterdam, The Netherlands, 458pp.

Hoddle, M.S. (ed.) (1998) *Innovation in Biological Control Research.* University of California, Berkeley, California, 245pp.

Hoffmann, M.P. and Frodsham, A.C. (1993) *Natural Enemies of Vegetable Insect Pests.* Cornell University, Cooperative Extension Publication, Ithaca, New York, New York, 63pp.

Hokkanen, H.M.T. (1985) Success in classical biological control. *CRC Critical Reviews in Plant Sciences* 3, 35–72.

Hokkanen, H.M.T. and Lynch, J.L. (1995) *Biological Control: Benefits and Risks.* OECD, Paris, France, 304pp.

Hong, L.W. (ed.) (2000) *Biological Control in the Tropics.* CAB International, Wallingford, UK, 155pp.

Hornby, D. (ed.) (1990) *Biological Control of Soil-borne Plant Pathogens.* CAB International, Wallingford, UK, 479pp.

Hoy, M.A. and Herzog, D.C. (eds) (1985) *Biological Control in Agricultural IPM Systems.* Academic Press, Orlando, Florida, 589pp.

Hoy, M.A., Cunningham, G.L. and Knutson, L. (eds) (1987) *Biological Control of Pests by Mites.* University of California, Berkeley, California, 185pp.

Hulme, M.A. (1982) *Biological Control in the Canadian Forestry Service.* Canadian Forestry Service, Hull, Quebec, 45pp.

Hunter, C.D. (1997) *Suppliers of Beneficial Organisms in North America.* California Environmental Protection Agency, Department of Pesticide Regulation, Sacramento, California, 32pp.

Hussey, N.W. and Scopes, N. (eds) (1985) *Biological Pest Control: the Glass House Experience.* Cornell University Press, Ithaca, New York, New York, 240pp.

Ignoffo, C.M. (ed.) (1988) *CRC Handbook of Natural Pesticides.* Volume 5, *Microbial Insecticides.* CRC Press, Boca Raton, Florida, 260pp.

Jackson, T.A. and Glare, T.R. (eds) (1992) *Use of Pathogens in Scarab Pest Management.* Intercept, Andover, UK, 298pp.

Jeffords, M.R. and Hodgins, A.S. (1995) *Pests Have Enemies Too: Teaching Young Scientists About Biological Control.* Illinois Natural History Survey, Champaign, Illinois, 64pp.

Jervis, M.A. and Kidd, N.A.C. (eds) (1996) *Insect Natural Enemies: Practical Approaches to Their Study and Evaluation.* Chapman and Hall, London, UK, 491pp.

Julien, M.H. and Griffiths, M.W. (eds) (1998) *Biological Control of Weeds: a World Catalogue of Agents and Their Target Weeds,* 4th edn. CAB International, Wallingford, UK, 223pp.

Julien, M. and White, G. (1997) *Biological Control of Weeds.* Australia Centre for International Agricultural Research, Canberra, Australia, 190pp.

Kauffman, W.C. and Nechols, J.E. (eds) (1992) *Selection Criteria and Ecological Consequences of Importing Natural Enemies.* Proceedings Thomas Say Publication in Entomology, 1, Entomological Society of America, Lanham, Maryland, 117pp.

Kelleher, J.S. and Hulme, M.A. (eds) (1984) *Biological Control Programmes Against Insects and Weeds In Canada, 1969–1980.* Commonwealth Agricultural Bureaux, Farnham Royal, UK, 410pp.

Kostiainen, T.S. and Hoy, M.A. (1996) *The Phytoseiidae as Biological Control Agents of Pest Mites and Insects: a Bibliography.* University of Florida, Agricultural Experiment Station, Gainesville, Florida, 355pp.

Kurstak, E. (ed.) (1982) *Microbial and Viral Pesticides.* Marcel Dekker, New York, New York, 720pp.

Laird, M. (ed.) (1981) *Biocontrol of Medical and Veterinary Pests.* Praeger, New York, New York, 235pp.

Laird, M., Lacey, L.A. and Davidson, E.W. (eds) (1990) *Safety of Microbial Insecticides.* CRC Press, Boca Raton, Florida, 259pp.

LaSalle, J. and Gauld, I.D. (1993) *Hymenoptera and Biodiversity.* CAB International, Wallingford, UK, 348pp.

Lindquist, E.E., Sabelis, M.W. and Bruin, J. (eds) (1996) *Eriophyoid Mites: Their Biology, Natural Enemies and Control.* Elsevier, Amsterdam, The Netherlands, 790pp.

Loomans, A.J.M. (1995) *Biological Control of Thrips Pests.* Wageningen Agricultural University, Wageningen, The Netherlands, 201pp.

Mackauer, M., Ehler, L.E. and Roland, J. (eds) (1989) *Critical Issues in Biological Control.* Intercept, Andover, UK, 330pp.

Mahr, D.L. and Ridgway, N.M. (1993) *Biological Control of Insects and Mites: an Introduction to Beneficial Natural Enemies and Their Use in Pest Management.* Cooperative Extension Publications, University of Wisconsin, Extension, Madison, Wisconsin, 91pp.

Malais, M. and Ravensberg, W.J. (1992) *Knowing and Recognizing: the Biology of Glasshouse Pests and Their Natural Enemies.* Koppert, Berkel en Rodenrijs, The Netherlands, 109pp.

Maramorosch, K. and Sherman, K.E. (eds) (1985) *Viral Insecticides for Biological Control.* Academic Press, Orlando, Florida, 809pp.

McClay, A.S. (1989) *Selection of Suitable Target Weeds for Classical Biological Control in Alberta.* Alberta Environmental Centre, Vegreville, Alberta, Canada, 97pp.

McClay, A.S. (1990) *Screening and Evaluation of Plant Diseases for Biological Control of Weeds.* Alberta Agriculture, Alberta, Canada, 57pp.

Minks, A.K. and Harrewijn, P. (eds) (1989) *World Crop Pests*, Vol. 2C: *Aphids, Their Biology, Natural Enemies and Control.* Elsevier, Amsterdam, The Netherlands, 322pp.

Mukerji, K.G. and Garg, K.L. (1988) *Biocontrol of Plant Diseases*, Vols 1 and 2. CRC Press, Boca Raton, Florida.

Mukerji, K.G., Chamola, B.P. and Upadhyay, R.K. (eds) (1999) *Biotechnological Approaches in Biocontrol of Plant Pathogens.* Plenum, London, UK, 255pp.

Navon, A. and Ascher, K.R.S. (eds) (2000) *Bioassays of Entomopathogenic Microbes and Nematodes.* CAB International, Wallingford, UK, 336pp.

Nechols, J.R., Andres, L.A., Beardsley, J.W., Goeden, R.D. and Jackson, C.G. (eds) (1995) *Biological Control in the Western United States.* University of California Division of Agriculture and Natural Resources, Publication 3361, Oakland, California, 356pp.

Nickle, W.R. (ed.) (1991) *Manual of Agricultural Nematology.* Marcel Dekker, New York, New York, 1035pp.

Noyes, J.S. and Hayat, M. (1994) *Oriental Mealybug Parasitoids of the Anagyrini (Hymenoptera: Encyrtidae): With a World Review of Encyrtidae Used in Classical Biological Control and an Index of Encyrtid Parasitoids of Mealybugs (Homoptera: Pseudococcidae).* CAB International, Wallingford, UK, 554pp.

Papavizas, G.C. (ed.) (1981) *Biological Control in Crop Production.* Allanheld, Osmun, Totowa, New Jersey, 461pp.

Patterson, R.S. and Rutz, D.A. (eds) (1986) *Biological Control of Muscoid Flies.* Misc. Publ. 61, Entomological Society of America, Lanham, Maryland, 174pp.

Pickett, C.H. and Bugg, R.L. (eds) (1998) *Enhancing Biological Control: Habitat Management to Promote Natural Enemies of Agricultural Pests.* University of California Press, Berkeley, California, 422pp.

Poinar, G.O. Jr and Jansson, H.-B. (eds) (1988) *Diseases of Nematodes.* CRC Press, Boca Raton, Florida, 2 volumes.

Poinar, G.O. Jr and Thomas, G.M. (1984) *Laboratory Guide to Insect Pathogens and Parasites.* Plenum Press, New York, New York, 392pp.

Powell, G.W. (1994) *Field Guide to the Biological Control of Weeds in British Columbia.* British Columbia Ministry of Forests, Victoria, British Columbia, Canada, 163pp.

Quicke, D.L.J. (1997) *Parasitic Wasps.* Chapman and Hall, London, UK, 470pp.

Raupp, M.J., van Driesche, R.G. and Davidson, J.A. (1993) *Biological Control of Insect and Mite Pests of Woody Landscape Plants: Concepts, Agents and Methods.* Maryland Cooperative Extension Service, College Park, Maryland, 39pp.

Rice Mahr, S.E., Mahr, D.L., and Wyman, J.A. (1993) *Biological Control of Insect Pests of Cabbage and Other Crucifers.* University of Wisconsin, Madison, Wisconsin, 54pp.

Ridgway, R.L., Hoffmann, M.P., Inscoe, M.N. and Glenister, C.S. (eds) (1998) *Mass-Reared Natural Enemies: Application, Regulation, and Needs.* Thomas Say Publications in Entomology 13, Entomological Society of America, Lanham, Maryland, 332pp.

Robinson, A.S. and Hooper, G.H.S. (eds) (1989) *Fruit Flies: Their Biology, Natural Enemies and Control,* Vols A and B, Elsevier, Amsterdam, The Netherlands.

Rosen, D. (ed.) (1990) *The Armored Scale Insects, Their Biology, Natural Enemies and Control,* Vol. 4B. Elsevier, Amsterdam, The Netherlands, 688pp.

Rosenthal, S.S., Maddox, D.M. and Brunetti, K. (1984) *Biological Methods of Weed Control.* Thomson Publications, Fresno, California, 88pp.

Rutz, D.A. and Patterson, R.S. (1990) *Biocontrol of Arthropods Affecting Livestock and Poultry.* Westview, Boulder, Colorado, 316pp.

Samways, M.J. (1981) *Biological Control of Pests and Weeds.* Arnold, London, UK, 57pp.

Sarazin, M. (1981–1991) *Insect Liberations in Canada.* Liberation Bulletin numbers 45–55, Agriculture and Agri-Food Canada, Research Branch, Ottawa, Ontario, Canada.

Sarazin, M. (1988–1995) *Biocontrol News.* Agriculture and Agri-Food Canada, Research Branch, Ottawa, Ontario, Vols 1–8.

Sarazin, M. (1992–2000) *Insect Liberations in Canada.* Agriculture and Agri-Food Canada, Research Branch. www.res2.agr.ca/ecorc/isbi/biocont/libhom.htm

Schaefer, P.W. (1983) *Natural Enemies and Host Plants of Species in the Epilachninae (Coleoptera: Coccinellidae): a World List.* University of Delaware, Agricultural Experiment Station, Newark, Delaware, 42pp.

Shaw, M.R. (1997) *Rearing Parasitic Hymenoptera.* Amateur Entomologists' Society, Orpington, UK, 45pp.

Steiner, M.Y. and Elliott, D.P. (1987) *Biological Pest Management for Interior Plantscapes.* Alberta Environmental Centre, Vegreville, Alberta, Canada, 32pp.

Stirling, G.R. (1991) *Biological Control of Plant Parasitic Nematodes: Progress, Problems and Prospects.* CAB International, Wallingford, UK, 282pp.

Tanada, Y. and Kaya, H.K. (1993) *Insect Pathology.* Academic Press, San Diego, California, 666pp.

Taylor, R.J. (1984) *Predation.* Chapman and Hall, London, UK, 166pp.

TeBeest, D.O. (ed.) (1991) *Microbial Control of Weeds.* Chapman and Hall, New York, New York, 284pp.

Tjamos, E.C., Papvizas, G.C. and Cook, R.J. (eds) (1992) *Biological Control of Plant Diseases.* Plenum Press, New York, New York, 462pp.

Toft, C.A., Aeschlimann, A. and Bolis, L. (eds) (1991) *Parasite–Host Associations: Coexistence or Conflict.* Oxford University Press, Oxford, UK, 384pp.

Van den Bosch, R., Messenger, P.S. and Gutierrez, A.P. (1982) *An Introduction to Biological Control.* Plenum Press, New York, New York, 247pp.

Van der Geest, L.P.S. and Evenhuis, H.H. (1991) *Tortricid Pests: Their Biology, Natural Enemies and Control.* Elsevier, Amsterdam, The Netherlands, 808pp.

Van Driesche, R.G. and Bellows, T.S. Jr (eds) (1992) *Steps in Classical Arthropod Biological Control.* Proceedings Thomas Say Publications in Entomology, 3, Entomological Society of America, Lanham, Maryland, 88pp.

Van Driesche, R.G. and Bellows, T.S. Jr (1996) *Biological Control.* Chapman and Hall, New York, New York, 539pp.

Van Lenteren, J.C., Minks, A.K. and de Ponti, O.M.B. (eds) (1992) *Biological Control and Integrated Crop Protection: Towards Environmentally Safer Agriculture.* Scientific Pudoc Publishers, Wageningen, The Netherlands, 239pp.

Vincent, C. and Coderre, D. (1992) *La Lutte Biologique.* G. Morin, Boucherville, Quebec, Canada, 671pp.

Waage, J.K. and Greathead, D. (eds) (1986) *Insect Parasitoids.* Academic Press, London, UK, 389pp.
Wajnberg, E. and Hassan, S.A. (eds) (1994) *Biological Control with Egg Parasitoids.* CAB International, Wallingford, UK, 286pp.
Wajnberg, E. and Vinson, S.B. (eds) (1991) *Trichogramma and Other Egg Parasitoids. Third International Symposium.* Les Colloques de L'INRA, 56, Paris, France, 246pp.
Waterhouse, D.F. (1994) *Biological Control of Weeds: Southeast Asian Prospects.* ACIAR, Canberra, Australia, 302pp.
Waterhouse, D.F. and Norris, K.R. (1987) *Biological Control: Pacific Prospects.* Inkata Press, Melbourne, Australia, 454pp.
Watson, A.K. (ed.) (1993) *Biological Control of Weeds Handbook.* Weed Science Society of America, Champaign, Illinois, 202pp.
Wilson, C.L. and Wisniewski, M.E. (eds) (1994) *Biological Control of Postharvest Diseases, Theory and Practice.* CRC Press, Boca Raton, Florida, 182pp.
Windels, C.E. and Lindow, S.E. (eds) (1985) *Biological Control on the Phylloplane.* American Phytopathological Society, St Paul, Minnesota, 169pp.
Wood, R.K.S. and Way, M.J. (1988) *Biological Control of Pests, Pathogens and Weeds: Developments and Prospects.* Royal Society, Vol. (Series B) 318, 111–376.

Appendix II: Canadian Suppliers of Biological Control Organisms

H.G. Philip

Introduction

The increased demand for biological control agents to manage insect pests since 1980 has led to a corresponding increase in their commercial production. The accompanying list gives details of Canadian commercial suppliers of over 50 different species. Addresses of suppliers in Mexico and the USA can be found at: http://www.cdpr.ca.govdocs/ipmnov/bensuppl.htm. The suppliers are listed in sections according to country (Canada, USA and Mexico) and each has a supplier number preceded with a country code: C = Canada; U = USA; M = Mexico. All the information on Mexican suppliers was obtained from Centro Nacional de Referencia de Control Biológico de la Dirección General de Sanidad Vegetal.

The name, address, telephone, facsimile and e-mail numbers are listed for each supplier along with a retail/wholesale notation. Under the retail/wholesale notation, there may be a brief note supplied by the company on their specialties and/or the services they provide. Suppliers belonging to the Association of National Bio-Control Producers (ANBP) are designated with 'ANBP member'. ANBP is an organization of companies and individuals whose goals are to enhance the standardization and the quality control of commercially available beneficial organisms, and the dissemination of accurate information on their use and handling.

Included are two separate indexes to suppliers. Both use scientific names because most organisms do not have common names. The first index lists beneficial organisms under 13 different categories in alphabetical order – Predatory Mites, Parasitic Nematodes, etc. The second index is a cross-reference to the scientific names of all the beneficial organisms listed in the publication. Each name is followed by the supplier numbers, e.g., C05, M31, U78, of companies supplying that organism. Use these numbers to locate the suppliers (addresses of Mexican and USA suppliers on the website). In addition to offering biological control organisms, some of the suppliers listed can provide consultation services on the use of these organisms alone or in an integrated pest management programme.

Importation of Biological Control Organisms

The Canadian Food Inspection Agency (CFIA) document *Import Requirements for Invertebrates and Microorganisms* (D-96-14e) (available at http://inspection.gc.ca/english/plaveg/protect/dir/d-96-14e.shtml) describes the current requirements for importation into Canada of certain living invertebrates (insects, mites, millipedes, terrestrial molluscs, nematodes) and microorganisms (bacteria, fungi, viruses), including those expressing novel traits introduced through biotechnology. An 'Application for Permit to Import (CFIA/ACIA 1274)' issued under the *Plant Protection Regulations* must be completed by every importer of organisms, including biological control agents, unless they are already approved for release. Other Acts and regulations may impose additional requirements. Permits can be issued for up to 3 years, depending on the organism and its application. For more information, contact you local CFIA office or the national Food Production and Inspection Branch, Animal and Plant Health Directorate, Plant Protection Division, 59 Camelot Drive, Nepean, Ontario, K1A 0Y9 (Tel.: 613-952-8000; Fax: 613-941-5671) or visit the Import Unit web site http://inspection.gc.ca/english/plaveg/oper/opere.shtml.

Acknowledgement

I am grateful to Charles D. Hunter, California Environmental Protection Agency, Department of Pesticide Regulation, for permission to reproduce part of his list of commercial suppliers of biological control organisms in North America.

Canada – Commercial Suppliers

C01 **Applied Bio-Nomics Ltd**
11074 West Saanich Road
Sidney, British Columbia V8L 5P5.
Canada
Tel.: (250) 656-2123 (Insectary); (604) 940-0290 (BC); (416) 793-000 (ONT)
Fax: (604) 656-3844
E-mail: bug@islandnet.com

Retail and wholesale.
Distributor list available to USA and Canada.
Free literature and price list. ANBP member.
Web site: www.highwaynorthdesign.com/applied/

C02 **Beneficial Insectary Canada**
60 Taggart Court, #1
Guelph, Ontario N1H 6H8
Canada
Tel.: (519) 763-8653
Fax: (519) 763-9103
E-mail: bi@insectary.com

Retail and wholesale.
Producing high-quality products.
Entomological staff available.
ANBP member.
Web site: www.beneficialinsectary.com

C03 **Better Yield Insects – Canada**
RR 3 1302 County Road 22
Belle River, Ontario N0R 1A0
Canada
Tel.: (800) 662-6562 (Toll Free – USA and Canada)
Fax: (519) 727-5989

Retail and wholesale.
Specializing in beneficial insects for 20 years.
Free consultation by telephone or fax.

C04 **BioBest Canada Ltd**
2020 Mersea Road #3, RR 4
Leamington, Ontario N8H 3V7
Canada
Tel.: (519) 322-2178
Fax: (519) 322-1271

Retail and wholesale.
Production of bumble bees and other beneficial organisms.
ANBP member.

C05 **Bio-Controle (Services) Inc.**
2600 Dalton
Foy, Quebec G1P 3S4
Canada
Tel.: (418) 653-3101; (418) 650-3709; (514) 528-9232 (Montreal)
Fax: (418) 653-3096
E-mail: biocontrole@accent.net

Retail and wholesale.
With purchase, information sheets on how to use Sainte *Trichogramma* and ladybird beetles successfully.
Both French and English spoken.

C06 **Coast Agri Ltd**
464 Riverside Road South RR#2
Abbotsford, British Columbia V2S 7N8
Canada
Tel.: (604) 864-9044
Fax: (604) 864-8418
E-mail: wporter@Coast-agri.com

Wholesale only.
Free informative catalogue available.
Consulting. ANBP member.

C07 **Halifax Seed Company Inc.**
P.O. Box 8026 Station A
5860 Kane Street
Halifax, Nova Scotia B3K 5L8
Canada
Tel.: (902) 454-7456, (902) 455-4364
Fax: (902) 455-5271
E-mail: hfxseed@sympatico.ns.ca

Retail and wholesale.
Providing beneficial organisms for commercial and home use in Atlantic Canada.

C08 **Koppert Canada Ltd**
3 Pullman Court
Scarborough, Ontario M1X 1E4
Canada
Tel.: (416) 291-0040; (800) 567-4195
Fax: (416) 291-0902
E-mail: kcanltd@aol.com

Retail and wholesale.
Free literature and pricing available upon request.
Web site: www.koppert.nl/e0216.shtml

C09 **Manbico Biological Ltd.**
Box 17, Group 242, RR2
Winnipeg, Manitoba R3C 2E6
Canada
Tel.: (204) 697-0863; Toll free (800) 665-2494
Fax: (204) 697-0887

Retail and wholesale.
Free catalogue, brochures and distributor list.

C10	**Natural Beginnings** PO Box 21036 Dartmouth, Nova Scotia B2W 6B2 Canada Tel.: (902) 435-4882 (customer service) Fax: (905) 382-4418 (customer inquiries) E-mail: natural@isisnet.com	Retail only. Free organic home gardening catalogues (Canada only). Beneficial organisms available from April–July only. School and environmental group fund raising programmes available.
C11	**Natural Insect Control** RR #2 Stevensville, Ontario L0S 1S0 Canada Tel.: (905) 382-2904 Fax: (905) 382-4418 E-mail: nic@niagara.com	Retail and wholesale. 48-page catalogue. Organic supplies. Ship worldwide. Large selection of beneficials. Technical telephone support. Bird and bat houses available. ANBP member. Web site: www.natural-insect-control.com
C12	**Nature's Alternative Insectary Ltd** Box 19 Dawson Road 1636 East Island Highway Nanoose Bay, British Columbia V0R 2R0 Canada Tel.: (250) 468-7912; (250) 468-7911 Fax: (250) 468-7912 E-mail: nai@number1inbugs.com	Retail and wholesale. Producer. Available year round. Weekly shipments within USA. ANBP and IOBC member. Web site: www.anbp.org/b-NAI.htm
C13	**Richters** 357 Highway 47 Goodwood, Ontario L0C 1A0 Canada Tel.: (905) 640-6677 Fax: (905) 640-6641 E-mail: info@richters.com	Retail only. Specializes in use of beneficials on commercial herb greenhouse and field crops. Provides advice to seed and plant customers. Web site: www.richters.com
C14	**Westgro Sales Inc.** 7333 Progress Way Delta, British Columbia V4G 1E7 Canada Tel.: (604) 940-0290 Fax: (604) 940-0258 E-mail: westgro@westgro.com	Retail and wholesale. Literature and price list available upon request. IPM consulting.

Beneficial organisms

Predatory mites

Galendromus annectans – for pest mites – U03, U05, U06, U19, U20, U23, U24, U32, U47, U78

Galendromus (= *Typhlodromus*) *helveolus* – for Persea mite on avocados – C09, C11, U03, U05, U06, U08, U19, U20, U23, U24, U31, U32, U47, U61, U78, U88, U92

Galendromus (= *Metaseiulus*, = *Typhlodromus*) *occidentalis* – western predatory mite for spider mites – C06, C07, C09, C11, C14, U02, U03, U04, U05, U06, U08, U12, U13, U19, U20, U22, U23, U24, U30, U31, U32, U40, U42, U43, U44, U48, U49, U51, U52, U61, U63, U66, U67, U73, U74, U78, U80, U82, U83, U88, U89, U92, U93, U95

Hypoaspis aculeifer – for fungus gnats and flower thrips – U56, U75

Hypoaspis miles – for fungus gnats and flower thrips – C01, C04, C06, C07, C08, C09, C11, C12, C14, U03, U06, U23, U24, U30, U31, U40, U42, U43, U44, U50, U51, U56, U64, U69, U75, U78, U88, U89, U92

Iphiseius (*Amblyseius*) *degenerans* – for western flower thrips and pest mites – C01, C04, C07, C08, C11, C12, C14, U05, U31, U50, U51, U56, U78, U88, U89, U92

Mesoseiulus (= *Phytoseiulus*) *longipes* – for spider mites – C07, C09, C11, C12, C14, U05, U06, U08, U13, U20, U24, U30, U31, U36, U42, U43, U44, U50, U51, U52, U67, U73, U74, U78, U88, U89, U92, U95

Neoseiulus (= *Amblyseius*, = *Phytoseiulus*) *barkeri* (= *mckenziei*) – for thrips – C09, C11, C14, U05, U06, U24, U51, U66, U80, U88

Neoseiulus (= *Amblyseius*) *californicus* – for spider mites – C04, C06, C07, C08, C09, C11, C14, U03, U05, U06, U08, U09, U13, U19, U20, U24, U29, U30, U31, U36, U37, U42, U43, U44, U50, U51, U52, U56, U66, U67, U69, U73, U74, U75, U78, U80, U88, U89, U92, U95

Neoseiulus (= *Amblyseius*) *cucumeris* – for thrips, cyclamen and broad mites – C01, C03, C04, C06, C07, C08, C09, C10, C11, C12, C14, U03, U05, U06, U09, U11, U13, U23, U24, U26, U29, U30, U31, U34, U37, U42, U43, U44, U50, U51, U52, U56, U64, U66, U67, U69, U73, U74, U75, U78, U80, U88, U89

Neoseiulus (= *Amblyseius*) *fallacis* – for European red and twospotted spider mites – C01, C07, C11, C12, C14, U05, U06, U09, U20, U24, U31, U37, U44, U50, U51, U78, U85, U88, U89, U92

Neoseiulus setulus – for cyclamen mites on strawberries – U06

Phytoseiulus macropilis – for spider mites – U05, U20, U30, U51, U89

Phytoseiulus persimilis – for spider mites – C01, C03, C04, C06, C07, C08, C09, C11, C12, C14, U03, U04, U05, U06, U08, U09, U11, U13, U20, U23, U24, U26, U27, U29, U30, U31, U32, U34, U36, U37, U40, U42, U43, U44, U50, U51, U52, U56, U57, U63, U64, U66, U67, U69, U73, U74, U75, U78, U86, U87, U88, U89, U92, U95

Pyemotes tritici – straw itch mite for ants and stored product pests – U24, U29, U72

Typhlodromus pyri – for various apple and other orchard mites – U44

Typhlodromus rickeri – for various orchard mites – U20, U78

Parasitic nematodes

Heterorhabditis bacteriophora (= *heliothidis*) – for manure flies, caterpillars, weevil larvae, and other soil-dwelling insects – C03, C09, C11, C12, U06, U09, U11, U17, U24, U31, U35, U36, U40, U42, U43, U44, U46, U51, U61, U64, U67, U68, U73, U74, U78, U80, U88, U89

Heterorhabditis megidis – for various soil-dwelling insects – C07, C08, C09, U06, U24, U40, U50, U56, U61, U75, U95

Steinernema (= *Neoaplectana*) *carpocapsae* – for caterpillars, beetle larvae, some flies, and other soil-dwelling insects – C03, C06, C07, C09, C10, C11, C12, C13, C14, U05, U06, U09, U11, U13, U17, U18, U21, U23, U24, U31, U39, U40, U42, U43, U44, U49, U52, U61, U63, U64, U67, U70, U73, U74, U78, U80, U88, U89, U91, U95

Steinernema (= *Neoaplectana*) *feltiae* (= *bibionis*) – for various soil-dwelling insects – C04, C07, C08, C11, U05, U06, U17, U24, U29, U30, U35, U40, U44, U46, U50, U51, U56, U64, U73, U75, U91

Steinernema (= *Neoaplectana*) *glaseri* – for soil-dwelling white grubs – U05, U06, U24, U40, U87

Steinernema riobravis – for maize earworm, mole crickets and the larvae of citrus weevils – U05, U06, U24, U40, U49, U91

Stored product pest parasites and predators

Anisopteromalus calandrae – a parasite for weevils – U15, U26

Bracon hebetor – a parasite for moth larvae – U06, U15, U24, U26, U72

Pyemotes tritici – straw itch mite, a predator for beetles and moths – U24, U29, U72

Xylocoris flavipes – warehouse pirate bug, a predator for various insects – U15, U24, U72

Aphid parasitoids and predators

Aphelinus abdominalis – a parasite – C04, C07, C08, C11, U29, U51, U56, U89

Aphidius colemani – a parasite – C04, C06, C07, C08, C11, C12, C14, U05, U23, U24, U29, U30, U31, U34, U39, U40, U42, U43, U50, U51, U52, U56, U69, U74, U75, U78, U88, U89

Aphidius ervi – a parasite – U56, U69

Aphidius matricariae – a parasite – C01, C07, C14, U05, U06, U24, U30, U31, U32, U43, U44, U51, U64, U67, U78, U88, U89, U92, U95

Aphidoletes aphidimyza – a predator – C01, C03, C04, C06, C07, C08, C11, C14 , U03, U05, U06, U09, U11, U13, U23, U24, U26, U29, U30, U31 U32, U39, U40, U42, U43, U44, U50, U51, U52, U56, U64, U66, U67, U73, U74, U75, U78, U88, U89, U92, U95

Chrysoperla (= *Chrysopa*) *carnea* – common green lacewing, a predator – C03, C04, C06, C07, C10, C11, C12, C14, M03, M07, M08, M09, M11, M12, M14, M17, M18, M20, M21, M24, M27, M29, M30, U01, U03, U04, U05, U06, U07, U11, U12, U19, U22, U23, U24, U26, U29, U30, U31, U32, U36, U39, U40, U42, U43, U44, U47, U49, U50, U51, U56, U58, U61, U63, U66, U67, U72, U78, U80, U82, U89, U90, U92, U94, U95

Chrysoperla (= *Chrysopa*) *comanche* – Comanche lacewing, a predator – C14, U04, U05, U06, U07, U12, U22, U23, U24, U26, U31, U40, U42, U44, U49, U50, U61, U80, U82, U89

Chrysoperla (= *Chrysopa*) *rufilabris* – a green lacewing, a predator – C02, C03, C05, C09, C11, C14, M31, U01, U03, U04, U05, U06, U07, U08, U09, U11, U12, U13, U15, U21, U22, U23, U24, U26, U30, U31, U36, U39, U40, U42, U43, U44, U47, U49, U50, U51, U52, U58, U61, U64, U67, U69, U72, U73, U74, U78, U80, U82, U87, U88, U89, U92, U94, U95

Coleomegilla maculata – pink spotted ladybird beetle, a predator – C05

Deraeocoris brevis – a true bug, a predator – C01, C07, C11, C14, U06, U31, U44, U78, U89, U92

Diaeretiella rapae – a parasite – U06

Harmonia axyridis – Asian multicoloured ladybird beetle, a predator – C01, C05, C08, C11, U31, U78, U89

Hippodamia convergens – convergent ladybird beetle, a predator – C03, C06, C07, C08, C09, C10, C11, C12, C13, C14, U01, U03, U04, U05, U06, U07, U09, U11, U12, U13, U15, U23, U24, U29, U30, U31, U36, U40, U42, U43, U44, U49, U50, U51, U52, U56, U58, U59, U60, U61, U63, U64, U66, U67, U73, U74, U75, U78, U82, U87, U88, U89, U90, U92, U94, U95

Lysiphlebus testaceipes – a parasite – U72

Macrolophus caliginosus – a predator – C07, C11

Orius insidiosus – a predator – C04, C06, C07, C08, C09, C11, C12, C14, U05, U06, U09, U13, U24, U30, U31, U34, U42, U43, U44, U50, U51, U56, U64, U67, U69, U73, U74, U75, U78, U80, U88, U89, U92, U95

Orius tristicolor – minute pirate bug, a predator – C03, C09, C10, U05, U06, U11, U23, U24, U30, U66, U78, U95

Whitefly parasitoids and predators

Chrysoperla (= *Chrysopa*) *carnea* – common green lacewing, a predator – C03, C04, C06, C07, C10, C11, C12, C14, M03, M07, M08, M09, M11, M12, M14, M17, M18, M20, M21, M24, M27, M29, M30, U01, U03, U04, U05, U06, U07, U11, U12, U19, U22, U23, U24, U26, U29, U30, U31, U32, U36, U39, U40, U42, U43, U44, U47, U49, U50, U51, U56, U58, U61, U63, U66, U67, U72, U78, U80, U82, U89, U90, U92, U94, U95

Chrysoperla (= *Chrysopa*) *comanche* – Comanche lacewing, a predator – C14, U04, U05, U06, U07, U12, U22, U23, U24, U26, U31, U40, U42, U44, U49, U50, U61, U80, U82, U89

Chrysoperla (= *Chrysopa*) *rufilabris* – a green lacewing, a predator – C02, C03, C05, C09, C11, C14, M31, U01, U03, U04, U05, U06, U07, U08, U09, U11, U12, U13, U15, U21, U22, U23, U24, U26, U30, U31, U36, U39, U40, U42, U43, U44, U47, U49, U50, U51, U52, U58, U61, U64, U67, U69, U72, U73, U74, U78, U80, U82, U87, U88, U89, U92, U94, U95

Delphastus pusillus – a predator – C01, C03, C04, C07, C09, C11, C12, C14, U03, U05, U06, U09, U11, U13, U19, U23, U24, U29, U31, U37, U39, U42, U43, U44, U50, U51, U52, U67, U73, U74, U78, U80, U88, U89, U92, U95

Deraeocoris brevis – a true bug, a predator – C01, C07, C11, C14, U06, U31, U44, U78, U89, U92

Encarsia deserti (= *luteola*) – a parasite for sweetpotato and silverleaf whiteflies – U03, U05, U43, U50, U88

Encarsia formosa – a parasite for greenhouse whitefly – C01, C03, C04, C06, C07, C08, C09, C10, C11, C12, C14, U03, U04, U05, U06, U07, U09, U11, U13, U21, U23, U24, U26, U29, U31, U32, U34, U36, U37, U39, U40, U42, U43, U44, U50, U51, U52, U56, U63, U64, U66, U67, U69, U72, U73, U74, U75, U78, U80, U87, U88, U89, U92, U95

Eretmocerus californicus – a parasite for sweetpotato and silverleaf whiteflies – C04, C07, C08, C09, C11, U03, U05, U06, U08, U24, U31, U34, U40, U43, U50, U51, U56, U64, U69, U75, U78, U88, U89

Macrolophus caliginosus – a predator – C07, C11

Parasitoids and predators for greenhouse pests

Aphelinus abdominalis – a parasite for aphids – C04, C07, C08, C11, U29, U51, U56, U89

Aphidoletes aphidimyza – a gall midge, a predator for aphids – C01, C03, C04, C06, C07, C08, C11, C14 , U03, U05, U06, U09, U11, U13, U23, U24, U26, U29, U30, U31 U32, U39, U40, U42, U43, U44, U50, U51, U52, U56, U64, U66, U67, U73, U74, U75, U78, U88, U89, U92, U95

Aphidius colemani – a parasite for aphids – C04, C06, C07, C08, C11, C12, C14, U05, U23, U24, U29, U30, U31, U34, U39, U40, U42, U43, U50, U51, U52, U56, U69, U74, U75, U78, U88, U89

Aphidius matricariae – a parasite for aphids – C01, C07, C14, U05, U06, U24, U30, U31, U32, U43, U44, U51, U64, U67, U78, U88, U89, U92, U95

Cryptolaemus montrouzieri – mealybug destroyer, a predator for various scales and mealybugs – C03, C04, C06, C07, C08,C09, C11, C12, C14, U03, U05, U06, U09, U11, U12, U13, U19, U23, U24, U26, U29, U30, U31, U32, U34, U39, U40, U42, U43, U44, U47, U49, U50, U51, U52, U56, U63, U64, U66, U67, U69, U71, U73, U74, U75, U78, U80, U82, U88, U89, U92, U95

Dacnusa sibirica – a parasite for leafminers – C04, C06, C07, C08, C11, C14, U03, U06, U24, U29, U31, U40, U43, U50, U51, U56, U66, U69, U75, U78, U89, U92

Delphastus pusillus – a predator for whiteflies – C01, C03, C04, C07, C09, C11, C12, C14, U03, U05, U06, U09, U11, U13, U19, U23, U24, U29, U31, U37, U39, U42, U43, U44, U50, U51, U52, U67, U73, U74, U78, U80, U88, U89, U92, U95

Diaeretiella rapae – a parasite for aphids – U06

Diglyphus isaea – a parasite for leafminers – C04, C06, C07, C08, C09, C11, C14, U03, U06, U24, U29, U31, U40, U42, U43, U44, U50, U51, U52, U56, U66, U69, U74, U75, U78, U88, U89, U92

Encarsia deserti (= *luteola*) – a parasite for sweetpotato and silverleaf whiteflies – U03, U05, U43, U50, U88

Encarsia formosa – a greenhouse whitefly parasite – C01, C03, C04, C06, C07, C08, C09, C10, C11, C12, C14, U03, U04, U05, U06, U07, U09, U11, U13, U21, U23, U24, U26, U29, U31, U32, U34, U36, U37, U39, U40, U42, U43, U44, U50, U51, U52, U56, U63, U64, U66, U67, U69, U72, U73, U74, U75, U78, U80, U87, U88, U89, U92, U95

Eretmocerus californicus – a parasite for sweetpotato and silverleaf whiteflies – C04, C07, C08, C09, C11, U03, U05, U06, U08, U24, U31, U34, U40, U43, U50, U51, U56, U64, U69, U75, U78, U88, U89

Feltiella acarisuga (=*Therodiplosis persicae*) – a gall midge, a predator for mites – C01, C04, C11, C12, C14, U31, U78, U89

Hippodamia convergens – convergent ladybird beetle, a general predator – C03, C06, C07, C08, C09, C10, C11, C12, C13, C14, U01, U03, U04, U05, U06, U07, U09, U11, U12, U13, U15, U23, U24, U29, U30, U31, U36, U40, U42, U43, U44, U49, U50, U51, U52, U56, U58, U59, U60, U61, U63, U64, U66, U67, U73, U74, U75, U78, U82, U87, U88, U89, U90, U92, U94, U95

Hypoaspis aculeifer – a predatory mite for fungus gnats and flower thrips – U56, U75

Hypoaspis miles – a predatory mite for fungus gnats and flower thrips – C01, C04, C06, C07, C08, C09, C11, C12, C14, U03, U06, U23, U24, U30, U31, U40, U42, U43, U44, U50, U51, U56, U64, U69, U75, U78, U88, U89, U92

Iphiseius (= *Amblyseius*) *degenerans* – a predatory mite for western flower thrips and pest mites – C01, C04, C07, C08, C11, C12, C14, U05, U31, U50, U51, U56, U78, U88, U89, U92

Lysiphlebus testaceipes – a parasite for aphids – U72

Orius insidiosus – a general predator – C04, C06, C07, C08, C09, C11, C12, C14, U05, U06,

U09, U13, U24, U30, U31, U34, U42, U43, U44, U50, U51, U56, U64, U67, U69, U73, U74, U75, U78, U80, U88, U89, U92, U95

Orius tristicolor – minute pirate bug, a general predator – C03, C09, C10, U05, U06, U11, U23, U24, U30, U66, U78, U95

Scolothrips sexmaculatus – sixspotted thrips, a predator for mites and thrips – U02, U03, U05, U19, U23, U48, U49, U78, U82, U83, U92

Stethorus punctillum – a predator for mites – C01, U89

Thripobius semiluteus – a parasite for thrips – C09, C11, U03, U06, U23, U24, U32, U44, U47, U51, U61, U67, U73, U78, U88, U89, U95

Scale and mealybug parasitoids and predators

Aphytis melinus – a parasite for red scale – C03, C09, C11, C12, C14, U03, U05, U06, U11, U12, U23, U24, U26, U30, U31, U32, U33, U38, U39, U42, U43, U44, U47, U49, U51, U64, U66, U71, U73, U78, U81, U82, U83, U88, U89, U92

Cryptolaemus montrouzieri – mealybug destroyer, a predator for various scales and mealybugs – C03, C04, C06, C07, C08, C09, C11, C12, C14, U03, U05, U06, U09, U11, U12, U13, U19, U23, U24, U26, U29, U30, U31, U32, U34, U39, U40, U42, U43, U44, U47, U49, U50, U51, U52, U56, U63, U64, U66, U67, U69, U71, U73, U74, U75, U78, U80, U82, U88, U89, U92, U95

Leptomastix dactylopii – a parasite for citrus mealybug – C04, C07, C08, C09, C11, C14, U06, U24, U30, U31, U43, U44, U50, U51, U56, U75, U88, U89

Metaphycus helvolus – a parasite for black scale – C03, C07, C09, C11, C14, U05, U06, U11, U24, U31, U33, U42, U43, U44, U50, U51, U66, U73, U78, U81, U88, U89, U92

Pseudaphycus angelicus – a parasite for mealybugs – U69

Rhyzobius (= *Lindorus*) *lophanthae* – a predator for various scales – C09, C11, C14, U05, U06, U24, U30, U31, U42, U43, U44, U50, U51, U64, U78, U88, U89, U92

Rhyzobius (= *Lindorus*) *ventralis* – a predator for various scales – U78

Insect egg parasitoids

Anagrus epos – for leafhoppers – U44

Anaphes iole – for lygus bugs – C05, C10, C11, U20, U51, U78, U88, U89, U92

Anastatus tenuipes – for brownbanded cockroaches – U76

Aprostocetus (= *Tetrastichus*) *hagenowii* – for American, smoky, brown, Australian, Oriental cockroaches – U76

Comperia merceti – for brownbanded cockroaches – U76

Trichogramma brassicae – for exposed eggs of various moths and pest butterflies – C02, C05, C06, C07, C09, C11, C12, C14, U04, U05, U06, U08, U09, U22, U23, U24, U31, U39, U44, U51, U52, U56, U61, U64, U67, U72, U73, U74, U75, U78, U87, U88, U92

Trichogramma evanescens – for exposed eggs of various moths and pest butterflies – C07, C09, C14, U05, U06, U08, U24, U50, U72

Trichogramma exiguum – for exposed eggs of various moths and pest butterflies – M16, M18, M24

Trichogramma minutum – minute egg parasite, primarily for exposed eggs of various moths and butterflies in orchards – C02, C03, C09, C11, C14, M13, M22, U01, U04, U05, U06, U07, U08, U09, U11, U12, U13, U19, U24, U29, U31, U36, U39, U40, U42, U43, U44, U49, U50, U51, U52, U58, U63, U64, U66, U67, U72, U73, U74, U78, U80, U82, U88, U89, U90, U92, U94, U95

Trichogramma platneri – primarily for exposed eggs of various moths and butterflies in orchards – C02, C07, C09, C11, C12, C14, U01, U03, U04, U05, U06, U07, U08, U12, U13, U19, U22, U23, U24, U26, U31, U32, U39, U42, U44, U47, U49, U50, U52, U61, U63, U64, U66, U67, U72, U73, U74, U78, U82, U88, U89, U92

Trichogramma pretiosum – primarily for exposed eggs of various moths and butterflies in vegetable and field crops – C02, C03, C09, C10, C11, C14, M01, M02, M03, M05, M06, M10, M11, M12, M13, M15, M17, M19, M20, M21, M22, M23, M25, M26, M27, M30, M31, M32, M33, U01, U03, U04, U05, U06, U07, U08, U09, U11, U12, U13, U15, U19, U21, U22, U24, U26, U29, U31, U32, U36, U39, U40, U42, U43, U44, U49, U50, U51, U52, U58, U61, U63, U64, U66, U67, U72, U73, U74, U78, U82, U87, U88, U89, U90, U92, U94, U95

Trichogrammatoidea bactrae – for exposed eggs of various moths and pest butterflies – C09, M04, U04, U05, U06, U23, U24, U31, U39, U44, U50, U61, U63, U72, U78, U92

Moth and butterfly larval parasitoids

Bracon hebetor – for moths in stored products – U06, U15, U24, U26, U72

Cotesia flavipes – for sugar cane borer – M05, U24
Cotesia melanoscelus – for gypsy moth – U24
Cotesia plutellae – for diamondback moth – U15, U26
Goniozus legneri – for navel orangeworm – U04, U08, U12, U19, U22, U25, U26, U32, U49, U61, U64, U73, U78, U82, U92
Pentalitomastix plethoricus – for navel orangeworm – U22, U25, U26, U61, U82

Filth fly parasitoids

Muscidifurax raptor – C05, C11, U01, U03, U04, U05, U06, U07, U08, U09, U22, U23, U24, U39, U44, U51, U58, U67, U72, U73, U95
Muscidifurax raptorellus – C11, C12, C14, U03, U04, U05, U06, U07, U08, U09, U10, U22, U23, U24, U31, U36, U39, U40, U42, U44, U49, U51, U58, U64, U67, U72, U73, U78, U84, U88, U89, U90
Muscidifurax raptoroides – M18
Muscidifurax zaraptor – C02, C03, C09, C10, C12, C14, U03, U04, U06, U07, U08, U09, U10, U11, U13, U21, U22, U23, U24, U31, U36, U39, U40, U42, U44, U52, U58, U63, U64, U66, U72, U73, U74, U78, U84, U88, U89, U90, U95
Nasonia vitripennis – C03, C09, C11, U03, U06, U11, U24, U66, U72, U95
Spalangia cameroni – U03, U08, U13, U23, U44, U49, U58, U61, U72, U74, U80, U84
Spalangia endius – C03, C12, C14, U01, U03, U05, U07, U08, U09, U11, U13, U15, U22, U31, U36, U39, U44, U58, U61, U63, U66, U72, U73, U74, U78, U84, U89, U95
Spalangia nigroaenea – U05, U07, U08, U13, U23, U39, U44, U49, U58, U61, U72, U73, U74, U80, U95

Other insect parasitoids

Aceratoneuromyia indica – for fruit fly larvae – M19
Bracon kirkpatricki – an external parasite for cotton boll weevil larvae – U24
Diachasmimorpha (= *Biosteres*, = *Opius*) *longicaudata* (= *longicaudatus*) – for fruit fly larvae – M19, U06
Pediobius foveolatus – for bean beetle – C09, U06, U24, U79, U89

General predators

Chrysoperla (= *Chrysopa*) *carnea* – common green lacewing – C03, C04, C06, C07, C10, C11, C12, C14, M03, M07, M08, M09, M11, M12, M14, M17, M18, M20, M21, M24, M27, M29, M30, U01, U03, U04, U05, U06, U07, U11, U12, U19, U22, U23, U24, U26, U29, U30, U31, U32, U36, U39, U40, U42, U43, U44, U47, U49, U50, U51, U56, U58, U61, U63, U66, U67, U72, U78, U80, U82, U89, U90, U92, U94, U95
Chrysoperla (= *Chrysopa*) *comanche* – Comanche lacewing – C14, U04, U05, U06, U07, U12, U22, U23, U24, U26, U31, U40, U42, U44, U49, U50, U61, U80, U82, U89
Chrysoperla (= *Chrysopa*) *rufilabris* – a green lacewing – C02, C03, C05, C09, C11, C14, M31, U01, U03, U04, U05, U06, U07, U08, U09, U11, U12, U13, U15, U21, U22, U23, U24, U26, U30, U31, U36, U39, U40, U42, U43, U44, U47, U49, U50, U51, U52, U58, U61, U64, U67, U69, U72, U73, U74, U78, U80, U82, U87, U88, U89, U92, U94, U95
Coleomegilla maculata – pink spotted ladybird beetle – C05
Cryptolaemus montrouzieri – mealybug destroyer, for scales and mealybugs – C03, C04, C06, C07, C08, C09, C11, C12, C14, U03, U05, U06, U09, U11, U12, U13, U19, U23, U24, U26, U29, U30, U31, U32, U34, U39, U40, U42, U43, U44, U47, U49, U50, U51, U52, U56, U63, U64, U66, U67, U69, U71, U73, U74, U75, U78, U80, U82, U88, U89, U92, U95
Delphastus pusillus – for whiteflies – C01, C03, C04, C07, C09, C11, C12, C14, U03, U05, U06, U09, U11, U13, U19, U23, U24, U29, U31, U37, U39, U42, U43, U44, U50, U51, U52, U67, U73, U74, U78, U80, U88, U89, U92, U95
Deraeocoris brevis – a true bug – C01, C07, C11, C14, U06, U31, U44, U78, U89, U92
Gambusia affinis – mosquito fish, for mosquitoes – U13, U28, U53, U66
Geocoris punctipes – a big-eyed bug – U15
Harmonia axyridis – Asian multi-coloured ladybird beetle – C01, C05, C08, C11, U31, U78, U89
Hippodamia convergens – convergent ladybird beetle – C03, C06, C07, C08, C09, C10, C11, C12, C13, C14, U01, U03, U04, U05, U06, U07, U09, U11, U12, U13, U15, U23, U24, U29, U30, U31, U36, U40, U42, U43, U44, U49, U50, U51, U52, U56, U58, U59, U60, U61, U63, U64, U66, U67, U73, U74, U75, U78, U82, U87, U88, U89, U90, U92, U94, U95
Macrolophus caliginosus – for aphids and whiteflies – C07, C11
Mantis religiosa – European mantid, a praying mantid – C03, U11

Orius insidiosus – insidious flower bug – C04, C06, C07, C08, C09, C11, C12, C14, U05, U06, U09, U13, U24, U30, U31, U34, U42, U43, U44, U50, U51, U56, U64, U67, U69, U73, U74, U75, U78, U80, U88, U89, U92, U95

Orius tristicolor – minute pirate bug – C03, C09, C10, U05, U06, U11, U23, U24, U30, U66, U78, U95

Podisus maculiventris – spined soldier bug – C11, U85, U89

Rumina decollata – decollate snail, for snails – C09, U03, U05, U06, U13, U23, U24, U32, U47, U52, U62, U65, U71, U73, U74, U78, U80, U81, U92

Scolothrips sexmaculatus – sixspotted thrips, for mites and pest thrips – U02, U03, U05, U19, U23, U48, U49, U78, U82, U83, U92

Stethorus picipes – for orchard mites – U78

Tenodera aridifolia sinensis – Chinese mantid, a praying mantid – C06, C09, C10, C11, U01, U05, U06, U09, U13, U24, U29, U31, U36, U42, U43, U44, U52, U63, U66, U67, U73, U74, U87, U88, U89, U90, U92, U94, U95

Xylocoris flavipes – warehouse pirate bug, for moths and beetles in stored grains – U15, U24, U72

Weed feeders

Aceria (= *Eriophyes*) *chondrillae* – for rush skeletonweed – U16

Agonopterix alstroemeriana – for poison hemlock – U16

Agrilus hyperici – for St John's wort (Klamath weed) – U14

Aphthona cyparissiae – for spurge – U16

Aphthona flava – for spurge – U16

Aphthona lacertosa – for spurge – U16

Aphthona nigriscutis – for spurge – U16

Apion fuscirostre – for Scotch broom – U14

Apion ulicis – for gorse – U14

Aplocera plagiata – for St John's wort (Klamath weed) – U16

Bangasternus orientalis – for yellow starthistle – U06, U14, U16, U26, U73

Brachypterolus pulicarious – for toadflax – U16

Cassida rubiginosa – for Canada and musk thistles – U16

Ceutorhynchus litura – for Canada thistle – U16

Chrysolina quadrigemina – Klamath weed beetle for St John's wort (Klamath weed) – U14, U16

Coleophora klimeschiella – for Russian thistle – U14

Coleophora parthenica – for Russian thistle – U14

Ctenopharyngodon idella – Chinese grass carp (white amur) for aquatic weeds – U45, U54, U55

Cystiphora schmidti – for rush skeletonweed – U16

Eustenopus villosus – for yellow starthistle – U26

Larinus planus – for Canada thistle – U16

Leucoptera spartifoliella – for Scotch broom – U14

Longitarsus jacobaeae – for tansy ragwort – U14, U16

Metzneria paucipunctella – for knapweed – U16

Microlarinus lareynii – puncturevine seed weevil – U14, U49, U61, U73

Microlarinus lypriformis – puncturevine stem weevil – U14, U49, U61, U73

Oberea erythrocephala – for leafy spurge – U16

Rhinocyllus conicus – different strains of weevil for Italian, milk and musk thistles – U14, U16

Spurgia esulae – for spurge – U16

Trichosirocalus horridus – for musk thistle – U16

Tyria jacobaeae – cinnabar moth for tansy ragwort – U14

Urophora affinis – for knapweed – U16

Urophora cardui – for Canada thistle – U16

Urophora quadrifasciata – for knapweed – U16

Urophora sirunaseva – for yellow starthistle – U14

Zeuxidiplosis giardi – for St John's wort (Klamath weed) – U14

Index of Scientific Names of Commercially Available Organisms

Aceratoneuromyia indica – M19

Aceria (= *Eriophyes*) *chondrillae* – U16

Agonopterix alstroemeriana – U16

Agrilus hyperici – U14

Amblyseius – (see *Iphiseius* and *Neoseiulus*)

Anagrus epos – U44

Anaphes iole – C05, C10, C11, U20, U51, U78, U88, U89, U92

Anastatus tenuipes – U76

Anisopteromalus calandrae – U15, U26

Aphelinus abdominalis – C04, C07, C08, C11, U29, U51, U56, U89

Aphidius colemani – C04, C06, C07, C08, C11, C12, C14, U05, U23, U24, U29, U30, U31, U34, U39, U40, U42, U43, U50, U51, U52, U56, U69, U74, U75, U78, U88, U89

Aphidius ervi – U56, U69

Aphidius matricariae – C01, C07, C14, U05, U06, U24, U30, U31, U32, U43, U44, U51, U64, U67, U78, U88, U89, U92, U95

Aphidoletes aphidimyza - C01, C03, C04, C06, C07, C08, C11, C14 , U03, U05, U06, U09, U11, U13, U23, U24, U26, U29, U30, U31 U32, U39, U40, U42, U43, U44, U50, U51, U52, U56, U64, U66, U67, U73, U74, U75, U78, U88, U89, U92, U95
Aphthona cyparissiae - U16
Aphthona flava - U16
Aphthona lacertosa - U16
Aphthona nigriscutis - U16
Aphytis melinus - C03, C09, C11, C12, C14, U03, U05, U06, U11, U12, U23, U24, U26, U30, U31, U32, U33, U38, U39, U42, U43, U44, U47, U49, U51, U64, U66, U71, U73, U78, U81, U82, U83, U88, U89, U92
Apion fuscirostre - U14
Apion ulicis - U14
Aplocera plagiata - U16
Aprostocetus (= *Tetrastichus*) *hagenowii* - U76
Bangasternus orientalis - U06, U14, U16, U26, U73
Biosteres - (see *Diachasmimorpha*)
Brachypterolus pulicarious - U16
Bracon hebetor - U06, U15, U24, U26, U72
Bracon kirkpatricki - U24
Cassida rubiginosa - U16
Ceutorhynchus litura - U16
Chrysolina quadrigemina - U14, U16
Chrysopa - (see *Chrysoperla*)
Chrysoperla (= *Chrysopa*) *carnea* - C03, C04, C06, C07, C10, C11, C12, C14, M03, M07, M08, M09, M11, M12, M14, M17, M18, M20, M21, M24, M27, M29, M30, U01, U03, U04, U05, U06, U07, U11, U12, U19, U22, U23, U24, U26, U29, U30, U31, U32, U36, U39, U40, U42, U43, U44, U47, U49, U50, U51, U56, U58, U61, U63, U66, U67, U72, U78, U80, U82, U89, U90, U92, U94, U95
Chrysoperla (= *Chrysopa*) *comanche* - C14, U04, U05, U06, U07, U12, U22, U23, U24, U26, U31, U40, U42, U44, U49, U50, U61, U80, U82, U89
Chrysoperla (= *Chrysopa*) *rufilabris* - C02, C03, C05, C09, C11, C14, M31, U01, U03, U04, U05, U06, U07, U08, U09, U11, U12, U13, U15, U21, U22, U23, U24, U26, U30, U31, U36, U39, U40, U42, U43, U44, U47, U49, U50, U51, U52, U58, U61, U64, U67, U69, U72, U73, U74, U78, U80, U82, U87, U88, U89, U92, U94, U95
Coleomegilla maculata - C05
Coleophora klimeschiella - U14
Coleophora parthenica - U14
Comperia merceti - U76
Cotesia flavipes - M05, U24
Cotesia melanoscelus - U24
Cotesia plutellae - U15, U26
Cryptolaemus montrouzieri - C03, C04, C06, C07, C08, C09, C11, C12, C14, U03, U05, U06, U09, U11, U12, U13, U19, U23, U24, U26, U29, U30, U31, U32, U34, U39, U40, U42, U43, U44, U47, U49, U50, U51, U52, U56, U63, U64, U66, U67, U69, U71, U73, U74, U75, U78, U80, U82, U88, U89, U92, U95
Ctenopharyngodon idella - U45, U54, U55
Cystiphora schmidti - U16
Dacnusa sibirica - C04, C06, C07, C08, C11, C14, U03, U06, U24, U29, U31, U40, U43, U50, U51, U56, U66, U69, U75, U78, U89, U92
Delphastus pusillus - C01, C03, C04, C07, C09, C11, C12, C14, U03, U05, U06, U09, U11, U13, U19, U23, U24, U29, U31, U37, U39, U42, U43, U44, U50, U51, U52, U67, U73, U74, U78, U80, U88, U89, U92, U95
Deraeocoris brevis - C01, C07, C11, C14, U06, U31, U44, U78, U89, U92
Diachasmimorpha longicaudata (= *Biosteres longicaudatus*, = *Opius longicaudatus*) - M19, U06
Diaeretiella rapae - U06
Diglyphus isaea - C04, C06, C07, C08, C09, C11, C14, U03, U06, U24, U29, U31, U40, U42, U43, U44, U50, U51, U52, U56, U66, U69, U74, U75, U78, U88, U89, U92
Encarsia deserti (= *luteola*) - U03, U05, U43, U50, U88
Encarsia formosa - C01, C03, C04, C06, C07, C08, C09, C10, C11, C12, C14, U03, U04, U05, U06, U07, U09, U11, U13, U21, U23, U24, U26, U29, U31, U32, U34, U36, U37, U39, U40, U42, U43, U44, U50, U51, U52, U56, U63, U64, U66, U67, U69, U72, U73, U74, U75, U78, U80, U87, U88, U89, U92, U95
Eretmocerus californicus - C04, C07, C08, C09, C11, U03, U05, U06, U08, U24, U31, U34, U40, U43, U50, U51, U56, U64, U69, U75, U78, U88, U89
Eriophyes - (see *Aceria*)
Eustenopus villosus - U26
Feltiella acarisuga (= *Therodiplosis persicae*) - C01, C04, C11, C12, C14, U31, U78, U89
Galendromus annectans - U03, U05, U06, U19, U20, U23, U24, U32, U47, U78
Galendromus (= *Typhlodromus*) *helveolus* - C09, C11, U03, U05, U06, U08, U19, U20, U23, U24, U31, U32, U47, U61, U78, U88, U92
Galendromus (= *Metaseiulus*, = *Typhlodromus*) *occidentalis* - C06, C07, C09, C11, C14, U02, U03, U04, U05, U06, U08, U12, U13, U19, U20, U22, U23, U24, U30, U31, U32, U40, U42, U43, U44, U48, U49, U51, U52, U61, U63, U66, U67, U73, U74, U78, U80, U82, U83, U88, U89, U92, U93, U95

Gambusia affinis - U13, U28, U53, U66

Geocoris punctipes - U15

Goniozus legneri - U04, U08, U12, U19, U22, U25, U26, U32, U49, U61, U64, U73, U78, U82, U92

Harmonia axyridis - C01, C05, C08, C11, U31, U78, U89

Heterorhabditis bacteriophora (= *heliothidis*) - C03, C09, C11, C12, U06, U09, U11, U17, U24, U31, U35, U36, U40, U42, U43, U44, U46, U51, U61, U64, U67, U68, U73, U74, U78, U80, U88, U89

Heterorhabditis megidis - C07, C08, C09, U06, U24, U40, U50, U56, U61, U75, U95

Hippodamia convergens - C03, C06, C07, C08, C09, C10, C11, C12, C13, C14, U01, U03, U04, U05, U06, U07, U09, U11, U12, U13, U15, U23, U24, U29, U30, U31, U36, U40, U42, U43, U44, U49, U50, U51, U52, U56, U58, U59, U60, U61, U63, U64, U66, U67, U73, U74, U75, U78, U82, U87, U88, U89, U90, U92, U94, U95

Hypoaspis aculeifer - U56, U75

Hypoaspis miles - C01, C04, C06, C07, C08, C09, C11, C12, C14, U03, U06, U23, U24, U30, U31, U40, U42, U43, U44, U50, U51, U56, U64, U69, U75, U78, U88, U89, U92

Iphiseius (= *Amblyseius*) *degenerans* - C01, C04, C07, C08, C11, C12, C14, U05, U31, U50, U51, U56, U78, U88, U89, U92

Larinus planus - U16

Leptomastix dactylopii - C04, C07, C08, C09, C11, C14, U06, U24, U30, U31, U43, U44, U50, U51, U56, U75, U88, U89

Leucoptera spartifoliella - U14

Lindorus - (see *Rhyzobius*)

Longitarsus jacobaeae - U14, U16

Lysiphlebus testaceipes - U72

Macrolophus caliginosus - C07, C11

Mantis religiosa - C03, U11

Mesoseiulus (= *Phytoseiulus*) *longipes* - C07, C09, C11, C12, C14, U05, U06, U08, U13, U20, U24, U30, U31, U36, U42, U43, U44, U50, U51, U52, U67, U73, U74, U78, U88, U89, U92, U95

Metaphycus helvolus - C03, C07, C09, C11, C14, U05, U06, U11, U24, U31, U33, U42, U43, U44, U50, U51, U66, U73, U78, U81, U88, U89, U92

Metaseiulus - (see *Galendromus*)

Metzneria paucipunctella - U16

Microlarinus lareynii - U14, U49, U61, U73

Microlarinus lypriformis - U14, U49, U61, U73

Muscidifurax raptor - C05, C11, U01, U03, U04, U05, U06, U07, U08, U09, U22, U23, U24, U39, U44, U51, U58, U67, U72, U73, U95

Muscidifurax raptorellus - C11, C12, C14, U03, U04, U05, U06, U07, U08, U09, U10, U22, U23, U24, U31, U36, U39, U40, U42, U44, U49, U51, U58, U64, U67, U72, U73, U78, U84, U88, U89, U90

Muscidifurax raptoroides - M18

Muscidifurax zaraptor - C02, C03, C09, C10, C12, C14, U03, U04, U06, U07, U08, U09, U10, U11, U13, U21, U22, U23, U24, U31, U36, U39, U40, U42, U44, U52, U58, U63, U64, U66, U72, U73, U74, U78, U84, U88, U89, U90, U95

Nasonia vitripennis - C03, C09, C11, U03, U06, U11, U24, U66, U72, U95

Neoaplectana - (see *Steinernema*)

Neoseiulus (= *Amblyseius*, = *Phytoseiulus*) *barkeri* (= *mckenziei*) - C09, C11, C14, U05, U06, U24, U51, U66, U80, U88

Neoseiulus (= *Amblyseius*) *californicus* - C04, C06, C07, C08, C09, C11, C14, U03, U05, U06, U08, U09, U13, U19, U20, U24, U29, U30, U31, U36, U37, U42, U43, U44, U50, U51, U52, U56, U66, U67, U69, U73, U74, U75, U78, U80, U88, U89, U92, U95

Neoseiulus (= *Amblyseius*) *cucumeris* - C01, C03, C04, C06, C07, C08, C09, C10, C11, C12, C14, U03, U05, U06, U09, U11, U13, U23, U24, U26, U29, U30, U31, U34, U37, U42, U43, U44, U50, U51, U52, U56, U64, U66, U67, U69, U73, U74, U75, U78, U80, U88, U89

Neoseiulus (= *Amblyseius*) *fallacis* - C01, C07, C11, C12, C14, U05, U06, U09, U20, U24, U31, U37, U44, U50, U51, U78, U85, U88, U89, U92

Neoseiulus setulus - U06

Oberea erythrocephala - U16

Opius - (see *Diachasmimorpha*)

Orius insidiosus - C04, C06, C07, C08, C09, C11, C12, C14, U05, U06, U09, U13, U24, U30, U31, U34, U42, U43, U44, U50, U51, U56, U64, U67, U69, U73, U74, U75, U78, U80, U88, U89, U92, U95

Orius tristicolor - C03, C09, C10, U05, U06, U11, U23, U24, U30, U66, U78, U95

Pediobius foveolatus - C09, U06, U24, U79, U89

Pentalitomastix plethoricus - U22, U25, U26, U61, U82

Phytoseiulus macropilis - U05, U20, U30, U51, U89

Phytoseiulus persimilis - C01, C03, C04, C06, C07, C08, C09, C11, C12, C14, U03, U04, U05, U06, U08, U09, U11, U13, U20, U23, U24, U26, U27, U29, U30, U31, U32, U34, U36, U37, U40, U42, U43, U44, U50, U51, U52, U56, U57, U63, U64, U66, U67, U69, U73, U74, U75, U78, U86, U87, U88, U89, U92, U95

Phytoseiulus – (see *Mesoseiulus* and *Neoseiulus*)
Podisus maculiventris – C11, U85, U89
Pseudaphycus angelicus – U69
Pyemotes tritici – U24, U29, U72
Rhinocyllus conicus – U14, U16
Rhyzobius (= *Lindorus*) *lophanthae* – C09, C11, C14, U05, U06, U24, U30, U31, U42, U43, U44, U50, U51, U64, U78, U88, U89, U92
Rhyzobius (= *Lindorus*) *ventralis* – U78
Rumina decollata – C09, U03, U05, U06, U13, U23, U24, U32, U47, U52, U62, U65, U71, U73, U74, U78, U80, U81, U92
Scolothrips sexmaculatus – U02, U03, U05, U19, U23, U48, U49, U78, U82, U83, U92
Spalangia cameroni – U03, U08, U13, U23, U44, U49, U58, U61, U72, U74, U80, U84
Spalangia endius – C03, C12, C14, U01, U03, U05, U07, U08, U09, U11, U13, U15, U22, U31, U36, U39, U44, U58, U61, U63, U66, U72, U73, U74, U78, U84, U89, U95
Spalangia nigroaenea – U05, U07, U08, U13, U23, U39, U44, U49, U58, U61, U72, U73, U74, U80, U95
Spurgia esulae – U16
Steinernema (= *Neoaplectana*) *carpocapsae* – C03, C06, C07, C09, C10, C11, C12, C13, C14, U05, U06, U09, U11, U13, U17, U18, U21, U23, U24, U31, U39, U40, U42, U43, U44, U49, U52, U61, U63, U64, U67, U70, U73, U74, U78, U80, U88, U89, U91, U95
Steinernema (= *Neoaplectana*) *feltiae* (= *bibionis*) – C04, C07, C08, C11, U05, U06, U17, U24, U29, U30, U35, U40, U44, U46, U50, U51, U56, U64, U73, U75, U91
Steinernema (= *Neoaplectana*) *glaseri* – U05, U06, U24, U40, U87
Steinernema riobravis – U05, U06, U24, U40, U49, U91
Stethorus picipes – U78
Stethorus punctillum – C01, U89
Tenodera aridifolia sinensis – C06, C09, C10, C11, U01, U05, U06, U09, U13, U24, U29, U31, U36, U42, U43, U44, U52, U63, U66, U67, U73, U74, U87, U88, U89, U90, U92, U94, U95
Therodiplosis – (see *Feltiella*)
Thripobius semiluteus – C09, C11, U03, U06, U23, U24, U32, U44, U47, U51, U61, U67, U73, U78, U88, U89, U95
Trichogramma brassicae – C02, C05, C06, C07, C09, C11, C12, C14, U04, U05, U06, U08, U09, U22, U23, U24, U31, U39, U44, U51, U52, U56, U61, U64, U67, U72, U73, U74, U75, U78, U87, U88, U92
Trichogramma evanescens – C07, C09, C14, U05, U06, U08, U24, U50, U72
Trichogramma exiguum – M16, M18, M24
Trichogramma minutum – C02, C03, C09, C11, C14, M13, M22, U01, U04, U05, U06, U07, U08, U09, U11, U12, U13, U19, U24, U29, U31, U36, U39, U40, U42, U43, U44, U49, U50, U51, U52, U58, U63, U64, U66, U67, U72, U73, U74, U78, U80, U82, U88, U89, U90, U92, U94, U95
Trichogramma platneri – C02, C07, C09, C11, C12, C14, U01, U03, U04, U05, U06, U07, U08, U12, U13, U19, U22, U23, U24, U26, U31, U32, U39, U42, U44, U47, U49, U50, U52, U61, U63, U64, U66, U67, U72, U73, U74, U78, U82, U88, U89, U92
Trichogramma pretiosum – C02, C03, C09, C10, C11, C14, M01, M02, M03, M05, M06, M10, M11, M12, M13, M15, M17, M19, M20, M21, M22, M23, M25, M26, M27, M30, M31, M32, M33, U01, U03, U04, U05, U06, U07, U08, U09, U11, U12, U13, U15, U19, U21, U22, U24, U26, U29, U31, U32, U36, U39, U40, U42, U43, U44, U49, U50, U51, U52, U58, U61, U63, U64, U66, U67, U72, U73, U74, U78, U82, U87, U88, U89, U90, U92, U94, U95
Trichogrammatoidea bactrae – C09, M04, U04, U05, U06, U23, U24, U31, U39, U44, U50, U61, U63, U72, U78, U92
Trichosirocalus horridus – U16
Typhlodromus pyri – U44
Typhlodromus rickeri – U20, U78
Typhlodromus – (see *Galendromus*)
Tyria jacobaeae – U14
Urophora affinis – U16
Urophora cardui – U16
Urophora quadrifasciata – U16
Urophora sirunaseva – U14
Xylocoris flavipes – U15, U24, U72
Zeuxidiplosis giardi – U14

Appendix III: Higher Classification for Family Names Cited in the Index

Families mentioned in the index are classified to Order, except for most bacteria (unassigned), viruses (unassigned), and imperfect fungi (Class). Virus classification follows Murphy *et al.* (1995) *Virus Taxonomy*, Springer-Verlag, Vienna, Austria.

Phylum Arthropoda

Family	Order
Acrididae	Orthoptera
Acrolepiidae	Lepidoptera
Agromyzidae	Diptera
Aleyrodidae	Hemiptera
Anthocoridae	Hemiptera
Anthomyiidae	Diptera
Aphelinidae	Hymenoptera
Aphididae	Hemiptera
Apidae	Hymenoptera
Apionidae	Coleoptera
Arctiidae	Lepidoptera
Blephariceridae	Diptera
Bombyliidae	Diptera
Braconidae	Hymenoptera
Buprestidae	Coleoptera
Cantharidae	Coleoptera
Carabidae	Coleoptera
Cecidomyiidae	Diptera
Cerambycidae	Coleoptera
Ceraphronidae	Hymenoptera
Ceratopogonidae	Diptera
Chaoboridae	Diptera
Chironomidae	Diptera
Chloropidae	Diptera
Chrysomelidae	Coleoptera
Chrysopidae	Neuroptera
Cicadellidae	Hemiptera
Cleridae	Coleoptera
Coccinellidae	Coleoptera
Cochylidae	Lepidoptera
Cosmopterygidae	Lepidoptera
Culicidae	Diptera
Curculionidae	Coleoptera
Cynipidae	Hymenoptera
Diapriidae	Hymenoptera
Diprionidae	Hymenoptera
Elateridae	Coleoptera
Encyrtidae	Hymenoptera
Ephydridae	Diptera
Eriophyidae	Acari
Erythraeidae	Acari
Eulophidae	Hymenoptera
Eupelmidae	Hymenoptera
Figitidae	Hymenoptera
Forficulidae	Dermaptera
Gelechiidae	Lepidoptera
Geometridae	Lepidoptera
Gracilariidae	Lepidoptera
Gryllidae	Orthoptera
Histeridae	Coleoptera
Hydropsychidae	Trichoptera
Hypoaspididae	Acari
Ichneumonidae	Hymenoptera
Laelapidae	Acari
Leptoceridae	Trichoptera
Lonchaeidae	Diptera
Lygaeidae	Hemiptera
Lymantriidae	Lepidoptera
Lyonetiidae	Lepidoptera
Macrochelidae	Acari
Megaspilidae	Hymenoptera
Mindaridae	Hemiptera
Miridae	Hemiptera
Momphidae	Lepidoptera
Muscidae	Diptera
Mycetophilidae	Diptera
Mymaridae	Hymenoptera
Nabidae	Hemiptera
Nitidulidae	Coleoptera
Noctuidae	Lepidoptera
Nymphalidae	Lepidoptera
Oecophoridae	Lepidoptera
Oestridae	Diptera
Pamphiliidae	Hymenoptera
Pemphigidae	Hemiptera
Pentatomidae	Hemiptera
Philodromidae	Araneae
Phymatidae	Hemiptera
Phytoseiidae	Acari
Platygastridae	Hymenoptera
Plutellidae	Lepidoptera
Psychodidae	Diptera
Psyllidae	Hemiptera
Pterdonchidae	Lepidoptera
Pteromalidae	Hymenoptera
Pterophoridae	Lepidoptera
Pyralidae	Lepidoptera
Reduviidae	Hemiptera
Rhizophagidae	Coleoptera

Sarcophagidae	Diptera
Scelionidae	Hymenoptera
Sciaridae	Diptera
Scolytidae	Coleoptera
Sesiidae	Lepidoptera
Simuliidae	Diptera
Sperchontidae	Acari
Sphingidae	Lepidoptera
Staphylinidae	Coleoptera
Stigmaeidae	Araneae
Syrphidae	Diptera
Tabanidae	Diptera
Tachinidae	Diptera
Tenebrionidae	Coleoptera
Tenthredinidae	Hymenoptera
Tephritidae	Diptera
Tetranychidae	Acari
Thomisidae	Araneae
Thripidae	Thysanoptera
Tortricidae	Lepidoptera
Trichogrammatidae	Hymenoptera
Uropodidae	Acari
Yponomeutidae	Lepidoptera

Phylum Chordata

Anatidae	Anseriformes
Ranidae	Anura
Salmonidae	Salmoniformes

Phylum Cnidaria

Hydridae	Hydrida

Phylum Mollusca

Physidae	Basommatophora

Phylum Nemata

Heterorhabditidae	Rhabditida
Mermithidae	Mermithida
Steinernematidae	Rhabditida
Tylenchidae	Tylenchida

Phylum Plantae

Aceraceae	Sapindales
Alstroemeriaceae	Liliales
Apiaceae	Apiales
Asteraceae	Asterales
Berberidaceae	Ranunculales
Betulaceae	Fagales
Boraginaceae	Solanales
Brassicaceae	Brassicales
Cannabinaceae	Rosales
Caprifoliaceae	Dipsacales
Caryophyllaceae	Caryophyllales
Ceratophyllaceae	Nymphaeales
Chenopodiaceae	Caryophyllales
Clusiaceae	Malpighiales
Convulvulaceae	Solanales
Curcubitaceae	Curcubitales
Cypetaceae	Cyperales
Elaeagnaceae	Rosales
Ericaceae	Ericales
Euphorbiaceae	Malpighiales
Fabaceae	Fabales
Fagaceae	Fagales
Haloragaceae	Saxifragales
Liliaceae	Liliales
Lythraceae	Myrtales
Malvaceae	Malvales
Nymphaeaceae	Nymphaeales
Pinaceae	Pinales
Poaceae	Poales
Pontederiaceae	Commelinales
Potamogetonaceae	Alismatales
Primulaceae	Ericales
Ranunculaceae	Ranunculales
Rhamnaceae	Rosales
Rosaceae	Rosales
Rubiaceae	Gentianales
Salicaceae	Malpighiales
Saxifragaceae	Saxifragales
Scrophulariaceae	Lamiales
Solanaceae	Solanales
Ulmaceae	Rosales
Vitaceae	Rosales

Phylum Platyhelminthes

Dugesiidae	Tricladida

Bacteria

Bacillus/Clostridium group	
CFB group	
Clostridiaceae	
Comamonadaceae	
Cytophagaceae	
Enterobacteriaceae	
Flavobacteriaceae	
Microbacteriaceae	
Micrococcaceae	
Moraxellaceae	
Pseudomonadaceae	
Rhizobiaceae	
Rickettsiaceae	Rickettsiales
Spingobacteria	
Streptomycetaceae	
Vibrionaceae	

Fungi

Acremonium	Hyphomycetes
Agaricaceae	Basidiomycetes
Albuginaceae	Oomycetes
Alternaria	Hyphomycetes
Ampelomyces	Coelomycetes
Amphisphaeriaceae	Ascomycetes
Ancylistaceae	Zygomycetes

Ascochyta	Coelomycetes
Aspergillus	Hyphomycetes
Atheliaceae	Basidiomycetes
Aureobasidium	Hyphomycetes
Baktoa	Entomophthorales
Beauveria	Hyphomycetes
Bondarzewiaceae	Basidiomycetes
Botrytis	Hyphomycetes
Candidaceae	Blastomycetes
Caudosporidae	Ascomycetes
Cercospora	Hyphomycetes
Cladosporium	Hyphomycetes
Coelomomycetaceae	Chytridiomycetes
Colletotrichum	Coelomycetes
Coniothyrium	Coelomycetes
Coriolaceae	Basidiomycetes
Cronartiaceae	Teliomycetes
Cryptococcaceae	Basidiomycetes
Culicinomyces	Hyphomycetes
Curvularia	Hyphomycetes
Darluca	Coelomycetes
Didymosphaeriaceae	Ascomycetes
Dilophospora	Coelomycetes
Diploceras	Hyphomycetes
Diplodia	Coelomycetes
Dothidiaceae	Ascomycetes
Drechslera	Hyphomycetes
Entomophthoraceae	Zygomycetes
Epicoccum	Hyphomycetes
Erysiphaceae	Ascomycetes
Exserohilum	Hyphomycetes
Fusarium	Hyphomycetes
Gliocladium	Hyphomycetes
Glomaceae	Zygomycetes
Harpellaceae	Trichomycetes
Helotiaceae	Ascomycetes
Hormonema	Hyphomycetes
Idriella	Hyphomycetes
Legeriomycetaceae	Trichomycetes
Leptosphaeriaceae	Ascomycetes
Melanconidaceae	Ascomycetes
Melanconium	Coelomycetes
Meruliaceae	Basidiomycetes
Metarhizium	Hyphomycetes
Microdochium	Hyphomycetes
Microsphaeropsis	Coelomycetes
Monilinia	Hyphomycetes
Monocillium	Hyphomycetes
Mucoraceae	Zygomycetes
Mycosphaerellaceae	Ascomycetes
Myrothecium	Hyphomycetes
Nectriaceae	Ascomycetes
Nidulariaceae	Basidiomycetes
Ophiostomataceae	Ascomycetes
Paecilomyces	Hyphomycetes
Penicillium	Hyphomycetes
Peniophoraceae	Basidiomycetes
Phaeotheca	Hyphomycetes
Phanerochaetaceae	Basidiomycetes
Phoma	Coelomycetes
Phomopsis	Coelomycetes
Phyllachoraceae	Ascomycetes
Phyllosticta	Coelomycetes
Plectosphaerella	Phyllachorales
Pleiochaeta	Hyphomycetes
Pleosporaceae	Ascomycetes
Pollaccia	Hyphomycetes
Polyporaceae	Basidiomycetes
Pothidieaceae	Ascomycetes
Pucciniaceae	Teliomycetes
Pucciniastraceae	Teliomycetes
Pyricularia	Hyphomycetes
Pythiaceae	Oomycetes
Rhizoctonia	Hyphomycetes
Saccharomycetaceae	Ascomycetes
Saprolegniaceae	Oomycetes
Schizophyllaceae	Basidiomycetes
Sclerosporaceae	Oomycetes
Sclerotiniaceae	Ascomycetes
Scytalidium	Hyphomycetes
Seimatosporium	Hyphomycetes
Septoria	Coelomycetes
Sporidesmium	Hyphomycetes
Sporobolomycetaceae	Basidiomycetes
Stachybotrys	Hyphomycetes
Stagonospora	Coelomycetes
Steccherinaceae	Basidiomycetes
Stemphylium	Hyphomycetes
Stilbella	Hyphomycetes
Synchytriaceae	Chytridiomycetes
Tolypocladium	Hyphomycetes
Trichocomaceae	Ascomycetes
Trichoderma	Hyphomycetes
Tricholomataceae	Basidiomycetes
Trichothecium	Hyphomycetes
Tuberculina	Hyphomycetes
Typhulaceae	Basidiomycetes
Ustilaginaceae	Basidiomycetes
Valsaceae	Ascomycetes
Venturiaceae	Ascomycetes
Verticillium	Hyphomycetes
Xylariaceae	Ascomycetes

Protozoa

Amblyopsoridae	Microsporida
Caudosporidae	Microsporida
Plasmodiidae	Eucoccidiida
Nosematidae	Microsporida
Pleistophoridae	Microsporida
Tetrahymenidae	Hymenostomatida
Thecamoebidae	Amoebida
Thelohaniidae	Microsporida
Tuzetiidae	Microsporida
Vampyrellidae	Aconchulinidae

Viruses
Baculoviridae
Bunyaviridae
Carlavirus
Closterovirus
Geminiviridae
Hypoviridae
Iridoviridae
Poxviridae
Reoviridae

Appendix IV: Contributors

Affolter, F.
CABI Bioscience Centre Switzerland
Rue des Grillons 1
CH-2800 Delémont
Switzerland

Babendrier, D.
Swiss Federal Research Station
for Agroecology and Agriculture
Reckenholzstr. 191
CH – 8046 Zürich
Switzerland

Bailey, K.
Agriculture and Agri-Food Canada
Saskatoon Research Centre
107 Science Place
Saskatoon, SK
Canada S7N 0X2

Bao, J.R.
United States Department of Agriculture
Agriculture Research Service
Rm 275 Bldg 011A BARC W.
Beltsville, MD 20705–2350
USA

Bardin, S.D.
Agriculture and Agri-Food Canada
Lethbridge Research Centre
5403 – 1st Avenue
Lethbridge, AB
Canada T1J 4B1

Beatty, P.H.
University of Alberta
Edmonton, AB
Canada, T6G 2E9

Bélanger, R.R.
Departement de phytologie – FSAA
Université Laval
Sainte-Foy QC
Canada G1K 7P4

Bernier, J.
Agriculture and Agri-Food Canada
Horticulture Research and Development Center
430 boulevard Gouin
Saint-Jean-sur-Richelieu, QC
Canada J3B 3E6

Bérubé, J.A.
Ressources Naturelles Canada
Service Canadien des Forêts
Centre de Foresterie des Laurentides
C.P. 3800, 1055 rue du P.E.P.S.
Sainte-Foy, QC
Canada G1V 4C7

Bissett, J.
Agriculture and Agri-Food Canada
Eastern Cereal and Oilseed Research Centre
K.W. Neatby Building
960 Carling Avenue
Ottawa, ON
Canada K1A 0C6

Boisvert, J.
Département de Chimie-Biologie
Université du Québec à Trois-Rivières
3351 boulevard des Forges
C.P. 500
Trois-Rivières, QC
Canada G9A 5H7

Boisvert, M.
Département de Chimie-Biologie
Université du Québec à Trois-Rivières
3351 boulevard des Forges
C.P. 500
Trois-Rivières, QC
Canada G9A 5H7

Boiteau, G.
Agriculture and Agri-Food Canada
Potato Research Centre
850 Lincoln Road
PO Box 20280

Fredericton, NB
Canada E3B 4Z7

Boivin, G.
Agriculture et Agroalimentaire Canada
Centre de recherches et de développement en horticulture
430 boulevard Gouin
Saint-Jean-sur-Richelieu, QC
Canada J3B 3E6

Boland, G.J.
Department of Environmental Biology
University of Guelph
Guelph, ON
Canada N1G 2W1

Boulter, J.I.
Department of Environmental Biology
University of Guelph
Guelph ON
Canada N1G 2W1

Bourchier, R.S.
Agriculture and Agri-Food Canada
Lethbridge Research Centre
5403 – 1st Avenue
Lethbridge, AB
Canada T1J 4B1

Boyetchko, S.M.
Agriculture and Agri-Food Canada
Saskatoon Research Centre
107 Science Place
Saskatoon, SK
Canada S7N 0X2

Braun, L.
Agriculture and Agri-Food Canada
Saskatoon Research Centre
107 Science Place
Saskatoon, SK
Canada S7N 0X2

Braun, M.P.
Agriculture and Agri-Food Canada
Saskatoon Research Centre
107 Science Place
Saskatoon, SK
Canada S7N 0X2

Broadbent, A.B.
Agriculture and Agri-Food Canada
Southern Crop Protection and Food Research Centre
1391 Sandford Street
London, ON
Canada N5V 4T3

Brockerhoff, E.G.
Forest Research
PO Box 29237
Fendalton, Christchurch
New Zealand

Butt, G.W.
Natural Resources Canada
Canadian Forest Service
PO Box 960
Corner Brook, NF
Canada A2H 6J3

Butts, R.A.
Agriculture and Agri-Food Canada
Lethbridge Research Centre
5403 – 1st Avenue
Lethbridge, AB
Canada T1J 4B1

Calpas, J.T.
Crop Diversification Centre South
Alberta Agriculture, Food and Rural Development
S.S. #4
Brooks, AB
Canada T1R 1E6

Carisse, O.
Agriculture et Agroalimentaire Canada
Centre de recherches et de développement en horticulture
430 boulevard Gouin
Saint-Jean-sur-Richelieu, QC
Canada J3B 3E6

Carl, K. (retired)
CABI Bioscience Centre Switzerland
Rue des Grillons 1
CH-2800 Delémont
Switzerland

Carney, V.
Southern Crop Protection and Food Research Centre
Agriculture and Agri-Food Canada
4902 Victoria Ave. N.
P.O. Box 6000
Vineland, ON
Canada L0R 2E0

Carter, N.
Department of Natural Resources and Energy
PO Box 6000
Fredericton, NB
Canada E3B 5H1

Cloutier, C.
Université Laval
Département de Biologie
Cité Universitaire
Québec, QC
Canada G1K 7P4

Colbo, M.H.
Department of Biology
Memorial University of Newfoundland
St John's, NF
Canada A1B 3X9

Conder, N.
Natural Resources Canada
Canadian Forest Service
506 W. Burnside Rd.
Victoria, BC
Canada V8Z 1M5

Conn, K.L.
Agriculture and Agri-food Canada
Southern Crop Protection and Food Research Centre
1391 Sandford Street
London, ON
Canada N5V 4T3

Corrigan, J.
Department of Environmental Biology
University of Guelph
Guelph, ON
Canada N1G 2W1

Cossentine, J.E.
Agriculture and Agri-Food Canada
Pacific Agri-Food Research Centre
4200 Hwy 97
Summerland, BC
Canada V0H 1Z0

Crowe, M.
Agriculture and Agri-Food Canada
Lethbridge Research Centre
5403 – 1st Avenue
Lethbridge, AB
Canada T1J 4B1

Cunningham, J.C. (retired)
Canadian Forest Service
Natural Resources Canada
PO Box 490
Sault Ste Marie, ON
Canada P6A 5M7

Darbyshire, S.
Eastern Cereal and Oilseed Research Centre
Agriculture and Agri-Food Canada
K.W. Neatby Building,
960 Carling Avenue
Ottawa, ON
Canada K1A 0C6

De Clerck-Floate, R.A.
Agriculture and Agri-Food Canada
Lethbridge Research Centre
5403 – 1st Avenue
Lethbridge, AB
Canada T1J 4B1

Digweed, S.C.
6020–104 Street
Edmonton, AB
Canada T6H 5S4

DiTommaso, A.
Department of Crop and Soil Sciences
Cornell University
Ithaca, NY 14853
USA

Dixon, P.L.
Agriculture and Agri-Food Canada
Atlantic Cool Climate Crop Research Centre
PO Box 39088
St John's, NF
Canada A1E 5Y7

Doane, J.F. (retired)
41 Simpson Crescent
Saskatoon, SK
Canada S7H 3C5

Dosdall, L.M.
Department of Agricultural, Food and Nutritional Science
4–16B Agriculture/Forestry Centre
University of Alberta
Edmonton, Alberta
Canada T6G 2P5

Dupont, S.
BioProducts Centre Inc.
The Atrium
101–111 Research Drive
Saskatoon SK
Canada S7N 3R2

Erb, S.
Agriculture and Agri-Food Canada
Lethbridge Research Centre
5403 – 1st Avenue
Lethbridge, AB
Canada T1J 4B1

Erlandson, M.A.
Agriculture and Agri-Food Canada
Saskatoon Research Centre
107 Science Place
Saskatoon, SK
Canada S7N 0X2

Ferguson, G.M.
Ministry of Agriculture, Food and Rural Affairs
Greenhouse and Processing Crops
Research Centre
Harrow, ON
Canada N0R 1G0

Fitzpatrick, S.M.
Agriculture and Agri-Food Canada
Pacific Agri-Food Research Centre
PO Box 1000
Agassiz, BC
Canada V0M 1A0

Floate, K.D.
Agriculture and Agri-Food Canada
Lethbridge Research Centre
5403 – 1st Avenue
Lethbridge, AB
Canada T1J 4B1

Foottit, R.G.
Agriculture and Agri-Food Canada
Eastern Cereal and Oilseed Research Centre
K.W. Neatby Building,
960 Carling Avenue
Ottawa, ON
Canada K1A 0C6

Frankenhuyzen, K. van
Natural Resources Canada
Canadian Forest Service
Great Lakes Forestry Centre
PO Box 490
Sault Ste Marie, ON
Canada P6A 5M7

Fry, K.M.
Crop and Plant Management
Alberta Research Council
PO Bag 4000
Vegreville, AB
Canada T9C 1T4

Gagnon, J.A.
Phytodata Inv.
Sherrington, QC
Canada J0L 2N0

Galloway, T.D.
Department of Entomology
The University of Manitoba
Winnipeg, MB
Canada, R3T 2N2

Gassmann, A.
CABI Bioscience Centre Switzerland
Rue des Grillons 1
CH-2800 Delémont
Switzerland

Gibson, G.A.P.
Agriculture and Agri-Food Canada
Eastern Cereal and Oilseed Research Centre
K.W. Neatby Building,
960 Carling Avenue
Ottawa, ON
Canada K1A 0C6

Gill, B.D.
Canadian Food Inspection Agency
Centre for Plant Quarantine Pests
Entomology Unit
K.W. Neatby Building,
960 Carling Avenue
Ottawa, ON
Canada K1A 0C6

Gill, J.J.
Agriculture and Agri-Food Canada
Food Research Program
93 Stone Road West
Guelph, ON
Canada N1G 5C9

Gillespie, D.R.
Agriculture and Agri-Food Canada
Pacific Agri-Food Research Centre
PO Box 1000
Agassiz, BC
Canada V0M 1A0

Goettel, M.S.
Agriculture and Agri-Food Canada
Lethbridge Research Centre
5403 – 1st Avenue
Lethbridge, AB
Canada T1J 4B1

Gracia-Garza, J.A.
Agriculture and Agri-Food Canada
Southern Crop Protection and Food Research Centre
PO Box 6000
4902 Victoria Ave N
Vineland Station, ON
Canada L0R 2E0

Green, S.
Agriculture and Agri-Food Canada
Saskatoon Research Centre
107 Science Place
Saskatoon SK
Canada S7N OX2

Hardman, J.M.
Agriculture and Agri-Food Canada
Atlantic Food and Horticulture Research Centre
32 Main Street
Kentville, NS
Canada B4N 1J5

Harris, P. (retired)
Agriculture and Agri-Food Canada
Lethbridge Research Centre
5403 – 1st Avenue
Lethbridge, AB
Canada T1J 4B1

Henderson, D.E.
E.S. Cropconsult Ltd.
3041 West 33rd Avenue
Vancouver, BC
Canada V6N 2G6

Heppner, D.G.
British Columbia Ministry of Forests
Vancouver Forest Region
2100 Labieux Rd.
Nanaimo, BC
Canada V9T 6E9

Hinz, H.L.
CABI Bioscience Centre Switzerland
Rue des Grillons 1
CH-2800 Delémont
Switzerland

Hoffmeister, T.S.
Zoologisches Institut, Oekologie
Christian-Albrechts-Universitaet Kiel
D-24098 Kiel
Germany

Holliday, N.J.
University of Manitoba
Department of Entomology
Winnipeg, MB
Canada R3T 2N2

Huang, H.C.
Agriculture and Agri-Food Canada
Lethbridge Research Centre
5403 – 1st Avenue
Lethbridge, AB
Canada T1J 4B1

Huber, J.T.
Natural Resources Canada
Canadian Forest Service c/o
K.W. Neatby Building,
960 Carling Avenue
Ottawa, ON
Canada K1A 0C6

Hueppelsheuser, T.
E.S. Cropconsult Ltd.
3041 West 33rd Avenue
Vancouver, BC
Canada V6N 2G6

Hulme, M.
Natural Resources Canada
Canadian Forest Service
506 West Burnside Road
Victoria, BC
Canada V8Z 1M5

Hunt, D.W.A.
Agriculture and Agri-Food Canada
Greenhouse and Processing Crops Research Centre
2585 Highway 20, E.
Harrow, ON
Canada N0R 1G0

Iranpour, M.
Department of Entomology
University of Manitoba
Winnipeg, MB
Canada R3T 2N2

Jabaji-Hare, S.H.
Department of Plant Sciences
McGill University, Macdonald Campus
21, 111 Lakeshore Road
Ste-Anne-de-Bellevue, QC
Canada H9X 3V9

Jarvis, W.R.
Greenhouse Crops Res. Centre
470 Thorn Ridge
Amherstburg, ON
Canada N9V 3X4

Jean, C.
Université Laval
Département de Biologie
Cité Universitaire
Québec, QC
Canada G1K 7P4

Jensen, K.I.M.
Agriculture and Agri-Food Canada
Atlantic Food and Horticulture Research Centre
32 Main Street
Kentville, NS
Canada B4N 1J5

Jensen, S.E.
University of Alberta
Edmonton, AB
Canada, T6G 2E9

Johnson, D.L.
Agriculture and Agri-Food Canada
Lethbridge Research Centre
5403 – 1st Avenue
Lethbridge, AB
Canada T1J 4B1

Kharbanda, P.D.
Alberta Research Council
PO Bag 4000
Vegreville, AB
Canada, T9C 1T4

Kenis, M.
CABI Bioscience Centre Switzerland
Rue des Grillons 1
CH-2800 Delémont
Switzerland

Kuhlmann, U.
CABI Bioscience Centre Switzerland
Rue des Grillons 1
CH-2800 Delémont
Switzerland

Lachance, S.
Recherche et Transfert de Technologie
Research and Technology Transfer
Alfred College
31 St-Paul Street, PO Box 580
Alfred, ON
Canada K0B 1A0

Laflamme, G.
Ressources Naturelles Canada
Service Canadien des Forêts
Centre de Foresterie des Laurentides
C.P. 3800 1055 rue du P.E.P.S.
Sainte-Foy, QC
Canada G1V 4C7

Langor, D.W.
Natural Resources Canada
Canadian Forest Service
Northern Forestry Centre
5320 – 122 Street
Edmonton, AB
Canada T6H 3S5

Lazarovits, G.
Agriculture and Agri-Food Canada
Southern Crop Protection and Food Research Centre
1391 Sandford Street
London, ON
Canada N5V 4T3

Li, S.Y.
Natural Resources Canada
Canadian Forest Service
Atlantic Forestry Centre
PO Box 960
Corner Brook, NF
Canada A2H 6J3

Lim, K.P.
15 Eastbourne #211
Brampton, ON
Canada L6T 3L9

Lindgren, C.J.
Manitoba Purple Loosestrife Project
Box 1160
Stonewall, MB
Canada R0C 2Z0

Lyons, D.B.
Natural Resources Canada
Canadian Forest Service
Great Lakes Forestry Centre
PO Box 490
Sault Ste Marie, ON
Canada P6A 5M7

Lysyk, T.J.
Agriculture and Agri-Food Canada
Lethbridge Research Centre
5403 – 1st Avenue
Lethbridge, AB
Canada T1J 4B1

Macey, D.E..
Natural Resources Canada
Canadian Forest Service
Pacific Forestry Centre
506 West Burnside Road
Victoria, BC
Canada V8Z 1M5

Mallett, K.I.
Natural Resources Canada
Canadian Forest Service
Northern Forestry Centre
5320 – 122 Street
Edmonton, AB
Canada T6H 3S5

MacRae, I.V.
Department of Entomology
University of Minnesota
NWROC
2900 University Avenue
Crookston MN 56716
USA

Mason, P.G.
Agriculture and Agri-Food Canada
Eastern Cereal and Oilseed Research Centre
K.W. Neatby Building,
960 Carling Avenue
Ottawa, ON
Canada K1A 0C6

McClay, A.
Alberta Research Council
P.O. Bag 4000
Vegreville AB
Canada T9C 1T4

Moeck, H.A. (retired)
4710 Sooke Road
Victoria, BC
Canada V9C 4B9

Mortensen, K. (retired)
Box 502
Balgonie, SK
Canada S0G 0E0

Moyer, J.
Agriculture and Agri-Food Canada
Lethbridge Research Centre
5403 – 1st Avenue
Lethbridge, AB
Canada, T1J 4B1

Nealis, V.G.
Natural Resources Canada
Canadian Forest Service
Pacific Forestry Centre
506 West Burnside Road
Victoria, BC
Canada V8Z 1M5

O'Hara, J.E.
Agriculture and Agri-Food Canada
Eastern Cereal and Oilseed Research Centre
K.W. Neatby Building,
960 Carling Avenue
Ottawa, ON
Canada K1A 0C6

Olfert, O.O.
Agriculture and Agri-Food Canada
Saskatoon Research Centre
107 Science Place
Saskatoon, SK
Canada S7N 0X2

Otvos, I.S.
Natural Resources Canada
Canadian Forest Service
Pacific Forestry Centre
506 W. Burnside Rd.
Victoria, BC
Canada V8Z 1M5

Parker, D.J.
Canadian Food Inspection Agency
Centre for Plant Quarantine Pests
Entomology Unit
K.W. Neatby Building,
960 Carling Avenue
Ottawa, ON
Canada K1A 0C6

Patterson, K.
Department of Environmental Sciences
Nova Scotia Agricultural College
PO Box 550
Truro, Nova Scotia
Canada B2N 5E3

Paulitz, T.C.
United States Department of Agriculture
Root Disease and Biological Control Research Unit
PO Box 646430
363 Johnson Hall
Washington State University
Pullman, WA 99164–6430
USA

Peschken, D.P. (retired)
2900 Rae St
Regina, SK
Canada S4S 1R5

Philion, V.
IRDA,
C.P. 480
St-Hyacinthe, QC
Canada J2S 7B8

Philip, H.G.
British Columbia Ministry of Agriculture and Food
1690 Powick Road
Kelowna, BC
Canada V1X 7G5

Prasad, R.P.
Natural Resources Canada
Canadian Forestry Service
Pacific Forestry Centre
506 West Burnside Road
Victoria, BC
Canada V8Z 1M5

Quednau, F.W. (retired)
Ressources Naturelles Canada
Service canadien des Forêts
Centre de Foresterie des Laurentides
C.P. 3800, 1055 rue du P.E.P.S.
Sainte-Foy, QC
Canada G1V 4C7

Raworth, D.A.
Agriculture and Agri-Food Canada
Pacific Agri-Food Research Centre
PO Box 1000
Agassiz, BC
Canada V0M 1A0

Reeleder, R.D.
Agriculture and Agri-Food Canada
Pest Management Research Centre
PO Box 186
Delhi, ON
Canada N4B 2W9

Ring, R.A.
Biology Department
University of Victoria
Victoria, BC
Canada V8W 3N5

Roy, M.
Laboratoire de diagnostic en phytoprotection MAPAQ
Complexe scientifique, D1.110
Sainte-Foy, QC
Canada G1P 3W8

Safranyik, L. (retired)
Natural Resources Canada
Canadian Forestry Service
Pacific Forestry Centre
506 West Burnside Road
Victoria, BC
Canada V8Z 1M5

Sampson, M.G.
Department of Environmental Sciences
Nova Scotia Agricultural College
PO Box 550
Truro NS
Canada B2N 5E3

Sarazin, M.J.
Agriculture and Agri-Food Canada
Eastern Cereal and Oilseed Research Centre
K.W. Neatby Building,
960 Carling Avenue
Ottawa, ON
Canada K1A 0C6

Schwarzländer, M.
Department of Plant, Soil and Entomological Sciences
College of Agriculture
University of Idaho
Moscow, ID 83844–2339
USA

Shamoun, S.
Natural Resources Canada
Canadian Forestry Service
Pacific Forestry Centre
506 West Burnside Road
Victoria, BC
Canada V8Z 1M5

Shepherd, R.F. (retired)
Natural Resources Canada
Canadian Forestry Service
Pacific Forestry Centre
506 West Burnside Road
Victoria, BC
Canada V8Z 1M5

Shipp, J.L.
Agriculture and Agri-Food Canada
Greenhouse and Processing Crops Research Centre
2585 Highway 20, E.
Harrow, ON
Canada N0R 1G0

Sholberg, P.
Agriculture and Agri-Food Canada
Pacific Agri-Food Research Centre
Box 4200, Hwy 97
Summerland, BC
Canada V0H 1Z0

Shore, T.L.
Natural Resources Canada
Canadian Forest Service
Pacific Forestry Centre
506 West Burnside Road
Victoria, BC
Canada V8Z 1M5

Smith, S.M.
University of Toronto
Forestry Department
33 Willcocks St.
Toronto, ON
Canada M5S 3B3

Sobhian, R. (retired)
European Biological Control Laboratory
USDA – ARS
Campus Internationale de Baillarguet
CS 90013 Montferrier sur Lez
34982 St Gely du Fesc, Cedex
France

Soltani, N.
Agriculture and Agri-food Canada
Southern Crop Protection and Food Research Centre
1391 Sandford Street
London, ON
Canada N5V 4T3

Soroka, J.J.
Agriculture and Agri-Food Canada
Saskatoon Research Centre
107 Science Place
Saskatoon, SK
Canada S7N 0X2

Spence, J.R.
Department of Biological Sciences
University of Alberta
Edmonton, AB
Canada T6G 2E3

Stewart-Wade, S.M.
Department of Crop Production
The University of Melbourne
Victoria 3010
Australia

Svircev, A.M.
Agriculture and Agri-Food Canada
Southern Crop Protection and Food Research Centre
PO Box 6000
4902 Victoria Ave N
Vineland Station, ON
Canada L0R 2E0

Sweeney, J.D.
Natural Resources Canada
Canadian Forest Service
Atlantic Forestry Centre
PO Box 4000
Fredericton, NB
Canada E3B 5P7

Teerling, C.
Agriculture and Agri-Food Canada
Southern Crop Protection and Food Research Centre
PO Box 6000
4902 Victoria Ave N
Vineland Station, ON
Canada L0R 2E0

Tenuta, M.
Agriculture and Agri-Food Canada
Southern Crop Protection and Food Research Centre
1391 Sandford Street
London, ON
Canada N5V 4T3

Teshler, I.B.
Department of Plant Science
McGill University MacDonald Campus
21,111 Lakeshore Road
Ste-Anne-de-Bellevue, QC
Canada H9X 3V9

Teshler, M.P.
Department of Plant Science
McGill University MacDonald Campus
21,111 Lakeshore Road
Ste-Anne-de-Bellevue, QC
Canada H9X 3V9

Tewari, J.P.
Department of Agricultural, Food, and Nutritional Science
4–10 Agriculture/Forestry Centre
University of Alberta
Edmonton, AB
Canada T6G 2P5

Thistlewood, H.M.A.
Agriculture and Agri-Food Canada
Pacific Agri-Food Research Centre
4200 Hwy 97
Summerland, BC
Canada V0H 1Z0

Thurston, G.S.
Natural Resources Canada
Canadian Forest Service
Atlantic Forestry Centre
PO Box 4000
Fredericton, NB
Canada E3B 5P7

Traquair, J.A.
Agriculture and Agri-Food Canada
Southern Crop Protection and Food Research Centre
1391 Sandford Street
London, ON
Canada N5V 4T3

Turgeon, J.J.
Natural Resources Canada
Canadian Forest Service
Great Lakes Forestry Service
PO Box 490
Sault Ste Marie, ON
Canada P6A 5M7

Turnock, W.J. (retired)
28 Vassar Road
Winnipeg, MB
Canada R3T 3M9

Utkhede, R.S.
Agriculture and Agri-Food Canada
Pacific Agri-Food Research Centre
PO Box 1000
Agassiz, BC
Canada V0M 1A0

Vincent, C.
Agriculture et Agroalimentaire Canada
Centre de recherches et de développement en horticulture
430 boulevard Gouin
Saint-Jean-sur-Richelieu, QC
Canada J3B 3E6

Watson, A.K.
Department of Plant Science
McGill University MacDonald Campus
21,111 Lakeshore Road
Ste-Anne-de-Bellevue, QC
Canada H9X 3V9

West, R.
PO Box 515
Portugal Cove, NF
Canada A0A 3K0

Whistlecraft, J.
Agriculture and Agri-Food Canada
Southern Crop Protection and Food Research Centre
1391 Sandford Street
London, ON
Canada N5V 4T3

White, D.J.
6346 112th Street
Edmonton, AB
Canada T6H 3J6

Whitney, H.S. (retired)
5033 Ayum Road
Sooke, BC
Canada V0S 1N0

Winchester, N.N.
Biology Department
University of Victoria
Victoria, BC
Canada V8W 3N5

Winder, R.S.
Natural Resources Canada
Canadian Forest Service
Pacific Forestry Centre
506 West Burnside Road
Victoria, BC
Canada V8Z 1M5

Yang, J.
Alberta Research Council
PO Bag 4000
Vegreville, AB
Canada T9C 1T4

Zhang, W.
Alberta Research Council
PO Bag 4000
Vegreville, AB
Canada T9C 1T4

Zhou, T.
Agriculture and Agri-Food Canada
Food Research Programme
93 Stone Road West
Guelph, ON
Canada N1G 5C9

Taxonomic Index

Each genus name is placed to family, where possible. For many genera of fungi only the class or superfamily is given because the perfect (teleomorph) stage, needed to classify a genus correctly to family, has not yet been associated with the corresponding anamorph (imperfect stage). Trinomials indicate subspecies unless otherwise indicated. Specific names for viruses follow Murphy, F.A., Fauquet, C.M., Bishop, D.H.L., Gabrial, S.A., Jarvis, A.W., Martelli, G.P., Mayo, M.A. and Summers, M.D. (eds) (1995) *Virus Taxonomy: Classification and Nomenclature of Viruses. Sixth Report of the International Committee on Taxonomy of Viruses*. Springer-Verlag, Vienna, Austria, 586pp.